C.THALER

ANNUAL REVIEW OF CELL BIOLOGY

ANNUAL REVIEW OF CELL BIOLOGY

VOLUME 5, 1989

GEORGE E. PALADE, *Editor*
Yale University School of Medicine

BRUCE M. ALBERTS, *Associate Editor*
University of California, San Francisco

JAMES A. SPUDICH, *Associate Editor*
Stanford University School of Medicine

ANNUAL REVIEWS INC 4139 EL CAMINO WAY P.O. BOX 10139 PALO ALTO, CALIFORNIA 94303-0897

ANNUAL REVIEWS INC.
Palo Alto, California, USA

International Standard Serial Number: 0743-4634
International Standard Book Number: 0-8243-3105-2

∞ The paper used in this publication meets the minimum requirements of American National Standard for Information Sciences—Permanence of Paper for Printed Library Materials, ANSI Z39.48-1984.

TYPESET BY AUP TYPESETTERS (GLASGOW) LTD., SCOTLAND
PRINTED AND BOUND IN THE UNITED STATES OF AMERICA

PREFACE

The recent history of Cell Biology has been characterized by rapid and functionally significant growth in a large number of unusually productive directions. The oldest and strongest among them has already led to the merging of the once distinct fields of Cell Biology and Molecular Biology into a continuous, common body of knowledge kept in active ferment by concepts and technologies derived from molecular genetics as well as from other sources.

The more we advance in developing this basic common body of knowledge, the more clearly we appreciate the remarkable versatility of the evolutionary process and the detailed and subtle richness of information contained within the cell's genome. Evolution has exploited minor changes as well as major rearrangements to create families or superfamilies of genes that encode for a variety of gene products with many common structural motifs yet diversified functions. The number of protein isoforms is continuously increasing, and the old notion that gene products are stable through the ontogeny of individual organisms is being revised, because of the established existence of multiple isoforms, products of single genes or gene families, differentially expressed at specific sites (organs or parts of organs), at specific times in ontogeny. It seems that interactions with the environment (even with microenvironments on homeostatic organisms) influence the expression of the genome to a larger extent than previously assumed. It is already well established that the genome includes instructions for the expression of specific sets of genes in specific cell types, but perhaps we should already search for instructions that control the expression of cell-type specific genes under specific environmental conditions in ontogeny or in pathological processes in the adult. In some of the latter, fetal gene products are known to be re-expressed.

For one reason or another, less attention is given to "junk DNA" at present than a few years ago, and less resistance is encountered by major projects, such as the sequencing of the human and other genomes of immediate interest. These genomes may prove to contain less "evolutionary garbage" than when the projects were first envisaged. Yet much more structural and chemical information is needed to understand in specific terms, rather than in general outline, how gene expression and genome replication are controlled. The cognate research areas are already, and will remain for some time, domains of critical importance in contemporary Cell Biology.

Periods of rapid scientific advances often generate illusions. Impressed somehow by the massive influx of new information, we may be inclined to believe that we are close to a full understanding of how cells are put together and carry out their functions. But often we discover that only a few years ago we were soaring over large areas of significant structures and activities we had blissfully ignored. And suddenly such areas become the equivalent of gold mining fields with the expected concentration of eager search forces, competition, frictions, claims, and counter claims. Protein traffic control in all cells, targeting, sorting, membrane traffic control in eukaryotic cells, and—connected with such operations—quality control of gene products appear to be among these recently opened gold mining fields. Signal transduction mechanisms represent a similar area, since so much remains to be uncovered between activated receptors and signal amplifiers on the distal side of the relevant chain of reactions and activated gene transcription or genome replication on the proximal end of the same chain. Equally attractive areas can be recognized in research fields dealing with the locomotor apparatus of the cell or the cell's interactions with its immediate environment, that is, with other cells and macromolecular components of the extracellular matrix. Each of these new and often unexpected developments is leading to spectacular advances in Cellular and Molecular Biology.

Yet there is still another area that may become in time even more fertile than those already mentioned. The genome and the basic organization at the cellular level is the same all the way from a fertilized egg to a fully mature organism that comprises a huge number of cells belonging to many highly differentiated cell types. All this complex but exquisitely controlled developmental process is guided by programmed genetic instructions that call for the expression of specific genes at specific sites and specific times in ontogeny. The basic mechanisms involving messengers, receptors, signal transduction, and other components of the relevant chains of reactants appear to be similar to those found in the cells of adult organisms, but the specific macromolecules that operate along these chains appear to be themselves differentiated. So far the best examples are found in terminal cell differentiation, especially in the immune system, that appears to have its own set of primary messengers—the interleukins—and its own set of receptors. We know little about the chain of reactants that connects activated receptors to the relevant programs of specific genes, but we know that the cognate gene products are cell-type specific and highly diversified. How many genes, we may ask, are involved in the construction of those elements of the immune system we have so far identified? And how many genes and gene products remain to be uncovered in the main cell types of the system or in their supporting cast represented by stromal cells, endothelial cells, and specialized epithelial cells?

It is worthwhile pointing out that work on the developmental history of the immune system has advanced faster than work on the development of any other complex organismic counterpart. This situation may be explained by some of the characteristics of the immune system: it consists of sortable populations of essentially free living cells, free because they have to patrol the organism in search of targets and interacting partners. And it relies on diffusible primary messengers for most cells interactions. Each of these characteristics has been used to generate tools and assays that have greatly facilitated research on the development of the entire system. The complexity of developmental interactions may be higher in other cases, especially in the central nervous system. Moreover, the uncovering of the guiding mechanisms may prove to be a much more arduous operation: the cells in case often form compact, cohesive parenchyma, and development seems to rely primarily on direct cell-cell interactions, bypassing the need for diffusible primary messengers. In this case (and undoubtedly many others), messengers as well as receptors appear to be integral membrane proteins, some of them related to cell receptors of the immune system as members of a large gene superfamily.

Starting from such considerations, we may ask: how many genes are involved in the construction of a sensory neuron that connects the periphery to the cerebellum? How many genes are needed to build up a cerebellum and how many to generate an entire central nervous system? The anticipated answers are: very large numbers. This may become, in fact, a recurrent theme: how many genes contribute to the development of a nephron and how many to that of an entire kidney, its specialized vasculature included? Perhaps, after all, a good fraction if not most of that DNA in the genome is functionally needed.

The Editorial Committee of the Annual Review of Cell Biology has tried to provide in this volume a useful sampling of current activities in many of the interesting and challenging research areas mentioned above. As the table of contents shows, the sampling includes topics belonging to developmental biology, cellular immunology, and plant cell biology, in addition to topics of general interest to cellular and molecular biologists. The readers will tell us, we hope, if they consider our attempts successful. Comments and suggestions coming from the reading public, including the membership of the American Society for Cell Biology, will be most welcome.

George E. Palade
Editor

Annual Review of Cell Biology
Volume 5, 1989

CONTENTS

(*continued*)

SOME RELATED ARTICLES IN OTHER *ANNUAL REVIEWS*

From the *Annual Review of Biochemistry*, Volume 58 (1989):

The Protein Kinase C Family: Heterogeneity and its Implications, U. Kikkawa, A. Kishimoto, Y. Nishizuka

Hemopoietic Cell Growth Factors and Their Receptors, N. A. Nicola

Structure and Biosynthesis of Prokaryotic Glycoproteins, J. Lechner and F. Wieland

DNA Conformation and Protein Binding, A. A. Travers

The Structure and Regulation of Protein Phosphatases, P. Cohen

The Heparin-Binding (Fibroblast) Growth Factor Family of Proteins, W. Burgess and T. Maciag

The Bacterial Photosynthetic Reaction Center as a Model for Membrane Proteins, D. C. Rees, H. Komiya, T. O. Yeates, J. P. Allen, and G. Feher

Phospholipid Biosynthesis in Yeast, G. M. Carman and S. A. Henry

Eukaryotic Transcriptional Regulatory Proteins, P. F. Johnson and S. L. McKnight

Glycosylation in the Nucleus and Cytoplasm, G. W. Hart, R. S. Haltiwanger, G. D. Holt, and W. G. Kelly

Topography of Membrane Proteins, M. L. Jennings

Molecular Mechanisms of Transcriptional Regulation in Yeast, K. Struhl

From the *Annual Review of Biophysics and Biophysical Chemistry*, Volume 18 (1989):

Biochemistry and Biophysics of Excitation-Contraction Coupling, S. Fleischer and M. Inui

Toward a Unified Model of Chromatin Folding, J. Widom

Structure and Function of the Red Blood Cell Anion Transport Protein, M. L. Jennings

Protein-Mediated Membrane Fusion, T. Stegmann, R. W. Doms, and A. Helenius

From the *Annual Review of Genetics*, Volume 23 (1989):

Prokaryotic Signal Transduction Mediated by Sensor and Regulator Protein Pairs, L. M. Albright, E. Huala, F. M. Ausubel

The Structure and Function of Telomeres, V. Zakian

(*continued*)

Maize Transposable Elements, A. Gierl, H. Saedler, P. A. Peterson

Alternative Splicing in the Control of Gene Expression, C. W. J. Smith, J. C. Patton, B. Nadal-Ginard

Homologous Recombination in Mammalian Cells, R. J. Bollag, A. S. Waldman, R. M. Liskay

Adenovirus E1A Protein: Paradigm Viral Transactivator, J. Flint, T. Shenk

From the *Annual Review of Immunology*, Volume 7 (1989):

The Cellular Basis of T-Cell Memory, J.-C. Cerottini and H. R. MacDonald

TH1 and TH2 Cells: Different Patterns of Lymphokine Secretion Lead to Different Functional Properties, T. R. Mosmann and R. L. Coffman

Cell Biology of Cytotoxic and Helper T-Cell Functions, A. Kupfer and S. J. Singer

Immunogenetics of Human Cell Surface Differentiation, W. J. Rettig and L. J. Old

The Biology of Cachectin/TNF- A Primary Mediator of the Host Response, B. Beutler and A. Cerami

From the *Annual Review of Microbiology*, Volume 43 (1989):

Role of the DNA/Membrane Complex in Prokaryotic DNA Replication, W. Firshein

How Antibiotic-Producing Organisms Avoid Suicide, E. Cundliffe

Genetics and Regulation of Bacterial Lipid Metabolism, T. Vanden Boom and J. E. Cronan, Jr.

Scrapie Prions, S. B. Prusiner

Transcription and Reverse Transcription of Retrotransposons, J. D. Boeke and V. G. Corces

Molecular Mechanisms of Immunoglobulin A Defense, N. K. Childers, M. G. Bruce, and J. R. McGhee

From the *Annual Review of Neuroscience*, Volume 13 (1990):

Patterned Activity, Synaptic Convergence, and the NMDA Receptor in Developing Visual Pathways, M. Constantine-Paton, H. T. Cline, and E. Debski

From the *Annual Review of Physiology*, Volume 51 (1989):

Structure and Regulation of Muscarinic Receptors, M. I. Schimerlik

Regulation of Tissue Plasminogen Activator Expression, R. D. Gerard and R. S. Meidell

Classes of Calcium-Activated Channels in Vertebrate Cells, B. P. Bean

The Calcium Pump of the Surface Membrane and of the Sarcoplasmic Reticulum, H. J. Schatzmann

External and Internal Signals for Epithelial Cell Surface Polarization, E. Rodriguez-Boulan and P. J. I. Salas

Biogenesis of Endogenous Plasma Membrane Proteins in Epithelial Cells, A. L. Hubbard, B. Stieger, and J. R. Bartles

Vesicle Recycling and Cell-Specific Function in Kidney Epithelial Cells, D. Brown

Development and Alteration of Polarity, M. Cereijido

Molecular Biological Approaches to Protein Sorting, M. G. Roth

From the *Annual Review of Plant Physiology and Plant Molecular Biology*, Volume 40 (1989):

The Uniqueness of Plant Mitochondria, R. Douce and M. Neuburger

Regulatory Interactions between Nuclear and Plastid Genomes, W. C. Taylor

Chloroplastic Precursors and Their Transport across the Envelope Membranes, K. Keegstra, L. J. Olson, and S. M. Theg

ANNUAL REVIEWS INC. is a nonprofit scientific publisher established to promote the advancement of the sciences. Beginning in 1932 with the *Annual Review of Biochemistry*, the Company has pursued as its principal function the publication of high quality, reasonably priced *Annual Review* volumes. The volumes are organized by Editors and Editorial Committees who invite qualified authors to contribute critical articles reviewing significant developments within each major discipline. The Editor-in-Chief invites those interested in serving as future Editorial Committee members to communicate directly with him. Annual Reviews Inc. is administered by a Board of Directors, whose members serve without compensation.

ANNUAL REVIEWS OF

Anthropology
Astronomy and Astrophysics
Biochemistry
Biophysics and Biophysical Chemistry
Cell Biology
Computer Science
Earth and Planetary Sciences
Ecology and Systematics
Energy
Entomology
Fluid Mechanics
Genetics
Immunology
Materials Science
Medicine
Microbiology
Neuroscience
Nuclear and Particle Science
Nutrition
Pharmacology and Toxicology
Physical Chemistry
Physiology
Phytopathology
Plant Physiology and Plant Molecular Biology
Psychology
Public Health
Sociology

SPECIAL PUBLICATIONS

Excitement and Fascination of Science, Vols. 1, 2, and 3

Intelligence and Affectivity, by Jean Piaget

A detachable order form/envelope is bound into the back of this volume.

Annu. Rev. Cell Biol. 1989. 5 : 1–23

CONTROL OF PROTEIN EXIT FROM THE ENDOPLASMIC RETICULUM

Hugh R. B. Pelham

MRC Laboratory of Molecular Biology, Hills Road, Cambridge CB2 2QH, England

---

CONTENTS

INTRODUCTION

The endoplasmic reticulum (ER) is the largest membrane-bound organelle in a typical eukaryotic cell. It consists of a continuous network of tubules and cisternae extending throughout the cytoplasm, with a total surface

0743–4634/89/1115–0001$02.00

area at least six times that of the plasma membrane. Most of the ER membrane is studded with ribosomes (thus forming the rough endoplasmic reticulum, or RER). Membrane and secretory proteins are synthesized on these ribosomes and enter the ER, which is the starting point for the secretory pathway; such proteins spend only a short time in this organelle before being transported, by a process of vesicle budding and fusion, to the Golgi apparatus and thence to the cell surface (Palade 1975). The ER membrane and the lumen that it encloses also contain a characteristic set of resident proteins that are involved in the processing of secretory proteins and in other metabolic functions such as phospholipid biosynthesis. Thus, export of newly synthesized polypeptides from the ER involves the separation of secretory proteins from resident proteins. In this article I discuss the mechanisms involved in this sorting process. Several other recent reviews have also covered this area (Pfeffer & Rothman 1987; Rothman 1987; Burgess & Kelly 1987; Rose & Doms 1988; Lodish 1988; Cutler 1988).

Outline of the Problem

Secretory and membrane proteins leave the ER in transport vesicles, which may bud from specialized ribosome-free regions of the ER membrane adjacent to the Golgi, termed transitional elements. They are delivered to the *cis* face of the Golgi apparatus and then continue their journey through the Golgi cisternae by means of repeated cycles of vesicle budding and fusion. The ER and the various regions of the Golgi differ in their content of oligosaccharide modifying enzymes, which allow the transport of glycoproteins through these compartments to be monitored by the sequential changes that occur to their oligosaccharide side chains (for reviews see Farquhar 1985; Kornfeld & Kornfeld 1985; Roth 1987). In general, glycoproteins found in the ER do not exhibit Golgi-specific modifications. This observation suggests that vesicular transport from ER to Golgi is both unidirectional and selective: most resident ER proteins do not reach the Golgi and once a secretory protein has reached the Golgi, it does not return to the ER.

Selective transport implies that individual proteins bear signals that control their movement. In principle, each step along the secretory pathway could be mediated by a specific transport signal and the destination of a protein determined by its lack of a signal for the next step. Alternatively, secretion could be the default pathway, driven by a nonselective process in which the contents of each compartment are transferred to the next; diversion from this pathway or retention in a specific compartment would then be signal-mediated (Pfeffer & Rothman 1987). There is now considerable evidence that a nonselective bulk flow to the cell surface does exist, and those signals that have been most clearly identified divert proteins from this flow: the mannose-6-phosphate marker directs soluble lysosomal enzymes to lysosomes (reviewed by von Figura & Hasilik 1986; Kornfeld & Mellman, this volume) and, as discussed below, the *C*-terminal tetrapeptide KDEL causes the retention of resident soluble proteins in the lumen of the ER (Munro & Pelham 1987).

PROTEIN EXPORT FROM THE ER

Nonselective Bulk Flow from the ER

The selectivity of export from the ER could be explained if this step was receptor-mediated. The most extreme version of this hypothesis predicts that every exported protein contains a transport signal and that disruption of this signal by mutation would prevent transport. A number of workers have searched for such signals by mutating the genes encoding secretory or membrane proteins and measuring the rate of transport of the altered proteins. Many mutant proteins do fail to be secreted, but it seems that in most cases this is a nonspecific effect: the proteins do not fold correctly, and malfolded proteins, in general, are slow to leave the ER, often because they form insoluble aggregates (reviewed by Rose & Doms 1988).

On the other hand, there are several cases where proteins that do not normally leave the ER are induced to do so by the deletion of certain amino acid sequences. Thus, removal of the *C* terminus from the resident ER Protein BiP (binding protein) (Munro & Pelham 1987), of the *N* terminus of the rotavirus protein VP7 (Poruchynsky et al 1985), or of the cytoplasmic tail of the adenovirus E19 protein, a transmembrane protein (Paabo et al 1987), causes each of them to enter the secretory pathway, even though they do not normally do so and therefore would not be expected to possess a transport signal. Similarly, expression of the prokaryotic protein β-lactamase in *Xenopus* oocytes also results in secretion although this enzyme is not expected to have signals for transport to the Golgi complex (Wiedmann et al 1984). Most strikingly, Wieland et al (1987) have shown that a synthetic tripeptide consisting of the glycosylation sequence Asn-Tyr-Thr, esterified to make it membrane-permeable, can enter cells, reach the ER, be glycosylated (thus making it membrane-impermeant), and then be rapidly secreted. The existence of appropriate modifications to the attached oligosaccharide demonstrates that at least some of the tripeptide passes through the Golgi complex on the normal secretory pathway. These results argue strongly in favor of a bulk flow model in which secretion is the default fate for a protein containing no specific signals.

The Rate of Bulk Flow

The halftime for exit of individual secretory or membrane proteins from the ER varies from about 15 min for viral glycoproteins (Quinn et al 1984; Copeland et al 1988) and some serum proteins to 2 h or more for other serum proteins (Fries et al 1984; Lodish et al 1983; Yeo et al 1985). The most rapidly secreted proteins reach the cell surface with a halftime of about 30 min. Is bulk flow sufficiently rapid to account for the export of these proteins? Wieland et al (1987) argued that it is: they estimated the halftime for secretion of the tripeptide from tissue culture cells to be about 10 min. However, this rate is not easy to measure, and the validity of this estimate has been questioned (Rose & Doms 1988). One problem is that the kinetics of glycosylation of the added tripeptide are distinctly biphasic, for unknown reasons, and examination of the published data suggests that halftimes of 25–50 min are equally possible.

There are other uncertainties: it is not clear whether the glycosylated tripeptide is truly a bulk marker, or whether it can bind to proteins within the secretory pathway. Also, its nonphysiological nature, while desirable for a bulk phase marker, raises doubts about the precise way in which it is metabolized by cells and the effects it might have on them. Further measurements using a different marker would clearly be useful. Nevertheless, the current data suggest that the bulk flow rate is close to the rate of transport of the most rapidly secreted proteins. If this is so, there is no compelling reason to postulate the existence of a positive transport signal for any protein.

Factors Affecting the Rate of Protein Export

Many proteins leave the ER more slowly than expected from the presumed bulk flow rate. The most likely explanation for this is that they interact with resident proteins in the ER, and are retarded in a manner analogous to chromatography (Pfeffer & Rothman 1987). A rapidly secreted protein would then be characterized by an absence of interactions rather than a dominant signal for transport. In principle, this can be tested by constructing fusions between rapidly and slowly exported proteins. For example, the soluble ER protein BiP, when its *C* terminus is removed, is secreted very slowly from COS cells ($t_{1/2}$ approx. 3 h; Munro & Pelham 1987); cathepsin D leaves the ER much more rapidly (Pelham 1988). A fusion protein containing both BiP and cathepsin D sequences leaves the ER at the slow rate characteristic of the truncated BiP (H. Pelham, unpublished observations), which is consistent with the idea that the low rate of export of truncated BiP is due to interactions with ER proteins rather than the lack of a "rapid transport" signal.

Membrane proteins seem to be particularly prone to hindrance. A large number of mutations have been introduced into such proteins as the VSV G protein and influenza hemagglutin (HA), and in many cases transport is impaired. Changes in the luminal, cytoplasmic, and transmembrane domains can all slow export and no general signal has been identified for either retention or transport (for extensive discussion of these experiments see Rose & Doms 1988). It seems that each protein is a special case. For example, VSV G, influenza HA, and the adenovirus E19 protein all have a single transmembrane segment and a small cytoplasmic tail. Removal of this tail has no effect on the rapid transport of HA (Doyle et al 1986), greatly slows VSV G export (Doms et al 1988), and promotes the export of E19, which normally resides in the ER (Paabo et al 1987).

PROTEIN FOLDING: THE ROLE OF RESIDENT ER PROTEINS

It has become apparent that the correct folding and assembly of secretory proteins is necessary for their efficient transport, presumably because unfolded proteins tend to bind to other proteins in the ER or form aggregates that are unable to enter transport vesicles (for discussion see Rose & Doms 1988; Lodish 1988; Helenius & Hurt, this volume). Folding of proteins in vitro is often a slow and inefficient process, but in vivo most proteins achieve their final conformation within a few minutes. This efficiency is due, at least in part, to the presence of several resident proteins of the ER that actively promote the folding process. In addition, these resident proteins probably contribute to the selectivity of transport: by binding (even weakly) to incompletely or incorrectly assembled proteins they may prevent their export.

Binding of Proteins to BiP

A major ER resident involved in protein assembly is the soluble protein BiP (binding protein; also known as the glucose-regulated protein GRP78). Increased synthesis of BiP is induced when abnormal proteins accumulate in the ER (Lee 1987; Kozutsumi et al 1988), and BiP preferentially associates with such proteins. Thus, it binds to mutant or malfolded forms of influenza HA (Gething et al 1986), to a derivative of SV40 T antigen that is fused to a signal sequence and thus enters the ER (Sharma et al 1985) and, in an in vitro translation-translocation system, to unoxidized (but not mature, disulphide-linked) prolactin and to unglycosylated, but not glycosylated, invertase (Kassenbrock et al 1988). Many proteins fold poorly or aggregate when their glycosylation is inhibited (e.g. Leavitt et al 1977), and in such cases association with BiP is frequently observed (Bole et al

1986; Gething et al 1986; Dorner et al 1987; Hendershot et al 1988). Binding of BiP to all these substrates is hydrophobic in nature, which suggests that they are recognized, at least in part, by the presence of exposed hydrophobic residues that are buried in the mature, properly folded protein.

Although many of the known substrates for BiP are aberrant proteins that remain in the ER, secretory proteins can also interact with BiP before they achieve their mature state. For example, BiP binds transiently to at least some immunoglobulin heavy chains prior to their association with light chains (Haas & Wabl 1983; Bole et al 1986), and to several human serum glycoproteins expressed in CHO cells (Dorner et al 1987). With the recent cloning of the yeast BiP gene, it has become possible to test the importance of BiP in the normal process of protein transport to the Golgi complex. Preliminary results indicate that BiP is essential for viability, and that within 15 min of warming a temperature sensitive BiP mutant to the nonpermissive temperature, import of proteins into the ER ceases (Rose & Misra 1989; L. Moran, personal communication; M. Rose, personal communication). These results suggest that BiP interacts with a variety of nascent proteins and may be required for some of them to complete their translocation into the ER.

Action of BiP

BiP is closely related to the cytoplasmic heat shock protein hsp70 and is likely to act in a similar way (Munro & Pelham 1986). Like hsp70, BiP binds tightly to ATP and ATP hydrolysis allows the protein to be released from at least some of the substrates to which it binds (Munro & Pelham 1986; Dorner et al 1987; Kassenbrock et al 1988). A simple working model for the function of these proteins is that they act as reversible detergents: they bind to hydrophobic surfaces, but at intervals use the energy of ATP hydrolysis to change their conformation to a nonbinding state (Lewis & Pelham 1985; Pelham 1986). This would have the effect of maintaining unfolded proteins in solution without sequestering them permanently in a nonfunctional complex; the proteins could thus avoid aggregation or precipitation but still be able to achieve their final tertiary and quaternary structures.

That such a detergent-like role could be useful is shown by studies of the effects of detergents on protein folding in vitro. For example, the enzyme rhodanese is composed of two domains with hydrophobic surfaces that normally interact with each other, but during refolding they tend to bind to other molecules and form nonproductive aggregates, so that very little active enzyme is obtained. Addition of the detergent lauryl maltoside weakens these nonproductive interactions and greatly increases the yield of correctly refolded enzyme (Tandon & Horowitz 1986).

Although the main function of BiP is probably to promote protein assembly, it may also serve to prevent export of misfolded proteins from the ER; such proteins remain associated with BiP until they either fold correctly or are degraded. It is not clear whether BiP is required for the retention of aberrant proteins, because most of them would probably aggregate and thus fail to be secreted even if BiP were absent. However, Dorner et al (1988) reported that the secretion of an overexpressed derivative of human tissue plasminogen activator was improved when the level of BiP was specifically reduced by the expression of anti-sense RNA. In this case, at least, it seems that interaction with BiP can limit the export of a secretory protein from the ER.

Binding of Proteins to Other ER Residents

Another ER resident that interacts with unfolded proteins is the enzyme protein disulphide isomerase (P.D.I.), which catalyses thiol oxidation and disulphide exchange reactions (Freedman 1984). P.D.I. is a remarkable protein with several distinct roles. Besides existing as a free monomer, it is also an essential subunit of prolyl 4–hydroxylase, an enzyme that catalyses the modification of prolyl residues in newly synthesized collagen (Pihlajaniemi et al 1987). Furthermore, it binds to the Asn-X-Ser/Thr acceptor sequence for *N*-linked glycosylation and is an important component of the oligosaccharide transferase (Geetha-Habib et al 1988). All forms of P.D.I. are soluble, or only loosely associated with the ER membrane, and are released from isolated microsomes when these are ruptured by incubation at high pH. P.D.I.-depleted microsomes can still import γ-gliadin (a nonglycosylated wheat storage protein), but the protein cannot achieve its correct disulphide-bonded state: reconstitution of microsomes in the presence of purified P.D.I. restores this function (Bulleid & Freedman 1988). P.D.I. can also be cross-linked to nascent immunoglobulin chains in vivo (Roth & Pierce 1987). These results indicate that P.D.I. interacts directly with newly synthesized secretory proteins and is required for their correct folding.

Other proteins may also help to lubricate the process of protein assembly in the ER and, perhaps at the same time, prevent premature export. Possible examples include a membrane protein termed CD-ω or TRAP, which associates transiently with partially assembled T-cell receptor-CD3 complexes (Alarcon et al 1988; Bonifacino et al 1988); the collagen-binding protein colligin (Hughes et al 1987; Saga et al 1987); and the abundant luminal protein GRP94, which is related to the cytoplasmic heat shock protein hsp90 and, like BiP, is synthesized at higher rates when aberrant proteins accumulate in the ER (Sorger & Pelham 1987; Lee 1987).

SORTING OF RESIDENT ER PROTEINS FROM SECRETED PROTEINS

Membrane Versus Soluble Proteins

Resident proteins of the ER must avoid export by the bulk flow pathway. For membrane proteins it is easy to imagine that residence results from the same features that slow the export of mutant proteins, in particular the tendency to form large aggregates. They may also be held in position by interactions between their cytoplasmic domains and other cellular structures. For example, the "crystalloid" ER that forms in cells expressing high levels of the enzyme HMG-CoA reductase consists of membrane tubules that interact extensively with each other, presumably via the cytoplasmic domains of integral membrane proteins (Anderson et al 1983). Similarly, the inner membrane of the nuclear envelope contains glycoproteins that are associated with the nuclear lamina (Senior & Gerace 1988). Major components of the RER, including the ribophorins and the signal peptidase complex, form an aggregate that resists disruption by nonionic detergents (Crimaudo et al 1987). These components should be further cross-linked in vivo as a result of their association with polysomes and are probably unable to enter transport vesicles.

Whether all ER membrane proteins are associated with some large structure remains to be seen. Few interactions have been studied in detail, although some progress has been made. For example, in adenovirus-infected cells the class I histocompatibility antigens remain in the ER because they are bound to the adenovirus E19 protein (Burgert & Kvist 1985; Andersson et al 1985; Severinsson & Peterson 1985). Retention of E19 in the ER is mediated by its short cytoplasmic tail (Paabo et al 1987), but the protein that interacts with this tail has not been identified. Another well-studied example is the VP7 polypeptide of the rotavirus SA11. Retention of this protein in the ER is mediated by the first 60 amino acids of the mature protein, which can form an amphipathic helix (Poruchynsky & Atkinson 1988). Surprisingly, the VP7 signal sequence is also required, even though it is cleaved from the mature protein: replacement with the HA signal leads to cleavage at the identical site, but also to secretion of VP7 as a soluble protein (Stirzaker et al 1987; Stirzaker & Both 1989). It seems that the formation of a membrane anchor requires the interaction of the VP7 signal sequence with the *N* terminus of the mature protein. Although the precise way in which VP7 becomes associated with the membrane is obscure, it is clear that it is this association that is responsible for its ER location.

For truly soluble luminal proteins, retention represents a much greater conceptual problem. If such proteins spend at least part of their time as diffusible entities of small size, they will inevitably tend to be transported by the bulk flow pathway. A special mechanism must therefore exist to prevent their loss from the cell. The following sections outline the evidence for the existence of soluble resident proteins in the ER and for a specific retention mechanism that keeps them there.

Existence and Function of Luminal ER Proteins

Table 1 lists some currently known luminal ER proteins, several of which have been independently discovered and named by different groups. Many of them are thought to bind to or act on newly synthesized secretory proteins: these include P.D.I., BiP, GRP94, and colligin, and also glucosidase II, which trims glucose residues from the oligosaccharide side chains of glycoproteins. There are a number of carboxyesterases that probably play a role in the detoxification of aromatic compounds; these include the gut esterase of *C. elegans*, the 60-kd liver esterases listed in Table 1, and other liver esterases such as egasyn (Brown et al 1987) and E1 (Harano et al 1988). Other proteins have less obvious functions: ERp72 is known only as an ER protein, while reticulin is a calcium-binding protein (as are the acidic proteins P.D.I., BiP and GRP94, at high calcium concentrations), and perhaps plays a role analogous to that of calsequestrin in the sarcoplasmic reticulum (Macer & Koch, 1988).

Solubility of the Luminal ER Proteins

The evidence that most of the proteins listed in Table 1 are not membrane-associated is of three main types. First, they are soluble in the absence of detergents, or in the presence of detergent concentrations that are too low to solubilize true integral membrane proteins (Strous & van Kerkhof 1989; Koch et al 1988; Bulleid & Freedman 1988; Bole et al 1986; Brown et al 1987). Second, glucosidase II, GRP94, P.D.I. and BiP have been localized by immunogold labeling at the EM level; they are not concentrated at the membrane but appear spread evenly across the lumen of the ER (Lucocq et al 1986; Koch et al 1988; Akagi et al 1988; J. Tooze, personal communication). Third, all of them apart from glucosidase II have been sequenced and they show no obvious transmembrane segments (see Table 1 for references). One report has suggested the presence of a single transmembrane segment in GRP94 (ERp99; Mazzarella & Green 1987). However, the 21-amino acid segment in question contains two charged residues and is part of an extended region of homology between GRP94 and the soluble protein hsp90. Given the other properties of GRP94, there seems no reason to believe that it spans the membrane (Koch et al 1988).

These arguments do not exclude the possibility that the proteins are attached to some immobile matrix in the ER lumen. This point has been

Table 1 Some luminal ER proteins

Protein	Other names	Species	*C* terminus	Refs.[e]
P.D.I.		rat	QKAV KDEL	1
	ERp59	mouse	QKAV KDEL	2
	Prolyl 4-hydroxylase[a]	human	QKAV KDEL	3
	T3BP[b]	cow	QKAV KDEL	4
	GSBP[c]	chick	QKAM KDEL	5
BiP	GRP78	rat	DTSE KDEL	6
		mouse	DTSE KDEL	7
		hamster	DTSE KDEL	8
		human	DTAE KDEL	9
		chick	EAAE KDEL	10
		Drosophila	DADL KDEL	11
		C. elegans[d]	DLDD KDEL	12
		C. elegans[d]	PSED HDEL	13
		tomato	EDDS HDEL	14
		S. cerevisiae	DYFE HDEL	15
		P. falciparum	EDVD SDEL	16
GRP94		hamster	STAE KDEL	17
	ERp99, endoplasmin	mouse	ESTE KDEL	18, 19
	hsp108	chick	STDV KDEL	20
Colligin	47 kd protein, gp46	rat	KM RDEL	21–23
ERp72		mouse	RSRT KEEL	24
Gut esterase		*C. elegans*	HSSN KDEL	25
60 kd esterase 1		rabbit	RETE HIEL	26
60 kd esterase 2		rabbit	HTEL	27
Auxin binding protein		*Zea mays*	FEAA KDEL	28
Reticulin	calregulin	rabbit	RRQA KDEL	29, 30
	CRP55	mouse	PAQA KDEL	31
Glucosidase II		mammals		32

[a] Prolyl 4-hydroxylase is a soluble $\alpha_2\beta_2$ tetramer; the β subunit is P.D.I.
[b] Thyroid hormone binding protein.
[c] Glycosylation site binding protein.
[d] *Caenorhabditis elegans* has two genes that encode BiP-like proteins.
[e] References: 1. Edman et al 1985, 2. Srinivasan et al 1988, 3. Pihlajaniemi et al 1987, 4. Yamauchi et al 1987, 5. Geetha-Habib et al 1988, 6. Munro & Pelham 1986, 7. Haas & Meo 1988, 8. Ting et al 1987, 9. Ting & Lee 1988, 10. Stoeckle et al 1988, 11. D. Rubin & K. Palter pers. commun., 12. Heschl & Baillie 1989, 13. M. Heschl & D. Baillie pers. commun., 14. A. Bennett pers. commun., 15. Rose & Misra 1989, 16. Kumar et al 1988, 17. Sorger & Pelham 1987, 18. Mazzarella & Green 1987, 19. Smith & Koch 1987, 20. Kulomaa et al 1986, 21. Kurkinen et al 1984, 22. Saga et al 1987, 23. D. Nandan, E. Zeuner pers. commun., 24. M. Green pers. commun., 25. J. McGhee pers. commun., 26. Korza & Ozols 1988, 27. Ozols 1988, 28. D. Klambt pers. commun., 29. Opas et al 1988, 30. Khanna et al 1987, 31. Macer & Koch 1988, M. Smith & G. Koch pers. commun., 32. Strous et al 1987.

addressed directly in the case of BiP. Ceriotti & Colman (1988) injected mRNA encoding rat BiP into one pole of *Xenopus* oocytes, and pulse-labeled the protein synthesized from it. They showed that BiP could diffuse (presumably within the ER) to the other side of the oocyte at a rate comparable to that of ovalbumin (a secretory protein), whereas ER membrane proteins labeled in the same way remained close to their site of synthesis. Despite the comparable mobilities of ovalbumin and BiP, only ovalbumin was secreted, as expected if rat BiP is recognized by the ER localization mechanism of the oocyte.

A Common Signal on Soluble ER Proteins

The first indication that a specific mechanism exists for the retention of these proteins was the observation that P.D.I. and BiP share a common *C* terminal tetrapeptide, KDEL (Munro & Pelham 1986). Subsequent cloning of luminal ER proteins from a wide range of species has shown that this sequence, or a closely related one such as HDEL, RDEL, or KEEL, is almost invariably present, whereas amino acids upstream of this tetrapeptide are not as closely related (Table 1). A possible exception is esterase E1 from rat liver, which has been reported to have SVL (or AVL) at the *C* terminus (Harano et al 1988). However, although it does not have the properties of a transmembrane protein, this esterase is not truly soluble and remains associated with microsomes under conditions that allow release of other luminal proteins (Harano et al 1988). Two 60-kd esterases from rabbit liver also have somewhat divergent sequences (HIEL and HTEL, Table 1); whether these are membrane-associated is unclear—at least one of them has been reported to be soluble only in the presence of detergents (Ozols 1988). Finally, a possible case of evolutionary divergence is represented by the presumptive BiP coding sequence from *Plasmodium falciparum*, which ends with SDEL (Kumar et al 1988). Despite these minor anomalies, it is clear from Table 1 that a similar *C* terminal sequence exists on a number of otherwise unrelated ER proteins, which suggests that this sequence may serve as a targeting signal.

Proof that the *C* terminal sequence is an essential part of the retention mechanism was provided by experiments in which altered proteins were expressed in COS cells. Removal of KDEL from BiP caused it to be secreted, albeit slowly. Conversely, addition of the sequence to lysozyme, a secretory protein, or cathepsin D, a lysosomal enzyme, caused these proteins to accumulate in the ER (Munro & Pelham 1987; Pelham 1988). Changing the KDEL sequence to KDAS or to KDELGL also resulted in secretion. These results strongly suggest that retention is mediated by a specific receptor that recognizes KDEL only when it is present at the extreme *C* terminus.

Function of KDEL: A Retrieval Mechanism

Although a simple way for KDEL-containing proteins to be retained in the ER would be for them to be permanently anchored to an immobilized membrane receptor or luminal matrix, the data argue against such a mechanism: the proteins are not membrane-associated, the presence or absence of KDEL does not affect the rate of BiP diffusion within the ER of *Xenopus* oocytes (Ceriotti & Colman 1988), and there is no candidate for a receptor of sufficient abundance to bind all the proteins stoichiometrically. Weak interactions with membrane proteins may slow their transport, but if the proteins are in solution even for part of the time, it seems inevitable that they will leave the ER in the general bulk flow.

A possible solution to this problem is for escaped resident proteins to be specifically and continuously retrieved from a subsequent compartment and returned to their correct location (Munro & Pelham 1987). Such a process would be entirely analogous to the sorting that occurs during endocytosis, or during delivery of lysosomal enzymes to their destination. The sorting compartment would have an internal environment (pH, salt concentration) that differs in some way from that of the ER. Binding of the KDEL sequence to a membrane receptor would occur in this environment, and the receptor/ligand complex would return to the ER by vesicular transport, where the KDEL-bearing protein would be released.

Evidence in favor of this recycling model comes from an experiment in which the KDEL sequence was attached to the lysosomal enzyme cathepsin D (Pelham 1988). Like other lysosomal enzymes, cathepsin D is modified by the addition of phosphate groups to mannose residues in its oligosaccharide side chains. This occurs in two steps: a phosphotransferase adds GlcNac-P to the mannose, and then the GlcNac residue is removed. There is good biochemical and kinetic evidence that the phosphotransferase resides not in the ER, but in a smooth-membraned compartment close to, but possibly distinct from, the *cis* Golgi; removal of the GlcNac residue probably occurs in a slightly later compartment (reviewed by Kornfeld & Kornfeld 1985; von Figura & Hasilik 1986).

The KDEL-containing cathepsin D was not transported to lysosomes, but accumulated in the ER as expected. However, it could still be labeled with ^{32}P and thus could reach the phosphotransferase. Labeling was blocked at low temperature, which supports the idea that transport out of the ER is necessary for the cathepsin D to reach the modifying enzyme (Pelham 1988). Removal of GlcNac from the phosphodiester was not observed. Thus it seems that luminal ER proteins can leave the ER, but do not penetrate far into the Golgi stack before returning. The latter conclusion is supported by studies of the oligosaccharides on GRP94, glucosidase II and colligin: these are of the high mannose type and do not appear to have been exposed to the Golgi mannosidase I, an enzyme regarded as a *cis*-Golgi marker (Lewis et al 1985; Strous et al 1987; Hughes et al 1987).

There is also an example of a lysosomal enzyme that is naturally held in the ER. Some of the β-glucuronidase of liver cells is found in a complex with a protein called egasyn, which has recently been shown to be one of the soluble esterases of the ER (Brown et al 1987). As with the KDEL-tagged cathepsin D, this ER form of β-glucuronidase has oligosaccharides containing phosphodiesters, but not phosphomonoesters (Mizuochi et al 1981). The simplest interpretation of this result is that egasyn is retained by the KDEL system and that, together with its associated β-glucuronidase, it can reach a compartment containing the phosphotransferase. At present, however, the *C* terminal sequence of egasyn is not known.

Likely Site of Sorting: The Salvage Compartment

The results described above suggest that the separation of resident ER proteins from secretory proteins occurs in a salvage compartment (Warren 1987) that contains the phosphotransferase but lacks N-acetylglucosaminidase and the *cis* Golgi marker mannosidase I (see Figure 1). For the retrieval model to be correct, this compartment must not be contiguous with the RER, because it has to have a distinct luminal environment that promotes the binding of KDEL to its receptor.

Interestingly, incubation of a macrophage cell line at 15°C, a temperature that is known to inhibit ER-Golgi transport, blocks N-acetylglucosaminidase action on newly synthesized lysosomal enzymes, but does not prevent the initial phosphotransferase step (Lazzarino & Gabel 1988). This suggests that exit from the salvage compartment is inhibited at this temperature, but entry to it is not. Saraste & Kuismanen (1984) found that incubation of Semliki Forest virus-infected BHK cells at 15° caused large vesicles containing viral glycoproteins to accumulate between ER and Golgi; after shifting to a higher temperature, these proteins were rapidly transferred to the Golgi stack, which suggests that the vesicles were derived from a normal intermediate of the secretory pathway. Low temperature incubation of other cell types also causes a proliferation of vesicles and tubules in this region (Saraste et al 1986; Tartakoff 1986), and it seems reasonable to speculate that the compartment that hypertrophies under these conditions is the salvage compartment.

Examination of cells incubated at normal growth temperatures has revealed a compartment between the transitional elements of the ER, where vesicle budding is often assumed to occur, and the *cis* face of the Golgi stack. This compartment is seen in electron micrographs as a

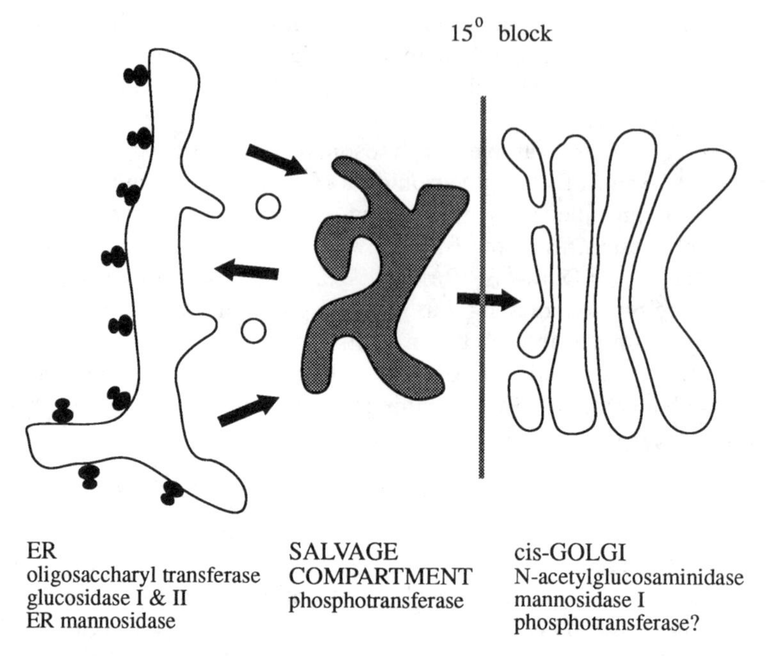

Figure 1 The figure depicts the transitional region of the ER, the salvage compartment (shaded) and the Golgi stack, and the proposed location of various marker enzymes is indicated. The phosphotransferase and N-acetylglucosaminidase act on lysosomal enzymes to generate the mannose-6-phosphate marker. The results of Lazzarino and Gabel (1988) suggest that phosphotransferase activity may be present in the *cis*-Golgi as well as in the salvage compartment.

collection of vesicles and/or tubules and is likely to be the normal form of the structure that is enlarged at 15°. It appears to be topologically distinct from both the Golgi stack and the ER and it lacks the ER marker glucose 6-phosphatase (Tooze et al 1984). In coronavirus-infected cells, the initial budding of viral particles is exclusively into this structure, and O-glycosylation of the viral E1 glycoprotein probably occurs there (Tooze et al 1988).

If this compartment is the site of sorting of luminal ER proteins, one might expect to detect such proteins in it, although if retrieval is efficient (and exit from the RER is slow), they may only be present at a low concentration. Immunoperoxidase localization of BiP showed staining of the ER and transitional elements, but not of the smooth vesicles between the transitional elements and Golgi (D. Bole, personal communication). However, glucosidase II has been detected in such structures by immunogold labeling, but not in the Golgi itself (Lucocq et al 1986). This supports the idea that soluble ER proteins are sorted in these vesicles, although it has not yet been shown that glucosidase II contains a KDEL sequence.

Recently, membrane proteins that are possible residents of the salvage compartment have been identified. Saraste et al (1987) described a 58 kd glycoprotein that by immunoelectron microscopy was located to the *cis* face of the Golgi and tubules and vesicles associated with it, and was also occasionally seen in the adjacent transitional portions of the RER. Schwiezer et al (1988) described a different, nonglycosylated protein that exists as a hexamer with a subunit molecular weight of 53 kd. This protein was found to be concentrated almost exclusively in smooth vesicles and tubules between ER and Golgi.

The identification of such marker proteins should allow a correlation to be made between morphological and cell fractionation studies. A low density vesicle fraction derived from VSV-infected cells with the characteristics expected for an intermediate between ER and Golgi has been described (Lodish et al 1987). However, these vesicles were not pure and their detailed biochemical characterization still represents a formidable challenge.

Retention of Soluble ER Proteins in Saccharomyces Cerevisiae

One way to identify important components of a biochemically intractable system is to make use of the powerful genetics of the yeast *S. cerevisiae*. This approach has identified a number of genes whose products are involved in the secretory process, including several that are required for ER to Golgi transport (Schekman 1985). Initial studies indicate that the sorting system for soluble ER proteins in yeast is similar to that in animal cells and thus genetic studies of this organism may be of general use in analyzing the mechanisms involved.

BiP in *S. cerevisiae* has the *C* terminal sequence HDEL (L. Moran, M.-J. Gething, personal communications; Rose & Misra 1989), and antibodies specific for this sequence detect at least two other proteins that cofractionate with the ER; thus HDEL, rather than KDEL, is probably the preferred ER signal in this organism (M. Lewis, K. Hardwick, J. Semenza, H. Pelham, unpublished observations). In support of this assumption, it was found that fusion of the last six amino acids of yeast BiP, via an appropriate spacer, to the secretory protein invertase resulted in retention of the invertase within the yeast cells (Pelham et al 1988). As with KDEL in animal cells, the HDEL sequence had to be at the extreme *C* terminus to function. KDEL itself it not an efficient retention signal in yeast, and conversely HDEL functions poorly in animal cells.

There is also evidence for recycling of HDEL-containing proteins from a post-ER compartment. Yeast glycoproteins such as invertase receive the same oligosaccharide side chains in the ER as do animal cell glycoproteins. Some of these are then modified by the addition of 100 or more mannose residues. This "outer chain" modification is regarded as a Golgi function because it is blocked at the appropriate temperature in *ts* mutants that are defective in transport from the ER (Schekman 1985). Invertase fusion proteins bearing the HDEL signal are still subject to this modification, as is at least one endogenous HDEL-containing glycoprotein, which suggests that such proteins can reach the Golgi and still be retrieved; however, kinetic studies indicate that outer chain glycosylation occurs much more slowly with HDEL-tagged proteins than with a similar protein lacking the HDEL sequence (Pelham et al 1988; K. Hardwick, N. Dean, H. Pelham, unpublished observations). Exposure of ER proteins to the modifying enzymes is probably a rare event; as with animal cells, retrieval may normally be from a pre-Golgi or very early Golgi compartment.

Mutants

Powerful selection and screening methods have been developed for the detection of rare mutant cells that secrete invertase fusion proteins. Use of these methods has allowed the isolation of mutants that fail to retain HDEL-tagged invertase in the ER (Pelham et al 1988). Many of the mutants also secrete their endogenous ER proteins, but do not mislocalize vacuolar proteins. They thus appear to be specifically deficient in the HDEL-dependent sorting pathway.

Mutants with a strong phenotype fall into two complementation groups and the corresponding genes have been cloned. One of these (termed *ERD1*, for ER retention defective) encodes a 43 kd membrane protein that is not required for normal growth or secretion of invertase (M. Lewis, J. Semenza, K. Hardwick, H. Pelham, unpublished observations). The precise functions of this protein and of the *ERD2* gene product remain to be determined, but they are obvious candidates for the putative HDEL receptor.

IMPLICATIONS OF BULK FLOW AND RETRIEVAL

The concept of bulk flow of the ER contents to the Golgi appears simple, but, as pointed out by Wieland et al (1987), it raises complex questions about the mechanisms involved. Nonselective transport implies a massive flux of vesicles: if the halftime for export is 15 min then an area of membrane roughly equivalent to the entire ER must reach the next compartment every 30 min (because the maximum rates of transport of membrane and soluble components are approximately equal, transport vesicles must have about the same surface-to-volume ratio as the ER). Since a typical tissue culture cell (BHK) has an ER surface area about six times that of the plasma membrane (Griffiths et al 1984), and the plasma membrane only doubles in area over a period of some 18 h the vast majority of the membrane lipid must be returned rapidly to the ER. This argument holds even if it is assumed that the bulk flow rate has been overestimated, and is closer to the slowest rate of protein export from the ER ($t_{1/2} \sim 3$ h).

The retrieval of lipid from the secretory pathway is accompanied by a significant concentration of the transported proteins. Semliki Forest virus glycoproteins are about eight-fold more concentrated in the Golgi than in the ER and a similar effect can be seen with rapidly secreted soluble proteins such as lysozyme (Quinn et al 1984; Munro & Pelham 1987). The first compartment in which an increased density of transported proteins can be detected is the putative salvage compartment (Saraste & Kuismanen 1984; Tooze et al 1984, 1988; Copeland et al 1988); once they reach the Golgi stack, proteins appear to be transported between cisternae without further concentration (Orci et al 1986). Thus it seems likely that most of the membrane lipid that leaves the ER is retrieved from the salvage compartment.

I have argued that KDEL-bearing proteins return from the salvage compartment to the ER by receptor-mediated vesicular transport. In principle, these vesicles could be the route by which membrane lipids return to the ER. But if this is so, it is hard to see how the vectorial bulk flow is maintained—proteins that can nonselectively enter vesicles would be transported equally well in each direction. The return flow, at least, must be selective in nature. By analogy with the process of endocytosis (Pearse & Bretscher 1981), selectivity could be achieved if the rare proteins that were destined to return to the ER (such as the KDEL receptor and its attached ligands) were concentrated in a small number of vesicles, thereby excluding most of the rest of the luminal and membrane proteins. This would leave a large amount of lipid that must somehow return to the ER without being accompanied by any luminal or membrane proteins, a mechanistic problem whose solution can, in the absence of any relevant data, only be imagined.

One ingenious solution was suggested by Wieland et al (1987), who proposed that the excess lipid returns not in vesicles but as single phospholipid molecules carried by proteins. A number of proteins that can transport lipid molecules have been identified (Spener & Wirtz 1985), and it is known that phospholipids can equilibrate between intracellular membranes with halftimes as short as 1 min (for review see Bishop & Bell

1988), but how the required directionality of lipid transport could be maintained is unclear. Another potential problem concerns the energy requirements: efficient transport would probably require energy, and if transfer of single lipid molecules is coupled to ATP hydrolysis, then at least one ATP molecule would be required for each. A rough calculation, assuming the halftime for transport from the ER is 20 min suggests that the amount of energy consumed by this process in a fibroblast would be comparable to the energy used for polypeptide synthesis from amino acids. This is a substantial but not impossible expense; however, in a specialized cell such as the hepatocyte, which has a much larger area of ER membrane, the energy required to maintain the same flow rate would be considerably greater.

Other mechanisms can easily be imagined. For example, some form of molecular sieve that allowed lipid vesicles to bud without incorporating any membrane or soluble proteins would solve the problem. A difficult task for the future is to design experiments that can test such hypotheses.

PROSPECTS

It is now generally accepted that a default pathway exists for the transport of secretory and membrane proteins from the ER to the Golgi. Although most membrane proteins of the RER may never enter this pathway, it seems very likely that some soluble ER proteins, as well as membrane lipids, do leave the RER and have to be returned to it in a selective manner. The mechanistic details of this process are obscure, but recent progress in this area is encouraging. Cell-free systems that show efficient transport from ER to Golgi have been developed from both yeast and animal cells, and the mechanics of vesicular transport are beginning to be unraveled through a combination of biochemistry and genetics (for reviews see Gruenberg & Howell, this volume; Bourne 1988). These approaches should allow models of the retrieval mechanism to be tested directly and lead to a much deeper understanding of the sorting processes that occur between the RER and the Golgi stack.

Acknowledgments

I would like to thank Mark Bretscher and Barbara Pearse for helpful criticisms of this review and the many colleagues at the MRC Laboratory of Molecular Biology with whom I have had stimulating discussions. I also thank all those who have communicated with me and allowed me to cite their unpublished results.

See note added in proof on p. 23.

Literature Cited

Akagi, S., Yamamoto, A., Yoshimori, T., Masaki, R., Ogawa, R., Tashiro, Y. 1988. Subcellular distribution of protein disulphide isomerase in rat hepatocytes and exocrine pancreatic cells. *P9.8.3* (*Abstr.*) *4th Int. Congr. Cell Biol., Montreal*

Alarcon, B., Berkhout, B., Breitmeyer, J., Terhorst, C. 1988. Assembly of the human T cell receptor-CD3 complex takes place in the endoplasmic reticulum and involves intermediary complexes between the CD3-$\gamma.\delta.\varepsilon$ core and single T cell receptor α and β chains. *J. Biol. Chem.* 263: 2953–61

Anderson, R. G. W., Orci, L., Brown, M. S., Garcia-Segura, L. M., Goldstein, J. L. 1983. Ultrastructural analysis of crystalloid endoplasmic reticulum in UT-1 cells and its disappearance in response to cholesterol. *J. Cell Sci.* 63: 1–20

Andersson, M., Paabo, S., Nilsson, T., Peterson, P. A. 1985. Impaired intracellular transport of class I MHC antigens as a possible means for adenoviruses to evade immune surveillance. *Cell* 43: 215–22

Bishop, W. R., Bell, R. M. 1988. Assembly of phospholipid into cellular membranes: biosynthesis, transmembrane movement and intracellular translocation. *Annu. Rev. Cell Biol.* 4: 579–610

Bole, D. G., Hendershot, L. M., Kearney, J. F. 1986. Posttranslational association of immunoglobulin heavy-chain binding-protein with nascent heavy chains in non-secreting and secreting hybridomas. *J. Cell Biol.* 102: 1558–66

Bonifacino, J. S., Lippincott-Schwartz, J., Chen, C., Antusch, D., Samelson, L. E., Klausner, R. D. 1988. Association and dissociation of the murine T cell receptor associated protein (TRAP): early events in the biosynthesis of a multisubunit receptor. *J. Biol. Chem.* 263: 8965–71

Bourne, H. R. 1988. Do GTPases direct membrane traffic in secretion? *Cell* 53: 669–71

Brown, J., Novak, E. K., Takeuchi, K., Moore, K., Medda, S., Swank, R. T. 1987. Lumenal location of the microsomal β-glucuronidase-egasyn complex. *J. Cell Biol.* 105: 1571–78

Bulleid, N. J., Freedman, R. B. 1988. Defective co-translational formation of disulphide bonds in protein disulphide-isomerase-deficient microsomes. *Nature* 335: 649–51

Burgert, H.-G., Kvist, S. 1985. An adenovirus type 2 glycoprotein blocks cell surface expression of human histocompatibility class I antigens. *Cell* 41: 987–97

Burgess, T. L., Kelly, R. B. 1987. Constitutive and regulated secretion of proteins. *Annu. Rev. Cell Biol.* 3: 243–93

Ceriotti, A., Colman, A. 1988. Binding to membrane proteins within the endoplasmic reticulum cannot explain the retention of the glucose-regulated protein GRP78 in *Xenopus* oocytes. *EMBO J.* 7: 633–38

Copeland, C. S., Zimmer, K.-P., Wagner, K. R., Healey, G. A., Mellman, I., Helenius, A. 1988. Folding, trimerization and transport are sequential events in the biogenesis of influenza virus haemagglutinin. *Cell* 53: 197–209

Crimaudo, C., Hortsch, M., Gausepohl, H., Meyer, D. I. 1987. Human ribophorins I and II: the primary structure and membrane topology of two highly conserved rough endoplasmic reticulum-specific glycoproteins. *EMBO J.* 6: 75–82

Cutler, D. F. 1988. The role of transport signals and retention signals in constitutive export from animal cells. *J. Cell Sci.* 91: 1–4

Doms, R. W., Ruusala, A., Machamer, C., Helenius, J., Helenius, A., Rose, J. K. 1988. Differential effects of mutations in three domains on folding, quaternary structure, and intracellular transport of vesicular stomatitis virus G protein. *J. Cell Biol.* 107: 89–99

Dorner, A. J., Bole, D. G., Kaufman, R. J. 1987. The relationship of N-linked glycosylation and heavy chain-binding protein association with the secretion of glycoproteins. *J. Cell Biol.* 105: 2665–74

Dorner, A. J., Krane, M. G., Kaufman, R. J. 1988. Reduction of endogenous GRP78 levels improves secretion of a heterologous protein in CHO cells. *Mol. Cell. Biol.* 8: 4063–70

Doyle, C., Sambrook, J., Gething, M.-J. 1986. Analysis of progressive deletions in the transmembrane and cytoplasmic domains of influenza hemagglutinin. *J. Cell Biol.* 103: 1193–1204

Edman, J. C., Ellis, L., Blacher, R. W., Roth, R. A., Rutter, W. J. 1985. Sequence of protein disulphide isomerase and implications of its relationship to thioredoxin. *Nature* 317: 267–70

Farquhar, M. G. 1985. Progress in unravelling pathways of Golgi traffic. *Annu. Rev. Cell Biol.* 1: 447–88

Freedman, R. B. 1984. Native disulphide bond formation in protein biosynthesis: evidence for the role of protein disulphide isomerase. *Trends Biochem. Sci.* 9: 438–41

Fries, E., Gustafsson, L., Peterson, P. A. 1984. Four secretory proteins synthesized by hepatocytes are transported from endoplasmic reticulum to Golgi complex at different rates. *EMBO J.* 3: 147–52

Geetha-Habib, M., Noiva, R., Kaplan, H. A., Lennarz, W. J. 1988. Glycosylation site binding protein, a component of oligosaccharyl transferase, is highly similar to three other 57 kd luminal proteins of the ER. *Cell* 1053–60

Gething, M.-J., McCammon, K., Sambrook, J. 1986. Expression of wild-type and mutant forms of influenza hemagglutinin: the role of protein folding in intracellular transport. *Cell* 46: 939–50

Griffiths, G., Warren, G., Quinn, P., Mathien-Costello, O., Hoppeler, H. 1984. Density of newly-synthesized plasma membrane proteins in intracellular membranes. I. Stereological studies. *J. Cell Biol.* 98: 2133–41

Haas, I., Meo, T. 1988. cDNA cloning of the immunoglobulin heavy chain binding protein. *Proc. Natl. Acad. Sci. USA* 85: 2250–54

Haas, I. G., Wabl, M. 1983. Immunoglobulin heavy chain binding protein. *Nature* 306: 387–89

Harano, T., Miyata, T., Lee, S., Aoyagi, H., Omura, T. 1988. Biosynthesis and localization of rat liver microsomal carboxyesterase E1. *J. Biochem. (Jpn.)* 103: 149–55

Hendershot, L. M., Ting, J., Lee, A. S. 1988. Identity of the immunoglobulin heavy-chain-binding protein with the 78,000-dalton glucose-regulated protein and the role of posttranslational modifications in its binding function. *Mol. Cell. Biol.* 8: 4250–56

Heschl, M. F. P., Baillie, D. L. 1989. Characterization of the hsp70 multigene family in *Caenorhabditis elegans*. *DNA*. In press

Hughes, R. C., Taylor, A., Sage, H., Hogan, B. L. M. 1987. Distinct patterns of glycosylation of colligin, a collagen-binding glycoprotein, and SPARC (osteonectin), a secreted Ca^{++}-binding glycoprotein: evidence for the localization of colligin in the endoplasmic reticulum. *Eur. J. Biochem.* 163: 57–63

Kassenbrock, C. K., Garcia, P. D., Walter, P., Kelly, R. B. 1988. Heavy chain binding protein recognizes aberrant polypeptides translated in vitro. *Nature* 333: 90–93

Khanna, N. C., Tokuda, M., Waisman, D. M. 1987. Comparison of calregulins from vertebrate livers. *Biochem. J.* 242: 245–51

Koch, G. L. E., Macer, D. R. J., Wooding, F. B. P. 1988. Endoplasmin is a reticuloplasmin. *J. Cell Sci.* 90: 485–91

Kornfeld, R., Kornfeld, S. 1985. Assembly of asparagine-linked oligosaccharides. *Annu. Rev. Biochem.* 54: 631–34

Korza, G., Ozols, J. 1988. Complete covalent structure of 60-kDa esterase isolated from 2,3,7,8-tetrachlorodibenzo-*p*-dioxin-induced rabbit liver microsomes. *J. Biol. Chem.* 263: 3486–95

Kozutsumi, Y., Segal, M., Normington, K., Gething, M.-J., Sambrook, J. 1988. The presence of malfolded proteins in the endoplasmic reticulum signals the induction of glucose-regulated proteins. *Nature* 332: 462–64

Kulomaa, M. S., Weigel, N. L., Kleinsek, D. A., Beattie, W. G., Conneely, O. M., et al. 1986. Amino acid sequence of a chicken heat shock protein derived from the complementary DNA nucleotide sequence. *Biochemistry* 25: 6244–51

Kumar, N., Syin, C., Carter, R., Quakyi, I., Miller, L. H. 1988. *Plasmodium falciparum* gene encoding a protein similar to the 78-kDa rat glucose-regulated stress protein. *Proc. Natl. Acad. Sci. USA* 85: 6277–81

Kurkinen, M., Taylor, A., Garrels, J., Hogan, B. L. M. 1984. Cell surface-associated proteins which bind native type IV collagen or gelatin. *J. Biol. Chem.* 259: 5915–22

Lazzarino, D. A., Gabel, C. A. 1988. Biosynthesis of the mannose 6-phosphate recognition marker in transport-impaired mouse lymphoma cells: demonstration of a two-step phosphorylation. *J. Biol. Chem.* 263: 10118–26

Leavitt, R., Schlesinger, S., Kornfeld, S. 1977. Impaired intracellular migration and altered solubility of nonglycosylated glycoproteins of vesicular stomatitis virus and sindbis virus. *J. Biol. Chem.* 252: 9018–23

Lee, A. S. 1987. Coordinated regulation of a set of genes by glucose and calcium ionophores in mammalian cells. *Trends Biochem. Sci.* 12: 20–23

Lewis, M. J., Pelham, H. R. B. 1985. Involvement of ATP in the nuclear and nucleolar functions of the 70-kd heat shock protein. *EMBO J.* 4: 3137–43

Lewis, M. J., Turco, S. J., Green, M. 1985. Structure and assembly of the endoplasmic reticulum: biosynthetic sorting of endoplasmic reticulum proteins. *J. Biol. Chem.* 260: 6926–28

Lodish, H. F. 1988. Transport of secretory and membrane glycoproteins from the rough endoplasmic reticulum to the Golgi. *J. Biol. Chem.* 263: 2107–10

Lodish, H. F., Kong, N., Hirami, S., Rasmussen, J. 1987. A vesicular intermediate in the transport of hepatoma secretory proteins from the rough endoplasmic reticulum to the Golgi complex. *J. Cell Biol.* 104: 761–67

Lodish, H. F., Kong, N., Snider, M., Strous, G. J. A. M. 1983. Hepatoma secretory proteins migrate from rough endoplasmic reticulum to Golgi at characteristic rates. *Nature* 304: 80–83

Lucocq, J. M., Brada, D., Roth, J. 1986. Immunolocalization of the oligosaccharide trimming enzyme glucosidase II. *J. Cell Biol.* 102: 2137–46

Macer, D. R. J., Koch, G. L. E. 1988. Identification of a set of calcium-binding proteins in reticuloplasm, the luminal content of the endoplasmic reticulum. *J. Cell Sci.* 91: 61–70

Mazzarella, R. A., Green, M. 1987. ERp99, an abundant, conserved glycoprotein of the endoplasmic reticulum, is homologous to the 90-kDa heat shock protein (hsp90) and the 94-kDa glucose regulated protein (GRP94). *J. Biol. Chem.* 260: 6926–31

Mizuochi, T., Nishimura, Y., Kato, K., Kobata, A. 1981. Comparative studies of asparagine-linked oligosaccharide structures of rat liver microsomal and lysosomal β-glucuronidases. *Arch. Biochem. Biophys.* 209: 298–303

Munro, S., Pelham, H. R. B. 1986. An hsp70-like protein in the ER: identity with the 78 kd glucose-regulated protein and immunoglobulin heavy chain binding protein. *Cell* 46: 291–300

Munro, S., Pelham, H. R. B. 1987. A C-terminal signal prevents secretion of luminal ER proteins. *Cell* 48: 899–907

Opas, M., Fliegel, L., Michalak, M. 1988. Reticulin, a 55-kDa Ca-binding protein is present in endoplasmic reticulum of many cell types. *P9.8.2* (*Abstr.*) *4th Int. Congr. Cell Biol., Montreal*

Orci, L., Glick, B. S., Rothman, J. E. 1986. A new type of coated vesicular carrier that appears not to contain clathrin: its possible role in protein transport within the Golgi stack. *Cell* 46: 171–84

Ozols, J. 1987. Isolation and characterization of a 60-kilodalton glycoprotein esterase from liver microsomal membranes. *J. Biol. Chem.* 262: 15316–21

Ozols, J. 1988. Covalent structure relationships in two microsomal luminal carboxyesterases. *J. Cell Biol.* 107: 772a (Abstr.)

Paabo, S., Bhat, B. M., Wold, W. S. M., Peterson, P. A. 1987. A short sequence in the COOH-terminus makes an adenovirus membrane glycoprotein a resident of the endoplasmic reticulum. *Cell* 50: 311–17

Palade, G. 1975. Intracellular aspects of the process of protein synthesis. *Science* 189: 347–58

Pearse, B. M. F., Bretscher, M. S. 1981. Membrane recycling by coated vesicles. *Annu. Rev. Biochem.* 50: 85–101

Pelham, H. R. B. 1986. Speculations on the functions of the major heat shock and glucose regulated proteins. *Cell* 46: 959–61

Pelham, H. R. B. 1988. Evidence that luminal ER proteins are sorted from secreted proteins in a post-ER compartment. *EMBO J.* 7: 913–18

Pelham, H. R. B., Hardwick, K. G., Lewis, M. J. 1988. Sorting of soluble ER proteins in yeast. *EMBO J.* 7: 1757–62

Pfeffer, S. R., Rothman, J. E. 1987. Biosynthetic protein transport and sorting by the endoplasmic reticulum and Golgi. *Annu. Rev. Biochem.* 56: 829–52

Pihlajaniemi, T., Helaakoski, T., Tasanen, K., Myllyla, R., Huhtala, M.-L. 1987. Molecular cloning of the β-subunit of human prolyl 4-hydroxylase. This subunit and the protein disulphide isomerase are products of the same gene. *EMBO J.* 6: 643–49

Poruchynsky, M. S., Atkinson, P. H. 1988. Primary sequence domains required for the retention of rotavirus VP7 in the endoplasmic reticulum. *J. Cell Biol.* 107: 1697–1706

Poruchynsky, M. S., Tyndall, C., Both, G. W., Sato, F., Bellamy, A. R., Atkinson, P. A. 1985. Deletions into an NH_2-terminal hydrophobic domain result in secretion of rotavirus VP7, a resident endoplasmic reticulum membrane glycoprotein. *J. Cell Biol.* 101: 2199–2209

Quinn, P., Griffiths, G., Warren, G. 1984. Density of newly-synthesized plasma membrane proteins in intracellular membranes. II. Biochemical studies. *J. Cell Biol.* 98: 2142–47

Rose, J. K., Doms, R. W. 1988. Regulation of protein export from the endoplasmic reticulum. *Annu. Rev. Cell Biol.* 4: 257–88

Rose, M. D., Misra, L. D. 1989. KAR2, a karyogamy gene, is the yeast homolog of mammalian BiP/GRP78. *Cell*. In press

Roth, J. 1987. Subcellular organization of glycosylation in mammalian cells. *Biochim. Biophys. Acta* 906: 405–36

Roth, M. A., Pierce, S. B. 1987. In vivo cross-linking of protein disulphide isomerase to immunoglobulins. *Biochemistry* 26: 4179–82

Rothman, J. E. 1987. Protein sorting by selective retention in the endoplasmic reticulum and Golgi stack. *Cell* 50: 521–22

Saga, S., Nagata, K., Chen, W.-T., Yamada, K. 1987. pH-dependent function, purification, and intracellular location of a major collagen-binding glycoprotein. *J. Cell Biol.* 105: 517–27

Saraste, J., Kuismanen, E. 1984. Pre- and post-Golgi vacuoles operate in the transport of Semliki Forest virus membrane glycoproteins to the cell surface. *Cell* 38: 535–49

Saraste, J., Palade, G. E., Farquhar, M. G. 1986. Temperature-sensitive steps in the transport of secretory proteins through the Golgi complex in exocrine pancreatic cells. *Proc. Natl. Acad. Sci. USA* 83: 6425–29

Saraste, J., Palade, G. E., Farquhar, M. G. 1987. Antibodies to rat pancreas Golgi subfractions: identification of a 58-kD *cis*-Golgi protein. *J. Cell Biol.* 105: 2021–30

Schekman, R. 1985. Protein localization and membrane traffic in yeast. *Annu. Rev. Cell Biol.* 1: 115–43

Schwiezer, A., Fransenm, J. A. M., Bachi, T., Ginsel, L., Hauri, H.-P. 1988. Identification, by a monoclonal antibody, of a 53-kD protein associated with a tubulo-vesicular compartment at the *cis*-side of the Golgi apparatus. *J. Cell Biol.* 107: 1643–53

Senior, A., Gerace, L. 1988. Integral membrane proteins specific to the inner nuclear membrane and associated with the nuclear lamina. *J. Cell Biol.* 107: 2029–36

Severinsson, L., Peterson, P. A. 1985. Abrogation of cell surface expression of human class I transplantation antigens by an adenovirus protein in *Xenopus laevis* oocytes. *J. Cell Biol.* 101: 540–47

Sharma, S., Rodgers, L., Brandsma, J., Gething, M.-J., Sambrook, J. 1985. SV40 T antigen and the exocytic pathway. *EMBO J.* 4: 1479–89

Smith, M. J., Koch, G. L. E. 1987. Isolation and identification of partial cDNA clones for endoplasmin, the major glycoprotein of mammalian endoplasmic reticulum. *J. Mol. Biol.* 194: 345–47

Sorger, P., Pelham, H. R. B. 1987. The glucose-regulated protein grp94 is related to heat shock protein hsp90. *J. Mol. Biol.* 194: 341–44

Spener, F., Wirtz, K. W. A. 1985. Intracellular lipid-binding and transfer proteins. *Chem. Phys. Lipids* 38: 1–222

Srinivasan, M., Mazzarella, R. A., Green, M. 1988. The cloning and expression of murine protein disulphide isomerase. *P9.8.8 (Abstr.) 4th Int. Congr. Cell Biol., Montreal*

Stirzaker, S. C., Both, G. W. 1989. The signal peptide of the rotavirus glycoprotein VP7 is essential for its retention in the endoplasmic reticulum as an integral membrane protein. *Cell* 56: 741–47

Stirzaker, S. C., Whitfield, P. L., Christie, D. L., Bellamy, A. R., Both, G. W. 1987. Processing of rotavirus glycoprotein VP7: implications for the retention of the protein in the endoplasmic reticulum. *J. Cell Biol.* 105: 2897–2903

Stoeckle, M. Y., Sugano, S., Hampe, A., Vashistha, A., Pellman, D., Hanafusa, H. 1988. 78-kilodalton glucose-regulated protein is induced in Rous sarcoma virus-transformed cells independently of glucose deprivation. *Mol. Cell. Biol.* 8: 2675–80

Strous, G. J., Van Kerkhof, P. 1989. Release of soluble resident as well as secretory proteins from HEPG2 cells by partial permeabilization of rough endoplasmic reticulum membranes. *Biochem. J.* 257: 159–63

Strous, G. J., Van Kerkhof, P., Brok, R., Roth, J., Brada, D. 1987. Glucosidase II, a protein of the endoplasmic reticulum with high mannose oligosaccharide chains and a rapid turnover. *J. Biol. Chem.* 262: 3620–25

Tandon, S., Horowitz, P. 1986. Detergent-assisted refolding of guanidinium chloride-denatured rhodanese: the effect of lauryl maltoside. *J. Biol. Chem.* 261: 15615–81

Tartakoff, A. M. 1986. Temperature and energy dependence of secretory protein transport in the exocrine pancreas. *EMBO J.* 5: 1477–82

Ting, J., Lee, A. S. 1988. Human gene encoding the 78,000-dalton glucose-regulated protein and its pseudogene: structure, conservation and regulation. *DNA* 7: 275–86

Ting, J., Wooden, S. K., Kriz, R., Kelleher, K., Kaufman, R. J., Lee, A. S. 1987. The nucleotide sequence encoding the hamster 78-kDa glucose-regulated protein (GRP78) and its conservation between hamster and rat. *Gene* 55: 147–52

Tooze, J., Tooze, S. A., Warren, G. 1984. Replication of coronavirus MHV-A59 in sac(−) cells: determination of the first site of budding of progeny virions. *Eur. J. Cell Biol.* 33: 281–93

Tooze, S. A., Tooze, J., Warren, G. 1988. Site of addition of N-acetyl-galactosamine to the E1 glycoprotein of mouse hepatitis virus-A59. *J. Cell Biol.* 106: 1475–87

von Figura, K., Hasilik, A. 1986. Lysosomal enzymes and their receptors. *Annu. Rev. Biochem.* 55: 167–93

Warren, G. 1987. Signals and salvage sequences. *Nature* 327: 17–18
Wiedmann, M., Huth, A., Rapoport, T. A. 1984. *Xenopus* oocytes can secrete bacterial β-lactamase. *Nature* 309: 637–39
Wieland, F. T., Gleason, M. L., Serafini, T. A., Rothman, J. E. 1987. The rate of bulk flow from the endoplasmic reticulum to the cell surface. *Cell* 50: 289–300
Yamauchi, K., Yamamoto, T., Hayashi, H., Koya, S., Takikawa, H., et al. 1987. Sequence of membrane-associated thyroid hormone binding protein from bovine liver: its identity with protein disulphide isomerase. *Biochem. Biophys. Res. Commun.* 146: 1485–92
Yeo, K.-T., Parent, J. B., Yeo, T.-K., Olden, K. 1985. Variability in transport rates of secretory glycoproteins through the endoplasmic reticulum and Golgi in human hepatoma cells. *J. Biol. Chem.* 260: 7896–7902

NOTE ADDED IN PROOF There have been several recent reports that the drug Brefeldin A is a specific inhibitor of ER-Golgi transport (Lippincott-Schwartz et al 1989; R. W. Doms et al, in press; J. B. Ulmer & G. E. Palade, in press). A further remarkable effect of this drug is to cause the reversible disruption of Golgi stacks and redistribution of at least some Golgi enzymes to the ER, a phenomenon that has been interpreted as evidence for a vesicle recycling pathway between the Golgi and the ER. It remains to be seen whether the drug reveals the existence of a normal pathway, or merely causes fusion of compartments that are normally distinct. Nevertheless, it should be a useful tool for the analysis of membrane traffic.

Doms, R. W., Russ, G., Yewdell, J. 1989. *J. Cell Biol.* In press
Lippincott-Schwartz, J., Yuan, L. C., Bonifacino, J. S., Klausner, R. D. 1989. *Cell* 56: 801–13
Ulmer, J. B., Palade, G. E. 1989. *Proc. Natl. Acad. Sci. USA* In press

Annu. Rev. Cell Biol. 1989. 5 : 25–50

ORIGIN AND EVOLUTION OF MITOCHONDRIAL DNA

Michael W. Gray

Department of Biochemistry, Dalhousie University, Halifax, Nova Scotia B3H 4H7, Canada

CONTENTS

INTRODUCTION

In contemplating how the mitochondrial genome originated and what has happened to it since, one is immediately confronted with the extraordinary variation in size, structure, organization, and mode of expression of mitochondrial DNA (mtDNA) in different eukaryotes (Gray 1982, 1988; Gray & Doolittle 1982; Wallace 1982; Sederoff 1984). This diversity has made it very difficult to discern the pathway(s) of mitochondrial genome evolution. Only within groups of relatively closely related organisms have comparative studies (and in particular sequence analysis) begun to reveal the mode and tempo of mtDNA evolution, permitting reasoned speculation about the mechanisms by which mtDNA divergence has occurred.

Underlying the present review are many questions: What is the evolutionary origin of the mitochondrial genome? What was its ancestral form?

0743–4634/89/1115–0025$02.00

What has happened to it since its appearance in the eukaryotic cell? By what mechanism(s) has mtDNA evolved in different eukaryotes? Did the range of mitochondrial genomes in contemporary eukaryotes originate in a single event, or have they had several, independent origins (i.e. is mtDNA monophyletic or polyphyletic)? Is it possible to decide between these two alternatives, and if so, how? Although much progress has been made in defining the structural and evolutionary basis of mtDNA diversity, we are still some way from being able to provide satisfactory answers to all but perhaps the first of these questions.

Several excellent reviews dealing with the genetics, molecular biology, and biogenesis of mitochondria have appeared recently (Tzagoloff & Myers 1986; Chomyn & Attardi 1987; Attardi & Schatz 1988); only limited reference to these topics, sufficient to frame the evolutionary perspective of the present review, will be made here. Considerable emphasis will, however, be placed on the nature and extent of mtDNA structural variation, as this is basic to an ultimate understanding of mtDNA evolution. Because we will take a broad look at the subject, a great deal of relevant data and many insights pertaining to intraspecific mtDNA divergence and intra-individual mtDNA variation [heteroplasmy; see Solignac et al (1987) and references therein] must be omitted. Much of this information has come from the increasing application of mtDNA analysis to questions of systematics and population genetics; again, detailed critical reviews of these topics may be found elsewhere (Wilson et al 1985; Avise et al 1987; Moritz et al 1987). Finally, in order to keep within reasonable bounds, this review will be selective in focus and will preferentially highlight topics from the most recent literature.

STRUCTURAL AND ORGANIZATIONAL FLUIDITY OF MITOCHONDRIAL DNA

Mitochondrial DNA has the same basic role in all eukaryotes that contain it: it encodes rRNA and tRNA components of a mitochondrial protein synthesizing system, whose function is to translate a small number of mtDNA-encoded mRNAs that specify essential polypeptide components of the mitochondrial electron transport chain (Tzagoloff & Myers 1986; Attardi & Schatz 1988). In animal mitochondria, the thirteen translation products (their genes indicated in parentheses) include seven subunits of NADH dehydrogenase (*ndh1,2,3,4,4L,5,6*), the apocytochrome *b* component of ubiquinol cytochrome *c* reductase (*cob*), three subunits of cytochrome *c* oxidase (*cox1,2,3*), and two subunits of ATP synthase (H^+-ATPase) (*atp6,8*). Not all of these genes are found in all mtDNAs: in fact, only *cox1* and *cob* and the genes encoding large subunit (LSU) and small subunit (SSU) rRNAs are common to all those mitochondrial genomes whose coding capacity has been completely defined. On the other hand, some mitochondrial genomes carry extra genes (e.g. for proteins associated with mitochondrial ribosomes or involved in mitochondrial RNA processing or intron transposition) (Chomyn & Attardi 1987).

Although conservative in basic genetic function, mtDNA differs radically in structure among and even within the four traditional eukaryotic kingdoms (Animalia, Plantae, Fungi and Protista) (Gray & Doolittle 1982; Gray 1982, 1988; Wallace 1982; Sederoff 1984; Clark-Walker 1985). Mitochondrial genome size varies over more than a 150-fold range, from a current low of 14.3 kbp in the nematode worm, *Ascaris suum* (Wolstenholme et al 1987) to an estimated high of 2400 kbp in the cucurbit, *Cucumis melo* (muskmelon) (Ward et al 1981). Most mtDNAs are circular, as deduced from their form when isolated or from the fact that they have circular genetic and/or physical maps. However, linear mtDNAs have also been identified in the animal *Hydra* (Warrior & Gall 1985), the fungi *Hansenula mrakii* (Wesolowski & Fukuhara 1981) and *Candida rhagii* (Kováč et al 1984), the ciliate protozoans *Paramecium* (Pritchard et al 1986) and *Tetrahymena* (Suyama et al 1985), and the green alga *Chlamydomonas reinhardtii* (Grant & Chiang 1980).

Animal Mitochondrial DNA

Initially, animal mtDNA was considered to be not only "an extreme example of genetic economy" (Attardi 1985), but a pardigm of structural stability. This perception was based on the finding that four vertebrate mtDNAs (human, mouse, bovine and *Xenopus laevis*) are very similar in size (16.5–17.6 kbp), have an invariant gene order, and are organized and expressed in an essentially identical manner. These mtDNAs mostly consist of structural genes, which are butt-joined one to another, with few or no spacer nucleotides between them. What little size variation exists among vertebrate mtDNAs is mainly confined to a 1–2 kbp noncoding region (the D-loop region), which contains the promoters of heavy (H) and light (L) strand transcription as well as the origin of H-strand replication (Clayton 1984).

More recently, this view of structural uniformity has been challenged by studies of additional animal mtDNAs, particularly nonvertebrate ones. Size variation now extends from 14.3 kbp in *A. suum* (Wolstenholme et al 1987) to as much as 32.1–39.3 kbp in *Plactopecten magellanicus* (sea scallop) (Snyder et al 1987). The smaller size of *Ascaris* mtDNA reflects the absence of the *atp8* gene found in vertebrate mtDNA, while expansion in some of the larger animal mtDNAs is due to localized sequence amplification, resulting in large (0.8–8.0 kbp) direct tandem duplications of both coding and noncoding portions of the genome (Moritz & Brown 1987; Bentzen et al 1988; Hyman et al 1988).

A different mitochondrial gene order has been found in each animal phylum studied to date (vertebrates, insects, echinoderms, nematodes, platyhelminths) (Clary & Wolstenholme 1984; Wolstenholme et al 1987; Jacobs et al 1988; Garey & Wolstenholme 1989). Even within phyla, variation in gene order may exist: e.g. between the mtDNAs of birds and other vertebrates (Yang & Zhou 1988), echinoid and asteroid echinoderms (Jacobs et al 1989), and orthopteran and dipteran insects (Haucke & Gellissen 1988). Nevertheless, vertebrate and invertebrate mtDNAs both maintain the extremely compact arrangement of genetic information first described in human and mouse mtDNA.

In several cases, gene order is seen to be related. Human mtDNA differs from that of sea urchin by only two transpositions in the order of protein and rRNA genes (Jacobs et al 1988), and from that of *Drosophila* by three inversions (Clary & Wolstenholme 1984). In other instances (e.g. *Ascaris* vs human) there is little evident relationship (Wolstenholme et al 1987). The positions of tRNA genes are much more variable. In both sea urchin (Jacobs et al 1988; Cantatore et al 1988) and starfish (Jacobs et al 1989), 15 of 22 tRNA genes are clustered together, whereas tRNA genes are more uniformly distributed throughout vertebrate, insect, and *A. suum* mtDNAs, although in different orders relative to each other and other genes.

Fungal Mitochondrial DNA

Among ascomycetes, mtDNA size varies from 17.6 kbp in *Schizosaccharomyces pombe* EF1 to 115 kbp in *Cochliobolus heterostrophus* (Wolf & Del Guidice 1988). Almost as great a variation (28–101 kbp) has been reported in a group of much more closely related yeasts (Hoeben & Clark-Walker 1986) and even within a single genus of basidiomycetes (36–121 kbp in *Suillus*; Bruns et al 1988). Features contributing to this large size variation have been extensively documented and are discussed in detail elsewhere (Clark-Walker 1985; Wolf & Giudice 1988). They include intergenic spacer length, presence or absence of optional introns, presence and number of short repetitive sequences, duplication of portions of the genome, presence of novel open reading frames (ORFs), and length polymorphisms within coding regions. Changes in the amount of sequence between and within coding regions (i.e. spacers and introns, respectively) are chiefly responsible for the wide intraspecies as well as interspecies variation in fungal mitochondrial genome size. Comparison of the mitochondrial genomes of *Torulopsis glabrata* (18.9 kbp) and *Saccharomyces cerevisiae* (68–81 kbp) graphically illustrates this point: although their coding regions share >80% sequence identity (see Clark-Walker et al 1983), these two genomes differ in size by 50–60 kbp, most of which is accounted for by A+T-rich intergenic DNA.

The smallest fungal mtDNAs, those of *S. pombe* and *T. glabrata*, are only slightly larger than typical animal mtDNAs. Not surprisingly, they also exhibit a compact arrangement of genetic material, although economy of organization is not taken to the extreme seen in animal mtDNA. Thus, whereas animal mtDNA entirely lacks spacers, genes in the smallest fungal mtDNAs are still separated by A+T-rich spacers of from a few to a few hundred bp in length (Lang et al 1983; Clark-Walker et al 1985b).

Intron content underlies major differences in the size of homologous genes in fungal mtDNA. In *S. cerevisiae*, *cox1* may contain up to seven introns, and as such span about 10 kbp, half the size of the entire *S. pombe* mitochondrial genome. Strain-dependent variation in intron number, indeed their complete dispensability (Séraphin et al 1987), characterizes fungal mitochondrial introns as optional. Although introns are reduced in number in the smallest fungal mitochondrial genomes, they are not entirely absent (Wolfe & Del Giudice 1988); in contrast, introns have not been found in animal mtDNAs.

Is intron variation a consequence of progressive intron loss from a larger, intron-containing progenitor, or have introns been progressively acquired starting with a small, intron-less mitochondrial genome? Clark-Walker et al (1983) have argued that because *S. cerevisiae* and *Neurospora crassa* mtDNAs share (at exactly the same positions) introns that are lacking in *T. glabrata* mtDNA, and because *S. cerevisiae* and *T. glabrata* are more closely related at the level of mtDNA sequence than either is to *N. crassa*, it is likely that introns have been selectively lost from *T. glabrata* mtDNA during its evolution, rather than selectively gained by *S. cerevisiae* mtDNA. However, the recognition of mitochondrial introns as mobile genetic elements (Dujon et al 1986), and the demonstration that duplicative transposition of a yeast mitochondrial intron is a highly site-specific event (Colleaux et al 1988), weaken this argument and make it likely that introns have both come and gone in the course of fungal mtDNA evolution. Recent evidence is, in fact, consistent with two independent acquisitions of an intron by an originally intron-less *cob* gene in *S. pombe* (Zimmer et al 1987). Other considerations have prompted speculation about the possibility of horizontal transfer of mitochondrial introns (Lang 1984; Dujon et al 1986; Michel & Dujon 1986; Wolf & Del Giudice 1987; Cummings & Domenico 1988).

Unique mitochondrial gene orders have been reported in every fungal genus examined, although conserved blocks of genes are evident in pairwise comparisons (Wolf & Del Giudice 1988). Of particular note is a five gene cluster that is common to the mtDNAs of *T. glabrata* and *Saccharomyces exiguus* (23.7 kbp), and that includes juxtaposed *atp6* and *atp9* genes in the same order and orientation as their homologues in the *Escherichia coli unc* operon. These two genes are dispersed in the larger yeast mtDNAs, which led Clark-Walker et al (1983) to suggest that the latter are descended from smaller ancestral forms bearing some resemblance to *T. glabrata*/*S. exiguus* mtDNAs.

Comparisons at intermediate levels of divergence have begun to reveal common or closely related patterns of fungal mtDNA organization. A common mitochondrial gene order is shared by three ascomycetes (two *Brettanomyces*, one *Eeniella*) whose mtDNAs range between 28.5 and 41.7 kbp (Hoeben & Clark-Walker 1986). Genes are also arrayed identically in four species of basidiomycetes (genus *Suillus*) having mtDNAs differing more than 3-fold in size (36–121 kbp) (Bruns et al 1988; Bruns & Palmer 1989). In both of these studies, similar mitochondrial gene orders, interconvertible by either one or two single-step rearrangements, were identified in related species or genera. Consideration of what appear to be single-step large scale rearrangements, coupled with intensive restriction site mapping of small size-conserved regions, permits the construction and evaluation of putative pathways of divergence, and has provided suggestive evidence for an expansion of the mitochondrial genome within *Suillus* (Bruns & Palmer 1989).

Plant Mitochondrial DNA

Plant mitochondrial genomes are by far the largest and most complex mtDNAs known (Pring & Lonsdale 1985; Newton 1988). The smallest plant mtDNA (208 kbp in *Brassica hirta*, white mustard; Palmer & Herbon 1987) is already larger than the largest fungal mtDNA (176 kbp, *Agaricus bitorquis*; Hintz et al 1985) and larger than most chloroplast DNAs (ctDNAs) (Palmer 1985a,b). Plant mitochondrial genome size varies over a 10-fold range (up to 2400 kbp in *C. melo*), and shows a 4-fold difference even within a single family (Ward et al 1981). Although a few novel genes have been described in plant mtDNA, and some plant mitochondrial genes contain introns, these two features do not contribute in a major way either to the large size or the size variation of plant mtDNA. Rather, most of the plant mitochondrial genome appears to consist of noncoding DNA. In contrast to fungal mtDNAs, putative spacer DNA is not A+T-rich in plant mtDNA, but has much the same base composition as the bulk mtDNA (ca. 47% G+C).

Genes in plant mtDNA are scattered and for the most part solitary; only in a few cases are coding sequences within close proximity to one another and either potentially or actually cotranscribed (Gray & Spencer 1983; Bland et al 1986; Wissinger et al 1988; Gualberto et al 1988). Gene maps have been published for the mtDNAs of maize (Dawson et al 1986), spinach (Stern & Palmer 1986), and turnip (*Brassica campestris*) (Makaroff & Palmer 1987); these are completely different from one another. The conserved juxtaposition of 18S and 5S rRNA genes (Huh & Gray 1982) is so far the only feature common to all angiosperm mtDNAs examined.

Other than this, species as closely related as wheat and maize (members of the same family) show no evident commonality in sequence arrangement. On the other hand, within species structural variation can be quite low (Palmer 1988).

A major step in understanding the physical organization of plant mtDNA has been the recognition that it contains directly repeated sequences that are recombinationally active (Palmer 1985a; Pring & Lonsdale 1985). This has led to the formulation of a multicircular genome model, in which the entire genetic complexity is represented in a circular molecule (the master chromosome) that can be resolved into a number of subgenomic circular molecules by intramolecular recombination (Palmer 1985a; Lonsdale et al 1988). In the simplest case, exemplified by the tripartite turnip mitochondrial genome, a 218 kbp master circle is in equilibrium with 135 and 83 kbp subgenomic circles generated by recombination between a single pair of 2 kbp direct repeats in the master chromosome (Palmer & Shields 1984). Direct evidence for the existence of the predicted circular forms has recently been obtained (Palmer 1988). As the number of recombinationally active repeats increases (as in the 570 kbp maize mtDNA), so do the possibilities for complex recombination (Lonsdale et al 1984). This model largely accounts for the physical heterogeneity displayed by isolated plant mtDNA (as seen by restriction and hybridization analysis; Bonen & Gray 1980; but see Palmer & Herbon 1987). Sublimons, substoichiometric restriction fragments presumed to result from infrequent recombination events between very short regions of sequence similarity (Small et al 1987), may also contribute to the physical heterodispersity of plant mtDNA. Judging from the results of sequence analysis (Gualberto et al 1988; Joyce et al 1988b; D. F. Spencer & M. W. Gray, unpublished), plant mtDNA appears to contain an abundance of short dispersed repeats, which are obvious candidates as foci for rearrangement.

Another hallmark of the plant mitochondrial genome is its acceptance of sequence information from other genomes, notably chloroplast (Stern & Lonsdale 1982) but also nuclear (Schuster & Brennicke 1987). Promiscuous chloroplast sequences are widely distributed in plant mtDNA, seemingly in random fashion both with respect to their variation in different plant species and the portion of the chloroplast genome incorporated (Stern & Palmer 1984). Sequential transfer of genetic information from chloroplast to mitochondrion in the course of evolution, and the persistence of these sequences in plant mtDNA over long periods of evolutionary time, has recently been reported (Nugent & Palmer 1988). Prominently represented among the promiscuous chloroplast sequences in plant mtDNA are tRNA genes, and there is increasing evidence that some of these are transcribed and matured and may actually function in translation in plant mitochondria (Maréchal et al 1987; P. B. M. Joyce & M. W. Gray, unpublished). Thus, the plant mitochondrial genome may be considered an evolutionary mosaic (Schuster & Brennicke 1988), with evidence for the acquisition of genetic information (and possibly even active genes) from several distinct sources in the course of evolution.

Protist Mitochondrial DNA

Considering the evolutionary diversity and antiquity of the unicellular eukaryotes, the true magnitude of which we are just now beginning to appreciate (Sogin et al 1989), it is a bit of an understatement to say that relatively little is known about protist mitochondrial genomes. Only within 5 of the 27 protistan (protoctist) phyla described by Margulis & Schwartz (1988) has mtDNA been characterized in any detail: Zoomastigina (genera *Crithidia*, *Leishmania*, *Trypanosoma*), Ciliophora (*Paramecium*, *Tetrahymena*), Chlorophyta (*Chlamydomonas*), Chytridiomycota (*Allomyces*, *Blastocladiella*), and Oomycota (*Achlya*).

To date, the most intensively studied protist mitochondrial genome has been the 20–40 kbp maxicircle DNA of the kinetoplastid (trypanosomid) protozoa (Benne 1985; Simpson 1986, 1987). In this genome, a relatively well conserved transcribed region (15–17 kbp) contains several of the classical mitochondrial genes (rRNA, *cox1,2,3*, *cob*, several *ndh*) and a few unidentified ORFs, but seemingly lacks *atp* or tRNA genes. The remainder of the mtDNA is an A+T-rich divergent (apparently nontranscribed) region containing a complex set of repetitive sequences (Muhich et al 1985). Coding information, although compactly organized (de la Cruz et al 1984), is not as condensed as in animal mtDNA. Maxicircle genes are conserved among the trypanosomids, but their order is entirely different from that seen in other organisms (Benne 1985). Consistent with the deep branching of the trypanosomid lineage (Sogin et al 1989), trypanosomid mitochondrial genes are less similar in sequence to the homologous yeast or human genes than the latter are to each other (de la Cruz et al 1984). Particularly unusual are the trypanosomid mitochondrial 9S (SSU) and 12S (LSU) rRNAs, which display only the barest resemblance to conventional rRNAs (Benne 1985; Simpson 1987). RNA editing, an intriguing phenomenon recently discovered in trypanosomid mitochondria, is reviewed elsewhere in this volume (see L. Simpson).

Unique features are also found in the linear mtDNAs of the ciliates *Paramecium* (41 kbp; Pritchard et al 1986) and *Tetrahymena* (55 kbp; Suyama et al 1985). *T. pyriformis* mtDNA contains duplicate LSU rRNA genes, located in 3 kbp sub-terminal inverted repeats and well separated from an internally localized, single copy SSU rRNA gene. The rRNA genes consist of two modules, encoding the two subunits (α and β) of split SSU and LSU rRNAs (Schnare et al 1986; Heinonen et al 1987); the α

cistrons specify the 5′-terminal 208 (SSU) or 280 (LSU) nucleotides of the respective rRNAs. A striking feature of the LSU gene is that the α module is rearranged relative to its expected position: in the direction of transcription, it is found downstream of the β coding sequence, and separated from it by a tRNA^{Leu} gene (Heinonen et al 1987). *Paramecium* mtDNA does not possess a terminal duplication, although its single LSU rRNA gene is still located at one end of the mtDNA and also specifies a split LSU rRNA (Seilhamer et al 1984). In this case, however, the α module is located conventionally, upstream of the β module in the direction of transcription. It appears that duplication of the mitochondrial LSU rRNA coding region took place in the *Tetrahymena* lineage after its divergence from *Paramecium*, and was accompanied by rearrangement of the LSU rRNA gene. Interestingly, one copy of the LSUα and tRNA^{Leu} genes is missing in several species of *Tetrahymena* (Morin & Cech 1988b). Other notable features of *Tetrahymena* mtDNA are (*a*) the presence of heterogeneous telomeric ends, composed of short (30–50 bp), species-specific tandem repeats (Morin & Cech 1986, 1988a), and (*b*) the apparent specification of fewer than the minimal number of tRNA genes required for mitochondrial protein synthesis (Suyama 1986). Ciliate mitochondrial protein genes are the most divergent known, more so even than their trypanosomid counterparts (Pritchard et al 1986); this is puzzling in view of the fact that (*a*) ciliates are supposed to have diverged later than trypanosomids in eukaryotic evolution (Sogin et al 1989), and (*b*) ciliate mitochondrial rRNA genes are much more conventional than their trypanosomid counterparts (Gray 1988).

At 15.8 kbp, the linear mtDNA of *Chlamydomonas reinhardtii* is the smallest protist mitochondrial genome described to date. It displays extensive physical and transcriptional linkage of genes, and in organization and mode of expression more closely resembles animal than plant mtDNA (Gray & Boer 1988). Notable features include (*a*) the apparent absence not only of any *atp* genes but of the otherwise ubiquitous *cox2* and *cox3* genes, (*b*) the presence of two novel ORFs, one of which specifies a reverse transcriptase-like polypeptide (Boer & Gray 1988a), (*c*) the presence of only three tRNA genes in 80% of the genome sequenced (Boer & Gray 1988c), and (*d*) a scrambled arrangement of rRNA coding modules for highly split SSU and LSU rRNAs (Boer & Gray 1988b).

Within the Chytridiomycota (phycomycetes), the size of the mtDNA ranges from 35.5 kbp in *Blastocladiella emersonii* to 56 kbp in *Allomyces macrogynus* (see Borkhardt & Olson 1986). Borkhardt et al (1987) reported pronounced differences in the restriction profiles of mtDNA from seven species of *Allomyces*, as well as of the (same size) mtDNAs from four isolates of *A. arbuscula*. In spite of this variation, available evidence suggests that gene order is the same in *B. emersonii*, *A. arbuscula*, and *A. macrogynus*

(Borkhardt et al 1988), with single SSU and LSU rRNA genes widely separated in the circular mtDNA, as in some fungi. In contrast, rRNA genes are localized within 10–12 kbp inverted repeats in the 50 kbp circular mitochondrial genomes of the oomycetes *Achlya ambisexualis* (Hudspeth et al 1983) and *A. klebsiana* (Boyd et al 1984). This arrangement, so far unique among mtDNAs, is found in most ctDNAs (Palmer 1985b); like the latter, *Achlya* mtDNA exists as isomeric forms that are generated by a flip-flop recombination mechanism involving the inverted repeats (Palmer 1985b).

RAPID VS SLOW EVOLUTION OF MITOCHONDRIAL DNA

A remarkable feature of mammalian mtDNA is its rapid rate of evolution (Brown 1985; Wilson et al 1985; Cantatore & Saccone 1987). Brown et al (1979, 1982) first reported that mtDNA in primates undergoes sequence divergence at a 5–10-fold higher rate than primate single copy nuclear DNA (nDNA), a finding subsequently extended to other mammals (Miyata et al 1982). Salient features of this accelerated rate of sequence change are (*a*) an exceptionally high proportion of silent (>90%) vs replacement (<10%) substitutions, (*b*) a high transition/transversion ratio (on the order of 10:1), and (*c*) a strong bias toward C↔T transitions in the L-strand (see Brown & Simpson 1982; Brown et al 1982). In pairwise comparisons, the transition/transversion ratio falls as the divergence times of the compared species increase, probably as a result of obliteration of the record of transitions by multiple substitutions at the same site (see Wilson et al 1985). Rates of sequence divergence for different functional portions of the mtDNA decrease in the order D-loop region > protein coding genes > rRNA and tRNA genes (Brown 1985). Although tRNA genes are changing in sequence at about one-half the rate of protein genes in primate mtDNA, they are evolving on the order of 100 times faster than their nuclear counterparts. This has been attributed to relaxed functional constraints on mitochondrial compared with nuclear tRNA genes (Brown et al 1982). Each of the 13 protein coding genes in mammalian mtDNA exhibits its own characteristic rate of change, which appears to be the same among mammals (Brown 1985). One striking exception is the *cox2* gene, which in the primate lineage has undergone at least a 5-fold acceleration in rate of divergence at the amino acid level (Brown & Simpson 1982; Cann et al 1984), concomitant with a parallel acceleration in the rate of change of cytochrome *c*, a nucleus-encoded partner protein.

At the present time there is some debate about whether other animal mtDNAs sustain the rapid rate of sequence evolution observed in mammalian mtDNA. Virtually equivalent rates of mtDNA and nDNA divergence have been reported for *Drosophila* (Powell et al 1986; Solignac et al

1986) and sea urchin (Vawter & Brown 1986), as well as an equivalent frequency of transitions and transversions in *Drosophila* mtDNA (Wolstenholme & Clary 1985). More recent work, however, indicates that the mtDNAs of Hawaiian *Drosophila* species evolve quickly and with a strong transition bias, but, intriguingly, do not become very different, i.e. mtDNA change saturates at a low ceiling (DeSalle et al 1987). One complicating factor in the case of *Drosophila* is the nucleotide composition bias of the mtDNA (74–80% A+T), which may place particular constraints on mtDNA sequence divergence. Through continuous selection for A/T nucleotides at all sites at which such selection is compatible with function (Wolstenholme & Clary 1985), an A+T-rich mtDNA may be prevented from becoming very different even at silent sites (i.e. the third position of codons) (DeSalle et al 1987). On the other hand, Vawter & Brown (1986) contend that the rapid rate of vertebrate mtDNA evolution is, in part, an artifact of a widely divergent rate of nuclear DNA evolution. To reconcile these divergent interpretations will require additional mtDNA and nDNA sequence comparisons of both closely and distinctly related species of both vertebrates and invertebrates.

The large amount of attention focused on rapid evolution of animal mtDNA has tended to foster the impression that mtDNA evolves this way in all eukaryotes. While some cases of rapid evolution of nonanimal mtDNA have been reported (e.g. in *Tetrahymena*; Morin & Cech 1988b), this phenomenon is by no means universal. In marked contrast to animal mtDNA, plant mtDNA diverges in sequence at an extremely slow rate. This difference was apparent in the earliest comparisons of protein coding (e.g. Bonen et al 1984) and rRNA (Chao et al 1984) genes, and has been verified and emphasized by subsequent work (Gwynn et al 1987; Sederoff 1987; Wolfe et al 1987; Palmer & Herbon 1988). Wolfe et al (1987) have estimated that angiosperm mtDNA evolves at least five times more slowly than nuclear sequences and even has a synonymous substitution rate at least 3-fold lower than ctDNA. In *Brassica* mtDNA this low rate extends over the entire genome, including noncoding as well as coding regions (Palmer & Herbon 1988). Thus, while animal mtDNA is one of the most rapidly evolving genomes known, plant mtDNA is one of the slowest!

MECHANISMS OF MITOCHONDRIAL GENOME DIVERGENCE

Animal Mitochondrial DNA

To account for the high rate of both point and length (Cann & Wilson 1983) mutations in animal mtDNA, an enhanced mutation pressure has been postulated (Brown et al 1982; Cann et al 1984; Wilson et al 1985). Factors that might contribute to such a pressure include (*a*) greater exposure of mtDNA to oxidative damage (e.g. superoxide; see Richter 1988); (*b*) a more error-prone system of DNA replication (poor fidelity at the level of dNTP selection and/or lack of editing); (*c*) absence or deficiency of DNA repair functions; (*d*) relative lack of a recombinational mechanism by which natural selection could eliminate mildly deleterious mutations (see Cann & Wilson 1983; Cann et al 1984), and (*e*) a high rate of turnover of mtDNA.

Mammalian mitochondria apparently lack both excision and recombination repair capacity (Clayton et al 1974; Lansman & Clayton 1975). However, enzymes implicated in DNA repair in other systems, such as uracil-DNA glycosylase and AP endonuclease, have been found in mammalian mitochondria (see Richter 1988; Tomkinson et al 1988). Moreover, the high fidelity of chick embryo pol-γ (mitochondrial DNA polymerase), attributable to a $3' \rightarrow 5'$ exonuclease (proofreading) activity (Kunkel & Soni 1988), seems at odds with the postulate of a high rate of error introduction during mtDNA replication. At this point, therefore, it is not clear to what extent the biochemistry of mtDNA replication and repair plays a role in its rapid evolution.

On the other hand, a role for relaxed selection and/or relaxed functional constraints seems indicated. In animal mitochondria an expanded codon recognition pattern means that a single tRNA species is able to decode all four codons in those quartets that specify a given amino acid. Third position codon changes are therefore effectively silent in these cases. In fact, an exceptionally high proportion of silent to replacement substitutions has been found in animal mitochondrial protein genes (Brown & Simpson 1982; Brown et al 1982), which supports the contention that the rapid evolution of mtDNA relative to nDNA is due only to silent changes and that amino acid altering substitutions accumulate in nDNA and mtDNA at comparable rates (Brown & Simpson 1982). Relaxed functional constraints have been invoked to account for the 10-fold greater (overall) rate of evolution of tRNA than protein coding genes in animal mtDNA (Brown et al 1982).

There is a general consensus that animal mtDNA does not undergo recombination (Wilson et al 1985; but see Horak et al 1974; Olivo et al 1983). On the other hand, although major rearrangements seem to have occurred relatively infrequently in the evolution of animal mtDNA, they obviously have occurred. It is not clear how rearrangement happens in such a tightly packed genome, or, equally, how it is tolerated. Recently, a role for tRNAs as agents of genomic change in animal mitochondria has been proposed (Moritz & Brown 1987; Cantatore et al 1987; Jacobs et al 1988). One model (in which tRNAs occasionally serve as illegitimate primers for DNA replication) predicts a relocation of tRNA genes adjacent to the replication origin, as is seen in sea urchin mtDNA (Jacobs et al 1988). However, in starfish mtDNA, an identical cluster of 13 tRNA genes is dissociated from the replication origin, which led Jacobs et al (1989) to propose that clustering of tRNA genes represents the ancestral situation and that evolutionary dispersal (rather than gathering together) of tRNA genes has taken place in the animal mitochondrial genome. Supporting this view is the presence of conserved juxtaposed clusters of tRNA genes in the mtDNAs of vertebrates, insects, echinoids, nematodes, and fungi (Jacobs et al 1989).

Fungal Mitochondrial DNA

Few studies of fungal mtDNA evolution have assessed the rapidity of sequence change (Weber et al 1986); instead, the focus has been almost exclusively on genome rearrangement. The *petite* mutation in *S. cerevisiae*, which involves the formation of altered, functionally defective mitochondrial genomes (Bernardi 1979), has provided the basis for models of how evolutionary rearrangement of yeast mitochondrial genomes may occur. In *S. cerevisiae*, excision of portions of the wild-type mtDNA is mediated by repetitive elements within the A+T-rich spacer regions, via a mechanism involving site-specific recombination events between homologous sequences (de Zamaroczy et al 1983); short G+C-rich clusters within the A+T-rich spacers seem to be particularly recombinogenic (Dieckmann & Gandy 1987; Zinn et al 1988). As in the case of plant mtDNAs, intramolecular recombination between direct repeats results in excision of the DNA segment between them as a circular molecule, which in *petite* mutants may then be amplified as a tandem array to generate a replication-competent (but respiratory-deficient) *petite* mitochondrial genome. Mitochondrial DNA in other fungi undergoes similar excision-amplification events characterized by intramolecular recombination between repeated sequences (De Vries et al 1981; Gross et al 1984; Turker et al 1987).

Features of yeast mtDNA that appear to contribute to evolutionary rearrangement include (*a*) a high rate of mtDNA recombination, (*b*) circularity of the mitochondrial genome, (*c*) a high proportion of intergenic, noncoding spacer sequences, and (*d*) the existence within spacers of short repeated sequences, dispersed throughout the genome (Clark-Walker & Miklos 1974). Elegant models of rearrangement, based on illegitimate recombination between short repetitive elements, have been elaborated by Clark-Walker & co-workers (Evans & Clark-Walker 1985; Clark-Walker et al 1985a). Direct evidence in support of the postulated rearrangement pathway comes from sequence analysis of novel junctions in rearranged recombinant molecules (G. D. Clark-Walker, personal communication).

Slonimski & co-workers (Kotylak et al 1985) have emphasized a key role for intron-encoded proteins in fungal mitochondrial genome evolution. Proteins encoded by yeast mitochondrial introns have been implicated in clean deletion of introns, duplicative intron transposition, and homologous recombination of exons with gene restructuring (Kotylak et al 1985).

Intron-encoded "nucleic acid wielding" proteins (Kotylak et al 1985) include putative or demonstrated reverse transcriptase (Michel & Lang 1985), DNA recombinase (Kotylak et al 1985), and DNA transposase (double-strand endonuclease) (Colleaux et al 1988) activities. Recently, evidence pointing to an independent evolution of structural and ORF regions of an intron has been reported (Mota & Collins 1988; Cummings & Domenico 1988). In one instance, highly homologous introns interrupt the *ndh1* genes of *Neurospora crassa* and *N. intermedia* Varkud at exactly the same position; however, these introns contain ORFs that differ both in sequence and location (Mota & Collins 1988).

Given the large size range of contemporary fungal mitochondrial genomes, can we make any conjecture about the likely size and organization of the ancestral form, and about the direction (large to small or vice versa) in which fungal mtDNA evolution has proceeded? Clark-Walker et al (1985a) suggest that the original yeast mtDNA may have had a size near the present median value (37 kbp), with smaller and larger yeast mtDNAs representing the evolutionary consequences of contraction and expansion processes, respectively. Similarities in size and gene arrangement are not, however, necessarily indicative of close evolutionary relationship (Clark-Walker et al 1987). For example, the mtDNAs of *T. glabrata* (18.9 kbp) and *S. exiguus* (23.7 kbp), although sharing considerable similarity in gene order (Clark-Walker et al 1983), are no more closely related to each other at the level of *cox2* gene sequence than each is to the larger (68–81 kbp) *S. cerevisiae* mitochondrial genome, which has a much different order and arrangement of genes (Clark-Walker et al 1985a). As the proportion of nonessential spacer sequence increases, we would expect that fungal mitochondrial genomes should become more prone to rearrangement by the mechanisms discussed above. There are observations both in support of (Hoeben & Clark-Walker 1986) and in conflict with (Bruns et al 1988) this expectation. Lability or resistance to rearrangement may depend on the extent to which the various features that promote evolutionary rearrangement of yeast mtDNA (outlined above) are present in other fungi.

We do not yet know how spacer sequences and their imbedded G+C-rich elements (Prunell & Bernardi 1977; Yin et al 1981) are introduced into (or deleted from) fungal mtDNAs and why these genomes are so AT-rich. As suggested by Wolfe & Del Giudice (1988), evolution toward A+T-rich genomes may reflect lack of a mechanism for repairing C to U delaminations (e.g. absence of a mitochondrial uracil N-glycosylase).

Plant Mitochondrial DNA

It is clear that plant mtDNA evolves rapidly in structure, but slowly in sequence (Palmer & Herbon 1988). It might be anticipated, therefore, that factors invoked to explain the rapid evolution of animal mtDNA, such as an enhanced mutation pressure, lack of recombination, and relaxed functional constraints (Cann et al 1984), might not be operating in plant mitochondria. We have no basis for judging the fidelity of DNA repair and the existence and efficiency of post-replication repair in plant mitochondria; there simply are no data that address these issues. Available information suggests that there is not a relaxed codon recognition pattern in plant mitochondria. Relative to their nuclear-encoded counterparts, plant mitochondrial rRNAs (Gray 1988) and tRNAs (Joyce et al 1988a) are much more highly conserved in primary sequence and secondary structure than are animal mitochondrial tRNAs and rRNAs. This might be an indication that translation is more constrained in plant than in animal mitochondria and less able to accommodate rapid change in protein coding sequences, even at nominally silent sites. However, an appeal to functional constraint cannot explain the slow rate of sequence change over the entire genome (noncoding as well as coding sequences; Palmer & Herbon 1988). Lonsdale et al (1988) have suggested that plant mtDNA exists as a panmictic population (due to fusion of mitochondria) in recombinational equilibrium, which suppresses nucleotide sequence change through processes such as copy correction.

The formation of novel mitochondrial genotypes as a result of somatic cell fusion provides direct evidence of mitochondrial fusion and active mtDNA recombination in plants (reviewed in Lonsdale 1987). Similarly, the discovery of rearranged (mosaic) regions in plant mtDNA (Dewey et al 1986) implicates recombination in the generation of evolutionary diversity. The structural determinants of rearrangement/recombination processes in plant mitochondria still have to be elucidated. Detailed comparative studies of rearranged regions may provide some clues (e.g. Joyce et al 1988b). Recently, this approach has provided evidence that models elaborated for the evolutionary rearrangement of yeast mtDNA (Evans & Clark-Walker 1985; Clark-Walker et al 1985a) may also apply to plant mtDNA (Small et al 1989). Even at this stage, it is evident that the plant mitochondrial genome is a highly plastic entity that is able to tolerate considerable structural variation with little or no effect on function. This implies that sequence context is largely immaterial to the expression of genetic information in plant mtDNA.

ORIGIN OF MITOCHONDRIAL DNA: THE ENDOSYMBIONT HYPOTHESIS

The endosymbiont hypothesis is generally regarded as the best explanation of the origin of the mitochondrial genome, as well as of the structural and functional complexity of the mitochondrion itself (Gray & Doolittle 1982; Gray 1983, 1988; Taylor 1987; but see Mikelsaar 1987). This hypothesis (see Margulis 1981) proposes that mitochondria originated in evolution as bacterial endosymbionts that were ultimately integrated into a host cell that provided the nuclear genome. Mitochondria remain semi-autonomous in the sense that they retain a distinctive genome that is replicated and expressed, but they are incapable of independent existence. According to the endosymbiont hypothesis, the contemporary mitochondrial genome is a bacterial remnant of the original endosymbiotic event(s) that created the ancestor of the eukaryotic cell in which it now resides. Most if not all of the genes now found in mtDNA are considered to represent genetic information retained from the original endosymbiont. Two assumptions implicit in this hypothesis are (*a*) that nuclear and mitochondrial genomes derive from distinctly separate lineages that enjoyed a long period of independent existence and evolutionary divergence before they were united within a single cell, and (*b*) that in the course of evolution there was massive transfer to the nucleus and/or loss of genetic information from the endosymbiont genome.

In posing the question, "Has the endosymbiont hypothesis been proven?", W. F. Doolittle and I argued, "If the evolutionary histories of nuclear genomes and one of the organellar genomes were known with certainty, and were with certainty different—that is, if the two could be shown to derive from genomic lineages which were phylogenetically distinct before the formation of the eucaryotic cell—then the endosymbiotic origin of that organelle could be taken as proven" (Gray & Doolittle 1982). In the case of mitochondria, there is now a satisfying concordance of structural, biochemical, and genetic evidence not only supporting an origin of this organelle from eubacteria, but, indeed, tracing this origin to a specific subdivision, the α-purple bacteria (Yang et al 1985; Villanueva et al 1985; John 1987; Cedergren et al 1988). Although the evolutionary origin of the nuclear genome is still uncertain, it is clearly distinct from that of the eubacterial genome (Woese 1987). Moreover, although the burden of evidence indicates that chloroplasts and mitochondria are both direct descendents of eubacteria, they have quite clearly arisen from separate and distinct phyla: cyanobacteria and α-purple bacteria, respectively (see Gray 1988). As emphasized by Doolittle (1980), this makes it impossible to maintain an autogenous evolutionary scenario for both organelles; in an autogenous scheme, the progenitor of the nuclear genome would have had to have been either a cyanobacterium or an α-purple bacterium, but quite obviously could not have been both.

TRACING THE PATHWAYS OF MITOCHONDRIAL GENOME EVOLUTION

Arguments for (Raven 1970; Dayhoff & Schwartz 1981; Stewart & Mattox 1984) and against (Cavalier-Smith 1987) multiple origins of mitochondria have been made. At least part of the vast structural diversity summarized in this review might be attributable to a polyphyletic origin of mitochondrial genomes; on the other hand, one of the strongest arguments in favor of a monophyletic origin is that the basic function of mtDNA (as reflected in the genes it encodes) is fundamentally the same in all eukaryotes. Assuming that contemporary mitochondrial genomes are the result of a massive trimming down of a much larger protomitochondrial genome, the fact that they contain much the same genetic information is hard to reconcile with a polyphyletic origin, since it would imply a process of gene loss and retention that has been highly convergent. It is not at all clear what the evolutionary forces operating in such a process would be (but see von Heijne 1986).

In the quest to explore the evolutionary history of mitochondria, and to distinguish between monophyletic and polyphyletic scenarios of mitochondrial origin, much emphasis is being placed on rRNA, the ultimate molecular chronometer (Woese 1987). The rationale for using rRNA sequence comparisons to establish global phylogenetic relationships among organisms and organelles has been discussed elsewhere (Gray et al 1984; Gray 1988; Cedergren et al 1988). One particular consideration is that the genes for SSU and LSU rRNAs are the only ones that are encoded by all organellar genomes as well as by nuclear and prokaryotic (eubacterial and archaebacterial) genomes. Thus, it is possible to use rRNA sequence information to construct phylogenetic trees that simultaneously reveal the evolutionary descent of nuclear and mitochondrial genomes. This allows one not only to determine relationships within each lineage, but to assess the degree of correspondence between nuclear and mitochondrial tree topologies (e.g. Morin & Cech 1988b). In the case of an early, monophyletic origin of mitochondria, nuclear and mitochondrial phylogenies are expected to be congruent, reflecting a parallel descent and evolution of the two genomes. In a polyphyletic scenario, major incongruities in nuclear and mitochondrial phylogenies may be anticipated, which reflects the fact that whereas nuclear genomes will have shared a common line of descent, mitochondrial genomes within these same eukaryotes will not.

It should be emphasized that there are a number of inherent difficulties in using rRNA sequence comparisons to determine mitochondrial phylogenies. Firstly, because of the extreme structural diversity displayed by homologous mitochondrial rRNAs (Gray 1988), one is constrained to use only the most conservative regions in such comparisons. This means that close-range relationships may not be well defined, because there are insufficient differences between the compared sequences. Secondly, because of widely differing rates of mtDNA sequence divergence in different eukaryotes (e.g. animals vs plants), the observed branching patterns may be subject to methodological artifacts (see Cedergren et al 1988). Thus, it is important to test for such artifacts and to interpret the results of rRNA phylogenetic analysis in the light of comparative information about mitochondrial genome organization and expression. That said, rRNA sequence comparisons currently provide the only generally applicable and objective method of establishing long-range evolutionary connections within the mitochondrial lineage.

CONCLUSIONS AND PERSPECTIVE: MITOCHONDRIAL DNA, AN EVOLUTIONARY MOSAIC?

In summing up this admittedly incomplete review of a very complex and controversial topic, we should perhaps return to some of the questions posed in the introduction.

WHAT IS THE EVOLUTIONARY ORIGIN OF THE MITOCHONDRIAL GENOME? At least in the case of rRNA genes, a definite answer can be given. Mitochondrial rRNA genes are of direct eubacterial ancestry (Spencer et al 1984; Gray et al 1984; Yang et al 1985; Cedergren et al 1988). To what extent this fact establishes the origin of the mitochondrial genome as a whole remains to be seen. Certainly, the eubacterial character of the mitochondrial respiratory chain, and its affiliation with one particular phylum of eubacteria (the same phylum to which the ancestry of mitochondrial rRNA genes has been traced), is evident (John 1987). Mitochondrially encoded genes for respiratory proteins are clear homologues of their eubacterial counterparts (Raitio et al 1987). Thus, an endosymbiotic origin of the mitochondrial genome is most consistent with the available evidence. It is true that, in contrast to rRNA genes, we do not have available a range of eubacterial and archaebacterial respiratory gene sequences for comparison with the range of available mitochondrial sequences. Nor do we know very much about genes that we presume were transferred to the nucleus as a consequence of a eubacterial endosymbiosis. Nuclear genes that we should particularly examine are those encoding mitochondrial ribosomal proteins, translation factors, and aminoacyl-tRNA synthetases—proteins having cytosolic homologues that are also encoded by nuclear genes.

WHAT WAS THE ANCESTRAL FORM OF THE MITOCHONDRIAL GENOME? This is particularly difficult to answer. The protomitochondrial genome may have been organized quite differently than contemporary eubacterial genomes; it may, for example, have contained introns and/or much more intergenic spacer sequence. It is reasonable to suppose that the original mitochondrial genome was a large, spaciously arrayed one, containing much noncoding sequence—perhaps not unlike present-day plant mitochondrial genomes. In such a genome considerable evolutionary restructuring could presumably occur with minimal effect on function. In this context we would regard the small compactly organized mtDNAs, such as that in animals, as derivative rather than ancestral forms. Certainly, it is difficult to accept that something resembling the highly condensed animal mitochondrial genome, an exquisite example of economy of organization and expression, could have existed as such at the very earliest stages of mtDNA evolution.

BY WHAT MECHANISM(S) HAS mtDNA EVOLVED IN DIFFERENT EUKARYOTES? As we have seen, comparative studies have started to reveal characteristic patterns of mitochondrial genome evolution, patterns that are quite distinct in the various eukaryotic phyla and that presumably account for much of the structural diversity we see among present-day mtDNAs. In some cases, both the mode and tempo of mtDNA evolution have become apparent as a result of careful, extensive studies of this kind. We can expect that as these studies are extended within a phylum, e.g. from vertebrate to invertebrate animals, the evolutionary basis for differences in mitochondrial genome arrangement and expression will become clearer. What will still be missing are the evolutionary connections between the mtDNAs in the major kingdoms (e.g. plants and animals); these connections may lie within the largely unexplored group of unicellular eukaryotes, the protists. Here again, answers can only come through judicious comparative studies. Equally, we are at a rather primitive stage in our understanding of the biochemistry of enzymes that act on mtDNA, and that surely have played a major role in setting the mitochondrial genome on different evolutionary pathways in different eukaryotes.

IS mtDNA MONOPHYLETIC OR POLYPHYLETIC? As outlined above, one way to address this question is through phylogenetic tree analysis. A recent application of this approach (Cedergren et al 1988) has uncovered a discordance with respect to the branching position of higher plants in nuclear and mitochondrial phylogenies, determined from separate SSU and LSU rRNA databases. Coupled with the fact that green algae (chlorophytes) and higher plants (metaphytes) do not appear to have shared a common mitochondrial ancestor as recently as they have shared a common nuclear ancestor (Gray & Boer 1988), this observation has raised the possibility that the rRNA genes of plant mitochondria have been derived more recently (in a separate endosymbiotic event) than the rRNA genes of other mitochondria (Gray et al 1989). An implication of this suggestion is that mtDNA may be an *evolutionary mosaic*, having derived (and lost) genetic information through various processes of lateral gene transfer in the course of its evolutionary history. Solid evidence that this has occurred in plant mtDNA already exists (Schuster & Brennicke 1988). This leads us to a new perspective on mtDNA evolution. Perhaps some of the genetic information in mtDNA (e.g. the genes encoding components of the respiratory chain) does derive from a single endosymbiotic event, whereas information encoding other components (e.g. of the mitochondrial transcription, translation, and/or RNA processing machinery) has been more labile in the course of evolution. More extensive phylogenetic analysis (involving mitochondrial protein coding as well as rRNA genes) may provide a critical test of this possibility. The concept that eukaryotic cells are evolutionary mosaics is now well established (Margulis 1981); the proposal that eukaryotic genomes are also, to varying extents, evolutionary mosaics, is one worth careful consideration.

ACKNOWLEDGMENTS

I am most grateful to colleagues around the world who provided reprints and preprints of unpublished data and observations. This information was invaluable in the preparation of this review. At the same time, I apologize to all those whose contributions to this topic could not be cited within the confines of this particular presentation. Finally, I greatly appreciate the assistance of Lisa Laskey in the preparation of the final manuscript. Support in the form of an Operating Grant from the Medical Research Council of Canada (MT-4124) and a Fellowship from the Canadian Institute for Advanced Research (Evolutionary Biology Program) is also gratefully acknowledged.

Literature Cited

Attardi, G. 1985. Animal mitochondrial DNA: an extreme example of genetic economy. *Int. Rev. Cytol.* 93: 93–145

Attardi, G., Schatz, 1988. Biogenesis of mitochondria. *Annu. Rev. Cell Biol.* 4: 289–333

Avise, J. C., Arnold, J., Ball, R. M., Bermingham, E., Lamb, T., et al. 1987. Intraspecific phylogeography: the mitochondrial DNA bridge between population genetics and systematics. *Annu. Rev. Ecol. Syst.* 18: 489–522

Benne, R. 1985. Mitochondrial genes in trypanosomes. *Trends Genet.* 1: 117–21

Bentzen, P., Leggett, W. C., Brown, G. G. 1988. Length and restriction site heteroplasmy in the mitochondrial DNA of American shad (*Alosa sapidissima*). *Genetics* 118: 509–18

Bernardi, G. 1979. The petite mutation in yeast. *Trends Biochem. Sci.* 4: 197–201

Bland, M. M., Levings, C. S. III, Matzinger, D. F. 1986. The tobacco mitochondrial ATPase subunit 9 gene is closely linked to an open reading frame for a ribosomal protein. *Mol. Gen. Genet.* 204: 8–16

Boer, P. H., Gray, M. W. 1988a. Genes encoding a subunit of respiratory NADH dehydrogenase (ND1) and a reverse transcriptase-like protein (RTL) are linked to ribosomal RNA gene pieces in *Chlamydomonas reinhardtii* mitochondrial DNA. *EMBO J.* 7: 3501–8

Boer, P. H., Gray, M. W. 1988b. Scrambled ribosomal RNA gene pieces in Chlamydomonas reinhardtii mitochondrial DNA. *Cell* 55: 399–411

Boer, P. H., Gray, M. W. 1988c. Transfer RNA genes and the genetic code in *Chlamydomonas reinhardtii* mitochondria. *Curr. Genet.* 14: 583–90

Bonen, L., Gray, M. W. 1980. Organization and expression of the mitochondrial genome of plants. I. The genes for wheat mitochondrial ribosomal and transfer RNA: evidence for an unusual arrangement. *Nucleic Acids Res.* 8: 319–35

Bonen, L., Boer, P. H., Gray, M. W. 1984. The wheat cytochrome oxidase subunit II gene has an intron insert and three radical amino acid changes relative to maize. *EMBO J.* 3: 2531–36

Borkhardt, B., Brown, T. A., Thim, P., Olson, L. W. 1988. The mitochondrial genome of the aquatic phycomycete *Allomyces macrogynus*. Physical mapping and mitochondrial DNA instability. *Curr. Genet.* 13: 41–47

Borkhardt, B., Olson, L. W. 1986. The mitochondrial genome of the aquatic phycomycete *Blastocladiella emersonii. Curr. Genet.* 11: 139–43

Borkhardt, B., Pedersen, M. B., Olson, L. W. 1987. Mitochondrial DNA in the aquatic fungus *Allomyces. Curr. Genet.* 12: 149–56

Boyd, D. A., Hobman, T. C., Gruenke, S. A., Klassen, G. R. 1984. Evolutionary stability of mitochondrial DNA organization in *Achlya. Can. J. Biochem. Cell Biol.* 62: 571–91

Brown, G. G., Simpson, M. V. 1982. Novel features of animal mtDNA evolution as shown by sequences of two rat cytochrome oxidase subunit II genes. *Proc. Natl. Acad. Sci. USA* 79: 3246–50

Brown, W. M. 1985. The mitochondrial genome of animals. In *Monographs in Evolutionary Biology: Molecular Evolutionary Genetics*, ed. R. J. MacIntyre, pp. 95–130. New York/London: Plenum

Brown, W. M., George, M. Jr., Wilson, A. C. 1979. Rapid evolution of animal mitochondrial DNA. *Proc. Natl. Acad. Sci. USA* 76: 1967–71

Brown, W. M., Prager, E. M., Wang, A., Wilson, A. C. 1982. Mitochondrial DNA sequences of primates: tempo and mode of evolution. *J. Mol. Evol.* 18: 225–39

Bruns, T. D., Palmer, J. D. 1989. Evolution of mushroom mitochondrial DNA: *Suillus* and related genera. *J. Mol. Evol.* 28: 349–62

Bruns, T. D., Palmer, J. D., Shumard, D. S., Grossman, L. I., Hudspeth, M. E. S. 1988. Mitochondrial DNAs of *Suillus*: three fold size change in molecules that share a common gene order. *Curr. Genet.* 13: 49–56

Cann, R. L., Wilson, A. C. 1983. Length mutations in human mitochondrial DNA. *Genetics* 104: 699–711

Cann, R. L., Brown, W. M., Wilson, A. C. 1984. Polymorphic sites and the mechanism of evolution in human mitochondrial DNA. *Genetics* 106: 479–99

Cantatore, P., Gadaleta, M. N., Roberti, M., Saccone, C., Wilson, A. C. 1987. Duplication and remoulding of tRNA genes during the evolutionary rearrangement of mitochondrial genomes. *Nature* 329: 853–55

Cantatore, P., Roberti, M., Rainaldi, G., Saccone, C., Gadaleta, M. N. 1988. Clustering of tRNA genes in *Paracentrotus lividus* mitochondrial DNA. *Curr. Genet.* 13: 91–96

Cantatore, P., Saccone, C. 1987. Organization, structure, and evolution of mammalian mitochondrial genes. *Int. Rev. Cytol.* 108: 149–208

Cavalier-Smith, T. 1987. The simultaneous symbiotic origin of mitochondria, chloroplasts, and microbodies. In *Endocytobiology III*, ed. J. J. Lee, J. F. Fredrick. *Ann. N.Y. Acad. Sci.* 503: 55–71

Cedergren, R., Gray, M. W., Abel, Y., Sankoff, D. 1988. The evolutionary relationships among known life forms. *J. Mol. Evol.* 28: 98–112

Chao, S., Sederoff, R., Levings, C. S. III. 1984. Nucleotide sequence and evolution of the 18S ribosomal RNA gene in maize mitochondria. *Nucleic Acids Res.* 12: 6629–44

Chomyn, A., Attardi, G. 1987. Mitochondrial gene products. *Curr. Top. Bioener.* 15: 295–329

Clark-Walker, G. D. 1985. Basis of diversity in mitochondrial DNAs. In *The Evolution of Genome Size*, ed. T. Cavalier-Smith, pp. 277–97. New York: Wiley

Clark-Walker, G. D., Evans, R. J., Hoeben, P., McArthur, C. R. 1985a. The basis of diversity in yeast mitochondrial DNAs. In *Achievements and Perspectives of Mitochondrial Research, Vol. II. Biogenesis*, ed. E. Quagliariello, E. C. Slater, F. Palmieri, C. Saccone, A. M. Kroon, pp. 71–78. Amsterdam/New York/Oxford: Elsevier

Clark-Walker, G. D., Hoeben, P., Plazinska, A., Smith, D. K., Wimmer, E. H. 1987. Application of mitochondrial DNA analysis to yeast systematics. In *The Expanding Realm of Yeast-Like Fungi*, ed. G. S. de Hoog, M. T. Smith, A. C. M. Weijman, pp. 259–66. Amsterdam: Elsevier

Clark-Walker, G. D., McArthur, C. R., Sriprakash, K. S. 1983. Order and orientation of genic sequences in circular mitochondrial DNA from *Saccharomyces exiguus*: implications for evolution of yeast mtDNAs. *J. Mol. Evol.* 19: 333–41

Clark-Walker, G. D., McArthur, C. R., Sriprakash, K. S. 1985b. Location of transcriptional control signals and transfer RNA sequences in *Torulopsis glabrata* mitochondrial DNA. *EMBO J.* 4: 465–73

Clark-Walker, G. D., Miklos, G. L. G. 1974. Mitochondrial genetics, circular DNA and the mechanism of the *petite* mutation in yeast. *Genet. Res., Camb.* 24: 43–57

Clary, D. O., Wolstenholme, D. R. 1984. The *Drosophila* mitochondrial genome. In *Oxford Surveys on Eukaryotic Genes*, ed. N. Maclean, Vol. 1, pp. 1–35. Oxford: Oxford Univ.

Clayton, D. A. 1984. Transcription of the mammalian mitochondrial genome. *Annu. Rev. Biochem.* 53: 573–94

Clayton, D. A., Doda, J. N., Friedberg, E. C. 1974. The absence of a pyrimidine dimer repair mechanism in mammalian mitochondria. *Proc. Natl. Acad. Sci. USA* 71: 2777–81

Colleaux, L., d'Auriol, L., Galibert, F., Dujon, B. 1988. Recognition and cleavage site of the intron-encoded *omega* transposase. *Proc. Natl. Acad. Sci. USA* 85: 6022–26

Cummings, D. J., Domenico, J. M. 1988. Sequence analysis of mitochondrial DNA from *Podospora anserina*. Pervasiveness of a class I intron in three separate genes. *J. Mol. Biol.* 204: 815–39

Dawson, A. J., Hodge, T. P., Issac, P. G., Leaver, C. J., Lonsdale, D. M. 1986. Location of the genes for cytochrome oxidase subunits I and II, apocytochrome *b*, α-subunit of the F_1 ATPase and the ribosomal RNA genes on the mitochondrial genome of maize (*Zea mays* L.). *Curr. Genet.* 10: 561–64

Dayhoff, M. O., Schwartz, R. M. 1981. Evidence on the origin of eukaryotic mitochondria from protein and nucleic acid sequences. In *Origins and Evolution of Eukaryotic Intracellular Organelles*, ed. J. F. Frederick. *Ann. N.Y. Acad. Sci.* 361: 92–104

de la Cruz, V. F., Neckelmann, N., Simpson, L. 1984. Sequences of six genes and several open reading frames in the kinetoplast maxicircle DNA of *Leishmania tarentolae*. *J. Biol. Chem.* 259: 15136–47

DeSalle, R., Freedman, T., Prager, E. M., Wilson, A. C. 1987. Tempo and mode of sequence evolution in mitochondrial DNA of Hawaiian *Drosophila*. *J. Mol. Evol.* 26: 157–64

Dewey, R. E., Levings, C. S. III, Timothy, D. H. 1986. Novel recombinations in the maize mitochondrial genome produce a unique transcriptional unit in the Texas male-sterile cytoplasm. *Cell* 44: 439–49

de Vries, H., de Jonge, J. C., van't Sant, P., Agsteribbe, E., Arnberg, A. 1981. A "stopper" mutant of *Neurospora crassa* containing two populations of aberrant mitochondrial DNA. *Curr. Genet.* 3: 205–11

de Zamaroczy, M., Faugeron-Fonty, G., Bernardi, G. 1983. Excision sequences in the mitochondrial genome of yeast. *Gene* 21: 193–202

Dieckmann, C. L., Gandy, B. 1987. Preferential recombination between GC clusters in yeast mitochondrial DNA. *EMBO J.* 6: 4197–4203

Doolittle, W. F. 1980. Revolutionary concepts in evolutionary cell biology. *Trends Biochem. Sci.* 5: 146–49

Dujon, B., Colleaux, L., Jacquier, A., Michel, F., Monteilhet, C. 1986. Mitochondrial introns as mobile genetic elements: the role of intron-encoded proteins. In *Extrachromosomal Elements in Lower Eukaryotes*, ed. R. B. Wickner, A. Hinnebusch, A. M. Lambowitz, I. C. Gunsalus, A. Hollaender, pp. 5–27. New York: Plenum

Evans, R. J., Clark-Walker, G. D. 1985. Elevated levels of petite formation in strains of *Saccharomyces cerevisiae* restored to respiratory competence. II: Organization of mitochondrial genomes in strains having high and moderate frequencies of petite mutant formation. *Genetics* 111: 403–32

Garey, J. R., Wolstenholme, D. R. 1989. Platyhelminth mitochondrial DNA: evidence for an early evolutionary origin of a $tRNA^{ser}$AGN that contains a dihydrouridine-arm replacement loop, and of serine-specifying AGA and AGG codons. *J. Mol. Evol.* 28: 374–87

Grant, D., Chiang, K.-S. 1980. Physical mapping and characterization of *Chlamydomonas* mitochondrial DNA molecules: their unique ends, sequence homogeneity, and conservation. *Plasmid* 4: 82–96

Gray, M. W. 1982. Mitochondrial genome diversity and the evolution of mitochondrial DNA. *Can. J. Biochem.* 60: 157–71

Gray, M. W. 1983. The bacterial ancestry of plastids and mitochondria. *BioScience* 33: 693–99

Gray, M. W. 1988. Organelle origins and ribosomal RNA. *Biochem. Cell Biol.* 66: 325–48

Gray, M. W., Boer, P. H. 1988. Organization and expression of algal (*Chlamydomonas reinhardtii*) mitochondrial DNA. *Philos. Trans. R. Soc. London Ser. B* 319: 135–47

Gray, M. W., Cedergren, R., Abel, Y., Sankoff, D. 1989. On the evolutionary origin of the plant mitochondrion and its genome. *Proc. Natl. Acad. Sci. USA* 86: 2267–71

Gray, M. W., Doolittle, W. F. 1982. Has the endosymbiont hypothesis been proven? *Microbiol. Rev.* 46: 1–42

Gray, M. W., Sankoff, D., Cedergren, R. J. 1984. On the evolutionary descent of organisms and organelles: a global phylogeny based on a highly conserved structural core in small subunit ribosomal RNA. *Nucleic Acids Res.* 12: 5837–52

Gray, M. W., Spencer, D. F. 1983. Wheat mitochondrial DNA encodes a eubacteria-like initiator methionine transfer RNA. *FEBS Lett.* 161: 323–27

Gross, S. R., Hsieh, T.-s., Levine, P. H. 1984. Intramolecular recombination as a source of mitochondrial chromosome heteromorphism in Neurospora. *Cell* 38: 233–39

Gualberto, J. M., Wintz, H., Weil, J.-H., Grienenberger, J.-M. 1988. The genes coding for subunit 3 of NADH dehydrogenase and for ribosomal protein S12 are present in the wheat and maize mitochondrial genomes and are co-transcribed. *Mol. Gen. Genet.* 215: 118–27

Gwynn, B., Dewey, R. E., Sederoof, R. R., Timothy, D. H., Levings, C. S. III. 1987. Sequence of the 18S–5S ribosomal gene region and the cytochrome oxidase II gene from mtDNA of *Zea diploperennis*. *Theor Appl. Genet.* 74: 781–88

Haucke, H.-R., Gellissen, G. 1988. Different mitochondrial gene orders among insects: exchanged tRNA gene positions in the COII/COIII region between an orthopteran and a dipteran species. *Curr. Genet.* 14: 471–76

Heinonen, T. Y. K., Schnare, M. N., Young, P. G., Gray, M. W. 1987. Rearranged coding segments, separated by a transfer RNA gene, specify the two parts of a discontinuous large subunit ribosomal RNA in *Tetrahymena pyriformis* mitochondria. *J. Biol. Chem.* 262: 2879–87

Hintz, W. E., Mohan, M., Anderson, J. B., Horgen, P. A. 1985. The mitochondrial DNAs of *Agaricus*: heterogeneity in *A. bitorquis* and homogeneity in *A. brunnescens*. *Curr. Genet.* 9: 127–32

Hoeben, P., Clark-Walker, G. D. 1986. An approach to yeast classification by mapping mitochondrial DNA from *Dekkera/Brettanomyces* and *Eeniella* genera. *Curr. Genet.* 10: 371–79

Horak, I., Coon, H. G., Dawid, I. B. 1974. Interspecific recombination of mitochondrial DNA molecules in hybrid somatic cells. *Proc. Natl. Acad. Sci. USA* 71: 1828–32

Hudspeth, M. E. S., Shumard, D. S., Bradford, C. J. R., Grossman, L. I. 1983. Organization of *Achlya* mtDNA: a population with two orientations and a large inverted repeat containing the rRNA genes. *Proc. Natl. Acad. Sci. USA* 80: 142–46

Huh, T. Y., Gray, M. W. 1982. Conservation of ribosomal RNA gene arrangement in the mitochondrial DNA of angiosperms. *Plant Mol. Biol.* 1: 245–49

Hyman, B. C., Beck, J. L., Weiss, K. C. 1988. Sequence amplification and gene rearrangement in parasitic nematode mitochondrial DNA. *Genetics* 120: 707–12

Jacobs, H. T., Elliot, D., Math, V. B., Farquharson, A. 1988. Nucleotide sequence and gene organization of sea urchin mitochondrial DNA. *J. Mol. Biol.* 202: 185–217

Jacobs, H. T., Asakawa, S., Araki, T., Miura, K.-i., Smith, M. J., Watanabe, K. 1989. Conserved tRNA gene cluster in starfish mitochondrial DNA. *Curr. Genet.* In press

John, P. 1987. *Paracoccus* as a free-living mitochondrion. In *Endocytobiology III*, ed. J. J. Lee, J. F. Frederick. *Ann. N.Y. Acad. Sci.* 503: 140–50

Joyce, P. B. M., Spencer, D. F., Bonen, L., Gray, M. W. 1988a. Genes for $tRNA^{Asp}$, $tRNA^{Pro}$, $tRNA^{Tyr}$ and two $tRNAs^{Ser}$ in wheat mitochondrial DNA. *Plant Mol. Biol.* 10: 251–62

Joyce, P. B. M., Spencer, D. F., Gray, M. W. 1988b. Multiple sequence rearrangements accompanying the duplication of a $tRNA^{Pro}$ gene in wheat mitochondrial DNA. *Plant Mol. Biol.* 11: 833–43

Kotylak, Z., Lazowska, J., Hawthorne, D. C., Slonimski, P. P. 1985. Intron encoded proteins of mitochondria: key elements of gene expression and genomic evolution. In *Achievements and Perspectives of Mitochondrial Research, Vol. II. Biogenesis*, ed. E. Quagliariello, E. C. Slater, F. Palmieri, C. Saccone, A. M. Kroon, pp. 1–20. Amsterdam/New York/Oxford: Elsevier

Kováč, L., Lazowska, J., Slonimski, P. P. 1984. A yeast with linear molecules of mitochondrial DNA. *Mol. Gen. Genet.* 197: 420–24

Kunkel, T. A., Soni, A. 1988. Exonucleolytic proofreading enhances the fidelity of DNA synthesis by chick embryo DNA polymerase-γ. *J. Biol. Chem.* 263: 4450–59

Lang, B. F. 1984. The mitochondrial genome of the fission yeast *Schizosaccharomyces pombe*: highly homologous introns are inserted at the same position of the otherwise less conserved *cox1* genes in *Schizosaccharomyces pombe* and *Aspergillus nidulans*. *EMBO J.* 3: 2129–36

Lang, B. F., Ahne, F., Distler, S., Trinkl, H., Kaudewitz, F., Wolf, K. 1983. Sequence of the mitochondrial DNA, arrangement of genes and processing of their transcripts in *Schizosaccharomyces pombe*. In *Mitochondria 1983. Nucleo-Mitochondrial Interactions*, ed. R. J. Schweyen, K. Wolf, F. Kaudewitz, pp. 313–29. Berlin/New York: de Gruyter

Lansman, R. A., Clayton, D. A. 1975. Selective nicking of mammalian mitochondrial DNA *in vivo*: photosensitization by incorporation of 5-bromodeoxyuridine. *J. Mol. Biol.* 99: 761–76

Lonsdale, D. M. 1987. The molecular

biology and genetic manipulation of the cytoplasm of higher plants. In *Genetic Engineering*, ed. P. W. J. Rigby, Vol. 6, pp. 47–102. London: Academic

Lonsdale, D. M., Brears, T., Hodge, T. P., Melville, S. E., Rottmann, W. H. 1988. The plant mitochondrial genome: homologous recombination as a mechanism for generating heterogeneity. *Philos. Trans. R. Soc. London Ser. B* 319: 149–63

Lonsdale, D. M., Hodge, T. P., Fauron, C. M.-R. 1984. The physical map and organization of the mitochondrial genome from the fertile cytoplasm of maize. *Nucleic Acids Res.* 12: 9249–61

Makaroff, C. A., Palmer, J. D. 1987. Extensive mitochondrial specific transcription of the *Brassica campestris* mitochondrial genome. *Nucleic Acids Res.* 15: 5141–56

Maréchal, L., Runeberg-Roos, P., Grienenberger, J. M., Colin, J., Weil, J. H., et al. 1987. Homology in the region containing a $tRNA^{Trp}$ gene and a (complete or partial) $tRNA^{Pro}$ gene in wheat mitochondrial and chloroplast genomes. *Curr. Genet.* 12: 91–98

Margulis, L. 1981. *Symbiosis in Cell Evolution*. San Francisco: Freeman

Margulis, L., Schwartz, K. V. 1988. *Five Kingdoms. An Illustrated Guide to the Phyla of Life on Earth*, pp. 75–149. New York: Freeman

Michel, F., Dujon, B. 1986. Genetic exchanges between bacteriophage T4 and filamentous fungi? *Cell* 46: 323

Michel, F., Lang, B. F. 1985. Mitochondrial class II introns encode proteins related to the reverse transcriptases of retroviruses. *Nature* 316: 641–43

Mikelsaar, R. 1987. A view of early cellular evolution. *J. Mol. Evol.* 25: 168–83

Miyata, T., Hayashida, H., Kikuno, R., Hasegawa, M., Kobayashi, M., Koike, K. 1982. Molecular clock of silent substitution: at least six-fold preponderance of silent changes in mitochondrial genes over those in nuclear genes. *J. Mol. Evol.* 19: 28–35

Morin, G. B., Cech, T. R. 1986. The telomeres of the linear mitochondrial DNA of Tetrahymena thermophila consist of 53 bp tandem repeats. *Cell* 46: 873–83

Morin, G. B., Cech, T. R. 1988a. Mitochondrial telomeres: surprising diversity of repeated telomeric DNA sequences among six species of Tetrahymena. *Cell* 52: 367–74

Morin, G. B., Cech, T. R. 1988b. Phylogenetic relationships and altered genome structures among *Tetrahymena* mitochondrial DNAs. *Nucleic Acids Res.* 16: 327–46

Moritz, C., Brown, W. M. 1987. Tandem duplications in animal mitochondrial DNAs: variation in incidence and gene content among lizards. *Proc. Natl. Acad. Sci. USA* 84: 7183–87

Moritz, C., Dowling, T. E., Brown, W. M. 1987. Evolution of animal mitochondrial DNA: relevance for population biology and systematics. *Annu. Rev. Ecol. Syst.* 18: 269–92

Mota, E. M., Collins, R. A. 1988. Independent evolution of structural and coding regions in a *Neurospora* mitochondrial intron. *Nature* 332: 654–56

Muhich, M. L., Neckelmann, N., Simpson, L. 1985. The divergent region of the *Leishmania tarentolae* kinetoplast maxicircle DNA contains a diverse set of repetitive sequences. *Nucleic Acids Res.* 13: 3241–60

Newton, K. J. 1988. Plant mitochondrial genomes: organization, expression and variation. *Annu. Rev. Plant Physiol. Plant Mol. Biol.* 39: 503–32

Nugent, J. M., Palmer, J. D. 1988. Location, identity, amount and serial entry of chloroplast DNA sequences in crucifer mitochondrial DNAs. *Curr. Genet.* 14: 501–9

Olivo, P. D., Van de Walle, M. J., Laipis, P. J., Hauswirth, W. W. 1983. Nucleotide sequence evidence for rapid genotypic shifts in the bovine mitochondrial DNA D-loop. *Nature* 306: 400–2

Palmer, J. D. 1985a. Evolution of chloroplast and mitochondrial DNA in plants and algae. In *Monographs in Evolutionary Biology: Molecular Evolutionary Genetics*, ed. R. J. MacIntyre, pp. 131–239. New York/London: Plenum

Palmer, J. D. 1985b. Comparative organization of chloroplast genomes. *Annu. Rev. Genet.* 19: 325–54

Palmer, J. D. 1988. Intraspecific variation and multicircularity in Brassica mitochondrial DNAs. *Genetics* 118: 341–51

Palmer, J. D., Herbon, L. A. 1987. Unicircular structure of the *Brassica hirta* mitochondrial genome. *Curr. Genet.* 11: 565–70

Palmer, J. D., Herbon, L. A. 1988. Plant mitochondrial DNA evolves rapidly in structure, but slowly in sequence. *J. Mol. Evol.* 28: 87–97

Palmer, J. D., Shields, C. R. 1984. Tripartite structure of the *Brassica campestris* mitochondrial genome. *Nature* 307: 437–40

Powell, J. R., Caccone, A., Amato, G. D., Yoon, C. 1986. Rates of nucleotide substitution in *Drosophila* mitochondrial DNA and nuclear DNA are similar. *Proc. Natl. Acad. Sci. USA* 83: 9090–93

Pring, D. R., Lonsdale, D. M. 1985. Molecular biology of higher plant mitochondrial DNA. *Int. Rev. Cytol.* 97: 1–46

Pritchard, A. E., Seilhamer, J. J., Cummings, D. J. 1986. *Paramecium* mitochondrial

DNA sequences and RNA transcripts for cytochrome oxidase subunit I, URF1, and three ORFs adjacent to the replication origin. *Gene* 44: 243–53

Prunell, A., Bernardi, G. 1977. The mitochondrial genome of wild-type yeast cells. VI. Genome organization. *J. Mol. Biol.* 110: 53–74

Raitio, M., Jalli, T., Saraste, M. 1987. Isolation and analysis of the genes for cytochrome *c* oxidase in *Paracoccus denitrificans*. *EMBO J.* 6: 2825–33

Raven, P. H. 1970. A multiple origin for plastids and mitochondria. *Science* 169: 641–46

Richter, C. 1988. Do mitochondrial DNA fragments promote cancer and aging? *FEBS Lett.* 241: 1–5

Schnare, M. N., Heinonen, T. Y. K., Young, P. G., Gray, M. W. 1986. A discontinuous small subunit ribosomal RNA in *Tetrahymena pyriformis* mitochondria. *J. Biol. Chem.* 261: 5187–93

Schuster, W., Brennicke, A., 1987. Plastid, nuclear and reverse transcriptase sequences in the mitochondrial genome of *Oenothera*: is genetic information transferred between organelles via RNA? *EMBO J.* 6: 2857–63

Schuster, W., Brennicke, A. 1988. Interorganellar sequence transfer: plant mitochondrial DNA is nuclear, is plastid, is mitochondrial. *Plant Sci. Lett.* 54: 1–10

Sederoff, R. R. 1984. Structural variation in mitochondrial DNA. *Adv. Genet.* 22: 1–108

Sederoff, R. R. 1987. Molecular mechanisms of mitochondrial-genome evolution in higher plants. *Amer. Natur.* 130: 530–45 (Suppl.)

Seilhamer, J. J., Gutell, R. R., Cummings, D. J. 1984. *Paramecium* mitochondrial genes. II. Large subunit rRNA gene sequence and microevolution. *J. Biol. Chem.* 259: 5173–81

Séraphin, B., Boulet, A., Simon, M., Faye, G. 1987. Construction of a yeast strain devoid of mitochondrial introns and its use to screen nuclear genes involved in mitochondrial splicing. *Proc. Natl. Acad. Sci. USA* 84: 6810–14

Simpson, L. 1986. Kinetoplast DNA in *Trypanosomid* flagellates. *Int. Rev. Cytol.* 99: 119–79

Simpson, L. 1987. The mitochondrial genome of kinetoplastid protozoa: genomic organization, transcription, replication, and evolution. *Annu. Rev. Microbiol.* 41: 363–82

Small, I. D., Isaac, P. G., Leaver, C. J. 1987. Stoichiometric differences in DNA molecules containing the *atpA* gene suggest mechanisms for the generation of mitochondrial genome diversity in maize. *EMBO J.* 6: 865–69

Small, I. D., Suffolk, R., Leaver, C. J. 1989. Evolution of plant mitochondrial genomes via sub-stoichiometric intermediates. *Cell.* In press

Snyder, M., Fraser, A. R., LaRoche, J., Gartner-Kepkay, K. E., Zouros, E. 1987. Atypical mitochondrial DNA from the deep-sea scallop *Placopecten magellanicus*. *Proc. Natl. Acad. Sci. USA* 84: 7595–99

Sogin, M. L., Gunderson, J. H., Elwood, H. J., Alonso, R. A., Peattie, D. A. 1989. Phylogenetic meaning of the kingdom concept: an unusual ribosomal RNA from *Giardia lamblia*. *Science* 243: 75–77

Solignac, M., Génermont, J., Monnerot, M., Mounolou, J.-C. 1987. Drosophila mitochondrial genetics: evolution of heteroplasmy through germ line cell divisions. *Genetics* 117: 687–96

Solignac, M., Monnerot, M., Mounolou, J.-C. 1986. Mitochondrial DNA evolution in the *melanogaster* species subgroup of *Drosophila*. *J. Mol. Evol.* 23: 31–40

Spencer, D. F., Schnare, M. N., Gray, M. W. 1984. Pronounced structural similarities between the small subunit ribosomal RNA genes of wheat mitochondria and *Escherichia coli*. *Proc. Natl. Acad. Sci. USA* 81: 493–97

Stern, D. B., Lonsdale, D. M. 1982. Mitochondrial and chloroplast genomes of maize have a 12-kilobase DNA sequence in common. *Nature* 299: 698–702

Stern, D. B., Palmer, J. D. 1984. Extensive and widespread homologies between mitochondrial DNA and chloroplast DNA in plants. *Proc. Natl. Acad. Sci. USA* 81: 1946–50

Stern, D. B., Palmer, J. D. 1986. Tripartite mitochondrial genome of spinach: physical structure, mitochondrial gene mapping, and locations of transposed chloroplast DNA sequences. *Nucleic Acids Res.* 14: 5651–66

Stewart, K. D., Mattox, K. R. 1984. The case for a polyphyletic origin of mitochondria: morphological and molecular comparisons. *J. Mol. Evol.* 21: 54–57

Suyama, Y. 1986. Two dimensional polyarcylamide gel electrophoresis analysis of *Tetrahymena* mitochondrial tRNA. *Curr. Genet.* 10: 411–20

Suyama, Y., Fukuhara, H., Sor, F. 1985. A fine restriction map of the linear mitochondrial DNA of *Tetrahymena pyriformis*: genome size, map locations of rRNA and tRNA genes, terminal inversion repeat, and restriction site polymorphism. *Curr. Genet.* 9: 479–93

Taylor, F. J. R. 1987. An overview of the

status of evolutionary cell symbiosis theories. In *Endocytobiology III*, ed. J. J. Lee, J. F. Fredrick. *Ann. N.Y. Acad. Sci.* 503: 1–16

Tomkinson, A. E., Bonk, R. T., Linn, S. 1988. Mitochondrial endonuclease activities specific for apurinic/apyrimidinic sites in DNA from mouse cells. *J. Biol. Chem.* 263: 12532–37

Turker, M. S., Domenico, J. M., Cummings, D. J. 1987. Excision-amplification of mitochondrial DNA during senescence in *Podospora anserina*. A potential role for an 11 base-pair consensus sequence in the excision process. *J. Mol. Biol.* 198: 171–85

Tzagoloff, A., Myers, A. M. 1986. Genetics of mitochondrial biogenesis. *Annu. Rev. Biochem.* 55: 249–85

Vawter, L., Brown, W. M. 1986. Nuclear and mitochondrial DNA comparisons reveal extreme rate variation in the molecular clock. *Science* 234: 194–96

Villanueva, E., Luehrsen, K. R., Gibson, J., Delihas, N., Fox, G. E. 1985. Phylogenetic origins of the plant mitochondrion based on a comparative analysis of 5S ribosomal RNA sequences. *J. Mol. Evol.* 22: 46–52

von Heijne, G. 1986. Why mitochondria need a genome. *FEBS Lett.* 198: 1–4

Wallace, D. C. 1982. Structure and evolution of organelle genomes. *Microbiol. Rev.* 46: 208–40

Ward, B. L., Anderson, R. S., Bendich, A. J. 1981. The mitochondrial genome is large and variable in a family of plants (Cucurbitaceae). *Cell* 25: 793–803

Warrior, R., Gall, J. 1985. The mitochondrial DNA of *Hydra attenuata* and *Hydra littoralis* consists of two linear molecules. *Arch. Sc. Genève* 38: 439–45

Weber, C. A., Hudspeth, M. E. S., Moore, G. P., Grossman, L. I. 1986. Analysis of the mitochondrial and nuclear genomes of two basidiomycetes, *Coprinus cinereus* and *Coprinus stercorarius*. *Curr. Genet.* 10: 515–25

Wesolowski, M., Fukuhara, H. 1981. Linear mitochondrial deoxyribonucleic acid from the yeast *Hansenula mrakii*. *Mol. Cell Biol.* 1: 387–93

Wilson, A. C., Cann, R. L., Carr, S. M., George, M., Gyllensten, U. B., et al. 1985. Mitochondrial DNA and two perspectives on evolutionary genetics. *Biol. J. Linn. Soc.* 26: 375–400

Wissinger, B., Hiesel, R., Schuster, W., Brennicke, A. 1988. The NADH-dehydrogenase subunit 5 gene in *Oenothera* mitochondria contains two introns and is co-transcribed with the 5S rRNA gene. *Mol. Gen. Genet.* 212: 56–65

Woese, C. R. 1987. Bacterial evolution. *Microbiol. Rev.* 51: 221–71

Wolf, K., Del Giudice, L. 1987. Horizontal gene transfer between mitochondrial genomes. *Endocyt. C. Res.* 4: 103–20

Wolf, K., Del Giudice, L. 1988. The variable mitochondrial genome of ascomycetes: organization, mutational alterations, and expression. *Adv. Genet.* 25: 185–308

Wolfe, K. H., Li, W.-H., Sharp, P. M. 1987. Rates of nucleotide substitution vary greatly among plant mitochondrial, chloroplast, and nuclear DNAs. *Proc. Natl. Acad. Sci. USA* 84: 9054–58

Wolstenholme, D. R., Clary, D. O. 1985. Sequence evolution of *Drosophila* mitochondrial DNA. *Genetics* 109: 725–44

Wolstenholme, D. R., Macfarlane, J. L., Okimoto, R., Clary, D. O., Wahleithner, J. A. 1987. Bizarre tRNAs inferred from DNA sequences of mitochondrial genomes of nematode worms. *Proc. Natl. Acad. Sci. USA* 84: 1324–28

Yang, D., Oyaizu, Y., Oyaizu, H., Olsen, G. J., Woese, C. R. 1985. Mitochondrial origins. *Proc. Natl. Acad. Sci. USA* 82: 4443–47

Yang, W., Zhou, X. 1988. rRNA genes are located far away from the D-loop region in Peking duck mitochondrial DNA. *Curr. Genet.* 13: 351–55

Yin, S., Heckman, J., RajBhandary, U. L. 1981. Highly conserved GC-rich palindromic DNA sequences flank tRNA genes in Neurospora crassa mitochondria. *Cell* 26: 325–32

Zimmer, M., Welser, F.,Oraler, G., Wolf, K. 1987. Distribution of mitochondrial introns in the species *Schizosaccharomyces pombe* and the origin of the group II intron in the gene encoding apocytochrome b. *Curr. Genet.* 12: 329–36

Zinn, A. R., Pohlman, J. K., Perlman, P. S., Butow, R. A. 1988. *In vivo* double-strand breaks occur at recombinogenic G+C-rich sequences in the yeast mitochondrial genome. *Proc. Natl. Acad. Sci. USA* 85: 2686–90

Annu. Rev. Cell Biol. 1989. 5 : 51–70

THE CHLOROPLAST CHROMOSOMES IN LAND PLANTS

Masahiro Sugiura

Center for Gene Research, Nagoya University, Nagoya 464-01, Japan

---

CONTENTS

INTRODUCTION

Chloroplasts are cytoplasmic organelles present in green plants, which contain the entire machinery for the process of photosynthesis and their own genetic system. Since the demonstration of a unique DNA species in chloroplasts (Sager & Ishida 1963), intensive studies of chloroplast chromosome structure and its gene expression have been made.

The chloroplast chromosome in most land plants consists of a single circular DNA molecule that ranges in size from 120 to 160 kilobase pairs (kbp) (Palmer 1985a). A characteristic feature is the presence of a large inverted repeat (IR), the segments of which are separated by a large and a small single-copy region (denoted as LSC and SSC, respectively). This arrangement results in a doubling of the rRNA genes and any other

0743–4634/89/1115–0051$02.00

genes included within the IRs. In addition, this arrangement is thought to increase the evolutionary stability of the chromosome (Palmer 1985b). The benefits, however, appear to be subtle, in as much as pea, broad bean, and pine chloroplast DNAs are exceptions to this pattern and lack IRs (Koller & Delius 1980; Palmer & Thompson 1981; Strauss et al 1988; Lidholm et al 1988). As another exception, chloroplast DNA of the unicellular alga *Euglena gracilis* contains three tandem repeats, each of which contains an rRNA gene cluster (Gray & Hallick 1978). The entire nucleotide sequences of chloroplast DNAs from tobacco (155,844 bp, Shinozaki et al 1986b); liverwort (121,024 bp, Ohyama et al 1986); and rice (134,525 bp, Hiratsuka et al 1989) have been determined. The chloroplast genes (including putative genes) found so far in vascular plants are presented in Table 1.

In this review I describe the structure of chloroplast genes found in land plants and note where interspecific variation has been observed (for comparison some unique chloroplast genes from algae are included). Although many of these genes have by now been sequenced in half a dozen or more plants, I shall limit references to the first one or two sequences reported when notable differences are not apparent in genes sequenced later. References omitted here should be readily available within the DNA sequence databases. In the final section I focus on the comparison of a monocot (rice) and a dicot (tobacco) chloroplast chromosome. Readers interested in gene expression are referred to the excellent review by Gruissem (1989). Other aspects of chloroplast genomes not covered here are discussed in several recent reviews (Palmer 1985a,b; Weil 1987; Rochaix 1987; Zurawski & Clegg 1987; Gray 1987; Mullet 1988; Sugiura 1989).

GENES FOR THE GENETIC APPARATUS

Ribosomal Genes

Chloroplasts contain 70S ribosomes that are distinct from their larger counterparts found in the cytoplasm. A 23S, a 5S, and a 4.5S rRNA are associated with the 50S subunit, and a 16S rRNA is associated with the 30S subunit. The 4.5S rRNA is homologous to the 3′ end of the 23S rRNA of prokaryotes (Edwards et al 1981). Cloning of the maize chloroplast rRNA genes (rDNAs) led to the discovery that these genes reside within the IRs (Bedbrook et al 1977). Sequencing of the maize (Schwarz & Kössel 1980; Edwards & Kössel 1981; Schwarz et al 1981b; Koch et al 1981) and tobacco (Takaiwa & Sugiura 1980; Tohdoh et al 1981; Takaiwa & Sugiura 1982a,b; Tohdoh & Sugiura 1982) rDNAs revealed a gene order of 16S, 23S, 4.5S, 5S plus an interspersion of tRNA genes within this gene cluster. In *Chlamydomonas reinhardii* the rDNA cluster consists of 16S, 7S, 3S, 23S, and 5S genes in this order (Rochaix & Malnoe 1978), and its 23S rRNA gene was the first split gene found in the chloroplast genome. The rDNA cluster of *Chlorella ellipsoidea* is split into two back-to-back operons: namely, operon 1, 16S rRNA-tRNA^{Ile} (GAU); and operon 2, tRNA^{Ala} (UGC)-23S rRNA-5S rRNA (Yamada & Shimaji 1986).

Chloroplast ribosomes contain about 60 ribosomal proteins, one-third of which are thought to be encoded by chloroplast DNA. Genes for chloroplast ribosomal proteins (*rpl* for large subunit proteins and *rps* for small subunit proteins) have been deduced through their homology with *E. coli* ribosomal protein genes (e.g. Sugita & Sugiura 1983; Subramanian et al 1983). The tobacco, liverwort, and rice chloroplast chromosomes each contain twenty different open reading frames (ORFs) potentially coding for polypeptides homologous to *E. coli* ribosomal proteins (Shinozaki et al 1986b; Ohto et al 1988; Ohyama et al 1986; Hiratsuka et al 1989). However, the typical angiosperm complement of ribosomal protein genes differs from that of liverwort in that *rpl*21 is absent but *rps*16, which contains an intron (Shinozaki et al 1986a), is present. The *rpl*23, *rpl*2, *rps*19, *rpl*22, *rps*3, *rpl*16, *rpl*14, *rps*8, *rpl*36, and *rps*11 sequences are clustered in this order; this arrangement corresponds to that of the homologous genes in the *E. coli* S10 and *spc* operons (Tanaka et al 1986). However legume chloroplast DNAs lack *rpl*22 (Spielmann et al 1988; Palmer et al 1988a).

The *rpl*2 gene generally contains an intron, but this intron is absent in some dicots (Zurawski et al 1984). The *rps*16 and *rpl*16 sequences also contain introns (Shinozaki et al 1986a; Tanaka et al 1986; Posno et al 1986). *Chlamydomonas rpl*16 contains no intron and is localized within the chloroplast DNA replication origin (Lou et al 1987). In *Euglena rpl*23, *rps*19 and *rps*3 contain multiple introns while no intron was found in *rpl*2 (Christopher et al 1988).

Most striking among chloroplast gene structures is *rps*12. This gene consists of three exons, with its 5′ exon (5′-*rps*12) located far away from the other two (3′-*rps*12) in all land plants studied to date (Fromm et al 1986; Torazawa et al 1986; Fukuzawa et al 1986; von Allmen & Stutz 1987). We have designated this gene structure a divided gene (Shinozaki et al 1986b). To produce the mature mRNA, *trans*-splicing between tobacco 5′-*rps*12 and 3′-*rps*12 transcripts occurs in vivo (Zaita et al 1987; Koller et al 1987; Hildebrand et al 1988). In contrast, *Euglena rps*12 is not split (Montandon & Stutz 1984).

In spinach, rice, and wheat, *rpl*23 is present as a pseudogene that bears an internal deletion. The spinach deletion, though short, splits the gene into two overlapping reading frames (Zurawski & Clegg 1987), and S1 nuclear experiments indicate that transcripts initiate both in front of and within this gene (Thomas et al 1988). The internally deleted *rpl*23 genes

Table 1 Chloroplast genes in vascular plants

Genes[a]	Products	Rice[b]	Tobacco[b]	Homology (%)[c]
Genes for the genetic apparatus				
23S rDNA	23S rRNA	2884	2810	94
16S rDNA	16S rRNA	1491	1489	97
5S rDNA	5S rRNA	121	121	97
4.5S rDNA	4.5S rRNA	95	103	83
*trn*A-UGC*	Ala-tRNA (UGC)	73	73	100
*trn*R-ACG	Arg-tRNA (ACG)	74	74	100
*trn*R-UCU	Arg-tRNA (UCU)	72	72	100
*trn*N-GUU	Asn-tRNA (GUU)	72	72	97
*trn*D-GUC	Asp-tRNA (GUC)	74	74	100
*trn*C-GCA	Cys-tRNA (GCA)	71	72	89
*trn*Q-UUG	Gln-tRNA (UUG)	72	72	96
*trn*E-UUC	Glu-tRNA (UUC)	73	73	96
*trn*G-GCC	Gly-tRNA (GCC)	71	71	82
*trn*G-UCC*	Gly-tRNA (UCC)	72	71	97
*trn*H-GUG	His-tRNA (GUG)	75	75	100
*trn*I-GAU*	Ile-tRNA (GAU)	72	72	100
*trn*I-CAU	Ile-tRNA (CAU)	74	74	92
*trn*L-UAA*	Leu-tRNA (UAA)	85	85	97
*trn*L-CAA	Leu-tRNA (CAA)	81	81	98
*trn*L-UAG	Leu-tRNA (UAG)	80	80	96
*trn*K-UUU*	Lys-tRNA (UUU)	72	72	97
*trn*fM-CAU	fMet-tRNA (CAU)	73	73	92
*trn*M-CAU	Met-tRNA (CAU)	73	73	97
*trn*F-GAA	Phe-tRNA (GAA)	73	73	97
*trn*P-UGG	Pro-tRNA (UGG)	74	74	99
*trn*S-GGA	Ser-tRNA (GGA)	87	87	97
*trn*S-UGA	Ser-tRNA (UGA)	88	92	91
*trn*S-GCU	Ser-tRNA (GCU)	88	88	94
*trn*T-GGU	Thr-tRNA (GGU)	72	72	96
*trn*T-UGU	Thr-tRNA (UGU)	73	73	93
*trn*W-CCA	Trp-tRNA (CCA)	74	74	99
*trn*Y-GUA	Tyr-tRNA (GUA)	84	84	99
*trn*V-GAC	Val-tRNA (GAC)	72	72	100
*trn*V-UAC*	Val-tRNA (UAC)	74	73	97
(*rps*2)	30S ribosomal protein CS2	236	236	79
(*rps*3)	CS3	239	218	63
(*rps*4)	CS4	201	201	80
*rps*7	CS7	156	155	85
(*rps*8)	CS8	136	134	75
(*rps*11)	CS11	143	138	68
(*rps*12*)	CS12	124	123	89
(*rps*14)	CS14	103	100	85
(*rps*15)	CS15	78	87	80
(*rps*16*)	CS16	86	86	90
(*rps*18)	CS18	163	101	70
*rps*19	CS19	93	92	68

Table 1 (*continued*)

Genes[a]	Products	Rice[b]	Tobacco[b]	Homology (%)[c]
rpl 2*	50S ribosomal protein CL2	273	274	90
(*rpl* 14)	CL14	123	123	83
(*rpl* 16*)	CL16	136	134	88
(*rpl* 20)	CL20	119	128	69
(*rpl* 22)	CL22	103	155	61
(*rpl* 23)	CL23	93	93	85
(*rpl* 33)	CL33	66	66	73
rpl 36	CL36	37	37	92
rpo A	RNA polymerase, subunit α	337	337	69
(*rpo* B)	β	1075	1070	81
(*rpo* C1*)	β'	682	687	78
(*rpo* C2)	β''	1514	1392	64
(*inf* A)	initiation factor 1	107	96	66
Genes for the photosynthetic apparatus				
rbc L	RuBisCO, large subunit	477	477	93
psa A	PSI, P700 apoprotein A1	750	750	96
psa B	P700 apoprotein A2	734	734	97
psa C	9 kDa protein	81	81	95
psb A	PSII, D1-protein	353	353	99
psb B	47 kDa protein	508	508	97
psb C	43 kDa protein	473	473	97
psb D	D2-protein	353	353	98
psb E	cytochrome b559 (9 kDa)	83	83	98
psb F	cytochrome b559 (4 kDa)	39	39	100
psb G	G-protein	246	284	82
psb H	10 kDa phosphoprotein	73	73	90
psb I	I-protein	36	52	97
psb K	K-protein	61	98	72
psb L	L-protein	38	38	100
pet A	b/f complex, cytochrome f	320	320	91
pet B*	cytochrome b6	215	215	99
pet D*	subunit IV	160	160	99
atp A	H^+-ATPase, subunit α	507	507	88
atp B	subunit β	498	498	92
atp E	subunit ε	137	133	73
atp F*	subunit I	180	184	79
atp H	subunit III	81	81	99
atp I	subunit IV	247	247	93
(*ndh* A*)	NADH dehydrogenase, ND1	362	364	76
(*ndh* B*)	ND2	510	387	96
(*ndh* C)	ND3	120	120	87
(*ndh* D)	ND4	500	509	82
(*ndh* E)	ND4L	101	101	83
(*ndh* F)	ND5	734	710	67

[a] * split genes, ( ) putative genes (products not detected).
[b] Numbers of base pairs (structural RNA gene) or amino acids (protein-coding gene).
[c] Calculated by the GENETYX program HOMOGAPN (tRNAs, rRNAs) and HOMOGAPP (proteins).

of wheat and rice are present at a secondary site and, at least in rice, an intact copy of the gene is still present in its normal position upstream of *rpl*2 (Moon et al 1988; Ogihara et al 1988; Hiratsuka et al 1989). It remains a puzzle why apparently defective copies of this gene should accumulate in chloroplasts, which for the most part maintain compact genomes.

Transfer RNA Genes

Chloroplast chromosomes apparently encode all of the tRNAs involved in chloroplast protein synthesis. In contrast to the clustered arrangement of the transfer RNA (*trn*) genes observed in *Euglena* (Hallick et al 1984), in land plant genomes they are scattered throughout the chromosome (e.g. Driesel et al 1979). The sequencing of numerous *trn* genes, beginning in 1981 (Schwarz et al 1981a,b; Kato et al 1981), enables us to draw generalizations about their structure. Although chloroplast tRNA sequences resemble those of prokaryotes, no tRNA genes code for their own 3′-CCA end; hence these must be added posttranscriptionally. Six tRNA genes harbor long single introns (0.5–2.5 kbp). The presence of introns in chloroplast tRNA genes was first demonstrated in maize *trn*I and *trn*A genes, which are located in the spacer separating the 16S and 23S rDNAs (Koch et al 1981). In contrast, unsplit *trn*I and *trn*A genes have so far been reported in all algae (e.g. Graf et al 1980). A split *trn*V-UAC gene was found in tobacco (Deno et al 1982). Interestingly, the *trn*G-UCC gene contains an intron in the D-stem region (Deno & Sugiura 1984), a feature unique, so far, to the chloroplast chromosome. The *trn*K gene has the longest intron (2.5 kbp) and contains an ORF in the intron, potentially encoding a 60 kd protein in tobacco (Sugita et al 1985) and mustard (Neuhaus & Link 1987).

The introns of these five tRNA genes can be folded into a secondary structure that is similar to the proposed structure for group II introns (Michel & Dujon 1983). However, introns of *trn*V, *trn*G, and *trn*K have conserved boundary sequences (5′-GTGYGNY, RYCNRYT(A)YNAY-3′), which resemble those of introns found in chloroplast protein genes (Hallick et al 1985; Sugita et al 1985). We have therefore classified these three introns together with those from split protein genes as a new group and designated the group III (Shinozaki et al 1986a). The *trn*L-UAA gene contains an intron between the first and second nucleotides of the anticodon (Steinmetz et al 1982). This intron can be folded into a structure similar to the postulated structure of the group I intron of the self-splicable rRNA precursor of *Tetrahymena* (Bonnard et al 1984). Therefore, the chloroplast introns can be classified into three distinct groups (Shinozaki et al 1986a; Sugiura et al 1987).

When all the tobacco chloroplast DNA fragments that hybridize to total chloroplast tRNAs were sequenced, thirty different tRNA genes were found (Wakasugi et al 1986). After the complete sequence of the tobacco chloroplast DNA had been determined, more tRNA genes were sought by computer, but none were found. Hence these 30 tRNA genes are all expressed, and they probably constitute all of the tRNA genes in the tobacco chloroplast (Shinozaki et al 1986b). All the tRNAs can form the cloverleaf structure and none has an abnormal form (Sugiura 1987). The rice chloroplast DNA contains the same set of tRNA sequences as in tobacco (Hiratsuka et al 1989). Thirty-one distinct tRNA genes were deduced from the entire DNA sequence of liverwort chloroplast DNA, a putative $tRNA^{Arg}$ (CCG) gene constituting the additional member (Ohyama et al 1986).

All possible codons are used in the sequences coding for polypeptides in chloroplasts (e.g. Wakasugi et al 1986). The minimum number of tRNA species required for translation of all codons had been thought to be 32. However, no tRNAs that recognize codons CUU/C(Leu), CCU/C(Pro), GCU/C(Ala), and CGC/A/G(Arg) by standard pairing rules have been found. If the two-out-of-three mechanism can operate in the chloroplast, as has been shown in an in vitro protein synthesizing system from *E. coli* (Samuelsson et al 1980), the single $tRNA^{Pro}$ (UGG), $tRNA^{Ala}$ (UGC), and $tRNA^{Arg}$ (ACG) species can read all four Pro, Ala, and Arg codons, respectively (note that these three tRNAs form only GC pairs in their first and second position codon pairings). Finally, if the gene for $tRNA^{Leu}$ (UAG) produces a tRNA with an unmodified U in the first position of the anticodon, this tRNA can read all four Leu codons (CUN) by U:N wobble. Since the bean, spinach, and soybean $tRNAs^{Leu}$ (UAG) do have an unmodified U in their anticodons ($UA^{m7}G$) (Pillay et al 1984), the 30 chloroplast tRNAs are probably sufficient to read all codons (Shinozaki et al 1986b).

Other Putative Genes

These genes for protein components of the chloroplast transcriptional and translational apparatus have all been found through homology with the corresponding *E. coli* proteins. Transcripts have been detected from most of these genes but few translation products have been identified. Therefore we must still refer to most of these as putative genes.

Chloroplast RNA polymerase appears to be encoded in the nuclear genome (Lerbs et al 1985). However, DNA sequences hybridizing with the α, β, and β' subunit genes of *E. coli* RNA polymerase were reported in the *Chlamydomonas* chloroplast chromosome (Watson & Surzycki 1983). Chloroplast DNA regions potentially coding for polypeptides similar to

E. coli RNA polymerase α, β, and β′ subunits (*rpo*A, *rpo*B, and *rpo*C) have been found in all completely sequenced chloroplast genomes, as well as in spinach (Sijben-Müller et al 1986; Ohme et al 1986; Shinozaki et al 1986b; Ohyama et al 1986; Hudson et al 1988; Hiratsuka et al 1989). However no sequences similar to a bacterial *rpo*D gene have been found. These findings suggest that chloroplasts synthesize three of their RNA polymerase subunits, but do not prove that these syntheses are exclusively organelle-encoded. Recently a polypeptide cross-reacting with an antiserum against a synthetic *rpo*A peptide was found in a maize chloroplast lysate (Ruf & Kössel 1988).

A putative gene for the protein synthesis initiation factor IF-1 (*inf*A) was found in spinach (Sijben-Müller et al 1986). A sequence similar to the *E. coli* EF-Tu gene (*tuf*A) has been found in *Euglena* (Montandon & Stutz 1984), but not in any land plant chloroplast DNAs sequenced to date.

GENES FOR THE PHOTOSYNTHETIC APPARATUS

Gene for the Large Subunit of RuBisCO

Ribulose-1,5-biphosphate carboxylase/oxygenase (RuBisCo) is the major stromal protein of chloroplasts and is composed of eight identical large subunits (LS) of 53 kd and eight identical small subunits (SS) of 12–14 kd. LS is encoded in the chloroplast DNA and SS is encoded in the nuclear DNA. The only exception to this pattern so far is a sea alga, *Olisthodiscus luteus*, whose SS gene is found in the chloroplast DNA (Reith & Cattolico, 1986).

Since its cloning and sequencing in maize (McIntosh et al 1980) the gene for LS (*rbc*L) has become the single most widely sequenced chloroplast gene, allowing sequence comparisons to be used to determine phylogenetic associations among plant species (e.g. Zurawski & Clegg 1987). In land plants and *Chlamydomonas*, the *rbc*L gene contains no introns, while nine introns have been found in this gene in *Euglena* (Koller et al 1984).

Genes for the Thylakoid Membrane System

Thylakoid membranes consist of five functionally distinct complexes: photosystem I (PSI), photosystem II (PSII), the light-harvesting chlorophyll protein complex (whose proteins are all nuclear-coded), the cytochrome b/f complex, and the ATP synthase complex.

At least three components of PSI are encoded in the chloroplast DNA. Genes for two subunits (A1 and A2) of P700 chlorophyll *a* apoprotein (*psa*A and *psa*B) were localized in the spinach chloroplast (Westhoff et al 1983a) and these genes were first sequenced from maize (Fish et al 1985). The *psa*A and *psa*B genes in land plants contain no introns, are situated tandemly, and bear about 45% similarity toward each other at the amino acid level. In *Chlamydomonas* the *psa*A gene is split into three exons scattered around the chloroplast DNA while the *psa*B gene is uninterrupted (Kück et al 1987). The three distantly separated exons of *psa*A constitute a functional gene, which probably operates by a *trans*-splicing mechanism. *Euglena psa*A and *psa*B contain multiple introns (Cushman et al 1988b). A gene for a 9 kd apoprotein for the iron-sulfur centers A and B (*psa*C) has also been located on the chloroplast SSC (Hayashida et al 1987; Høj et al 1987). This gene was identified by first comparing the N-terminal amino acid sequence of the corresponding spinach 9 kd polypeptide with the sequence of tobacco chloroplast DNA and then detecting the appropriate transcripts in tobacco.

At least 11 components of PSII are encoded in the chloroplast chromosome. The gene (*psb*A) for the 32 kd protein Q_B (or D1 protein) was the first PSII component gene sequenced in spinach and *Nicotiana debneyi* (Zurawski et al 1982a). This gene has been studied intensively in a number of plants because the 32 kd protein binds to herbicides such as atrazine. *Psb*A genes isolated from herbicide-resistant mutants have point mutations resulting in substitution from serine to glycine or alanine at codon 264 of the 32 kd protein (e.g. Hirschberg & McIntosh 1983; Goloubinoff et al 1984). Though not split in land plants, the *psb*A genes of both *Euglena* and *Chlamydomonas* are interrupted by four introns (Karabin et al 1984; Erickson et al 1984).

The genes for the 47 kd and 43 kd proteins (*psb*B and *psb*C) were first localized in the spinach chloroplast DNA (Westhoff et al 1983b) and the *psb*B sequence was determined (Morris & Herrmann 1984). The D2 protein or 32 kd-like protein is also encoded in the chloroplast DNA (*psb*D). Sequencing of the *psb*C and *psb*D genes from land plants has revealed that *psb*D overlaps *psb*C by 50 bp (e.g. Alt et al 1984; Holschuh et al 1984).

The genes for the 9 kd and 4 kd subunits of cytochrome b-559 (*psb*E and *psb*F) were first sequenced from spinach (Herrmann et al 1984) and contain no introns. However *Euglena psb*E and *psb*F contain introns (Cushman et al 1988a). An ORF on the maize chloroplast DNA was identified as the gene for the 24 kd G-protein (*psb*G) by using antibodies against synthetic oligopeptides deduced from the DNA sequence for Western blotting (Steinmetz et al 1986). The gene for the 10 kd phosphoprotein (*psb*H) was identified by comparing the N-terminal amino acid sequence of the spinach peptide (Farchaus & Dilley 1986) with the entire sequence of tobacco chloroplast DNA (Shinozaki et al 1986b).

Two ORFs of 52 and 98 codons have been found between *trn*S and *trn*Q in the tobacco chloroplast DNA (Deno & Sugiura 1983). ORF52 was suggested as the gene for the 6 kd protein (*psb*I) by using an antibody against a synthetic peptide (Kato et al 1987) and later identified by determination of the partial amino acid sequence of spinach 4.8 kd protein (Ikeuchi & Inoue 1988). ORF98 was identified as the gene for the 2 kd protein by comparison with a partial amino acid sequence (Murata et al 1988). The N-terminal amino acid sequences of wheat 3.2 kd PSII protein (Webber et al 1988) and the related spinach 5 kd PSII protein (Ikeuchi et al 1989) matched the polypeptide encoded by ORF38 located just downstream of *psb*E/*psb*F; hence ORF38 was identified as the gene for the small polypeptide associated with PSII preparations (*psb*L).

The cytochrome b/f complex consists of six components, of which cytochrome f (*pet*A), cytochrome b6 (*pet*B), and subunit IV (*pet*D) are coded for by the chloroplast chromosome. *Pet*A was first sequenced in pea (Willey et al 1984). *Pet*B and *pet*D are located close together (Alt et al 1983a) and were first sequenced in pea and spinach (Phillips & Gray 1984; Heinemeyer et al 1984). The *pet*B and *pet*D genes are clustered with *psb*B and *psb*H; both *pet*B and *pet*D contain single introns with short first exons (6–8 bp).

Six genes for ATP synthase subunits (α, β, ε, I, III, and IV) are present in the chloroplast DNA. Genes for the β and ε subunits (*atp*B and *atp*E) are located upstream from the *rbc*L gene (e.g. Westhoff et al 1981) and were first sequenced from maize and spinach (Krebbers et al 1982, Zurawski et al 1982b). The *atp*B and *atp*E genes in most plants overlap by 4 bp, but in pea and sweet potato these two genes do not overlap (Zurawski et al 1986; Kobayashi et al 1987). The gene for the α subunit (*atp*A) is located far away from the *atp*B/*atp*E cluster (e.g. Westhoff et al 1981; Howe et al 1983) and was initially sequenced in tobacco (Deno et al 1983). The genes for the IV, III, and I subunits (*atp*I, *atp*H, and *atp*F) are clustered in this order just before *atp*A (e.g. Howe et al 1982; Alt et al 1983b; Hennig & Herrmann 1986; Cozens et al 1986). The *atp*F genes contain single introns (Bird et al 1985). The predicted amino acid sequences of these six subunits show homology with their counterparts in *E. coli*.

Putative ndh *Genes*

Six chloroplast DNA sequences whose predicted amino acid sequences resemble those of components (ND1, 2, 3, 4, 4L, and 5) of the respiratory chain NADH dehydrogenase from human mitochondria have been found in tobacco (Shinozaki et al 1986b). As these sequences are highly expressed, they are likely to be the genes for components of a chloroplast NADH dehydrogenase (*ndh*A, B, C, D, E, and F) (Matsubayashi et al 1987). The tobacco *ndh*A and *ndh*B genes contain single introns. Similar sequences have also been found in the chloroplast DNA of several other land plants

(Meng et al 1986; Ohyama et al 1986; vom Stein & Hachtel 1988; Schantz & Bogorad 1988; Hiratsuka et al 1989). These observations suggest the existence of a respiratory chain in chloroplasts, in agreement with earlier physiological studies in *Chlamydomonas* (Bennoun 1982; Godde 1982).

COMPARISON OF MONOCOT AND DICOT CHLOROPLAST CHROMOSOMES

The evolution and organization of chloroplast chromosome structure in plants have been studied by comparative DNA sequence mapping, which makes use of heterologous DNA hybridization (e.g. Palmer 1985a; Palmer et al 1988b). The determination of corresponding DNA sequences from two or more chloroplast chromosomes has enabled us more accurately to identify the rearrangements that have occurred during evolution (e.g. Quigley & Weil 1985; Howe 1985; Zurawski & Clegg 1987; Howe et al 1988; vom Stein & Hachtel 1988; Bowman et al 1988).

As the complete nucleotide sequences of rice (a monocot) and tobacco (a dicot) chloroplast DNAs are available, an overall comparison has been made between them (Hiratsuka et al 1989). The rice chromosome is about 14% smaller than the tobacco chromosome (Figure 1). The length of the rice SSC is about two-thirds as long as that of the tobacco SSC, a difference largely accounted for by the absence of 4.8 kbp of DNA found in tobacco between *rps*15 and *trn*N. In rice, ORF1244 found in tobacco is missing and *rps*15 is contained within the IRs (hence two copies per genome). The rice IRs also have expanded into a sequence corresponding to the LSC of tobacco and contain *trn*H and *rps*19 from the LSC of tobacco. This expansion is not simple because *trn*H is located between *rpl*2 and *rps*19. The gene order *rpl*23, *rpl*2, *rps*19, *rpl*22, and so on is colinear in both dicot chloroplasts and the *E. coli* S10 operon, which suggests that this constitutes the ancestral gene order, and that the insertion of *trn*H into the cluster was a more recent event. A similar expansion of IRs is observed in wheat (Bowman et al 1988). Despite the apparent expansion of the IRs into both directions, the rice IR is about 18% shorter in length. Apparent deletions in rice relative to tobacco account for this discrepancy, the largest of which occurs near the junction with the LSC. At this site 6 kbp of a sequence corresponding in tobacco to ORF1708 and ORF581 are absent. The length of the rice LSC is about 6% shorter than that of tobacco. Numerous short apparent deletions exist in the LSC as compared with the tobacco LSC. In the region between *rbc*L and ORF36, the rice sequence is 1.4 kbp shorter than tobacco and has very little homology with tobacco.

Extensive rearrangements are evident within the rice LSC. These

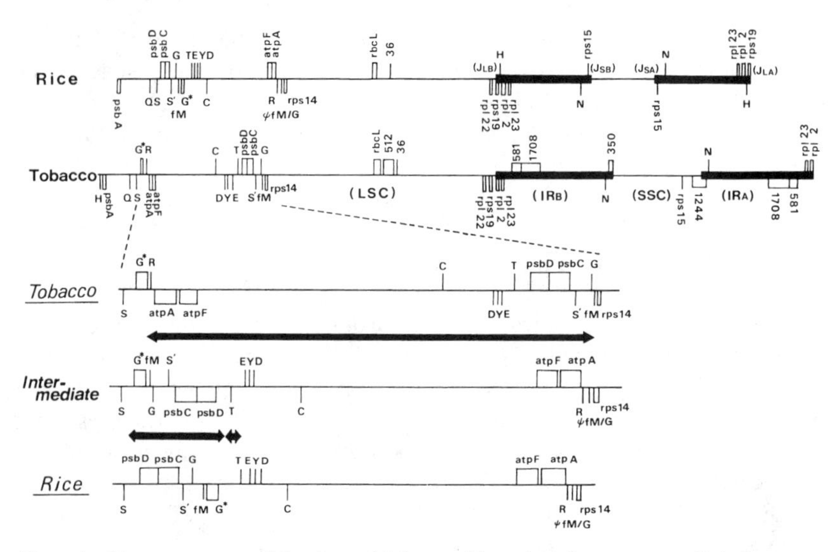

Figure 1 Linear gene map of the rice and tobacco chloroplast chromosomes. Only the genes and their neighbors discussed in this review are shown. Expanded maps (*bottom 3 lines*) show the evolution of an extensively rearranged part of the rice chloroplast DNA from an ancestral form in Tobacco, through an intermediate form to the present form in Rice. Bold arrows represent inversions. Genes shown above lines are transcribed rightward and those below lines, leftward. Letters indicate *trn* genes and numerals indicate ORFs.

rearrangements are probably confined within the grass family, since other monocots, such as orchids, retain a tobacco-like gene order (Palmer et al 1988b). However a gene arrangement similar to rice was reported for the corresponding region of the wheat LSC. It has been suggested that the rearrangement of the wheat LSC with respect to the tobacco (and spinach) LSC is the result of three inversions (Quigley & Weil 1985; Howe 1985; Howe et al 1988). Starting from a tobacco-like ancestral form, a large inversion is thought to have occurred encompassing 28 kbp, its endpoints lying between *trn*G* (UCC) and *trn*R (UCU) at one end and between *rps*14 and *trn*fM at the other end (see Figure 1). This led to the intermediate form. Two further inversions [the region between *psb*D and *trn*G* (UCC), and the short region containing *trn*T] subsequently gave rise to the present form observed in rice (and wheat) chloroplasts. Copies of a short, repeated sequence have been described near the endpoints of the largest inversion in wheat, and it was proposed that the repeat mediated the inversion via homologous recombination (Howe 1985). The endpoints of the second and third inversions in wheat are not known to be associated with repeated sequences. However the endpoints of all three inversions lie adjacent to at least one tRNA gene, which suggests that tRNA genes might be involved in rearrangements of chloroplast DNA (Howe et al 1988).

A pseudogene to *trn*fM (between *rps*14 and *trn*R) has been noted in wheat (Howe 1985) and maize (Rodermel et al 1987). A close comparison of this sequence with other rice tRNA genes shows that it is chimeric, deriving its 5′ sequence from *trn*fMet and its 3′ sequence from *trn*G* (UCC) (Hiratsuka et al 1989). This chimeric tRNA pseudogene overlaps the endpoints of the largest inversion. A model invoking illegitimate intermolecular recombination between tRNA genes was proposed that accounts simultaneously for the origin of this pseudogene, the large inversion, and the creation of repeated sequences near the endpoints (Hiratsuka et al 1989).

CONCLUSIONS

Thirty-four genes for RNA components and 55 genes for polypeptides (including putative genes) have so far been found in the chloroplast genome of vascular plants, but dozens of ORFs still remain to be identified. Most of the putative genes (*ndh*A-F, *rpo*A-C, *inf*A, and most of *rpl*/*rps*) described through homology with the corresponding genes of other organisms are transcribed in chloroplasts, but their translation products remain to be isolated and characterized.

The chloroplast chromosome shows both prokaryotic and eukaryotic features. Chloroplast genes coding for components of the genetic system and subunits of the ATP synthase have substantial sequence homology with their prokaryotic counterparts. On the other hand, at least sixteen chloroplast genes contain introns, which are typical in eukaryotic genes. Splicing and processing of precursor RNAs are often catalyzed by protein-RNA complexes. RNA molecules are thought not to be imported into chloroplasts from the cytoplasm, which suggests that the RNA components, if any, should be encoded in the chloroplast chromosome. No genes for RNAs other than tRNAs and rRNAs have so far been identified.

In some cases coding sequences have sequences associated nearby that resemble regulatory sequences such as promoters and terminators from *E. coli*. However, assumptions should be avoided. Only a few chloroplast promoters have been experimentally identified (e.g. Link 1984; Gruissem & Zurawski 1985). Furthermore, a class of chloroplast genes was shown to require no upstream promoter elements for transcription (Gruissem et al 1986), and some hair-pin structures in the chloroplast DNA have been shown to serve in transcript processing rather than termination (Stern &

Gruissem 1987). It is essential to use in vitro transcription systems and mutational analyses for an accurate analysis of the regulatory sequences on the chloroplast chromosome.

The chloroplast chromosomes in vascular plants are similar in size, gene content, and gene order. Moreover the sequences of the chloroplast's genes are highly conserved (see Table 1). This indicates that the chloroplast genome is a slowly evolving chromosome (Palmer 1985a). The endosymbiotic theory, which proposes that chloroplasts were derived from an ancestral photosynthetic prokaryote related cyanobacteria, has been supported in part by comparisons between chloroplast and cyanobacterial *rrn* operons and several other genes. This suggests that ancestral cyanobacteria had introns in their genomes, which are still retained today in chloroplast genomes, but have been lost in present day eubacteria (Tomioka & Sugiura 1984). The fact that several archaebacterial genes have been shown to contain introns supports this idea.

ACKNOWLEDGMENTS

I am grateful to Dr. H. Shimada for preparing the table and figure, and Dr. R. Whittier for critical reading of the manuscript.

Literature Cited

Alt, J., Morris, J., Westhoff, P., Herrmann, R. G. 1984. Nucleotide sequence of the clustered genes for the 44 kd chlorophyll *a* apoprotein and the "32 kd"-like protein of the photosystem II reaction center in the spinach plastid chromosome. *Curr. Genet.* 8: 597–606

Alt, J., Westhoff, P., Sears, B. B., Nelson, N., Hurt, E., et al. 1983a. Genes and transcripts for the polypeptides of the cytochrome b6/f complex from spinach thylakoid membranes. *EMBO J.* 2: 979–86

Alt, J., Winter, P., Sebald, W., Moser, J. G., Schedel, R., et al. 1983b. Localization and nucleotide sequence of the gene for the ATP synthase proteolipid subunit on the spinach plastid chromosome. *Curr. Genet.* 7: 129–38

Bedbrook, J. R., Kolodner, R., Bogorad, L. 1977. Zea mays chloroplast ribosomal RNA genes are part of a 22,000 base pair inverted repeat. *Cell* 11: 739–49

Bennoun, P. 1982. Evidence for a respiratory chain in the chloroplast. *Proc. Natl. Acad. Sci. USA* 79: 4352–56

Bird, C. R., Koller, B., Auffret, A. D., Huttly, A. K., Howe, C. J., et al. 1985. The wheat chloroplast gene for CF_0 subunit of ATP synthase contains a long intron. *EMBO J.* 4: 1381–88

Bonnard, G., Michel, F., Weil, J. H., Steinmetz, A. 1984. Nucleotide sequence of the split $tRNA^{Leu}$ UAA gene from *Vicia faba* chloroplasts: evidence for structural homologies of the chloroplast $tRNA^{Leu}$ intron with the intron from the autosplicable *Tetrahymena* ribosomal RNA precursor. *Mol. Gen. Genet.* 194: 330–36

Bowman, C. M., Barker, R. F., Dyer, T. A. 1988. In wheat ctDNA, segments of ribosomal protein genes are dispersed repeats, probably conserved by nonreciprocal recombination. *Curr. Genet.* 14: 127–36

Christopher, D. A., Cushman, J. C., Price, C. A., Hallick, R. B. 1988. Organization of ribosomal protein genes *rpl*23, *rpl*2, *rps*19, *rpl*22, and *rps*3 on the *Euglena gracilis* chloroplast genome. *Curr. Genet.* 14: 275–86

Cozens, A. L., Walker, J. E., Phillips, A. L., Huttly, A. K., Gray, J. C. 1986. A sixth subunit of ATP synthase, an F_0 component, is encoded in the pea chloroplast genome. *EMBO J.* 5: 217–22

Cushman, J. C., Christopher, D. A., Little,

M. C., Hallick, R. B., Price, C. A. 1988a. Organization of the *psb*E, *psb*F, *orf*38, and *orf*42 gene loci on the *Euglena gracilis* chloroplast genome. *Curr. Genet.* 13: 173–80

Cushman, J. C., Hallick, R. B., Price, C. A. 1988b. The two genes for the P_{700} chlorophyll *a* apoproteins on the *Euglena gracilis* chloroplast genome contain multiple introns. *Curr. Genet.* 13: 159–71

Deno, H., Kato, A., Shinozaki, K., Sugiura, M. 1982. Nucleotide sequences of tobacco chloroplast genes for elongator $tRNA^{Met}$ and $tRNA^{Val}$ (UAC): the $tRNA^{Val}$ (UAC) gene contains a long intron. *Nucleic Acids Res.* 10: 7511–20

Deno, H., Shinozaki, K., Sugiura, M. 1983. Nucleotide sequence of tobacco chloroplast gene for the α subunit of proton-translocating ATPase. *Nucleic Acids Res.* 11: 2185–91

Deno, H., Sugiura, M. 1983. The nucleotide sequences of $tRNA^{Ser}$ (GCU) and $tRNA^{Gln}$ (UUG) genes from tobacco chloroplasts. *Nucleic Acids Res.* 11: 8407–14

Deno, H., Sugiura, M. 1984. Chloroplast tide sequences of $tRNA^{Ser}$ (GCU) and $tRNA^{Gln}$ (UUG) genes from tobacco chloroplast genes for $tRNA^{Gly}$ (UCC) and $tRNA^{Arg}$ (UCU). *Proc. Natl. Acad. Sci. USA* 81: 405–8

Driesel, A. J., Crouse, E. J., Gordon, K., Bohnert, H. J., Herrmann, R. G., et al. 1979. Fractionation and identification of spinach chloroplast transfer RNAs and mapping of their genes on the restriction map of chloroplast DNA. *Gene* 6: 285–306

Edwards, K., Bedbrook, J., Dyer, T., Kössel, H. 1981. 4.5S rRNA from *Zea mays* chloroplasts shows structural homology with the 3′-end of prokaryotic 23S rRNA. *Biochem. Int.* 2: 533–38

Edwards, K., Kössel, H. 1981. The rRNA operon from *Zea mays* chloroplasts: Nucleotide sequence of 23S rDNA and its homology with *E. coli* 23S rDNA. *Nucleic Acids Res.* 9: 2853–69

Erickson, J. M., Rahire, M., Rochaix, J. D. 1984. *Chlamydomonas reinhardii* gene for the 32,000 mol. wt. protein of photosystem II contains four large introns and is located entirely within the chloroplast inverted repeat. *EMBO J.* 3: 2753–62

Farchaus, J., Dilley, R. A. 1986. Purification and partial sequence of the Mr 10,000 phosphoprotein from spinach thylakoids. *Arch. Biochem. Biophys.* 244: 94–101

Fish, L. E., Kück, U., Bogorad, L. 1985. Two partially homologous adjacent light-inducible maize chloroplast genes encoding polypeptides of the P700 chlorophyll *a*-protein complex of photosystem I. *J. Biol. Chem.* 260: 1413–21

Fromm, H., Edelman, M., Koller, B., Goloubinoff, P., Galun, E. 1986. The enigma of the gene coding for ribosomal protein S12 in the chloroplast of *Nicotiana*. *Nucleic Acids Res.* 14: 883–98

Fukuzawa, H., Kohchi, T., Shirai, H., Ohyama, K., Umesono, K., et al. 1986. Coding sequences for chloroplast ribosomal protein S12 from liverwort, *Marchantia polymorpha*, are separated far apart on the different DNA strands. *FEBS Lett.* 198: 11–15

Godde, D. 1982. Evidence for a membrane bound NADH-plastoquinone-oxidoreductase in *Chlamydomonas reinhardii* CW-15. *Arch. Microbiol.* 131: 197–202

Goloubinoff, P., Edelman, M., Hallick, R. B. 1984. Chloroplast-coded atrazine resistance in *Solanum nigrum*: *psb*A loci from susceptible and resistant biotypes are isogenic except for a single codon change. *Nucleic Acids Res.* 12: 9489–96

Graf, L., Kössel, H., Stutz, E. 1980. Sequencing of 16S-23S spacer in a ribosomal RNA operon of *Euglena gracilis* chloroplast DNA reveals two tRNA genes. *Nature* 286: 908–10

Gray, J. C. 1987. Genetics and synthesis of chloroplast membrane proteins. In *Photosynthesis*, ed. J. Amesz, pp. 319–42. Amsterdam: Elsevier

Gray, P. W., Hallick, R. B. 1978. Physical mapping of the *Euglena gracilis* chloroplast DNA and ribosomal RNA gene region. *Biochem.* 17: 284–89

Gruissem, W. 1989. Chloroplast RNA: transcription and processing. In *Biochemistry of Plants*, ed. A. Marcus. pp. 151. San Diego: Academic

Gruissem, W., Elsner-Menzel, C., Latshaw, S., Narita, J. O., Schaffer, M. A., et al. 1986. A subpopulation of spinach chloroplast tRNA genes does not require upstream promoter elements for transcription. *Nucleic Acids Res.* 14: 7541–56

Gruissem, W., Zurawski, G. 1985. Identification and mutational analysis of the promoter for a spinach chloroplast transfer RNA gene. *EMBO J.* 4: 1637–44

Hallick, R. B., Gingrich, J. C., Johanningmeier, U., Passavant, C. W. 1985. Introns in *Euglena* and *Nicotiana* chloroplast protein genes. In *Molecular Form and Function of the Plant Genome*, ed. L. van Vloten-Doting, G. S. P. Groot, T. C. Hall, pp. 211–31. New York: Plenum

Hallick, R. B., Hollingsworth, M. J., Nickoloff, J. A. 1984. Transfer RNA genes of *Euglena gracilis* chloroplast DNA. *Plant Mol. Biol.* 3: 169–75

Hayashida, N., Matsubayashi, T., Shinozaki, K., Sugiura, M., Inoue, K., et al. 1987. The gene for the 9 kd polypeptide, a possible apoprotein for the iron-sulfur centers A and B of the photosystem I complex, in tobacco chloroplast DNA. *Curr. Genet*. 12: 247–50

Hennig, J., Herrmann, R. G. 1986. Chloroplast ATP synthase of spinach contains nine nonidentical subunit species, six of which are encoded by plastid chromosomes in two operons in a phylogenetically conserved arrangement. *Mol. Gen. Genet*. 203: 117–28

Herrmann, R. G., Alt, J., Schiller, B., Widger, W. R., Cramer, W. A. 1984. Nucleotide sequence of the gene for apocytochrome b-559 on the spinach plastid chromosome: Implications for the structure of the membrane protein. *FEBS Lett*. 176: 239–44

Heinemeyer, W., Alt, J., Herrmann, R. G., 1984. Nucleotide sequence of the clustered genes for apocytochrome b6 and subunit 4 of the cytochrome b/f complex in the spinach plastid chromosome. *Curr. Genet*. 8: 543–49

Hildebrand, M., Hallick, R. B., Passavant, C. W., Bourque, D. P. 1988. Trans-splicing in chloroplasts: The *rps*12 loci of *Nicotiana tabacum*. *Proc. Natl. Acad. Sci. USA* 85: 372–76

Hiratsuka, J., Shimada, H., Whittier, R. F., Ishibashi, T., Sakamoto, M., et al. 1989. The complete sequence of the rice (*Oryza sativa*) chloroplast genome. *Mol. Gen. Genet*. In press

Hirschberg, J., McIntosh, L. 1983. Molecular basis of herbicide resistance in *Amaranthus hybridus*. *Science* 222: 1346–49

Holschuh, K., Bottomley, W., Whitfeld, P. R. 1984. Structure of the spinach chloroplast genes for the D2 and 44 kd reaction center proteins of photosystem II and for $tRNA^{Ser}$ (UGA). *Nucleic Acids Res*. 12: 8819–34

Høj, P. B., Svendsen, I., Scheller, H. V., Moller, B. L. 1987. Identification of a chloroplast-encoded 9-kDa polypeptide as a 2[4Fe-4S] protein carrying centers A and B of photosystem I. *J. Biol. Chem*. 262: 12676–84

Howe, C. J. 1985. The endpoint of an inversion in wheat chloroplast DNA are associated with short repeated sequences containing homology to *att*-lambda. *Curr. Genet*. 10: 139–45

Howe, C. J., Auffret, A. D., Doherty, A., Bowman, C. M., Dyer, T. A., et al. 1982. Location and nucleotide sequence of the gene for the proton-translocating subunit of wheat chloroplast ATP synthase. *Proc. Natl. Acad. Sci. USA* 79: 6903–7

Howe, C. J., Barker, R. F., Bowman, C. M., Dyer, T. A. 1988. Common features of three inversions in wheat chloroplast DNA. *Curr. Genet*. 13: 343–49

Howe, C. J., Bowman, C. M., Dyer, T. A., Gray, J. C. 1983. The genes for the alpha and proton-translocating subunits of wheat chloroplast ATP synthase are close together on the same strand of chloroplast DNA. *Mol. Gen. Genet*. 190: 51–55

Hudson, G. S., Holton, T. A., Whitfeld, P. R., Bottomley, W. 1988. Spinach chloroplast *rpo*BC genes encode three subunits of the chloroplast RNA polymerase. *J. Mol. Biol*. 200: 639–54

Ikeuchi, M., Inoue, Y. 1988. A new photosystem II reaction center component (4.8 kDa protein) encoded by chloroplast genome. *FEBS Lett*. 241: 99–104

Ikeuchi, M., Takio, K., Inoue, Y. 1989. N-terminal sequencing of photosystem II low-molecular-mass proteins: 5 and 4.1 kDa components of the O_2-evolving core complex from higher plants. *FEBS Lett*. 242: 263–69

Karabin, G. D., Farley, M., Hallick, R. B. 1984. Chloroplast gene for M_r 32,000 polypeptide of photosystem II in *Euglena gracilis* is interrupted by four introns with conserved boundary sequences. *Nucleic Acids Res*. 12: 5801–12

Kato, A., Shimada, H., Kusuda, M., Sugiura, M. 1981. The nucleotide sequences of two $tRNA^{Asn}$ genes from tobacco chloroplasts. *Nucleic Acids Res*. 9: 5601–7

Kato, K., Sayer, R. T., Bogorad, L. 1987. Expression of the PSII-I gene in maize chloroplasts. *Proc. Ann. Meet. Jpn. Soc. Plant Physiol*. Urawa. pp. 208 (in Japanese)

Kobayashi, K., Nakamura, K., Asahi, T. 1987. CF_1 ATPase β- and ε-subunit genes are separated in the sweet potato chloroplast genome. *Nucleic Acids Res*. 15: 7177

Koch, W., Edwards, K., Kössel, H. 1981. Sequencing of the 16S-23S spacer in a ribosomal RNA operon of Zea mays chloroplast DNA reveals two split tRNA genes. *Cell* 25: 203–13

Koller, B., Delius, H. 1980. *Vicia faba* chloroplast DNA has only one set of ribosomal RNA genes as shown by partial denaturation mapping and R-loop analysis. *Mol. Gen. Genet*. 178: 261–69

Koller, B., Gingrich, J. C., Stiegler, G. L., Farley, M. A., Delius, H., et al. 1984. Nine introns with conserved boundary sequences in the *Euglena gracilis* chloroplast ribulose-1,5-bisphosphate carboxylase gene. *Cell* 36: 545–53

Koller, B., Fromm, H., Galun, E., Edelman, M. 1987. Evidence for in vivo *trans*-

splicing of pre-mRNAs in tobacco chloroplasts. *Cell* 48: 111–19

Krebbers, E. T., Larrinua, I. M., McIntosh, L., Bogorad, L. 1982. The maize chloroplast genes for the β and ε subunits of the photosynthetic coupling factor CF1 are fused. *Nucleic Acids Res.* 10: 4985–5002

Kück, U., Choquet, Y., Schneider, M., Dron, M., Bennoun, P. 1987. Structural and transcriptional analysis of two homologous genes for the P700 chlorophyll *a*-apoproteins in *Chlamydomonas reinhardii*: evidence for *in vivo trans*-splicing. *EMBO J.* 6: 2185–95

Lerbs, S., Bräutigam, E., Parthier, B. 1985. Polypeptides of DNA-dependent RNA polymerase of spinach chloroplast: characterization by antibody-linked polymerase assay and determination of sites of synthesis. *EMBO J.* 4: 1661–66

Lidholm, J., Szmidt, A. E., Hällgren, J.-E., Gustafsson, P. 1988. The chloroplast genomes of conifers lack one of the rRNA-encoding inverted repeats. *Mol. Gen. Genet.* 212: 6–10

Link, G. 1984. DNA sequence requirements for the accurate transcription of a protein-coding plastid gene in a plastid *in vitro* system from mustard (*Sinapis alba* L.). *EMBO J.* 3: 1697–704

Lou, J. K., Wu, M., Chang, C. H., Cuticchia, A. J. 1987. Localization of a r-protein gene within the chloroplast DNA replication origin of *Chlamydomonas*. *Curr. Genet.* 11: 537–41

Matsubayashi, T., Wakasugi, T., Shinozaki, K., Shinozaki, Y. K., Zaita, N., et al. 1987. Six chloroplast genes (*ndh*A to F) homologous to human mitochondrial genes encoding components of the respiratory-chain NADH dehydrogenase are actively expressed: Determination of the splice-sites in *ndh*A and *ndh*B pre-mRNAs. *Mol. Gen. Genet.* 210: 385–93

McIntosh, L., Poulsen, C., Bogorad, L. 1980. Chloroplast gene sequence for the large subunit of ribulose bisphosphate carboxylase of maize. *Nature* 288: 556–60

Meng, B. Y., Matsubayashi, T., Wakasugi, T., Shinozaki, K., Sugiura, M., et al. 1986. Ubiquity of the genes for components of a NADH dehydrogenase in higher plant chloroplast genomes. *Plant Sci.* 47: 181–84

Michel, F., Dujon, B. 1983. Conservation of RNA secondary structures in two intron families including mitochondrial-, chloroplast- and nuclear-encoded members. *EMBO J.* 2: 33–38

Montandon, P.-E., Stutz, E. 1984. The genes for the ribosomal proteins S12 and S7 are clustered with the gene for the EF-Tu protein on the chloroplast genome of *Euglena gracilis*. *Nucleic Acids Res.* 12: 2851–59

Morris, J., Herrmann, R. G. 1984. Nucleotide sequence of the gene for the P680 chlorophyll *a* apoprotein of the photosystem II reaction center from spinach. *Nucleic Acids Res.* 12: 2837–50

Moon, E., Kao, T.-H., Wu, R. 1988. Rice mitochondrial genome contains a rearranged chloroplast gene cluster. *Mol. Gen. Genet.* 213: 247–53

Mullet, J. E. 1988. Chloroplast development and gene expression. *Annu. Rev. Plant Physiol.* 39: 475–502

Murata, N., Miyao, M., Hayashida, N., Hidaka, T., Sugiura, M. 1988. Identification of a new gene in the chloroplast genome encoding a low-molecular-mass polypeptide of photosystem II complex. *FEBS Lett.* 235: 283–88

Neuhaus, H., Link, G. 1987. The chloroplast $tRNA^{Lys}$ (UUU) gene from mustard (*Sinapis alba*) contains a class II intron potentially coding for a maturase-related polypeptide. *Curr. Genet.* 11: 251–57

Ogihara, Y., Terachi, T., Sasakuma, T. 1988. Intramolecular recombination of chloroplast genome mediated by short direct-repeat sequences in wheat species. *Proc. Natl. Acad. Sci. USA* 85: 8573–77

Ohme, M., Tanaka, M., Chunwongse, J., Shinozaki, K., Sugiura, M. 1986. A tobacco chloroplast DNA sequence possibly coding for a polypeptide similar to *E. coli* RNA polymerase β subunit. *FEBS Lett.* 200: 87–90

Ohto, C., Torazawa, K., Tanaka, M., Shinozaki, K., Sugiura, M. 1988. Transcription of ten ribosomal protein genes from tobacco chloroplasts: a compilation of ribosomal protein genes found in the tobacco chloroplast genome. *Plant Mol. Biol.* 11: 589–600

Ohyama, K., Fukuzawa, H., Kohchi, T., Shirai, H., Sano, T., et al. 1986. Chloroplast gene organization deduced from complete sequence of liverwort *Marchantia polymorpha* chloroplast DNA. *Nature* 322: 572–74

Palmer, J. D. 1985a. Comparative organization of chloroplast genomes. *Annu. Rev. Genet.* 19: 325–54

Palmer, J. D. 1985b. Evolution of chloroplast and mitochondrial DNA in plants and algae. In *Monographs in Evolutionary Biology*: *Molecular Evolutionary Genetics*, ed. R. J. Macintyre, pp. 131–240. New York: Plenum

Palmer, J. D., Osorio, B., Thompson, W. F. 1988a. Evolutionary significance of inversions in legume chloroplast DNAs. *Curr. Genet.* 14: 65–74

Palmer, J. D., Jansen, R. K., Michaels, H. J., Chase, M. W., Manhart, J. R. 1988b. Chloroplast DNA variation and plant phylogeny. *Ann. M. Bot. Gard.* 75: 1180–218

Palmer, J. D., Thompson, W. F. 1981. Rearrangements in the chloroplast genomes of mung bean and pea. *Proc. Natl. Acad. Sci. USA* 78: 5533–37

Phillips, A. L., Gray, J. C. 1984. Location and nucleotide sequence of the gene for the 15.2 kDa polypeptide of the cytochrome b-f complex from pea chloroplast. *Mol. Gen. Genet.* 194: 477–84

Pillay, D. T. N., Guillemaut, G., Weil, J. H. 1984. Nucleotide sequences of three soybean chloroplast $tRNAs^{Leu}$ and re-examination of bean chloroplast $tRNA_2^{Leu}$ sequence. *Nucleic Acids Res.* 12: 2997–3001

Posno, M., van Vliet, A., Groot, G. S. P. 1986. The gene for *Spirodela oligorhiza* chloroplast ribosomal protein homologous to *E. coli* ribosomal protein L16 is split by a large intron near its 5′ end: structure and expression. *Nucleic Acids Res.* 14: 3181–95

Quigley, F., Weil, J. H. 1985. Organization and sequence of five tRNA genes and of an unidentified reading frame in the wheat chloroplast genome: evidence for gene rearrangements during the evolution of chloroplast genomes. *Curr. Genet.* 9: 495–503

Reith, M., Cattolico, R. A. 1986. Inverted repeat of *Olisthodiscus luteus* chloroplast DNA contains genes for both subunits of ribulose-1,5-bisphosphate carboxylase and the 32,000-dalton Q_B protein: Phylogenetic implications. *Proc. Natl. Acad. Sci. USA* 83: 8599–603

Rochaix, J.-D. 1987. Molecular genetics of chloroplasts and mitochondria in the unicellular alga *Chlamydomonas. FEMS Microbiol. Rev.* 46: 13–34

Rochaix, J. D., Malnoe, P. 1978. Anatomy of the chloroplast ribosomal DNA of *Chlamydomonas reinhardii. Cell* 15: 661–70

Rodermel, S., Orlin, P., Bogorad, L. 1987. The transcription termination region between two convergently-transcribed photoregulated operons in the maize plastid chromosome contains *rps*14, *trn*R(UCU) and a putative *trnf*M pseudogene. *Nucleic Acids Res.* 15: 5493

Ruf, M., Kössel, H. 1988. Structure and expression of the gene coding for the α-subunit of DNA-dependent RNA polymerase from the chloroplast genome of *Zea mays. Nucleic Acids Res.* 16: 5741–54

Sager, R., Ishida, M. R. 1963. Chloroplast DNA in *Chlamydomonas. Proc. Natl. Acad. Sci. USA* 50: 725–30

Samuelsson, T., Elias, P., Lusting, F., Axberg, T., Fölsch, G., et al. 1980. Aberrations of the classic codon reading scheme during protein synthesis in vitro. *J. Biol. Chem.* 255: 4583–88

Schantz, R., Bogorad, L. 1988. Maize chloroplasts genes *ndh*D, *ndh*E, and *psa*C. Sequences, transcripts and transcript pools. *Plant Mol. Biol.* 11: 239–47

Schwarz, Z., Jolly, S. O., Steinmetz, A. A., Bogorad, L. 1981a. Overlapping divergent genes in the maize chloroplast chromosome and in vitro transcription of the gene for $tRNA^{His}$. *Proc. Natl. Acad. Sci. USA* 78: 3423–27

Schwarz, Z., Kössel, H. 1980. The primary structure of 16S rDNA from *Zea mays* chloroplast is homologous to *E. coli* 16S rRNA. *Nature* 283: 739–42

Schwarz, Z., Kössel, H., Schwarz, E., Bogorad, L. 1981b. A gene coding for $tRNA^{Val}$ is located near 5′ terminus of 16S rRNA gene in *Zea mays* chloroplast genome. *Proc. Natl. Acad. Sci. USA* 78: 4748–52

Shinozaki, K., Deno, H., Sugita, M., Kuramitsu, S., Sugiura, M. 1986a. Intron in the gene for the ribosomal proteins S16 of tobacco chloroplast and its conserved boundary sequences. *Mol. Gen. Genet.* 202: 1–5

Shinozaki, K., Ohme, M., Tanaka, M., Wakasugi, T., Hayashida, N., et al. 1986b. The complete nucleotide sequence of the tobacco chloroplast genome: its gene organization and expression. *EMBO J.* 5: 2043–49

Sijben-Müller, G., Hallick, R., Alt, J., Westhoff, P., Herrmann, R. G. 1986. Spinach plastid genes coding for initiation factor IF-1, ribosomal protein S11 and RNA polymerase α-subunit. *Nucleic Acids Res.* 14: 1029–44

Spielmann, A., Roux, E., von Allmen, J.-M., Stutz, E. 1988. The soybean chloroplast genome: complete sequence of the *rps*19 gene, including flanking parts containing exon 2 of *rpl*2 (upstream), but lacking *rpl*22 (downstream). *Nucleic Acids Res.* 16: 1199

Steinmetz, A. A., Castroviejo, M., Sayre, R. T., Bogorad, L. 1986. Protein PSII-G: An additional component of photosystem II identified through its plastid gene in maize. *J. Biol. Chem.* 261: 2485–88

Steinmetz, A., Gubbins, E. J., Bogorad, L. 1982. The anticodon of the maize chloroplast gene for $tRNA^{Leu}$ UAA is split by a large intron. *Nucleic Acids Res.* 10: 3027–37

Stern, D. B., Gruissem, W. 1987. Control of plastid gene expression: 3′ inverted repeats act as mRNA processing and stabilizing elements, but do not terminate transcription. *Cell* 51: 1145–57

Strauss, S. H., Palmer, J. D., Howe, G. T., Doerksen, A. H. 1988. Chloroplast genomes of two conifers lack a large inverted repeat and are extensively rearranged. *Proc. Natl. Acad. Sci. USA* 85: 3898–902

Subramanian, A. R., Steinmetz, A., Bogorad L. 1983. Maize chloroplast DNA encodes a protein sequence homologous to the bacterial ribosome assembly protein S4. *Nucleic Acids Res.* 11: 5277–86

Sugita, M., Shinozaki, K., Sugiura, M. 1985. Tobacco chloroplast $tRNA^{Lys}$ (UUU) gene contains a 2.5-kilobase-pair intron: An open reading frame and a conserved boundary sequence in the intron. *Proc. Natl. Acad. Sci. USA* 82: 3557–61

Sugita, M., Sugiura, M. 1983. Putative gene of tobacco chloroplast coding for ribosomal protein similar to *E. coli* ribosomal protein S19. *Nucleic Acids Res.* 11: 1913–18

Sugiura, M. 1987. Structure and function of the tobacco chloroplast genome. *Bot. Mag. Tokyo* 100: 407–36

Sugiura, M. 1989. The chloroplast genome. In *Biochemistry of Plants*, ed. A. Marcus, 15: 133–50. San Diego: Academic

Sugiura, M., Shinozaki, K., Tanaka, M., Hayashida, N., Wakasugi, T., et al. 1987. Split genes and *cis/trans* splicing in tobacco chloroplasts. In *Plant Molecular Biology*, ed. D. von Wettstein, N. H. Chua, pp. 65–76. New York: Plenum

Takaiwa, F., Sugiura, M. 1980. The nucleotide sequence of 4.5S and 5S ribosomal RNA genes from tobacco chloroplasts. *Mol. Gen. Genet.* 180: 1–4

Takaiwa, F., Sugiura, M. 1982a. Nucleotide sequence of the 16S-23S spacer region in an rRNA gene cluster from tobacco chloroplast DNA. *Nucleic Acids Res.* 10: 2665–76

Takaiwa, F., Sugiura, M. 1982b. The complete nucleotide sequence of a 23S ribosomal gene from tobacco chloroplasts. *Eur. J. Biochem.* 124: 13–19

Tanaka, M., Wakasugi, T., Sugita, M., Shinozaki, K., Sugiura, M. 1986. Genes for the eight ribosomal proteins are clustered on the chloroplast genome of tobacco (*Nicotiana tabacum*): Similarity to the S10 and *spc* operons of *Escherichia coli*. *Proc. Natl. Acad. Sci. USA* 83: 6030–34

Thomas, F., Massenet, O., Dorne, A. M., Briat, J. F., Mache, R. 1988. Expression of the *rpl*23, *rpl*2 and *rps*19 genes in spinach chloroplasts. *Nucleic Acids. Res.* 16: 2461–72

Tohdoh, N., Shinozaki, K., Sugiura, M. 1981. Sequence of a putative promoter region for the rRNA genes of tobacco chloroplast DNA. *Nucleic Acids Res.* 9: 5399–406

Tohdoh, N., Sugiura, M. 1982. The complete nucleotide sequence of a 16S ribosomal RNA gene from tobacco chloroplast. *Gene* 17: 213–18

Tomioka, N., Sugiura, M. 1984. Nucleotide sequence of the 16S-23S spacer region in the *rrn*A operon from a blue-green alga, *Anacystis nidulans*. *Mol. Gen. Genet.* 193; 427–30

Torazawa, K., Hayashida, N., Obokata, J., Shinozaki, K., Sugiura, M. 1986. The 5′ part of the gene for ribosomal protein S12 is located 30 kbp downstream from its 3′ part in tobacco chloroplast genome. *Nucleic Acids Res.* 14: 3143

vom Stein, J., Hachtel, W. 1988. Chloroplast DNA differences in the genus *Oenothera* subsection *Munzia*: a short direct repeat resembling the lambda chromosomal attachment site occurs as a deletion/insertion within an intron of an NADH-dehydrogenase gene. *Curr. Genet.* 13: 191–97

von Allmen, J.-M., Stutz, E. 1987. Complete sequence of 'divided' *rps*12 (r-protein S12) and *rps*7 (r-protein S7) gene in soybean chloroplast DNA. *Nucleic Acids Res.* 15: 2387

Wakasugi, T., Ohme, M., Shinozaki, K., Sugiura, M. 1986. Structures of tobacco chloroplast genes for $tRNA^{Ile}$ (CAU), $tRNA^{Leu}$ (CAA), $tRNA^{Cys}$ (GCA), $tRNA^{Ser}$ (UGA) and $tRNA^{Thr}$ (GGU): a compilation of tRNA genes from tobacco chloroplasts. *Plant Mol. Biol.* 7: 385–92

Watson, J. C., Surzycki, S. J. 1983. Both the chloroplast and nuclear genomes of *Chlamydomonas reinhardii* share homology with *Escherichia coli* genes for transcriptional and translational components. *Curr. Genet.* 7: 201–10

Webber, A. N., Hird, S. M., Packman, L. C., Dyer, T. A., Gray, J. C. 1989. A photosystem II polypeptide is encoded by an open reading frame cotranscribed with genes for cytochrome *b*-559 in wheat chloroplast DNA. *Plant Mol. Biol.* 12: 141–51

Weil, J. H. 1987. Organization and expression of the chloroplast genome. *Plant Sci.* 49: 149–57

Westhoff, P., Alt, J., Nelson, N., Bottomley, W., Bunemann, H., et al. 1983a. Genes and transcripts for the P_{700} chlorophyll *a* apoprotein and subunit 2 of the photosystem I reaction center complex from spinach thylakoid membranes. *Plant Mol. Biol.* 2: 95–107

Westhoff, P., Alt, J., Herrmann, R. G.

1983b. Localization of the genes for the two chlorophyll *a*-conjugated polypeptides (mol. wt. 51 and 44 kd) of the photosystem II reaction center on the spinach plastid chromosome. *EMBO J.* 2: 2229–37

Westhoff, P., Nelson, N., Bünemann, H., Herrmann, R. G. 1981. Localization of genes for coupling factor subunits on the spinach plastid chromosome. *Curr Genet.* 4: 109–20

Willey, D. L., Auffret, A. D., Gray, J. C. 1984. Structure and topology of cytochrome *f* in pea chloroplast membranes. *Cell* 36: 555–62

Yamada, T., Shimaji, M. 1986. Peculiar feature of the organization of rRNA genes of the *Chlorella* chloroplast DNA. *Nucleic Acids Res.* 14: 3827–39

Zaita, N., Torazawa, K., Shinozaki, K., Sugiura, M. 1987. *Trans* splicing in vivo: joining of transcripts from the 'divided' gene for ribosomal protein S12 in the chloroplasts of tobacco. *FEBS Lett.* 210: 153–56

Zurawski, G., Bohnert, H. J., Whitfeld, P. R., Bottomley, W. 1982a. Nucleotide sequence of the gene for the M_r 32,000 thylakoid membrane protein from *Spinacia oleracea* and *Nicotiana debneyi* predicts a totally conserved primary translation product of M_r 38,950. *Proc. Natl. Acad. Sci. USA* 79: 7699–703

Zurawski, G., Bottomley, W., Whitfeld, P. R. 1982b. Structure of the genes for the β and ε subunits of spinach chloroplast ATPase indicates a dicistronic mRNA and an overlapping translation stop/start signal. *Proc. Natl. Acad. Sci. USA* 79: 6260–64

Zurawski, G., Bottomley, W., Whitfeld, P. R. 1984. Junctions of the large single copy region and the inverted repeats in *Spinacia oleracea* and *Nicotiana debneyi* chloroplast DNA: sequence of the genes for $tRNA^{His}$ and the ribosomal proteins S19 and L2. *Nucleic Acids Res.* 12: 6547–58

Zurawski, G., Bottomley, W., Whitfeld, P. R. 1986. Sequence of the genes for the β and ε subunits of ATP synthase from pea chloroplasts. *Nucleic Acids Res.* 14: 3974

Zurawski, G., Clegg, M. T. 1987. Evolution of higher-plant chloroplast DNA-encoded genes: implications for structure-function and phylogenetic studies. *Annu. Rev. Plant Physiol.* 38: 391–418

Annu. Rev. Cell Biol. 1989. 5 : 71–92

TENASCIN: AN EXTRACELLULAR MATRIX PROTEIN PROMINENT IN SPECIALIZED EMBRYONIC TISSUES AND TUMORS

Harold P. Erickson

Department of Cell Biology, Duke University Medical Center, Durham, North Carolina 27710

Mario A. Bourdon

La Jolla Cancer Research Foundation, La Jolla, California 92037

CONTENTS

INTRODUCTION—HISTORY OF DISCOVERIES AND NAMES

Tenascin is a large oligomeric glycoprotein of the extracellular matrix (ECM). It is synthesized at specific times and locations during embryonic

0743–4634/89/1115–0071$02.00

development; it is present in very restricted locations in adult tissues; and it is prominently expressed in a variety of tumors. The protein is present in some apparently stable locations, but is especially prominent during development and growth of embryonic tissues and tumors.

Tenascin was discovered independently by several laboratories, each laboratory giving it a different name: glioma mesenchymal extracellular matrix (GMEM) protein, myotendinous antigen, hexabrachion protein, cytotactin, and J1. The history of these discoveries and a number of important earlier observations are reviewed by Erickson & Lightner (1988). Other recent reviews are by Chiquet (1989) and Ekblom & Aufderheide (1989). We will briefly mention the early discoveries most relevant to the themes of this review.

One of the earliest discoveries was that of the GMEM protein by Bourdon et al (1983, 1984, 1985). They immunized mice with whole human glioblastoma cells, and isolated a monoclonal antibody that reacted with a novel component in the ECM of most gliomas, and also with many other human tumors. At approximately the same time Chiquet & Fambrough (1984a,b) obtained a monoclonal antibody against an ECM molecule in chicken embryos. This antibody (M1) demonstrated a distribution quite different from fibronectin, laminin, or other ECM proteins. It stained dense connective tissues (tendons, ligaments, bones, cartilage) and smooth muscle. These authors did not associate the protein with tumors, but Chiquet-Ehrismann et al (1986), using a polyclonal antibody against chicken myotendinous antigen, demonstrated that this EMC protein was absent in adult rat mammary gland, but was very prominent in the stroma of chemically induced mammary carcinomas. They proposed the name tenascin (which was not meant to imply tenacity, but combined the original word tendon and the association with growth, nascent).

Erickson & Iglesias (1984) discovered the molecule as a contaminant of cell surface fibronectin preparations. They used electron microscopy to demonstrate the elaborate six-armed structure of the disulfide-bonded oligomer, and termed the molecule a hexabrachion. Vaughan et al (1987) and Erickson & Taylor (1987) showed that the myotendinous antigen was the chicken hexabrachion and the latter authors also showed that GMEM was the human hexabrachion. The demonstrated homology of GMEM and the myotendinous antigen confirmed the link between embryonic and tumorigenic expression demonstrated by Chiquet-Ehrismann et al (1986).

Tenascin was discovered as a major ECM protein of embryonic brain of mouse and chicken and given the names J1 (Kruse et al 1985) and cytotactin (Grumet et al 1985). These proteins are now known to be tenascin of mouse and chicken (Erickson & Taylor 1987; Hoffman et al 1988; Faissner et al 1988). Gulcher et al (1986) identified a large glycoprotein of human brain, which is now known to be tenascin (Gulcher et al 1989).

BIOCHEMICAL CHARACTERIZATION

Figure 1 shows the structure of a tenascin molecule observed by rotary shadowing electron microscopy. The molecule was called a hexabrachion in reference to the six long, thin arms (Erickson & Iglesias 1984). The characteristic features of the hexabrachion are (*a*) a terminal knob on each arm, (*b*) a thick distal segment, (*c*) a thin proximal segment, (*d*) a T-junction where three arms are joined to form a trimer, and (*e*) a central knob, where two trimers are joined to form the hexamer. The protein runs as an oligomer on nonreducing gels, which implies that the connections at the T-junction and at the central knob involve disulfide bonds, probably

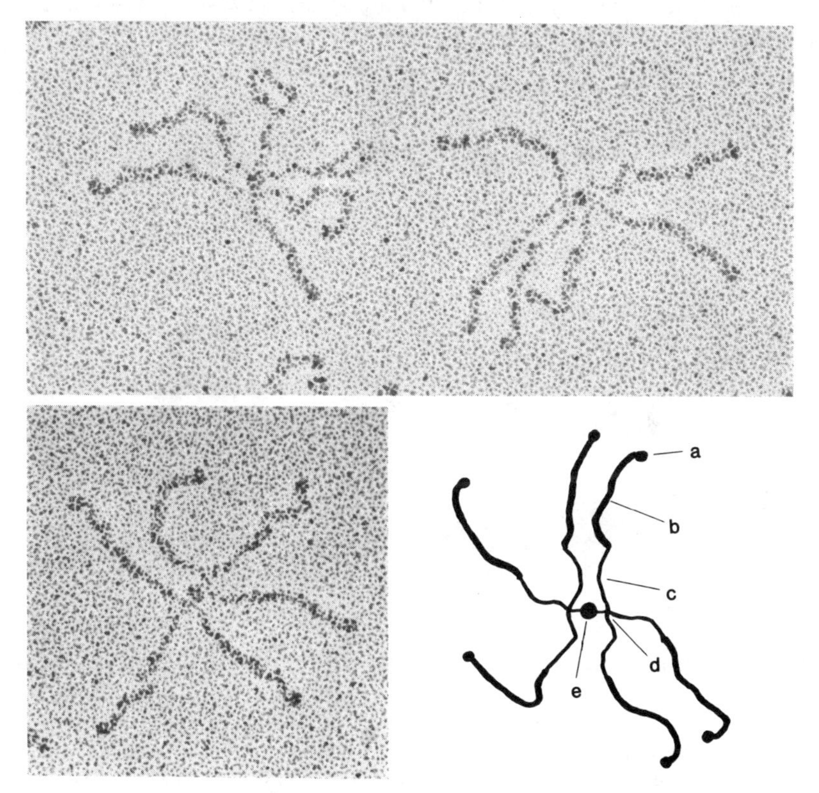

Figure 1 Rotary shadowed electron micrograph of hexabrachions, 250,000X. The lettered features in the diagram (reprinted from Erickson & Iglesias 1984) are identified in the text.

of two distinct types. Although the hexabrachion is the usual form of the molecule, trimers are also found as separate units and probably represent a subassembly. The disulfide bonding and 3-D structure are discussed further by Taylor et al (1989) and by Erickson & Lightner (1988).

The subunit composition as visualized by SDS gels is shown in Figure 2 and molecular weights are summarized in Table 1. The absolute values stated for the subunit masses have varied considerably from one laboratory to another and have also changed as the mass of standards has been refined. The mass of nonreduced human hexabrachions, determined both by agarose gel electrophoresis in SDS and by sedimentation equilibrium of native protein, is 1,900 kd, approximately six times the estimated mass of the subunit (Taylor et al 1989). This confirms that each arm corresponds to a single subunit and lends credibility to the subunit masses in Table 1.

Tenascin from human, chicken, and rat cell cultures shows a characteristic pattern of a large subunit form and two smaller subunit forms (Figure 2), as first described by Chiquet & Fambrough (1984b). These authors demonstrated that all three bands of chicken tenascin were stained by their

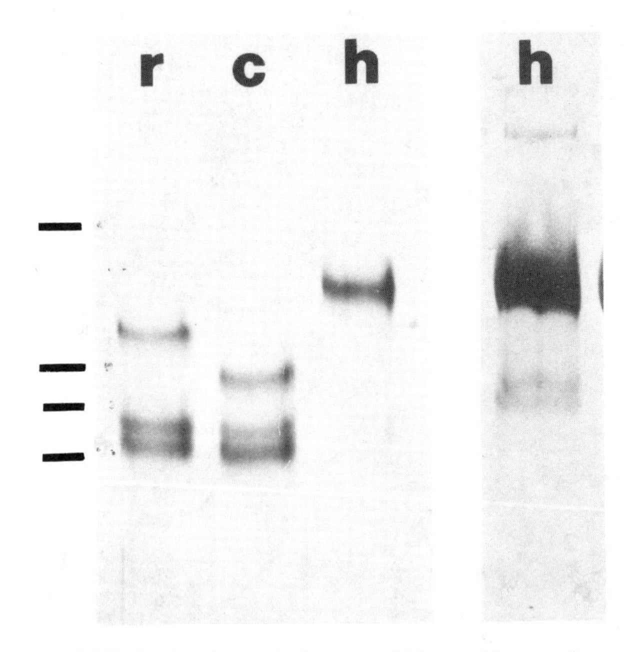

Figure 2 SDS-PAGE of reduced tenascin from rat, chicken and human. A separate sample of human tenascin at higher concentration shows the weak doublet at 220–230 kd (*right lane*). The marks at the left indicate molecular weight markers: myosin 212 kd; laminin 225 kd; fibronectin 250 kd; laminin 400 kd.

Table 1 Biochemical data for tenascin[a]

		Human	Chicken	Rat
Subunit mass	large subunit	320 kd	250 kd	280 kd
(by SDS-PAGE, reducing)	small subunits	230, 220 kd	230, 220 kd	230, 220 kd
Native hexabrachion mass				
Agarose gel electrophoresis	large H × B[b]	1,900 kd	1,540 kd	
	small	—	1,280 kd	
Sedimentation equilibrium	large	1,900 kd	—	
Average arm length	large	87 nm	72 nm	82 nm
	small		<66 nm	68 nm
Sedimentation coefficient	large	14 S	13 S	13 S
	small		12 S	11 S
Extinction coefficient				
(A_{277} 1 mg/ml, 1 cm)		0.97	—	—

[a] From Taylor et al (1989) and (rat tenascin) I. Aukhil et al, in preparation.
[b] H × B = hexabrachion.

monoclonal antibody M1, and they had virtually identical peptide maps. They found no evidence that the smaller peptides were produced by proteolysis. The pattern is remarkable in that the smaller subunit forms usually appear as a pair of equally intense bands, separated by about 10 kd in all three species (the small subunits are weak but still detectable in tenascin from most human cell lines). This 10 kd corresponds in size to one fibronectin type III domain. Sequence data discussed below suggests that these different bands correspond to subunit forms with different numbers of FN-III-type domains, the result of alternative RNA splicing. Tenascin purified from tissue homogenates can show a more complex pattern of bands of higher and lower M_r (Erickson & Taylor 1987; Taylor et al 1989).

The arms of the human hexabrachion are longer than those of the chicken, and correlate well with the difference in subunit mass (Erickson & Iglesias 1984). On a finer level, both chicken and rat hexabrachions can be separated into fractions containing predominantly large or small subunit forms (Taylor et al 1989; I. Aukhil et al, in preparation). This separation and a separation using a monoclonal antibody specific for the large subunit form (Lightner et al 1989b) suggest that the different subunit forms are predominantly segregated into large hexabrachions, which have long arms corresponding to the large subunit, and small hexabrachions with short arms. Hexabrachions with a mixture of long and short arms are also found (H. P. Erickson, unpublished observations). Both long and short arms have connection at the T-junction and a normal terminal knob, consistent with the evidence from DNA cloning that the different sized subunits are produced by alternative RNA splicing.

The splicing of this segment to give the larger subunit may be developmentally important. Hoffman et al (1988) reported that the fraction of larger subunit increased with age in embryonic development of brain. Aufderheide & Ekblom (1988) showed a similar and more absolute transition in development of the intestine in the mouse. Extracts of intestine from 15- to 18-day embryos showed almost exclusively the smaller subunit forms, while extracts from adult intestine, as well as extracts from fibroblast cultures, showed equal portions of the larger and the smaller subunit forms.

A significant glycosylation of tenascin has been demonstrated by labeling with ^{3}H-glucosamine (Chiquet & Fambrough 1984b; Bourdon et al 1985), and by a faster migration on SDS gels following digestion with endoglycosidase F (about a 20 kd loss, Vaughan et al 1987), or with neuraminidase (about a 10 kd loss, Taylor et al 1989). Grumet et al (1985) reported that tenascin isolated from embryonic chicken brain had the HNK-1 antigenic determinant, and Hoffman et al (1988) provided evidence that a small fraction of tenascin subunits isolated from brain had a covalently bound chondroitin sulfate chain. Glycosaminoglycans have not been found on tenascin from cell cultures. The subunit molecular weights given in Table 1 are about 30% higher than the peptide molecular weights predicted from sequence data, which is consistent with a large carbohydrate content.

AMINO ACID SEQUENCE AND STRUCTURAL IMPLICATIONS

The majority of the amino acid sequence of chicken tenascin has been deduced from cDNA clones by Jones et al (1988) and Pearson et al (1988). The remainder of the sequence of chicken tenascin, including the structure of three splicing variants, has recently been determined by Spring et al (1989). The sequence of two variants of chicken tenascin has also been completed by Jones et al (1989). A large part of the sequence of human tenascin, which shows two splicing variants, has been determined by Gulcher et al (1989). A striking feature of the sequence is the multiple repeats of small segments, each probably folding into an independent domain. There are three types of domains, with sequence similarity to domains already identified in other proteins. Figure 3 shows a model in which these domains are matched to the structure of the hexabrachion.

The amino terminus corresponds to the central nodule of the hexabrachion. The first 150 amino acids include a short region of alpha helix with four heptad repeats that could form a triple-stranded coiled coil. These are flanked by cysteines that could bond three arms together to

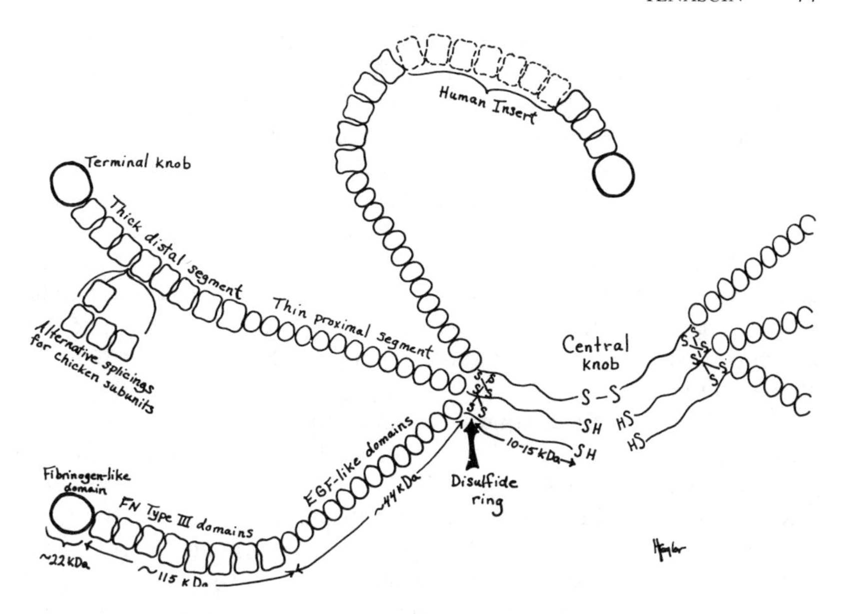

Figure 3 A diagram of the hexabrachion structure, based on the sequence data for alternatively spliced forms (Jones et al, 1988, 1989; Pearson et al 1988; Gulcher et al 1989; Spring et al 1989). The central region is enlarged relative to the arms to show the proposed disulfide bonding. We thank Hope Taylor for this drawing (from Taylor et al 1989).

make a trimer, and there is a more proximal cysteine that could bond two trimers to make the hexabrachion.

The arm continues with 13 repeats of a 31-amino acid segment with sequences similar to the EGF-like domains found in several other ECM proteins. The EGF-like domains of tenascin are remarkable in their high degree of similarity to each other.

There follows a string of 8 to 15 domains, the number depending on mRNA splicing variants that are similar to the Type-III domains of fibronectin. These FN-III domains, about 90 amino acids each with virtually no cysteine, show a similarity to each other and to the FN-III domains of fibronectin at the level of 26–40% identity. Corresponding domains from chicken and human tenascin show a sequence similarity of 80% if conservative substitutions are accounted for (Gulcher et al 1989). The third FN-III domain in both chicken and human tenascin has an RGD sequence. The flanking amino acids are different in tenascin and fibronectin, but the RGD is in essentially the same position in the two proteins.

The first five and the last three FN-III domains are present in all clones so far examined: the shortest forms of both the chicken and human protein contain these eight domains. Jones et al (1989) and Spring et al (1989) have identified clones that give three alternatively spliced variants of the chicken protein. In addition to the shortest form, either one or three additional FN-III domains can be inserted after FN-III-5. Gulcher et al (1989) have identified clones indicating two alternatively spliced variants of the human protein: the short form with eight FN-III domains, and one with seven additional FN-III domains inserted after FN-III-5. This much larger insert in the human protein contains four domains that are highly similar to each other, 80% identical in amino acid sequence. These four domains each contain two or three identically placed sites for N-linked glycosylation.

The carboxyl-terminal sequence is similar to the carboxyl-terminal segment of fibrinogen β and γ chains, specifically the region that forms the globular domains of the fibrinogen D-nodule. These domains are involved in longitudinal and diagonal contacts in polymerization of fibrin (Erickson & Fowler 1983), which suggests a possible role in protein-protein association for this domain of the hexabrachion.

The sequence information can now be fit into a model of the hexabrachion that explains the structural features seen by electron microscopy. The domain model is shown in Figure 3. The amino-terminal segment makes up the central nodule and T-junction that joins the arms of the hexabrachion first into trimers and then into the hexamer. The 13 EGF-like domains form the thin proximal segment, and the 8 to 15 FN-III domains form the thick distal segment of the arm. Finally, the terminal knob of the hexabrachion arm corresponds to the fibrinogen-like domain.

The alternative splicing can explain the triplet of bands seen in tenascin of chicken, rat, and human (Figure 2). The lowest band, designated 220 kd for all three species, corresponds to the shortest arm, with eight FN-III domains. The next band is the variant with one additional FN-III domain (the clone for this form has not yet been identified for the human but the bands on SDS gels suggest that this species must exist). Finally, the 250 kd chicken and the 320 kd human bands correspond to the species with three and seven additional FN-III domains. The 280 kd band in the rat protein suggests an insert of five additional FN-III domains in this subunit form.

Alternative splicing is also demonstrated by the identification of multiple mRNA species in both chicken (Jones et al 1988, 1989) and human (Gulcher et al 1989) cells. Southern blot analyses indicated a single tenascin gene in both species. Alternatively spliced domains are also present in fibronectin, and may provide unique functions for a subset of fibronectin molecules that contain them (Schwartzbauer et al 1983; Humphries et al

1986). A unique function for the alternative domains of tenascin is suggested by the fact that they are missing from the tenascin secreted in the earliest stages of development and incorporated later (Hoffman et al 1988; Aufderheide & Ekblom 1988).

SOURCES AND PURIFICATION OF TENASCIN

Tenascin is secreted by fibroblasts and glial cells in tissue culture. One of the best sources of human tenascin is the glioma cell line U-251 MG. Other human glioma lines such as U-373 also secrete large amounts of tenascin. Media collected from confluent cultures of U-251 MG and harvested after three days contains 5–20 μg/ml tenascin (Lightner et al 1989). Primary cultures of chicken embryo fibroblasts produce 2–5 μg/ml tenascin, while human fibroblasts produce about tenfold less (Erickson & Lightner 1988). The U-251 MG cell line has the additional advantage that it can be grown in 1% fetal calf serum, or for periods of three days in serum free medium.

A direct, one-step purification can be achieved by immunoaffinity, preferably using a monoclonal antibody as originally described by Chiquet & Fambrough (1984b). The M1 antibody against chicken tenascin can be obtained from the Developmental Studies Hybridoma Bank, The Johns Hopkins University School of Medicine, Biophysics Bldg., Rm 311, 725 N. Wolfe St., Baltimore MD 21205. Monoclonal antibodies to the human protein are not yet commercially available. The antibody is coupled to agarose (Sepharose 2-B is preferable because its large pore size gives maximum access of the hexabrachions to the bound antibody). Cell supernatant or tissue homogenate is passed over the column, the column is washed with 0.5 M NaCl, and the protein is eluted with buffer at pH 11, or with 2 M sodium thiocyanate, or 4 M urea. The eluted protein is typically at a concentration of 20–100 μg/ml.

The anion exchange purification of Slemp et al (1988) can provide an additional level of purification following immunoaffinity elution. Protein in 4 M urea, diluted to bring the NaCl concentration to 0.25 M or less, is pumped through a Mono Q column (from Pharmacia; a DEAE cellulose column could also be used with adjustment of salt). The tenascin binds to the column, is freed of urea by washing with 0.2 M NaCl, and eluted at high concentration by a step to 0.6 M NaCl. This procedure has the additional advantage of removing proteoglycans and a form of laminin that can contaminate tenascin (Slemp et al 1988).

Tenascin can also be purified from cell culture supernatants by a combination of gel filtration, anion exchange chromatography, and gradient sedimentation (Slemp et al 1988; I. Aukhil et al, in preparation). The purification involves four steps: (*a*) ammonium sulfate precipitation (37%

saturation); (*b*) chromatography on Sephacryl-500 HR (Pharmacia) (tenascin elutes just behind the void volume, ahead of fibronectin and most other proteins); (*c*) anion exchange chromatography on a Mono Q column (tenascin elutes between 0.32 and 0.40 M NaCl); and (*d*) a second step of Sephacryl chromatography, or glycerol gradient sedimentation (hexabrachions sediment at 12–14 S, separated from most contaminants). Three to five mg of tenascin can be prepared from 400 ml of U-251 MG cell supernatant in two days.

Tenascin can also be purified from tissue homogenates; embryonic chicken tissues have been used most extensively. Brain, gizzard, and wings from 11- to 14-day chick embryos are all relatively rich in tenascin. The protein can be extracted by homogenizing the tissues in phosphate buffered saline (Grumet et al 1985), but a better extraction may be obtained in saline buffered at pH 11 (Erickson & Taylor 1987; Lightner et al 1989a). Vaughan et al (1987) used 1 M NaCl pH 8 to extract tenascin from the sternal cartilages of 17-day chick embryos. In most reports, tenascin is purified from the tissue extract by immunoaffinity chromatography. Vaughan et al obtained a reasonably pure preparation of chick tenascin from their cartilage extract by a single step of anion exchange chromatography on DEAE-Trisacryl. In recent work, Hoffman et al (1988) have used immunoaffinity as a primary step in purifying their protein from brain extract, but have added steps of anion exchange chromatography on DE52, density gradient centrifugation on CsCl, and gel filtration on Sephacryl S-500 to separate a proteoglycan that they found associated with tenascin in tissues and homogenates.

BINDING OF TENASCIN TO OTHER ECM MOLECULES

Several laboratories have explored the possibility that tenascin may bind to other molecules of the ECM. There is general agreement for tenascin binding to only one molecule, chondroitin sulfate proteoglycan (CSPG). In the original isolation of tenascin by immunoaffinity chromatography, Chiquet & Fambrough (1984b) noted that a CSPG secreted by muscle cell cultures bound to the tenascin on the antibody column and eluted with tenascin at pH 11. Vaughan et al (1987) noted similarly that a CSPG copurified with tenascin from cartilage. A specific CSPG from chicken brain has been characterized in more detail by Hoffman & Edelman (1987), Tan et al (1987), and Hoffman et al (1988). Tenascin and the brain CSPG are produced by different cell types, glial cells and neurons, respectively. They are frequently but not always colocalized in embryonic tissues, which suggests a possible coordinate interaction of different cell types and ECM molecules during morphogenesis (Hoffman et al 1988). In these studies tenascin was reported to bind to the 280 kd core protein of the proteoglycan. It will be important to determine whether the tenascin-binding CSPGs from muscle cell culture (Chiquet & Fambrough, 1984a) and cartilage (Vaughan et al 1987) are related to the molecule isolated from brain, and whether the glycosaminoglycan chains play any role in binding tenascin.

Fibronectin does not bind to tenascin during immunoaffinity purification (Chiquet & Fambrough 1984b). Lightner et al (1988), and M. A. Bourdon (unpublished observations) found no binding of tenascin to fibronectin in ELISA. In contrast, Chiquet-Ehrismann et al (1988) and Hoffman et al (1988) both reported binding of tenascin to fibronectin. However, the assays used in these two studies did not give quantitative information on the affinity of the interaction, which could be quite weak. It should be noted that binding of fibronectin to tenascin has not been demonstrated in solution (Lightner et al 1988). Indeed, if binding were significant, the bivalent fibronectin and the multivalent hexabrachions should form a precipitate, which is not observed at protein concentrations up to 200 μg/ml (H. P. Erickson, unpublished).

Lightner et al (1988) used an ELISA to examine binding of soluble tenascin to a variety of ECM molecules immobilized on plastic. Significant binding was observed to the monoclonal antibody to tenascin, a positive control, but no binding was observed to collagens I, III, IV, and V, gelatin, fibronectin, or laminin. Thus so far the only clearly identified ECM molecule that binds to tenascin is a chondroitin sulfate proteoglycan.

INTERACTION OF TENASCIN WITH CELLS

An interaction of tenascin with the cell surface has been observed by several laboratories that used a cell attachment assay as an indication of binding (Chiquet-Ehrismann et al 1988; Friedlander et al 1988; Bourdon & Ruoslahti 1989). Two important characteristics of the cell binding or attachment to tenascin are (*a*) not all cells are capable of adhering to tenascin-coated surfaces, and (*b*) cells do not flatten and spread on a tenascin substrate as they do on fibronectin or laminin. Cells attached to tenascin maintain a rounded to spindle or branching morphology depending on the cell line. Moreover, cell attachment to tenascin is much weaker than to fibronectin. The distinctive morphology of cells on tenascin, the lack of spreading, and the weakness of the attachment, suggest a role quite different from other adhesion molecules. Studies from several laboratories have demonstrated that tenascin can modulate cell behavior on fibronectin coated surfaces (Tan et al 1987; Chiquet-Ehrismann et al 1988). How this is accomplished is uncertain, but may well involve several mechanisms.

One mechanism by which cells interact with tenascin is RGD dependent (Friedlander et al 1988; Bourdon & Ruoslahti, 1989). Cell attachment to tenascin has been demonstrated for a number of cell types, including tumor cells, fibroblasts, and endothelial cells. This interaction is inhibited by RGD-containing peptides at peptide concentrations a hundredfold lower than needed to inhibit adhesion to fibronectin (Bourdon & Ruoslahti, 1989). This adhesion likely involves the RGD sequence already identified in chicken and human tenascin (Jones et al 1988; Gulcher et al 1989). Preliminary results indicate that the sensitivity of tenascin-cell interaction reflects, at least in part, a higher receptor affinity for the RGD peptides (M. A. Bourdon, unpublished observations). Friedlander et al (1988) have shown that fragments of tenascin, which may include the RGD sequence, have a twentyfold higher cell adhesion activity than intact tenascin.

Bourdon & Ruoslahti (1989) have recently identified and isolated an integrin receptor that binds tenascin in an RGD-dependent manner. The cell surface receptor, which was eluted from a tenascin affinity column by RGD peptide, is a heterodimer composed of a 145 kd α-subunit and a 125 kd β subunit. The β subunit of the tenascin receptor is indistinguishable by immunoblot analysis from the β_1 subunit shared by all members of the fibronectin receptor subgroup of integrins. The α subunit of the tenascin receptor is similar to that of the larger fibronectin receptor α_5 subunit in its composition of a large and small peptide linked by disulfide bond. However, the identity of the α subunit has not been established and may or may not be a new subunit unique for the tenascin receptor. By SDS–PAGE and immunoblot analysis, the α subunit clearly is not a fibronectin or vitronectin receptor α subunit. In addition, the tenascin receptor does not appear to bind fibronectin, vitronectin, or laminin.

The results of Chiquet-Ehrismann et al (1988) and Friedlander et al (1988) suggest that a non-RGD-dependent mechanism may also play a role in mediating cell-tenascin interactions. What receptors may be involved in these interactions are as yet undefined, but may include cell surface or matrix proteoglycan (Hoffman et al 1988). The expression of several different cell surface receptors to tenascin may provide multiple ways of responding to tenascin or of modulating those responses discussed below.

Several studies have reported that cell adhesion to tenascin was not detectable using conventional assays (Chiquet-Ehrismann et al 1986; Erickson & Taylor 1987; Slemp et al 1988). Lotz et al (1987) used a sensitive and quantitative centrifugation assay, which showed a transient binding of glioma cells to tenascin. The initial binding at 4°C was weaker than to fibronectin, and, in contrast to fibronectin on which cells spread and strengthened their attachment, adhesion to tenascin was completely lost after 15 min. The initial binding is consistent with the presence of a tenascin receptor on the glial cells, but the weakness of this binding and its eventual loss suggests that tenascin is not a cell adhesion molecule in the way we think of fibronectin adhesion. The weakness of adhesion to tenascin may mean that (*a*) cell receptors for tenascin are too few in number to promote the strong, and physiologically complex adhesion seen with fibronectin, and (*b*) that cell adhesion may not be the primary role of these receptors. A role in signal transduction is an attractive possibility for the tenascin receptors, with a weak adhesion to tenascin coated plastic simply reflecting an affinity of tenascin for its cell surface receptor(s).

Cell substrate adhesion requires that the adhesion molecule either form or bind to a substrate. Fibronectin forms its own matrix of covalently cross-linked strands and also binds to collagen. Thus adhesion to fibronectin can attach cells to the fibronectin matrix or to collagen. As discussed above, tenascin does not appear to bind to fibronectin, laminin or collagens, nor is there any evidence that it is cross-linked to itself to form a matrix. Thus it seems unlikely that tenascin mediates cell substrate adhesion. It might be noted that the structure of the hexabrachion, in particular its large size and its hexavalency, is ideally suited to mediate a direct cell-cell adhesion, in particular in nervous tissue and smooth muscle where cell surfaces are closer together than the length of a hexabrachion arm. However, the weakness of adhesion for neurons and glia (Friedlander et al 1988; Lotz et al 1987) again argues that cell-cell adhesion is probably not the primary purpose of the interaction between tenascin and its cell receptors.

Cell motility and migration are closely related to and dependent on cell adhesion. Tan et al (1987) studied migration of neural crest cells on substrates composed of alternating strips of fibronectin and tenascin. The cells showed a strong preference for migration on fibronectin. The few cells that wandered onto the tenascin strips assumed a rounded morphology. Mackie et al (1988b) made very similar observations in frog and rat embryos, and both groups postulated that tenascin might somehow modulate cell migration on fibronectin or other substrates. This possibility is also suggested by experiments of Chiquet-Ehrismann et al (1988) that show that tenascin can inhibit cell adhesion to fibronectin.

Tenascin involvement in cell migration has been demonstrated in vivo by antibody inhibition. Antibodies to tenascin were found to inhibit migration of granule cells in organ explants of developing cerebellum (Chuong et al 1987). Tenascin has been observed along pathways of neural crest cell migration (Tan et al 1987; Epperlein et al 1988; Mackie et al

1988b). Bronner-Fraser (1988) confirmed this localization and focused more on the head region of early chick embryos. The injection of anti-tenascin antibodies into the head region of early (pre 9-somite) embryos led to abnomalities in neural crest cell migration and neural tube development. Similar injection into the trunk region had no effect on development. Bronner-Fraser offered several possible ways in which tenascin might play its role in neural crest cell migration, including restricting cell migration to the more permissive fibronectin tracks.

Cells grown on tenascin substrates can behave quite differently from those grown on fibronectin or plastic. Chiquet-Ehrismann et al (1986) observed that primary epithelial tumor cells cultured on tenascin had substantially enhanced growth, compared to cultures on collagen, fibronectin, or laminin. Mackie et al (1987b) found that cultures of chicken embryo wing bud produced more cartilage nodules when grown on tenascin, compared to uncoated plastic. It is attractive to think that stimulatory or hormonal action can be transmitted from the substrate-bound tenascin and affect the growth of the cultured cells. It must be noted, however, that the tenascin substrate was always the least adhesive, and it is possible that some of these effects on cell growth are primarily a consequence of altered adhesion to the substrate.

INDUCTION OF TENASCIN SECRETION IN MESENCHYMAL CELLS

Chiquet-Ehrismann et al (1986) noted that the fibroblasts of adult rat mammary gland, although capable of synthesizing tenascin when grown in tissue culture, were apparently repressed for tenascin synthesis in vivo: in normal adult mammary gland no tenascin was detectable by immunofluorescence. The tenascin that accumulates in carcinomas does not appear to be synthesized by the epithelial carcinoma cells. A wide range of human carcinoma cell lines (breast and ovarian) have been assayed for tenascin secretion and none produced a detectable amount of tenascin (Bourdon et al 1983; Erickson & Lightner 1988). Mammary carcinoma and squamous cell carcinoma were devoid of tenascin staining in the islands of carcinoma cells, but tenascin was prominent in the surrounding connective tissue (Mackie et al 1987a; Lightner et al 1989a). Inaguma et al (1988) demonstrated tenascin synthesis in mesenchyme surrounding heterologous transplanted breast carcinoma, clearly separate from the donor carcinoma cells. Thus the transformed epithelial cells appear to induce the fibroblasts of the underlying connective tissue to synthesize tenascin. Similar inductive effects are observed in embryonic development (discussed below) as growing epithelia appear to induce mesenchyme condensation and tenascin synthesis. It is reasonable to postulate some soluble growth factor(s) as mediating the induction.

Fibroblasts and glial cells appear to be the primary cells that synthesize tenascin. The absence of tenascin in most normal adult tissues implies that synthesis is turned off in most fibroblasts. The prominent accumulation near certain classes of growing epithelial cells (transformed, embryonic, and in healing wounds) suggests that these epithelial cells (and in other locations some other cell types) secrete one or more growth factors that stimulate the fibroblasts of adjacent connective tissue to synthesize and secrete tenascin. Important experimental evidence for such induction was obtained by Aufderheide & Ekblom (1988), who found that gut mesenchyme from 13-day mouse embryos, when cultured alone or with B16-F1 melanoma cells, expressed only trace amounts of tenascin. However, these same explants, when cultured with an epithelial cell line MDCK, secreted substantial amounts of tenascin. This cell line appears to secrete some growth factors that stimulate the mesenchyme cells to secrete tenascin. The factors have not yet been identified.

Pearson et al (1988) reported that primary cultures of chicken embryo fibroblasts secreted large amounts of tenascin when grown in 10% fetal calf serum (which contains a multitude of growth factors), but very little when grown in 1% serum. Addition of TGF-β to cultures growing in 0.3% serum resulted in a significant increase in secretion of both tenascin and fibronectin. This was perhaps a general enhancement of the cells' secretory activity. Interestingly, serum enhanced secretion of tenascin with a much smaller effect on fibronectin secretion. There are no doubt many growth factors that can affect secretion of tenascin.

LOCALIZATION OF TENASCIN IN TISSUES

The distribution of tenascin is widespread in embryonic development, but much more restricted than fibronectin and laminin. The distribution was first described by Chiquet & Fambrough (1984a) and in more detail, in particular for neural tissue, by Crossin et al (1986). These references and the review of Erickson & Lightner (1988) should be consulted for discussion of the diverse tissues expressing tenascin. We will focus here on some recent specialized studies and attempt some general statements.

Tenascin is expressed at several locations and times in neural development (Crossin et al 1986; Chuong et al 1987; Bronner-Fraser 1988; Mackie et al 1988b; Riou et al 1988; Steindler et al 1989). It first appears during gastrulation and is found along the neural tube and notochord. Shortly thereafter it is found in pathways for neural crest cell migration, in particular in the anterior portion of somites. Subsequently tenascin is expressed in the developing brain, particularly within the external granular layer and molecular layer, and along the radial fibers of Bergman glia. Evidence for a role in cell migration is discussed under cell adhesion. Tenascin expression generally declines as the brain matures. Steindler et al (1989) reported a transient appearance of tenascin within the somatosensory cortex of rat 24–48 hours after birth, at the time of innervation by sensory neurons from vibrissae. Tenascin appeared at the boundaries of prospective barrel fields, associated with glial and neuronal cell membranes, and at glial end feet on blood vessels. After a brief expression, tenascin rapidly disappeared. This result suggests that the transient expression of tenascin, possibly by altering cell adhesion, plays some role in organizing the structure of these neural fields.

In the peripheral nervous system tenascin has been identified in the perineureum and within the nerve near the nodes of Ranvier (ffrench-Constant et al 1986; Rieger et al 1986). Maier & Mayne (1987) described a very restricted localization of tenascin in the outer spindle capsule, but not around the muscle fibers, of adult chicks. Sanes et al (1986) observed that tenascin was absent from adult muscle tissue of the rat, but could be induced by experimental manipulation. If the nerve to the muscle was cut, crushed, or paralyzed, tenascin accumulated prominently in the ECM around the motor endplate. Tenascin staining disappeared when the muscle was reinnervated.

Tenascin is expressed in restricted locations in embryonic mesenchyme, frequently in response to induction by epithelium, and often at regions of mesenchyme condensation. Tenascin was observed selectively in the condensing mesenchyme near budding epithelium of developing mammary gland, vibrissae, and teeth (Chiquet-Ehrismann et al 1986; Thesleff et al 1987). In embryonic kidney a part of the mesenchyme condenses and develops into the S-shaped bodies of kidney tubule epithelium (Aufderheide et al 1987). Tenascin appeared in the mesenchyme surrounding these bodies, but only after these were identified by keratin markers as having epithelial character. When tubulogenesis was blocked in organ culture, tenascin expression was inhibited, consistent with the induction hypothesis. In their study of mammary gland development, Inaguma et al (1988) observed that tenascin was induced in the stroma near certain specific segments of the budding epithelium. In a study of tenascin in early amphibian development, Riou et al (1988) manipulated induction experimentally. Ectoderm isolated at early gastrula normally differentiated into epidermal vesicles devoid of tenascin. If, however, the ectoderm were induced to form a neural tube, either by a brief association with dorsal blastopore lip, or artificially by the lectin concanavalin A, the mesenchyme surrounding the induced neural tube structures showed prominent staining for tenascin. Continued presence of epithelium was not required for this induction.

Not all epithelial growth is accompanied by tenascin synthesis. Chiquet-Ehrismann et al (1986) and Inaguma et al (1988) noted that the stromal tissue of the adult rat mammary gland showed no tenascin even during pregnancy, when epithelial growth and differentiation are occurring. The latter study also noted that tenascin was restricted to certain segments of the developing epithelial buds and tubes in embryonic mammary gland. There are a number of epithelial structures in embryonic development that have little or no associated tenascin (Chiquet & Fambrough 1984a; Crossin et al 1986).

Although frequently associated with epithelial growth or differentiation from mesenchyme, tenascin is also prominent in several nonepithelial tissues. It is perhaps most conspicuous in developing dense connective tissues including tendons and ligaments, embryonic cartilage, and the periosteal layer of developing bone (Chiquet & Fabrough 1984a; Crossin et al 1986; Vaughan et al 1987; Mackie et al 1987b; Tan et al 1987; Thesleff et al 1987, 1988). Tenascin could be extracted in large quantity from embryonic chicken cartilage, which suggests a distribution throughout the cartilage matrix (Vaughan et al 1987), but tenascin appears predominantly in the perichondrium by immunohistochemistry (Mackie et al 1987b). Another example of tenascin associated with nonepithelial tissues is its prominence in smooth muscle of arterial walls and in gizzard.

Given the association of tenascin with tissue growth and differentiation, it is not surprising to find tenascin prominently expressed during wound healing. Mackie et al (1988a) and Lightner et al (1989a) showed that tenascin is present in a thin zone beneath the dermal-epidermal junction, but is absent from most of the dermis. In healing wounds Mackie et al (1988a) observed extensive accumulation of tenascin in the granulation tissue. Tenascin was prominent in the dermis under the advancing epithelium, but it was also found more generally in the deeper regions of the wound, far from the surface and in advance of reepithelialization. Thus the massive accumulation of tenascin in the wounded dermis may be partly induced by the growing epidermis, but is probably also stimulated by other cells, perhaps macrophages or leukocytes, in the granulation tissue. Perhaps as remarkable as its expression during healing is the disappearance of tenascin following reepithelialization and contraction of the wound.

Most studies of tenascin by immunohistochemistry have used light microscope techniques, often at low magnification. At least three groups have extended the studies to the electron microscope level. Sanes et al (1986) and Mackie et al (1988a), using a lactoperoxidase technique, reported a weak diffuse staining in the ECM, but noted that stain accumulated prominently around the outside of collagen fibrils and at the basement membrane. Lightner et al (1989a) noted a similar stain pattern with the lactoperoxidase technique, but found a different distribution using gold bead second antibody for immunolocalization. The gold bead label showed no association of tenascin to collagen fibrils or at the basement membrane. Rather the gold beads labeled small patches about the size of single hexabrachions between, but not attached to, the collagen fibrils of the papillary dermis. Lightner et al noted previous evidence that the peroxidase reaction product can diffuse from the site of the antigen, and suggested that the apparent staining of basement membranes and collagen fibrils in this technique was due to adventitious accumulation of the diffused reaction product on these prominent surfaces. In the immuno-gold technique the hexabrachions appear scattered in the ECM as single molecules and small clusters, not attached to any obvious structure. It might be noted that proteoglycans are not well stained, so a network of hexabrachions bound to CSPGs might be the structure keeping the hexabrachions in place.

Tenascin is prominent in both mesenchymal tumors and carcinomas, including gliomas, fibrosarcomas, osteosarcomas, melanomas, Wilms tumor, mammary and lung carcinoma, and squamous cell carcinoma (Bourdon et al 1983; Chiquet-Ehrismann et al 1986; Mackie et al 1987a; Lightner et al 1989a). It is remarkable that tenascin is most prominent in the most anaplastic tumors. For example, it is almost always found in glioblastoma multiforme, but infrequently in the more differentiated astrocytomas. Similarly, tenascin is prominent in malignant breast carcinomas but rarely in benign lesions. Tenascin may, therefore, be a marker for undifferentiated tumors (Bourdon et al 1983; Mackie et al 1987a).

The prominence of tenascin in tumors has potential medical implications for diagnosis and therapy. Bourdon et al (1984), Bullard et al (1986), and Mackie et al (1987a) have explored the potential use of monoclonal antibodies to image or identify tumors. The possible therapeutic use of radiolabeled antibodies to target a large dose of radiation to the tenascin containing tumors has been explored by Bourdon et al (1984), Blasberg et al (1987), and Lee et al (1988a,b). This has seemed a particularly attractive approach for gliomas, because the prominence of tenascin in the tumor is in contrast to its absence from normal adult brain.

Although the distribution of tenascin is quite restricted, much more so than fibronectin, the locations in which it is found are quite diverse. Erickson & Lightner (1988) concluded that tissues as diverse as brain, gizzard, and cartilage have so little in common that one cannot suggest a single function for tenascin in all locations. Rather, the diversity suggests that tenascin may be a multifunctional protein, with only a specific subset of functions being utilized in each tissue. The modular design of the hexabrachion arm (Figure 3) underscores this possibility of multiple independent functions. The FN-III domains vary considerably in amino acid sequence, so each of these, as well as the fibrinogen-like terminal knob, could have an independent cell surface receptor or ECM ligand. So far, one cell surface receptor, an integrin, has been identified (Bourdon & Ruoslahti, 1989), with intriguing hints of other cell receptors and ECM ligands.

The most striking and consistent theme that emerges from studies so far is the prominence of tenascin during embryonic development, wound healing, and tumor growth. Tenascin is a highly specialized matrix product of mesenchymal and glial cells, and these cells in turn are responsive to growth factors from growing epithelium. The expression of tenascin correlates with cell proliferation and migration, growth and remodeling of the extracellular matrix, and patterning in normal development. Similar processes are likely taking place in the abnormal growth and vascular proliferation associated with tumors. The biology of tenascin is clearly going to be complex, but its prominence in development and disease promise to make research on this molecule most rewarding.

Acknowledgments

This work was supported by National Institutes of Health grants CA-47056 (HPE) and CA-45586 (MAB).

Literature Cited

Aufderheide, E., Chiquet-Ehrismann, R., Ekblom, P. 1987. Epithelial-mesenchymal interactions in the developing kidney lead to expression of tenascin in the mesenchyme. *J. Cell. Biol.* 105: 599–608

Aufderheide, E., Ekblom, P. 1988. Tenascin during gut development: Appearance in the mesenchyme, shift in molecular forms, and dependence on epithelial-mesenchymal interactions. *J. Cell Biol.* 107: 2341–49

Blasberg, R. G., Nakagawa, H., Bourdon, M. A., Groothuis, D. R. 1987. Regional localization of a glioma-associated antigen defined by monoclonal antibody 81C6 in vivo: kinetics and implications for diagnosis and therapy. *Cancer Res.* 47: 4432–43

Bourdon, M. A., Coleman, R. E., Blasberg, R. G., Groothius, D. R., Bigner, D. D. 1984. Monoclonal antibody localization in subcutaneous and intracranial human glioma zenografts: paired-label and imaging analysis. *Anticancer Res.* 4: 133–40

Bourdon, M. A., Matthews, T. J., Pizzo, S. V., Bigner, D. D. 1985. Immunochemical and biochemical characterization of a glioma-associated extracellular matrix glycoprotein. *J. Cell Biochem.* 28: 183–95

Bourdon, M. A., Ruoslahti, E. 1989. Tenascin mediates cell attachment through an RGD-dependent receptor. *J. Cell Biol.* 108: 1149–55

Bourdon, M. A., Wikstrand, C. J., Furthmayer, H., Matthews, T. J., Bigner, D. D. 1983. Human glioma-mesenchymal extracellular matrix antigen defined by monoclonal antibody. *Cancer Res.* 43: 2796–2805

Bronner-Fraser, M. 1988. Distribution and function of tenascin during cranial neural crest development in the chick. *J. Neurosci. Res.* 21: 135–47

Bullard, D. E., Adams, C. J., Coleman, R. E., Bigner, D. D. 1986. In vivo imaging of intracranial human glioma zenografts comparing specific with nonspecific radiolabeled monoclonal antibodies. *J. Neurosurg.* 64: 257–62

Chiquet, M. 1989. Tenascin/J1/Cytotactin: the potential function hexabrachion proteins in neural development. *Dev. Neurosci.* In press

Chiquet, M., Fambrough, D. M. 1984a. Chick myotendinous antigen. I. A monoclonal antibody as a marker for tendon and muscle morphogenesis. *J. Cell Biol.* 98: 1926–36

Chiquet, M., Fambrough, D. M. 1984b. Chick myotendinous antigen. II. A novel extracellular glycoprotein complex consisting of large disulfide-linked subunits. *J. Cell Biol.* 98: 1937–46

Chiquet-Ehrismann, R., Kalla, P., Pearson, C. A., Beck, K., Chiquet, M. 1988. Tenascin interferes with fibronectin action. *Cell* 53: 383–90

Chiquet-Ehrismann, R., Mackie, E. J., Pearson, C. A., Sakakura, T. 1986. Tenascin: an extracellular matrix protein involved in tissue interactions during fetal development and oncogenesis. *Cell* 47: 131–39

Chuong, C. M., Crossin, K. L., Edelman, G. M. 1987. Sequential expression and differential function of multiple adhesion molecules during the formation of cerebellar cortical layers. *J. Cell Biol.* 104: 331–42

Crossin, K. L., Hoffman, S., Grumet, M., Thiery, J. P., Edelman, G. M. 1986. Site-restricted expression of cytotactin during development of the chicken embryo. *J. Cell. Biol.* 102: 1917–30

Ekblom, P., Aufderheide, E. 1989. Stimulation of tenascin expression in mesenchyme by epithelial-mesenchymal interactions. *Int. J. Dev. Biol.* In press

Epperlein, H.-H., Halfter, W., Tucker, R. P. 1988. The distribution of fibronectin and tenascin along migratory pathways of the neural crest in the trunk of amphibian embryos. *Development* 103: 743–56

Erickson, H. P., Fowler, W. E. 1983. Electron microscopy of fibrinogen, its plasmic fragments and small polymers. *Ann. N.Y. Acad. Sci.* 408: 146–63

Erickson, H. P., Iglesias, J. L. 1984. A six-armed oligomer isolated from cell surface fibronectin preparations. *Nature* 311: 267–69

Erickson, H. P., Lightner, V. A. 1988. Hexabrachion protein (tenascin, cytotactin, brachionectin) in connective tissues, embryonic brain, and tumors. In *Advances in Cell Biology*, Vol. 2: 55–90, ed. K. R. Miller. London: JAI

Erickson, H. P., Taylor, H. C. 1987. Hexabrachion proteins in embryonic chicken tissues and human tumors. *J. Cell Biol.* 105: 1387–94

Faissner, A., Kruse, J., Chiquet-Ehrismann, R., Mackie, E. 1988. The high-molecular-weight J1 glycoproteins are immunochemically related to tenascin. *Differentiation* 37: 104–114

ffrench-Constant, C., Miller, R. H., Kruse, J., Schachner, M., Raff, M. C. 1986. Molecular specialization of astrocyte processes at nodes of Ranvier in rat optic nerve. *J. Cell Biol.* 102: 844–52

Friedlander, D. R., Hoffman, S., Edelman, G. M. 1988. Functional mapping of cytotactin: Proteolytic fragments active in cell-substrate adhesion. *J. Cell Biol.* 107: 2329–40

Grumet, M., Hoffman, S., Crossin, K. L., Edelman, G. M. 1985. Cytotactin, an extracellular matrix protein of neural and non-neural tissues that mediates glia-neuron interaction. *Proc. Natl. Acad. Sci. USA* 82: 8075–79

Gulcher, J. R., Marton, L. S., Stefansson, K. 1986. Two large glycosylated polypeptides found in myelinating oligodendrocytes but not in myelin. *Proc. Natl. Acad. Sci. USA* 83: 2118–22

Gulcher, J. R., Nies, D. E., Marton, L. S., Stefansson, K. 1989. An alternatively spliced region of the human hexabrachion contains a novel repeat of potential N-glycosylation sites. *Proc. Natl. Acad. Sci. USA* 86: 1588–92

Hoffman, S., Crossin, K. L., Edelman, G. M. 1988. Molecular forms, binding functions, and developmental expression patterns of cytotactin and cytotactin-binding proteoglycan, an interactive pair of extracellular matrix molecules. *J. Cell Biol.* 106: 519–32

Hoffman, S., Edelman, G. M. 1987. A proteoglycan with HNK-1 antigenic determinants is a neuron-associated ligand for cytoactin. *Proc. Natl. Acad. Sci. USA* 84: 2523–27

Humphries, M. J., Akiyama, S. K., Komoriya, A., Olden, K., Yamada, K. M. 1986. Identification of an alternatively spliced site in human plasma fibronectin that mediates cell type-specific adhesion. *J. Cell Biol.* 103: 2637–47

Inaguma, Y., Kusakabe, M., Mackie, E. J., Pearson, C. A., Chiquet-Ehrismann, R., Sakakura, T. 1988. Epithelial induction of stromal tenascin in the mouse mammary gland: from embryogenesis to carcinogenesis. *Dev. Biol.* 128: 245–55

Jones, F. S., Burgoon, M. P., Hoffman, S., Crossin, K. L., Cunningham, B. A., Edelman, G. M. 1988. A cDNA clone for cytotactin contains sequences similar to epidermal growth factor-like repeats and segments of fibronectin and fibrinogen. *Proc. Natl. Acad. Sci. USA* 85: 2186–90

Jones, F. S., Hoffman, S., Cunningham, B. A., Edelman, G. M. 1989. A detailed structural model of cytotactin: protein homologies, alternative RNA splicing, and binding regions. *Proc. Natl. Acad. Sci. USA* 86: 1905–9

Kruse, J., Keilhauer, G., Faissner, A., Timpl, R., Schachner, M. 1985. The J1 glycoprotein—a novel nervous system cell adhesion molecule of the L2/HNK-1 family. *Nature* 316: 146–48

Lee, Y., Bullard, D. E., Humphrey, P. A., Colapinto, E. V., Friedman, H. S., et al. 1988a. Treatment of intracranial human glioma xenografts with 131I-labeled anti-tenascin monoclonal antibody 81C6. *Cancer. Res.* 48: 2904–10

Lee, Y. S., Bullard, D. E., Zalutsky, M. R., Coleman, R. E., Wikstrand, C. J., et al. 1988b. Therapeutic efficacy of antiglioma mesenchymal extracellular matrix 131I-radiolabeled murine monoclonal antibody in a human glioma xenograft model. *Cancer Res.* 48: 559–66

Lightner, V. A., Rajagopalan, P., Erickson, H. P. 1988. Hexabrachion/tenascin does not bind to fibronectin, laminin or collagen. *J. Cell Biol.* 107: 600a

Lightner, V. A., Gumkowski, F., Bigner, D. D., Erickson, H. P. 1989a. Tenascin/hexabrachion in human skin: biochemical identification and localization by light and electron microscopy. *J. Cell Biol.* In press

Lightner, V. A., Pegram, C., Bigner, D., Erickson, H. P. 1989b. Monoclonal antibody 127 recognizes a subpopulation of chicken hexabrachion/tenascin. In "Structure, interactions and assembly of cytoskeletal and extracellular proteins", ed. P. M. Bailey. Berlin: Springer Verlag. In press

Lotz, M. M., Burdsal, C. A., Erickson, H. P., McClay, D. R. 1987. Comparison of the adhesion of NIL fibroblasts and glioma cells to fibronectin and hexabrachion (tenascin, cytotactin). *J. Cell Biol.* 105: 138a (Abstr.)

Mackie, E. J., Chiquet-Ehrismann, R., Pearson, C. A., Inaguma, Y., Taya, K., et al. 1987a. Tenascin is a stromal marker for epithelial malignancy in the mammary gland. *Proc. Natl. Acad. Sci. USA* 84: 4621–25

Mackie, E. J., Thesleff, I., Chiquet-Ehrismann, R. 1987b. Tenascin is associated with chondrogenic and osteogenic differentiation in vivo and promotes chondrogenesis in vitro. *J. Cell Biol.* 105: 2569–79

Mackie, E. J., Halfter, W., Liverani, D. 1988a. Induction of tenascin in healing wounds. *J. Cell Biol.* 107: 2757–67

Mackie, E. J., Tucker, R. P., Halfter, W., Chiquet-Ehrismann, R., Epperlein, H. H. 1988b. The distribution of tenascin coincides with pathways of neural crest cell migration. *Development* 102: 237–50

Maier, A., Mayne, R. 1987. Distribution of connective tissue proteins in chick muscle spindles as revealed by monoclonal antibodies: a unique distribution of brachionectin/tenascin. *Am. J. Anat.* 180: 226–36

Pearson, C. A., Pearson, D., Shibahara, S., Hofsteenge, J., Chiquet-Ehrismann, R. 1988. Tenascin: cDNA cloning and induction by TGF-beta. *EMBO J.* 7: 2677–81

Rieger, F., Daniloff, J. K., Pincon-Raymond, M., Crossin, K. L., Grumet, M., Edelman, G. M. 1986. Neuronal cell adhesion molecules and cytotactin are colocalized at the node of Ranvier. *J. Cell Biol.* 103: 379–91

Riou, J.-F., Shi, D.-L., Chiquet, M., Boucaut, J.-C. 1988. Expression of tenascin in response to neural induction in amphibian embryos. *Development* 104: 511–24

Sanes, J. R., Schachner, M., Covault, J. 1986. Expression of several adhesive macromolecules (N-CAM, L1, J1, NILE, uvomorulin, laminin, fibronectin, and a heparan sulfate proteoglycan) in embryonic, adult, and denervated adult skeletal muscle. *J. Cell Biol.* 102: 420–31

Schwarzbauer, J. E., Tamkun, J. W., Lemischka, I. R., Hynes, R. O. 1983. Three different fibronectin mRNAs arise by alternative splicing within the coding region. *Cell* 35: 421–31

Slemp, C. A., Lightner, V. A., Briscoe, G., Erickson, H. P. 1988. Separation of tenascin/hexabrachion, with no cell attachment activity, from a variant form of laminin, with strong fibroblast attachment activity. *J. Cell Biol.* 107: 232a

Spring, J., Beck, K., Chiquet-Ehrismann, R. 1989. Complete tenascin sequence: multidomain structure of three splicing variants and of recombinant tenascin fragments. *Cell.* Submitted

Steindler, D. A., Cooper, N. G., Faissner, A., Schachner, M. 1989. Boundaries defined by adhesion molecules during development of the cerebral cortex: the J1/tenascin glycoprotein in the mouse somatosensory cortical barrel field. *Dev. Biol.* 131: 234–60

Tan, S. S., Crossin, K. L., Hoffman, S., Edelman, G. M. 1987. Asymmetric expression in somites of cytotactin and its proteoglycan ligand is correlated with neural crest cell distribution. *Proc. Natl. Acad. Sci. USA* 84: 7977–81

Taylor, H. C., Lightner, V. A., McCasslin,

D. A., Beyer, W. F., Briscoe, G., Erickson, H. P. 1989. Biochemical characterization of tenascin/hexabrachion proteins. *J. Cell. Biochem.* In press

Thesleff, I., Mackie, E., Vainio, S., Chiquet-Ehrismann, R. 1987. Changes in the distribution of tenascin during tooth development. *Development* 101; 289–96

Thesleff, I., Kantomaa, T., Mackie, E., Chiquet-Ehrismann, R. 1988. Immunohistochemical localization of the matrix glycoprotein tenascin in the skull of the growing rat. *Archs. Oral Biol.* 33: 383–90

Vaughan, L., Huber, S., Chiquet, M., Winterhalter, K. H. 1987. A major, six-armed glycoprotein from embryonic cartilage. *EMBO J.* 6: 349–53

Annu. Rev. Cell Biol. 1989. 5 : 93–117

GROWTH FACTORS IN EARLY EMBRYOGENESIS

Malcolm Whitman and D. A. Melton

7 Divinity Avenue, Department of Biochemistry and Molecular Biology, Harvard University, Cambridge, Massachusetts 02138

CONTENTS

CELL-CELL INTERACTIONS IN DEVELOPMENT: THE SEARCH FOR MORPHOGENS

Over sixty years ago, Spemann & Mangold (1924) strikingly demonstrated that a small piece of embryonic tissue could organize an entire second body axis following transplantation to an ectopic site in a host embryo. Host cells that would have formed epidermis instead made muscle, notochord, and neural tissue, patterned into recognizable organ structures. The fate of host embryonic cells was clearly being shifted from one developmental pathway to another by experimental manipulation of their position relative to other embryonic tissues. Since that time, similar demonstrations in a wide variety of developing organisms have established the principle that the development of embryonic pattern is dependent upon intercellular interactions between the constituent parts of the developing

0743–4634/89/1115–0093$02.00

embryo. Through these interactions, the developmental fate of cells or tissues is specified with respect to their position within the embryo. The molecules postulated to mediate such cellular interactions have been termed morphogens; their identification is an essential step in understanding the molecular basis of pattern formation.

Although much information has been accumulated concerning the molecular mechanisms of intercellular communication in somatic cells (see Cold Spring Harbor Symposia on Quantitative Biology Vol. LIII 1989 for review), it is not yet clear to what extent these same mechanisms may be involved in intercellular signaling in early development. The identification of peptide growth factors in the early embryo and the recognition that these factors can have a broad range of effects on both the proliferation and differentiation of cultured cells (Sporn & Roberts 1987) support the possibility that these factors play a regulatory role during embryogenesis. This chapter will emphasize recent work demonstrating that peptide growth factors can specify the fate of embryonic cells, and hence may function as morphogens within the early embryo. If known peptide growth factors indeed function as morphogens in vivo, then understanding the molecular basis of morphogenesis need no longer be focused on the search for developmentally specific intercellular signaling molecules. It might rather be focused on understanding how the cellular transduction apparatus interprets the binding of a given extracellular ligand as a morphogenetic rather than a mitogenic signal.

Morphogenetic cellular interactions have been identified during the early development of many vertebrate and invertebrate metazoans (Slack 1983; Nieuwkoop et al 1985). Such interactions have been best characterized, however, in embryos in which surgical excision and recombination of cells or tissues may most easily be used to study the effect of cell and tissue interactions on developmental fate. These early interactions have been defined primarily through the study of chicken and amphibian embryos that are amenable to surgical manipulation from their earliest stages. In these organisms, the best characterized class of morphogenetic intercellular interactions are embryonic inductions: the action of an inducing tissue on a responding tissue that alters the developmental fate of the responding tissue.

Spemann & Mangold (1924) Spemann (1938) identified a region at the dorsal blastopore lip of amphibian gastrulae that can induce the formation of a second dorsal axis when grafted onto the ventral region of a host embryo. Further transplantation/recombination experiments identified several distinct events in early amphibian embryonic patterning: the induction of mesoderm from primordial ectoderm by endoderm (mesodermal induction), followed by the induction of neural tissue from primordial ectoderm by chordamesoderm (neural induction). These early patterning events are then followed by an ongoing series of inductive interactions throughout organogenesis that include development of mouth, heart, lung, eye, kidney, and pancreas (Jacobson & Sater 1988). A tissue that is induced at one developmental stage may itself become an inducer at a later developmental stage (e.g. amphibian chordamesoderm), and thus each successive embryonic induction is dependent on a preceding series of inductive decisions (Wessells 1977; Jacobson & Sater 1988). Although embryonic inductions have been most extensively characterized in amphibians because of the relative ease of surgical manipulation in these organisms, similar manipulations in numerous other organisms ranging from nematodes (Preiss & Thomson 1987) to mammals (Wessells 1977) have identified cellular interactions necessary for correct specification of cell fates.

Once the importance of embryonic inductions in early development had been established, a search for the chemical basis of these interactions soon began. Identification of a variety of agents that could mimic the function of Spemann's dorsal "organizer" appeared to establish an assay system in which endogenous morphogens might be identified. Attempts to purify the neuralizing agent in Spemann's organizer, however, revealed that many nonspecific agents could mimic the action of the organizer. These included pH shock, dead tissues from a broad range of sources, as well as chemical agents such as ethanol (Holtfreter 1934; Brachet 1950). This discovery raised the possibility that the biological specificity in these interactions resided almost entirely in the responding tissue and, thus, that the search for specific intercellular morphogens would prove fruitless. Therefore the search for the chemical basis of the organizer generated a great deal of theoretical speculation concerning the nature of inductive interactions, but did not lead to the identification of the molecular signals involved in this phenomenon (reviewed in Hamburger 1988).

Studies of induction of mesoderm in amphibians offered somewhat more hope for the identification of specific factors. Nieuwkoop (1969) utilized a convenient assay system for mesodermal induction in which ectodermal tissue from the animal pole of the amphibian embryo (the animal cap) was dissected and combined in a "sandwich" with endoderm or other putative inducing substances. Nieuwkoop and his colleagues showed that endoderm produces a signal that induces mesoderm from adjacent prospective ectodermal tissue (Figure 1). The inducing function of endoderm could be mimicked by a limited set of heterologous tissues and factors (reviewed in Neiuwkoop 1985; Gurdon 1987; Tiedemann 1966). Most of these inducers, however, were found in tissues (e.g. guinea pig bone marrow, chicken embryos, fish swim bladder) that bore no obvious relation to amphibian

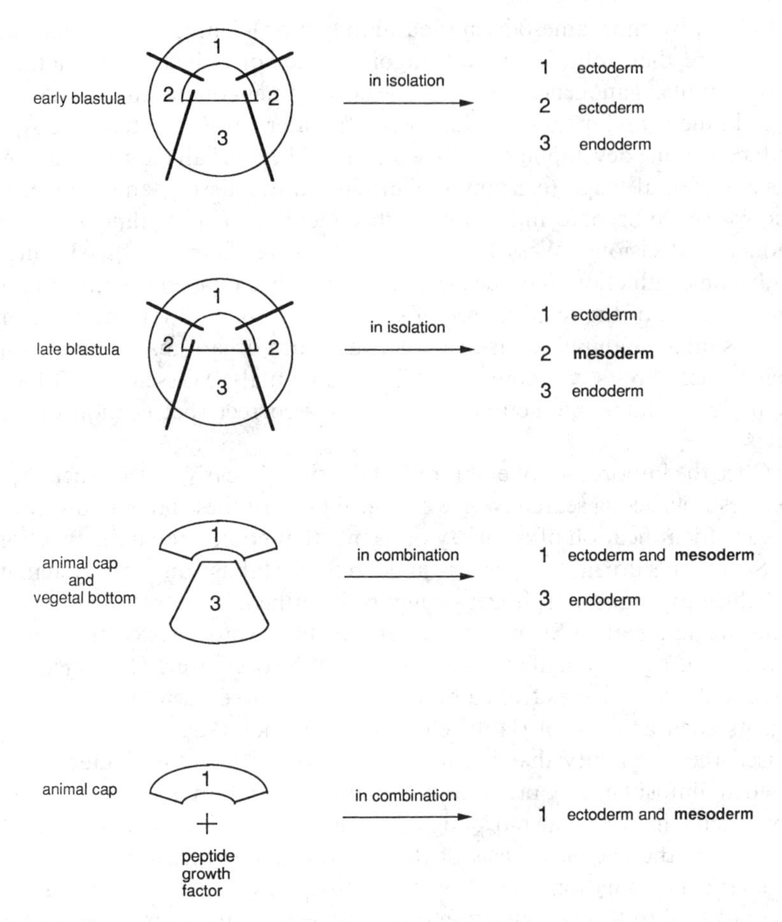

Figure 1 Schematic representation of mesodermal induction in isolation/recombination experiments. Animal (*1*), equatorial (*2*), or vegetal (*3*) tissue is dissected from blastulae and allowed to differentiate in culture. Mesoderm is formed from the equatorial region of late blastulae (*second panel from top*), animal cap tissue recombined with vegetal tissue (*third panel from top*), or animal cap tissue cultured in the presence of mesoderm inducing growth factors (*bottom panel*).

endoderm, the natural inducing tissue. In addition, these agents function as inducers only in the form of insoluble pellets, which makes further molecular analysis of their interaction with ectodermal tissue difficult. In contrast, both mesodermal and neural induction occur when inducing and responding tissues are separated by nucleopore filters, which indicates that soluble factors are likely to play a role in the natural inductive process.

(Grunz & Tacke 1986; Saxen et al 1976; Toivonen et al 1976; Gurdon 1989).

The study of soluble factors as mesodermal inducers blossomed with the identification and purification by Smith (1987) (Smith et al 1988) of a soluble peptide from medium conditioned by a *Xenopus* tissue culture cell line that acts as a potent inducer of mesoderm when incubated with animal cap tissue. With the identification of soluble inducing factors and of molecular markers for mesodermal tissue, the coupling of a developmental stimulus (inducer) to a developmental response (mesodermal differentiation) has become amenable to analysis at the cellular and biochemical levels. In this assay system, a number of growth factor or growth-factor-like polypeptides have now been identified that can induce mesoderm in animal caps (Slack et al 1987; Godsave et al 1988; Rosa et al 1988; Kimelman & Kirschner 1987).

These observations have focused new attention on the function of growth factors in development. There is ample evidence that such factors can regulate both cell proliferation and differentiation in somatic cells, and it seems likely that they might play a similar role in embryonic cells. The identification of growth factors as putative inducers, however, provides the first indication that these factors may play more than simply a supportive role in the development of embryonic tissues; they may function as morphogens. We first review the general properties of growth factors, which suggest they may function in early embryogenesis, and then return to recent work that indicates growth factors may act as morphogens in early embryonic induction.

IDENTIFICATION AND CLASSIFICATION OF PEPTIDE GROWTH FACTORS

Peptide growth factors were initially characterized using either cultured cell systems (PDGF, TGF) or in vivo assays (EGF) as small, soluble polypeptides capable of regulating cellular physiology (Ross et al 1986; Carpenter & Cohen 1979; Froesch et al 1985; Gospodarowicz et al 1987; Sporn et al 1986). The purification and sequencing of these factors and the genes encoding them has permitted their grouping into a relatively small number of families (Table 1) (Mercola & Stiles 1988). Within each growth factor family, members are closely related structurally, and, in some but not all instances, may interact with similar or identical receptors. In a few cases, however, family members may have distinct or even opposing biological effects; for example the TGFβ-related molecules, activin and inhibin, have antagonistic effects on both the differentiation of erythroid

Table 1 Peptide growth factor families

Family	Members
Epidermal growth factor (EGF)	EGF
	Transforming growth factor-alpha (TGFα)
	notch
	delta
	lin-12
Transforming growth factor-Beta (TGFβ)	TGFβ_{1-5}
	Xenopus tissue culture mesoderm-inducing factor (XTC-MIF)
	Inhibin-A
	Inhibin-B
	Activin A
	Activin-AB
	Mullerian inhibiting substance (MIS)
	Bone morphogenetic protein 2A (BMP-2A)
	Bone morphogenetic protein 3 (BMP-3)
	Decapentaplegic (*dpp*)
	Vg-1
Heparin binding growth factor (HBGF)	acidic fibroblast growth factor (aFGF)
	basic fibroblast growth factor (bFGF)
	embryonal carcinoma derived growth factor (ECDGF)
	int-2
Platelet derived growth factor (PDGF)	PDGF-A
	PDGF-B
	PDGF-AB
Insulin-like growth factor (IGF)	IGF-I
	IGF-II (Somatomedin-C)

Growth factors and their homologues discussed in this review are listed under family headings. Family groupings are generally based on sequence similarity, with the exception of XTC-MIF, which has been classed as TGFβ-like based on its biochemical and antigenic properties (Smith et al 1988; Rosa et al 1988). Developmentally important genes encoding products with significant sequence similarity to peptide growth factors are indicated in italics. (Adapted with permission from Mercola & Stiles 1988).

cells (Yu et al 1987) and on the release of pituitary follicle stimulating hormone (Vale et al 1986; Ling et al 1986; reviewed in Massagué 1987). More commonly, growth factor family members may have similar or indistinguishable effects; for example acidic and basic FGF have virtually indistinguishable effects on proliferation or differentiation of a variety of cell types (reviewed in Gospodarowicz et al 1987).

In general, the peptide growth factors have been defined as such primarily by their ability to regulate the proliferation of cultured cells. It is now recognized, however, that peptide growth factors are multifunctional and may trigger a broad range of cellular responses (reviewed by Sporn &

Roberts 1987), including effects on extracellular matrix formation, cell motility, cell survival, secretion, and the expression and stabilization of differentiated characteristics. The actions of growth factors may therefore be usefully subdivided into two general categories: effects on cell proliferation and effects on cell phenotype other than proliferation. These categories often overlap; regulation of division often alters other aspects of cell phenotype. Nevertheless, these categories are useful in that they distinguish two aspects of embryonic morphogenesis: differential rates of tissue proliferation and the organization of differentiated cell types into characteristic structures and organs.

Effects on Cultured Cells: Proliferation

The ability of peptide growth factors to stimulate cell proliferation is clearly of potential importance during development. Growth factors are generally specific for distinct tissue types, for example bFGF is a potent mitogen for cells of mesodermal origin, particularly epithelia (reviewed in Gospodarowicz et al 1987), while the interleukins are mitogenic primarily for lymphocytes (reviewed in O'Garra 1988). In the case of TGFβ, positive or negative effects on cell proliferation may be observed, depending on the cell type or cell state examined. TGFβ inhibits the proliferation of many cell types, including fibroblasts (Sporn et al 1986), but it can also stimulate growth of fibroblasts when they have been cultured in soft agar suspension (Roberts et al 1981).

Localized cell proliferation at the site of organ formation is a common phenomenon, and microinjection of growth factors into developing tissues or organs can concomitantly stimulate local cell proliferation and trigger premature developmental events (reviewed in Carpenter & Cohen 1979). These phenomena are observed relatively late in development, however, at a stage when cultured cells isolated from the embryo are highly dependent on serum factors for continued proliferation. The importance of embryonic growth factors for regulation of cell proliferation during early embryogenesis is less clear. In amphibians and other oviparous vertebrates and invertebrates (Zalokar 1976; Newport & Kirschner 1982), early embryonic cell divisions proceed without discernible G_1 or G_2 phases of the cell cycle (Graham & Morgan 1966), and blastomeres isolated from early amphibian embryos continue to divide through the first 17 cleavage divisions in the absence of cell-cell contact in a simple salt solution (Winklbauer 1986). Through the mid-blastula transition (approximately 10–12 cleavages), these cleavages are synchronous, even among isolated blastomeres (Hara 1977). Although these observations do not rule out the possibility of autocrine mitogenic stimulation, they indicate that in these early embryos cell division is under the control of an internal clock rather than external cues. It is only relatively late in amphibian development that localized regulation of cell proliferation begins to become evident (Fischberg 1949).

Blastomeres isolated from early mammalian embryos have normal G_1 and G_2 phases, but also appear to be able to divide to form blastocysts in the absence of exogenous growth factors (Biggers 1987; Tarkowski & Wrobleska 1967). These early blastomeres express transcripts for several growth factors (Rappolee et al 1988), and embryonal carcinoma cells secrete a number of factors with mitogenic activity (Heath & Isacke 1984; Heath & Shi 1986; Rizzino & Bowen-Pope 1985). Autocrine or paracrine growth factor stimulation of cell proliferation may therefore play some role in maintaining the proliferation of early mammalian blastomeres. In addition, as Rappolee et al (1988) have pointed out, factors secreted by the blastocyst at the time of implantation may also be directed at maternal tissues, stimulating uterine proliferation and angiogenesis necessary for support of the developing embryo.

Differentiation/Dedifferentiation

Examples of the regulation of the differentiated state of both primary and established cells in culture have now been identified for a number of growth factor families (Sporn & Roberts 1987; Gospodarowicz et al 1987; Froesch et al 1985; Sporn et al 1986). In many cases growth factors can either stimulate or inhibit cell differentiation, depending on the cell type or cell state examined. For example, TGFβ stimulates the differentiation of bronchial epithelial cells and inhibits terminal differentiation of B cells and adipogenesis of 3T3 fibroblasts (Ignotz & Massagué 1985; Sporn et al 1986). It can also enhance the expression of a differentiated cartilaginous phenotype in primitive mesenchymal cells, but suppresses expression of cartilage markers in differentiated cells of mesenchymal origin (Rosen et al 1988). FGF can delay differentiation of myoblasts (Gospodarowicz et al 1976), but stimulates differentiation or stabilizes the differentiated characteristics of a variety of cell types, including chondorocytes, endothelial cells, and a rat pheochromocytoma line (PC12) (reviewed in Gospodarowicz et al 1987).

Effects of growth factors on cell differentiation, both positive and negative, may in some cases be coupled to regulation of cell proliferation and in others not. PDGF potently inhibits the differentiation of oligodendrocyte/type 2 astrocyte progenitor cells by maintaining their continued proliferation (Noble et al 1988; Raff et al 1988); inhibition of B-cell proliferation by TGFβ appears to inhibit the expression of differentiated characteristics (Kehrl et al 1985). The enhancement of glial cell differentiation by FGF is associated with stimulated mitogenesis (Morrison et al 1986), while TGFβ appears to stabilize the differentiated state of kidney epithelial cells by inhibiting cell proliferation (Fine et al 1985). In contrast, in the case of stimulation of neurite outgrowth in PC12 cells by FGF (Togari et al 1983) and inhibition of 3T3 cell differentiation to adipocytes by TGFβ (Ignotz & Massagué 1985), growth factor effects on cell differentiation do not appear to be coupled to effects on cell proliferation.

In addition to initiating or maintaining cellular differentiation, peptide growth factors can regulate function in a number of ways. Both FGF and TGFβ can affect the formation of the extracellular matrix (ECM) (reviewed by Folkman & Klagsbrun 1987; Gospodarowicz et al 1987; Sporn et al 1987), and this activity may mediate some of these factors' effects on cell proliferation and differentiation. Both in cultured cells and upon injection in vivo (Roberts et al 1986), TGFβ can modify the ECM by enhancing formation of collagen and fibronectin (Ignotz & Massagué 1986), increasing secretion of protease inhibitors, and inhibiting secretion of cellular proteases (Laiho et al 1986). TGFβ can also modify other types of secretory phenomena: for instance, the responses of ovarian granulosa cells to FSH is strongly potentiated by pretreatment with TGFβ (Ying et al 1986). Similarly, FGF can regulate secretion by pituitary cells (reviewed in Baird et al 1986). Cell movement may also be regulated by peptide growth factors. EGF stimulates the nondirected movement of keratinocytes (Barrandon & Green 1987), while PDGF and TGFβ are both chemotactic for fibroblasts and other cell types (Grotendorst et al 1981; Seppa et al 1982; Postlethwaite et al 1987; Wahl et al 1987).

Cell proliferation, cell differentiation, ECM formation, secretion, and cell movement are all critical events in the morphogenesis of the embryo, and therefore may be important targets for regulation by growth factors during development. Although many cellular functions have been identified as potential targets for growth factor regulation, it is clear that the effect of a growth factor is highly dependent on the nature of the target cell. Thus it is difficult to extrapolate from studies of cultured cells or adult animals to specific developmental events; a range of possible functions for a given growth factor can be defined in such alternative systems, but this does not suffice to specify the effect of this factor on an embryonal target cell. Growth factors must therefore be associated with specific developmental events as a prerequisite for assessing their function in the regulation of these events.

Identification and Localization in Developing Tissues

Peptide growth factors or their mRNAs have been detected from the earliest stages of embryogenesis in both mammalian and nonmammalian organisms. Transcripts for PDGF have been detected in unfertilized eggs of both mouse and frog (Rappolee et al 1988; Mercola et al 1988), transcripts for TGFα and TGFβ are expressed in preimplantation mouse embryos (Rappolee et al 1988), and both FGF transcript and protein have been identified in frog oocytes and early embryos (Kimelman et al 1988; Slack & Isaacs 1989). Embryonal carcinoma cells derived from early mouse blastomeres secrete PDGF-like and FGF-line activities, and express IGF-II transcripts (Heath & Isacke 1984; Heath & Shi 1986; Rizzino & Bowen-Pope 1985). As development proceeds and organogenesis begins, the full range of identified growth factors becomes detectable in fetal tissues (Mercola & Stiles 1988).

For growth factors to play a truly morphogenetic role in early embryogenesis, not only must they be present in early embryos, they must also be distributed in a spatially heterogeneous fashion. In the frog, a maternal mRNA Vg-1, which has significant sequence similarity to TGFβ, is localized to the vegetal pole early in oogenesis (Melton 1987; Ysraeli & Melton 1988; Weeks & Melton 1987). The protein encoded by this mRNA is localized in vegetal cells during the early cleavage and blastula stages (D. Tannahill & D. A. Melton, in preparation). This localization is consistent with the proposed role of this molecule as a signaling molecule generated in the vegetally derived endoderm. A transcript encoding an EGF-homologous polypeptide is found to be localized to ectodermal tissue in early sea urchin embryos (Hursh et al 1987) and a member of the FGF gene family, the proto-oncogene int-2, is expressed both in extra-embryonic mouse endoderm and in migrating primordial mesoderm during gastrulation (Wilkinson et al 1988). A specific function during embryogenesis has not yet been demonstrated for any of these developmentally localized growth factor homologues.

Studies of localization of growth factors in developing tissues have been focused primarily on later development. TGFβ has been localized immunohistochemically in 11–18-day mouse embryos in mesodermally derived tissues (Heine et al 1988). TGFβ expression was correlated with a variety of histogenetic and morphogenetic events, including formation of vertebrae, bones, meninges, and hair follicles. TGFβ1 transcripts appear to occur in an overlapping but nonidentical distribution to that reported for TGFβ protein (Lehnert & Akhurst 1988); they are found in the mesenchymal tissues of bone and several internal organs, but are also expressed in epithelia overlying mesenchymal tissues in which TGFβ protein was found at high levels. The authors therefore suggest that in some tissues TGFβ may act as a paracrine factor, diffusing from its site of synthesis (epithelia) to its site of action (mesenchyme), while in other sites (e.g. bone), its synthesis and action are within the same tissue. Although this latter distribution might reflect an autocrine function for this factor, there is no evidence to exclude the possibility that distinct cell populations serve as source and target for TGFβ in bone mesenchyme. Extracellular matrix formation appears to play a significant role in epithelio-mesenchymal inductive interactions (Grobstein 1967; Wessells 1977), and therefore the identification of TGFβ in epithelio-mesenchymal tissue during organogenesis suggests that effects of TGFβ on the ECM (see above) might mediate inductive interactions in these tissues. In 10–16-day rat embryos IGF-II transcripts were found to be expressed in mesodermal tissue with a pattern of expression similar to that described for TGFβ (Beck et al 1987; Stylianapolou et al 1988). TGFβ and IGF-II can act synergistically to stimulate anchorage independent growth in cultured fibroblasts, which raises the possibility that their coexpression in embryonic mesoderm reflects a functional interaction between them (Massagué et al 1985).

Identification of growth factors and their transcripts during embryogenesis has established possible times and sites of action, but not the actual target tissues for these factors. The observation that TGFβ protein may be concentrated in tissues ajacent to but distinct from those in which the gene is expressed highlights the need to identify target as well as source tissues for embryonal growth factors. Relatively little is known about the presence or localization of growth factor receptors during early development. EGF receptors have been identified in post-11-day mouse embryonic tissue as well as in 4-day extra-embryonic trophectoderm (Adamson & Meek 1984). IGF-II receptors have been identified in murine blastocysts and later embryos (Smith et al 1987), but have not been localized within embryonic tissues. TGFβ receptors appear to be ubiquitously expressed on cells in culture (Wakefield et al 1987), but it is not certain whether the same is true for early embryonic cells. Further characterization of the timing and location of expression of growth factor receptors during development will clarify the likely target tissues for embryonal growth factors.

Genetic Evidence for Growth Factor Function in Development

Investigation of the developmental expression of growth factors, like the study of growth factor effects on cultured cells, helps to define the possible range of action of growth factors in development. Direct evidence that growth factors participate in the specification of developmental fate, however, requires experimental manipulation of growth factors in embryonic tissues. Developmental genetics has provided one approach to such manipulations; many of the genes found to be important in the regulation of developmentally important events have been found to encode genes with significant structural or functional similarities to peptide growth factors or their receptors. The *Drosophila* homeotic genes *notch* and *delta*, which participate in the differential specification of epidermal and neurogenic tissue, share domains of sequence similarity with EGF (Wharton et al 1985; Kidd et al 1986; Kopycinski et al 1988) as does the nematode homeotic gene *lin-12* (Greenwald 1985). Similarly, the *Drosophila* gene complex *decapentaplegic* (*dpp*), which functions both in embryonic dorsal-ventral specification and in imaginal disk morphogenesis, is homologous to TGFβ as well as to the developmentally localized frog mRNA Vg-1 (Padgett et al 1987). The segment polarity gene *wingless* is the *Drosophila* homolog of the mammalian proto-oncogene int-1 (Rijsewik et al 1987), the product of which has been proposed be a secreted factor localized in mouse embryos to the developing neural tube (Shackleford & Varmus, 1987; Wilkinson et al 1987). Other developmentally important genes appear to encode peptide receptors: The *Drosophila* gene *torpedo*, which participates in the establishment of the embryonic dorsal-ventral axis (Schupbach 1987), is the *Drosophila* homolog of mammalian EGF receptor (Price & Schubach 1989), and the homeotic gene *sevenless* bears structural homology to a number of membrane tyrosine kinase growth factor receptors (Hafen et al 1987).

Mutations in the developmental genes described above clearly involve changes in cell fate. In some instances a very restricted class of cells may be respecified; for example the *Drosophila* mutation *sevenless* transforms the R7 photoreceptor neuron of the ommatidium into a nonneuronal cell (Tomlinson & Ready 1986). In other cases whole tissues may be transformed; mutations at the *notch* locus result in replacement of the embryonic epidermis by hypertrophied neural tissue (Lehmann et al 1983). Some of these genes appear to affect only the cells in which they are expressed (Wharton et al 1985; Tomlinson & Ready 1987), others appear to also act on cells or tissues adjacent to those in which they are expressed (Spencer 1984; Posakony 1987; Reinke & Zipursky 1988). Because products of these genes have been characterized primarily on the basis of sequence similarities to known signaling molecules, there is no direct biochemical evidence that mechanisms of action of these gene products are similar to those characterized for peptide growth factors. Nevertheless, the sequence similarities between developmentally important genes and growth factors and their receptors strongly suggest functional similarities between them.

GROWTH FACTORS AND MESODERM INDUCTION

Direct evidence that growth factors can specify cell fate has come from studies of the mesoderm induction in amphibians. During normal amphibian embryogenesis, a signal emanating from the vegetal region of the embryo specifies cells in the equatorial region to become mesoderm (Nieuwkoop 1985; Gurdon 1987; Slack 1983). During gastrulation, many desendants of these mesodermally specified blastomeres invaginate at the blastopore lip, and soon afterwards begin to differentiate into mesodermal tissues and structures. Those located most dorsally along the anterior-posterior axis form the notochord and somites. This dorsal mesoderm induces the formation of the neural tube from overlying ectoderm and will ultimately differentiate into much of the musculature and connective tissue of the tadpole. Those located more laterally and ventrally, the lateral plate and ventral mesenchyme, will form much of the viscera, including heart, blood, and kidneys. The final patterning of mesodermal tissues involves differentiation of specific cell types (e.g. muscle), organization of these cells into organ structures, and organization of these structures along the dorsal-ventral and anterior-posterior axes.

An assay that reproduced the phenomenon of mesodermal induction was developed by Nieuwkoop (1969), who found that endoderm could induce formation of mesoderm from explanted animal pole ectoderm, which in the absence of endoderm would form epidermis. Induction of mesoderm by this technique generates mesodermal cell types (e.g. muscle, notochord), usually organized into discrete blocks of tissue. Later experiments (Boterenbrood & Nieuwkoop 1973; Dale et al 1985; Dale & Slack 1987) have demonstrated that dorsal vegetal blastomeres induce more dorsal mesodermal tissues than do ventral vegetal blastomeres, which indicate that a dorsal-ventral distinction can be reproduced in this assay system.

Induction by Growth Factors

A number of soluble factors, some derived from *Xenopus* and some from other organisms, have been identified that can mimic the inductive effects of vegetal blastomeres: they shift cells from the animal pole blastomeres from an epidermal differentiation pathway to a mesodermal pathway. These include a number of purified growth factors such as both bovine and frog FGF, porcine TGFβ-2, mouse ECDGF, and a TGFβ-like molecule from medium conditioned by cultured *Xenopus* cells (XTC-MIF) (Slack et al 1987; Smith et al 1988; Kimelman & Kirschner 1987; Rosa et al 1988; Kimelman et al 1988; Slack & Isaacs 1989). TGFβ has also been reported to act synergistically with FGF in increasing the quantity of muscle actin induced by the latter by $\sim$10-fold (Kimelman & Kirschner 1987). Godsave et al (1988) have argued that mesoderm inducing factors fall into two distinct classes: heparin binding growth factors, and a second class of factors similar to *Xenopus* XTC-MIF, which may be related to TGFβ (Rosa et al 1988; Smith et al 1988). There is no compelling evidence, however, that the list of possible inducing factors is complete.

In the original studies of mesoderm induction by vegetal tissue, identification of induced tissues depended on histological analysis of differentiated structures. As emphasized by Gurdon in a recent review (1987), the identification of molecular markers for induction has proven extremely useful as a more sensitive, quantitative assay for induction. RNA (Gurdon et al 1984, 1985) and antibody (Kintner & Brockes 1984; Dale et al 1985; Godsave et al 1988; Symes et al 1988) probes for muscle tissue have been of particular value. Molecular assays for early markers of notochord (Smith & Watt 1985; Zanetti et al 1985; LaFlamme et al 1988), epidermal (Jones 1985; Symes et al 1988; Jonas et al 1985), and neural differentiation (Jacobson & Rutishauser 1986; Kintner & Melton 1987) have also been developed. Mesoderm induction in animal caps by peptide growth factors can now be assessed by a variety of criteria: elongation/extension of animal caps characteristic of gastrulating marginal zone tissue (Keller et al 1985; Symes & Smith 1987), muscle actin gene transcription (Gurdon et al 1985), appearance of muscle specific antigens (e.g. Symes et al 1988; Godsave et al 1988), as well as histological identification of muscle, notochord, mesenchyme, and blood.

Induction of mesoderm by peptide growth factors does not involve any detectable change in the proliferation of embryonic blastomeres (Symes & Smith 1987; Slack et al 1987), nor is there any obvious change in the timing of cellular differentiation; epidermal differentiation is suppressed in animal pole and mesodermal differentiation occurs instead (Grunz et al 1975). Recently it was shown that the suppression of epidermal characteristics and the onset of mesodermal differentiation are separable processes during mesodermal induction. When mesoderm inducing factor from *Xenopus* cultured cells (XTC-MIF) is added to isolated animal pole blastomeres, expression of epidermal markers is suppressed, but mesodermal markers are not expressed (Symes et al 1988). Only when the blastomeres are allowed to aggregate after XTC-MIF treatment does differentiation of mesodermal tissue occur. Therefore mesodermal differentiation, but not the suppression of epidermal differentiation by XTC-MIF, appears to be dependent on cell-cell contact. Even when endoderm is used as an inducer, expression of mesodermal characteristics in animal pole blastomeres is facilitated by association with other animal pole blastomeres. This "community effect" (Heasman et al 1984; Gurdon 1987) indicates that mesodermal induction is a multistep process, and that cellular interactions other than addition of a soluble inducer are necessary for the expression of mesodermal characteristics.

The likelihood that the observed effects of growth factors on animal pole tissue are relevant to the natural inductive process has been increased by the discovery that FGF mRNA and protein are present in both eggs and early embryos (Kimelman et al 1988; Slack & Isaacs 1989); the protein appears to be present in quantities consistent with its proposed role as an inducing factor. However, FGF has not directly been shown to be localized to or secreted by endodermal tissue. Although heparin appears to inhibit the ability of endoderm to induce formation of mesoderm across a filter, it is not clear that this inhibition is due to the ability of heparin to bind FGF (Slack et al 1987). The hypothesis that FGF actually functions as an endoderm-secreted mesodermal inducer therefore remains unproven, but the available evidence strongly supports this possibility.

Neither TGFβ-1, TGFβ-2, nor the TGFβ-like inducer XTC-MIF has been identified in early frog embryos. A maternal RNA with significant homology to the TGFβ family (Weeks & Melton 1987), Vg-1, is translated throughout early embryogenesis (D. Tannahill & D. A. Melton, in preparation). Both Vg-1 mRNA (Melton 1987; Ysraeli & Melton 1988) and embryonically synthesized Vg-1 protein (D. Tannahill & D. A. Melton, in preparation) appear to be localized to the vegetal pole, from which endoderm, the source of the natural inducing signal, arises. Recently, an mRNA encoding a new member of the TGFβ family, TGFβ-5, has been identified in frog oocytes and embryos (Kondaiah et al, submitted). Neither Vg-1 protein or TGFβ-5, however, has been directly demonstrated to function as a mesodermal inducer. The role of TGFβ-like molecules in the natural inductive process therefore remains uncertain.

Although putative mesoderm inducers have been identified, they do not in themselves suffice to explain the patterning of mesodermal structures. As several authors (Slack 1983; Jacobson & Slater 1988; Cooke & Smith 1989) have emphasized, the endodermal inductive signal may serve to establish a "morphogenetic field" rather than to specify individual cell fates. In this interpretation, the initial signal that specifies mesoderm may serve to establish a field of pluripotential mesodermal cells in the equatorial region of the embryo; subsequent intercellular interactions are then necessary to specify individual tissue and cell types (e.g. muscle, notochord) within this region. A successful molecular description of mesodermal induction would therefore identify and characterize several distinct events in the patterning of mesodermal tissue. The proposed pluripotential mesodermal state has not yet been sufficiently defined experimentally to allow investigation of the role peptide growth factors might play in its establishment. It is possible that the suppression of epidermal characteristics observed in isolated XTC-MIF-treated blastomeres (see above) reflects such a pluripotential mesodermal state, but it has not been demonstrated that such a state exists even transiently in vivo. Thus, in spite of the recognition that FGF and TGFβ-like molecules can shift explanted embryonic blastomeres from an epidermal differentiation pathway to a mesodermal one, the point at which these factors may act on mesodermal patterning in the intact embryo is unclear.

The transmission of inducing signals through the responding tissue may in itself be a complex phenomenon. Several experiments indicate that cells induced by either the natural inducing signal or artificial inducers may themselves acquire inductive potency. This phenomenon, known as homoiogenic induction, was originally observed by Mangold & Spemann (1927) as the ability of notoplate tissue to induce new notoplate from host tissue at an ectopic site. More recently, Kirihara & Sasaki (1981) found that the natural inductive signal was not transmitted through layers of ectodermal tissue that were no longer competent to respond to it. These experiments do not rule out the possibility that a passive block to the inductive signal develops in noncompetent tissue, but suggests the possibility that propagation of the inductive signal through responding tissue is an active, cellular process rather than a passive, extracellular one. Cooke et al (1987) have found that tissue induced by XTC-MIF can in turn induce mesodermal structures in cells no longer competent to respond to XTC-MIF. It would therefore seem that an inducing molecule distinct from XTC-MIF is produced by XTC-MIF induced cells that can, in turn, induce mesodermal structures in cells no longer competent to respond to XTC-MIF.

Patterning of Mesoderm

The basis for the organization of mesoderm along the dorsal-ventral and anterior-posterior axes presents an additional problem in interpreting the roles of growth factors in induction. Notochord and muscle are characteristic dorsal mesodermal structures, while blood and mesenchyme are characteristic ventral tissues. Dorsal vegetal tissue induces more dorsal structures from competent ectoderm than does ventral vegetal tissue (Boterenbrood & Nieuwkoop 1973; Dale et al 1985; Dale & Slack 1987), which indicates that endodermal blastomeres differ in their inducing capacity. FGF appears to induce primarily ventral mesodermal tissue at low concentrations, and more dorsal structures at higher concentration, which raises the possibility that a simple dorsal-ventral gradient of FGF might organize mesoderm along this axis. However, FGF can induce notochord only very rarely, even at very high concentrations (Slack et al 1987; Slack & Isaacs 1989). It is therefore uncertain whether FGF alone is sufficient to generate the full range of dorsal mesodermal structures.

Alternatively, FGF concentration might be fixed along the dorsal-ventral axis, and a second, synergistically acting factor, such as a TGFβ-like molecule, might provide the dorsalizing signal. Slack and collaborators have argued for the existence of a distinct dorsalizing signal, based primarily on their observation of the ability of dorsal mesoderm to dorsalize more ventrally derived mesoderm (Dale & Slack 1987). These experiments do not, however, directly demonstrate the existence of a distinct dorsalizing signal, and therefore the molecular basis for dorsal-ventral organization remains to be established.

Different cell fates along the anterior-posterior axis may also be established by signals emanating from the endoderm (Cooke 1985). Differences in mesodermal tissues are most commonly discussed as dorsal-ventral, i.e. notochord as more dorsal than muscle that is, in turn, more dorsal than blood. While this is certainly an accurate description of the anatomical relations at the post-gastrula stage, it may not accurately reflect the developmental mechanism used to set up mesodermal axes. In particular, it may be the case that the anterior-posterior differences are established by peptide growth factor signals from the endoderm (A. Ruiz i Altaba & D. A. Melton, in preparation). Investigation of this possibility has been limited by a lack of markers for anterior-posterior differences in mesoderm. Differences in dorsal versus ventral mesoderm can be assayed by cell phenotype (e.g. notochord vs blood); an anterior somite may be histologically indistinguishable from a posterior one, but may be equally well distinguished from a posterior somite in terms of how it was developmentally specified. The recent identification of gene markers showing anterior-posterior differences in mesodermal tissues (Condie & Harland 1987; Ruiz i Altaba & Melton 1989) makes it possible to address this issue. Thus, we may find out that mesoderm is not only specified by cell type or dorsal-ventral value, but also by its anterior-posterior position.

Finally, we note that the development of mesodermal structure depends not only on the specification axes but also, more locally, on the organization of differentiated cell types into discrete tissues. For example, somitic muscle and notochord are first observed as discrete, self-contained blocks of tissue. This may occur either because cells induced to form these tissues possess some capacity for migration and selective adhesion, or that induction occurs cooperatively in discrete blocks of competent cells. Muscle tissue can form in discrete blocks in the absence of detectable cell movement, which suggests that in this instance induction is a cooperative process among a localized set of cells, perhaps reflecting an inductive community effect (Gurdon 1987). In other cases, cell movement and cell-type specific cell adhesion molecules may participate in the formation of discrete tissue structures from regions of induced tissue (Trinkaus 1984; McClay & Ettensohn 1987).

Although molecular markers can be used to model the induction of mesoderm as a relatively simple stimulus-response coupling, the considerations discussed above illustrate that the identification of growth factors as possible mesodermal inducers has not in itself sufficed to explain the morphogenesis of mesodermal structures. One current practical limitation to interpretation of investigations of tissue differentiation and patterning within mesoderm may be the lack of an appropriate range of molecular markers. Muscle specific markers have been the primary quantitative molecular marker of mesodermal induction used to date. When using markers for a single-tissue type, it is difficult to distinguish changes in the intensity of an inductive response from changes in dorsal-ventral specification. Consideration of axial specification of mesodermal tissues (e.g mesenchyme, notochord, blood) at the molecular level is thus problematic. In addition, it is quite possible that many of the secondary morphogenetic interactions responsible for mesodermal patterning proceed by mechanisms other than soluble intercellular factors. Gap junctions, cell-surface proteins, and extracellular matrix proteins may all mediate the transmission of signals important in the correct patterning of mesodermal structures. Nevertheless, as additional tools become available, the identification of peptide growth factors as mesodermal inducers may provide a framework within which the more complicated issues of mesodermal morphogenesis may be characterized in molecular terms.

Growth Factors and Developmental Specificity

The regional specification of the early embryo depends on a continuing series of intercellular interactions triggering specific developmental decisions. The specificity of a given developmental event can be resolved into two distinct components: the specificity of the intercellular signal and the potency of the target tissue to respond appropriately. This type of interaction is illustrated by the elegant transplantation experiments of Saunders et al (1957): when chick prospective thigh mesoderm (a posterior, proximal tissue) is transplanted under the apical ectodermal ridge of the growing chick wing (an anterior, distal region), the grafted tissue is induced to form a foot (a posterior, distal structure), while arm mesoderm at the same site generates wing tissue. In this case, an inductive signal specifies the formation of a distal (rather than proximal) structure, but the competence of the target tissue (thigh mesoderm) determines its differentiation as a posterior structure (foot) rather than an anterior one (wing). Thus an inductive signal in chick wing bud can elicit distal limb structures whose anterior-posterior identity is determined by the prior developmental history of the induced tissue.

The study of peptide growth factors has revealed numerous cases in which response to a growth factor is dependent on the state or type of target cell (see above). The identification of growth factors as developmental signaling molecules may allow the analysis of developmental competence in molecular terms: how does a given growth factor elicit a developmentally specific response from a competent target cell. When added to a variety of somatic cell types, FGF stimulates cell division, but when added to embryonic ectoderm, it triggers mesodermal differentiation without modulating cell proliferation. The molecular basis on which different cell types interpret FGF as a mitogenic or differentiative signal may lie at any of a number of points. In embryonic blastomeres, FGF binding may generate a membrane or cytosolic messenger distinct from those mediating FGF stimulated mitogenesis in fibroblasts. Alternatively, the same membrane/cytosolic second messengers may be generated in response to FGF in different cell types, and the biological specificity of the response would in this case lie in the presence of different specific targets, e.g. transcription factors for those messengers. The recent observation that the viral oncogene polyoma middle T can, like FGF, induce mesoderm from prospective ectoderm suggests that intracellular signals interpreted as mitogenic in fibroblasts may be interpreted as differentiation signals in embryonic blastomeres (Whitman & Melton 1989). Tigges et al (1989) have recently proposed that IL-2 can stimulate either proliferation or differentiation of B-lymphocytes through identical receptor/signal transduction pathways: differential interpretation of the membrane/cytosolic messenger would, in this case, presumably reside in the competence of the target cell nucleus.

FUTURE PROSPECTS

The response of early ectoderm to FGF provides the best example of the specific response of an embryonic target cell to a multipotent growth factor signal. Studies of cultured cells suggest that the cellular responses to a given growth factor may be highly cell type or cell-state dependent. Mapping the pathways of cellular responsiveness to growth factors from the membrane to the nucleus in different cell types should begin to identify the molecular basis for cell type and cell-state specific responses. The identification of growth factors as intercellular signals during development provides a basis for the application of the biochemistry of signal transduction to specific developmental events, so that the establishment of embryonic pattern may begin to be understood in molecular terms.

ACKNOWLEDGMENT

The authors would like to thank Charles Jennings and Ariel Ruiz i Altaba for critical reading of the manuscript. M.W. was supported by post-doctoral fellowship from the Jane Coffin Childs Memorial Fund for Medical Research; D.M. was supported by a grant from the National Institutes of Health.

Literature Cited

Adamson, E. D., Meek, J. 1984. The ontogeny of epidermal growth factor during development. *Dev. Biol.* 103: 62–70

Baird, A., Esch, F., Mormede, P., Ueno, N., Ling, N., et al. 1986. Molecular characterization of fibroblast growth factor: distribution and biological activities in various tissues. *Recent Prog. Horm. Res.* 42: 143–98

Barrandon, Y., Green, H. 1987. Cell migration is essential for sustained growth of keratinocyte colonies: the roles of transforming growth factor-alpha and epidermal growth factor. *Cell* 50: 1131–37

Beck, F., Samani, N. J., Penschow, J. D., Thorley, B., Tregear, G. W., Coghlan, J. P. 1987. Histochemical localization of IGF-I and IGF-II in the developing rat embryo. *Development* 101: 175–84

Biggers, J. D. 1971. New observations on the nutrition of the mammalian oocyte and preimplantation embryo. In *Biology of the Blastocyst*, ed. R. J. Blandau, pp. 319–28. Chicago: Univ. Chicago

Boterenbrood, E. C., Nieuwkoop, P. D. 1973. The formation of the mesoderm in Urodelean amphibians. V. Its regional induction by the endoderm. *Wilhelm Roux' Arch. Entwicklungsmech. Org.* 173: 319–32

Brachet, J. 1950. *Chemical Embryology*, pp. 345–423. New York: Interscience 2nd ed.

Carpenter, G., Cohen, S. 1979. Epidermal growth factor. *Annu. Rev. Biochem.* 48: 193–216

Cold Spring Harbor Symp. Quantitative Biology. 1989. *Molecular Biology of Signal Transduction*. Vol. LIII. 556 pp.

Condie, B. G., Harland, R. M. 1987. Posterior expression of a homeobox gene in early *Xenopus* embryos. *Development* 101: 93–105

Cooke, J. 1985. The system specifying body position in the early development of *Xenopus*, and its response to early perturbations. *J. Embryol. Exp. Morphol. Suppl.* 89: 69–87

Cooke, J., Smith, J. C. 1989. Gastrulation and larval pattern in *Xenopus* after blastocoelic injection of a *Xenopus* derived inducing factor: experiments testing models for the normal organization of mesoderm. *Dev. Biol.* 131: 383–400

Cooke, J., Smith, J. C., Smith, E. J., Yaqoob, M. 1987. The organization of mesodermal pattern in *Xenopus laevis*: experiments using a *Xenopus* mesoderm inducing factor. *Development* 101: 893–908

Dale, L., Slack, J. M. W. 1987. Regional specification within the mesoderm of early embryos of *Xenopus laevis*. *Development* 100: 279–95

Dale, L., Smith, J. C., Slack, J. M. W. 1985. Mesoderm induction in *Xenopus laevis*: a quantitative study using a cell lineage label and tissue specific antibodies. *J. Embryol. Exp. Morphol. Suppl.* 89: 289–312

Fine, L. G., Holley, R. W., Nasri, H., Badie-Dezfooley, B. 1985. BSC-1 growth inhibitor transforms a mitogenic stimulus into a hypertrophic stimulus for renal proximal tubular cells: Relationship to Na^+/H^+ antiport activity. *Proc. Nat. Acad. Sci. USA* 82: 6163

Fischberg, M. 1949. Experimentalle auslosung von heteroploidie durch kaltebehandlung der eier von *Triton alpestris* aus verschidenen populationen. *Genetica* 24: 213–329

Folkman, J., Klagsbrun, M. 1987. Angiogenic factors. *Science* 235: 442–46

Froesch, E. R., Schmid, C., Schwander, J., Zapf, J. 1985. Actions of insulin-like growth factors. *Annu. Rev. Physiol.* 47: 443–67

Godsave, S. F., Isaacs, H. V., Slack, J. M. W. 1988. Mesoderm inducing factors: a small class of molecules. *Development* 102: 555–66

Gospodarowicz, D., Ferrera, N., Schweigerer, L., Neufeld, G. 1987. Structural characterization and biological functions of fibroblast growth factor. *Endocr. Rev.* 8: 95–113

Gospodarowicz, D., Weseman, J., Moran, J., Linstrom, J. 1976. Effect of fibroblast

growth factor on the division and fusion of myoblasts. *J. Cell. Biol.* 70: 395
Graham, C. F., Morgan, R. W. 1966. Changes in the cell cycle during early amphibian development. *Dev. Biol.* 14: 439
Greenwald, I. 1985. lin-12, a nematode homeotic gene, is homologous to a set of mammalian proteins that includes epidermal growth factor. *Cell* 43: 583–90
Grobstein, C. 1967. The problem of the chemical nature of embryonic inducers. *Ciba Found. Symp. Cell Differentiation*, ed. A. V. S. de Reuck, J. Knight. pp. 131–36. London: Churchill
Grotendorst, G. R., Seppa, H. E., Kleinman, H. K., Martin, G. 1981. Attachment of smooth muscle cells to collagen and their migration toward PDGF. *Proc. Nat. Acad. Sci. USA* 78: 3669–72
Grunz, H., Multier-Lajous, A. M., Herbst, R., Arkenberg, G. 1975. The differentiation of isolated amphibian ectodern with and without treatment with an inductor. *Wilhelm Roux' Arch. Dev. Biol.* 178: 277–84
Grunz, H., Tacke, L. 1986. The inducing capacity of the presumptive endoderm studied by transfilter experiments. *Wilhelm Roux' Arch. Dev. Biol.* 195: 467–73
Gurdon, J. B. 1987. Embryonic induction-molecular prospects. *Development* 99: 285–306
Gurdon, J. B. 1989. The localization of an inductive response. *Development* 105: 27–33
Gurdon, J. B., Brennan, S., Fairman, S., Mohun, T. J. 1984. Transcription of muscle specific actin genes in early *Xenopus* development: nuclear transplantation and cell dissociation. *Cell* 38: 691–700
Gurdon, J. B., Fairman, S., Mohun, T. J., Brennan, S. 1985. Activation of muscle specific actin genes by an induction between animal and vegetal cells of a blastula. *Cell* 41: 913–22
Hafen, E., Basler, K., Edstroem, J. E., Rubin, G. 1987. *Sevenless*, a cell specific homeotic gene of *Drosophila*, encodes a putative transmembrane receptor with a tyrosine kinase domain. *Science* 236: 55–63
Hamburger, V. 1988. *The Heritage of Experimental Embryology: Hans Spemann and the Organizer*. Oxford/New York: Oxford 240 pp.
Hara, K. 1977. The cleavage pattern of the axolotl egg studied by cinematography and cell counting. *Wilhelm Roux' Arch. Dev. Biol.* 181: 73–87
Heasman, J., Wylie, C. C., Hausen, P., Smith, J. C. 1984. Fates and states of determination of single vegetal pole blastomeres. *Cell* 37: 185–94
Heath, J. K., Isacke, C. M. 1984. PC13 embryonal carcinoma-derived growth factor. *EMBO J.* 3: 1957–62
Heath, J. K., Shi, W. K. 1986. Developmentally regulated expression of insulin-like growth factors by differentiated murine teratocarcinomas and extraembryonic mesoderm. *J. Embryol. Exp. Morphol.* 95: 193–212
Heine, U. I., Munoz, E. F., Flanders, K. C., Ellingsworth, L. R., Lam, H. Y., et al. 1988. Role of transforming growth factor beta in the development of the mouse embryo. *J. Cell Biol.* 105: 2861–76
Holtfreter, J. 1934. Der einfluss thermischer, merchanischer, und chemischer eingriffe auf die induzierfahigkeit von Triton-keimteilen. *Wilhelm Roux' Arch. Entwicklungsmech. Org.* 132: 225–306
Hursh, D. A., Andrews, M. E., Raff, R. A. 1987. A sea urchin gene encodes a polypeptide homologous to epidermal growth factor. *Science* 238: 1487–90
Ignotz, R. A., Massagué, J. 1985. Type β transforming growth factor controls the adipogenic differentiation of 3T3 fibroblasts. *Proc. Nat. Acad. Sci. USA* 82: 8530
Ignotz, R. A., Massagué, J. 1986. Transforming growth factor-β stimulates the expression of fibronectin and collagen and their incorporation into the extracellular matrix. *J. Biol. Chem.* 261: 4337–45
Jacobson, A. G., Sater, A. K. 1988. Features of embryonic induction. *Development* 104: 341–59
Jacobson, M., Rutishauser, U. 1986. Induction of neural cell adhesion molecule (N-CAM) in *Xenopus* embryos. *Dev. Biol.* 116: 524–31
Jonas, E., Sargent, T. D., Dawid, I. B. 1985. Epidermal keratin gene expression in embryos of *Xenopus laevis*. *Proc. Nat. Acad. Sci. USA* 82: 5413–17
Jones, E. A. 1985. Epidermal development in *Xenopus laevis*: the definition of a monoclonal antibody to an epidermal marker. *J. Cell Sci. Suppl.* 89: 155–166
Keller, R. E., Danilchik, M., Gimlich, R., Shih, J. 1985. The function and mechanism of convergent extension during gastrulation of *Xenopus laevis*. *J. Embryol. Exp. Morphol. Suppl.* 89: 185–209
Kidd, S., Kelley, M. W., Young, M. W. 1986. Sequence of the *Notch* locus: relationship of the encoded protein to mammalian clotting and growth factors. *Mol. Cell Biol.* 6: 3094–3108
Kimelman, D., Abraham, J. A., Haaparanta, T., Palisi, T. M., Kirschner, M. W. 1988. The presence of fibroblast growth

factor in the frog egg: its role as a natural mesodermal inducer. *Science* 242: 1053–56

Kimelman, D., Kirschner, M. W. 1987. Synergistic induction of mesoderm by FGF and TGFβ and the identification of an mRNA coding for FGF in the early *Xenopus* embryo. *Cell* 51: 869–77

Kintner, C., Brockes, J. 1984. Monoclonal antibodies against blastemal cells derived from dedifferentiating muscle in newt limb regeneration. *Nature* 308: 67–69

Kintner, C., Melton, D. A. 1987. Expression of *Xenopus* N-CAM in ectoderm is an early response to neural induction. *Development* 99: 311–20

Kopczinski, C. C., Alton, A. K., Fechtel, K., Kooh, P. J., Muscavitch, M. A. T. 1988. *Delta*, a *Drosophila* neurogenic gene, is transcriptionally complex and encodes a protein related to blood coagulation factors and epidermal growth factor of vertebrates. *Genes Dev.* 2: 1723–35

Kurihara, K., Sasaki, N. 1981. Transmission of homiogenetic induction in presumptive ectoderm of newt embryo. *Dev. Growth Differ.* 23: 361–69

LaFlamme, S. E., Jamrich, M., Richter, K., Sargent, T. D., Dawid, I. 1988. *Xenopus* EndoB is a keratin preferentially expressed in the embryonic notochord. *Genes Dev.* 2: 853–62

Laiho, M., Saksela, O., Andreasen, P. A., Keski-Oja, J. 1986. Enhanced production and extracellular deposition of the endothelial-type plasminogen activator inhibitor in cultured human lung fibroblasts by transforming growth factor beta. *J. Cell. Biol.* 103: 2403–10

Lehmann, R., Jiminez, F., Dietrich, U., Campos-Ortega, J. A. 1983. On the phenotype and development of mutants of early neurogenesis in *Drosophila melanogaster*. *Wilhelm Roux' Arch. Dev. Biol.* 192: 62–74

Lehnert, S. A., Akhurst, R. J. 1988. Embryonic expression pattern of TGF beta type-1 RNA suggests both paracrine and autocrine mechanisms of action. *Development* 104: 263–73

Ling, N., Ying, S. Y., Ueno, N., Shimasaki, S., Esch, F., et al. 1986. Pituitary FSH is released by a heterodimer of the β subunits from the two forms of inhibin. *Nature* 321: 779–82

Mangold, O., Spemann, H. 1927. Uber Induktion von Medullarplatte durch Medullarplatte im jungren Keim, ein Beispeil homoogenetischer oder assimilatorischer Induktion. *Wilhlem Roux' Arch. Entwicklungsmech. Org.* 111: 341–422

Massagué, J., Kelly, B., Mottola, C. 1985. Stimulation by insulin-like growth factors is required for cellular transformation by TGFβ. *J. Biol Chem.* 260: 4551–54

Massagué, J. 1987. The TGF-β family of growth and differentiation factors. *Cell* 49: 437–38

McClay, D. R., Ettensohn, C. A. 1987. Cell adhesion in morphogenesis. *Annu. Rev. Cell Biol.* 3: 319–45

Melton, D. A. 1987. Translocation of maternal mRNA to the vegetal pole of *Xenopus* oocytes. *Nature* 328: 80–82

Mercola, M., Melton, D. A., Stiles, C. D. 1988. Platelet-derived growth factor A chain is maternally encoded in *Xenopus* embryos. *Science* 241: 1223

Mercola, M., Stiles, C. D. 1988. Growth factor superfamilies and mammalian embryogenesis. *Development* 102: 451–60

Morrison, R. S., De Veillis, J., Lee, Y. L., Bradshaw, R. A. 1986. Hormones and growth factors induce the synthesis of glial fibrillary acidic protein in rat brain astrocytes. *J. Neurosci. Res.* 14: 167

Newport, J., Kirschner, M. 1982. A major developmental transition in early *Xenopus* embryos: I. Characterization and timing of cellular changes at the midblastula stage. *Cell* 30: 675–86

Nieuwkoop, P. D. 1969. The formation of mesoderm in urodelean amphibians. I. Induction by the endoderm. *Wilhelm Roux' Arch. Entwicklungsmech.* 162: 341–73

Nieuwkoop, P. D. 1985. Inductive interactions in early amphibian development and their general nature. *J. Embryol. Exp. Morphol. Suppl.* 89: 333–47

Nieuwkoop, P. D., Johnen, A. G., Albers, B. 1985. *The Epigenetic Nature of Early Chordate Development*. Cambridge: Cambridge Univ. 373 pp.

Noble, M., Murray, K., Stroobant, P., Waterfield, M. D., Riddle, P. 1988. Platelet derived growth factor promotes division and motility and inhibits premature differentiation of the oligodendrocyte/type 2 astrocyte progenitor cell. *Nature* 333: 560–62

O'Garra, A., Umland, S., DeFrance, T., Christiansen, J. 1988. B-cell factors are pleiotropic. *Immunol. Today* 9: 45

Padgett, R. W., St. Johnston, D., Gelbart, W. M. 1987. A transcript from a *Drosophila* pattern gene predicts a protein homologous to the TGF-β gene family. *Nature* 325: 81–84

Posakony, L. 1987. *The role of the decapentaplegic gene complex in the development of imaginal disks in Drosophila melanogaster*. PhD thesis. Harvard. 196 pp.

Postlethwaite, A. E., Keski-Oja, J., Moses,

H. L., Kang, A. H. 1987. Stimulation of the chemotactic migration of human fibroblasts by TGF-β. *J. Exp. Med.* 165: 251–56

Preiss, J. R., Thomson, J. N. 1987. Cellular interactions in early *C. elegans* embryos. *Cell* 48: 241–50

Price, J. V., Clifford, R. J., Schüpback, T. 1989. The maternal ventralizing locus *torpedo* is allelic to *faint little ball*, an embryonic lethal, and encodes the *Drosophila* EGF receptor homolog. *Cell* 56: 1085–92

Rappolee, D. A., Brenner, C. A., Schultz, R., Mark, D., Werb, Z. 1988. Developmental expression of PDGF, TGF-alpha, and TGF-β genes in preimplantation mouse embryos. *Science* 241: 1823–25

Raff, M. C., Lillien, L. E., Richardson, W. D., Burne, J. F., Noble, M. D. 1988. Platelet derived growth factor from astrocytes drives the clock that times oligodendrocyte development in culture. *Nature* 333: 562–64

Reinke, R., Zipursky, S. L. 1988. Cell-cell interaction in the *Drosophila* retina: the *bride of sevenless* gene is required in photoreceptor cell R8 for R7 development. *Cell* 55: 321–30

Rijsewik, F., Schuermann, M., Wagenaar, E., Parren, P., Weigel, D., Nusse, R. 1987. The *Drosophila* homolog of the mouse mammary oncogene *int-1* is identical to the segment polarity gene *wingless*. *Cell* 50: 649–57

Rizzino, A., Bowen-Pope, D. F. 1985. Production of PDGF like factors by embryonal carcinoma cells and response to PDGF by endoderm-like cells. *Dev. Biol.* 110: 15–22

Roberts, A. B., Anzano, M. A., Lamb, L. C., Smith, J. M., Sporn, M. B. 1981. New class of transforming growth factors potentiated by epidermal growth factor: isolation from non-neoplastic tissues. *Proc. Nat. Acad. Sci. USA* 78: 5339

Roberts, A. B., Sporn, M. B., Assoian, R. K., Smith, J. M., Roche, S., et al. 1986. Transforming growth factor type-beta: rapid induction of fibrosis and angiogenesis in vivo and stimulation of collagen formation in vitro. *Proc. Nat. Acad. Sci. USA* 83: 4167–71

Rosa, F., Roberts, A. B., Danielpour, D., Dart, L. L., Sporn, M. B., Dawid, I. B. 1988. Mesoderm induction in amphibians: the role of TGF-β_2 like factors. *Science* 239: 783–85

Rosen, D. M., Stempien, S. A., Thompson, A. Y., Seyedin, P. R. 1988. Transforming growth factor beta modulates the expression of osteoblast and chondroblast phenotypes in vitro. *J. Cell Physiol* 134: 337–46

Ross, R., Raines, E. W., Bowen-Pope, D. F. 1986. The biology of platelet derived growth factor. *Cell* 46: 155–69

Ruiz i Altaba, A., Melton, D. A. 1989. Bimodal and graded expression of the *Xenopus* homeobox gene Xhox3 during embryonic development. *Development* 106: 173–83

Saunders, J. W., Cairns, J. M., Gasseling, M. T. 1957. The role of the apical ridge in the differentiation of the morphological structure and inductive specificity of limb parts in the chick. *J. Morphol.* 101: 57–88

Saxen, L., Lehtonen, E., Karkinen-Jaaskelainen, M., Nordling, S., Wartiovarra, J. 1976. Are morphogenetic tissue interactions mediated by transmissible signal substances or through cell contacts? *Nature* 259: 662–63

Schupbach, T. 1987. Germ line and soma cooperate during oogenesis to establish the dorsoventral pattern of egg shell and embryo in *Drosophila melanogaster*. *Cell* 49: 699–707

Seppa, H., Grotendorst, G., Seppa, S., Schiffmann, E., Martin, G. R. 1982. Platelet derived growth factor is chemotactic for fibroblasts. *J. Cell Biol.* 92: 584–88

Shackleford, G. M., Varmus, H. E. 1987. Expression of the proto-oncogene int-1 is restricted to postmeiotic male germ cells and the neural tube of midgestational embryos. *Cell* 50: 89–95

Slack, J. M. W. 1983. *From Egg to Embryo: Determinative Events in Early Development*. Cambridge/London: Cambridge Univ. 241 pp.

Slack, J. M. W., Darlington, B. G., Heath, J. K., Godsave, S. F. 1987. Mesoderm induction in the early *Xenopus* embryos by heparin-binding growth factors. *Nature* 326: 197–200

Slack, J. M. W., Isaacs, H. V. 1989. Presence of fibroblast growth factor in the early *Xenopus* embryo. *Development* 105: 147–54

Smith, E. P., Sadler, T. W., D'Ercole, A. J. 1987. Somatomedins/insulin-like growth factors, their receptors and binding proteins are present during mouse embryogenesis. *Development* 101: 73–82

Smith, J. C. 1987. A mesoderm inducing factor is produced by a *Xenopus* cell line. *Development* 99: 3–14

Smith, J. C., Watt, F. 1985. Biochemical specificity of *Xenopus* notochord. *Differentiation* 29: 109–15

Smith, J. C., Yaqoob, M., Symes, K. 1988. Purification, partial characterization and biological effects of the XTC mesoderm-inducing factor. *Development* 103: 591–600

Spemann, H. ed. 1938. *Embryonic Development and Induction.* New Haven: Yale Univ.

Spemann, H., Mangold, H. 1924. Uber induction von embryoanlagen durch implantation atrfremder organis atoren. *Wilhelm Roux' Arch. Entwicklungsmech. Org.* 100: 599–638

Spencer, F. 1984. *The decapentaplegic gene complex and adult pattern formation in Drosophila.* PhD thesis. Harvard. 181 pp.

Sporn, M. B., Roberts, A. B. 1987. Peptide growth factors are multifunctional. *Nature* 332: 217–19

Sporn, M. B., Roberts, A. B., Wakefield, L. M., Assoian, R. K. 1986. Transforming growth factor-β: biological function and chemical structure. *Science* 233: 532–34

Sporn, M. B., Roberts, A. B., Wakefield, L. M., de Crombrugge, B. 1987. Some recent advances in the chemistry and biology of transforming growth factor-beta. *J. Cell Biol.* 105: 1039–45

Stylianopolou, F., Efstratiadis, A., Herbert, J., Pintar, J. 1988. Pattern of the insulin-like growth factor II gene expression during rat embryogenesis. *Development* 103: 497–506

Symes, K., Smith, J. C. 1987. Gastrulation movements provide an early marker of mesoderm induction in *Xenopus laevis. Development* 101: 339–49

Symes, K., Yaqoob, M., Smith, J. 1988. Mesoderm induction in *Xenopus laevis*: responding cells must be in contact for mesoderm formation but suppression of epidermal differentiation can occur in single cells. *Development* 104: 609–18

Tarkowski, A. K., Wrobleska, J. 1967. Development of blastomeres of mouse eggs isolated at the 4 and 8-cell stage. *J. Embryol. Exp. Morphol.* 18: 155–80

Tiedemann, H. 1966. The molecular basis of differentiation in early development of amphibian embryos. *Curr. Topics Dev. Biol.* 1: 85–112

Tigges, M. A., Casey, L. S., Koshland, M. E. 1989. Mechanism of interleukin-2 signaling: mediation of different outcomes by a single receptor and transduction pathway. *Science* 243: 781–86

Togari, A., Dickens, G., Huzuya, H., Guroff, G. 1983. The effect of fibroblast growth factor on PC-12 cells. *J. Neurosci.* 5: 307

Toivonen, S., Tarin, D., Saxen, L. 1976. The transmission of morphogenetic signals from amphibian mesoderm to ectoderm in primary induction. *Differentiation* 5: 49–55

Tomlinson, A., Ready, D. F. 1986. *Sevenless*: a cell specific homeotic mutation of the *Drosophila* eye. *Science* 231: 400

Tomlinson, A., Ready, D. F. 1987. Cell fate in the *Drosophila* ommatidium. *Dev. Biol.* 123: 264–75

Trinkaus, J. B. 1984. *Cells into organs: the forces that shape the embryo.* Englewood Cliffs, NJ: Prentice-Hall. 543 pp. 2nd ed.

Vale, W., Rivier, J., Vaughan, J., McClintock, R., Corrigan, A., et al. 1986. Purification and characterization of an FSH releasing protein from porcine ovarian follicular fluid. *Nature* 321: 776–79

Wahl, S. M., Hunt, D. A., Wakefield, L. M. McCartney-Francis, N., Wahl, L. M., et al. 1987. Transforming growth factor beta induces monocyte chemotaxis and growth factor production. *Proc. Nat. Acad. Sci. USA* 84: 5788–92

Wakefield, L. M., Smith, D. M., Masui, T., Harris, C. C., Sporn, M. B. 1987. Distribution and modulation of the cellular receptor for TGF-β. *J. Cell Biol.* 105: 965–75

Weeks, D. L., Melton, D. A. 1987. A maternal messenger RNA localized to the vegetal hemisphere in *Xenopus* eggs codes for a growth factor related to TGF-β. *Cell* 51: 861–67

Wessels, N. K. 1977. *Tissue Interactions and Development.* Menlo Park: Benjamin. 276 pp.

Wharton, K., Johansen, K. M., Xu, T., Artavanis-Tsakonas, S. 1985. Nucleotide sequence from the neurogenic locus *Notch* implies a gene product that shares homology with proteins containing EGF-like repeats. *Cell* 43: 567–81

Whitman, M., Melton, D. A. 1989. Induction of mesoderm by a viral oncogene in early *Xenopus* embryos. *Science* 244: 803–6

Wilkinson, D. G., Bailes, J. A., McMahon, A. P. 1987. Expression of the proto-oncogene int-1 is restricted to specific neural cells in the developing mouse embryo. *Cell* 50: 79–88

Wilkinson, D. G., Peters, G., Dickson, C., McMahon, A. P. 1988. Expression of the FGF-related protooncogene int-2 during gastrulation and neurulation in the mouse. *EMBO J.* 7: 691–95

Winklbauer, R. 1986. Cell proliferation in the ectoderm of the *Xenopus* embryo: development of substratum requirements for cytokinesis. *Dev. Biol.* 118: 70–81

Ying, S., Becker, A., Ling, N., Ueno, N., Guillemin, R. 1986. Inhibin and Beta type transforming growth factor (TGFβ) have opposite modulating effects on the FSH-induced aromatase activity of cultured rat granulosa cells. *Biochim. Biophys. Res. Comm.* 136: 969

Ysraeli, J., Melton, D. A. 1988. The maternal

mRNA Vg1 is correctly localized following injection into *Xenopus* oocytes. *Nature* 336: 592–95

Yu, J., Shao, L., Lemas, V., Yu, A. L., Vaughan, J., et al. 1987. Importance of FSH-releasing protein and inhibin in erythrodifferentiation. *Nature* 330: 765–69

Zalokar, M. 1976. Autoradiographic study of protein and RNA formation during early development in *Drosophila* eggs. *Dev. Biol.* 52: 31–42

Zanetti, M., Ratcliffe, A., Watt, F. M. 1985. Two subpopulations of differentiated keratinocytes identified with a monoclonal antibody to keratan sulphate. *J. Cell. Biol.* 101: 53–59

Annu. Rev. Cell Biol. 1989. 5 : 119–51

DYNEIN STRUCTURE AND FUNCTION

Mary E. Porter

Department of Cell Biology and Neuroanatomy, University of Minnesota, Minneapolis, Minnesota 55455

Kenneth A. Johnson

Department of Biochemistry, Microbiology, Molecular, and Cell Biology, The Pennsylvania State University, University Park, Pennsylvania 16802

CONTENTS

INTRODUCTION

Microtubules participate in a wide variety of cellular functions such as the maintenance of cell shape, particle and organelle movements, the separation of chromosomes on the mitotic spindle, and the beating of eukaryotic cilia and flagella. To accomplish these diverse functions, microtubules interact with a large number of microtubule associated proteins (MAPs). One class of MAPs is the microtubule motors, mechanochemical enzymes that convert the energy derived from nucleotide binding and

0743–4634/89/1115–0119$02.00

hydrolysis into the movement of cellular objects relative to the microtubule surface. The intrinsic structural polarity of the microtubule (due to the asymmetric arrangement of tubulin dimers) has made it possible to identify two distinct families of motors that move in opposite directions, the kinesins and the dyneins (Vale et al 1985b; Sale & Satir 1977; Paschal & Vallee 1987). The dynein family is the subject of this review.

Dyneins were first described as the inner and outer arm structures present on the outer doublet microtubules of ciliary and flagellar axonemes (Gibbons 1965). These arms are the ATPases responsible for generating the sliding movements between microtubules that underly ciliary and flagellar motility. The nine outer doublet microtubules bearing the dynein arms are held in a ring by interdoublet linkages termed nexin links. In most cases, this ring surrounds a central core of two singlet microtubules and associated radial spokes to form the characteristic 9+2 structure of the axoneme. These structures constrain and coordinate the dynein-induced sliding of microtubules to generate the flagellar waveform (Gibbons 1981).

Axonemal dyneins are the best characterized microtubule motors for several reasons. First and foremost, the axoneme is a stable and highly ordered structure with easily observed motility. While the axoneme is normally enclosed in an extension of the cell membrane, it can easily be detached from the cell surface and purified for biochemical analysis. Second, in several systems it is possible to remove the cell membrane and reactivate the movement of the axoneme by the addition of ATP in the appropriate buffer. Such functional models have made possible the selective extraction and readdition of axonemal components to reconstitute motility. Finally, the loss of axoneme function is rarely lethal to an organism, and thus there exist a number of genetic mutations that affect axoneme structure and motility (Luck 1984). The combined biochemical, structural, functional, and genetic analyses of the past thirty years have therefore provided a fairly good understanding of the structure and composition of the axonemal dyneins.

The frequent observations of other forms of ATP-dependent, microtubule-based motility have naturally led to a search for dynein-like motors outside the axoneme. Historically, these cytoplasmic dyneins have been most intensively studied in sea urchin eggs, and several investigators have demonstrated that egg dyneins share enzymatic and immunological properties with the axonemal dyneins (reviewed in Pratt 1989). Renewed interest in cytoplasmic dyneins has come from two developments. First, in recent years several investigators have developed in vitro models for organelle transport and begun to characterize the associated motor activities (reviewed in Vale 1987). Second, improvements in light microscopy have made it possible to see individual microtubules and organelles (Inoué 1981;

Allen et al 1981) and to observe their movements in crude cell extracts (Brady et al 1982; Allen et al 1985; Vale et al 1985a,b). This led to the development of in vitro assays for microtubule translocators and the discovery of kinesin (Vale et al 1985a). Shortly thereafter, it was recognized that kinesin was a unidirectional motor that moved objects toward the "plus" ends of microtubules. A second activity, distinct from kinesin, was responsible for movement toward the "minus" end, the retrograde motor (Vale et al 1985b). (The two ends of a microtubule are defined as plus and minus based on their relative rates of assembly.) The pharmacological properties and polarity of the retrograde motor were similar to those of axonemal dyneins (Vale et al 1985b). The retrograde motor was subsequently identified as cytoplasmic dynein (Pascal & Vallee 1987), and cytoplasmic dynein has since been shown to be an enzyme distinct from its axonemal cousins (Paschal et al 1987b; Lye et al 1987; Shpetner et al 1988). Since cytoplasmic microtubules are usually organized with their minus ends at the cell center or centrosome and their plus ends at the cell periphery (Euteneuer & McIntosh 1981), objects driven by cytoplasmic dynein should be directed toward the centrosome.

AXONEMAL DYNEINS

Composition and Structure

Axonemal dyneins have been most extensively studied in three experimental systems; the ciliated protozoan *Tetrahymena*, the unicellular green alga *Chlamydomonas*, and the sea urchin spermatozoan. Comparison of dyneins from these and other sources has revealed an overall similarity in structure and composition, with apparently only minor species differences (reviewed in Marchese-Ragona & Johnson 1989). Comparison of outer arm and inner arm dyneins has, on the other hand, revealed a greater structural and functional diversity.

PURIFICATION AND POLYPEPTIDE COMPOSITION

Outer arms The dynein arms and their ATPase activities can be selectively solubilized from the axoneme either by brief exposure to buffers containing high salt (0.5–0.6 M NaCl or KCl) or by overnight dialysis against low ionic strength buffers plus EDTA (Gibbons 1965). The dynein containing extracts are then further purified by a combination of ion exchange chromatography and/or sucrose density gradient centrifugation. Since the dynein arms and their associated subunits are large particles, which sediment between 10–30S, the identification of dynein polypeptides has been primarily by the cosedimentation of components with ATPase activity. In *Chlamydomonas*, these polypeptides also coincide with the polypeptides that are missing in axonemes prepared from mutants that lack the outer dynein arm (Huang et al 1979). All outer arm dyneins examined thus far contain 2–3 heavy chains of 400–500 kd (named α, β, γ), two or more intermediate chains of 55–125 kd, and a cluster of 4–8 light chains around 20 kd (see Figure 1).

Inner arms The inner dynein arms have not been as widely studied as the outer dynein arms, but analysis of mutants in *Chlamydomonas* that lack inner dynein arms has indicated that they are even more complicated than outer arms (Huang et al 1979; Brokaw & Kamiya 1987; Piperno 1988). The inner arm subunits are constructed along the same basic plan as the outer dynein arm with their own set of heavy, intermediate, and light chains (see Figure 1) (Piperno & Luck 1981; Mitchell & Rosenbaum 1985; Goodenough et al 1987; Piperno 1988). Comparisons in other species

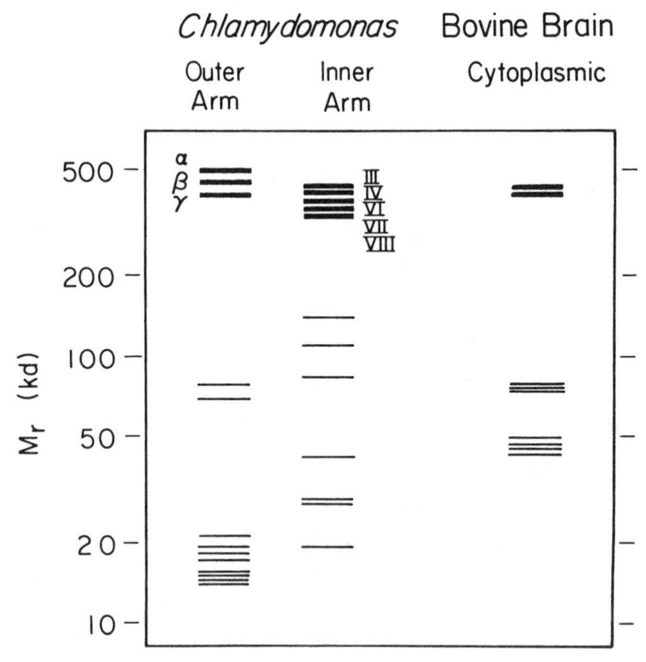

Figure 1 Polypeptide composition of axonemal and cytoplasmic dyneins. The polypeptide compositions of the outer and inner arm dyneins from *Chlamydomonas* flagella and the cytoplasmic dynein from bovine brain are shown diagrammatically on a theoretical gel system that resolves components between 10–500 kd. For details see the following references: *Chlamydomonas* dyneins (Piperno & Luck 1979, 1981; Huang et al 1979; Luck & Piperno 1989; King & Witman 1989); bovine brain cytoplasmic dynein (Paschal et al 1987b; Vallee et al 1988).

support the notion that inner and outer arm polypeptides are different (Porter & Johnson 1983a; Warner et al 1985; Sale et al 1989). Furthermore, antibodies to outer arm polypeptides that cross-react across species fail to cross-react with inner arm polypeptides (Ogawa et al 1980, 1982; Piperno 1984, 1988).

STRUCTURE OF THE DYNEIN ARM IN VITRO

Outer arm Scanning transmission electron microscopy (STEM) and quick-freeze deep-etch replicas of isolated dyneins have revealed that the outer arms are actually arranged in the form of a bouquet, with two or three globular heads connected by slender stems to a common base (Johnson & Wall 1983; Goodenough & Heuser 1984) (see Figure 2). The mass of the isolated arm (1250–1960 kd) can be correlated with the number of heavy chains and the number of heads (Johnson & Wall 1983; Witman et al 1983; Sale et al 1985). More recent images have revealed substructure to the dynein bouquet; these include slender stalks extending from each head, asymmetries in the shape of the heads, and the presence of smaller globular domains at the base of the stems (Sale et al 1985; Goodenough et al 1987; Toyoshima 1987a). In *Chlamydomonas* the masses of the heads as measured by STEM are too small to accommodate the full lengths of the

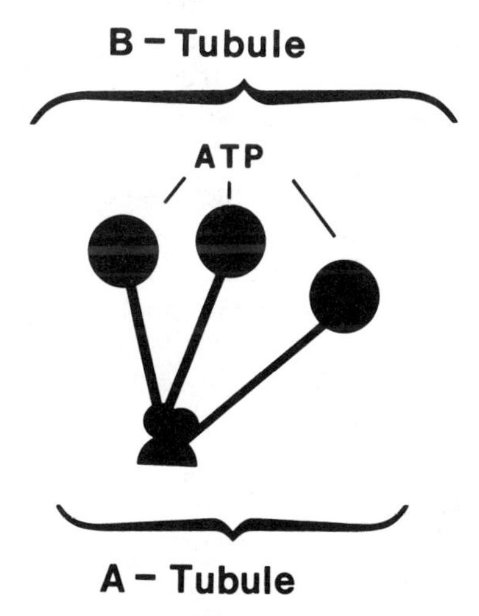

Figure 2 The outer dynein arm from *Tetrahymena*. The bouquet structure of the outer dynein arm in vitro is shown schematically. The three globular heads contain the sites of ATP binding and hydrolysis and form the ATP-sensitive attachment sites to the B-tubule. The base of the arm forms the ATP-insensitive attachment site to the A-tubule (Johnson 1985).

dynein heavy chains (Witman et al 1983; King & Witman 1987). Thus, each heavy chain forms a single head and part of its associated stem (King & Witman 1989), a conclusion substantiated more definitively by proteolytic digestion (see below). The particles at the base of the stems have been correlated with the presence of intermediate chains (Sale et al 1985). More detailed mapping on the location of the different intermediate and light chains as well as specific domains of the heavy chains is currently underway in several laboratories.

Inner arms Analysis of inner arm dynein subunits from *Chlamydomonas* has demonstrated the presence of a single two-headed particle and two distinct single-headed particles (Goodenough et al 1987). Each particle reiterates the head, stem, base motif described above for the outer arm dyneins, but the relationship of the isolated particles to the inner arm structures seen in situ is presently unknown. Analysis of the 14S dyneins from *Tetrahymena* by STEM has also revealed the presence of two single-headed dynein particles (Marchese-Ragona et al 1988). These may be the *Tetrahymena* equivalents of the *Chlamydomonas* inner arm subunits, but the 14S dyneins have not been definitively identified as inner arm components in this organism.

STRUCTURES OF THE DYNEIN ARM IN SITU

Outer arm While there is consensus that the outer dynein arms in solution display the bouquet configuration (Johnson & Wall 1983; Goodenough & Heuser 1984), it is still uncertain how these two- or three-headed bouquets are arranged in the axoneme. The extended configuration seen in vitro is ~50 nm across (Johnson & Wall 1983; Toyoshima 1987a), and must be compacted in some fashion to give the characteristic 24 nm repeat of the outer dynein arm seen in situ (Amos et al 1976).

Most images of the dynein arm obtained by negative stain or thin section electron microscopy are consistent with a model in which the base of the bouquet anchors the dynein arm to the A-tubule and the globular heads reach out across the interdoublet space to make contact with B-subfiber (reviewed in Johnson 1985, see Figure 2). In views of the dynein arm obtained by the analysis of quick-freeze, deep-etch replicas (Goodenough & Heuser 1982, 1984, 1989), the globular heads do not make direct contact with the B-subfiber. Instead, attachment occurs via a structure named the B-link that is thought to be composed of the slender stalks extending from each head (Goodenough & Heuser 1989).

Inner arms The increased complexity in inner arm polypeptide composition (5–6 heavy chains vs 2–3 heavy chains in outer arms) has been matched by a corresponding increase in structural diversity. Early thin section analysis suggested that the structure and spacing of the inner arms differed from that of the outer arms (Huang et al 1979). More recently, quick-freeze deep-etch replicas have indicated that the inner arms repeat in groups of three distinct subunits every 96 nm (Goodenough & Heuser 1985, 1989). Examination of inner arm mutants that are missing subsets of inner arm polypeptides (Brokaw & Kamiya 1987; Piperno 1988; Luck & Piperno 1989) will be an important next step in sorting out inner arm complexity.

FUNCTION OF POLYPEPTIDE SUBUNITS The large size and complexity of the dynein molecule suggest that there are several structural domains that may be involved in the assembly and function of the arms in flagellar motility. Considerable effort has therefore been devoted to the analysis of the function of the polypeptide subunits.

Heavy chains The dynein heavy chains contain the sites of ATP binding and hydrolysis, as evidenced by the photoincorporation of radioactive ATP analogs (Pfister et al 1984, 1985). Vanadate-mediated photolysis also indicates that the ATP binding sites are located within the central 100 kd region of each heavy chain (Lee-Eiford et al 1986; Gibbons et al 1987). Vanadate is a phosphate analog that inhibits dynein's ATPase activity by binding to a dynein-ADP intermediate to form a stable dead end complex (Shimizu & Johnson 1983a). UV irradiation of the dynein ADP-vanadate complex results in the quantitative cleavage of the heavy chain into two smaller fragments simultaneously with the loss of ATPase activity (Lee-Eiford et al 1986; Gibbons et al 1987). This site of cleavage (the V_1 site) is thought to be the region normally occupied by the γ phosphate of ATP. UV irradiation of the dynein heavy chains in the presence of Mn^{++} and oligomeric vanadate, but without nucleotide, results in the cleavage of the dynein heavy chain at a second site (the V_2 site), $\sim$100 kd away from the V_1 site (Tang & Gibbons 1987). The V_2 site is thought to represent the region normally occupied by the α or β phosphate of ATP. The sensitivity to vanadate-mediated photolysis is a property that has been conserved among the heavy chains from both the outer and inner dynein arms (Gibbons & Gibbons 1987; Sale et al 1989), as well as the cytoplasmic dyneins (see below).

The smaller sizes of the vanadate cleavage fragments ($\sim$200 kd) have made it possible to obtain more reliable estimates of the molecular masses of the dynein heavy chains (400–500 kd) (Lee-Eiford et al 1986; Gibbons et al 1987), and to construct one-dimensional maps that identify the positions of the amino and carboxy termini, ATP binding sites, phosphorylation sites, protease sensitive domains, and epitopes recognized by monoclonal antibodies (King & Witman 1988, 1989; Mocz et al 1988; see Figure

3). A tentative model has been proposed that correlates the one-dimensional maps of the heavy chains with the bouquet configuration seen by electron microscopy (Figure 2) (Mocz et al 1988; King & Witman 1989). The globular heads seen in EM are thought to represent a protease resistant central core of the heavy chains and include both the ATPase sites and the sites for ATP-sensitive binding to the B-tubule (Ogawa 1973; Toyoshima 1987a,b; Ow et al 1987; Mocz et al 1988; King & Witman 1988). The protease sensitive domains on either end of the heavy chains presumably correlate with the stems that extend to the base and interact with each other and the intermediate chains to form the structural attachment site to the A tubule (Bell & Gibbons 1982; Toyoshima 1987b; King & Witman 1989) (see Figures 2 and 3).

The α heavy chain of outer arm dyneins and a subset of the inner arm

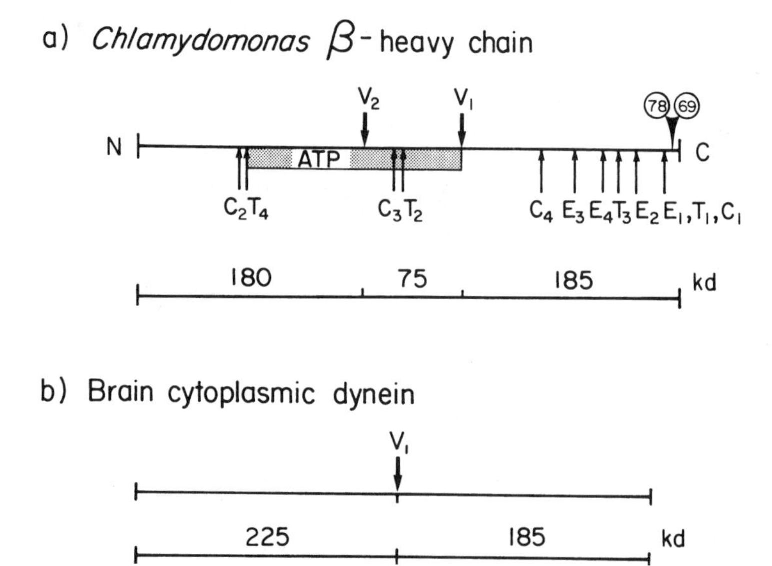

Figure 3 One dimensional maps of the structure of the dynein heavy chains. (*a*) The β heavy chain of outer arm dynein from *Chlamydomonas* is drawn to scale, and the relative positions of the amino and carboxy termini, the ATP binding site, the vanadate-UV cleavage sites (V_1 and V_2), and the proteolytically sensitive sites (C = chymotrypsin, E = elastase, and T = trypsin) are indicated. The intermediate chains (78 kd and 69 kd) are placed near the carboxy terminus of the β-heavy chain based on the solution studies of Pfister & Witman (1984) and Mitchell & Rosenbaum (1986), and the work of King & Witman (1989). Redrawn after Figure 12 in King & Witman (1988) and Figure 12 in King & Witman (1989). (*b*) The heavy chain of bovine brain cytoplasmic dynein and the relative position of the V_1 cleavage site are indicated. See Paschal et al (1987b) for details.

heavy chains are also substrates for phosphorylation (Piperno & Luck 1981; King & Witman 1989). Changes in the phosphorylation state of some of these sites have been correlated with changes in axonemal motility (Tash & Means 1988), but their precise roles are still unknown (see below).

Intermediate and light chains Solution studies have indicated that each heavy chain is tightly associated with at least one intermediate or light chain (Piperno & Luck 1979, 1981; Pfister et al 1982; Pfister & Witman 1984; Mitchell & Rosenbaum 1986; Tang et al 1982). The intermediate chains may mediate interactions between the heavy chains to form the A-tubule binding site (Mitchell & Rosenbaum 1986; Sale et al 1985; King & Witman 1989), but their other functions are largely unknown. Some of the light chains are also substrates for phosphorylation (Piperno & Luck 1981; Tash & Means 1988; Hamasaki et al 1989) and may be involved in the regulation of motility.

The Dynein Crossbridge Cycle

The premise of most models of the dynein crossbridge cycle is that the dynein arm on one A-tubule of an outer doublet must reach across the interdoublet space and interact transiently with the adjacent B-subfiber in an ATP-sensitive fashion (reviewed in Gibbons 1981). One of the puzzles in early studies of axoneme structure was that this attachment to the B-subfiber was never observed, either in the presence or absence of ATP. However, Gibbons & Gibbons (1974) later demonstrated that it was possible to "catch" the attached state of the dynein arm by the rapid dilution of ATP in solutions of reactivated sperm flagella, which produce rigid waves analogous to rigor mortis in muscle. Preservation of the attached state for electron microscopy was extremely sensitive to both buffer and fixation conditions (Gibbons 1975; Zanetti et al 1979). Addition of low concentrations ($\sim 1\ \mu M$) of MgATP resulted in the relaxation of the rigor waveforms (Gibbons & Gibbons 1974) and the detachment of the dynein arms from the B-subfiber (Takahashi & Tonomura 1978; Mitchell & Warner 1980; Satir et al 1981). These observations set the stage for a series of studies aimed at probing the microtubule properties of different dynein preparations and reconstituting the dynein crossbridge cycle in vitro.

MICROTUBULE BINDING PROPERTIES

Outer arms Reconstitution studies clearly established that each dynein arm contains two distinct microtubule binding sites that can be distinguished by their appearance in cross section and their sensitivity to the presence of ATP (Haimo et al 1979; Warner & McIlvain 1982; Porter & Johnson 1983a). The thinner A-end of the dynein arm (corresponding to the structural site normally bound to the A-tubule) binds in an ATP-insensitive fashion, whereas the broader B-end (corresponding to the force producing site normally bound to the B-subfiber) binds in an ATP-sensitive mode. Outer arm dyneins are also capable of binding purified singlet microtubules in both modes with the same characteristic 24 nm repeat observed in situ (Haimo et al 1979; Satir et al 1981; Porter & Johnson 1983a; Haimo & Fenton 1984). These data indicate that the microtubule binding sites of the outer arm recognize the tubulin surface lattice and do not require the presence of other axonemal structures for binding. The mechanism that specifies the unique positions of the dynein arms in the axoneme is, however, unknown.

Inner arms Early attempts to rebind inner arm dyneins to extracted axonemes were largely unsuccessful (Gibbons 1965), but more recently Warner et al (1985) have succeeded in rebinding the 14S *Tetrahymena* ATPases at the positions expected for the inner dynein arm. 14S dynein is also capable of binding singlet microtubules in the absence, but not the presence of microtubule associated proteins, but no obvious periodicity has been observed (Porter & Johnson 1983a; Marchese-Ragona et al 1988).

THE CROSSBRIDGE CYCLE IN VITRO The study of the dynein crossbridge cycle in vitro depends on the ability to distinguish between the two types of microtubule binding sites, the structural site and the force producing site. With *Tetrahymena* outer arm dynein it is possible to establish conditions in which greater than 90% of the dynein is bound by its force producing site, and to measure both the rate and extent of crossbridge association and dissociation, using stopped flow light scattering techniques (Porter & Johnson 1983a,b). This has opened the way for a detailed kinetic analysis of the crossbridge cycle in vitro (see Johnson 1985 for review).

Kinetic analysis Kinetic analysis of the dynein-microtubule crossbridge cycle has established the following reaction pathway:

$$\begin{array}{ccccccc}
\text{D-MT+ATP} & \xrightarrow{k_1} & \text{D-MT-ATP} & \text{---} & \text{D-MT-ADP-P}_i & \xrightarrow{k_5} & \text{D-MT+ADP+P}_i \\
| & & \downarrow k_d & & \uparrow k_4 & & | \\
\text{D+ATP} & \text{---} & \text{D-ATP} & \xrightarrow{k_2} & \text{D-ADP-P}_i & \xrightarrow{k_3} & \text{D+ADP+P}_i
\end{array}$$

Addition of ATP to the dynein-microtubule complex results in the rapid dissociation ($k_d > 1000\ s^{-1}$) of the dynein from the microtubule immediately following ATP binding ($k_1 = 4.7 \times 10^6\ M^{-1}\ s^{-1}$) (Porter & Johnson 1983b). Direct measurement of ATP binding and hydrolysis by chemical quench flow methods demonstrated the ATP hydrolysis proceeds more slowly ($k_2 + k_{-2} = 60\ s^{-1}$) on the dissociated dynein crossbridge (Johnson 1983). The rate limiting step at steady state is the release of products (ADP and P_i) from dynein's active site ($k_3 = 8\ s^{-1}$). If the concentration of microtubules is high enough, microtubules can bind to the dynein-ADP-P_i intermediate ($k_4 = 1.2–6 \times 10^4\ s^{-1}$) and stimulate the release of these products from the enzyme's active site ($k_5 = 33\ s^{-1}$) (Omoto & Johnson 1986). This release of products is thought to be coupled to a conformational change in the dynein arm that results in the net movement of microtubules. The crossbridge cycle restarts with the binding of another molecule of ATP. These studies demonstrate that the key steps of crossbridge attachment and detachment are coupled to the energy associated with the binding and release of nucleotide, and that ATP hydrolysis simply converts the dynein arm between two states with different affinities for the microtubule.

Multiple heads Kinetic analysis has also established the presence of three ATP binding sites per dynein arm (Johnson 1983), in good agreement with the observed number of heads by STEM (Johnson & Wall 1983) and the observed number of heavy chains (Porter & Johnson 1983a; Toyoshima 1987a). Titration experiments have further revealed that three molecules of ATP are required to dissociate a dynein arm from the microtubule, one for each head (Shimizu & Johnson 1983b). These data favor a model in which all three heads interact with microtubule to form the ATP-sensitive binding site (see Figure 2).

Inhibitors The dynein crossbridge cycle can be inhibited by a number of nucleotide and phosphate analogs that mimic ATP or ADP·P_i intermediates (see Penningroth 1989 for recent review). These analogs have been used as probes for the different conformational states of the dynein arm in situ (see below), and as agents to test for the presence of dynein-like ATPases in other systems (see Table 1). However, kinetic analysis indicates that many of these inhibitors are actually poor analogs with respect to axonemal dynein. For example, the nonhydrolyzable ATP analog, AMPPNP, has been used to determine if crossbridge dissociation and flagellar relaxation require ATP hydrolysis, with conflicting results (see Penningroth 1989). More recent work has established that AMPPNP can induce the dissociation of dynein from mictotubules in solution, but it is approximately 100,000-fold less effective than ATP (Chilcote & Johnson 1989). Erythro-9-[3-(2-hydroxynonyl)]Adenine (EHNA) is the substrate analog of adenosine that has also been used as a dynein inhibitor (see Table 1). Analysis by steady state techniques indicates that EHNA acts as a weak, mixed inhibitor of dynein's ATPase activity at low ATP concentrations, but the mechanism of inhibition is still unknown (Penningroth 1989). These studies indicate that these analogs should be used with caution.

The phosphate analog vanadate is, on the other hand, a potent inhibitor of the dynein's ATPase activity and ciliary and flagellar motility (Gibbons

Table 1 Properties of axonemal and cytoplasmic dyneins

Property	Axonemal	Cytoplasmic
Microtubule binding sites		
ATP-sensitive	+	+
ATP-insensitive	+	−
ATPase sites		
Heavy chains	+	+
Sensitive to V_i-UV cleavage	+	+
Phosphorylation sites (in vivo)		
Heavy chains	+	?
Intermediate chains	−	?
Light chains	+	NA
MgATPase activity		
Substrate preference		
Soluble	ATP ≫ CTP	CTP ≫ ATP
Microtubule-activated	ATP	ATP
Microtubule binding ($M^{-1}\ s^{-1}$)[a]	1.2–6 × 10^4	4.7 × 10^5
Inhibitors		
Vanadate-K_i	<1 μM	5–10 μM
EHNA-K_i[b]	<1 mM	<1 mM
Microtubule motility		
Substrate preference	MgATP	MgATP, CaATP
Inhibitors		
Vanadate-10 μM	−	±
EHNA-1 mM[c]	−	−
NEM-1 mM	−	−
Vmax		
in vitro	4–8 μm/s	1–2 μm/s
in situ	14–16 μm/s	ND
Polarity	minus-end	minus-end

NA = not applicable, ND = not determined.

[a] The rate constants for microtubule binding were determined by measuring the initial slopes of the curves describing microtubule activation of the dynein ATPase activity and assuming the presence of one active site per head (Omoto & Johnson 1986; Shpetner et al 1988).

[b] ATPase activity measured in the presence of 0.6 M salt.

[c] Motility assayed in the presence of 0.1 mM ATP.

References: axonemal dynein (Gibbons 1965; Omoto & Johnson 1986; Vale & Toyoshima 1988; Fox & Sale 1987; Sale & Fox 1988; Piperno & Luck 1981; Penningroth 1989), and cytoplasmic dynein (Paschal et al 1987b; Lye et al 1987; Paschal & Vallee 1987; Shpetner et al 1988).

et al 1978). Detailed kinetic analysis of the mechanism of vanadate inhibition indicates that vanadate binds rapidly to the dynein-ADP intermediate after a single turnover of ATP hydrolysis to form a stable dead end complex (Shimizu & Johnson 1983a). This intermediate is the complex that mediates the cleavage of the dynein heavy chains in the presence of UV light (Gibbons et al 1987). ADP and vanadate can mimic an effect of ATP by inducing the slow quantitative dissociation of dynein from microtubules in solution (Shimizu & Johnson 1983b). This sensitivity to vanadate has become diagnostic for dynein polypeptides in vitro (see below and Table 1), but, given the phosphate based metabolism of the cell, it is not clear that vanadate inhibition is selective for dynein in vivo.

THE DYNEIN CROSSBRIDGE CYCLE IN SITU Analysis of the dynein crossbridge cycle in situ has utilized two independent strategies. One approach has been to compare the steady state ATPase activities of soluble and axonemal bound dynein under different experimental conditions, and the other has been to correlate morphological changes in the conformation of the dynein arm with different stages of the ATPase cycle.

Axonemal ATPase activity The rebinding of soluble dynein to its original location on the A-tubule (i.e. via its ATP-insensitive binding site) has been correlated with an increase in total ATPase activity (Takahashi & Tonomura 1978; Gibbons & Fronk 1979; Warner & McIlvain 1986). This effect has been interpreted as the result of the high local concentration of microtubule binding sites available to the B-end (the ATP-sensitive end) of the bound dynein arm (Gibbons & Fronk 1979). Other measurements on reactivated flagellar models have indicated that the axonemal ATPase activity is elevated under conditions in which motility is stimulated and depressed when motility is inhibited (Brokaw & Benedict 1968; Gibbons & Gibbons 1972). These data support the hypothesis that microtubule activation of dynein's ATPase is closely coupled to force generation.

ATP-induced conformational changes in the dynein arm Most models of crossbridge behavior predict some change in the position of the dynein arm on the A-tubule relative to the surface lattice of the opposing B-subfiber. Examination of sliding axonemes by electron microscopy has demonstrated that force generation is unidirectional. The dynein arms on doublet n "push" the adjacent n + 1 doublet toward the tip or plus end of the axoneme as they "walk" toward the minus end (Sale & Satir 1977; Fox & Sale 1987).

The conformation of the dynein arm in situ has been examined by a wide variety of techniques. The basic strategy has been to lock the dynein arm into different stages of the crossbridge cycle by manipulating the nucleotide levels in a preparation. However, the different methodologies employed to observe these changes in conformation have generated significantly different images and models for movement. The two major points of disagreement appear to be the conformation of the dynein arm in the absence of ATP (the rigor image) and the nature of nucleotide-induced changes in conformation during the crossbridge cycle (see Goodenough & Heuser 1989; Satir 1989).

In the absence of ATP, thin sectioned or negatively stained material shows the globular head(s) of dynein arm attached directly to the adjacent B-subfiber, giving the classic rigor image (Gibbons 1975; reviewed in Satir 1989). In the presence of ATP, or ATP plus vanadate, the arm adopts a relaxed configuration in which it is detached from the B-subfiber and tilts proximally towards the base of the axoneme. Addition of the nonhydrolyzable ATP analog AMPPCP generates a third conformation in which ~50% of the arms are tilted proximally as in the presence of ATP but remain attached to the B-subfiber (Avolio et al 1984). This new orientation has been interpreted as analogous to the transient attached state of the dynein·ADP·P intermediate that is believed to occur during the active crossbridge cycle (Avolio et al 1984). However, evidence for the equivalence of these two states is lacking.

Images of axonemes in quick-freeze deep-etch replicas have given a different impression (reviewed in Goodenough & Heuser 1989). In the absence of ATP, the dynein heads make contact with the B-subfiber by means of the B-link described above (Goodenough & Heuser 1982, 1984, 1989). In the presence of ATP the heads shift proximally, but the B-link remains attached to the B-subfiber (Goodenough & Heuser 1982, 1984, 1989; Sale et al 1985). The rigor configuration is never observed in preparations that were actively motile at the time of freezing, nor have the heads ever been observed to make direct contact with the B-subfiber (Goodenough & Heuser 1989). These images are therefore difficult to reconcile with biochemical data showing crossbridge attachment and detachment as the basis for movement.

Kinetic analysis of the dynein crossbridge cycle in vitro has demonstrated, however, that the lifetime of the attached state in the presence of ATP would be extremely brief (Porter & Johnson 1983b). Analysis of flagellar mechanics also predicts that the number of attached crossbridges in motile axonemes would be relatively low (Brokaw 1975; Brokaw & Johnson 1989). Thus, the failure to observe the B-subfiber bound intermediate in motile axonemes, while disappointing, may not be surprising. The major unresolved discrepancies are therefore the difference in the rigor images obtained with different methods and the function of the B-links. The development of alternative imaging strategies such as cryoelectron microscopy and three-dimensional reconstruction of the axoneme may be the ultimate solution to this dilemma (Murray 1987).

Regulation of Dynein Activity and Axonemal Motility

The three-dimensional bending waveforms of cilia and flagella are obviously more complex than the sliding of two microtubules described in most models of the dynein crossbridge cycle. The mechanisms that convert microtubule sliding into axoneme beating are not well understood, but it is clear that other structural components in the axoneme must be considered. One must not overlook, however, other possible functions of the dynein arms.

THE FUNCTION OF THE DYNEIN ARMS IN AXONEMAL MOTILITY

Outer arms The ability to selectively extract and reintroduce the dynein arms has been an important experimental tool: first, to demonstrate that the arms contain the sites of ATPase activity (Gibbons 1965), and later, to define the functional unit as the complete outer arm (Gibbons & Gibbons 1979; Mitchell & Warner 1980). More recently, reconstitution studies have examined the quantitative relationship between dynein arm activity and axonemal motility.

Selective removal of the outer row of dynein arms, either by chemical extraction or by mutation, reduces flagellar beat frequency to 30–50% of normal levels, but does not significantly alter the flagellar waveform (Gibbons & Gibbons 1973; Mitchell & Rosenbaum 1985; Brokaw & Kamiya 1987; Fox & Sale 1987; Kamiya 1988). Direct measurements of microtubule sliding in disintegrating axonemes indicate that sliding velocities are reduced in the absence of outer arms (Hata et al 1980; Okagaki & Kamiya 1986; Fox & Sale 1987; Kamiya et al 1989). Readdition of native or wild-type extracts containing outer arms can restore beat frequency (Gibbons & Gibbons 1979; Sakikibara & Kamiya 1989) and sliding velocities (Yano & Miki-Noumura 1981) to normal or wild-type levels. These observations have often been cited as evidence that the functions of the inner and outer arms are equivalent, and that the outer arms simply add more power for flagellar motility (Gibbons & Gibbons 1973). Recent analyses of the outer arm in vitro and the inner arm in situ have indicated, however, that this is an oversimplification.

The in vitro motility assay used in the identification of kinesin and cytoplasmic dynein has since been adapted for the analysis of axonemal dyneins (Paschal et al 1987a; Vale & Toyoshima 1988). Outer arm dynein from both *Tetrahymena* cilia and sea urchin sperm flagella will adsorb to a glass surface and move microtubules over that surface in an ATP-dependent fashion (Vale & Toyoshima 1988; Paschal et al 1987a). The direction of movement is the same as that observed in situ; dynein pushes the microtubule toward its plus end as it walks toward the minus end (Vale

& Toyoshima 1988). A single-headed dynein complex will also support microtubule gliding at rates even faster than the intact outer arm (Sale & Fox 1988). These results indicate that a single head is sufficient to support motility and begs the question of the function of multiple heads in the outer arm.

Inner arms The structural complexity of the inner dynein arms appears to reflect additional functional diversity. The only mutant that lacks all the inner arm polypeptides is paralyzed (Huang et al 1979), whereas mutants that lack subsets of inner arm polypeptides are motile (Brokaw & Kamiya 1987; Piperno 1988; Kamiya et al 1989). Analysis of their flagellar waveforms indicates that, unlike the outer arm mutants, the beat frequency of the partial inner arm mutants is almost normal but the pattern of the waveform is altered (Brokaw & Kamiya 1987; Kamiya et al 1989). Direct measurement of microtubule sliding in disintegrating axonemes confirms that the sliding velocities of the inner arm mutants are ~80% of the wild-type rates (Kamiya et al 1989). These observations suggest that inner and outer arms have different functions in the generation of axonemal motility.

Analysis of the 14S ATPase of *Tetrahymena* by in vitro motility assays indicates a potential inner arm function. The 14S ATPase will support the translocation of microtubules, but the microtubules undergo a clockwise rotation as they move toward their plus ends (Vale & Toyoshima 1988). If the 14S dynein is an inner arm subunit, this rotation may be related to either the axonemal twist or the central pair rotation observed in some three-dimensional waveforms (Omoto & Kung 1979).

REGULATION BY ACCESSORY STRUCTURES The complexity and diversity of axonemal waveforms necessitates many levels of regulatory activity. At the most basic level, the activity of the dynein arms must be locally activated both along the length of the outer doublets and around the axoneme in order to produce discrete regions of sliding. The recognition that all dynein arms push in the same direction (Sale & Satir 1977) led to the proposal for a switching mechanism that regulates dynein arm activity (Satir & Sale 1977). The "switch" hypothesis suggests that to produce a bend, dynein arms on one side of an axoneme must be activated to slide, whereas dynein arms on the other side must be inhibited. As the first bend is then propagated down the axoneme, the pattern of dynein arm activity must switch and initiate a new bend in the opposite direction (Satir 1985).

Since the flagellar beat cycle can be as high as 50–100 Hz, the switch must operate with a response time of less than 10 msec (Satir 1985). For this reason it has been argued that the switch or oscillator must involve a mechanical feedback loop, such as the model of curvature control (Brokaw

1971). Bending of outer doublets could change their relative spacing or introduce distortions in the microtubule surface lattice that would selectively activate or inhibit the dynein arms (Douglas 1975). Several models based on curvature control have been computer simulated with increasing sophistication over the years (reviewed in Brokaw 1989). These models can generate and propagate waves, but they do not fully account for the observed behavior of flagellar waveforms under a wide variety of experimental conditions (Brokaw 1985, 1989).

An alternative approach is to determine the minimum structure required to generate flagellar motility (see Figure 4). Simple oscillatory bending has

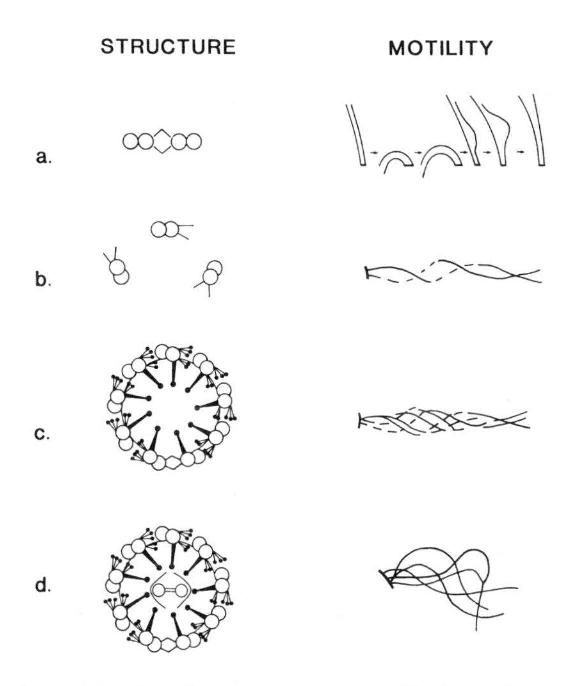

Figure 4 Structure and motility of axoneme components. The increasing sophistication in flagellar motility associated with the increased complexity of axoneme structure is shown diagrammatically. The axoneme structure is shown in cross section on the left, and the pattern of motility is shown on the right. (*a*) Two outer doublet microtubules and their associated linkages in frayed axonemes of *Chlamydomonas* flagella (redrawn from Figure 5 in Kamiya & Okagaki 1986). (*b*) The 3+0 axonemes of gregarine sperm flagella beating at low frequencies (1.5 Hz) (Prensier et al 1979). (*c*) The higher frequency helical waveforms observed in 9+0 axonemes of the Asian horseshoe crab sperm flagella (Ishijima et al 1988). (*d*) The asymmetric, planar waveforms observed with the 9+2 axonemes in the sperm flagella of the American horseshoe crab (Ishijima et al 1988).

been observed between two outer doublet microtubules tethered at their base in frayed axonemes (Kamiya & Okagaki 1986). These observations suggest that the basic oscillator must reside in the dynein arms and/or the microtubules themselves. Propagation of a bend requires at least three microtubules, as evidenced by the low frequency (1.5 Hz), helical waveforms generated by the 3+0 axonemes of gregarine sperm flagella (Prensier et al 1979). This is consistent with the model of opposing sets of arm activity. More effective forward progression is achieved by 9+0 axonemes that can beat at higher frequencies (Gibbons et al 1985; Ishijima et al 1988). These observations demonstrate that the dynein arms, the microtubules, and the associated linkages are the basic components required to initiate and propagate the more primitive helical waveforms.

The evolution of more sophisticated planar waveforms is associated with appearance of structural asymmetries in the axoneme. These include the radial spoke/central pair complex as well as specialized structures on individual outer doublets (Gibbons 1961; Hoops & Witman 1983). The function of the radial spoke/central pair complex appears to be twofold. One is to help to establish a planar form of beat (Ishijima et al 1988), and the second is to provide additional coordination of the mechanism of active sliding required for generation of asymmetric waveforms (see Figure 4). Evidence for the latter function comes from the analysis of bypass suppressor mutations in *Chlamydomonas* (Brokaw & Luck 1985). *Chlamydomonas* axonemes that lack radial spokes or central pair structures are normally paralyzed (Witman et al 1978), but it is possible to bypass these defects with second-site mutations that restore motility without restoring the missing radial spokes or central pair structures (Huang et al 1982). Analysis of the waveforms of the different mutant combinations indicates that the only significant difference is the inability of the strains lacking normal radial spokes or central pair to generate asymmetric waveforms (Brokaw et al 1982; Brokaw & Luck 1985). Thus, the radial spoke/central pair complex can be thought of as a modifier of the switching mechanism.

The bypass suppressor mutations themselves identify a second control system in the axoneme (Huang et al 1982). This control system may normally inactivate the dynein arms in the absence of a signal from the radial spoke/central pair complex, and lead to paralysis in the radial spoke/central pair mutants. The second site suppressor mutants short circuit this inhibitory pathway and allow dynein arm activity in the absence of a normal signal from the radial spoke/central pair complex. Consistent with this hypothesis, the suppressor mutations are associated with a number of defects in axonemal polypeptides, which include structural defects in the β heavy chain of the outer arm dynein (Huang et al 1982) and defects in phosphorylation of inner arm components (Luck & Piperno 1989).

These observations suggest that there are domains within the dynein arms that can regulate the sliding process.

REGULATION BY CALCIUM AND cAMP Regulation of axonemal motility also involves second order control mechanisms that allow the cell to generate transient or long lasting changes in flagellar waveforms in response to external stimuli. The ability to induce these changes in vitro indicates that the responsive elements are built into the structure of the axoneme. The elucidation of these regulatory pathways may provide new insights into the basic mechanisms of flagellar motility.

Regulation by calcium The ability of cilia and flagella to transiently alter their waveforms in response to changes in their environment is primarily mediated by Ca^{++}. External stimuli, either natural or artificial, that raise intracellular calcium, can produce ciliary arrest, reversal, or activation depending on cell type. Furthermore, there is evidence for graded responses to changes in calcium concentrations suggestive of multiple pathways (Naitoh & Kaneko 1972; Omoto & Brokaw 1985; Brokaw 1987). In most cases, the regulatory mechanism appears to be intrinsic to the structure of the axoneme, as reactivated cell models can mimic the different waveforms in response to changes in the calcium concentration of the reactivation medium; but in some instances, demonstration of Ca^{++} sensitivity requires readdition of soluble components or modification of extraction conditions (Brokaw & Nagayama 1985). Not surprisingly there is evidence for both soluble and tightly bound calcium binding proteins, including calmodulin (see Otter 1989 for recent review).

The pathways by which Ca^{++} mediates its effects on axonemal motility are not fully defined, but there are at least two distinct mechanisms that have been suggested thus far. The first model is an enzymatic mechanism that predicts that calcium binding to calmodulin (or a related calcium binding protein) results in the activation of an enzyme that leads to a change in the phosphorylation state of key axonemal polypeptides. For example, both vertebrate and invertebrate sperm flagella contain a calmodulin-dependent phosphatase, calcineurin, which will dephosphorylate several axonemal polypeptides, including the regulatory subunit of the cAMP dependent protein kinase, axokinin (Tash & Means 1988; Tash et al 1988). Thus, calcium may mediate its effect both directly and indirectly by antagonizing protein phosphorylation in the axoneme (Tash et al 1988).

What axonemal structures might be the substrates of such a phosphorylation/dephosphorylation cycle? Two likely candidates include the radial spoke/central pair complex and the dynein arms. The absence or loss of the radial spoke/central pair complex can be correlated with both the absence of calcium sensitivity and the inability to generate asymmetric waveforms (Gibbons et al 1985; Brokaw & Luck 1985; Hosokawa & Miki-Noumura 1987). Thus, Ca^{++}-induced changes in asymmetry might be mediated via the radial spoke/central pair complex. Alternatively, Ca^{++} might modify the phosphorylation state of the dynein arms (Tash & Means 1988; Hamasaki et al 1989), but there is no evidence yet to indicate that calcium modifies dynein activity, as assayed by microtubule sliding (Walter & Satir 1979). Since both dynein and radial spoke polypeptides are substrates for phosphorylation in vivo (Piperno & Luck 1981; Piperno et al 1981; Huang et al 1981), it will be worthwhile to determine if these polypeptides are also substrates for dephosphorylation by Ca^{++}/calmodulin/calcineurin.

The second model for Ca^{++}/calmodulin regulation is a structural mechanism mediated by calcium-induced changes in outer doublet curvature (reviewed in Otter 1989). This model is based on the observation that a significant fraction of axonemal calmodulin is tightly bound at a stoichiometry of ~1 mol of calmodulin per 60 mol of tubulin and that this ratio is unchanged in mutants that lack dynein arms, radial spokes, or central pairs (Otter 1989). Ca^{++} binding to the tightly bound calmodulin in the outer doublets might then mediate a change in the flexural rigidity or curvature of the microtubules themselves (Miki-Noumura & Kamiya 1979) that results in a biased baseline that generates asymmetric waveforms independent of sliding mechanisms (Eshel & Brokaw 1987; Okuno & Brokaw 1981). The significance of such a structural mechanism remains to be determined.

Regulation by cAMP and protein phosphorylation The effects of cyclic nucleotides on axonemal motility are just as varied and diverse as calcium, but in general cAMP dependent phosphorylation appears to define the long term level of activity in the axoneme. For example, the initiation and maintenance of motility in sperm flagella is associated with cAMP dependent phosphorylation (reviewed in Brokaw 1987). cAMP dependent phosphorylation also plays a role in sensory adaptation in other systems. For instance, in both protozoan ciliates and metazoan epithelial cilia, cAMP not only stimulates the beat frequency but can override the Ca^{++} reversal or arrest response (Nakaoka & Ooi 1985; Stephens & Stommel 1989). The relevant substrates for cAMP dependent phosphorylation vary in the different systems, but they include dynein polypeptides (Tash & Means 1988; Hamasaki et al 1989), axokinin, the regulatory subunit (RII) of the cAMP dependent protein kinase (Tash et al 1986; Tash & Means 1988), and several previously unidentified polypeptides (see Brokaw 1977; Stephens & Stommel 1989). The phosphorylated form of axokinin is also a substrate of calcineurin, the calmodulin dependent protein phosphatase described above (Tash & Means 1988), which suggests a tight interplay between the two systems of regulation.

CYTOPLASMIC DYNEINS

Historical Background

Nonaxonemal dynein ATPase activity was first described in extracts of unfertilized sea urchin eggs over twenty years ago (Weisenberg & Taylor 1968). The "egg dynein" ATPase sedimented at 13S on sucrose density gradients and was characterized by a broad pH optimum, a strict requirement for divalent cations, and a high nucleotide specificity (ATP $\gg$ NTP) (Weisenberg & Taylor 1968). Later work established that the egg dynein ATPase activity copurified with high molecular weight polypeptides and was sensitive to inhibition by sodium vanadate (Pratt 1980). Egg dynein was subsequently purified by a number of different protocols which yielded slightly different preparations with subtle differences in polypeptide composition, sensitivity to ATPase inhibitors, and microtubule binding properties (reviewed in Pratt 1989; Lye et al 1989). At least some of these egg dynein polypeptides were antigenically related to axonemal dynein polypeptides, but all the available antibodies did not cross-react with all the different preparations (Piperno 1984; Asai 1986; Hisinaga et al 1987; Pratt 1989; Porter et al 1988). A significant fraction of the egg dynein polypeptides were, however, related to a precursor pool for ciliogenesis (Stephens 1977; Asai 1986; Porter et al 1988).

This complexity stimulated the search for dynein-like polypeptides in cells and tissues that do not make ciliary precursors. Early efforts met with limited success (reviewed in Pratt 1989; Lye et al 1989). The advent of in vitro motility assays for microtubule translocators, however, made possible the identification of dynein-like molecules with clear mechanochemical activity (Lye et al 1987; Paschal et al 1987b).

Cytoplasmic dynein microtubule translocators have now been identified in bovine brain (Paschal et al 1987b), the nematode *C. elegans* (Lye et al 1987), rat liver and testis (Collins & Vallee 1988; Neely & Boekelheide 1988), tissue culture cells (Lye et al 1989), the giant amoeba *Reticulomyxa* (Euteneuer et al 1988), and *Drosophila* embryos (T. S. Hays, personal communication). Reexamination of the sea urchin system indicates that the egg contains multiple dynein isoforms, some of which may be related to the cytoplasmic dynein translocators described elsewhere (Porter et al 1988; Grissom et al 1988; Pratt 1989). However, for the purpose of the following discussion we focus primarily on cytoplasmic dyneins identified by means of in vitro motility assays.

Composition and Structure

PURIFICATION AND POLYPEPTIDE COMPOSITION Cytoplasmic dynein translocators have been purified by means of their affinity for microtubules in the absence of ATP (Lye et al 1987; Paschal et al 1987b). Microtubules prepared from crude cell extracts are enriched for cytoplasmic dynein by the depletion of nucleotides, and the dynein is subsequently released from the microtubules by extraction with ATP. The dynein containing extract is then further purified by sucrose density gradient centrifugation to yield a 20S particle (Paschal et al 1987b; Lye et al 1987).

All cytoplasmic dyneins described thus far contain 1–2 heavy chain polypeptides >400 kd (depending on the gel system used for analysis), and a cluster of intermediate chain polypeptides between 50 and 150 kd (see Figure 1). The heavy chains contain the primary sites of ATP binding and hydrolysis, as evidenced by their susceptibility to vanadate-mediated UV cleavage (Lye et al 1987, 1989; Paschal et al 1987b). V_1 cleavage occurs approximately midway through the heavy chain to generate two fragments that differ only slightly in size from their axonemal counterparts (Lye et al 1987, 1989; Paschal et al 1987b). Cleavage of the cytoplasmic dynein heavy chains can also be correlated with the loss of the ATPase activity (Porter et al 1988) and the inhibition of motility (Lye et al 1987). These results suggest that the ATP binding site is conserved both in structure and location between cytoplasmic and axonemal dynein isoforms. Comparison of cytoplasmic dyneins from different species and tissues indicates some variability in intermediate chain composition (Neely & Boekelheide 1988; C. A. Collins, personal communication). This may reflect tissue specific functions, but the role of the intermediate chains is presently unknown.

STRUCTURE IN VITRO Examination of brain cytoplasmic dynein by STEM (Vallee et al 1988) and quick-freeze deep-etch (Steuer et al 1988) indicates that cytoplasmic dynein is a two-headed molecule of ~1220 kd, similar in overall shape to the axonemal dyneins. The sizes of the heads (327 kd) are slightly smaller than the sizes of the heavy chain polypeptides (410 kd), which gives rise to the hypothesis that the heavy chains form both the heads and part of the stems (Vallee et al 1988). The shape of the two heads in quick-freeze deep-etch replicas appears the same, as contrasted with images of ciliary and flagellar dyneins that indicate structural asymmetries (Steuer et al 1988; Sale et al 1985). By analogy with axonemal dynein, the two-head domains are thought to contain both the microtubule binding sites and the sites for ATP hydrolysis. A third globular domain is occasionally observed at the base of the stalks connecting the two heads (Vallee et al 1988). To be consistent with the mass analysis obtained by STEM, this third domain is proposed to contain the intermediate chain polypeptides and may be the site for attachment to other cellular objects in vivo (Vallee et al 1989).

None of the cytoplasmic dynein translocators has yet been visualized in situ. Several antibody localization studies in sea urchin embryos have indicated the presence of dynein related antigens in either the mitotic spindle, embryonic cilia, or vesicular components of the cytoplasm (see Pratt 1989 for review), but given the diversity of egg dynein isoforms, these results should be interpreted with caution. The recent availability of antibody probes to other cytoplasmic dynein translocators may soon reveal the subcellular distribution of these motors.

The Cytoplasmic Dynein Crossbridge Cycle

MICROTUBULE BINDING PROPERTIES Cytoplasmic dynein binds microtubules in the absence of nucleotide (Paschal et al 1987b; Lye et al 1987), and dissociation of cytoplasmic dynein from microtubules is specific for MgATP (Shpetner et al 1988). Nonhydrolyzable ATP analogs will also induce dissociation, but are less effective than ATP (Shpetner et al 1988) as has been observed previously with axonemal dyneins (Chilcote & Johnson 1989). No ATP insensitive microtubule binding site has been described.

ATPase ACTIVITY Cytoplasmic dyneins exhibit an ATPase activity that is similar to axonemal and egg dynein activity in some respects, but significantly different in others (see Table 1). Like other dyneins, cytoplasmic dynein has a requirement for divalent cations (Shpetner et al 1988), and is sensitive to inhibition by vanadate, EHNA, and the sulfhydryl reagent, N-ethyl maleimide (NEM) (Lye et al 1987; Shpetner et al 1988; Neely & Boekelheide 1988). However, cytoplasmic dynein differs from other dyneins in its substrate specificity (Shpetner et al 1988), and its relative insensitivity to activation by nonionic detergents (Lye et al 1987; Shpetner et al 1988). The ability of cytoplasmic dyneins to hydrolyze pyrimidine nucleotides is probably of little physiological significance, but it may be a useful tool for sorting out dynein isoforms in complex cells and tissues such as testis (Collins & Vallee 1988) and sea urchin eggs (Grissom et al 1988).

The specific ATPase activities of the cytoplasmic dyneins are ~tenfold lower than those reported for both axonemal dyneins and egg dyneins (Paschal et al 1987b; Lye et al 1987; Shpetner et al 1988; Neely & Boekelheide 1988). Addition of purified tubulin in the form of assembled microtubules stimulates cytoplasmic dynein's ATPase activity five to sevenfold (Paschal et al 1987b; Shpetner et al 1988; Neely & Boekelheide 1988; Collins & Vallee 1988). The low basal ATPase activity allows one to detect microtubule activation at relatively low tubulin concentrations

($K_m \simeq 0.5$ mg/ml tubulin) as compared to axonemal dyneins ($K_m > 25$ mg/ml tubulin), however, the rates of microtubule binding by the two enzymes differ only by a factor of ten (see Table 1). The high basal ATPase activity of axonemal dyneins may reflect the lack of constraints for efficient utilization of ATP in a system that is always "on."

RECONSTITUTION OF MOTILITY IN VITRO Cytoplasmic dynein activity in vitro was first detected as the retrograde movement of latex beads in kinesin-depleted axoplasmic supernatants (Vale et al 1985b). The identification of cytoplasmic dynein as the retrograde motor came following purification of the enzyme from mammalian brain (Paschal et al 1987b; Paschal & Vallee 1987). Two puzzles remained, however; the observations that the movement mediated by cytoplasmic dynein prepared from *C. elegans* was of the opposite polarity (Lye et al 1987) and that cytoplasmic dynein prepared from *Reticulomyxa* supported bidirectional movement (Euteneuer et al 1988). Reexamination of the *C. elegans* cytoplasmic dynein with alternative assays subsequently established that it, too, is a retrograde motor (Lye et al 1989). The *Reticulomyxa* dynein, however, continues to be an exception to the rule that all dyneins move toward the minus end (Euteneuer et al 1988). Whether this reflects the presence of additional factors in the reconstituted system responsible for movement in the plus direction or the existence of a complex regulatory mechanism that specifies dynein polarity remains to be established.

The Function of Cytoplasmic Dynein

RECONSTITUTION OF MOTILITY IN SITU The evidence that cytoplasmic dynein functions as an organelle motor or microtubule translocator in vivo has been somewhat indirect. Dynein-like polypeptides have been described in semi-purified preparations of organelles (Gilbert & Sloboda 1986; Pratt 1986). Furthermore, purified cytoplasmic dynein supports the minus end directed movement of objects over microtubules at the same speed and with the same pharmacological properties as has been observed for the retrograde movement of organelles in vivo (Smith 1988) and in vitro (Vale et al 1985b; Koonce & Schliwa 1986; Dabora & Sheetz 1988; Haimo & Fenton 1988), but not all retrograde organelle movements in situ are sensitive to dynein inhibitors (discussed in Bloom et al 1989). This discrepancy may reflect the presence of additional factors that regulate organelle movements in different systems.

More direct evidence for cytoplasmic dynein function in retrograde organelle movements comes from recent reconstitution studies. Semi-purified organelle fractions from chick embryo fibroblasts will translocate on isolated microtubules, but unlike axoplasmic preparations, these organelles move primarily toward the minus end (Dabora & Sheetz 1988). High salt treatment of semi-purified organelle fractions completely abolishes their capacity for movement in in vitro assays (Schroer et al 1989). Addition of purified cytoplasmic dynein will not restore motility, but readdition of cytosolic extracts containing cytoplasmic dynein will rescue the motility of the salt-washed organelles. UV irradiation of the cytosol extract in the presence of ATP and vanadate (followed by norepinephrine treatment to reduce the vanadate) completely abolishes the capacity of the extract to restore motility. However, readdition of purified cytoplasmic dynein to the UV-cleaved cytosol restores motility. These data suggest that cytoplasmic dynein is required but not sufficient to drive retrograde organelle movement (Schroer et al 1989). Identification of the cytosolic factors that mediate cytoplasmic dynein rescue of organelle movement will be an important next step toward our understanding of the mechanism of organelle movement.

REGULATION OF CYTOPLASMIC DYNEIN ACTIVITY In vitro models of organelle movements (reviewed in Vale 1987) indicate that some systems are sensitive to external regulation that alters the direction of movement. For instance, in axons, individual organelles usually move unidirectionally (Brady et al 1982; Allen et al 1985), but under some treatments, such as a cold block or axon ligation, individual organelles can be observed to reverse their direction (Smith 1988). Analysis of the inhibitor sensitivity of the two directions indicates that the organelles appear to be switching between two distinct motor activities (Smith 1988). Other models of bidirectional movement such as pigment granules indicate that the direction of movement can be regulated by modulating the levels of cAMP dependent phosphorylation (Rozdzial & Haimo 1986; Haimo & Fenton 1988) or the calcium concentration of the medium (McNiven & Ward 1988). The regulatory pathways and motor molecules in these systems have yet to be identified, but, by analogy with axonemal motility, the regulatory targets may be the motor molecules themselves or a switching mechanism that specifies which motors are active.

FUTURE STUDIES

The large size and complexity of the dynein molecules means that any analysis of structure and function will be a long term effort. Considerable progress has been made in mapping functional domains of the axonemal dynein heavy chains, but we still know very little about the nature of the microtubule binding sites, the function of intermediate and light chains, and the effects of phosphorylation on dynein activity. Recent progress in cloning some of the structural genes of dynein polypeptides (Williams et al 1986; Wilkerson et al 1988; Foltz & Asai 1988; Mitchell & Warner 1989) suggests that some primary sequence information may be available soon. Comparison of these data with sequence information from nucleotide and microtubule binding domains in other proteins may shed more light on the significant structural domains in both axonemal and cytoplasmic dyneins.

Better understanding of the roles played by the individual subunits in axonemal dynein should come from the continued study of mutant strains of *Chlamydomonas* with defects in flagellar motility. Mutations that alter the assembly and function of dynein arms have been mapped to several genetic loci (Huang et al 1979; Kamiya 1988; Luck & Piperno 1989; Kamiya et al 1989). However, the corresponding mutant gene products have not yet been identified. As cloned genes encoding different dynein polypeptides become available, it will be possible to map these DNAs with respect to the genetically identified loci by restriction fragment length polymorphism (RFLP) analysis (Ranum et al 1988). In this way the polypeptide gene product of a given locus may be identified, and its function(s) can be inferred from mutant phenotype(s). Isolation of new mutant alleles at the locus may then identify additional functions of the polypeptide. Similar approaches should be possible with cytoplasmic dyneins in organisms such as *C. elegans* and *Drosophila*.

Definition of the conformational changes during crossbridge action must await higher resolution structural information. Current efforts to use frozen-hydrated specimens (Murray 1987) should more easily help to define the structure of the dynein-microtubule rigor complex. Changes in the conformation observed with nucleotide binding must be interpreted within the framework of the biochemical and kinetic data that establish two important points: (*a*) the connections from the heads to the base of the dynein molecule are flexible, and (*b*) the short lifetimes of the force producing states of dynein attached to the microtubule may preclude observation of these states. For these reasons, it has been exceedingly difficult to define the conformational basis for force production with any ATPase.

Regulation of dynein activity and axonemal motility by cAMP dependent phosphorylation and calcium has been a complex subject for many years. Recent progress in sorting out some of the regulatory pathways and their interconnections suggests that there may soon be a general framework in which to assimilate the various data (Tash & Means 1988). One must not oversimplify the subject, however. In *Chlamydomonas*, for example, calcium modifies the shape of the flagellar waveform, the direction of movement, and the beat frequency, and the relevant calcium concentration varies in each case (Omoto & Brokaw 1985). Clearly there are multiple pathways that must be considered in any discussion of regulation.

Although cytoplasmic dyneins were first described over twenty years ago, they have gained respectability only recently. Progress in the development of in vitro models for microtubule and organelle motility has now established both cytoplasmic dynein and kinesin as mechanochemical enzymes of opposite polarity with the appropriate characteristics for organelle motors. Current efforts are now focused on identifying the receptors that mediate the binding of these motors to their respective targets and on elucidating the pathways that regulate organelle motility. At the present time, it is uncertain if individual organelles determine their destinations by the selection of unidirectional motors or by the regulation of a bidirectional complex. Finally, other forms of microtubule-based motility such as chromosome movements on the mitotic spindle or changes in cell shape may be additional sites of cytoplasmic dynein or kinesin activity. The development of specific probes for these polypeptides will undoubtedly provide additional information about their subcellular locations and cellular functions.

ACKNOWLEDGMENTS

We wish to extend our thanks to all our colleagues who provided reprints and preprints. Thanks also to J. R. McIntosh, T. S. Hays, and P. Wilson for their thoughtful reading of the manuscript, to C. Inouye and G. Anderson for their cheerful preparation of the many drafts, and to D. Lorenz and L. Harwood for their skillful artwork. Grants from the National Science Foundation (to M.E.P. and S. K. Dutcher) and the National Institutes of Health (to K.A.J.) have supported our work on dyneins.

Literature Cited

Allen, R. D., Allen, N. S., Travis, J. L. 1981. Video enhanced differential interference contrast (AVEC-DIC) microscopy: a new method capable of analyzing microtubule-related movement in the reticulopodia of *Allogromia laticollaris*. *Cell Motil.* 1: 191–302

Allen, R. D., Weiss, D. G., Hayden, J. H., Brown, D. T., Fujiwake, H., Simpson, M. 1985. Gliding movement of and bidirectional organelle transport along single native microtubules from squid axoplasm. Evidence for an active role of microtubules in cytoplasmic transport. *J. Cell Biol.* 100: 1736–52

Amos, L. A., Linck, R. W., Klug, A. 1976. Molecular structure of flagellar microtubules. See Goldman et al 1976, pp. 847–67

Asai, D. J. 1986. An antiserum to the sea urchin 20S egg dynein reacts with embryonic ciliary dynein but it does not react with the mitotic apparatus. *Dev. Biol.* 118: 416–24

Avolio, J., Lebduska, S., Satir, P. 1984. Dynein arm substructure and the orientation of arm-microtubule attachments. *J. Mol. Biol.* 173: 389–401

Bell, C. W., Gibbons, I. R. 1982. Structure of the dynein-1 outer arm in sea urchin sperm flagella II. Analysis by proteolytic cleavage. *J. Biol. Chem.* 257: 516–22

Bloom, G. S., Wagner, M. C., Pfister, K. K., Leopold, P. L., Brady, S. T. 1989. Involvement of microtubules and kinesin in the fast axonal transport of membrane-bounded organelles. See Warner & McIntosh 1989, pp. 321–33

Brady, S. T., Lasek, R. J., Allen, R. D. 1982. Fast axonal transport in extruded axoplasm from squid giant axon. *Science* 218: 1129–31

Brokaw, C. J. 1971. Bend propagation by a sliding filament model for flagella. *J. Exp. Biol.* 55: 289–304

Brokaw, C. J. 1975. Cross bridge behavior in a sliding filament model for flagella. See Inoué & Stephens 1975, pp. 165–79

Brokaw, C. J. 1985. Computer simulation of flagellar movement. VI. Simple curvature controlled models are incompletely specified. *Biophys. J.* 48: 633–42

Brokaw, C. J. 1987. Regulation of sperm flagella motility by calcium and cAMP-dependent phosphorylation. *J. Cell. Biochem.* 35: 175–84

Brokaw, C. J. 1989. Operation and regulation of the flagellar oscillator. See Warner et al 1989, pp. 267–79

Brokaw, C. J., Benedict, B. 1968. Mechanochemical coupling in flagella. 1. Movement dependent dephosphorylation of ATP. *Arch. Biochem. Biophys.* 125: 770–78

Brokaw, C. J., Johnson, K. A. 1989. Perspective: dynein-induced microtubule sliding and force generation. See Warner et al 1989, pp. 191–98

Brokaw, C. J., Kamiya, R. 1987. Bending patterns of *Chlamydomonas* flagella IV. Mutants with defects in inner and outer dynein arms indicate differences in dynein arm function. *Cell Motil. Cytoskel.* 8: 68–75

Brokaw, C. J., Luck, D. J. L. 1985. Bending patterns of *Chlamydomonas* flagella III. A radial spoke deficient mutant and a central pair deficient mutant. *Cell Motil.* 5: 195–208

Brokaw, C. J., Luck, D. J. L., Huang, B. 1982. Analysis of the movement of *Chlamydomonas* flagella: The function of the radial spoke system is revealed by comparison of wild type and mutant flagella. *J. Cell Biol.* 92: 722–32

Brokaw, C. J., Nagayama, S. M. 1985. Modulation of the asymmetry of sea urchin sperm flagellar binding by calmodulin. *J. Cell Biol.* 100: 1875–83

Chilcote, T. J., Johnson, K. A. 1989. Microtubule-dynein cross-bridge cycle and the kinetics of 5′-adenylyl imidodiphosphate (AMPPNP) binding. See Warner et al 1989, pp. 235–43

Collins, C. A., Vallee, R. B. 1988. Abundant cytoplasmic dynein related to brain dynein (MAPIC) in rat liver and testis but not in sea urchin eggs. *Cell Motil. Cytoskel.* 11: 195–96

Dabora, S. L., Sheetz, M. P. 1988. Cultured cell extracts support organelle movement on microtubules in vitro. *Cell Motil. Cytoskel.* 10: 482–95

Douglas, G. J. 1975. Sliding filaments in sperm flagella. *J. Theor. Biol.* 53: 247–52

Eshel, D., Brokaw, C. J. 1987. New evidence for a "biased baseline" mechanism for calcium regulated asymmetry of flagellar bending. *Cell Motil. Cytoskel.* 7: 160–68

Euteneuer, U., Koonce, M. P., Pfister, K. K., Schliwa, M. 1988. An ATPase with properties expected for the organelle motor of the giant amoeba, *Reticulomyxa*. *Nature* 332: 176–78

Euteneuer, U., McIntosh, J. R. 1981. Polarity of some motility-related microtubules. *Proc. Natl. Acad. Sci. USA* 78: 372–76

Foltz, K. R., Asai, D. J. 1988. Developmental expression and responses to deciliation of the A_β heavy chain of sea urchin dynein. *J. Cell Biol.* 107: 246a

Fox, L. A., Sale, W. S. 1987. Direction of force generated by the inner row of dynein arms on flagellar microtubules. *J. Cell Biol.* 105: 1781–87

Gibbons, B. H., Baccetti, B., Gibbons, I. R. 1985. Live and reactivated motility in the 9+0 flagellum of *Anguilla* sperm. *Cell Motil.* 5: 333–50

Gibbons, B. H., Gibbons, I. R. 1972. Flagellar movement and adenosine triphosphatase activity in sea urchin sperm extracted with Triton X-100. *J. Cell Biol.* 54: 75–97

Gibbons, B. H., Gibbons, I. R. 1973. The effect of partial extraction of dynein arms on the movement of reactivated sea urchin sperm. *J. Cell Sci.* 13: 337–57

Gibbons, B. H., Gibbons, I. R. 1974. Properties of flagellar "rigor waves" produced by abrupt removal of adenosine triphosphate from actively swimming sea urchin sperm. *J. Cell Biol.* 63: 970–85

Gibbons, B. H., Gibbons, I. R. 1979. Relationship between the latent adenosine triphosphatase state of dynein 1 and its ability to recombine functionally with KCl extracted sea urchin sperm flagella. *J. Biol. Chem.* 254: 197–207

Gibbons, B. H., Gibbons, I. R. 1987. Vanadate sensitized cleavage of dynein heavy chains by 365 nm irradiation of demembranated sperm flagella and its effect on the flagellar motility. *J. Biol. Chem.* 262: 8354–59

Gibbons, I. R. 1961. The relationship between fine structure and direction of beat in gill cilia of a lamellibranch mollusc. *J. Biophys. Biochem. Cytol.* 11: 179–205

Gibbons, I. R. 1965. Chemical dissection of the cilia. *Arch. Biol.* (*Liège*) 76: 317–52

Gibbons, I. R. 1975. The molecular basis of flagellar movement in sea urchin sper-

matozoa. See Inoué & Stephens 1975, pp. 207–32
Gibbons, I. R. 1981. Cilia and flagella of eucaryotes. *J. Cell Biol.* 91: 107s–24s
Gibbons, I. R., Cosson, M. P., Evans, J. A., Gibbons, B. H., Houck, B., et al. 1978. Potent inhibition of dynein adenosine-triphosphatase and of the motility of cilia and sperm flagella by vanadate. *Proc. Natl. Acad. Sci. USA* 75: 2220–24
Gibbons, I. R., Fronk, E. 1979. A latent adenosine triphosphatase form of dynein 1 from sea urchin sperm flagella. *J. Biol. Chem.* 254: 187–96
Gibbons, I. R., Lee-Eiford, A., Mocz, G., Phillipson, C. A., Tang, W.-J. Y., Gibbons, B. H. 1987. Photosensitized cleavage of dynein heavy chains. *J. Biol. Chem.* 262: 2780–86
Gilbert, S. P., Sloboda, R. D. 1986. Identification of a MAP-2-like ATP binding protein associated with axoplasmic vesicles that translocate on isolated microtubules. *J. Cell Biol.* 103: 947–56
Goldman, R. D., Pollard, T. D., Rosenbaum, J. L., eds. 1976. *Cell Motility*, Vol. 3, pp. 841–1373. New York: Cold Spring Harbor
Goodenough, U. W., Gebhart, B., Mermall, V., Mitchell, D. R., Heuser, J. E. 1987. High pressure liquid chromatography fractionation of *Chlamydomonas* dynein extracts and characterization of inner arm dynein subunits. *J. Mol. Biol.* 194: 481–94
Goodenough, U. W., Heuser, J. E. 1982. Substructure of the outer dynein arm. *J. Cell Biol.* 95: 798–815
Goodenough, U. W., Heuser, J. E. 1984. Structural comparison of purified dynein proteins with in situ dynein arms. *J. Mol. Biol.* 180: 1083–1118
Goodenough, U. W., Heuser, J. E. 1985. Substructure of inner dynein arms, radial spokes, and the central pair projection complex. *J. Cell Biol.* 100: 2008–18
Goodenough, U. W., Heuser, J. E. 1989. Structure of the soluble and in situ ciliary dyneins visualized by quick-freeze deep-etch microscopy. See Warner et al 1989, pp. 121–40
Grissom, P. M., Porter, M. E., McIntosh, J. R. 1988. Isoforms of cytoplasmic dynein in sea urchin eggs. *J. Cell Biol.* 107: 246a
Haimo, L. T., Fenton, R. D. 1984. Microtubule crossbridging by *Chlamydomonas* dynein. *Cell Motil.* 4: 371–85
Haimo, L. T., Fenton, R. D. 1988. Dynein mediated retrograde organelle movements in melanophores. *J. Cell Biol.* 107: 245a
Haimo, L. T., Telzer, B. R., Rosenbaum, J. L. 1979. Dynein binds to and crossbridges cytoplasmic microtubules. *Proc. Natl. Acad. Sci. USA* 76: 5759–63
Hamasaki, T., Murtaugh, T. J., Satir, B. H., Satir, P. 1989. In vitro phosphorylation of *Paramecium* axonemes and permeabilized cells. *Cell Motil. Cytoskel.* 12: 1–11
Hata, H., Yano, Y., Mohri, T., Mohri, H., Miki-Noumura, T. 1980. ATP-driven tubule extrusion from axonemes without outer dynein arms of sea urchin sperm flagella. *J. Cell Sci.* 41: 331–40
Hisanaga, S.-I., Tanaka, T., Masaki, T., Sakai, H., Mabuchi, I., Hiramoto, Y. 1987. Localization of sea urchin egg cytoplasmic dynein in mitotic apparatus studied by using a monoclonal antibody against sea urchin sperm flagellar 21S dynein. *Cell Motil. Cytoskel.* 7: 97–109
Hoops, H. J., Witman, G. B. 1983. Outer doublet heterogeneity reveals structural polarity related to beat direction in *Chlamydomonas* flagella. *J. Cell Biol.* 97: 902–8
Hosokawa, Y., Miki-Noumura, T. 1987. Bending motions of *Chlamydomonas* axonemes after extrusion of central-pair microtubules. *J. Cell Biol.* 105: 1297–1301
Huang, B., Piperno, G., Luck, D. J. L. 1979. Paralyzed flagella mutants of *Chlamydomonas reinhardtii* defective for axonemal doublet microtubule arms. *J. Biol. Chem.* 254: 3091–99
Huang, B., Piperno, G., Ramanis, Z., Luck, D. J. L. 1981. Radial spokes of *Chlamydomonas* flagella: Genetic analysis of assembly and function. *J. Cell Biol.* 88: 80–88
Huang, B., Ramanis, Z., Luck, D. J. L. 1982. Suppressor mutations in *Chlamydomonas* reveal a regulatory mechanism for flagellar function. *Cell* 28: 115–24
Inoué, S. 1981. Video image processing greatly enhances contrast, quality, and speed in polarization-based microscopy. *J. Cell Biol.* 89: 346–56
Inoué, S., Stephens, R. E., eds. 1975. *Molecules and Cell Movement*. New York: Raven. 450 pp.
Ishijima, S., Sekiguchi, K., Hiramoto, Y. 1988. Comparative study of the beat patterns of American and Asian horseshoe crab sperm: Evidence for a role of the central pair complex in forming planar waveforms in flagella. *Cell Motil. Cytoskel.* 9: 264–70
Johnson, K. A. 1983. The pathway of ATP hydrolysis by dynein. Kinetics of a presteady state burst. *J. Biol. Chem.* 258: 13825–32
Johnson, K. A. 1985. Pathway of the microtubule-dynein ATPase and the structure of dynein. *Annu. Rev. Biophys. Biophys. Chem.* 14: 161–88
Johnson, K. A., Wall, J. S. 1983. Structure

and molecular weight of the dynein ATPase. *J. Cell Biol.* 96: 669–78

Kamiya, R. 1988. Mutations at twelve independent loci result in absence of outer dynein arms in *Chlamydomonas reinhardtii. J. Cell Biol.* 107: 2253–58

Kamiya, R., Kurimoto, E., Sakakibara, H., Okagaki, T. 1989. A genetic approach to the function of inner and outer arm dynein. See Warner et al 1989, pp. 209–18

Kamiya, R., Okagaki, T. 1986. Cyclical bending of two outer-doublet microtubules in frayed axonemes of *Chlamydomonas. Cell Motil Cytoskel.* 6: 580–85

King, S. M., Witman, G. B. 1987. Structure of the α and β heavy chains of the outer arm dynein from *Chlamydomonas* flagella. Masses of chains and sites of ultraviolet-induced vanadate dependent cleavage. *J. Biol. Chem.* 262: 17596–17604

King, S. M., Witman, G. B. 1988. Structure of the α and β heavy chains of the outer arm dynein from *Chlamydomonas* flagella. Location of epitopes and protease-sensitive sites. *J. Biol. Chem.* 263: 9244–55

King, S. M., Witman, G. B. 1989. Molecular structure of *Chlamydomonas* outer arm dynein. See Warner et al 1989, pp. 61–75

Koonce, M. P., Schliwa, M. 1986. Reactivation of organelle movements along the cytoskeletal framework of a giant freshwater amoeba. *J. Cell Biol.* 103: 605–12

Lee-Eiford, A., Ow, R. A., Gibbons, I. R. 1986. Specific cleavage of dynein heavy chains by ultraviolet irradiation in the presence of ATP and vanadate. *J. Biol. Chem.* 261: 2337–42

Luck, D. J. L. 1984. Genetic and biochemical dissection of the eucaryotic flagellum. *J. Cell Biol.* 98: 789–94

Luck, D. J. L., Piperno, G. 1989. Dynein arm mutants of *Chlamydomonas.* See Warner et al 1989, pp. 49–60

Lye, R. J., Pfarr, C. M., Porter, M. E. 1989. Cytoplasmic dynein and microtubule translocators. See Warner & McIntosh 1989, pp. 141–54

Lye, R. J., Porter, M. E., Scholey, J. M., McIntosh, J. R. 1987. Identification of a microtubule based cytoplasmic motor in the nematode *C. elegans. Cell* 51: 309–18

Marchese-Ragona, S. P., Johnson, K. A. 1989. Structure and subunit organization of the soluble dynein particle. See Warner et al 1989, pp. 37–48

Marchese-Ragona, S. P., Wall, J. S., Johnson, K. A. 1988. Structure and mass analysis of 14S dynein obtained from *Tetrahymena* cilia. *J. Cell Biol.* 106: 127–32

McNiven, M. A., Ward, J. B. 1988. Calcium regulation of pigment transport in vitro. *J. Cell Biol.* 106: 111–25

Miki-Noumura, T., Kamiya, R. 1979. Conformational change in the outer doublet microtubules from sea urchin sperm flagella. *J. Cell Biol.* 81: 355–60

Mitchell, D. R., Rosenbaum, J. L. 1985. A motile *Chlamydomonas* flagellar mutant that lacks outer dynein arms. *J. Cell Biol.* 100: 1228–34

Mitchell, D. R., Rosenbaum, J. L. 1986. Protein-protein interactions in the 18S ATPase of *Chlamydomonas* outer dynein arms. *Cell Motil. Cytoskel.* 6: 510–20

Mitchell, D. R., Warner, F. D. 1980. Interaction of dynein arms with B subfibers of *Tetrahymena* cilia: Quantitation of the effects of magnesium and adenosine triphosphate. *J. Cell Biol.* 87: 84–97

Mitchell, D. R., Warner, F. D. 1989. Protein-protein interactions in dynein function. See Warner et al 1989, pp. 141–53

Mocz, G., Tang, W.-J. Y., Gibbons, I. R. 1988. A map of photolytic and tryptic cleavage sites on the β heavy chain of dynein ATPase from sea urchin sperm flagella. *J. Cell Biol.* 106: 1607–14

Murray, J. M. 1987. Electron microscopy of frozen hydrated eukaryotic flagella. *J. Ultrastruc. Res.* 95: 196–206

Naitoh, Y., Kaneko, H. 1972. Reactivated Triton-extracted models of *Paramecium*: Modification of ciliary movement by calcium ions. *Science* 176: 523–24

Nakaoka, Y., Ooi, H. 1985. Regulation of ciliary reversal in Triton-extracted *Paramecium* by calcium and cyclic adenosine monophosphate. *J. Cell Sci.* 77: 185–95

Neely, D. M., Boekelheide, K. 1988. Sertoli cell processes have axoplasmic features: An ordered microtubule distribution and an abundant high molecular weight microtubule associated protein (Cytoplasmic dynein). *J. Cell Biol.* 107: 1767–76

Ogawa, K. 1973. Studies on flagellar ATPase from sea urchin spermatozoa. II. Effect of trypsin digestion on the enzyme. *Biochim. Biophys. Acta* 293: 514–25

Ogawa, K., Negishi, S., Obika, M. 1980. Dynein 1 from rainbow trout spermatozoa. Immunological similarity between trout and sea urchin dynein 1. *Arch. Bioch. Biophys.* 203: 196–203

Ogawa, K., Negishi, S., Obika, M. 1982. Immunological dissimilarity in protein component (Dynein 1) between outer and inner arms within sea urchin sperm axonemes. *J. Cell Biol.* 92: 706–13

Okagaki, T., Kamiya, R. 1986. Microtubule sliding in mutant *Chlamydomonas* devoid of outer dynein arms. *J. Cell Biol.* 103: 1895–1902

Okuno, M., Brokaw, C. J. 1981. Calcium-induced change in form of demembranated sea urchin sperm flagella immobilized by vanadate. *Cell Motil.* 1: 349–62

Omoto, C. K., Brokaw, D. J. 1985. Bending patterns of *Chlamydomonas* flagella: II. Calcium effects on reactivated *Chlamydomonas* flagella. *Cell Motil.* 5: 53–60

Omoto, C. K., Johnson, K. A. 1986. Activation of the dynein adenosine triphosphatase by microtubules. *Biochem.* 25: 419–27

Omoto, C. K., Kung, C. 1979. The pair of central tubules rotates during ciliary beat in *Paramecium*. *Nature* 279: 532–34

Otter, T. 1989. Calmodulin and the control of flagellar movement. See Warner et al 1989, pp. 281–98

Ow, R. A., Tang, W.-J. Y., Mocz, G., Gibbons, I. R. 1987. Tryptic digestion of dynein 1 in low salt medium: origin and properties of fragment A. *J. Biol. Chem.* 262: 3409–14

Paschal, B. M., King, S. M., Moss, A. G., Collins, C. A., Vallee, R. B., Witman, G. B. 1987a. Isolated flagellar outer arm dynein translocates brain microtubules in vitro. *Nature* 330: 672–74

Paschal, B. M., Shpetner, H. S., Vallee, R. B. 1987b. MAP 1C is a microtubule-activated ATPase which translocates microtubules in vitro and has dynein-like properties. *J. Cell Biol.* 105: 1273–82

Paschal, B. M., Vallee, R. B. 1987. Retrograde transport by the microtubule associated protein MAP 1C. *Nature* 330: 181–83

Penningroth, S. M. 1989. Chemical inhibitors of the dynein adenosine triphosphatases. See Warner et al 1989, pp. 167–79

Pfister, K. K., Fay, R. B., Witman, G. B. 1982. Purification and polypeptide composition of dynein ATPases from *Chlamydomonas* flagella. *Cell Motil.* 2: 525–47

Pfister, K. K., Haley, B. E., Witman, G. B. 1984. The photoaffinity probe 8-azidoadenosine-5′-triphosphate selectively labels the heavy chain of *Chlamydomonas* 12S dynein. *J. Biol. Chem.* 259: 8499–8504

Pfister, K. K., Haley, B. E., Witman, G. B. 1985. Labeling of *Chlamydomonas* 18S dynein polypeptides by 8-azidoadenosine 5′-triphosphate, a photoaffinity analog of ATP. *J. Biol. Chem.* 260: 12844–50

Pfister, K. K., Witman, G. B. 1984. Subfractionation of *Chlamydomonas* 18S dynein into two unique subunits containing ATPase activity. *J. Biol. Chem.* 259: 12072–80

Piperno, G. 1984. Monoclonal antibodies to dynein subunits reveal the existence of cytoplasmic antigens in sea urchin egg. *J. Cell Biol.* 98: 1842–50

Piperno, G. 1988. Isolation of a sixth dynein subunit adenosine triphosphatase. *J. Cell Biol.* 106: 133–40

Piperno, G., Huang, B., Ramanis, Z., Luck, D. J. L. 1981. Radial spokes of *Chlamydomonas* flagella: Polypeptide composition and phosphorylation of stalk components. *J. Cell Biol.* 88: 73–79

Piperno, G., Luck, D. J. L. 1979. Axonemal adenosine triphosphatases from flagella of *Chlamydomonas reinhardtii*. *J. Biol. Chem.* 254: 3084–90

Piperno, G., Luck, D. J. L. 1981. Inner arm dyneins from flagella of *Chlamydomonas reinhardtii*. *Cell* 27: 331–40

Porter, M. E., Grissom, P. M., Scholey, J. M., Salmon, E. D., McIntosh, J. R. 1988. Dynein isoforms in sea urchin eggs. *J. Biol. Chem.* 263: 6759–71

Porter, M. E., Johnson, K. A. 1983a. Characterization of the ATP-sensitive binding of *Tetrahymena* 30S dynein to bovine brain microtubules. *J. Biol. Chem.* 258: 6575–81

Porter, M. E., Johnson, K. A. 1983b. Transient state kinetic analysis of the ATP-induced dissociation of the dynein-microtubule complex. *J. Biol. Chem.* 258: 6582–87

Pratt, M. M. 1980. The identification of a dynein ATPase in unfertilized sea urchin eggs. *Devel. Biol.* 74: 364–78

Pratt, M. M. 1986. Stable complexes of axoplasmic vesicles and microtubules: protein composition and ATPase activity. *J. Cell Biol.* 103: 957–68

Pratt, M. M. 1989. Dyneins in sea urchin eggs and nerve tissue. See Warner & McIntosh 1989, pp. 125–40

Prensier, G., Vivier, E., Goldstein, S., Schrevel, J. 1979. Motile flagellum with a "3+0" ultrastructure. *Science* 207: 1493–94

Ranum, L. P. W., Thompson, M. D., Schloss, J. A., Lefebvre, P. A., Silflow, C. D. 1988. Mapping flagellar genes in *Chlamydomonas* using restriction fragment length polymorphisms. *Genetics* 120: 109–22

Rozdzial, M. M., Haimo, L. T. 1986. Bidirectional pigment granule movements of melanophores are regulated by protein phosphorylation and dephosphorylation. *Cell* 47: 1061–70

Sakakibara, H., Kamiya, R. 1989. Functional recombination of outer dynein arms with outer arm missing flagellar axonemes of a *Chlamydomonas* mutant. *J. Cell Sci.* 92: 77–83

Sale, W. S., Fox, L. A. 1988. Isolated β-heavy chain subunit of dynein translocates microtubules in vitro. *J. Cell Biol.* 107: 1793–97

Sale, W. S., Fox, L. A., Milgram, S. L. 1989. Composition and organization of the inner row dynein arms. See Warner et al 1989, pp. 89–102

Sale, W. S., Goodenough, U. W., Heuser, J. E. 1985. The substructure of isolated and in situ outer dynein arms of sea urchin sperm flagella. *J. Cell Biol.* 101: 1400–12

Sale, W. S., Satir, P. 1977. The direction of active sliding of microtubules in *Tetrahymena* cilia. *Proc. Natl. Acad. Sci. USA* 74: 2045–49

Satir, P. 1985. Switching mechanisms in the control of ciliary motility. *Mod. Cell Biol.* 4: 1–46

Satir, P. 1989. Structural analysis of the dynein cross-bridge cycle. See Warner et al 1989, pp. 219–34

Satir, P., Sale, W. S. 1977. Tails of *Tetrahymena*. *J. Protozool.* 24: 498–501

Satir, P., Wais-Steider, J., Lebduska, S., Nasr, A., Avolio, J. 1981. The mechanochemical cycle of the dynein arm. *Cell Motil.* 1: 303–27

Schroer, T. A., Steuer, E. R., Sheetz, M. P. 1989. Cytoplasmic dynein is a minus-end-directed motor for membraneous organelles. *Cell* 56: 937–46

Shimizu, T., Johnson, K. A. 1983a. Presteady state kinetic analysis of vanadate-induced inhibition of the dynein ATPase. *J. Biol. Chem.* 258: 13833–40

Shimizu, T., Johnson, K. A. 1983b. Kinetic evidence for multiple dynein ATPase sites. *J. Biol. Chem.* 258: 13841–46

Shpetner, H. S., Paschal, B. M., Vallee, R. B. 1988. Characterization of the microtubule-activated ATPase of brain cytoplasmic dynein (MAP 1C). *J. Cell Biol.* 107: 1001–9

Smith, R. S. 1988. Studies on the mechanism of the reversal of rapid organelle transport in myelinated axons of *Xenopus laevis*. *Cell Motil. Cytoskel.* 10: 296–308

Stephens, R. E. 1977. Differential protein synthesis and utilization during cilia formation in sea urchin embryos. *Dev. Biol.* 61: 311–29

Stephens, R. E., Stommel, E. W. 1989. Role of cyclic adenosine monophosphate in ciliary and flagellar motility. See Warner et al 1989, pp. 299–316

Steuer, E. R., Heuser, J. R., Sheetz, M. P. 1988. Cytoplasmic dynein and ciliary outer arm dynein: A structural comparison. *Cell Motil. Cytoskel.* 11: 200–1 (Abstr.)

Takahashi, M., Tonomura, Y. 1978. Binding of 30S dynein with the B-tubule of the outer doublet of axonemes from *Tetrahymena pyriformis* and adenosine triphosphate-induced dissociation of the complex. *J. Biochem.* 84: 1339–55

Tang, W.-J. Y., Bell, C. W., Sale, W. S., Gibbons, I. R. 1982. Structure of the dynein-1 outer arm in sea urchin sperm flagella. I. Analysis by separation of subunits. *J. Biol. Chem.* 257: 508–15

Tang, W.-J. Y., Gibbons, I. R. 1987. Photosensitized cleavage of dynein heavy chains. Cleavage at the V2 site by irradiation at 365 nm in the presence of oligovanadate. *J. Biol. Chem.* 263: 17728–34

Tash, J. S., Hidaka, H., Means, A. R. 1986. Axokinin phosphorylation by cAMP-dependent protein kinase is sufficient for activation of sperm motility. *J. Cell Biol.* 103: 649–55

Tash, J. S., Krinks, M., Patel, J., Means, R. L., Klee, C. B., Means, A. R. 1988. Identification, characterization, and functional correlation of calmodulin dependent protein phosphatase in sperm. *J. Cell Biol.* 106: 1625–33

Tash, J. S., Means, A. R. 1988. Interplay between cAMP- and Ca^{2+}-calmodulin (CAM)-dependent pathways regulating flagellar motility. *Cell Motil. Cytoskel.* 11: 216–17 (Abstr.)

Toyoshima, Y. Y. 1987a. Chymotryptic digestion of *Tetrahymena* 22S dynein I. Decomposition of three headed 22S dynein to one-headed and two headed particles. *J. Cell Biol.* 105: 887–95

Toyoshima, Y. Y. 1987b. Chymotryptic digestion of *Tetrahymena* 22S dynein II. Pathway of degradation of 22S dynein heavy chains. *J. Cell Biol.* 105: 897–901

Vale, R. D. 1987. Intracellular transport using microtubule based motors. *Annu. Rev. Cell Biol.* 3: 374–78

Vale, R. D., Reese, T. S., Sheetz, M. P. 1985a. Identification of a novel class of force generating proteins (kinesin) involved in microtubule-based motility. *Cell* 41: 39–50

Vale, R. D., Schnapp, B. J., Mitchison, T., Steuer, E., Reese, T. S., Sheetz, M. P. 1985b. Different axoplasmic proteins generate movement in opposite directions along microtubules in vitro. *Cell* 43: 623–32

Vale, R. D., Toyoshima, Y. Y. 1988. Rotation and translocation of microtubules in vitro induced by dyneins from *Tetrahymena* cilia. *Cell* 52: 459–69

Vallee, R. B., Paschal, B. M., Shpetner, H. S. 1989. Characterization of microtubule associated protein (MAP) 1C as the motor for retrograde transport and its identification as dynein. See Warner & McIntosh 1989, pp. 211–22

Vallee, R. B., Wall, J. S., Paschal, B. M., Shpetner, H. S. 1988. Microtubule-associated protein 1C from brain is a two headed cytosolic dynein. *Nature* 332: 561–63

Walter, M. F., Satir, P. 1979. Calcium does not inhibit active sliding of microtubules from mussell gill cilia. *Nature* 278: 69–70

Warner, F. D., McIlvain, J. H. 1982. Binding stoichiometry of 21S dynein A and B subfiber microtubules. *Cell Motil.* 2: 429–43

Warner, F. D., McIlvain, J. H. 1986. Kinetic properties of microtubule activated 13S and 21S dynein ATPases. *J. Cell Sci.* 83: 251–67

Warner, F. D., McIntosh, J. R., eds. 1989. Kinesin, dynein, and microtubule dynamics. In *Cell Movement*, Vol. 2. New York: Liss. 478 pp.

Warner, F. D., Perreault, J. G., McIlvain, J. H. 1985. Rebinding of *Tetrahymena* 13S and 21S dynein ATPases to extracted doublet microtubules. The inner row and outer row dynein arms. *J. Cell Sci.* 77: 263–87

Warner, F. D., Satir, P., Gibbons, I. R., eds. 1989. The Dynein ATPases. In *Cell Movement*, Vol. 1. New York: Liss. 337 pp.

Weisenberg, R. C., Taylor, E. W. 1968. Studies on ATPase activity of sea urchin eggs and isolated mitotic apparatus. *Exp. Cell Res.* 53: 372–84

Wilkerson, C., Piperno, G., Luck, D. 1988. Isolation and characterization of a genomic sequence encoding an outer arm heavy chain subunit. *J. Cell Biol.* 107: 247a

Williams, B. D., Mitchell, D. R., Rosenbaum, J. R. 1986. Molecular cloning and expression of flagellar radial spoke and dynein genes of *Chlamydomonas*. *J. Cell Biol.* 103: 1–11

Witman, G. B., Johnson, K. A., Pfister, K. K., Wall, J. S. 1983. Fine structure and molecular weight of the outer arm dyneins of *Chlamydomonas*. *J. Submicrosc. Cytol.* 15: 193–98

Witman, G. B., Plummer, J., Sander, G. 1978. *Chlamydomonas* flagellar mutants lacking radial spokes and central tubules. *J. Cell Biol.* 76: 729–47

Yano, Y., Miki-Noumura, T. 1981. Recovery of sliding ability in arm depleted flagellar axonemes after recombination with extracted dynein 1. *J. Cell Sci.* 48: 223–39

Zanetti, N. C., Mitchell, D. R., Warner, F. D. 1979. Effects of divalent cations on dynein cross bridging and ciliary microtubule sliding. *J. Cell Biol.* 80: 573–88

Annu. Rev. Cell Biol. 1989. 5 : 153–80

COMMUNICATION BETWEEN MITOCHONDRIA AND THE NUCLEUS IN REGULATION OF CYTOCHROME GENES IN THE YEAST *SACCHAROMYCES CEREVISIAE*

Susan L. Forsburg and Leonard Guarente

Department of Biology, Massachusetts Institute of Technology, Cambridge, Massachusetts 02139

CONTENTS

INTRODUCTION

The mitochondrion is an essential organelle used by the cell for respiration, as well as to sequester numerous metabolic pathways. Located within this organelle are the enzymes of the citric acid cycle, enzymes required in pathways for the biosynthesis of numerous amino acids and the heme

0743–4634/89/1115–0153$02.00

biosynthetic enzymes. The electron transport chain that drives energy production lies in the inner membrane. The mitochondria contain their own, limited genome that encodes a small fraction of the proteins necessary for electron transport and oxidative phosphorylation. The rest are encoded in the nucleus, translated in the cytoplasm, and imported into the mitochondria. (For additional recent reviews on various aspects of mitochondrial biogenesis, see Attardi & Schatz 1988; Tzagaloff & Myers 1986).

The yeast *Saccharomyces cerevisiae*, because it is readily amenable to microbiological and genetic manipulation, is especially suitable for study of regulatory processes in cell biology. It is a facultative anaerobe, and unlike most eukaryotes can survive the loss of part or all of its mitochondrial DNA (Dujon 1981). Numerous genes in the nucleus required for mitochondrial function have been identified through the isolation and characterization of petite (*pet*) mutations (e.g. Ebner et al 1973; Sherman 1963; Burkl et al 1976; Tzagaloff & Myers 1986). The petite cells are unable to grow on a nonfermentable carbon source because they are respiratory-deficient. All told, genes that are targets for petite mutations include not only cytochromes and other proteins directly involved in electron transport and oxidative phosphorylation, but also genes involved in the expression of the mitochondrial genome. Among the latter are genes encoding the mitochondrial RNA polymerase (Lustig et al 1982a), maturases needed for splicing the type II introns in mitochondrial transcripts (McGraw & Tzagaloff 1983; Faye & Simon, 1983; Pillar et al 1983), and mitochondrial ribosomal proteins (Myers et al 1987). Ribosomal RNA and tRNAs that decode unusual mitochondrial codons are encoded in the mitochondrial genome (Dujon 1981; Tzagaloff & Myers 1986). Nuclear genes may also play other roles such as in the addition of uridine bases to mitochondrial transcripts in trypanosomes (Feagin et al 1987, 1988; Benne et al 1986).

Nuclear petites are distinguished from several types of mitochondrial mutants in which the mitochondrial genome itself is deficient. Mutants in the mitochondrial genome are termed *mit*$^-$, in which one of the mitochondrially encoded proteins is mutated ρ^-, in which mitochondrial DNA is partially deleted, or ρ^0, in which the mitochondrial DNA is entirely deleted (Dujon 1981). All these mutants are respiratory-deficient, and ρ^- and ρ^0 cells are deficient in mitochondrial protein synthesis.

The essential function of the electron transport chain in the mitochondria is oxidative phosphorylation, in which the passage of electrons from one component to another (with molecular oxygen the terminal accepter) harvests energy to produce ATP. The electrons are passed via the heme prosthetic groups of cytochromes. The cytochromes can exist as single polypeptides or as members of larger complexes. The cytochrome b-c_1 complex, including 9 subunits (Ljungdahl et al 1986), passes electrons to cytochrome *c*, which in turn transfers them to the cytochrome *c* oxidase complex, which also contains 9 different proteins as subunits (Power et al 1984). Of these polypeptides, the mitochondrial genome encodes only cytochrome *b* and three subunits of cytochrome *c* oxidase. Also encoded in the mitochondria are two of nine subunits of the ATPase complex (Dujon 1981). The rest of the components in the electron transport chain are encoded by nuclear genes.

This arrangement presents the cell with a technical problem. It must coordinate expression of the nuclear genes with the status of the mitochondria and with the growth conditions of the cell. The cell must balance production of proteins encoded in two very different genomes to allow assembly of all the components in the inner membrane. An additional complexity arises because several proteins are encoded by duplicated genes. For example, in the yeast *Saccharomyces cerevisiae* considered in this review there are two subtly different forms of cytochrome *c* encoded respectively by the genes *CYC1* and *CYC7*. The cell must integrate the expression of these gene families into the overall pattern and thus fine-tune its respiratory response.

In this review, we will discuss how this balance is achieved by using, as our principal model, the transcriptional regulation of the nuclear-encoded iso-1 cytochrome *c* gene. First, we will present an overview of signal transduction as it pertains to the communication between mitochondria and nucleus and to the sensing of the growth conditions of the cell. We will briefly discuss regulation of respiratory genes that are duplicated. Then we will describe in detail the transcriptional activators that provide this regulation to nuclear genes, and lastly, present the implications of these studies for eukaryotic gene expression in general.

SIGNAL TRANSDUCTION

It was generally accepted that nucleic acids—either DNA or RNA—do not pass the mitochondrial membrane (Dujon 1981). However recent experiments show that RNA, as part of a ribonucleoprotein complex, is imported into mitochondria (Chang & Clayton 1987). Nonetheless, it is safe to assume that, in most cases, the currency of exchange between these two organelles must come at the level of proteins or of metabolites that can diffuse or be transported across the membranes. (For reviews on how nuclear-encoded proteins are imported into mitochondria, see Attardi & Schatz 1988; van Loon et al 1988).

Mitochondria to Nucleus

Several findings suggest that the mitochondrion can communicate with the nucleus. First, respiratory deficiency induced by drugs specifically affecting mitochondrial functions (Siemens et al 1980) has been shown to affect levels of cytochrome *c* expression in the cell nucleus. More recently, Parikh et al (1987) showed that the status of the mitochondrial genome itself can affect expression of nuclear genes. These authors used subtractive cDNA hybridization to determine whether there is any difference in the expression of nuclear genes when phenotypically identical cells of different mitochondrial genotypes are compared. Three strains of identical nuclear genotypes were compared: they varied only in their mitochondrial DNA. One was mit^-, one was ρ^-, and one was ρ^0. Clones corresponding to two classes of RNA were isolated. Levels of Class I RNAs were increased in the three respiratory deficient strains relative to wildtype (ρ^+). Class II RNAs were increased in the ρ^- or ρ^0 strains, relative to the mit^- strain. Oddly enough, of those RNAs that could be definitely identified, one was a transcript derived from the spacer region between the 5S rRNA and 35S rRNA genes (enhanced in the ρ^0 strain), and one was derived from a region of the native yeast plasmid 2μ (from the ρ^- strain). It is unclear whether the enhancements reflect actual rates of transcription or increased RNA stability. Probes to nuclear genes known to encode mitochondrial subunits, such as *CYC1*, *COX6* (subunit VI of cytochrome *c* oxidase), or the F_1 ATPase showed, in most cases, no difference in expression in the three different mitochondrial backgrounds. The exception was one of the several transcripts of *COX6*, which appeared to be enhanced in the ρ^- and ρ^0 cells.

These results suggest that a direct way exists for the nucleus to monitor the mitochondrial genome. Perhaps the transcripts thus identified indicated the cell's attempt to compensate for the loss of its respiratory competency. Parikh et al (1987) suggest two models for a possible mechanism. In one, there is simply a mitochondrially encoded protein that acts as messenger. When mitochondrial protein synthesis is shut down (in a ρ^- or ρ^0 mutant), this signalling molecule is absent. Alternatively, a nuclear protein could monitor mitochondrial DNA levels by entering the mitochondria and binding to the DNA. If the mitochondrial DNA were deleted, then the import of this protein might be inhibited by some novel mechanism, thereby partitioning the protein mainly to the cell nucleus where its presence could influence nuclear transcription or RNA stability.

A second signal that emanates from the mitochondria to influence transcription in the nucleus is the cytochrome cofactor itself, heme. Strains mutant in *HEM1* (δ-aminolevulinate synthase) may be grown without heme if they are given a source of unsaturated fatty acids and the membrane sterol, ergosterol (Gollub et al 1977; Mattoon et al 1979; Guarente & Mason 1983). Such cultures showed a 100-fold reduction in expression of a *CYC1-lacZ* reporter gene (Guarente & Mason 1983). Heme could be added exogenously and shown to be an inducer of the transcription of the *CYC1* gene (Guarente & Mason 1983). Further, this induction occurred via an upstream activation site (UAS) about 300 base pairs away from the start of transcription. Earlier work implicated heme in the transcription and translation of catalase T (*CTT1*; Hortner et al 1982; Richter et al 1980; Woloszauk et al 1980; Hamilton et al 1982). Heme was also shown to be required for the accumulation of the nuclear-encoded subunits of cytochrome oxidase (Saltzgaber-Muller & Schatz, 1978; Woodrow & Schatz 1979; Lustig et al 1982b; Gollub & Dayan 1985) and, indeed, for transcription of the *COX* genes that encode them (Trueblood et al 1988; Kloekner-Gruissem et al 1987).

The immediate precursor to heme, missing the Fe atom, is not an inducer, since ferrochelatase mutants are blocked in the Fe addition step of heme synthesis and do not have detectable levels of *CYC1* expression (L. Guarente, unpublished observation). Heme also represses a class of genes that includes the gene *HEM13* (coproporphyrinogen oxidase; Zagorec & Labbe-Bois 1986) and the gene *ANB1* (function unknown; Lowry & Lieber 1986; Lowry & Zitomer 1984). Repression of *HEM13* indicates that end product repression of the heme pathway occurs at a late step in heme biosynthesis, presumably because the pathway bifurcates at an early intermediate, prophobilogen, which is needed for synthesis of sirroheme. The latter is used as a cofactor by sulfite reductase, needed in the synthesis of methionine (Tait 1978).

How does the heme signal exert its effects on transcription of nuclear genes? The major activator of genes that respond to heme is the HAP1 protein (Guarente et al 1984) which is allelic to the CYP1 gene product (Clavilier et al 1976; Verdiere et al 1986). *CYP1* is a locus in which dominant mutations were isolated that increased expression of the *CYC7* gene (Verdiere et al 1985). Loss of function mutations in *HAP1* were isolated because they reduce expression from UAS1, one of the two independent UAS elements of the *CYC1* gene (Guarente et al 1984). The identification of these mutants was facilitated by a *CYC1-lacZ* reporter gene under the control of UAS1. The recessive alleles of *HAP1* reduce expression of *CYC7* (iso-2 cytochrome *c*; Prezant et al 1987), *CTT1* (Winkler et al 1988), and *CYT1* (cytochrome *c1*; Schneider 1989), and probably other genes involved in electron transport that are induced by heme. Curiously, *CYC7* is regulated both positively and negatively by heme, via *HAP1* and *ROX1*, respectively (Prezant et al 1987; Wright & Zitomer 1984; Zitomer et al 1987). How heme exits the mitochondria is not known. In the next section we consider how heme levels are integrated into a signal transduction pathway involving O_2, and in the section on transcriptional regulators of cytochrome genes we discuss the molecular basis for the activation of HAP1-regulated genes by heme. Table 1 identifies those genes that we will discuss in detail throughout this review.

Other genes involved in regulation by heme include the *ROX* genes (Lowry & Lieber 1986; Lowry & Zitomer 1984; Lowry & Zitomer 1988), which are genetically defined as negative regulators that prevent expression of anaerobically induced genes, such as *ANB1* (Lowry & Leiber 1986). Transcription of the *ROX1* repressor gene itself was shown to be activated by heme, although it is not known whether this activation requires *HAP1* (Lowry & Zitomer 1988). Likewise, the *REO1* gene represses the *COX5B* gene (subunit V_B of cytochrome *c* oxidase) under aerobic conditions (Trueblood et al 1988). Expression of this minor subunit of the oxidase is thus induced under anaerobic conditions.

Nucleus to Mitochondria

Many nuclear petites have been isolated that define genes required for expression of the mitochondrial genome; these gene products provide communication in the other direction, from nucleus to mitochondrion. Those mutants with broad effects include those in general transcription or translation factors such as the nucleus-encoded mitochondrial RNA polymerase (Lustig et al 1982a) or the mitochondrial ribosomal proteins (Fearon & Mason 1988; Partaledis & Mason 1988). Other types of mutants also exist. For example, the nuclear *PET111* gene is specifically required for translation of the mitochondrial *oxi2* gene, encoding subunit III of

Table 1 Gene names and protein products[a]

Gene	Product
ANB1	anaerobically regulated gene, of unknown function
COX4	subunit IV of cytochrome *c* oxidase
COX5A	subunit V_A of cytochrome *c* oxidase (major isoform)
COX5B	subunit V_B of cytochrome *c* oxidase (minor isoform)
CYC1	iso-1 cytochrome *c*
CYC7	iso-2 cytochrome *c*
CYT1	cytochrome *c1*
HAP1	Heme Activator Protein; positive transcriptional regulator of UAS1 of *CYC1* and other respiratory genes
HAP2,3,4	Heme Activator Protein; positive transcriptional regulators of UAS2 of *CYC1* and other respiratory genes
HEM1	δ-aminolevulinate synthase; a heme biosynthetic enzyme
ROX1	Regulation by oxygen; negative transcriptional regulator of *ANB1* and *CYC7*

[a] Gene names and protein products of those loci discussed in detail throughout this review. References for loci appear in the text.

cytochrome *c* oxidase (Poutre & Fox 1987). The *PET494* gene is required for the translation of the mitochondrial *oxi3* gene, which encodes subunit I (Costanzo & Fox 1986, 1988). Likewise, the *CBP6* gene is required for translation of cytochrome *b* (*COB*; Dieckmann & Tzagaloff 1985), as are *CBS1* and *CBS2* (Roedel 1986; Roedel & Fox 1987; Roedel et al 1985). Presumably these proteins produced in the nucleus act as regulators to provide an exact connection between the growth conditions of the cell (as recognized in the nucleus) and the mitochondria. As will be discussed below, one would expect these gene products to provide regulation in response to the same signals that regulate expression of the many nuclear cytochrome *c* oxidase subunits.

Numerous mitochondrial genes contain introns, and splicing of a given intron can be specifically affected by nuclear genes. For example, *CBP2* (McGraw & Tzagaloff 1983; Pillar et al 1983) is required for splicing of introns 3 and 5 in *COB*. A gene specifically affecting splicing of subunit I of cytochrome *c* oxidase has been defined (Faye & Simon 1983). Dominant mutations in a nuclear gene, *NAM2*, suppress a defective COB intron-encoded maturase (Labouesse et al 1985, 1987). Gene disruptions in *NAM2* are defective in splicing of *COB*. Surprisingly, the *NAM2* gene product, as well as a nuclear gene of *Neurospora* involved in mitochondrial splicing, encode mitochondrial tRNA synthetases (Herbert et al 1988; Akins & Lambowitz 1987).

Response to Physiologic Signals

HEME/OXYGEN There are two principal signals that inform the cell that it is capable of, or requires, respiration: oxygen and carbon source. Since oxidative phosphorylation requires O_2 as an electron acceptor, the presence of oxygen, i.e. an aerobic growth condition, is clearly required for respiratory growth (Somlo & Fukuhara, 1965). But how can the cell register the presence of oxygen and respond accordingly?

Several enzymes late in the heme biosynthetic pathway are oxygenases. As a result, anaerobic cells do not make heme, but accumulate porphyrins instead (Mattoon et al 1979). Therefore heme levels reflect oxygen levels. Heme's role as a mediator of oxygen induction is demonstrated by *CYC1* induction under anaerobic conditions when heme or heme analogues are added to the media (Pfeifer 1988; Mason et al 1984). Similar experiments showed that *ROX1* expression also responds to heme rather than to O_2 directly (Lowry & Zitomer 1988). Since heme synthesis occurs in the mitochondria, the cell can, in principle, integrate the levels of oxygen in the environment with the rate of consumption of oxygen in electron transport. However, heme may not be the only mediator of the presence of oxygen. Recent experiments indicate that oxygen and not heme induces expression of *PET494* and show that this induction occurs at the level of translation (Marykwas & Fox 1989).

CARBON The second signal that affects genes encoding mitochondrial proteins is carbon source. Catabolite or glucose repression ensures that the cell preferentially utilizes the simple sugar glucose. Numerous genes have been implicated in carbon control in the study of other catabolic pathways. Mutations in any of several genes termed *SNF* prevent induction of *SUC2* (invertase) or *GAL* (galactose catabolic) genes when cells are shifted out of glucose (Carlson et al 1981; Neigeborn & Carlson 1984; Ciriacy 1977). The *SNF1* gene product has been shown to be a ser/thr kinase (Celenza & Carlson 1986). Suppressors of *snf*$^-$ mutations that restored *SUC2* gene expression were isolated and termed *ssn* (for suppressors of *snf*; Carlson et al 1984; Neigeborn & Carlson 1987). Interestingly, the gene encoding *SSN6* is allelic to a locus that was termed *CYC8*; mutations in *CYC8* were isolated because they increase the usually low level of *CYC7* expression (Trumbly 1986; Rothstein & Sherman 1980). This finding suggests a possible link between glucose repression of carbon source catabolic pathways and mitochondrial proteins.

Catabolite repression of genes encoding mitochondrial proteins has been evident for some time (Ephrussi & Slonimski 1950; Polakis et al 1965; Ibrahim et al 1973; Ciriacy 1977). Some genes known to be regulated by carbon source at the transcriptional level are *CYC1* (Zitomer et al 1979; Guarente et al 1984; Laz et al 1984), *CYC7* (Prezant et al 1987; Zitomer et al 1987), the catalase genes (Winkler et al 1988; Hortner et al 1982), *COX5A* (Trueblood et al 1988), and *PET494* (Marykwas & Fox 1989). Catabolite repression of enzymes early in the heme biosynthetic pathway has also been reported (Mahler & Lin 1978; Mattoon et al 1979; Labbe-Bois et al 1983), which suggests that the heme and carbon source signals may intersect. Mutations have been reported that specifically release mitochondrial functions from glucose repression, as well as mutations that affect both mitochondrial functions and carbon source catabolic pathways (Boker-Schmitt et al 1982; Michels & Romanoski 1980; Rothstein & Sherman 1980; Ciriacy 1978; Szekely & Montgomery 1984).

The fact that catabolite repression pathways for carbon source catabolic genes and mitochondrial genes are not identical is also demonstrated by mutations in the HAP2/3/4 transcriptional activation system for cytochrome genes. Mutations in these three *HAP* genes were identified by virtue of a reduction in expression of a *CYC1-lacZ* reporter driven by the second *CYC1* upstream activation sequence, UAS2 (Guarente et al 1984; Pinkham et al 1987; S. L. Forsburg & L. Guarente, in preparation). Mutations in *HAP2, HAP3*, or *HAP4* prevent activation of many other nuclear cytochrome genes including *COX4* (subunit IV of cytochrome *c* oxidase; Schneider 1989), *CYT1* (cytochrome *c1*; Schneider 1989), *COX5A* (subunit V_A of cytochrome *c* oxidase; Trueblood et al 1988), as well as *HEM1* (δ-aminolevulinate synthase; Keng & Guarente 1987). *hap2*$^-$, *hap3*$^-$, and *hap4*$^-$ mutants, however, do not reduce expression of genes encoding carbon catabolic pathways (Guarente et al 1984; S. L. Forsburg & L. Guarente, in preparation).

COORDINATION BETWEEN NUCLEUS AND MITOCHONDRIA

We can begin to see an integrated circuit that coordinates expression of the nuclear and mitochondrial genes, best illustrated in the case of cytochrome oxidase (Figure 1). The physiological signals of oxygen, working through heme and carbon source are monitored by the effects of regulatory genes such as *HAP1/2/3/4, ROX1, REO1*, and probably others.

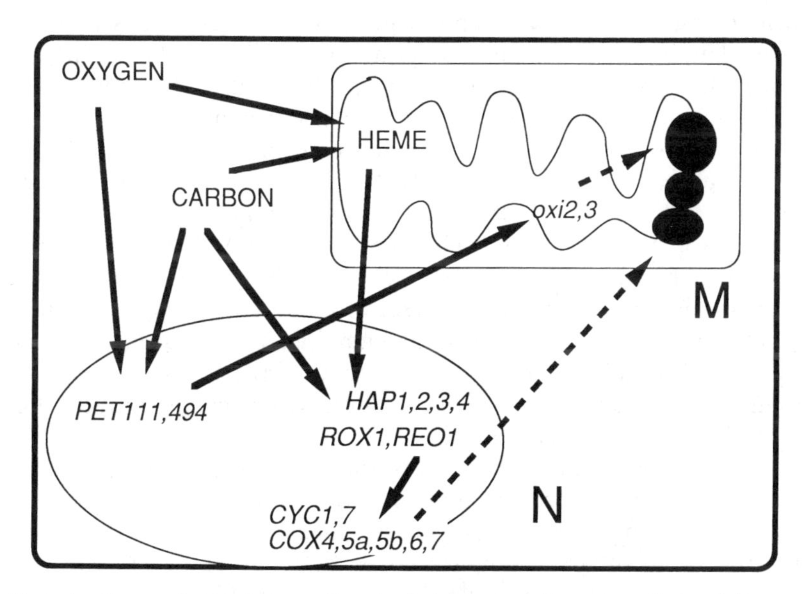

Figure 1 Communication between the mitochondrion and the nucleus. Some of the regulatory pathways providing communication between nucleus (N) and mitochondria (M) are diagrammed here. Oxygen influences the level of heme, which in turn influences activity or expression of various regulators, including *HAP1*. Oxygen also acts upon *PET494* in a heme-independent manner. Carbon source independently influences activity of the regulators *HAP2/3/4*; it can also act independently of the *HAP2/3/4* system and as well as affect heme synthesis. Solid lines indicate regulatory effects; dashed lines indicate localization of structural proteins.

Thus a particular level of expression of nuclear genes encoding oxidase subunits is established. Other nuclear genes responding to these signals include *PET494* and *PET111*, which are needed for translation of the genes *oxi3* and *oxi2* of the mitochondrial genome. In this way, the nuclear and mitochondrial oxidase genes are coordinated. To complete the circuit, the nucleus can in some way monitor the quality of the mitochondrial genome and can be influenced by mitochondrial metabolites, such as heme. This simple picture does have some complications. For instance, while *COX4* and *COX5A* are under control of HAP2/3/4, the *PET494* gene, which conveys the same physiologic signals to the mitochondrial genes, is not. Therefore the details of the picture presented in Figure 1 remain to be spelled out.

DUPLICATED GENES

There are now numerous examples in yeast in which specific enzymes or structural proteins are encoded by duplicated genes that differ slightly in their coding sequence. There appear to be two general reasons why cells have related genes that encode isoforms with similar functions. The first is that the products of the two genes are targeted to different cellular compartments by differences at their amino termini. This mechanism allows cells to partition various enzymes between the mitochondria and the cytoplasm, including certain tRNA synthetases (Myers & Tzagaloff 1985; Pape & Tzagaloff 1985; Pape et al 1985) and citrate synthase (Kim et al 1986). An alternative mechanism is used by cells to achieve partitioning of histidinyl tRNA synthetase and a tRNA modification enzyme. In the former case, there is a single gene with transcripts that differ at their 5′ ends such that one contains a mitochondrial signal sequence and one does not (Natsoulis et al 1986). In the latter case, the *MOD5* gene encodes a single tRNA modification enzyme localized to both compartments (Martin & Hopper 1982; Dihanich et al 1987).

A second reason that cells have isoforms is that it allows regulation based on subtle differences in protein activities, since the genes encoding different isoforms can be expressed under different physiologic conditions. The two forms of cytochrome *c* encoded by *CYC1* and *CYC7* are 78% identical at the amino acid level (Montgomery et al 1980). Likewise, there are two forms of subunit V of cytochrome *c* oxidase, encoded by the *COX5A* and *COX5B* genes; these isoforms are 66% identical (Cumsky et al 1987). Under normal aerobic growth conditions, *CYC1* and *COX5A* produce the vast majority of cytochrome *c* and subunit V of cytochrome *c* oxidase, respectively (Sherman & Stewart 1971; Cumsky et al 1985). Both of these genes are activated by heme via the HAP1 protein. Under anaerobic conditions these genes are turned off while *CYC7* and *COX5B* continue to be expressed. As mentioned previously, the latter genes are repressed under aerobic conditions by the *ROX* and *REO* loci. Another pair of genes showing differential regulation by oxygen is *ANB1*, expressed under anaerobic conditions, and *TR1*, expressed under aerobic conditions. Transcripts of these two genes hybridize to the same probe, but their functions are not known (Lowry & Zitomer 1984).

Presumably the minor isoforms of cytochrome *c* and subunit V of cytochrome *c* oxidase are more efficient under weakly aerobic conditions than the major forms, although physical evidence for this assertion is still lacking (Trueblood et al 1988). Consistent with such a role, the apo-protein of the iso-2 cytochrome *c* expressed under anaerobic conditions, unlike that of iso-1, is stable in the absence of its heme cofactor (Matner & Sherman 1982). This stability may poise the cell for the transition from anaerobic to aerobic growth.

TRANSCRIPTIONAL REGULATORS OF CYTOCHROME GENES

Regulators of transcription have been identified in many eukaryotic systems. In animal cells, these proteins were first identified as factors from extracts that would bind to a regulatory site of a particular promoter. Regulatory sites are genetically defined as DNA regions that, when deleted or substituted, will alter or abolish the expression from the promoter. Such sites include binding sequences for transcriptional activators, which are termed UASs (upstream activation sites) in yeast, and distal promoter elements or enhancers in higher cells. Proteins that bind to UASs or enhancers are themselves regulated in response to physiologic signals in the cell and thus convey to their target genes a characteristic pattern of control. A second critical element found in most eukaryotic promoters is the promoter-proximal TATA box. Gene-specific activators bound at UASs and basic transcriptional factors bound at the TATA box cooperate to bring about the initiation of transcription by RNA polymerase II.

Numerous mammalian proteins that bind to regulatory sites have been purified and found to activate transcription in vitro. Further, by making oligonucleotide probes that encode defined regions of the purified proteins, it has been possible to isolate genes encoding many of these factors (Johnson & McKnight 1989). In yeast, the identification of regulatory proteins has been genetic, by the isolation of mutations that alter gene expression in vivo (Guarente 1987). The regulator proteins that have been so defined include the products of the GAL4 and GCN4 genes. Mutations in these genes reduce expression of galactose catabolic enzymes and many amino acid biosynthetic enzymes, respectively (Johnston 1987a; Hinnebusch 1988; see Table 1).

HAP1: A Heme-Responsive DNA Binding Protein

As mentioned above, the *CYC1* promoter was shown to contain two functional UASs, termed UAS1 (which is activated by HAP1) and UAS2 (which is activated by HAP2/3/4), as well as three functional TATA boxes (Guarente et al 1984; Hahn et al 1985). Below, we discuss the relationships between these HAP proteins and the heme and carbon signals (Figure 2). We also describe features of these transcriptional activators that help provide new insights into eukaryotic gene control.

In order to gain insight into the molecular function of the HAP1 activator, we cloned the gene and carried out DNA mobility shift assays using extracts from strains that were *hap1*$^-$ or *HAP1*$^+$, or contained *HAP1* on a multicopy plasmid, or contained truncated *HAP1*. In this way, we

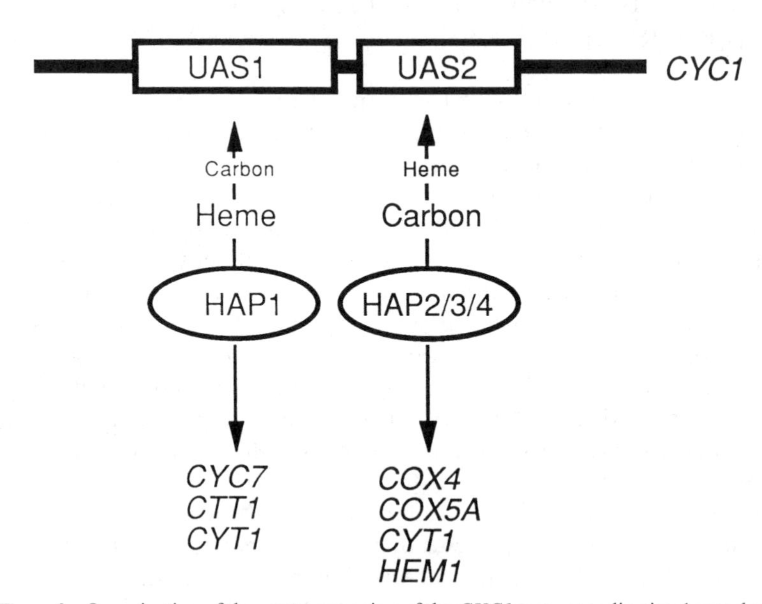

Figure 2 Organization of the upstream region of the *CYC1* gene, encoding iso-1 cytochrome *c*. *CYC1* is regulated by two independent systems. The HAP1 system responds principally to heme, and has been shown to regulate *CYC7*, *CTT1*, and *CYT1* expression. The HAP2/3/4 system responds to carbon source, such that cells deficient in any one of the three are petite, have reduced levels of all cytochromes, and are unable to grow on a nonfermentable carbon source. HAP2/3/4 have also been shown to regulate *COX4, COX5A*, *CYT1*, and *HEM1* expression.

showed that the *HAP1* protein binds to UAS1 or to the genetically defined UAS of the *CYC7* gene (Pfeifer et al 1987a,b). In addition, this binding was heme-dependent in vitro, which suggests that heme is an effector that interacts directly with the HAP1 protein. This suggestion was strengthened by the finding that HAP1 derivatives synthesized in a wheat germ translation system bind to DNA in a heme-dependent manner (Pfeifer et al 1989).

The native HAP1 protein contains 1383 amino acids. Interestingly, while a HAP1 derivative bearing amino acid residues 1–445 required heme for binding in vitro, shorter truncations (amino acids 1–245 or 1–148) did not (Pfeifer et al 1989). Further, an internal deletion of residues 245–445 resulted in a heme-independent constitutive phenotype in vivo. We conclude from these findings that HAP1 contains an internal repression sequence that masks DNA binding in the absence of heme. The region of HAP1 between residues 245 and 445 contains the sequence (Lys/Arg)-Cys-Pro-(Val/Ile)-Asp-His repeated seven times. Because this repeat contains precisely spaced cys and his, residues that are implicated in the binding of heme in other hemoproteins, we have proposed that this internal part of HAP1 binds heme and that this binding unmasks the ability of the protein to bind DNA. Whether masking is intramolecular or is mediated by the binding to HAP1 of an additional cellular protein is not known at present.

MULTI-SEQUENCE RECOGNITION It was noted that the UAS1 and *CYC7* sequences bound by HAP1 displayed no similarity (Pfeifer et al 1987b). Further, the contacts made by HAP1 at the two sites, as determined by methylation interference assays, are different. Since these observations were made, three additional HAP1 binding sites have been identified, one in *CTT1* (catalase T; Winkler et al 1988), one in *CYT1* (cytochrome *c1*; Schneider 1989), and a second site in UAS1 (Pfeifer 1988). These three new sites show clear similarity to UAS1 and no obvious similarity to *CYC7*. The relative affinities of UAS1 and *CYC7* for HAP1 are comparable in vitro binding assays (Pfeifer et al 1987b). Two interesting questions arise from these observations. First, how does HAP1 recognize two different sequences? Second, does multi-sequence recognition play some role in gene control?

By examining the DNA binding properties of mutant HAP1 derivatives in vitro, we were able to conclude that the same domain in HAP1 is required to bind to the two sequence motifs (Pfeifer et al 1989). Like GAL4 and many other candidate yeast regulatory proteins, HAP1 contains a cysteine-rich zinc finger at its amino terminus (amino acids 63 to 99). Mutation of the conserved cys64 or cys83 to any of several residues knocked out binding to both UAS1 and *CYC7*. Further carboxyl terminal deletion mutations showed that a fragment containing residues 1–148 would bind efficiently to either sequence (Pfeifer et al 1989). However, within this single domain, non-identical sets of amino acids recognize the two motifs. Mutant derivatives of *HAP1* exist that selectively abolish binding to one site or the other. One such mutation is *HAP1-18* (*CYP1-18*; Pfeifer et al 1987b), which changes ser63 to arg (Pfeifer et al 1989). This mutation reduces *CYC1* expression by eliminating HAP1 binding to UAS1 while increasing HAP1 activity at the *CYC7* gene. A second class of mutations, consisting of carboxyl terminal deletions ending at residue 115 or 99, selectively eliminate HAP1 binding to *CYC7* while leaving activity at UAS1 (Pfeifer et al 1989).

In order to gain further insight into the role of residue 63 of HAP1 in site recognition, we replaced this residue with 15 different amino acids and examined the effects of the changes on binding to UAS1 and *CYC7* (K.-S. Kim & L. Guarente, in preparation). The only residues that allowed a normal level of binding to UAS1 were ser, thr, and cys, which suggests that this residue might be involved in hydrogen bonding to a base in UAS1. Residues that allowed normal binding to *CYC7* were these three amino acids plus arg, ile, and met. Many other changes eliminated binding to both sites. In the latter case we do not know if the lack of binding in vitro simply reflects a failure of the altered side chain to make an appropriate DNA contact, or whether the mutation has changed the conformation of the adjacent zinc finger.

A genetic method similar to that used to show that the GAL4 cys-rich region binds zinc was used to demonstrate that the cysteine-rich region is a zinc finger. In the GAL4 instance, mutation of pro26 to leu reduces the affinity of the finger for zinc and allows the dependence to be demonstrated readily, since the mutant may be rescued by increased concentrations of zinc in the growth media (Johnston 1987b). When the corresponding change was made in the HAP1 protein, the DNA binding domain of the mutant protein was shown to require zinc for binding in vitro (K.-S. Kim & L. Guarente, in preparation).

A second question of interest relates to the possible role of multi-sequence recognition in gene control. Several observations lead us to propose that HAP1 has different properties when bound at UAS1 than when bound at *CYC7*. First, synthetic oligonucleotides encoding the UAS1/HAP1-binding site drive a much higher level of expression of a *lacZ* reporter in vivo in response to HAP1 than oligonucleotides encoding the HAP1-binding site from *CYC7*. Second, internal deletions in HAP1 (for example, removal of amino acids 247 to 1308) increase its activity at UAS1 but decrease its activity at *CYC7* in vivo. The domain of HAP1 that activates transcription is an acidic region at the carboxyl terminus of the protein (residues 1308 to 1483), far from the DNA binding and heme regulatory regions. From these observations, we devised a model by which the DNA sequence of *CYC7* acts like an allosteric effector to cause the DNA binding domain of HAP1 to mask its activation domain. We suggest that the internal deletions reduce expression at *CYC7* by bringing the DNA binding and activation domains into closer proximity, thereby favoring the masking interaction.

The *HAP1-18* mutation does not increase the affinity of HAP1 for the *CYC7* site, but it increases the activity of the HAP1 protein at *CYC7*. We find that *HAP1-18* still needs the carboxy terminal activation domain of the protein for this activity. The simplest interpretation is that the *HAP1-18* mutation in the DNA binding domain of the HAP1 allows the activation domain to be more effective at *CYC7*. According to the above model, the increase in the activity of *HAP1-18* at *CYC7* could be the result of reducing the intramolecular masking that normally reduces HAP1 activity.

When we replaced the HAP1 activation domain with that of GAL4, the HAP1-GAL4 fusion was found to have roughly similar activity at UAS1 and *CYC7*. This contrasts with the results obtained with the native HAP1 protein, which normally activates *CYC1* (by binding to UAS1) much more than it activates *CYC7*. This result therefore supports the hypothesis that the activity of the native HAP1 protein is down-regulated by the *CYC7* site.

HAP2, HAP3, HAP4: A Heteromeric Regulatory Complex

The major form of control at UAS2 of *CYC1* is a 50-fold induction when cells are shifted from glucose to lactate (Guarente et al 1984). The UAS2 site was shown by linker scanning analysis to span some 65 nucleotides and to contain two critical regions (Forsburg & Guarente 1988). One region, responsible for the carbon response of this UAS, is similar to the CCAAT box element found in many mammalian promoters. The wild-type UAS2 contains the sequence CCAAC, and a mutation called *UP1*, which increases the activity of UAS2 10-fold, changes this sequence to CCAAT (Guarente et al 1984). The second UAS2 region is separated by a 7-base spacer from the CCAAT box and has no obvious features.

Using DNA mobility shift gel assays, we identified a DNA binding complex at UAS2, formation of which was dependent upon the *UP1* mutation and enhanced in extracts from lactate grown cells (Olesen et al 1987). By testing extracts from cells expressing size variants of HAP2 or HAP3, we showed that the mobility of the complex shifted with each size variant, and therefore that the complex contained both proteins. Extracts from strains mutant in *HAP2* or *HAP3* had no binding activity. Activity could be recovered by mixing together the extracts prepared from the two mutant strains; thus, HAP2 and HAP3 bind to UAS2 in an interdependent manner (Olesen et al 1987). Footprinting analysis showed that the HAP complex bound to precisely those sequences defined by linker scanning analysis as critical for carbon response (Olesen et al 1987; Forsburg & Guarente 1988). In order to determine whether HAP2 and HAP3 formed a complex apart from the DNA, we fractionated an extract containing bifunctional fusions of HAP2 to lexA, and of HAP3 to β-galactosidase, over four successive columns. UAS2 binding activity (UP1 mutant), lexA-HAP2, and HAP3-β-galactosidase cofractionated at each step, showing that the two HAP proteins are in a complex in solution (Hahn & Guarente 1988).

The *HAP2* and *HAP3* genes were cloned in order to help solve the puzzle of why multiple gene products are required to activate transcription at a single site (Pinkham & Guarente 1985; Hahn et al 1988). Sequence analysis of the *HAP2* gene predicted a 265 residue protein with several interesting features (Pinkham et al 1987). The carboxyl terminal one-third is highly basic, with 28% lys or arg residues, while the rest of the protein contains only about 6% lys or arg. There is also a poly-glutamine tract between residues 120 and 133. By deletional analysis we have found that a short region of the protein encompassing roughly the amino terminal half of the basic domain is active in vivo. Sequence analysis of the *HAP3* gene indicated a 144-residue protein (Hahn et al 1988). Interestingly, there is a 3 kb antisense RNA transcript containing no sizable open reading frame that almost completely overlaps the sense RNA. The 3 kb RNA may be eliminated without impairing HAP3 function. A possible role for this antisense-transcript in regulating *HAP3* expression has yet to be shown.

In a more recent search for additional mutations that reduce expression of a UAS2-driven *CYC1-lacZ* fusion (using the UP1 mutant form of UAS2), additional alleles of *HAP2* and *HAP3* were generated, along with two mutations in a third gene, *HAP4* (S. L. Forsburg & L. Guarente, in preparation). Like *hap2*$^-$ and *hap3*$^-$ mutants, the *hap4*$^-$ mutants failed to grow on nonfermentable carbon sources and displayed no UAS2 binding activity. DNA mobility shift analysis with extracts containing HAP4 size variants showed that this protein is a third subunit of the UAS2 binding complex. Inspection of the *HAP4* sequence reveals a 554-residue protein with a highly acidic region toward the carboxyl terminus. There is no such domain in either HAP2 or HAP3. Deletion of this acidic region abolishes the function of the protein in vivo, and replacing this domain with the activation domain of the yeast GAL4 regulator restores function. These data suggest that HAP4 has two functional domains: a carboxy terminal region that is the principal activation domain for the entire HAP2/3/4 complex, and an amino terminal region that is required for assembly of the DNA binding complex (S. L. Forsburg & L. Guarente, in preparation). The role of HAP4 is thus analogous to that of vp-16 of herpes simplex virus, which forms a complex with cellular DNA binding proteins and provides an acidic activation domain to the complexes (Gerster & Roeder 1988; Preston et al 1988; Treizenberg et al 1988; McKnight et al 1987; O'Hare et al 1988).

HAP4 differs from HAP2 and HAP3 in one other important respect. While *HAP2* and *HAP3* are constitutively transcribed, *HAP4* transcription is induced when cells are shifted from glucose to lactate (S. L. Forsburg & L. Guarente, in preparation). Thus in a sense, HAP4 is the regulatory subunit of the complex. How much of the regulation at UAS2 is due to changes in HAP4 levels remains to be determined.

WHY HETEROMERIC ACTIVATION COMPLEXES? Since discovery of the HAP2/3/4 regulatory complex, it has become clear that regulatory protein complexes are a feature of many eukaryotic transcriptional activators. In a particularly interesting example, the mammalian factor AP1 was found to form a complex with the oncogene product *fos*. This interaction occurs when the *fos* gene is induced by serum growth factors and results in a *fos*-AP1 heteromer that has altered binding properties at the AP1 DNA site (Halazonetis et al 1988; Kouzarides & Ziff 1988; Nakabeppu et al 1988; Chiu et al 1988).

What could be the *raison d'etre* of a cellular complex such as HAP2/3/4? Any transcriptional activator must have the ability to bind to a specific DNA sequence, activate transcriptions, and respond to regulatory signals. In many cases these three properties all reside in a single polypeptide. By dividing them into separate polypeptide chains, the cell can employ different combinations of subunits to gain an added range of control from a given repertoire of proteins. For example, we envision that there are genes that are regulated by a subset of the HAP2, HAP3, and HAP4 polypeptides. Consistent with this view, *hap2*$^-$ and *hap3*$^-$ mutants have a poor ability to utilize ammonia as a nitrogen source, while *hap4*$^-$ mutants are like wild-type in this respect (J. T. Olesen, personal communication). We imagine that there are genes involved in ammonia utilization that are activated by a complex that includes HAP2 and HAP3 but not HAP4. There might even be a HAP4 counterpart in such a complex that is regulated by nitrogen source rather than carbon source.

As discussed above, there is some evidence that the various functions of the HAP2/3/4 complex are distributed among the individual subunits. HAP4 apparently provides the major activation domain for the complex. There is recent genetic evidence that the functional core region of HAP2 is involved in recognition of UAS2 (J. T. Olesen, personal communication). We do not believe that this region is sufficient for recognition, however, since *hap3*⁻ or *hap4*⁻ mutant extracts do not bind. In one model, the core of HAP2 recognizes one of the two critical regions in UAS2. The amino terminus of HAP4 may contact the other region. In this view, HAP3 would be a linchpin for the complex, holding the parts together. This mechanism would allow other HAP4 counterparts that interact with HAP2 and HAP3 to dictate binding to different classes of sequence elements encoding UASs for other gene sets.

TRANSCRIPTIONAL MECHANISMS ARE CONSERVED IN EUKARYOTES

In the past year, a great deal of new information has appeared indicating that the fundamental process of transcription and its regulation are conserved in mechanistic detail in all eukaryotes. Several different kinds of experiments, to be summarized below, illustrate that this conservation includes the basic transcriptional machinery as well as specific activators of transcription that act at a distance. In addition, the mechanism by which activators stimulate the transcription machinery, although not well understood, has also been conserved in all eukaryotes.

The first indication that there is conservation in the basic transcription machinery came with the sequences of the large subunit of RNA polymerase from yeast, *Drosophila*, and mammals. All three proteins had substantial homology throughout their sequences. This homology extended to the beta-prime subunit of *E. coli* RNA polymerase. However, unlike the subunits of the bacterial enzyme, the three eukaryotic polypeptides all had a heptapeptide repeat at their carboxyl terminus: the mammalian large subunit has the sequence Tyr-Ser-Pro-Thr-Ser-Pro-Ser repeated 52 times while the yeast large subunit has exactly the same sequence repeated 26 times (Allison et al 1985, 1988; Cordon et al 1985; Sweetser et al 1987).

Other transcription factors are essential for RNA polymerase to initiate transcription adjacent to a TATA box, the essential proximal element in eukaryotic promoters. These factors have been separated chromatographically into fractions TFIIA, TFIIB, TFIID, and TFIIE (Sawadago & Roeder 1985a; Dignam et al 1983). The TFIID fraction contains a factor that binds directly to the TATA box (Sawadago & Roeder 1985b; Nakajima et al 1988). The binding of TFIID initiates an ordered pathway of assembly of the other factors along with RNA polymerase to build the active initiation complex (Buratowski et al 1989).

Recent experiments have shown that yeast has a factor that will substitute for the mammalian TFIID in an in vitro transcription reaction. Using this transcription assay the protein has been purified and identified as a 25-kd monomer that will bind specifically to TATA box elements of various eukaryotic promoters (Buratowski et al 1988; Cavallini et al 1988). The key transcription factor TFIID is therefore functionally conserved in eukaryotes that range from yeast to mammals. More recently the transcription factor TFIIA from yeast was likewise shown to substitute for the mammalian TFIIA (S. Hahn & L. Guarente, in preparation). The conclusion from these types of experiments is that there is functional conservation in eukaryotes in the protein-protein and protein-DNA interactions of many, if not all, of the basic transcription factors.

In addition to the TATA box element, eukaryotic promoters have other elements that stimulate transcription initiation. As discussed in the previous section, sites called UASs or enhancers are bound by specific proteins that activate transcription and that themselves interface with physiologic signals to the cell. Transcriptional activators contain domains that bind specific DNA sequences and, in many cases, separate domains that activate transcription. The activation domains for numerous activators, including several yeast activators, consist of regions very rich in acidic amino acids (Hope & Struhl 1986; Ma & Ptashne 1987; Giniger & Ptashne 1987).

Interestingly, the yeast activator GAL4 has been shown to activate transcription in higher cells, while mammalian activators such as steroid hormone receptors will function in a hormone responsive manner in yeast (Guarente 1988; Kakidani & Ptashne 1988; Webster et al 1988; Fischer et al 1988; Ma et al 1988; Schena & Yamamoto 1988; Metzger et al 1988). Thus the mechanism of activation, which must involve some form of interaction between activation domains and the basic transcription machinery, has been conserved in eukaryotes that range from yeast to mammals.

Further, since hormonal regulation of the steroid receptors is preserved in yeast, factors governing this form of control must also be conserved. Recent data suggest that one of these factors could be the heat shock factor hsp90, which binds to the receptor in the absence of hormone, thereby inhibiting DNA binding (Pratt et al 1988). The hormone binding domain thus acts on an internal repression sequence in the absence of hormone (Picard et al 1988). The cellular machinery needed to mediate hormonal induction, which could include hsp90, must be conserved from yeast to mammals. While yeast has no proteins that are obvious counterparts of the hormone receptors, the activator HAP1 shares some of the features of this control mechanism. As detailed above, an internal sequence in the protein blocks DNA binding in the absence of the inducer, heme

(Pfeifer et al 1989). Whether this region of HAP1 actually binds heme or interacts with the yeast HSP90 protein in the absence of heme has not yet been determined.

Beyond this conservation in the basic transcription factors and in the activation mechanism, there is conservation in the transcriptional activators themselves and the DNA sequences that they recognize. For example, the mammalian transcriptional activator AP1, which is the cellular homologue of the mammalian oncogene *jun*, binds to the DNA sequence TGACTCA (Angel et al 1988; Lee et al 1987; Maki et al 1987). The yeast activator GCN4 binds to the same DNA sequence and actually has significant sequence similarity in its DNA binding domain to *jun* (Vogt et al 1987).

In a second example, the yeast HAP2/3/4 complex binds to UAS2 of *CYC1*, a key part of which is the sequence CCAAT. This is also an important sequence element in many mammalian promoters (Gluzman 1985; Breathnach & Chambon 1981; McKnight & Tjian 1986). HeLa cells contain several factors that each bind to a subset of CCAAT sites in a variety of mammalian promoters that differ in their flanking sequences (Dorn et al 1988; Santoro et al 1988; Chodosh et al 1988a). As determined by methylation interference footprinting, one of these factors, CP1, makes identical major groove contacts to the adenovirus major late promoter to those made by the HAP2/3/4 complex at UAS2. A further analogy between the HAP2/3/4 complex and CP1 is that the latter also appears to be a heteromeric complex (Chodosh et al 1988a). CP1 may be separated chromatographically into two fractions that must be combined to give binding activity. In the most striking example of the similarity between CP1 and the HAP complex, the CP1A chromatographic fraction will restore activity to an extract from a *hap3*$^-$ mutant, and CP1B will restore binding activity to an extract from a *hap2*$^-$ mutant (Chodosh et al 1988b). Therefore, CP1A contains a functional equivalent of HAP3 while CP1B contains a functional equivalent of HAP2. This means that during more than a billion years of evolution the ability of individual subunits of the heteromeric complex to bind to one another, and the ability of the complex so formed to recognize the CCAAT box, have been conserved. In this experiment, both *hap2*$^-$ and *hap3*$^-$ extracts would provide HAP4, therefore making it impossible to determine which HeLa fraction might contain a counterpart of this subunit.

The above example of the CCAAT binding complex and *jun*/AP1 indicate that fundamental rules governing the interaction of specific gene activators with their binding sites were put in place in an ancestral eukaryote and have remained invariant. In the example of the CCAAT complex, not only the interaction of the factor with the DNA but also the protein-

protein interactions in the complex have been conserved over evolution. The one feature of these activators that has changed is the way in which they interact with and respond to physiologic signals. While the yeast HAP2/3/4 complex is tightly regulated by carbon source, the mammalian CCAAT factors are apparently constitutive activators. Likewise, while GCN4 activates in response to amino acid starvation, AP1 is a constitutive factor that can be further activated by forming a complex with the oncogene product *fos* in response to serum factors.

One interesting link between the HAP2/3/4 complex and GCN4 is that they are both global regulators in a yeast cell. The HAP2/3/4 complex regulates many genes that encode cytochromes, heme biosynthetic enzymes, and other functions involved in respiration. GCN4 regulates many amino acid biosynthetic pathways (Hinnebusch 1988). Other more specialized but well-studied yeast activators such as GAL4 and HAP1 do not yet have identified counterparts in higher cells.

One is led to speculate that a relatively small number of transcriptional activators that serve critical functions in the cell have been retained over evolution. A possible explanation could be that if a single DNA binding protein recognizes a relatively large number of sites sprinkled around the genome, the cell is unlikely to tolerate a change in the protein that alters its DNA interaction. This constraint is similar to that affecting the isoacceptor tRNAs and codons, making the genetic code invariant. More specialized gene activators, which bind to only a few sites, may have more freedom to diverge. Of course, an alternative proposal is that counterparts of the yeast global activators have been found in higher cells because they are more abundant, and have therefore been among the first DNA binding proteins identified biochemically. Future work may uncover many more factors that are evolutionarily related to yeast regulators.

SUMMARY AND PERSPECTIVE

Since the early studies on gene regulation in the *lac* operon of *E. coli*, it has been clear that transcriptional regulators are linked to specific signal transduction pathways that govern control of their target genes. Transcriptional activators of the yeast nuclear cytochrome genes are no exception. Oxygen and carbon sources serve as key regulators of these transcriptional activators. In this system, an additional layer of complexity is added by the need for communication between two different cellular compartments, the nucleus and the mitochondria. This need is met, in part, by signals that emanate from the mitochondria, such as heme, and transmit their effects to the nucleus. The nucleus, in turn, encodes specific regulators of mitochondrial gene expression and components of the mitochondrial RNA processing machinery that are imported into that organelle. At least some of these nuclear gene products are likewise regulated by the oxygen and carbon source signals. Thus communication between the two compartments is achieved and coordinated regulation of mitochondrial biogenesis occurs. Our understanding of how the nucleus and mitochondria communicate is still fragmentary, however. We can look forward to an exciting period of findings on this unique interaction between two separate genetic systems housed in the same cell.

Literature Cited

Akins, R. A., Lambowitz, A. M. 1987. A protein required for splicing Group I introns in *Neurospora* mitochondria is mitochondrial tyrosyl-tRNA synthase or a derivative thereof. *Cell* 50: 331–45

Allison, L. A., Moyles, M., Shales, M., Ingles, C. J. 1985. Extensive homology among the largest subunits of eukaryotic and prokaryotic RNA polymerase. *Cell* 42: 599–610

Allison, L. A., Wang, J. K.-C., Fitzpatrick, V. D., Moyle, M., Ingles, C. J. 1988. The C-terminal domain of the largest subunit of RNA polymerase II of *Saccharomyces cerevisiae*, *Drosophila melanogaster*, and mammals: a conserved structure of essential function. *Mol. Cell. Biol.* 8: 321–29

Angel, P., Allegretto, E. A., Okino, S., Hattori, K., Boyle, W. J. 1988. Oncogene jun encodes a sequence specific trans-activator similar to AP-1. *Nature* 332: 166–71

Attardi, G., Schatz, G. 1988. Biogenesis of mitochondria in yeast. *Annu. Rev. Cell Biol.* 4: 289–333

Benne, R., van den Burg, J., Brackenjoff, J., Sloof, P., vanBoom, J. H., et al. 1986. Major transcription of the frame shifted coxII gene I from trypanosome mitochondria contains four nucleotides that are not encoded in the DNA. *Cell* 46: 819–46

Boker-Schmitt, E., Francisci, S., Schweyen, R. J. 1982. Mutations releasing mitochondrial biogenesis from glucose repression in *Saccharomyces cerevisiae*. *J. Bact.* 151: 303–10

Breathnach, R., Chambon, P. 1981. Organization and expression of eukaryotic split genes coding for proteins. *Annu. Rev. Biochem.* 50: 349–53

Buratowski, S., Hahn, S., Guarente, L., Sharp, P. A. 1989. Five intermediate complexes in transcription initiation by RNA polymerase II. *Cell* 56: 549–61

Buratowski, S., Hahn, S., Sharp, P. A., Guarente, L. 1988. Function of a yeast TATA element binding protein in a mammalian transcripion system. *Nature* 334: 37–42

Burkl, G., Denmer, W., Holzer, H., Schweizer, E. 1976. Temperature sensitive nuclear petite mutants of *Saccharomyces cerevisiae*. In *Genetics, Biogenesis and bioenergetics of mitochondria*, ed. W. Bandlow et al. pp. 39. Berlin: de Gruyter

Carlson, M., Osmond, B. C., Botstein, D. 1981. Mutants of yeast defective in sucrose utilization. *Genetics* 98: 25–40

Carlson, M., Osmond, B. C., Neigeborn, L., Botstein, D. 1984. A suppressor of *snf1* mutations confers constitutive high level invertase synthesis in yeast. *Genetics* 107: 19–32

Cavallini, B., Huet, J., Plassat, J.-L., Sentenac, A., Egly, J. M., Chambon, P. 1988. A yeast activity can substitute for the HeLa cell TATA box factor. *Nature* 334: 77–80

Celenza, J. L., Carlson, M. 1986. A yeast gene that is essential for release from glucose repression encodes a protein kinase. *Science* 233: 1175–80

Chang, D. D., Clayton, D. A. 1987. A mammalian mitochondrial RNA processing activity contains nucleus-encoded RNA. *Science* 235: 1178–84

Chiu, R., Boyle, W. J., Meek, J., Smeal, T., Hunter, T., Karin, M. 1988. The c-Fos protein interacts with c-Jun/AP-1 to stimulate transcription of AP-1 responsive genes. *Cell* 54: 541–52

Chodosh, L. A., Baldwin, A. S., Carthew, R. W., Sharp, P. A. 1988a. Human CCAAT binding proteins have heterologous subunits. *Cell* 53: 11–24

Chodosh, L. A., Olesen, J., Hahn, S., Baldwin, A. S., Guarente, L., Sharp, P. A. 1988b. A yeast and human CCAAT binding protein have heterologous subunits that are functionally interchangeable. *Cell* 53: 25–35

Ciriacy, M. 1977. Isolation and characterization of yeast mutants defective in intermediary carbon metabolism and in catabolite derepression. *Mol. Gen. Genet.* 129: 329–35

Ciriacy, M. 1978. A yeast mutant with glucose-resistant form of mitochondria enzymes. *Mol. Gen. Genet.* 154: 213–20

Clavilier, L., Aubert, G. P., Somlo, M., Slonimski, P. P. 1976. Reseau d'interactions entre des genes non lies: regulation synergique ou antagoniste de la synthese de l'iso-1-cytochrome *c*, de l'iso-2-cytochrome *c*, et du cytochrome b2. *Biochimie* 58: 155–72

Cordon, J. L., Cadena, D. L., Ahearn, J. M., Dahmus, M. E. 1985. A unique structure at the carboxyl terminal of the largest subunit of the eukaryotic RNA polymerase II. *Proc. Natl. Acad. Sci. USA* 82: 7934–38

Costanzo, M. C., Fox, T. D. 1986. Product of *Saccharomyces cerevisiae* nuclear gene *PET494* activates translation of a specific mitochondrial mRNA. *Mol. Cell. Biol.* 6: 3694–703

Costanzo, M. C., Fox, T. D. 1988. Specific translational activation by nuclear gene product occurs in the 5′ untranslated leader of a yeast mitochondrial mRNA. *Proc. Natl. Acad. Sci. USA* 88: 2677–681

Cumsky, M. G., Ko, C., Trueblood, C. E., Poyton, R. O. 1985. Two nonidentical forms of subunit V are functional in yeast cytochrome *c* oxidase. *Proc. Natl. Acad. Sci. USA* 82: 2235–39

Cumsky, M. G., Trueblood, C. E., Ko, C., Poyton, R. O. 1987. Structural analysis of 2 genes encoding divergent forms of cytochrome *c* oxidase subunit V. *Mol. Cell. Biol.* 7: 3511–19

Dieckmann, C. L., Tzagaloff, A. 1985. Assembly of the mitochondrial membrane system. *CBP6*, a yeast nuclear gene necessary for synthesis of cytochrome *b*. *J. Biol. Chem.* 260: 1513–20

Dihanich, M. E., Najarian, D., Clark, R., Gillman, E. C., Martin, N. C., Hopper, A. K. 1987. Isolation and characterization of MOD5, a gene required for isopentylation of cytopasmic and mitochondrial tRNAs of *Saccharomyces cerevisiae*. *Mol. Cell. Biol.* 7: 177–84

Dignam, J. D., Martin, P. L., Shastry, B. S., Roeder, R. G. 1983. Eukaryotic gene transcription with purified components. *Meth. Enzymol.* 104: 582–98

Dorn, A., Bellekens, J., Stauls, A., Benoist, C., Mathis, D. 1988. A multiplicity of CCAAT-binding proteins. *Cell* 50: 863–72

Dujon, B. 1981. Mitochondrial genetics and functions. In *Molecular Biology of the Yeast Saccharomyces*, eds. J. S. Strathern, E. W. Jones, J. R. Broach, pp. 505–651. Cold Spring Harbor: Cold Spring Harbor Lab.

Ebner, E., Mennucci, L., Schatz, G. 1973. Mitochondrial assembly in respiration-deficient mutants of *Saccharomyces cerevisiae*: I. Effect of nuclear mutations on mitochondrial protein synthesis. *J. Biol. Chem.* 248: 5360–68

Ephrussi, B., Slonimski, P. P. 1950. La synthese adaptive des cytochromes chez la levure de boulangerie. *Biochim. Biophys. Acta* 6: 256–67

Faye, G., Simon, M. 1983. Analysis of a yeast nuclear gene involved in the maturation of mitochondrial premessenger RNA of the cytochrome *c* oxidase subunit I. *Cell* 32: 77–87

Feagin, J. E., Jasmer, D. P., Stuart, K. 1987. Developmentally regulated addition of nucleotides within the apocytochrome *b* transcripts in *Trypanosoma brucei*. *Cell* 49: 337–45

Feagin, J. E., Shaw, J. M., Simpson, L., Stuart, K. 1988. Creation of AUG initiation codons by addition of uridines within cytochrome *b* transcripts of kinetoplastids. *Proc. Natl. Acad. Sci. USA* 85: 539–43

Fearon, K., Mason, T. L. 1988. Structure and regulation of a nuclear gene in *Saccharomyces cerevisiae* that specifies MRP7, a protein of the large subunit of the mitochondrial ribosome. *Mol. Cell. Biol.* 8: 3636–46

Fischer, J. A., Giniger, E., Maniatis, T., Ptashne, M. 1988. GAL4 activates transcription in *Drosophila*. *Nature* 332: 853–56

Forsburg, S. L., Guarente L. 1988. Mutational analysis of upstream activation site 2 of the *Saccharomyces cerevisiae CYC1* gene: a *HAP2-HAP3* responsive site. *Mol. Cell. Biol.* 8: 647–54

Gerster, T., Roeder, R. G. 1988. A herpes virus trans-activating protein interacts with transcriptional factor OTF-1 and other cellular proteins. *Proc. Natl. Acad. Sci. USA* 85: 6347–51

Giniger, E., Ptashne, M. 1987. Transcription in yeast is activated by a putative amphipathic a-helix linked to a DNA binding domain. *Nature* 330: 670–72

Gluzman, Y. 1985. In *Eukaryotic transcription: the role of cis and trans-acting elements in initiation*, ed. Y. Gluzman. Cold Spring Harbor: Cold Spring Harbor Lab. 200 pp.

Gollub, E. G., Dayan, J. 1985. Regulation by heme of synthesis of cytochrome *c* oxidase subunits V and VII in yeast. *Bioch. Biophys. Res. Comm.* 1128: 1447–54

Gollub, E. G., Liu, K. P., Dayan, J.,

Aldersbergh, M., Sprinson, D. B. 1977. Yeast mutants deficient in heme biosynthesis and a heme mutant additionally blocked in cyclization of 2,3 oxido squalene. *J. Biol. Chem.* 252: 2846–54

Guarente, L. 1987. Yeast Regulatory Proteins. *Annu. Rev. Genet.* 21: 425–52

Guarente, L. 1988. UASs and enhancers: common mechanisms of transcriptional activation in yeast and mammals. *Cell* 52: 303–5

Guarente, L., Lalonde, B., Gifford, P., Alani, E. 1984. Distinctly regulated tandem upstream activation sites mediate catabolite repression of the *CYC1* gene of *S. cerevisiae*. *Cell* 36: 503–11

Guarente, L., Mason, T. 1983. Heme regulates transcription of the *CYC1* gene in *S. cerevisiae* via an upstream activation site. *Cell* 32: 1279–86

Hahn, S., Guarente, L. 1988. Yeast HAP2 and HAP3: transcriptional activators in a heteromeric complex. *Science* 240: 317–21

Hahn, S., Hoar, E., Guarente, L. 1985, Each of three "TATA elements" specifies a subset of the transcription initiation sites at the *CYC1* promoter of *Saccharomyces cerevisiae*. *Proc. Natl. Acad. Sci. USA* 82: 8562–66

Hahn, S., Pinkham, J., Wei, R., Miller, R., Guarente, L. 1988. The *HAP3* regulatory locus of *Saccharomyces cerevisiae* encodes divergent overlapping transcripts. *Mol. Cell. Biol.* 8: 655–63

Halazonetis, T. D., Georgopoulos, K., Greenberg, M. E., Leder, P. 1988. *c*-Jun dimerizes with itself or with *c*-Fos, forming complexes of different DNA binding affinity. *Cell* 55: 917–24

Hamilton, B., Hogbauer, R., Ruis, R. 1982. Translational control of catalase synthesis by heme in the yeast *Saccharomyces cerevisiae*. *Proc. Natl. Acad. Sci. USA* 179: 7609–13

Herbert, C. J., Labouesse, M., Dujardin, G., Slonimski, P. P. 1988. The *NAM2* proteins from *S. cerevisiae* and *S. douglassi* are mitochondrial leucyl-tRNA synthetases, and are involved in mRNA splicing. *EMBO J.* 7: 473–83

Hinnebusch, A. G. 1988. Mechanisms of gene regulation in the general control of amino acid biosynthesis in *Saccharomyces cerevisiae*. *Microb. Rev.* 52: 248–73

Hope, I., Struhl, K. 1986. Functional dissection of a eukaryotic transcriptional activator protein, GCN4 in yeast. *Cell* 46: 885–94

Hortner, H., Ammerer, G., Hartter, E., Hamilton, B., Rytka, J., et al. 1982. Regulation of the synthesis of catalases and iso-1 cytochrome *c* in *Saccharomyces cerevisiae* by glucose, oxygen and heme. *Eur. J. Biochem.* 128: 179–84

Ibrahim, N. G., Stuchell, R. N., Beattie, D. S. 1973. Formation of yeast mitochondrial membranes. I. Effects of glucose on mitochondrial protein synthesis. *Eur. J. Biochem.* 36: 519–27

Johnson, P., McKnight, S. L. 1989. Eukaryotic transcriptional regulatory proteins. *Annu. Rev. Bioch.* 58: In press

Johnston, M. 1987a. A model fungal gene regulatory mechanism: the *GAL* genes of *Saccharomyces cerevisiae*. *Microb. Rev.* 51: 458–76

Johnston, M. 1987b. Genetic evidence that zinc is an essential cofactor in the DNA binding domain of GAL4 protein. *Nature* 328: 353–55

Kakidani, H., Ptashne, M. 1988. GAL4 activates gene expression in mammalian cells. *Cell* 52: 161–67

Keng, T., Guarente, L. 1987. Multiple regulatory systems result in constitutive expression of the yeast *HEM1* gene. *Proc. Natl. Acad. Sci. USA* 84: 9113–17

Kim, K.-S., Rosenkrantz, M. S., Guarente, L. 1986. *Saccharomyces cerevisiae* contains two functional citrate synthase genes. *Mol. Cell. Biol.* 6: 1936–42

Kloekner-Gruissem, B., McEwan, J. M., Poyton, R. O. 1987. Nuclear functions required for cytochrome *c* oxidase biogenesis in *S. cerevisiae:* multiple trans-acting nuclear genes exert specific effects on expression of each of the cytochrome *c* oxidase subunits encoded in mitochondrial DNA. *Curr. Genet.* 12: 311–22

Kouzarides, T., Ziff, E. 1988. The role of the leucine zipper in the fos-jun interaction. *Nature* 336: 646–51

Labbe-Bois, R., Urban-Grimal, D., Volland, C., Camadro, J.-M., Dehoux, P. 1983. About the regulation of Protoheme synthesis in the yeast *Saccharomyces cerevisiae*. In *Mitochondria 1983*. pp. 523–34. Berlin: de Gruyter

Labouesse, M., Dujardin, G., Slonimski, P. P. 1985. The yeast nuclear gene *NAM2* is essential for mitochondrial DNA integrity and can cure a mitochondrial RNA maturase deficiency. *Cell* 41: 133–43

Labouesse, M., Herbert, C. J., Dujardin, G., Slonimski, P. P. 1987. Three suppressor mutations which are a mitochondrial RNA maturase deficiency occur at the same codon in the open reading frame of the nuclear *NAM2*. *EMBO J.* 713–21

Lalonde, B., Arcangioli, B., Guarente L. 1986. A single *Saccharomyces cerevisiae* upstream activation site (UAS1) has two distinct regions essential for its activity. *Mol. Cell. Biol.* 6: 4690–96

Laz, T. M., Pietras, D. F., Sherman, F. 1984.

Differential regulation of duplicated iso-cytochrome *c* genes in yeast. *Proc. Natl. Acad. Sci. USA* 81: 3375

Lee, W., Mitchell, P., Tjian, R. 1987. Purified transcription factor AP-1 interacts with TPA-inducible enhancer elements. *Cell* 49: 741–52

Ljungdahl, P. O., Pennoyer, J. D., Trumpower, B. L. 1986. Purification of cytochrome bc_1 complexes from phylogenetically diverse species by a single method. *Meth. Enzymol.* 126: 181–91

Lowry, C. V., Lieber, R. H. 1986. Negative regulation of the *Saccharomyces cerevisiae ANB1* gene by heme as mediated by *ROX1* gene product. *Mol. Cell. Biol.* 6: 4145–48

Lowry, C. V., Zitomer, R. S. 1984. Oxygen regulation of anaerobic and aerobic genes mediated by common factor in yeast. *Proc. Natl. Acad. Sci. USA* 81: 6129–33

Lowry, C. V., Zitomer, R. S. 1988. *ROX1* encodes a heme-induced repression factor regulating *ANB1* and *CYC7* of *Saccharomyces cerevisiae*. *Mol. Cell. Biol.* 4651–58

Lustig, A., Levens, D., Rabinowitz, M. 1982a. The biogenesis and regulation of the yeast mitochondrial RNA polymerase. *J. Biol. Chem.* 247: 5800–8

Lustig, A., Padmanabon, G., Rabinowitz, M. 1982b. Regulation of the nuclear encoded peptides of yeast cytochrome *c* oxidase. *Bioch.* 21: 309–16

Ma, J., Ptashne, M. 1987. Deletion analysis of GAL4 defines two transcriptional activating segments. *Cell* 48: 847–53

Ma, J., Przibilla, E., Hu, J., Bogorad, L., Ptashne, M. 1988. Yeast activators stimulate plant gene expression. *Nature* 334: 631–33

Mahler, H. R., Lin, C. C. 1978. Molecular events during the release of w-aminolevulinate dehydratase for catabolite repression. *J. Bacteriol.* 135: 54–61

Maki, Y., Bos, T. J., Davis, C., Starbuck, M., Vogt, P. D. 1987. Avian sarcoma virus 17 carries the *jun* oncogene. *Proc. Natl. Acad. Sci. USA* 84: 2848–52

Martin, N. C., Hopper, A. K. 1982. Isopentylation of both cytoplasmic and mitochondrial tRNA is affected by a single nuclear mutation. *J. Biol. Chem.* 277: 10562–65

Marykwas, D. L., Fox, T. D. 1989. Control of the *Saccharomyces cerevisiae* regulatory gene PET494: Transcriptional repression by glucose and translational induction by oxygen. *Mol. Cell. Biol.* 9: 484–91

Mason, T., Little, H. N., van Sickle, C., Guarente, L. 1984. The role of heme and oxygen in the regulation of mitochondrial cytochromes. In *Nucleo-cytoplasmic interactions*, eds. K. Wolf, R. Schweyen, F. Kanudewitz. Berlin: de Gruyter. 185 pp.

Matner, R. R., Sherman, F. 1982. Differential accumulation of cytochromes *c* in processing mutants in yeast. *J. Biol. Chem.* 257: 9811

Mattoon, J. R., Lancashire, W. E., Sanders, H. K., Carvajal, E., Malamud, D. R., et al. 1979. Oxygen and catabolite regulation of hemoprotein biosynthesis in yeast. In *Biochemical and Clinical Aspects of Oxygen*, ed. W. J. Caughey, pp. 421–35. New York: Academic

McGraw, P., Tzagaloff, A. 1983. Assembly of the mitochondrial membrane system. Characterization of a yeast nuclear gene involved in the processing of the cytochrome b pre-mRNA. *J. Biol. Chem.* 258: 9459–68

McKnight, J. L. C., Kristie, T. M., Roizman, B. 1987. Binding of the virion protein mediating a gene induction in Herpes Simplex Virus-1 infected cells to its *cis* site requires cellular proteins. *Proc. Natl. Acad. Sci. USA* 84: 7061–65

McKnight, S. L., Tjian, R. 1986. Transcriptional selectivity of viral genes in mammalian cells. *Cell* 46: 795–805

Metzger, D., White, J. H., Chambon, P. 1988. The human oestrogen receptor functions in yeast. *Nature* 334: 31–36

Michels, C. A., Romanoski, A. 1980. Pleiotropic glucose-repression resistant mutants of *S. carlsbergensis*. *J. Bacteriol.* 143: 674–79

Montgomery, D. L., Leung, D. W., Smith, M., Shalit, P., Faye, G., Hall, B. 1980. Isolation and sequence of the gene for iso-2 cytochrome *c* in *S. cerevisiae*. *Proc. Natl. Acad. Sci. USA* 77: 541–45

Myers, A. M., Crevellone, M. D., Tzagaloff, A. 1987. Assembly of the mitochondrial membrane system. MRP1 and MRP2, two yeast nuclear genes coding for mitochondrial ribosomal proteins. *J. Biol. Chem.* 2626: 2288–97

Myers, A. M., Tzagaloff, A. 1985. MSW: a yeast gene coding for the mitochondrial tryptophanyl tRNA synthetase. *J. Biol. Chem.* 360: 15371–77

Nakajima, M., Horikoshi, M., Roeder, R. G. 1988. Factors involved in specific transcription by mammalian RNA polymerase II: Purification, genetic specificity, and TATA box promoter interactions at TFIID. *Mol. Cell. Biol.* 8: 4028–40

Nakabeppu, Y., Ryder, K., Nathans, D. 1988. DNA binding activities of three murine Jun proteins: stimulation by Fos. *Cell* 55: 907–15

Natsoulis, G., Hilger, F., Fink, G. R. 1986. The HTS1 gene encodes both the cytoplasmic and mitochondrial histidine tRNA synthetases of *S. cerevisiae*. *Cell* 46: 235–43

Neigeborn, L., Carlson, M. 1984. Genes affecting regulation of *SUC2* gene expression by glucose repression in *Saccharomyces cerevisiae*. *Genetics* 108: 845–58

Neigeborn, L., Carlson, M. 1987. Mutations causing constitutive invertase synthesis in yeast: genetic interactions with *snf* mutations. *Genetics* 115: 247–53

O'Hare, P., Goding, C. R., Haigh, A. 1988. Direct combinatorial interactions between a herpes simplex virus regulatory protein and a cellular octamer binding factor mediates virus immediate early gene expression. *EMBO J.* 7: 4231–38

Olesen, J. T., Hahn, S., Guarente, L. 1987. Yeast HAP2 and HAP3 activators both bind to the *CYC1* upstream activation site UAS2 in an interdependent manner. *Cell* 51: 953–61

Pape, L. K., Koerner, T. J., Tzagaloff, A. 1985. Characterization of a nuclear gene (MST1) coding for the mitochondrial threonyl tRNA synthetase. *J. Biol. Chem.* 260: 15632–70

Pape, L. K., Tzagaloff, A. 1985. Cloning and characterization of the gene for the yeast cytoplasmic threonyl tRNA synthetase. *Nucl. Acids Res.* 13: 6171–83

Parikh, V. S., Morgan, M. M., Scott, R., Clements, L. S., Butow, R. A. 1987. The mitochondrial genotype can influence nuclear gene expression in yeast. *Science* 235: 576–80

Partaledis, J. A., Mason, T. L. 1988. Structure and regulation of a nuclear gene in *Saccharomyces cerevisiae* that specifies MRP13, a protein of the small subunit of the mitochondrial ribosome. *Mol. Cell. Biol.* 8: 3647–60

Pfeifer, K. 1988. *Coordinate regulation of cytochrome c genes by heme and HAP1*. PhD thesis. MIT. 250 pp.

Pfeifer, K., Arcangioli, B., Guarente, L. 1987a. Yeast HAP1 activator competes with the factor RC2 for binding to the upstream activation site UAS1 of the *CYC1* gene. *Cell* 49: 9–18

Pfeifer, K., Kim, K.-S., Kogan, S., Guarente, L. 1989. Functional dissection and sequence of the yeast HAP1 activator. *Cell* 56: 291–301

Pfeifer, K., Prezant, T., Guarente, L. 1987b. Yeast HAP1 activator binds to two upstream sites of different sequences. *Cell* 49: 19–27

Picard, D., Salser, S. J., Yamamoto, K. R. 1988. A movable and regulable inactivation function within the steroid binding domain of the glucocorticoid receptor. *Cell* 54: 1073–80

Pillar, T., Lang, B. F., Steinberger, I., Vogt, B., Kaudewitz, F. 1983. Expression of the split gene cob in yeast mitochondrial DNA. Nuclear mutations specifically block the excision of different introns from its primary transcript. *J. Biol. Chem.* 258: 7954–59

Pinkham, J. L., Guarente, L. 1985. Cloning and molecular analysis of the *HAP2* locus, a global regulator of respiratory genes in *Saccharomyces cerevisiae*. *Mol. Cell. Biol.* 5: 3410–16

Pinkham, J. L., Olesen, J. T., Guarente, L. 1987. Sequence and nuclear localization of the *Saccharomyces cerevisiae* HAP2 protein, a transcriptional activator. *Mol. Cell. Biol.* 7: 578–87

Polakis, E. S., Bartley, W., Meek, G. A. 1965. Changes in the activity of respiratory enzymes during the aerobic growth of yeast on different carbon sources. *Bioch. J.* 97: 298–302.

Poutre, C. G., Fox, T. D. 1987. *PET111* a *Saccharomyces cerevisiae* nuclear gene required for translation of the mitochondrial mRNA encoding cytochrome *c* oxidase subunit II. *Genetics* 115: 637–42

Power, S. D., Lochrie, M. A., Sevarino, K. A., Patterson, T. E., Poyton, R. O. 1984. The nuclear coded subunits of yeast cytochrome *c* oxidase. I. Fractionation of the holoenzyme into chemically pure polypeptides and the identification of two new subunits using solvent extraction and reverse phase high pressure liquid chromatography. *J. Biol. Chem.* 259: 6564–70

Pratt, W., Jolly D., Pratt, D., Hollenberg, S., Giguere, V., et al. 1988. A region in the steroid binding domain determines function of the non-DNA binding, 9S glucocorticoid receptor complex. *J. Biol. Chem.* 263: 267–73

Preston, C. M., Frame, M. C., Campbell, M. E. M. 1988. A complex formed between cell components and an HSV structural polypeptide binds to a viral immunoglobulin early gene regulatory sequence. *Cell* 52: 425–34

Prezant, T., Pfeifer, K., Guarente, L. 1987. Organization of the regulatory region of the yeast *CYC7* gene: multiple factors are involved in regulation. *Mol. Cell. Biol.* 7: 3252–59

Richter, K., Ammerer, G., Hartter, E., Ruis, H. 1980. The effect of δ-aminolevulinate on catalase T mRNA levels in δ-aminolevulinate synthase deficient mutants of *Saccharomyces cerevisiae*. *J. Biol. Chem.* 255: 8019–22

Roedel, G. 1986. Two yeast nuclear genes, *CBS1* and *CBS2*, are required for translation of mitochondrial transcripts bearing the 5′ untranslated COB leader. *Curr. Genet.* 11: 41–45

Roedel, G., Fox, T. D. 1987. The yeast nuclear gene *CBS1* is required for translation of mitochondrial mRNAs bearing the cob 5′ untranslated leader. *Mol. Gen. Genet.* 206: 45–50

Roedel, G., Korte, A., Kaudewitz, F. 1985. Mitochondria suppression of a yeast nuclear mutation which affects the translation of the mitochondrial apo-cytochrome *b* transcript. *Curr. Genet.* 9: 641–48

Rothstein, R., Sherman, F. 1980. Genes affecting expression of cytochrome *c* in yeast: genetic mapping and genetic interactions. *Genetics* 94: 871

Saltzgaber-Muller, J. S., Schatz, G. 1978. Heme is necessary for accumulation and assembly of cytochrome *c* oxidase subunits in *Saccharomyces cerevisiae. J. Biol. Chem.* 253: 305–10

Santoro, C., Mermod, M., Andrews, P. C., Tjian, R. 1988. A family of human CCAAT-box binding proteins active in transcription and DNA replication: cloning and expression of multiple cDNAs. *Nature* 334: 218–24

Sawadago, M., Roeder, R. G. 1985a. Factors involved in specific transcription by human RNA polymerase II: Analysis by a rapid and quantitative *in vitro* assay. *Proc. Natl. Acad. Sci. USA* 82: 4394–98

Sawadago, M., Roeder, R. G. 1985b. Interaction of a gene specific transcription factor with the adenovirus major late promoter upstream of the TATA box region. *Cell* 43: 165–75

Schena, M., Yamamoto, K. R. 1988. Mammalian glucocorticoid receptor derivates enhance transcription in yeast. *Science* 241: 965–68

Schneider, J. C. 1989. *Mechanism of coordinate induction of cytochrome genes in Saccharomyces cerevisiae.* PhD thesis. MIT. 208 pp.

Sherman, F. 1963. Respiration deficient mutants of yeast. I: Genetics. *Genetics* 48: 375

Sherman, F., Stewart, J. 1971. Genetics and biosynthesis of cytochrome *c. Annu. Rev. Genet.* 5: 257

Siemens, T. V., Nichols, D. L., Zitomer, R. S. 1980. The effect of mitochondrial functions of the synthesis of yeast cytochrome *c. J. Bacteriol.* 142: 499–507

Somlo, M., Fukuhara, H. 1965. On the necessity of molecular oxygen for the synthesis of respiratory enzymes in yeast. *Bioch. Biophys. Res. Comm.* 19: 587–91

Sweetser, D., Nonet, M., Young, R. 1987. Prokaryotic and eukaryotic RNA polymerase have homologous core subunits. *Proc. Natl. Acad. Sci. USA* 84: 1192–96

Szekely, E., Montgomery, D. L. 1984. Glucose represses transcription of *Saccharomyces cerevisiae* nuclear genes that encode mitochondrial subunits. *Mol. Cell. Biol.* 4: 939–46

Tait, G. H. 1978. The biosynthesis and degradation of heme. In *Heme and Hemoproteins*, eds. F. DeMateis, W. N. Aldridge. pp. 1–48. New York: Springer Verlag

Triezenberg, S. J., LaMarco, K. L., McKnight, S. L. 1988. Evidence of a DNA : protein interaction that mediates HSV 1 immediate early gene activation by vp16. *Genes Dev.* 2: 730–42

Trueblood, C. E., Wright, R. M., Poyton, R. O. 1988. Differential regulation of the two genes encoding *Saccharomyces cerevisiae* cytochrome *c* oxidase subunit V by heme and *HAP2* and *REO1* genes. *Mol. Cell. Biol.* 8: 4537–40

Trumbly, R. S. 1986. Isolation of *Saccharomyces cerevisiae* mutants constitutive for invertase synthesis. *J. Bacteriol* 166: 1123–27

Tzagaloff, A., Myers, A. M. 1986. Genetics of mitochondrial biogenesis. *Annu. Rev. Biochem.* 55: 249–88

van Loon, A. P. G. M., Eilers, M., Baker, A., Verner, K. 1988. Transport proteins into yeast mitochondria. *J. Cell Bioch.* 36: 59–71

Verdiere, J., Creusot, F., Guarente, L. 1986. The overproducing CYP1 mutation and the underproducing hap1 mutations are alleles of the same gene which regulates in *trans* the expression of the structural genes encoding iso-cytochromes *c. Curr. Genet.* 10: 339–42

Verdiere, J., Creusot, F., Guerineau, M. 1985. Regulation of the expression of the iso-2 cytochrome *c* in *S. cerevisiae*: cloning of the positive regulatory gene *CYP1* and identification of the region of its target sequence on the structural gene *CYP3. Mol. Gen. Genet.* 199: 524–33

Vogt, P. K., Box, T. J., Doolittle, R. F. 1987. Homology between the DNA binding domain of the GCN4 regulatory protein of yeast and the carboxy-terminal region of a protein coded for by the oncogene *jun. Proc. Natl. Acad. Sci. USA* 84: 3316–19

Webster, N., Jin, J. R., Green, S., Hollis, M., Chambon, P. 1988. The yeast UAS_G is a transcriptional enhancer in human HeLa cells in the presence of the GAL4 transactivator. *Cell* 52: 169–78

Winkler, H., Adam, G., Mattes, E., Schanz, M., Hartig, A., Ruis, H. 1988. Coordinate control of synthesis of mitochondria and non-mitochondria hemoproteins a binding site of the HAP1 (CYP1) protein in UAS region of the yeast catalase T (CTT) gene. *EMBO J.* 7: 1799–804

Woloszauk, W., Sprinson, D. B., Ruis, H. 1980. Relation of heme to catalase T apoprotein synthesis in yeast. *J. Biol. Chem.* 255: 2624

Woodrow, G., Schatz, G. 1979. The role of oxygen in the biosynthesis of cytochrome *c* oxidase of yeast mitochondria. *J. Biol. Chem.* 254: 6088–93

Wright, C. F., Zitomer, R. S. 1984. A positive regulatory site and a negative regulatory site control the expression of the *Saccharomyces cerevisiae CYC7* gene. *Mol. Cell. Biol.* 4: 2023–30

Zagorec, M., Labbe-Bois, R. 1986. Negative control of yeast coproporphyrinogen oxidase synthesis by heme and oxygen. *J. Biol. Chem.* 261: 2506–9

Zitomer, R. S., Montgomery, D. L., Nichols, D. L., Hall, B. D. 1979. Transcriptional regulation of the yeast cytochrome *c* gene. *Proc. Natl. Acad. Sci. USA* 76: 3627–31

Zitomer, R. S., Sellers, J. W., McCarter, D. W., Hastings, G. A., Wick, P., Lowry, C. V. 1987. Elements involved in oxygen regulation of the *Saccharomyces cerevisiae* CYC7 gene. *Mol. Cell. Biol.* 7: 2212–20

Annu. Rev. Cell Biol. 1989. 5 : 181–96

MOLECULAR GENETICS OF DEVELOPMENT STUDIED IN THE TRANSGENIC MOUSE[1]

Heiner Westphal

Laboratory of Molecular Genetics, National Institute of Child Health and Human Development, National Institutes of Health, Bethesda, Maryland 20892

Peter Gruss

Department of Molecular Cell Biology, Max-Planck Institute of Biophysical Chemistry, 3400 Göttingen, West Germany

CONTENTS

INTRODUCTION

Transgenic technology allows the insertion of defined genetic material into the germline of mice and other organisms. Of all disciplines of the life sciences few stand to profit more from this than developmental biology. Important insights into the molecular genetics of development can be gained either by overproducing factors that regulate or interfere with development in specific target organs of the growing organism, or by mutating resident chromosomal genes that regulate development. While virtually every new phenotype observed in a transgenic mouse also involves changes in some aspect of development, we will deal selectively with broad changes in pattern formation and in organogenesis initiated by genes that have been inserted or changed in the mouse genome. This field of research is very much at its beginning, hence our review will be short. Nonetheless, we hope to elicit in our readers an interest in the enormous potential of these experiments for the understanding of the molecular genetics of mammalian development.

EXPERIMENTAL DESIGNS

A transgenic mouse is characterized by the presence of foreign DNA sequences inserted in the genome by laboratory techniques. A recent review details this approach (Jaenisch 1988). Most commonly, a transcription cassette is constructed and injected into a pronucleus of a one-cell embryo so that it can integrate in the host chromatin and become part of every cell of the growing organism. This includes the germ line, so that the original transgenic mouse will become founder of a transgenic line. Transgenes appear to integrate randomly by illegitimate recombination events. The spatial and temporal control of expression is influenced by *cis*-acting regulatory elements incorporated in the construct or present at the integration site.

An alternative method to introduce new genetic material into mice involves the use of embryonic stem (ES) cells. DNA can be introduced into ES cells in culture, the desired transformants can be selected and placed in recipient blastocysts (see Figure 1). The pluripotent ES cells are capable of colonizing all or most of the tissues of the resulting offspring (reviewed by Bradley & Robertson 1986). If they participate in the formation of germ line chimeras, their specific traits can be propagated. Since this methodology allows a great number of cells to be subjected to selective steps in culture, it holds a great deal of promise for producing mutations at defined loci either by conventional or insertional mutagenesis or by gene replacement via homologous recombination. Initial work focused on a model locus, the gene encoding HPRT or hypoxanthine-guanosine phosphoribosyl transferase. HPRT-deficient mice were produced from ES cells that carried either spontaneous mutations or retroviral inserts in the target gene (Hooper et al 1987; Kuehn et al 1987). The *hprt* gene is located on

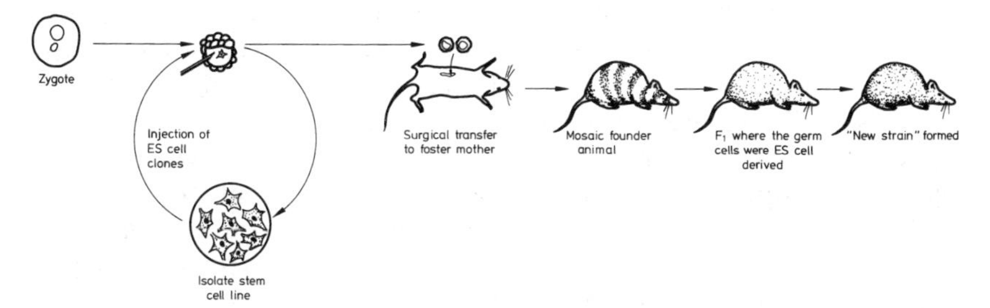

Figure 1 Schematic representation of the generation of transgenic mice using embryonic stem (ES) cells. Pluripotent ES cells can be grown in culture from cells of the inner cell mass of mouse blastocysts (Evans & Kaufman 1981; Martin, 1981; Bradley et al, 1984). Since large numbers of these cells can be transformed with DNA in culture and subjected to selection, even rare events such as homologous recombination can be detected and the respective cell clone can be isolated. Upon introduction of preselected ES cells into blastocysts and surgical transfer of blastocysts into a foster mother, chimeric animals can be obtained. Provided the ES cell has maintained its pluripotency, transmission into the germ line is possible. Animals can be bred to generate homozygous mice.

the X chromosome, hence only one mutant copy is needed in male cells to yield a phenotype that can easily be selected. The initial results using the *hprt* gene as a model established the feasibility of introducing changes in resident chromosomal genes of the mouse. However, most genes of interest are present as diploid copies for which no positive selection procedures exist. We will describe below two different approaches that screen for ES cells that have undergone homologous recombination events in one or another of such nonselectable diploid genes.

PROTO-ONCOGENES AND ONCOGENES

Cellular proto-oncogenes or viral oncogenes are primarily discussed in the context of cancer and not in the context of embryogenesis. However there is a rapidly growing body of information that assigns cellular proto-oncogenes a physiologic role in development. The temporal and spatial control of expression, as much as their nature and quantity, determines the outcome of their action. Viral oncogenes or a nonphysiologic dose of cellular proto-oncogenes interfere with development in characteristic ways that can be assessed in the transgenic mouse.

Several laboratories have begun to examine the effects of targeting the expression of defined viral oncogenes or their cellular equivalents in preselected tissues or organs, notably the eye lens, the pancreas, and the mammary gland (most recently reviewed by Cory & Adams 1988). Inasmuch as these studies are focused on oncogenesis, they will not be discussed here. However, this type of experiment can also register the interaction of a specific oncogene or proto-oncogene product with unknown factors that define a specific cell phenotype in the developing organism. The nature and composition of such factors change during differentiation. Therefore, one may predict that the phenotype of transgenic mice expressing a given oncogene will reflect not only the nature of the target tissue but also its developmental stage at the time of initial exposure to the oncogene product. One of our laboratories has developed a system to test this prediction.

With the help of the αT transgene (a promoter sequence of the murine αA-crystallin gene fused to sequences encoding for the tumor antigens of SV40), we have been able to direct malignant growth to the lens of transgenic mice (Mahon et al 1987). The lens has a simple architecture. At day E11 of gestation, posterior epithelia of the embryonic mouse lens begin to differentiate into crystallin producing fibers that are destined to form the lens nucleus. T. Nakamura et al (in preparation) have recently compared two very different phenotypes of mouse lines carrying the αT transgene. Line αT1 accumulates SV40 tumor antigens (TAG) in E11 lens epithelia prior to primary fiber differentiation. Progeny of this line develop fast growing lens tumors comprised of poorly differentiated cells that are largely deficient in lens crystallin synthesis. By contrast, individuals of line αT2 accumulate TAG at day E12/13 in differentiating fiber cells. Their lens tumors grow very slowly and consist of cells that retain some characteristics of differentiation, including the ability to synthesize significant amounts of crystallins. These observations led us to propose a model that is illustrated in Figure 2. According to this model, the oncogene products immortalize some basic properties of the initial target cell in which they accumulate. In αT1, the target cells are the embryonic lens epithelia. We postulate that these cells are not yet fully committed to differentiation into fiber cells and hence, they produce no or very little β and γ crystallins. Their growth regulation has a higher degree of plasticity, which makes them proliferate fast under the influence of the oncogene product. By contrast, the developing lens fiber cells of αT2 approach terminal differentiation and produce β and γ crystallins. These cells proliferate more slowly in the presence of TAG and maintain their program of β and γ crystallin synthesis. Thus it appears that a powerful transforming activity can be modified by the state of differentiation of the cell.

A quite different approach, designed to assess the role of a proto-oncogene in development, has recently been formulated by Mansour et al (1988), who altered the *int*-2 gene of ES cells by insertional mutagenesis. *Int*-2 is a proto-oncogene that displays a highly restricted pattern of expression during mouse organogenesis (Wilkinson et al 1988), which

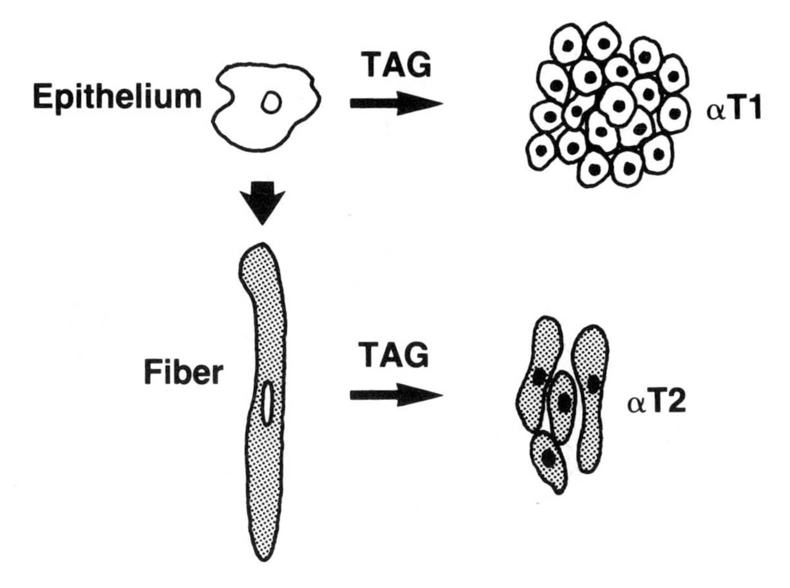

Figure 2 Model of SV40 tumor antigen (TAG) mediated transformation of lens epithelia and fibers in the transgenic mouse. Posterior lens epithelia differentiate (arrowhead) into fibers which begin to synthesize β and γ crystallins (shaded). In αT1 mice, TAG mediated transformation of undifferentiated epithelia leads to rapid proliferation. In αT2 mice, differentiated fiber cells undergo transformation and proliferate slowly. Both the αT1 and αT2 lens cells contain nuclear TAG (indicated by black color).

suggests important functions in mammalian development. Mansour et al developed an elegant method that enabled them to enhance the selection for cells that have undergone homologous recombination at the *int*-2 locus. The bacterial *neo* gene, which confers neomycin resistance, was fused to a promoter active in ES cells. This construct was inserted such that it disrupted the coding sequence of one of the *int*-2 exons. In addition, a herpes virus gene encoding thymidine kinase (HSV-TK) was linked to one end of the linear DNA fragment to serve as a negative selectable marker. Upon introduction of this DNA construct by electroporation, a method that insures the integration of a limited number of gene copies (Doetschman et al 1988), the drug G418 was added to eliminate all cells that did not integrate and express the *neo* gene. Surviving cells were treated with the nucleoside analog gancyclovir (GANC). This drug was originally developed as an antiviral agent because it is cytotoxic for cells containing functional herpes virus thymidine kinase. In an illegitimate recombination event the entire linear DNA fragment is integrated and HSV-TK is expressed, killing the cell in the presence of GANC. During homologous recombination, however, the DNA sequence encoding HSV-TK is lost, and the cell survives GANC treatment. The method enhances the selection at least 2000–fold for the rare homologous combination event. It is hoped, of course, that the ES cell containing the *int*-mutation is able to participate in the formation of germ-line chimeras so that mice expressing the mutant trait can be derived.

The example of mice carrying dominant, spontaneous mutations in the c-*kit* proto-oncogene shows that this approach can lead to interesting developmental phenotypes. Recently c-*kit* has been identified as the product of the *W* locus (Geissler et al 1988). White coat color, sterility, and anemia are characteristic for *W* locus mutants. These phenotypes have been attributed to failure of stem cell populations to migrate and to proliferate properly. The ability to produce defined mutations via homologous recombination will allow the role of proto-oncogenes in mammalian development to be probed in a systematic way. For example, the murine homolog of the c-*mos* proto-oncogene has been linked to meiotic maturation in frog and mouse oocytes (Mutter et al 1988; Sagata et al 1988; R. Paules et al, in preparation). Targeting the chromosomal *mos* gene in the mouse germ line could reveal similar and additional functions of *mos* in mammalian development.

HOMEOBOX, PAIRED BOX, AND FINGER STRUCTURE ENCODING GENES

The assumption that structural conservation is indicative of functional conservation prompted many researchers to screen mammalian genomes for the presence of sequences first discovered in developmental control genes of *Drosophila* (McGinnis et al 1984a,b; Scott & Weiner 1984). To date three gene families classified by common sequence motifs have been identified in the mouse genome. These include the homeobox (*Hox*), paired box (*Pax*), and finger structure (*Zpf*) encoding genes (reviewed by Dressler & Gruss 1988; Holland & Hogan 1988).

First indications that these genes may indeed be involved in mammalian developmental control were obtained by studying their temporal and spatial pattern of expression. The homeobox containing genes are expressed in a tissue and region specific manner, the paired box containing gene, *Pax* 1, is expressed along the entire rostro-caudal axis in segmented structure, and a finger containing gene, *Mkr2/Zpf2*, is expressed specifically in neurons (see Holland & Hogan 1988; Dressler & Gruss 1988, for review).

Investigation of a mouse mutant, undulated (*un*), strongly supported the view that these conserved genes play a role in development. *Un* exhibits abnormalities of the vertebral column beginning during embryonic development (Grüneberg 1954) and the adult has distortions along the entire rostro caudal axis. *Pax* 1 is expressed in the structures affected by the mutation (Deutsch et al 1988; Balling et al 1988). Thus, it was rewarding to find that in *un* mice there is a glycine for serine substitution in the most conserved region of the paired box of *Pax* 1, which suggests that this mutation is causing the developmental defect (Balling et al 1988). Although the identification of *un* as a gene controlling mammalian development was made possible by analyzing a preexisting mouse mutant, it is unlikely that mutants for all the members of the above mentioned gene families will be found. Thus alternative approaches, including the generation of dominant gain of function or loss of function mutations, must be explored.

Gain of Function Mutants

A temporally or spatially misdirected expression during embryogenesis may result in phenotypic alterations and consequently help us to define the role of genes involved in developmental control. Using a large genomic fragment including the putative control signals and the coding region of *Hox* 1.4, Wolgemuth & collaborators (1989) generated several lines of transgenic mice. The rationale was that, by insertion of one or more additional copies of the gene, a finely tuned control mechanism could be thrown out of balance in the tissue in which this gene is normally expressed. Indeed, all transgenic animals overexpressed the *Hox* 1.4 gene in testis as well as in the intestinal tract. In midgestation embryos the level of transgene transcripts was at least fivefold higher, particularly in the gut. Interestingly, a phenotype identified as megacolon developed most likely as a consequence of the activity of the transgene. Mouse mutants exhibiting the megacolon phenotype show a deficiency of myenteric ganglion cells in the colon. These represent neural crest derivatives that migrate into the developing gut (Figure 3).

In a related type of experiment, Balling et al (1989) used a DNA construct in which the β-actin promoter was placed upstream of the *Hox* 1.1 coding region in order to express the *Hox* 1.1 gene at ectopic sites. Several founder animals were obtained. One of them was used to breed transgenic offspring. The transgenic offspring remained small, did not feed and died soon after birth most likely due to a phenotype known as cleft-palate. These mice show a number of craniofacial abnormalities, including open eyes at birth, non-fused pinnae, and cleft secondary palate. This phenotype resembles the effects observed after application of retinoic acid during pregnancy (retinoic acid embryopathy). Because of the similarity in phenotypes, it has been suggested that cranial neural crest cells that participate

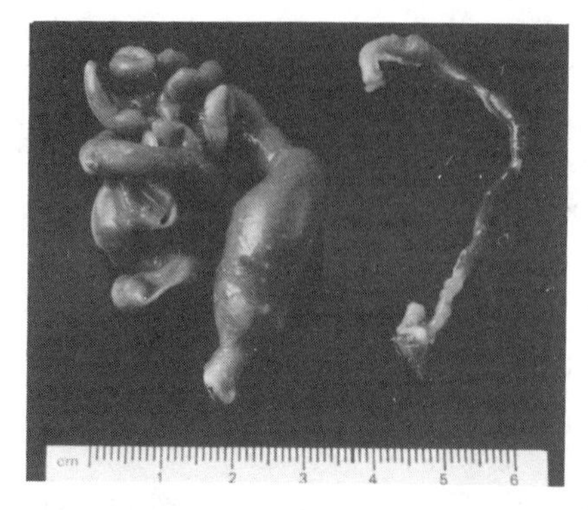

Figure 3 Mouse mutants exhibiting the megacolon phenotype and showing a deficiency of myenteric ganglion cells in the colon. (With permission, *Nature* 337: 464–67 copyright © 1989. Macmillan Ltd.)

in the formation of the distorted structures may represent a target of *Hox* 1.1 ectopic expression (Balling et al 1989). Again, the phenotype could be correlated with the activity of the transgene. In parallel experiments using the histocompatibility gene H2 promoter to activate the *Hox* 1.1 gene, no transgenic offspring have been obtained (R. Balling et al, personal communication), which suggests lethal effects due to the ectopic expression of *Hox* 1.1 during embryogenesis.

Loss of Function Mutants

Certain mutant gene products can exert dominant negative effects on development by competing with the wild-type product. Resulting phenotypes in which gene function is blocked at the protein level were first described as antimorphs (for review, see Herskowitz 1987). The possibility to supply mutant gene products via transgene insertion and thereby interfere with development in the mouse was realized by Stacey et al (1988). These authors aimed at generating a mouse model of a human genetic disease, *osteogenesis imperfecta* type II, which is associated with substitutions of single glycine residues in $\alpha 1$(I) collagen. A mutant collagen gene was inserted in the germ line of transgenic mice. As little as 10% mutant gene expression in these mice can disrupt normal collagen function, presumably by interfering with the formation of the multimeric structure of genuine collagen. The resulting dominant lethal phenotype with its severe skeletal abnormalities resembles the human disease. In another example of dominant negative gene regulation on the protein level, mutant forms of a viral transcriptional activator have recently been shown to interfere with the action of their wild-type counterpart (Friedman et al 1988).

By analogy, the role of genes encoding putative transcription factors, such as homeobox containing genes, could be probed by expression of a nonfunctional equivalent gene. The products of these genes may be involved in a hierarchy of transcriptional regulation, which ultimately establishes the embryo pattern. Expression of a mutant form of one of the members in the hierarchy should lead to a disturbance in pattern formation (see Levine & Hoey 1988 for review). Similarly, transgenes encoding mutant products of a wide variety of mouse genes may eventually be expressed in the embryo in order to assess the function of their wild-type corollaries during development.

Dominant loss of function can also be induced by antisense RNA. RNA complementary to messenger RNA has been shown to repress the expression of specific genes in mammalian cells (Weintraub et al 1985; Melton 1985; Izant & Weintraub 1985) and in *Drosophila* embryos where it induces a specific mutant phenotype (Rosenberg et al 1985). Thus, antisense expression in transgenic mice provides a tool to functionally inactivate a target gene and thereby possibly induce a mutant phenotype. This approach has been used by Katsuki et al (1988) to convert normal behavior to shiverer (*shi*) by antisense expression in transgenic mice. Mice homozygous for the mutation show a violent shiver when disturbed. This phenotype is due to a deficiency in myelin basic protein (mbp) (Roach et al 1983; Kimura et al 1985; Molineaux et al 1986). Katsuki and coworkers (1988) consequently inserted a myelin basic protein cDNA into an expression vector in the reverse orientation. This expression vector promotes transcription from the mbp gene promoter, carries a splicing signal from the rabbit β globin gene, and a polyadenylation signal from SV40. Several transgenic offspring of a founder transgenic mouse were converted from the normal to the *shi* phenotype. It was demonstrated that antisense messenger RNA was expressed in these mutant mice, which allowed the conclusion that the phenotypic alterations probably were the result of expression of antisense mbp RNA. In a related type of experiment Strickland & coworkers (1988) showed that the synthesis of tissue plasminogen activator can be blocked by injection of cognate antisense RNAs into maturing oocytes.

Finally, loss of function of developmental control genes can be achieved using targeted mutagenesis. We have already discussed this approach in the context of proto-oncogenes. One of our laboratories has targeted the *Hox* 1.1 gene of the mouse ES cell. An oligonucleotide linker, 20-base pairs long, was inserted into the coding region of the second exon of *Hox* 1.1 (Kessel et al 1987) causing a frame-shift mutation (Zimmer & Gruss 1989). The construct was placed by microinjection (Capecchi 1980) directly into the nucleus of ES cells. After a few growth cycles, the cells were screened with the help of the PCR, or polymerase chain reaction (Saiki et al 1985). This highly sensitive technique can detect an event of homologous recombination even if it occurs in only one of a pool of cells (Kim & Smithies 1988). For the PCR reaction, an oligonucleotide identical to the one we inserted in the *Hox* 1.1 gene was used as primer 1. Primer 2 is an oligonucleotide that is complementary to a sequence located directly upstream of the target site of homologous recombination (Figure 4). Amplification of the sequence between primer 1 and primer 2 leading to a fragment of the expected size will occur only in the case of a homologous recombination event. Using this protocol, 1 of 150 injected cells was shown to carry a homologously recombined mutated allele. This remarkably high frequency readily allowed the subcloning of the respective cell. In the future, we foresee a variation of this protocol. Instead of the introduction of the oligonucleotide, the insertion could equally well be designed to produce a single point exchange, an approach that should also be suitable for somatic gene therapy. Currently, many laboratories are attempting to utilize ES cell lines in order to generate chimeric mice. So far, several groups have succeeded in producing animals that are mosaics and thus descended from both the foster blastocyst and the ES cells introduced. Although previous experiments have demonstrated that ES cells trans-

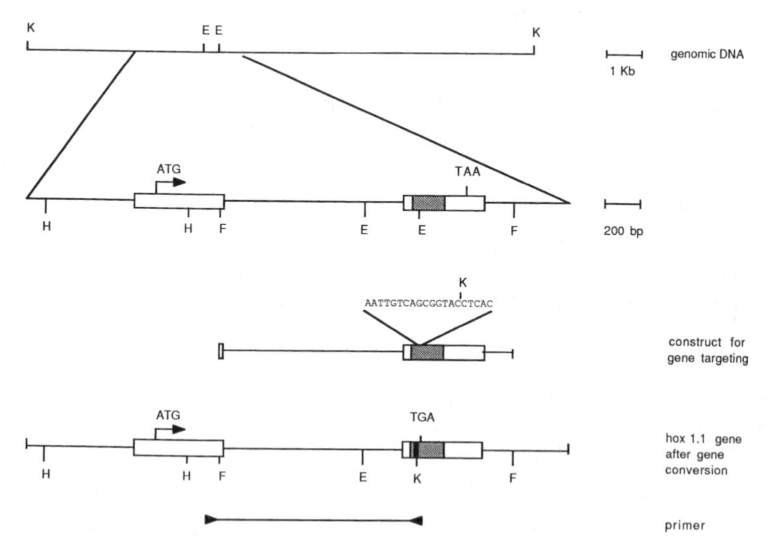

Figure 4 Schematic representation of the mutagenesis of the *Hox* 1.1 gene and its analysis by PCR. A 20 bp oligonucleotide was inserted into the Eco RI (E) site of a Fsp I (F) fragment and subsequently used for microinjection into ES cells as a construct for gene targeting. The successful homologous recombination event was monitored using PCR; as primer 1 the 20 bp oligonucleotide, and as primer 2 a sequence specific for the *Hox* 1.1 site located upstream of the Fsp I fragment was used (Zimmer & Gruss 1989).

fected in vitro can be transmitted through the germ line, there is so far only one report on germ line transmission of ES cells carrying a gene altered by homologous recombination (Thompson et al, 1989). Based on previous experience, however, it can only be a question of time until this promising method will enable researchers to introduce mutations into a variety of genes of the animal and to analyze their effect on developmental control.

INSERTIONAL MUTAGENESIS

Insertion of transgenes in chromosomal DNA can cause mutations if functions of essential genes are disrupted. Typically, these mutations are recessive and result in noticeable phenotypes with an estimated frequency of 5–10% (Jaenisch 1988). They often interfere with development and, in severe cases, cause embryonic lethality. One well-characterized mouse mutant, Mov 13, carries a retroviral insert in the first intron of the α1(I)collagen gene. This mutant is being utilized to investigate the role of collagen in development (reviewed by Jaenisch 1988).

Sometimes transgene insertion brings about defects that are allelic with known spontaneous mutants. Examples include *ld* or limb deformity and *dt* or *dystonia musculorum.* The *ld* mutation in mice homozygous for transgene insertion affects genes involved in the regulation of organogenesis. Development of transgenic *ld* limb buds is impaired as early as day E10 of embryogenesis. Kidney development is also affected. Using the transgene fortuitously integrated in *ld*, genomic sequences were cloned and used to characterize *ld* cDNAs. Transcription of the normal *ld* gene begins as early as the egg cylinder stage, and transcripts are also found in adult tissues. One of these transcripts contains an open reading frame for a protein of 1468 amino acids. Transcription is quantitatively and qualitatively altered in *ld* mutants, as would be expected from an insertion event that interrupts *ld* gene function (Woychik et al 1985 and in preparation).

Homozygous individuals of the transgenic mouse line allelic to *dt* lose limb coordination shortly after birth and die young. The histological picture shows a specific loss of dorsal root sensory axons along the spinal cord. The fact that the transgene interrupting the *dt* locus contains a *lacZ* transcriptional cassette was utilized to obtain stained preparations of the neural tube, which showed β-galactosidase activity beginning at day E9.5 in the rostral-most region, and involving the entire length of the neural tube by day E13.5. Thus it is likely that the transgene has come under the control of *cis*-acting regulatory sequences of the *dt* locus (Kothary et al 1988). With the help of the inserted *lacZ* sequence, it should therefore be possible to characterize the *dt* gene at the molecular level.

Following work in other systems, Allen et al (1988) have utilized the excellent properties of *lacZ* as an in situ tracer of gene activity in their design of a general method aimed at detecting genes that are involved in mammalian organogenesis and pattern formation. They microinjected into mouse embryos a *lacZ* transcription cassette that lacks active *cis*-acting elements conferring strong spatial or temporal control of expression. If this cassette happens to integrate at a site controlled by a strong promoter that is active in specific cell systems of the developing organism, *lacZ* will also be expressed in these cells from the allele in which it integrated. Indeed, sections obtained from transgenic embryos revealed unique patterns of β-galactosidase staining that were symmetrically arranged, and affected organs and tissues that differed from one integration event to the other. The result is best explained by postulating that the transgene integrated into one of a number of chromosomal regions actively expressed during development. The *lacZ* reporter gene thus came under the control of hitherto unknown genes involved in the regulation of pattern formation. Such genes can now be traced with the help of the integrated *lacZ* sequence.

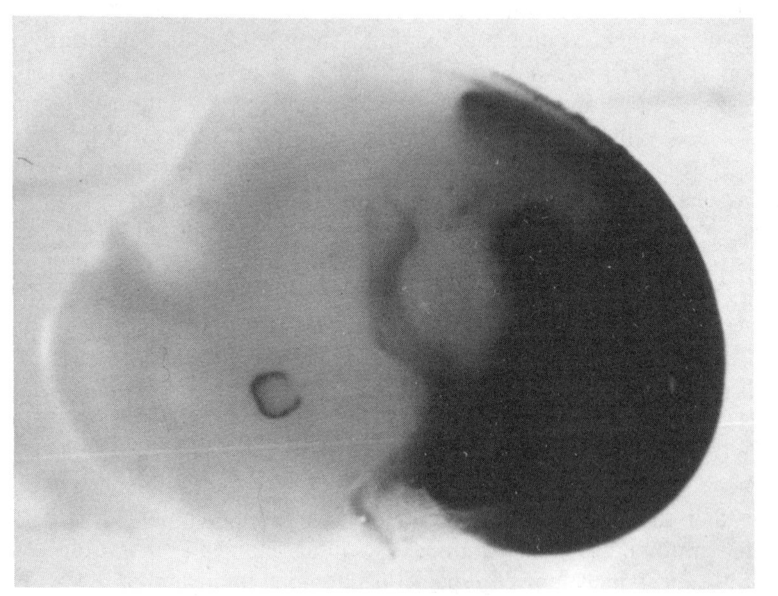

Figure 5 Expression of a *Hox* 1.1 promoter-*lacZ* fusion gene in a transgenic day E11.5 embryo. Transgenic mice were generated using a *Hox* 1.1 promoter-*lacZ* fusion gene. Day E11.5 embryos were fixed in 1% formaldehyde, 0.2% glutaraldehyde, 0.02% NP40, 1 × PBS, washed twice in PBS and stained in 1 mg/ml X-Gal, 5 mM $K_3Fe(CN)_6$, 5 mM $K_4Fe(CN)_6$, 2 mM $MgCl_2$, 1 × PBS. Expression of the transgene is detectable in spinal ganglia, neural tube, mesoderm, and epidermis.

One of our laboratories has used the β-galactosidase enzymatic activity as a reporter in order to map the tissue-specific control elements of a *Hox* 1.1 homeobox containing gene. In these experiments the coding region of *Hox* 1.1 was replaced by the coding region of *lacZ*, which placed it under the transcriptional and post-transcriptional regulation of *Hox* 1.1 control elements (Figure 5). This construct was introduced into zygotes and transgenic mouse lines were established and bred. Analysis of the resulting embryos by using the conventional X-gal stain, revealed the presence of β-galactosidase activity. As shown in Figure 5, a distinct spatial expression pattern was obtained in the central nervous system, as well as in sclerotomes and the mesodermal component of several organs (A. Püschel & P. Gruss, in preparation). The overall temporal and spatial expression pattern was similar if not identical to the expression of the endogenous *Hox* 1.1 gene as studied by in situ analysis (Mahon et al 1988).

TRANSGENES AND GENOMIC IMPRINTING

Both classical genetic experiments (Searle & Beechey 1978; Searle & Beechey 1985; Cattanach & Kirk 1985; Cattanach 1986; Solter 1987) and nuclear transfer experiments have demonstrated that a genomic contribution from both the mother and the father is essential for successful development (McGrath & Solter 1984; Surani et al 1984; McGrath & Solter 1986; Mann & Lovell-Badge 1984; Renard & Babinet 1986; Surani et al 1986a,b; Surani 1987; Monk 1987). These data further revealed that the paternal genome seems to be more important for development of extraembryonic tissues, while the maternal genome plays a greater role for embryonic development. Since the maternal and paternal genetic contributions do not function equivalently during development, differential genome imprinting must occur during female and male gametogenesis.

If a transgene integrates into a chromosomal locus subject to genomic imprinting, it can be used as a molecular marker to examine the molecular details of this phenomenon. Since in appropriate crosses the transgene is inherited either from the mother or from the father, it is subject to imprinting during maternal or paternal gametogenesis. It was demonstrated that methylation of a given transgene is accurately switched between the maternal and the paternal pattern (Reik et al 1987; Sapienza et al 1987). Moreover, expression studies using the transgene RSV-*myc* as a probe revealed that the expression is strictly dictated by its paternal origin (Swain et al 1987). Only the transgenic alleles inherited from the male parent were expressed, while the same transgene inherited from the female parent was not. Interestingly, the transgene inherited from the father was under-methylated. Thus there is a correlation between the degree of methylation and the activity or inactivity of a gene. These elegant experiments clearly document differential imprinting imposed on the genetic material during gametogenesis, and the molecular details involved are under investigation.

CONCLUSIONS

Our review points out various approaches toward defining genes that regulate mammalian genes and ways to modify their action in the transgenic mouse. In addition, we discuss experiments that probe the interrelationship between development and oncogenesis. The power of transgenic technology has already brought fascinating results and is about to provide us with entirely new insights into the molecular genetics of mammalian development.

ACKNOWLEDGMENTS

We thank Dr. Kathleen Mahon and Dr. Corrinne Lobe for valuable comments and Mrs. Kathy Shoobridge for typing the manuscript. P. G. acknowledges financial support by the Max-Planck Society and the Deutsche Forschungsgemeinschaft.

Literature Cited

Allen, N. D., Cran, D. G., Barton, S. C., Hettle, S., Reik, W., et al. 1988. Transgenes as probes for active chromosomal domains in mouse development. *Nature* 333: 852–55

Balling, R., Deutsch, U., Gruss, P. 1988. Undulated, a mutation affecting the development of the mouse skeleton, has a point mutation in the paired box or *Pax* 1. *Cell* 55: 531–35

Balling, R., Mutter, G., Gruss, P., Kessel, M. 1989. Cranofacial abnormalities induced by ectopic expression of the homeobox gene *Hox* 1.1 in transgenic mice. *Cell*. In press

Bradley, A., Evans, M., Kaufman, M. H., Robertson, E. 1984. Formation of germline chimeras from embryo-derived teratocarcinoma cell lines. *Nature* 309: 255–56

Bradley, A., Robertson, E. 1986. Embryo-derived stem cells; a tool for elucidating the developmental genetics of the mouse. *Curr. Top. Dev. Biol.* 20: 357–71

Capecchi, M. R. 1980. High efficiency transformation by direct microinjection of DNA into cultured mammalian cells. *Cell* 22: 479–88

Cattanach, B. M. 1986. Parental origin effects in mice. *J. Embryol Exp. Morph.* 97: 137–50 (Suppl.)

Cattanach, B. M., Kirk, M. 1985. Differential activity of maternally and paternally derived chromosome regions in mice. *Nature* 315: 496–98

Cory, S., Adams, J. M. 1988. Transgenic mice and oncogenesis. *Annu. Rev. Immunol.* 6: 25–48

Deutsch, U., Dressler, G. R., Gruss, P. 1988. *Pax* 1, a member of a paired box homologous murine gene family, is expressed in segmented structures during development. *Cell* 53: 617–25

Doetschman, T., Maeda, N., Smithies, O. 1988. Targeted mutation of the *Hprt* gene in mouse embryonic stem cells. *Proc. Natl. Acad. Sci. USA* 85: 8583–87

Dressler, G. R., Gruss, P. 1988. Do multigene families regulate vertebrate development? *Trends Genet.* 4: 214–19

Evans, M. J., Kaufman, M. H. 1981. Establishment in culture of pluriopotential cells from mouse embryos. *Nature* 292: 154–56

Friedman, A. D., Triezenberg, S. J., McKnight, S. L. 1988. Expression of a truncated viral *trans*-activator selectively impedes lytic infection by its cognate virus. *Nature* 335: 452–54

Geissler, E. N., Ryan, M. A., Housman, D. E. 1988. The dominant-white spotting (W) locus of the mouse encodes the *c-kit* proto-oncogene. *Cell* 55: 185–92

Grüneberg, H. 1954. Genetical studies on the skeleton of the mouse. XII. The development of undulated. *J. Genet.* 52: 441–55

Herskowitz, I. 1987. Functional inactivation of genes by dominant negative mutations. *Nature* 329: 219–22

Holland, P. W. H., Hogan, B. L. M. 1988. Expression of homeobox genes during mouse development: a review. *Genes Dev.* 2: 773–82

Hooper, M., Hardy, K., Handyside, A., Hunter, S., Monk, M. 1987. HPRT-deficient (Lesch-Nyhan) mouse embryos derived from germline colonization by cultured cells. *Nature* 326: 292–95

Izant, J. G., Weintraub, H. 1985. Constitutive and conditional suppression of exogenous and endogenous genes by anti-sense RNA. *Science* 229: 345–52

Jaenisch, R. 1988. Transgenic Animals. *Science* 240: 1468–74

Katsuki, M., Sato, M., Kimura, M., Yokoyama, M., Kobayashi, K., et al. 1988. Conversion of normal behavior to shiverer by myelin basic proteins antisense cDNA in transgenic mice. *Science* 241: 593–95

Kessel, M., Schulze, F., Fibi, M., Gruss, P. 1987. Primary structure and nuclear localization of a murine homeodomain protein. *Proc. Natl. Acad. Sci. USA* 84: 5305–10

Kim, H. S., Smithies, O. 1988. Recombinant fragment assay for gene targeting based on the polymerase chain reaction. *Nucleic Acids Res.* 16: 8887–903

Kimura, M., Inoko, H., Katsuki, M., Ando, A., Sato, T., et al. 1985. Molecular genetic analysis of myelin-deficient mice: shiverer mutant mice show deletion in gene(s) coding for myelin basic protein. *J. Neurochem.* 44: 692–96

Kothary, R., Clapoff, S., Brown, A., Campbell, R., Peterson, et al. 1988. A transgene containing *lacZ* inserted into the *dystonia* locus is expressed in neural tube. *Nature* 335: 435–37

Kuehn, M. R., Bradley, A., Robertson, E. J., Evans, M. J. 1987. A potential animal model for Lesch-Nyhan syndrome through introduction of HPRT mutations into mice. *Nature* 326: 295–98

Levine, M., Hoey, T. 1988. Homeobox proteins as sequence-specific transcription factors. *Cell* 55: 537–40

Mahon, K. A., Chepelinsky, A. B., Khillan, J. S., Overbeek, P. A., Piatigorsky, J., et al. 1987. Oncogenesis of the lens in transgenic mice. *Science* 235: 1622–28

Mahon, K. A., Westphal, H., Gruss, P. 1989. Expression of mouse homeobox gene *Hox* 1.1 during mouse embryogenesis. *Development* 104: 187–95 (Suppl.)

Mann, J. R., Lovell-Badge, R. H. 1984. Inviability of parthogenones is determined by pronuclei, not egg cytoplasm. *Nature* 310: 66–67

Mansour, S. L., Thomas, K. R., Capecchi, M. R. 1988. Disruption of the proto-oncogene *int*-2 in mouse embryo-derived stem cells: a general strategy for targeting mutations to non-selectable genes. *Nature* 336: 348–52

Martin, G. R. 1981. Isolation of a pluripotent cell line from early mouse embryos cultured in medium conditioned by teratocarcinoma stem cells. *Proc. Natl. Acad. Sci. USA* 78: 7634–38

McGinnis, W., Garber, R. L., Wirz, J., Kuroiwa, A., Gehring, W. J. 1984a. A homologous protein-coding sequence in *Drosophila* homeotic genes and its conservation in other metazoans. *Cell* 37: 403–8

McGinnis, W., Levine, M. S., Hafen, E., Kuroiwa, A., Gehring, W. J. 1984b. A conserved DNA sequence in homeotic genes of the *Drosophila* Antennapedia and bithorax complexes. *Nature* 308: 428–33

McGrath, J., Solter, D. 1984. Completion of mouse embryogenesis requires both the maternal and paternal genomes. *Cell* 37: 179–83

McGrath, J., Solter, D. 1986. Nucleocytoplasmic interactions in the mouse embryo. *J. Embryol. Exp. Morphol.* 97: 277–89

Melton, D. A. 1985. Injected anti-sense RNAs specifically block messenger RNA translation *in vivo*. *Proc. Natl. Acad. Sci. USA*. 82: 144–48

Molineaux, G. S., Engh, H., deFerra, F., Hudson, L., Lazzarini, R. A. 1986. Recombination within the myelin basic protein gene created the dysmyelinating shiverer mouse mutation. *Proc. Natl. Acad. Sci USA* 83: 7542–46

Monk, M. 1987. Genomic imprinting. Memories of mother and father. *Nature* 328: 203–4

Mutter, G. L., Grills, G. S., Wolgemuth, D. J. 1988. Evidence for the involvement of the proto-oncogene *c-mos* in mammalian meiotic maturation and possibly very early embryogenesis. *EMBO J.* 7: 683–89

Reik, W., Collick, A., Norris, M. L., Barton, S. C., Surani, M. A. 1987. Genomic imprinting determines methylation of parental alleles in transgenic mice. *Nature* 328: 248–51

Renard, J. P., Babinet, C. 1986. Identification of a paternal developmental effect on the cytoplasm of one-cell-stage mouse embryos. *Proc. Natl. Acad. Sci. USA* 83: 6883–86

Roach, A., Boylan, K., Horvath, S., Prusiner, S. B., Hood, L. E. 1983. Characterization of cloned cDNA representing rat myelin basic protein: absence of expression in brain of shiverer mutant mice. *Cell* 34: 799–806

Rosenberg, U. B., Preiss, A., Seifert, E., Jäckle, H., Knipple, D. C. 1985. Production of phenocopies by Krüppel antisense RNA injection into *Drosophila* embryos. *Nature* 313: 703–6

Sagata, N., Oskarsson, M., Copeland, T., Brumbaugh, J., Vande Woude, G. F. 1988. Function of *c-mos* proto-oncogene product in meiotic maturation in *Xenopus* oocytes. *Nature* 335: 519–25

Saiki, R. K., Scharf, S., Falcona, F., Mullis, K. B., Horn, G. T., et al. 1985. Enzymatic amplification of beta-globin genomic sequences and restriction site analysis for diagnosis of sickle cell anemia. *Science* 230: 1350–54

Sapienza, C., Peterson, A. C., Rossant, J., Balling, R. 1987. Degree of methylation of transgenes is dependent on gametes of origin. *Nature* 328: 251–54

Scott, M. D., Weiner, A. J. 1984. Structural relationships among genes that control development; sequence homology between the Antennapedia, Ultrabithorax, and fushi tarazu loci of *Drosophila. Proc. Natl. Acad. Sci. USA* 81: 4115–19

Searle, A. G., Beechey, C. V. 1978. Complementation studies with mouse translocations. *Cytogenet. Cell Genet.* 20: 282–303

Searle, A. G., Beechey, C. V. 1985. *Aneuploidy: etiology and mechanisms*, eds. V. L. Dellarco, P. E. Voylek, A. Hollaender, 3: 23–27. New York: Plenum. pp. 363–76

Solter, D. 1987. Inertia of the embryonic genome in mammals. *Trends Genet.* 3: 23–27

Stacey, A., Bateman, J., Choi, T., Mascara, T., Cole, W., et al. 1988. Perinatal lethal *osteogenesis imperfecta* in transgenic mice bearing an engineered mutant pro-α1 (I) collagen gene. *Nature* 332: 131–36

Strickland, S., Huarte, J., Belin, D., Vassalli, A., Rickles, F. J., et al. 1988. Antisense RNA directed against the 3′ noncoding region prevents dormant mRNA activation in mouse oocytes. *Science* 241: 680–84

Surani, M. A., Barton, S. C., Norris, M. L. 1984. Development of reconstituted mouse eggs suggests imprinting of the genome during gametogenesis. *Nature* 308: 548–50

Surani, M. A., Barton, S. C., Norris, M. L. 1986a. Nuclear transplantation in the mouse: heritable differences between parental genomes after activation of the embryonic genome. *Cell* 45: 127–36

Surani, M. A. 1987. *Experimental Approaches to Mammalian Embryonic Development*, eds. J. Rossant, R. A. Pederson. pp. 401–36. New York: Cambridge

Surani, M. A., Reik, W. Norris, M. L., Barton, S. C. 1986b. Influence of germline modifications of homologous chromosomes on mouse development. *J. Embryol. Exp. Morph.* 97: 123–36 (Suppl.)

Swain, J. L., Stewart, T. A., Leder, P. 1987. Parental legacy determines methylation and expression of an autosomal transgene: a molecular mechanism for parental imprinting. *Cell* 50: 719–27

Thompson, S., Clarke, A. R., Pow, A. M., Hooper, M. L., Melton, D. W. 1989. Germ line transmission and expression of a corrected *HPRT* gene produced by gene targeting in embryonic stem cells. *Cell* 56: 313–21

Weintraub, H., Izant, J. G., Harland, R. M. 1985. Anti-sense RNA as a molecular tool for genetic analysis. *Trends Genet.* 1: 22–25

Wilkinson, D. G., Peters, G., Dickson, C., McMahon, A. P. 1988. Expression of the FGF-related proto-oncogene *int*-2 during gastrulation and neurulation in the mouse. *EMBO J.* 7: 691–95

Wolgemuth, D. L., Behringer, R. R., Mostoller, M. D., Brinster, R. L., Palmiter, R. D. 1989. Transgenic mice overexpressing the mouse homeobox containing gene *Hox* 1.4 exhibit abnormal gut development. *Nature* 337: 464–67

Woychik, R. P., Stewart, T. A., Davis, L. G., D'Eustachio, P., Leder P. 1985. An inherited limb deformity created by insertional mutagenesis in a transgenic mouse. *Nature* 318: 36–40

Zimmer, A., Gruss, P. 1989. Production of chimaeric mice containing embryonic stem (ES) cells carrying a homeobox *Hox* 1.1 allele mutated by homologous recombination. *Nature* 338: 150–53

Annu. Rev. Cell Biol. 1989. 5 : 197–245

INITIATION OF EUKARYOTIC DNA REPLICATION IN VITRO

Bruce Stillman

Cold Spring Harbor Laboratory, P.O. Box 100, Cold Spring Harbor, New York 11724

CONTENTS

PROLOGUE

The aim of modern cell biology is to understand how cells work at the molecular level. A prime goal is to seek a complete understanding of the biochemistry of cell growth and division, including an integration of the growth regulatory pathways with the processes they regulate. These regulated events include the mechanisms of maintaining a cell in a quiescent (nonreplicating) state, of exit from that state to a state of cell proliferation, and of the events that are specifically required during cell reproduction such as DNA replication, mitosis, and cytokinesis. Indeed, a detailed description at the biochemical level of exactly how eukaryotic cells grow and divide is essential if the regulatory pathways are to be fully appreciated.

0743–4634/89/1115–0197$02.00

For example, DNA synthesis is a commonly used assay for studying the effects of growth factors and pharmacological reagents on cellular metabolism, but any interpretation of such results necessarily suffers from our incomplete understanding of how DNA is replicated and how this replication is controlled. Further studies on the mechanism and regulation of DNA replication in eukaryotic cells should therefore contribute significantly to bridging the gap between the growth regulatory pathways and the critical regulated events in cellular proliferation.

Studies on DNA replication and its regulation in eukaryotic cells have been limited in part by the lack of useful genetic approaches in higher eukaryotic cells, but also by the feeling that the mechanisms of DNA replication in eukaryotic cells will merely reflect what we have already learned from studying bacteria and their phages. This view ignores the unique and interesting problems specific to eukaryotic cells and, perhaps, incorrectly assumes that mechanisms will be the same. Recently, much progress has been made, predominantly through the use of cell-free systems that allow the biochemistry of DNA replication to be dissected. In this review, I summarize what has been learned from studying some of these cell-free replication systems and attempt to place this knowledge in the broader context of cell proliferation and its control. Various aspects of eukaryotic DNA replication have been discussed in previous general reviews, most notably by Challberg & Kelly (1982) and Campbell (1986), and in reviews of the more specialized topics to be mentioned.

ADENOVIRUS DNA REPLICATION

The viruses that grow in mammalian cells have provided relatively simple model systems for understanding eukaryotic biology, just as bacteriophage systems have done in prokaryotic biology. The human DNA virus group, the adenoviruses, have played a pioneering and significant role in the development of many aspects of eukaryotic molecular biology; true to form, they have also provided the first cell-free system to study eukaryotic DNA replication. The mechanism of adenovirus DNA replication is unusual and does not reflect the processes that replicate the host cell's chromosomes. Nonetheless, studies of adenovirus DNA replication have yielded a novel mechanism for initiation of DNA replication and revealed a dual role for certain site-specific DNA binding proteins in DNA replication and transcription. The replication of adenovirus DNA has been reviewed frequently (Kelly et al 1988; Challberg & Kelly 1982; Stillman 1983; Kelly 1984).

Human adenoviruses contain linear, double-stranded DNA genomes approximately 36,000 bp in length and an inverted, terminally repeated

DNA sequence of 102–162 bp at each molecular end of the chromosome. A striking feature is that the virus chromosome isolated from purified virus particles contains a 55,000 M_r protein, called terminal protein (TP), covalently linked to the end of each DNA strand; this linkage is a phosphodiester bond between the 5′-deoxycytidine residue in the DNA and a serine residue in the protein (Robinson et al 1973; Robinson & Bellett 1974; Rekosh et al 1977; Stillman et al 1981; Desiderio & Kelly 1981; Challberg & Kelly 1981; Smart & Stillman 1982). The discovery of this protein covalently attached to the 5′ end of each DNA strand directly led to the suggestion that the protein was involved in priming DNA replication at the replication origin (Robinson & Bellett 1974; Rekosh et al 1977). Since all DNA polymerases synthesize DNA in the 5′ to 3′ direction, but require a primer for polymerization to commence, this model represents a novel solution to the recognized problem of how the very 5′ end of the linear chromosome can be duplicated.

The general mechanism of adenovirus DNA replication had been elucidated prior to the development of a cell-free replication system (Figure 1, see previous reviews for references). Both the origins of DNA replication and the termini of DNA synthesis were mapped by pulse labeling and electron microscopic techniques to the two ends of the chromosome. These observations, coupled with the high proportion of single-strand DNA found in the replicating fraction of adenovirus DNA in infected cells, demonstrated that adenovirus DNA replication begins at either end (randomly) of the chromosome and proceeds unidirectionally via a strand displacement mechanism (Figure 1). This means that one strand (the leading strand) is synthesized continuously and that the nontemplate strand is first displaced as a single strand (type I replication) and subsequently copied by a similar unidirectional mechanism (type II replication). Considerable evidence obtained from experiments in vivo suggests that the inverted, repeated sequences at each end of the displaced strand pair with each other to regenerate a double-strand replication origin for replication via type II intermediates (Stow 1981, see Figure 1). Recently Leegwater et al (1988) have provided evidence that the regenerated duplex origin on type II replication intermediates initiates replication by a mechanism similar to initiation of type I replication.

The development by Challberg & Kelly (1979a,b) of a cell-free extract derived from adenovirus infected cells that could faithfully initiate and complete one round of replication on an exogenous adenovirus DNA template was a watershed event in eukaryotic DNA replication studies. The template for replication was the virus DNA that had been extracted from the virion particle with the terminal protein intact; removal of this protein by protease digestion eliminated template activity. Various studies

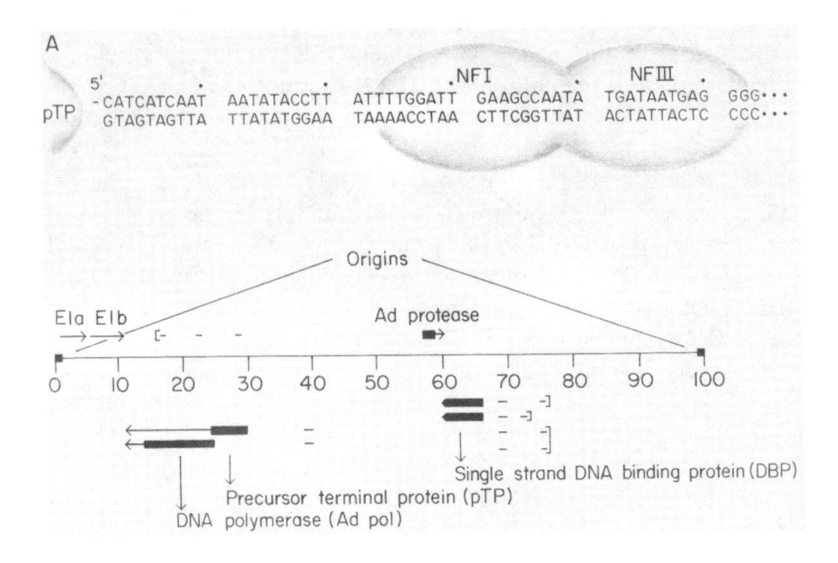

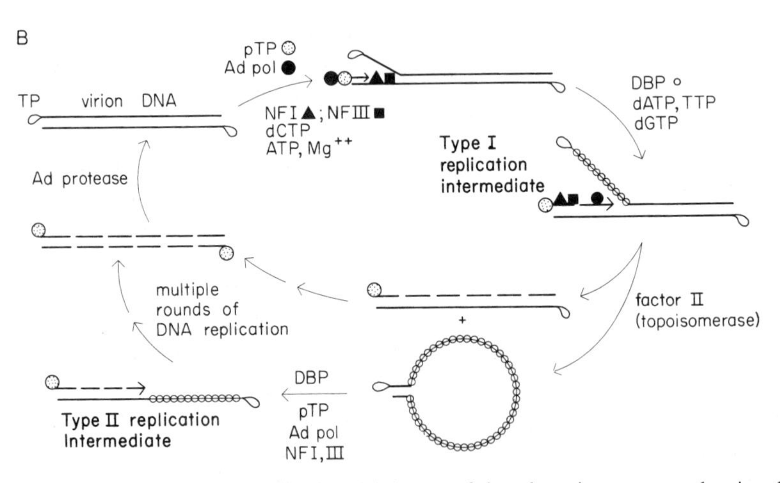

Figure 1 Adenovirus DNA replication. (a) A map of the adenovirus genome showing the three virus encoded replication proteins and an expanded region of the origins of replication. (b) Schematic representation of the replication of adenovirus DNA. Part of this figure was modified from Stillman (1983).

demonstrated that the overall mechanism of DNA replication in vitro reflected the mechanism deduced from studies in vivo. It thus became possible to pursue a detailed understanding of the replication of this virus DNA and obtain new insights into cellular macromolecular metabolism.

Virus-Encoded Replication Proteins

Adenovirus encodes three proteins that are directly required for the replication of its DNA, although this was not known in 1979. These proteins are the single-strand DNA binding protein (DBP), the precursor terminal protein (pTP), and the adenovirus-encoded DNA polymerase.

SINGLE STRAND DNA BINDING PROTEIN (DBP) The DBP was one of the first recognized adenovirus-encoded proteins synthesized during the early phase of virus infection, predominantly because it is produced in relatively large amounts (van der Vliet & Levine 1973). The protein binds quantitatively to single-strand DNA cellulose, but not to double-strand DNA cellulose and thus was relatively easy to purify. The purified protein is extensively phosphorylated on serine and threonine residues in the amino-terminal region of the molecule, but phosphorylation does not alter its DNA binding properties (Klein et al 1979; Linne & Philipson 1980; Jeng et al 1977). Indeed, a proteolytic fragment derived from the carboxy-terminal two-thirds of DBP lacks phosphate modification and can bind DNA and support adenovirus DNA replication in vitro (Ariga et al 1980; Friefeld et al 1983a). DBP binds cooperatively to single-strand DNA (van der Vliet & Levine 1973; van der Vliet et al 1978; Fowlkes et al 1979; Schecter et al 1980), with approximately one molecule per nine-eleven nucleotides. The DBP-DNA complex exists in a rigid, extended confirmation (van Amerongen et al 1987). Furthermore, DBP does not appear to lower the melting temperature of DNA (van der Vliet et al 1978; Fowlkes et al 1979). In many respects, the DBP resembles the classic single-strand DNA binding proteins such as the gene 32 protein from bacteriophage T4.

The function of DBP in DNA replication was first indicated by genetic studies. Temperature-sensitive mutants in the gene encoding DBP (E2A region of the virus genome) resulted in a defect in DNA replication at the nonpermissive temperature in vivo (Ensinger & Ginsberg 1972; van der Vliet & Sussenbach 1975; Ginsberg et al 1977; van Bergen & van der Vliet 1983). Subsequently, DBP purified from wild-type infected cells was shown to complement defective extracts made from temperature-sensitive mutant infected cells (Kaplan et al 1979; Friefeld et al 1983a; Ostrove et al 1983; Prelich & Stillman 1986). Most importantly, the DBP is essential for adenovirus DNA replication reconstituted in vitro with purified proteins (Ikeda et al 1981; Nagata et al 1983b; Rosenfeld et al 1987; Leegwater et al 1988; Stillman et al 1982a). DBP clearly functions during the elongation stage, probably by binding to and stabilizing the displaced nontemplate strand, but also by increasing the processivity of the adenovirus polymerase (Lindenbaum et al 1986). Although DBP is not absolutely required for initiation of DNA replication (as measured by the formation of the pTP-dCMP initiation complex, see below), it stimulates pTP-dCMP complex formation about three-fold (Nagata et al 1982). On the other hand, extracts prepared from cells infected with viruses carrying a temperature-sensitive DBP are fully active for pTP-dCMP complex formation, but are defective for subsequent strand elongation, even for synthesis of the first 26 bases of the nascent strand (Friefeld et al 1983a; Prelich & Stillman 1986). The defective DBPs display reduced binding to single-strand DNA. These results suggest that DBP may be bifunctional; the stimulation of initiation by DBP may involve its interaction with another initiation protein, whereas a separate elongation activity seems to require the single-strand DNA binding function.

Sequencing of the genes encoding DBPs from various human virus serotypes (Kruijer et al 1983b; Quinn & Kitchingman 1984; Kitchingman 1985; Vos et al 1988) and of mutant virus genes encoding replication defective DPBs (Kruijer et al 1983a; Prelich & Stillman 1986) has revealed highly conserved domains that are required for DBP function. One interesting region, containing two temperature-sensitive mutations that alter replication and single-strand DNA binding properties, contains a region (Histidine.X.Cysteine.X_8.Cysteine.X.Histidine) that has some similarity to the metal ion binding motif that may facilitate DNA binding (Vos et al 1988), either by directly interacting with the DNA or by promoting protein-protein cooperativity in DNA binding. This observation further strengthens the similarity between DBP and the phage T4 gene 32 protein, which has been demonstrated to bind a Zn^{2+} ion (Giedroc et al 1986), and to have a region with similarities to the zinc finger that is required for cooperative binding to the DNA (Giedroc et al 1987). The crystallization of DBP may aid in structure-function studies (Tsernoglou et al 1984).

The similarity between the adenovirus DBP and prokaryotic single-strand DNA binding proteins (SSBs) such as the gene 32 protein (and even other eukaryotic viral DBPs, such as the HSV ICP8 protein) prompted numerous searches for similar DBPs in mammalian and yeast cells (see Chase & Williams 1986), that were fueled by the assumption that eukaryotic cell SSBs would bind tightly to single-strand DNA, but not to double-strand DNA. This assumption however, may prove incorrect, as described below, a human cell single-strand DNA binding protein required for SV40 DNA replication has properties different from the viral DBPs or prokaryotic SSBs.

PRECURSOR TERMINAL PROTEIN (TP) As described above, the discovery of a 55,000-dalton protein that was covalently linked to the 5′ ends of adenovirus virion-derived DNA led to the speculation that this protein might function in DNA replication (Rekosh et al 1977). However, at that time it was not even known if the terminal protein (TP) was encoded by the virus, and its function remained enigmatic. It was known that replicating DNA isolated from infected cells had a protein covalently attached to the 5′ end of all nascent DNA strands (Coombs et al 1979; Stillman & Bellett 1979; van Weilink et al 1979; Robinson et al 1979) and that all nascent DNA strands replicated in vitro contained a covalently linked protein. However, it was surprising to find this protein had an apparent mass of 80–87 kd (Challberg et al 1980). The 80 kd protein, like the 55-kd TP protein, was linked to the 5′ end of the DNA by a serine-phosphodiester bond, which suggested that the 80-kd protein was a precursor to the 55-kd TP. The 80-kd protein was also shown to be present on intracellular DNA (Challberg & Kelly 1981). Subsequent studies (Stillman et al 1981; Challberg & Kelly 1981; Binger et al 1982) demonstrated conclusively that the 80-kd protein is a precursor to the terminal protein (pTP) and that it is encoded by the adenovirus genome from a transcription unit that also encodes the adenovirus DBP (and the virus DNA polymerase). The 80-kd protein is processed during virion assembly to the 55-kd form by a virus encoded protease (Stillman et al 1981; Challberg & Kelly 1981).

These results strongly implicated the involvement of pTP in initiation of DNA replication. This was demonstrated by a label transfer experiment in which ^{32}P-label in dCTP was transferred to the 80-kd pTP during DNA replication in vitro (Challberg et al 1980) and more directly by the in vitro formation of the proposed priming complex; a covalent complex between pTP and dCMP in a reaction that depended upon adenovirus template DNA linked to terminal protein, ATP, $MgCl_2$, and dCTP (Enomoto et al 1981; Lichy et al 1981; Pincus et al 1981; Stillman 1981; Challberg et al 1982; Ikeda et al 1982; Tamanoi & Stillman 1982). The pTP was purified in a functional form and was shown to be tightly complexed with a 140-kd DNA polymerase activity (Enomoto et al 1981). The DNA polymerase (Ad pol) and pTP could be separated by sedimentation in glycerol gradients containing urea, and by using this technique both pTP and Ad pol were shown to be necessary (but not sufficient) for reconstitution of adenovirus DNA replication in vitro and for formation of the pTP-dCMP initiation complex (Lichy et al 1982).

ADENOVIRUS DNA POLYMERASE (AD POL) Fractions containing the 140-kd protein that copurified with pTP also contained a DNA polymerase activity that was distinguishable from the known cellular polymerases α,

β and γ (Enomoto et al 1981; Lichy et al 1982). The 140-kd protein was identified as an adenovirus-encoded protein using a different approach. Replication extracts prepared from cells infected at the nonpermissive temperature with temperature-sensitive mutants of adenovirus that were defective for DNA replication in vivo failed to synthesize the pTP-dCMP initiation complex in vitro (Stillman et al 1982a; Friefeld et al 1983b; Ostrove et al 1983; van Bergen & van der Vliet 1983). The 140-kd DNA polymerase, purified using complementation of these defective extracts as an assay, reversed the defect in vitro, whereas purified pTP failed to complement. The Ad5*ts*149 temperature-sensitive mutant did not render the 140-kd DNA polymerase activity thermolabile in vitro. Initiation of DNA replication, however, was more sensitive to inactivation by addition of urea, which suggests that the Ad5*ts*149 mutation might destabilize the interaction of polymerase with pTP or another initiation factor.

The pTP and Ad pol proteins are encoded by the E2B region of the virus genome, whereas the DBP is encoded by the E2A region (Stillman et al 1982a). The three adenovirus replication proteins are encoded from the same transcriptional unit and thus may be coordinately regulated at the level of transcription throughout the virus infection cycle. Recently it was recognized that both pTP and Ad pol open reading frames extend into an upstream exon that donates the same three amino acids to the amino-terminus of the proteins (Pettit et al 1988; Shu et al 1987, 1988). These amino-terminal amino acids are required for protein expression, but it is not clear whether they provide the same function in both proteins.

The adenovirus polymerase has an intrinsic 3′ to 5′ exonuclease activity that is probably used for proof-reading during chain elongation (Field et al 1984). Interestingly, polymerase activity on primed single-strand synthetic templates is specifically stimulated approximately 100-fold by DBP (but not by the *E. coli* single-strand DNA binding protein, SSB) and this stimulation is further augmented another 3–4-fold by inclusion of ATP in the reaction (Field et al 1984; Lindenbaum et al 1986). The DBP stimulates the processivity of the Ad pol (the number of nucleotides that a single polymerase molecule adds before dissociating from a growing 3′ DNA end), and products of up to 40 kd (approximately the length of the adenovirus genome) can be synthesized. A similar increase in processivity is conferred upon the herpes simplex virus (HSV) DNA polymerase by the HSV single-strand DNA binding protein ICP8 (O'Donnell et al 1987). Mutations in the gene encoding ICP8 lead to altered drug sensitivity to DNA polymerase inhibitors (Chiou et al 1985), which suggests that the mechanism of elongation of DNA synthesis in HSV infected cells may parallel that on the adenovirus genome. Sequencing these mutations may be informative, because ICP8 shares limited sequence similarity with the proliferating cell nuclear antigen (PCNA), which increases the processivity of polymerase δ (Prelich et al 1987b; Matsumoto et al 1987; Almendral et al 1987) (see below).

The primary structure of the adenovirus polymerase has been compared to the amino acid sequences of a large number of prokaryotic and eukaryotic DNA polymerases. There is surprising sequence similarity between the polymerases from diverse sources such as bacteriophages T4, T7, and ϕ29, the eukaryotic viruses such as adenovirus and HSV, and cellular polymerase α from mammalian cells and the analogous polymerase I from yeast (for example, see Wang et al 1989). Perhaps more surprising is the fact that none of these polymerases has homology with the predominant replicative polymerase from *E. coli*, DNA polymerase III. Conjecture based on evolution might predict the existance of a eukaryotic cell DNA polymerase with some similarity to the *E. coli* pol III. Candidates were the mammalian polymerase δ and yeast polymerase III enzymes, but these latter polymerases appear to be biochemically more related to the polymerase α class. Recent sequence analysis of the DNA polymerase III from *S. cerevisiae* demonstrates that it is homologous to the polymerase α class of polymerases (Boulet et al 1989). For a detailed discussion on DNA polymerases, see Burgers 1989.

The biochemistry of the adenovirus replication proteins has been studied in detail and there are ongoing studies to examine their cell biology. Both the pTP and Ad pol can be expressed in a functional form after plasmid transfection of expression vectors (Shu et al 1987; Pettit et al 1988) or by infection with recombinant vaccinia viruses (Stunnenberg et al 1988). Both proteins localize to the nucleus (Green et al 1981; Sasaguri et al 1987), but a recent study indicates that the nuclear localization of the adenovirus polymerase is facilitated by its interaction with pTP (Zhao & Padmanabhan 1988). Thus pTP and Ad pol may enter the nucleus as a complex that is preformed in the cytoplasm, such a complex would provide a useful model for study of the intercellular transport of proteins that are destined to work together in the nucleus.

Cellular Replication Proteins

NUCLEAR FACTOR I The three adenovirus-encoded replication proteins are essential but not sufficient for replication of an adenovirus DNA-protein complex in vitro. Using reconstitution of replication as an assay, three cellular factors that are required for efficient adenovirus DNA replication have been identified and characterized. The first of these, called nuclear factor I (NFI), was purified based upon its ability to stimulate the formation of the pTP-dCMP complex (Nagata et al 1982). NFI stimulates pTP-dCMP complex formation even in the absence of DBP, but it becomes essential for this reaction in the presence of DBP (Nagata et al 1982; de Vries et al 1985). NFI binds specifically to a sequence within the origin of DNA replication (Nagata et al 1983a; Rawlins et al 1984a; Leegwater et al 1985; Diffley & Stillman 1986; Rosenfeld & Kelly 1986; Rosenfeld et al 1987; de Vries et al 1987) and to other virus and cellular DNA sequences (Gronostajski et al 1984; Rawlins et al 1984a; Henninghausen et al 1985; Leegwater et al 1985; Nowock et al 1985; Shaul et al 1986; Jeang et al 1987; Miksicek et al 1987; Oikarinen et al 1987). Most of these cellular and virus binding sites are associated with transcription promoter or enhancer elements, implicating NFI as a transcription factor. The consensus NFI recognition sequence obtained from these studies and others (Rawlins et al 1984a; Gronostajski et al 1984; Gronostajski 1986, 1987; de Vries et al 1987) is TGG($^{A}/_{C}$)N_5GCCAA, a partially palindromic sequence with a variable central pentamer. When NFI binding sites are placed directly upstream of the TATA box in the adenovirus major late promoter, transcription in vitro increases (Gronostajski et al 1988).

A number of different cellular proteins can bind to the adenovirus NFI binding site and stimulate replication (Nagata et al 1982; Diffley & Stillman 1986; Rosenfeld & Kelly 1986). The NFI proteins identified by Rosenfeld and Kelly are particularly interesting because they are identical to a closely related family of proteins each of which functions as the mammalian transcription factor CTF (Jones et al 1987; Santoro et al 1988). CTF binds to CCAAT recognition sites in promoters to activate transcription. Since CTF expressed in and purified from *E. coli* can activate adenovirus DNA replication in vitro, it is clear that the CTF/NFI protein has a duel replication and transcription activity (Santoro et al 1988). It is not known, however, if the protein acts via the same type of mechanism in both situations. A number of other CCAAT binding proteins have been described in mammalian cells (for example, see Dorn et al 1987; Chodosh et al 1988), but it is not clear if these can also function at the adenovirus origin to stimulate replication.

NUCLEAR FACTOR III Another cellular, site specific DNA binding protein, called Nuclear Factor III, was identified as a factor that stimulates initiation of adenovirus DNA replication and binds to a site adjacent to the NFI binding site (Pruijn et al 1986; Rosenfeld et al 1987; O'Neill & Kelly 1988; NFIII has also been named ORP-C). The protein interacts with sequences within the adenovirus terminally repeated sequence (Pruijn et al 1988) that are similar to promoter and enhancer elements in mammalian and virus chromosomes (Pruijn et al 1987; O'Neill & Kelly 1988). Indeed, NFIII is probably the same as the cellular enhancer and promoter binding factor called Octamer transcription factor (OTF-1) (or Octamer binding protein OBP-100), because it binds to a conserved octamer sequence (see Sturm et al 1987 and references therein) and purified OTF-1 can functionally substitute for NFIII (O'Neill et al 1988). A cDNA encoding OBP-100/OTF-1 has recently been cloned and sequenced and shown to encode the ubiquitiously expressed protein that binds both to the octamer sequence and an unrelated TAATGARAT sequence (R = either A or G) first identified in herpes simplex virus promoters (Sturm et al 1988). The corresponding gene, called oct-1, is related to the oct-2 gene encoding a B cell specific transcription factor, and to genes encoding a pituitary transcription factor and a *C. elegans* developmental switch protein. Each of these proteins contains a homeobox and an upstream conserved domain, collectively called the POU domain (see Herr et al 1988). The POU specific region distinguishes these proteins from other homeobox proteins because the entire POU domain, not just the homeobox, is necessary for DNA binding.

The function of NFIII in replication initiation is not as clear as the function of NFI. Initiation with purified proteins can occur in the absence of NFIII when NFI is present as the only cellular site specific DNA binding protein (Guggenheimer et al 1984). Initiation in the absence of NFIII has also been reported in the presence of the three purified virus-encoded proteins, purified NFI, and a crude cellular fraction derived from HeLa cells that does not appear to contain NFIII, since NFIII did not substitute for this fraction (Pruijn et al 1986; Rosenfeld et al 1987). Under these circumstances, the NFIII binding site was not absolutely required for replication initiation, but could stimulate initiation. In apparent contrast, O'Neill et al (1988) described conditions in which initiation of replication was dependent upon NFIII and, moreover, NFIII substituted for the crude cellular factor. Thus, although the role of NFIII in DNA replication needs to be resolved, it is clear that NFIII does stimulate, to varying extents in different experiments, the initiation of adenovirus replication.

NUCLEAR FACTOR II A third cellular protein called Nuclear Factor II (NFII) was identified as a factor required to complete elongation of adenovirus DNA replication when full-length virus templates were used; however, NFII was not required for initiation or elongation of shorter templates (Guggenheimer et al 1983a,b; Nagata et al 1983b). NFII copurified with a type I topoisomerase activity and the purified HeLa or calf thymus topoisomerase 1, but not *E. coli* topoisomerase 1, could functionally substitute for NFII. Because the eukaryotic topoisomerases but not the *E. coli* enzyme can remove positive supercoils in DNA, these results suggest that the role of NFII is in removing positive superhelical tension induced in the DNA template ahead of the replication fork. However, NFII has a native molecular mass between 25 and 45 kd and copurified with proteins of 14.5 and 15.5 kd (Nagata et al 1983b), which suggests that NFII is not the native HeLa cell topoisomerase I. It is still not clear what relationship NFII has to the much larger, well-characterized type I topoisomerase, nor is it clear why a topoisomerase should be required for replication of a linear DNA molecule. One possibility is that the two ends of the chromosome interact during replication, which creates some torsional constraint.

The Origin of DNA Replication

Comparative sequence analysis of the terminally repeated sequences present in adenoviruses from different serotypes and diverse species has revealed heterogeneity in the length of the repeat, but has also pointed to highly conserved sequences adjacent to the molecular end of the chromosome, the site of initiation (Tolun et al 1979; Shinagawa et al 1983; Aleström et al 1982; Stillman et al 1982b). But this analysis did not clearly define the sequences required for initiation. A breakthrough came when it was demonstrated that origin sequences cloned into bacterial plasmids could support initiation provided that the plasmids were subsequently linearized such that the origin sequences were at the end of the DNA molecules (Tamanoi & Stillman 1982; van Bergen et al 1983). In addition to the protein requirements for initiation of replication on the adenovirus DNA-protein complex (i.e. DBP, pTP, Adpol, NFI, NFIII), initiation of replication on linearized plasmid DNAs requires a cellular protein called factor pL (Guggenheimer et al 1984a–c). This protein is a $5' \rightarrow 3'$ exonuclease that appears to remove sequences from the $5'$ end of the non-template stand and exposes the $3'$ dG residue in the template strand, thereby allowing formation of the pTP-dCMP complex (Kenny & Hurwitz 1988; Kenny et al 1988). Since factor pL is not required for initiation when the adenovirus DNA-protein complex is used as a template, it seems that the TP or pTP (on the $5'$ ends of virion and intracellular DNA respectively) functions to aid in unwinding the origin.

Linear plasmids containing the adenovirus origin sequences have been used to define the sequences necessary for initiation of replication in vitro (Enns et al 1983; Tamanoi & Stillman 1983; van Bergen et al 1983; Challberg & Rawlins 1984; de Vries et al 1985; Guggenheimer et al 1984; Lally et al 1984; Rawlins et al 1984b; Adhya et al 1986; Wides et al 1987). These experiments led to the following observations (Figure 1). The terminal C·G base pair is essential, however the adjacent pairs (2–8) can accommodate point mutations, but not insertions or deletions. The highly conserved A·T rich sequence (base pairs 9–18) is required for initiation and the terminal 18 bp that includes them constitute the core origin sequence, although initiation on this minimal origin is very inefficient. The NFI binding site (base pairs 25–39) is essential for efficient initiation, whereas the presence of the adjacent NFIII binding site (base pairs 39–50) is not essential, but stimulates replication to varying amounts, depending upon the assay. The NF-1 binding site can be replaced by cellular NFI binding sites and can be inverted, but must lie adjacent to the highly conserved A·T rich sequence with correct spacing. The highly conserved A·T rich region and the sequences between this region and the NF-I binding site interact specifically with the pTP·Adpol protein complex (Rijnders et al 1983; Kenny & Hurwitz 1988). Thus initiation may occur by the formation of a pTP·Adpol·NFI complex that recognizes specific DNA sequences and places the active site of pTP (Smart & Stillman 1982) in proximity to the 3′ dG residue on the template strand. This general view of the origin structure has been confirmed by studying the replication of adenovirus mini chromosomes in vivo (Hay & McDougal 1986; Hay et al 1984; Hay 1985a,b; Wang & Pearson 1985; Bernstein et al 1986).

These biochemical studies on adenovirus DNA replication have uncovered a novel and interesting mechanism for initiation of DNA replication that also operates in the replication of other chromosomes, most notably the *B. subtitis* bacteriophage ϕ29 chromosome (Salas 1983). Furthermore, an interesting relationship between transcription factors and origin binding proteins has been unearthed that may be a common theme in virus and perhaps cellular DNA replication (DePamphilis 1988).

SIMIAN VIRUS 40 DNA REPLICATION

From the time of its disquieting discovery as a contaminant in a poliomyelitis virus vaccine, simian virus 40 (SV40) has been the subject of intensive study, primarily because it can induce tumors in newborn hamsters and is therefore oncogenic, but more recently because these continuing investigations on the interaction between SV40 and its host cell have provided valuable insight into mammalian cell biology. This is particularly so for studies on eukaryotic DNA replication; SV40 has long provided fundamental information on the mechanism of DNA replication in mammalian cells. A valuable feature is that the SV40 virus DNA present in infected cells exists in a chromatin structure that resembles the chromatin structure of the host cell chromosomes, and thus studies on SV40 DNA replication can address issues such as chromatin (rather than just DNA) replication. A distinguishing feature of eukaryotic genome reproduction is that both the DNA and its associated proteins must be duplicated each S phase, but little is known about the latter process.

The general mechanism of SV40 DNA replication was discovered by studying the structure of replication intermediates isolated from infected cells and recent detailed reviews have described these studies (DePamphilis & Bradley 1986; Kelly et al 1988). Another feature that makes SV40 so attractive is its apparent simplicity. The virus genome is a duplex circular DNA chromosome of 5243 base pairs and contains a single origin of DNA replication. On an individual molecule, replication occurs bidirectionally from the origin, creating replication intermediates that resemble the θ structures first described for the *E. coli* chromosome (Cairns 1963). Termination of DNA replication occurs when the two replication forks meet each other and the daughter molecules segregate into monomer circles. This general scheme for DNA replication resembles the process assumed to occur from a single replication origin in a cellular chromosome. A trivial difference for chromosomal replication is that termination would occur when two forks that originated from adjacent replication origins collide, rather than collision of forks originating from a single sequence, as is the case for SV40. Of the six proteins encoded by the SV40 chromosome, only four are essential for growth in cell culture. Three of these are virion structural proteins and another one, the SV40 large tumor (T) antigen, plays an essential role in virus DNA replication. Thus all but one of the proteins that replicate SV40 DNA are encoded by the host cell and knowing their identity should shed light on cellular DNA replication. In this section, I will review recent studies on SV40 DNA replication and primarily concentrate on new information obtained from studies with cell free replication systems. The reader is referred to previous reviews (DePamphilis & Bradley 1986; Kelly et al 1988) that provide a basis for this discussion.

Cell-Free Replication Systems

Initial studies on SV40 DNA replication in vitro utilized virus chromosomes isolated from infected monkey cells as templates to complete elongation of DNA replication, but there was no convincing evidence for initiation occurring on these templates (DePamphilis & Bradley 1986). Reports of T antigen-dependent cell free replication of plasmid DNAs containing a functional SV40 origin of DNA replication, of plasmids containing cellular Alu repeated sequences, and of the cellular c-*myc* protein substituting for T antigen in these replication systems have appeared over the last few years (Ariga & Sugano 1983; Ariga 1984; Ariga et al 1985; Iguchi-Ariga et al 1987). However, these results have not been repeated in several laboratories and must therefore await independent confirmation (Li & Kelly 1985; Gutierrez et al 1988).

A clear breakthrough in the quest to understand mammalian DNA replication came with the development by Li & Kelly (1984) of conditions for preparing soluble extracts from SV40 infected or uninfected monkey cells that would support initiation and complete replication of plasmid DNAs containing the SV40 origin of replication. When uninfected cell extracts were used, replication was only observed when purified SV40 T antigen was added to the reaction mix. Thus the development of rapid immunoaffinity techniques for purification of SV40 T antigen contributed to this success (Dixon & Nathans 1985; Simanis & Lane 1985). More recently, the over production of T antigen in adenovirus (Stillman & Gluzman 1985) or baculovirus vectors (Lanford 1988; Murphy et al 1988; O'Reilly & Miller 1988) has enabled large quantities of purified T antigen to be obtained. Soon after the initial report, several laboratories extended these results by showing that cell-free extracts from competent human HeLa or 293 cells could support SV40 DNA replication in the presence of T antigen (Li & Kelly 1985; Stillman & Gluzman 1985; Wobbe et al 1985); cells that have the advantage of growing in suspension culture, so that large quantities of cellular extracts can be isolated. To date there are no significant differences between the human and monkey cell-free systems.

The replication of SV40 origin containing plasmids in vitro was shown by a variety of techniques to start at the replication origin and proceed, in a semi-conservative manner, bidirectionally around the molecule at rates comparable to those observed in vivo (Li & Kelly 1984, 1985; Stillman & Gluzman 1985; Wobbe et al 1985; Stillman et al 1986; Decker et al 1987). Furthermore, replication in vitro was predominantly via Cairns θ-type replication intermediates that yielded covalently closed, but relaxed circular DNA molecules. Interestingly, the capacity of extracts prepared from cells of different species reflected the ability of those cells to replicate SV40 DNA in vivo (Li & Kelly 1985). Additionally, replication in vitro was shown to proceed with higher fidelity than DNA synthesis with purified DNA polymerase α/DNA primase complex, which suggested that some proofreading of nucleotide incorporation was occurring (Hauser et al 1988; Roberts & Kunkel 1988). Thus this vigorous cell free system has provided an experimental boost for a detailed biochemical investigation into eukaryotic DNA replication. Moreover, conditions have been described that allow the replicating DNA to be assembled into a chromatin structure that is similar to the chromatin structure of cellular chromosomes (Stillman 1986). In this system, chromatin assembly occurs preferentially, and perhaps exclusively, on replicating DNA, thereby reproducing the events that occur during the S phase of the cell cycle.

The Origin of DNA Replication

The SV40 origin of DNA replication lies within a complicated noncoding region of the viral genome and overlaps with the divergent promoters for transcription of genes that are expressed early and late in the infection cycle. The structure of the replication origin was initially determined by analysis of viable SV40 variants or by site directed mutagenesis of the origin region followed by transfection of the plasmid DNAs into SV40 infected cells, which provided T antigens in *trans*, or monkey COS cells, a cell line expressing SV40 T antigen constituitively (Gluzman 1981). These initial experiments demonstrated that sequences required for DNA replication could be subdivided into two different classes, those that are essential and those that, under some circumstances, increase the amount of DNA replication. Detailed reviews of the results obtained in vivo have been presented previously (DePamphilis & Bradley 1986; Kelly et al 1988) and are summarized here. The sequences that stimulate DNA replication in vivo, but which are not essential, flank the essential core origin and correspond to the transcriptional regulatory sequences such as the SV40 enhancer sequences and the binding sites for the cellular protein Sp1. Both the enhancer sequences and the Sp1 binding sites function independently of each other, and the extent of stimulation by these auxiliary sequences in vivo is variable, depending upon the precise assay used. In the standard in vitro assays, however, it is clear that they have no effect on SV40 DNA replication (Stillman et al 1985; Li et al 1986; Smale & Tjian 1986).

More clear-cut evidence for transcription regulatory elements that are associated with the core replication origin and required for replication in vivo has been obtained in several other replication systems. These include the related papovavirus, polyomavirus (deVilliers et al 1984), bovine papilloma virus (Lusky & Botchan 1984, 1986; Stenlund et al 1987; Mecsas & Sugden 1987), Epstein-Barr virus (Reisman & Sugden 1986; Mecsas & Sugden 1987), adenovirus (see above), the Drosophilia amplification controlling element found within the chorion gene clusters that are amplified by DNA replication at a specific stage during larval development (Spradling & Orr-Weaver 1987), the Tetrahymena ribosomal DNA clusters (rDNA clusters, Larson et al 1986), and the origin of replication and transcriptional silencer at the HMR locus of the yeast *S. cerevisiae* (Brand et al 1987). In all these cases, the function of the transcription controlling elements that are required for replication in vivo is unclear. For example, a cell-free replication system from mouse cells is capable of replicating plasmids containing the polyomavirus origin, but in stark contrast to the in vivo situation where the auxiliary sequences (enhancer) are essential, the auxiliary sequences have no effect in vitro, and the core origin sequence is sufficient (Murakami et al 1986a; Prives et al 1987). This result, and a similar observation for SV40 (Stillman et al 1985; Li et al 1986) suggest that transcription per se is not required for the formation of a primer for initiation of DNA replication in vivo. It is more likely that the *cis*-acting transcription elements and their *trans*-acting binding proteins either activate the origin region by opening up the chromatin structure or that transcription near the origin facilitates sequence recognition by initiation proteins. Recently a scenario of the latter type has been described for transcription activation at the *E. coli ori* C replication origin (Baker & Kornberg 1988).

The SV40 core origin, without transcription regulatory elements, is necessary and sufficient for DNA replication in vitro (Stillman et al 1985; Li et al 1986; Smale & Tjian 1986; Dean et al 1987a), but to date, only naked plasmid DNA has been utilized as a template. Thus any flanking sequence effects on replication that act via changes in the chromatin structure would have been missed. The core origin is only 64 bp (nucleotides 5211 to 31, see Figure 2; Deb et al 1986a), and it contains a binding site for T antigen (called site II). Immediately adjacent to this core origin sequence is an additional binding site for SV40 T antigen (site I, Tjian 1978). T antigen binding site I is predominantly required for autoregulation of transcription of the T antigen gene (Rio et al 1980). This sequence is not essential for replication, but its presence stimulates initiation of replication approximately twofold, both in vivo and in vitro (Stillman et al 1985; DeLucia et al 1986; Li et al 1986).

The structure of the minimal SV40 core origin has recently been defined in great detail, primarily through the efforts of Tegtmeyer and his colleagues. The core origin consists of three domains that have been defined by site directed mutagenesis and are highly conserved in all primate papovavirus genomes (Deb et al 1986a). The first domain, located between T antigen binding sites I and II, consists of one-half of a partially palindromic sequence that is perfectly conserved in other papovaviruses that replicate in primate cells (nucleotides 5211–5220, Figure 2; Deb et al 1986a). Correct spacing between this essential "palindrome domain" and T antigen binding site II is important, but the sequence of the spacer between them is not. It has been reported that a cellular protein (called MCF), associated with DNA polymerase α, binds specifically to the palindrome domain, both to its double-strand and to one of its single strands (Traut & Fanning 1988; Fanning et al 1988). The ability of MCF to bind mutant origin sequences correlates with the replication activity of those DNAs, which suggests that MCF may function in DNA replication, but a direct demonstration of such a role has not yet surfaced. It is noteworthy that this conserved, essential domain corresponds to the region that is locally unwound by SV40 T antigen in the presence of ATP (see below). A second origin domain is the 20 base pair A·T rich sequence on the other side of T antigen site II (nucleotides 12–31, Figure 2). This region is essential for SV40 DNA replication in vitro (Stillman et al 1985; Deb et al 1986b; Gerard & Gluzman 1986; Li et al 1986; Hertz et al 1987); it

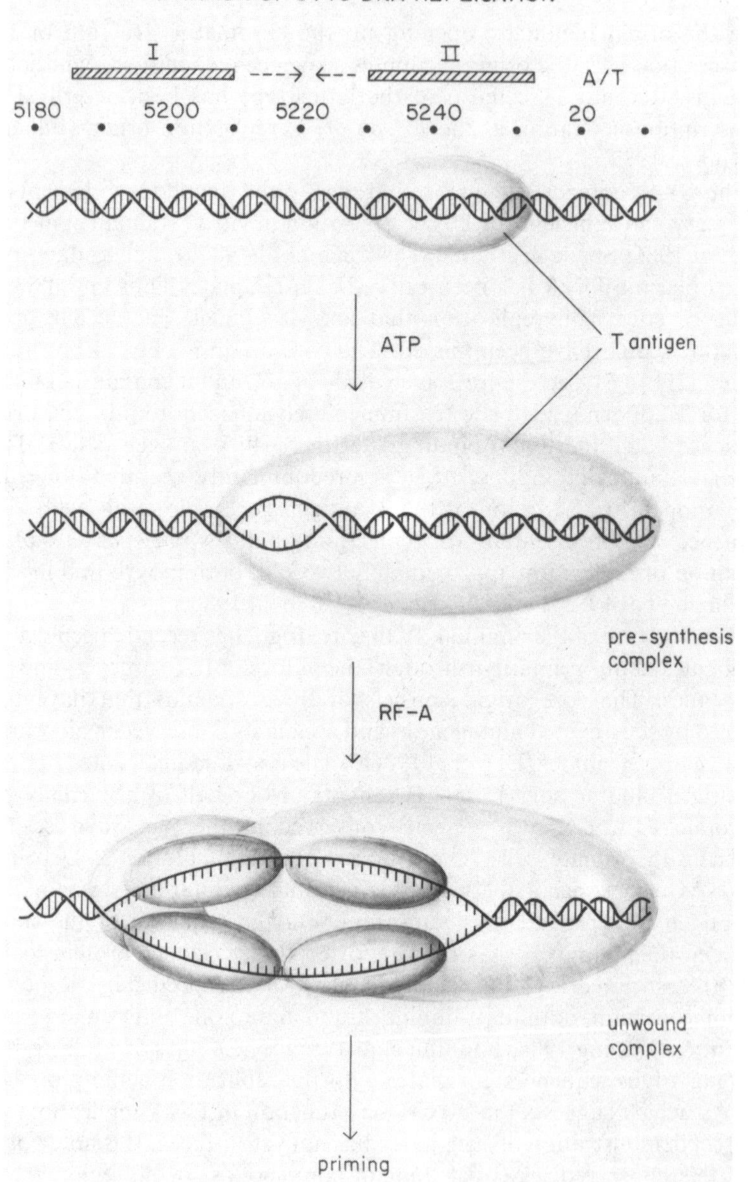

Figure 2 A model for the initiation of SV40 DNA replication. Top panel shows the structure of the minimal core origin of replication, with T antigen bound to site II (-ATP). The middle panel shows T antigen bound in the presence of ATP and the locally unwound region of the core origin. The bottom panel shows the unwound complex following addition of the cellular protein, RF-A. See text for details.

influences the binding of SV40 T antigen to site II (Müller et al 1987), is structurally altered upon T antigen binding to site II in the presence of ATP (Borowiec & Hurwitz 1988b), and is a center for DNA bending (Deb et al 1986b). Interestingly, a bent DNA sequence can function as a replication auxiliary sequence in the yeast origin of replication ARS1 (Snyder et al 1986; Williams et al 1988).

A cellular protein (called LOB) binds to the A·T rich domain from nucleotides 13 to 35 and induces DNA bending of the origin DNA (Baur & Knippers 1988). Although there is no evidence that that LOB protein is involved in DNA replication, it binds to the A·T rich region in the presence of SV40 T antigen and together these proteins produce an extended DNase I footprint that mimics the T antigen footprint seen in the presence of ATP (Borowiec & Hurwitz 1988b; see below). The LOB binding site overlaps with a region of the SV40 origin that inhibits SV40 DNA replication in vitro when added as a competitor sequence (Yamaguchi & DePamphilis 1986). As for the MCF protein, it nevertheless remains to be shown that the LOB protein functions in DNA replication.

The third and central feature of the core origin is the SV40 T antigen binding site II, which contains four repeats of the sequence 5′-GAGGC-3′ arranged in a 27 bp palindrome (Tjian 1978; DiMaio & Nathans 1980; DeLucia et al 1983; Tegtmeyer et al 1983; Tenen et al 1983; Jones et al 1984; Gottlieb et al 1985; Deb et al 1987). T antigen binds to this sequence in the absence of ATP and forms a variety of oligomeric structures, ranging up to tetramers (see Figure 2; Mastrangelo et al 1985; Dean et al 1987c), however binding in the presence of ATP is substantially different. In the presence of ATP and at 37°C (conditions for replication in vitro), T antigen protects a much extended region that now encompasses the entire core origin from DNase I digestion (Dean et al 1987c; Deb & Tegtmeyer 1987; Borowiec & Hurwitz 1988b; see Figure 2). Moreover, under these conditions, T antigen forms a double hexamer structure at the origin and causes the local unwinding of approximately eight base pairs in one-half of the palindrome domain plus a structural deformation of the A·T rich region (Dean et al 1987c; Borowiec & Hurwitz 1988a; Mastrangelo et al 1989; Figure 2). A similar ATP dependent local unwinding (or melting) of origin sequences that flank the origin-specific protein binding sites is observed during the initial step in initiation of bidirectional replication at the origin of replication on both the *E. coli* chromosome (*ori* C) and the bacteriophage λ chromosome (Bramhill & Kornberg 1988a,b; Schnos et al 1988). It is therefore probable that proteins and DNA sequences that combine to locally unwind immediately adjacent DNA are a common feature of origins of bidirectional replication in other virus DNAs, and perhaps in cellular chromosomes as well.

SV40 T Antigen

The large tumor antigen (T antigen) encoded by the SV40 gene expressed early in infection is an extraordinary protein by any measure. This multifunctional phosphoprotein of 708 amino acids was one of the first oncogene products to be characterized in detail and plays multiple roles during the productive infection of cells by SV40. In addition to its role in replication of SV40 DNA, which is discussed below in some detail, T antigen also autoregulates its own synthesis by a classical repressor/operator mechanism, induces quiescent cells to enter the proliferative cycle, activates cellular DNA and ribosomal RNA synthesis, and binds to a number of interesting cellular proteins that have been implicated in growth control, namely the protein encoded by the recessive retinoblastoma oncogene, the cellular p53 protein, the cellular transcription factor AP2, and possibly DNA polymerase α/DNA primase complex (Smale & Tjian 1986; Mitchell et al 1987; deCaprio et al 1988; see reviews by Rigby & Lane 1983; DePamphilis & Bradley 1986). Of prime importance for this discussion is the function of T antigen in DNA replication and how it interacts with the host cell replication machinery.

One function of T antigen that is not directly involved in the actual synthesis of DNA, but nevertheless is important for SV40 DNA replication, is the ability of T antigen to induce quiescent cells to proliferate (i.e. transit from G0 to G1; Tjian et al 1978; Mueller et al 1978). This property is shared by the adenovirus E1A oncogene product and is of considerable general interest. Both proteins interact with the cellular retinoblastoma recessive oncogene-encoded protein, a protein implicated in cellular growth control (Whyte et al 1988; de Caprio et al 1988). Both SV40 and adenovirus intervene in this growth stimulatory process at a relatively late stage in the transition from a resting state to actively dividing cells because both will induce DNA synthesis in a temperature-sensitive rat cell line blocked in G1 at the nonpermissive temperature, whereas 10% serum fails (Ninomiya-Tsuji et al 1987). T antigen and E1A are nuclear localized proteins and thus must act inside the cell, whereas serum stimulates the G0 to G1 transition via extracellular signals and growth factor receptors. Thus, T antigen and E1A may target a critical regulatory step in the mitogenic pathway that is far downstream of growth factor receptors and their second messengers. Understanding the regulation of G0 to G1 transition and how T antigen and E1A intervene will be an important achievement in biological research.

SV40 T antigen has at least two direct roles in SV40 DNA replication and both functions can be separated by mutagenesis or partial proteolysis from the oncogenic functions. The first, discussed briefly above, is the ability to bind specifically to the minimal core origin of replication (Tjian 1978). Extensive mutagenesis of T antigen has demonstrated that the region of T antigen required for site specific origin binding is a short 114 amino acid region (amino acids 132–246 from the amino-terminus; Gluzman & Ahrens 1982; Kalderon & Smith 1984; Manos & Gluzman 1984, 1985; Stringer 1982; Huber et al 1985; Peden & Pipas 1985; Stillman et al 1985; Cole et al 1986; Paucha et al 1986; Gish & Botchan 1987; Strauss et al 1987; Simmons 1986, 1988; D. McVey & Y. Gluzman, personal communication). This origin binding domain, which recognizes the pentameric sequence 5′-GAGGC-3′ (see above), does not contain a recognizable DNA binding motif and does not include the putative zinc finger motif at amino acids 302–320. Mutations in the zinc binding motif nevertheless inactivate the replication function of SV40 T antigen (Loeber et al 1989). It is possible that this putative zinc finger is involved in oligomerization of T antigen at the origin or that it increases T antigen single-strand DNA binding affinity, in a manner analogous to the zinc dependent cooperative binding of the gene 32 protein of phage T4 to single-strand DNA (Giedroc et al 1987; see adenovirus section for discussion).

As described above, T antigen binding to site II is influenced by the presence of ATP (Figure 2). The *E. coli ori* C specific, DNA binding protein, dna A also is influenced by ATP and may be a paradigm for a subset of the functions performed by T antigen. The dna A protein can exist in two states, an inactive ADP bound form or an active ATP bound form, the latter of which is able to induce local unwinding at *ori* C (Bramhill & Kornberg 1988a). The membrane phospholipid cardiolipin (diphosphatidylglycerol) promotes exchange of the ADP for ATP, thereby allowing dna A to function at *ori* C (Sekimizu & Kornberg 1988). These effects are probably mediated by membrane attachment of the dna A protein and perhaps indicate that compartmentalization of dna A is critical to its function and regulation (Sekimizu et al 1988; Yung & Kornberg 1988). It is worth speculating that an analogous phenomenon may occur with T antigen. A subfraction of T antigen is associated with a nuclear structure commonly referred to as the nuclear matrix (Covey et al 1984; Hinzpeter & Deppert 1987; Schirmbeck & Deppert 1987). This association occurs concomitant with the onset of virus DNA replication in infected cells and is temperature-sensitive in transformed cells containing the replication defective *ts*A58 mutant. However, there is no definitive evidence that nuclear matrix association is required for DNA replication in vivo.

Another more direct method of regulating DNA replication is by phosphorylation and there is clear evidence that this occurs for SV40 T antigen. T antigen contains two clusters of phosphorylated serine and threonine residues, one of which is adjacent to the DNA binding domain (Ser 106,

111, 112, 123, and Thr 124) and the other near the carboxy-terminus (Ser 639, 676, 677, 679, and Thr 701) (Scheidtmann et al 1982). Dephosphorylation of T antigen with alkaline phosphatase increases the ability of T antigen to bind to site II and its ability to initiate DNA replication in vitro but has no effect on the DNA helicase or ATPase activity (Simmons et al 1986; Grässer et al 1987; Mohr et al 1987; Klausing et al 1988). The systematic mutagenesis of each of these phosphorylated amino acids to non-phosphorylatable residues has been more revealing (Schneider & Fanning 1988). Mutations in Ser 123 and Thr 124 disabled T antigen so that it could not support DNA replication in vivo and interestingly, the Thr 124 mutant was able to bind to T antigen binding site I, but not to site II within the core origin. In contrast, mutants in Ser 677 and 679 supported more DNA replication in vivo than did wild-type. The Ser 677 mutant was able to bind site II, but not site I and the Ser 679 mutant bound to site II better than wild-type. Thus phosphorylation of residues outside the DNA binding domain can dramatically affect the ability of T antigen to bind to, and discriminate between, two functionally distinct DNA sites and may explain the functionally distinct subclasses of T antigen in infected cells (Schirmbeck & Deppert 1988). This example should be noted when considering how phosphorylation of cellular DNA binding proteins may regulate specific gene expression, or perhaps even cellular DNA replication.

A second function of SV40 T antigen in DNA replication is its DNA helicase activity and the associated ATPase and single-strand binding activities. The helicase function maps to the carboxy-terminal half of the protein. T antigen was long recognized as an ATPase that did not depend upon exogenous DNA, but was stimulated by the addition of poly (dT) (Giacherio & Hagar 1979; Tjian & Robbins 1979). Studies of mutant phenotypes demonstrated that the ATPase activity is required for DNA replication (Huber et al 1985; Stillman et al 1985; Cole et al 1986; Farber et al 1987), and monoclonal antibodies that block the ATPase activity also block replication in vitro (Clark et al 1981; Stahl et al 1985; Smale & Tjian 1986; Wielkowski et al 1987). Surprisingly, in some of these studies, monoclonal antibodies inhibited replication in a cell-free system that was capable of elongation, but not initiation of replication. This observation, coupled with the observation that T antigen appears to be bound to origin distal regions of replicating molecules (Stahl & Knippers 1983; Tack & DePamphilis 1983; Tack & Proctor 1987), led Stahl et al (1986) to the important discovery that T antigen is also a DNA helicase capable of unwinding duplex DNA into its single-strand components. The DNA helicase moves in the 3′ to 5′ direction on the strand to which it is bound, implicating it as a helicase that moves on the leading strand template (Goetz et al 1988; Wiekowski et al 1988). This is in contrast to the direction of movement of the major replicative helicase in *E. coli*, the dna B helicase that moves in the 5′ to 3′ direction and constitutes part of the primosome (LeBowitz & McMacken 1986).

By taking stock of these observations, the emerging view of the initial events in SV40 DNA replication is that T antigen binds to site II and in the presence of ATP, forms an oligomeric structure capable of binding to an extended region of the origin and locally unwinding the early palindrome region (Figure 2). In the presence of the *E. coli* single-strand DNA binding protein (SSB) to stabilize unwound single strands and topoisomerase I to relieve torsional strain in the circular molecule, T antigen can extensively unwind plasmid DNAs containing the origin of replication, even in the absence of DNA synthesis (Dean et al 1987b; Dodson et al 1987; Wold et al 1987; Goetz et al 1988; Wiekowski et al 1988). A protein from human cells can substitute for *E. coli* SSB in this reaction (see below). After deproteinization, the product of this reaction is a highly negatively supercoiled, underwound circular molecule called form U that resembles a similar product formed by unwinding plasmids containing the *E. coli ori* C region in the presence of the dna A (origin binding protein), dna B (helicase), dna C (which associates with dna B), SSB (single-strand DNA binding protein) and gyrase (topoisomerase) proteins (see Bramhill & Kornberg 1988a). Thus T antigen fulfills the role of three *E. coli* proteins (dna A, B, and C) in a similar unwinding reaction. It is probable that extensive unwinding of the template does not normally occur prior to the initiation of DNA synthesis and that unwinding and DNA synthesis at the replication fork are normally linked events (Dröge et al 1985).

Cellular Replication Proteins

As we have seen, SV40 T antigen plays a central role in the initiation and perhaps elongation of DNA replication from the SV40 origin, but it does not do everything! Local unwinding of the origin, the synthesis of both strands at the replication fork, and termination and segregation require cellular proteins that probably fulfill similar roles in the uninfected host cell. Consequently, the identification and characterization of these cellular proteins may pave the way for comprehending the mechanism and regulation of cellular replication events.

RF-A, A SINGLE STRAND DNA BINDING PROTEIN The identification of cellular proteins involved in SV40 DNA replication in vitro has predominantly relied upon the direct fractionation of the cellular extracts into multiple components, with subsequent purification of the active factors. Alternatively, when previously known cellular proteins were suspected to be involved (e.g. polymerase α and topoisomerases), antibodies directed against these proteins have proven useful. A cellular factor required, along with SV40 T antigen, for the initiation of DNA replication in vitro has been identified as a multi-subunit protein called RF-A (Wobbe et al 1987; Wold & Kelly 1988; Fairman & Stillman 1988; note that this protein has the alternative names of RP-A or HeLa SSB). The protein consists of three tightly complexed subunits with molecular weights of 70,000 (70 kd), 34,000 (34 kd), and 11,000 (11 kd) and binds to both single- and double-strand DNA, but preferentially to single strands (note that the 76 and 72 kd HeLa SSB described by Wobbe et al is identical to the 70, 34 and 11-kd protein described in other laboratories). RF-A can substitute for the *E. coli* SSB in the unwinding reaction that yields form U (Wobbe et al 1987; Wold & Kelly 1988; T. Tsurimoto, M. P. Fairman & B. Stillman, submitted for publication), but RF-A must be doing more than binding to single-strand DNA during replication because *E. coli* SSB will not substitute for RF-A for DNA replication.

The steps involving RF-A in initiation of replication are of interest because RF-A may interact with cellular proteins that recognize cellular origins of DNA replication. Furthermore, extracts prepared from human cells that had been separated by centrifugal elutriation into cells in the G1, S, and G2 phases of the cell cycle revealed that G1 extracts were inefficient in supporting SV40 origin specific replication, whereas extracts from both S phase and G2 cells had significantly higher specific activities (Roberts & D'Urso 1988). This difference was also reflected in the SV40 origin specific unwinding reaction (form U production), since G1 extracts were essentially inactive, but S phase and G2 extracts were active. Since RF-A and topoisomerase I were the only cellular proteins known to be required for form U production, both RF-A and topoisomerase were added to G1 extracts, however neither increased the activity of these extracts. These results suggest that RF-A functions with other cellular components in a key cell cycle regulated step in the initiation of DNA replication, although other possibilities exist.

Inspection of the kinetics of DNA synthesis in the SV40 replication reaction reveals a lag of 10 to 15 min before any synthesis occurs; thereafter vigorous DNA synthesis ensues over the next one to two hr (Stillman et al 1986; Wobbe et al 1986). The time lag can be eliminated if T antigen and a subset of cellular proteins are preincubated with the template DNA, whereupon immediate DNA replication occurs (Stillman et al 1986; Wobbe et al 1986; Fairman et al 1987; Wold et al 1987; Fairman et al 1988). Recently it was recognized that part of this presynthesis reaction requires cellular proteins to overcome the inhibitory action of another cellular protein, which inhibits origin specific unwinding (T. Tsurimoto et al, submitted for publication). If the inhibitor is removed by biochemical fractionation of the extracts, then the time lag can be overcome by preincubation with T antigen, RF-A, and topoisomerase I with the template DNA. Since addition of RF-A does not activate G1 extracts to support efficient SV40 DNA replication (Roberts & D'Urso 1988), the cell cycle results may be explained by the effect of inhibitors on the unwinding reaction. It is clear that future studies must address the roles played by RF-A and these inhibitors in cellular DNA synthesis so as to determine if they are involved in a cell cycle regulated reaction in vivo.

POLYMERASE α/PRIMASE COMPLEX The DNA polymerase α/DNA primase complex from mammalian cells has long been implicated in cellular DNA replication (reviewed by Burgers 1989) and is clearly required for SV40 DNA replication. The animal cell complex is highly conserved in structure and contains polypeptides of 180 kd (the catalytic subunit), 70 kd and two subunits of 58 and 48 kd that comprise the DNA primase subunits. The *S. cerevisiae* analogue, DNA polymerase I, has a similar polypeptide composition and mutants in the gene encoding the catalytic subunit are defective in DNA replication (reviewed by Burgers 1989). A cDNA for the human polymerase α catalytic subunit has been isolated and its sequence reveals strong homology with polymerases from diverse sources such as the bacteriophage T4, herpes virus, adenovirus, vaccinia virus, and yeast (Wang et al 1989). The polymerase α activity and expression of the gene appear not to be cell cycle regulated, but gene expression and consequently enzyme activity are dramatically increased when quiescent cells are stimulated to proliferate (Wahl et al 1988).

The involvement of polymerase α/primase in SV40 DNA replication in vitro was initially investigated by depleting the complex from crude extracts by immunoaffinity chromatography, demonstrating that the flow-through was inactive, and then reconstituting replication with purified complex (Murakami et al 1986b). It was of considerable interest that replication could only be restored with purified DNA polymerase α/primase complex derived from primate (including human) cells that are permissive for SV40 DNA replication, but not with the analogous enzymes from the nonpermissive cells of murine or bovine origin. Moreover, addition of purified human polymerase α/primase to extracts made from the nonpermissive murine FM3A cells enabled these previously inactive extracts to support SV40 DNA replication. These results suggested that the host cell specificity of SV40 replication was determined, at least in part, by the interaction of the virus T antigen or origin DNA with the host cell polymerase α/primase complex. Elegant confirmation of these results was obtained when replication from the murine polyomavirus origin, dependent upon purified polyomavirus T antigen, could occur in normally nonpermissive human HeLa cell extracts only upon the addition of purified murine polymerase α/primase complex (Murakumi et al 1986a).

These results took on added significance when an interaction between SV40 T antigen and polymerase α was suggested by coprecipitation experiments (Smale & Tjian 1986). A small fraction of DNA polymerase α activity was selectively retained on a T antigen affinity column when crude extracts, but not purified DNA polymerase α/primase complex were used, which suggests that T-antigen and polymerase α/primase complex interact either directly, or via a third cellular component. A subset of anti-T antigen monoclonal antibodies blocked this interaction (Smale & Tjian 1986) and it was recognized that this subset of antibodies also blocked the binding of T antigen to the murine nuclear phosphoprotein, p53 (Gannon & Lane 1987). The cellular p53 protein from murine cells was able to block the association of T antigen with polymerase α in crude extracts in vitro (Gannon & Lane 1987), or block SV40 origin dependent DNA replication in vivo (Braithwaite et al 1987). These results suggest that p53 from murine cells may interact with T antigen and form a tight complex that would prevent the T antigen from functioning in DNA replication. A liberal interpretation of these results is that p53 might be directly involved in the regulation of DNA replication. The p53 protein is present in elevated levels in transformed and tumor derived cells and has been implicated in growth control (Mercer et al 1982, 1984 and references therein). An intriguing observation provides more fuel to this speculation about the role of p53 in replication and may provide an experimental approach to examine this possibility. A monoclonal antibody that recognizes human p53 was shown to cross react with oligomeric forms of SV40 T antigen, but not with monomeric forms (Leppard & Crawford 1984). It is possible that p53 and the oligomeric form of T antigen are structurally related because they bind the same cellular protein. If this is so, it should be possible to identify this hypothetical cellular component by immunological techniques and test whether it is involved in SV40 DNA replication and whether it binds T antigen and p53. DNA polymerase α would be a good candidate for this protein. The above results suggest that the polymerase α/primase complex may recognize the unwound region at the replication origin by binding to T antigen, RF-A, both of these, or via some other intermediate protein.

Protein-free, single-strand DNA derived from both strands of the SV40 replication origin and cloned into the bacteriophage M13 genome have been used as template DNAs for purified polymerase α/primase complex from monkey cells, or for purified primase from murine cells (Tseng & Ahlem 1984; Yamaguchi et al 1985). These experiments revealed synthesis of RNA primers at specific sites within the same region of the palindrome domain that is locally unwound by T antigen in vitro. Thus purified primase can recognize the sequence that is locally exposed by T antigen as single strands within the core origin. However more work will be required to determine the significance of these results especially since the primer sites do not precisely coincide with the sites suggested to be used in vivo (Hay & DePamphilis 1982).

PROLIFERATING CELL NUCLEAR ANTIGEN (PCNA) The initiation of DNA replication at the SV40 origin involves a number of distinct stages including origin recognition by T antigen, local unwinding of the duplex DNA, the formation of a multiprotein complex that includes T antigen and the RF-A protein, and finally synthesis of the first nascent strand at the origin, probably by the polymerase α/primase complex. Elongation of DNA replication is perhaps more complicated because a replication fork must be established on each side of the origin that is capable of synthesizing both the leading (continuously synthesized) and lagging (discontinuously synthesized) strands. Studies using the SV40 cell free replication system have begun to dissect these events and determine how they might be regulated.

A 36,000 dalton replication factor (initially called replication factor B, or RF-B) was identified as a cellular protein that is essential for complete DNA replication in vitro (Prelich et al 1987a). In contrast to other previously described proteins, this factor was not required for initiation of replication, but was required at a later stage (discussed in detail below). When this replication factor was purified, its biochemical and physical properties indicated that it was similar to three previously and independently characterized proteins suspected to be involved in cell growth or division. Indeed, it is now known that this replication factor is identical to these known proteins (Prelich et al 1987a,b; Bravo et al 1987) and a brief description of each follows. A more detailed review of PCNA has recently appeared (Matthews 1989).

The first protein was initially identified as a human auto-antigen using sera from a number of patients with systemic lupus erythematosus and was found to be present in cells in culture or in tissues that were actively undergoing growth and division (Miyachi et al 1978). Thus, this antigen, identified as a 36,000 (36 kd) protein, was called the proliferating cell nuclear antigen (PCNA). At approximately the same time, a protein called cyclin was observed to be preferentially synthesized in proliferating cells and in a cell cycle dependent manner (Bravo & Celis 1980; Bravo et al 1981). The identity of cyclin and PCNA was recognized a number of years later (Mathews et al 1984) and because the name cyclin has been used for another cell cycle regulated protein involved in mitotic control (see chapter by Cross, Roberts & Weintraub, this volume), the name PCNA will be used henceforth.

The third protein equated with the replication factor was then a newly identified protein that stimulated the processivity of a relatively unknown mammalian DNA polymerase called polymerase δ (Tan et al 1986; Bravo et al 1987; Prelich et al 1987b). This result raised the distinct possibility that in addition to polymerase α, polymerase δ was required for SV40 DNA replication. Polymerase δ is one of four known cellular DNA polymerases in animal cells and has a functional analogue in the yeast *S. cerevisiae* called polymerase III (Bauer et al 1988; for a detailed review of DNA polymerases, see Burgers 1989; note also that there may be two polymerase δ enzymes, a PCNA dependent enzyme, and a PCNA independent enzyme; Wong et al 1989). Furthermore, a yeast protein that increases the processivity of yeast polymerase III has also been identified and is the yeast PCNA (Burgers 1988; Bauer & Burgers 1988). Neither polymerase δ, nor yeast polymerase III has an associated DNA primase activity, but in the presence of a suitable primer and PCNA, they copy long stretches of template in a highly processive manner (Lee et al 1984; Tan et al 1986; Prelich et al 1987b; Bauer and Burgers 1988; Burgers 1989).

The role of PCNA in SV40 DNA replication was investigated by comparing the structure of replication products formed in reactions with or without added PCNA (Prelich & Stillman 1988). These results demonstrated that initiation of DNA replication and the formation of the first nascent strands at the replication origin occurred normally in the absence of PCNA, but subsequent elongation was aberrant. In contrast to the normal sequence of events, in which bidirectional replication occurs on both leading and lagging strands, in the absence of PCNA leading strand replication did not occur and lagging strand synthesis was abnormal. The lagging strand products remained short (Okazaki fragment size) and were displaced from the template DNA. These results suggested two things. First, that PCNA is normally required for the coordinated synthesis of both leading and lagging strands at a replication fork and second, that DNA synthesized at the replication origin is not automatically elongated, but must be switched to an elongation mode of DNA synthesis following initiation. These results also suggest a model for the replication fork, based upon the model for the replication fork in phage T4 (Sinha et al 1980; Alberts et al 1982) and *E. coli* (Kornberg 1982) DNA replication. Because PCNA is a processivity factor for polymerase δ, it is possible that polymerase δ is the leading strand polymerase. Because polymerase α contains an associated DNA primase needed for Okazaki fragment synthesis, it is possible that it is the lagging strand polymerase. A model based upon these ideas is presented in Figure 3, but it should be noted that to date there is only indirect evidence demonstrating a function for polymerase δ in SV40 DNA replication. Recently the *S. cerevisiae* CDC2 gene was shown to

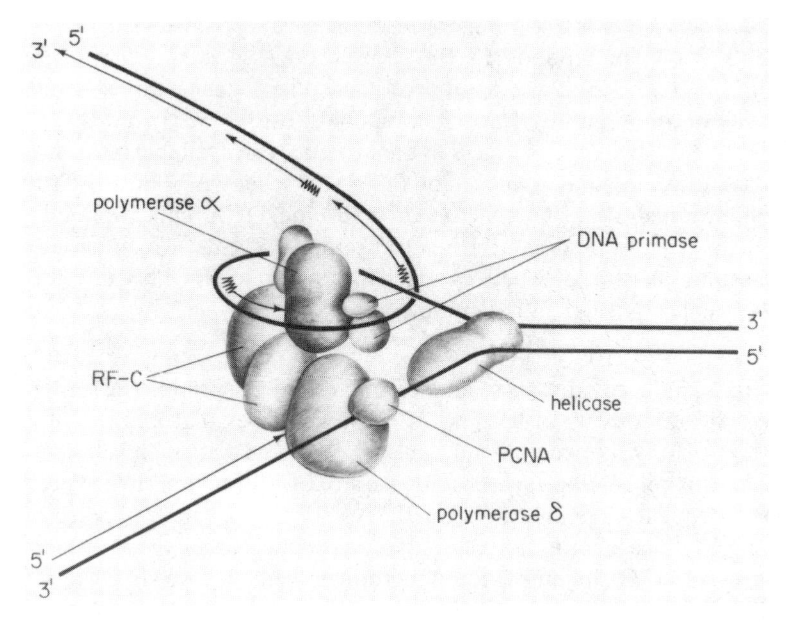

Figure 3 Elongation of SV40 DNA replication. Proposed model of a eukaryotic replication fork showing replication factors RF-C, PCNA and the two asymmetric DNA polymerases α and δ synthesizing the lagging and leading strands respectively. During SV40 DNA replication, the DNA helicase at the fork may be SV40 T antigen (see text for details).

encode the catalytic subunit of DNA polymerase III (polymerase δ, Sitney et al 1989; Boulet et al 1989). Mutations in this gene arrest cells in the S phase of the cell cycle, which provides strong genetic evidence for the role of two DNA polymerases in DNA replication.

PCNA seems to have a role in cellular DNA replication because a fraction of this protein is associated with replicating DNA (Bravo & McDonald-Bravo 1985; Celis & Celis 1985). Moreover, antibodies against PCNA inhibit DNA synthesis in isolated nuclei and in frog embryos (Wong et al 1987; Zuber et al 1989), and PCNA is essential for cell growth (Jaskulski et al 1988). PCNA can be detected by immunofluorescence using two different protocols and these studies have revealed two populations of the nuclear protein. The first, utilizing methanol fixation, detects a subset of nuclear PCNA that is only detected during the S phase of the cell cycle and that colocalizes with sites of DNA replication (Bravo & McDonald-Bravo 1985; Celis & Celis 1985; Madsen & Celis 1985). The second, utilizing formaldehyde fixation, detects all the PCNA in the nucleus (Bravo & McDonald-Bravo 1987). These two populations can be physically separated by selective extraction of cells with Triton detergent; the soluble fraction corresponds to the bulk of intranuclear PCNA, whereas the insoluble fraction corresponds to the replication associated PCNA. Even though PCNA is synthesized in a cell cycle dependent manner, it has a half-life of about 20 hr (Bravo & McDonald-Bravo 1987), and therefore appears to be available in excess in the nucleus. The available information to date indicates that a subset of PCNA is sequestered into a Triton-insoluble structure at sites of active chromosomal DNA replication. PCNA is not detectably modified by posttranslational events, which raises the interesting question of what regulates this apparent assembly into a presumed replicative elongation complex.

REPLICATION FACTOR C, RF-C A second elongation factor called RF-C has recently been identified and characterized (Tsurimoto & Stillman 1989). RF-C consists of a complex of polypeptides that migrate in two clusters on SDS-polyacrylamide gels in the 37–41 kd and 100–140 kd range. This replication factor is not required for initiation, but is needed at a subsequent stage. Indeed, when the type of product analysis performed in the presence or absence of PCNA was repeated in the presence or absence of RF-C, similar replication products were observed in the absence of RF-C as has been observed in the absence of PCNA (Tsurimoto & Stillman 1989; T. Tsurimoto et al, submitted for publication). Thus it appears that multiple replication factors are required to set up an elongation complex for coordinated leading and lagging strand synthesis following initiation of replication at the origin. The precise role played by RF-C is not known, but one possibility is that it provides a link between the two polymerases at the replication fork and enables them to synthesize DNA in a coordinated manner (see Figure 3). Recently it was suggested that PCNA overcomes the effect of an inhibitor of elongation during SV40 DNA replication in a reaction dependent upon a number of unidentified activator fractions (Lee et al 1988). It is possible that RF-C is one of these activators, but it is not yet clear how the inhibitor would fit into the scheme of replication outlined above.

TOPOISOMERASES I AND II The roles in DNA replication of the two known cellular topoisomerases, types I and II, were known from studies on SV40 DNA replication in vivo (Sundin & Varshavsky 1980, 1981; Weaver et al 1985; Snapka 1986; Richter et al 1987; Champoux 1988; Snapka et al 1988) and from genetic studies on the yeast *S. cerevisiae* and *S. pombe* (DiNardo et al 1984; Thrash et al 1984; Uemura & Yanagida 1984; Holm et al 1985). The consensus of these results is that topoisomerase I is not essential for cell viability and therefore for DNA replication, but it probably functions normally as a swivel enzyme to relieve torsional strain in the DNA caused by unwinding the parental DNA strands at the replication fork. On the other hand, topoisomerase II is essential and is required for the segregation of daughter molecules following termination of DNA replication. Topoisomerase II can also act like topoisomerase I as a "swivelase" during replication fork movement.

These topoisomerases I and II functions can also be observed during SV40 DNA replication in vitro. If topoisomerase I is inhibited by the specific inhibitor camptothecin, normal DNA synthesis occurs (Yang et al 1987). In contrast, if topoisomerase II function is blocked by the addition of the specific inhibitor VM26, DNA replication is not inhibited, but the replicated daughter molecules are not separated from each other and remained intertwined. Similar results were obtained when topoisomerase I and topoisomerase II were immuno-depleted from the extracts prior to DNA replication (Yang et al 1987). In this case, DNA synthesis was dependent upon addition of either topoisomerase I or II, but only the type II enzyme would decatenate the daughter molecules. Not surprisingly, therefore, topoisomerases are required for the complete reconstitution of DNA replication when using fractionated cell extracts (Wobbe et al 1987; Wold & Kelly 1988; Tsurimoto & Stillman 1989).

CELLULAR DNA REPLICATION

Eukaryotic chromosomes, unlike their prokaryotic counterparts and the relatively simple virus genomes, contain multiple origins of DNA replication that cooperate to replicate the large amount of DNA in the long linear chromosomes once per cell cycle. The origins must remain silent in the G1, G2, and M phases of the cell cycle, yet be collectively activated in the S phase. The nature of replication origins in mammalian cells remains a mystery because they have been refractory to analysis by techniques so effectively used in the unicellular eukaryotes and for mammalian DNA viruses. In the yeast *Saccharomyces cerevisiae*, chromosomal origins of replication can be excised from the chromosomes and placed onto artificially constructed plasmids that replicate in yeast once per cell cycle (for excellent reviews of yeast DNA replication, see Umek et al 1989; Newlon 1988). These replication origins, called autonomously replicating sequences (ARS) clearly delineate the place where DNA replication begins and they function only once per S phase. Several reports of cell-free systems that replicate ARS plasmid DNAs in vitro have not proved to be reproducible or suitable for further analysis (reviewed by Jong & Scott 1985) and thus little progress on yeast DNA replication has been made using biochemical approaches. Fortunately, the powerful genetics of the yeast system has enabled the identification of replication proteins such as DNA polymerases and DNA primase (Burgers 1988; Newlon 1988).

In striking contrast to the situation in unicellular eukaryotes, amphibian eggs offer little hope for a simple genetic analysis, but have provided a very interesting cell-free replication system that may yield fundamental information about the regulation of replication. The early divisions in *Xenopus laevis* embryos occur in rapid succession with minimal G1 and G2 phases of the cell cycle, and DNA replication and cell division occur without significant gene expression (Newport & Kirschner 1982). Upon reaching the midblastula transition, replication and cell division slow down and gene expression commences. The activated egg contains many proteins that are required for DNA replication and that can replicate exogenously added nuclei or naked DNA by a semiconservative mechanism that is regulated in step with the normal cell cycle (Harland & Laskey 1980). One surprising observation is that there is little sequence specificity for initiation of DNA replication, since all DNAs replicate, be they bacterial plasmid, phage DNA, or complicated chromosomes (Méchali & Kearsey 1984). This observation does not imply that initiation on any given DNA occurs randomly. It is entirely possible that each DNA molecule would contain short sequences that resemble the host cell origins because origins of replication may indeed be relatively simple, as appears to be the case in *S. cerevisiae* (Palzkill & Newlon 1988). Alternatively, during the early stages of embryogenesis, the sequence specificity of replication origins may be relaxed, as occurs for polyomavirus DNA replication in one-cell mouse embryos (Martínez-Salas et al 1988).

Lohka & Masui (1983, 1984) first demonstrated that cell-free extracts from activated *Rana pipiens* eggs could assemble demembranated *Xenopus laevis* sperm nuclei containing highly condensed chromosomes into a structure resembling interphase nuclei. Moreover, these reconstituted nuclei supported DNA replication, evidenced by the incorporation of ^{3}H-thymidine into DNA. Subsequently, it was demonstrated that similar cell free extracts from *Xenopus* eggs could replicate demembranated sperm nuclei (Blow & Laskey 1986) or bacteriophage λ DNA (Newport 1987) by a process that was preceded by decondensation of the chromosomes and their assembly into nuclear structures. The cell-free replication of sperm nuclear DNA was shown to be semiconservative and, under some circumstances, rereplication was observed (Blow & Laskey 1986; Hutchison et al 1987). The cell-free extract consists of a soluble fraction that is capable of decondensing chromatin and presumably provides replication proteins, and a vesicular fraction that contributes the nuclear envelope membranes (Sheenan et al 1988). Both fractions are required for correct nuclear assembly, formation of nuclear pores, and for DNA replication. Since initiation of DNA replication requires an intact nucleus, it is possible that initiation only occurs when DNA is formed into a specialized nuclear structure. Alternatively, the nucleus might simply need to concentrate replication proteins before initiation can occur.

In this cell-free system, complete DNA replication occurred in some of the reconstituted nuclei, but not in others present in the same reaction, which demonstrated that each nucleus acts as an independent integrated unit (Blow & Watson 1987). Furthermore, within any one nucleus once DNA replication began, it nearly always went to completion because the haploid (n) content of DNA was quantitatively converted to a diploid (2n) DNA content even when neighboring nuclei had not begun to synthesize DNA (Blow & Watson 1987). Under ideal conditions, which corresponded to conditions that maintain the nuclei intact, replication doubled the DNA content and then ceased, thus reflecting the process of a single S phase. However, if the nuclei were deliberately permeabilized following one round of DNA replication, another round of replication followed in some of the nuclei (Blow & Laskey 1988). Interestingly, if the extracts were made so that they continue to synthesize proteins from endogenous mRNA and ribosomes, multiple rounds of replication occurred. In the presence of the protein synthesis inhibitor cycloheximide in the same extracts, only one round of replication ensued. The cycloheximide block could be overcome by the addition of an extract containing the maturating (mitosis) promoting factor (MPF), which induces nuclear envelope breakdown (Blow & Laskey 1988). Thus the protein synthesis requirement may only reflect the synthesis of active MPF. These interesting experiments have led Blow & Laskey (1988) to propose a model for control of replication in the cell cycle in which a critical factor, called licensing factor, can only enter the nucleus when the nuclear envelope is permeabilized or broken down, as it is in mitosis; in this model licensing factor promotes initiation of DNA replication, but once it is used it is inactivated. This model is consistent with the classic cell fusion experiments of Rao & Johnson (1970) who demonstrated that G2 nuclei do not replicate in heterokaryons containing G1 or S phase nuclei until the entire cell undergoes mitosis. But this model does not explain how G1 nuclei, which they would predict contains licensing factor, remain inactive for replication. Clearly other activating events must occur in G1 prior to S phase. The *Xenopus* egg cell-free system should be valuable for examining these questions, as well as providing a means to test if replication factors such as PCNA, RF-A, RF-C, and the various DNA polymerases identified in other systems function in cell chromosomal DNA replication.

EPILOGUE

In this discussion of three cell-free replication systems that reproduce the various mechanisms of DNA replication that occur in the nucleus of eukaryotic cells, one is struck by the apparent diversity by which DNA is replicated and how studies on DNA replication often fuse with other areas of investigation. The novel mechanism of protein priming of adenovirus DNA replication, also used in the prokaryotic world by bacteriophage ϕ29, involves recruitment of cellular proteins that normally function as transcription factors in the host cell. Similarly, transcription elements function as auxiliary regions for the replication of papovavirus, papillomavirus, Epstein-Barr virus, and some cellular origins in *Tetrahymena* and *Drosophila*. It is not yet clear if the majority of replication origins within eukaryotic chromosomes will be analogous to these examples.

In contrast to the protein priming mechanism for initiation of adenovirus DNA replication, SV40 DNA replication appears to start by a local unwinding of the duplex DNA. Most probably the first primer for DNA synthesis is formed by the polymerase α associated DNA primase. The mechanism of local unwinding of the SV40 origin by T antigen, ATP, and the cellular protein RF-A clearly parallels similar events at the *E. coli* and phage λ replication origins, which demonstrates that this mechanism of initiation has been conserved during evolution and suggests a possible mechanism for initiation at the multiple origins present in cellular chromosomes. Following initiation at the SV40 origin, the cellular replication apparatus involving replication factors such as PCNA, RF-C, topoisomerases, and possibly two DNA polymerases, α and δ, function to complete replication of the chromosomes. It is very likely that these factors, and others, perform a similar function in cellular DNA replication. Cell-free systems like the *Xenopus* egg extract model should be able to confirm this notion. Indeed, it may be possible to establish a chromosome replication system from mammalian cells that is similar to the *Xenopus* egg system, which would allow direct studies on the mammalian replication factors and their cell cycle regulation.

Soon all the cellular proteins that are required to replicate SV40 DNA in vitro will be identified. However, this is not the final goal of such studies, but merely the start of perhaps more interesting studies. As noted in the introduction, these factors should be an interesting starting point for investigating the regulatory pathways that determine cell proliferation and cell cycle control. Furthermore, if the replication factor RF-A functions in initiation at cellular replication origins, then identifying the cellular proteins with which it interacts may yield origin specific binding proteins analogous to T antigen.

Finally, one must consider how general the studies to date are by investigating other model systems. Because of space limitations I have not discussed the extensive work demonstrating transcription induced priming of mitochondrial DNA replication in vitro (see Chang & Clayton 1989 and references therein) and will only briefly mention three other virus systems that may prove valuable for future studies. Recent work on herpes simplex virus (HSV) DNA replication has led to the identification of seven virus-encoded proteins that are required for HSV DNA replication in vivo (Wu et al 1988; Weller et al 1988). These proteins include the previously characterized DNA polymerase and the single-strand DNA binding protein (ICP8), but also a newly discovered origin binding protein (Elias & Lehman 1988; Olivo et al 1988; Koff & Tegtmeyer 1988), and possibly a virus-induced DNA helicase (Crute et al 1988; McGeoch et al 1988), and DNA primase (Holmes et al 1988). Since all the HSV replication proteins were identified by genetic means and thus the genes encoding them are in hand, their over-expression in various vector systems should facilitate further functional studies and make the HSV system an attractive model for studying the mechanism of DNA synthesis. More work, however, is required to understand the mechanism of HSV DNA replication in infected cells.

Two eukaryotic viruses that may provide insight into the mechanism of DNA replication control, particularly cell cycle control, are the bovine papillomavirus (BPV) and Epstein-Barr virus (EBV). A recent comprehensive review of these model systems has appeared (Mecsas & Sugden 1987). Both viruses can exist in a latent state where the virus genome is maintained as an extra-chromosomal plasmid in the nucleus of infected cells. These circular DNAs replicate in synchrony with the host cell chromosomes and under normal circumstances replicate once per cell cycle. Interestingly, both virus DNAs contain a single origin of replication in such latently infected cells and like SV40, only contribute one (in the case of EBV) or possibly two (for BPV) virus-encoded replication proteins. Thus the cellular replication machinery must be responsible for the bulk of replication of these virus DNAs. These are presumably the same proteins that replicate SV40 DNA. If this is true, then understanding the essential differences between SV40, which replicates many times within a single S phase, and BPV and EBV, which replicate once per cell cycle will be illuminating. Clearly the differences must lie with the proteins involved in the initiation of replication. Therefore it will be interesting to understand the functions of the EBV nuclear antigen (EBNA-1), which is required for both transcription control and DNA replication and binds to two sites within the EBV origin, one a transcriptional enhancer sequence necessary for DNA replication. Similarly, we need to understand how the BPV, M, and R replication proteins work (M functions to regulate DNA replication and transcription, whereas R seems to be a positive replication factor). These virus-encoded proteins must interact with cellular proteins to activate the replication origin at a specific time in the cell cycle (Roberts &

Weintraub 1986; Berg et al 1986). Understanding this mechanism and how it is controlled may point to a similar mechanism of activating replication origins in uninfected eukaryotic cells. With these expectations, the field of eukaryotic DNA replication has an exciting future.

ACKNOWLEDGMENTS

I wish to thank many colleagues for helpful discussions and copies of published and unpublished papers. I am particularly grateful to Susan Smith, Ian Mohr, John Diffley, Tom Melendy, Toshiki Tsurimoto, and Winship Herr for critically reading draft manuscripts. I am indebted to Barbara Weinkauff for typing many drafts and Jim Duffy for his excellent artwork. Work in the author's laboratory was supported by grants from the National Institutes of Health (CA13106 and AI20460).

Literature Cited

Adhya, S., Schneidman, P. S., Hurwitz, J. 1986. Reconstruction of adenovirus replication origins with a human nuclear factor I binding site. *J. Biol. Chem.* 261: 3339–46

Alberts, B. M., Barry, J., Bedinger, P., Formosa, T., Jongeneel, C. V., Kreuzer, K. N. 1982. Studies on DNA replication in the bacteriophage T4 in vitro system. *Cold Spring Harbor Symp. Quant. Biol.* 47: 655–68

Aleström, P., Stenlund, A., Li, P., Pettersson, U. 1982. A common sequence in the inverted terminal repetitions of human and avian adenoviruses. *Gene* 18: 193–97

Almendral, J. M., Huebsch, D., Blundell, P. A., Macdonald-Bravo, H., Bravo, R. 1987. Cloning and sequence of the human nuclear protein cyclin: Homology with DNA-binding proteins. *Proc. Natl. Acad. Sci. USA* 84: 1575–79

Ariga, H. 1984. Replication of cloned DNA containing the Alu family sequence during cell extract-promoting simian virus 40 DNA synthesis. *Mol. Cell. Biol.* 4: 1476–82

Ariga, H., Klein, H., Levine, A. J., Horwitz, M. S. 1980. A cleavage product of the adenovirus DNA binding protein is active in DNA replication in vitro. *Virology* 101: 307–10

Ariga, H., Sugano, S. 1983. Initiation of simian virus 40 DNA replication in vitro. *J. Virol.* 48: 481–91

Ariga, H., Tsuchihaski, A., Naruto, M., Yamada, M. 1985. Cloned mouse DNA fragments can replicate in a simian virus 40 T antigen-dependent system in vivo and in vitro. *Mol. Cell. Biol.* 5: 563–68

Baker, T. A., Kornberg, A. 1988. Transcriptional activation of initiation of replication from the *E. coli* chromosomal origin: an RNA-DNA hybrid near *ori* C. *Cell* 55: 113–23

Bauer, G. A., Burgers, P. M. J. 1988. The yeast analog of mammalian cyclin/proliferating-cell nuclear antigen interacts with mammalian DNA polymerase δ. *Proc. Natl. Acad. Sci. USA* 85: 7506–10

Bauer, G. A., Heller, H. M., Burgers, P. M. J. 1988. DNA polymerase III from *Saccharomyces cerevisiae* 1 purificaton and characterization. *J. Biol. Chem.* 263: 917–24

Baur, C., Knippers, R. 1988. Protein-induced bending of the simian virus 40 origin of replication. *J. Mol. Biol.* 203: 1009–19

Berg, L., Stenlund, A., Botchan, M. R. 1986. Repression of bovine papilloma virus replication is mediated by a virally encoded trans-acting factor. *Cell* 46: 753–62

Bernstein, J. A., Porter, J. M., Challberg, M. D. 1986. Template requirements for in vivo replication of adenovirus DNA. *Mol. Cell. Biol.* 6: 2115–24

Binger, M. H., Flint, S. J., Rekosh, D. M. 1982. Expression of the gene encoding the adenovirus DNA terminal protein inproductively-infected and transformed cells. *J. Virol.* 42: 488–501

Blow, J. J., Laskey, R. A. 1966. Initiation of DNA replication in nuclei and purified DNA by a cell-free extract of *Xenopus* eggs. *Cell* 47: 577–87

Blow, J. J., Laskey, R. A. 1988. A role for the nuclear envelope in controlling DNA replication within the cell cycle. *Nature* 332: 546–48

Blow, J. J., Watson, J. V. 1987. Nuclei act as independent and integrated units of replication in a *Xenopus* cell-free DNA replication system. *EMBO J.* 6: 1997–2002

Borowiec, J. A., Hurwitz, J. 1988a. Localized melting and structural changes in the SV40 origin of replication induced by T antigen. *EMBO J.* 7: 3149–58

Borowiec, J. A., Hurwitz, J. 1988b. ATP stimulates the binding of simian virus 40 (SV40) large tumor antigen to the SV40 origin of replication. *Proc. Natl. Acad. Sci. USA* 85: 64–68

Boulet, A., Simon, M., Faye, G., Bauer, G. A., Burgess, P. M. J. 1989. Structure and function of the *Saccharomyces cerevisiae* CDC2 gene encoding the large subunit of DNA polymerase III. *EMBO J.* 8: 1849–54

Braithwaite, A. W., Sturzbecher, H.-W., Addison, C., Palmer, C., Rudge, K., Jenkins, J. R. 1987. Mouse p53 inhibits SV40 origin-dependent DNA replication. *Nature* 329: 458–60

Bramhill, D., Kornberg, A. 1988a. A model for initiation at origins of DNA replication. *Cell* 54: 915–18

Bramhill, D., Kornberg, A. 1988b. Duplex opening by dnaA protein at novel sequences in initiation of replication at the origin of the *E. coli* chromosome. *Cell* 52: 743–55

Brand, A. H., Micklem, G., Nasmyth, K. 1987. A yeast silencer contains sequences that can promote autonomous plasmid replication and transcriptional activation. *Cell* 51: 709–19

Bravo, R., Celis, J. E. 1980. A search for differential polypeptide synthesis throughout the cell cycle of HeLa cells. *J. Cell. Biol.* 84: 795–802

Bravo, R., Fey, S. J., Bellatin, J., Mose Larsen, M., Arevalo, J., Celis, J. E. 1981. Identification of a nuclear and of a cytoplasmic polypeptide whose relative proportions are sensitive to changes in the rate of cell proliferation. *Exp. Cell. Res.* 136: 311–19

Bravo, R., Frank, R., Blundell, P. A., Macdonald-Bravo, H. 1987. Cyclin/PCNA is the auxiliary protein of DNA polymerase-δ. *Nature* 326: 515–17

Bravo, R., Macdonald-Bravo, H. 1985. Changes in the nuclear distribution of cyclin (PCNA) but not its synthesis depend on DNA replication. *EMBO J.* 4: 655–61

Bravo, R., Macdonald-Bravo, H. 1987. Existence of two populations of cyclin/proliferating cell nuclear antigen during the cell cycle: Association with DNA replication sites. *J. Cell Biol.* 105: 1549–54

Burgers, P. M. J. 1988. Mammalian cyclin/PCNA (DNA polymerase δ auxiliary protein) stimulates processive DNA synthesis by yeast DNA polymerase III. *Nucleic Acids Res.* 16: 6297–306

Burgers, P. M. J. 1989. Eukaryotic DNA polymerases α and δ: conserved properties and interactions from yeast to mammalian cells. *Prog. Nucleic Acid Res. Mol. Biol.* In press

Cairns, J. 1963. The bacterial chromosome and its manner of replication as seen by autoradiography. *J. Mol. Biol.* 6: 208–13

Campbell, J. L. 1986. Eukaryotic DNA replication. *Annu. Rev. Biochem.* 55: 733–71

Celis, J. E., Celis, A. 1985. Cell cycle-dependent variations in the distribution of the nuclear protein cyclin (proliferating cell nuclear antigen) in cultured cells: Subdivision of S phase. *Proc. Natl. Acad. Sci. USA* 82: 3262–66

Challberg, M. D., Desiderio, S. V., Kelly, T. J. Jr. 1980. Adenovirus DNA replication in vitro: Characterization of a protein covalently linked to nascent DNA strands. *Proc. Natl. Acad. Sci. USA* 77: 5105–109

Challberg, M. D., Kelly, T. J. Jr. 1979a. Adenovirus DNA replication in vitro: origin and direction of daughter strand synthesis. *J. Mol. Biol.* 135: 999–1012

Challberg, M. D., Kelly, T. J. Jr. 1979b. Adenovirus DNA replication in vitro. *Proc. Natl. Acad. Sci. USA* 76: 655–59

Challberg, M. D., Kelly, T. J. Jr. 1981. Processing of the adenovirus terminal protein. *J. Virol.* 38: 272–77

Challberg, M. D., Kelly, T. J. Jr. 1982. Eukaryotic DNA replication: viral and plasmid model systems. *Annu. Rev. Biochem.* 51: 901–34

Challberg, M. D., Ostrove, J. M., Kelly, T. J. Jr. 1982. Initiation of adenovirus DNA replication: detection of covalent complexes between nucleotide and the 80 Kd terminal protein. *J. Virol.* 41: 265–70

Challberg, M. D., Rawlins, D. R. 1984. Template requirement for the initiation of adenovirus DNA replication. *Proc. Natl. Acad. Sci. USA* 81: 100–4

Champoux, J. 1988. Topoisomerase I is preferentially associated with isolated replicating simian virus 40 molecules after treatment of infected cells with camptotheoin. *J. Virol.* 62: 3675–83

Chang, D. D., Clayton, D. A. 1989. Mouse RNAase MRP RNA is encoded by a nuclear gene and contains a decamer sequence complementary to a conserved region of mitochondrial RNA substrate. *Cell* 56: 131–39

Chase, J. W., Williams, K. R. 1986. Single-stranded DNA binding proteins required for DNA replication. *Annu. Rev. Biochem.* 55: 103–36

Chiou, H. C., Weller, S. K., Coen, D. M. 1985. Mutations in the Herpes Simplex virus major DNA binding protein gene leading to altered sensitivity to DNA polymerase. *J. Virol.* 145: 213–26

Chodosh, L. A., Baldwin, A. S., Carthew, R. W., Sharp, P. A. 1988. Human CCAAT-binding proteins have heterologous subunits. *Cell* 53: 11–24

Clark, R., Lane, D. P., Tjian, R. 1981. Use of monoclonal antibodies as probes of Simian virus 40 T antigen ATPase activity. *J. Biol. Chem.* 256: 11854–58

Cole, C. M., Tornow, J., Clark, R., Tjian, R. 1986. Properties of the simian virus 40 (SV40) large T antigens encoded by SV40 mutants with deletions in gene A. *J. Virol.* 57: 539–46

Coombs, D. H., Robinson, A. J., Bodnar, J. W., Jones, C. J., Pearson, G. D. 1978. Detection of DNA-protein complexes: The adenovirus DNA-terminal protein and Hela DNA-protein complexes. *Cold Spring Harbor Symp. Quant. Biol.* 43: 741–53

Covey, L., Choi, Y., Prives, C. 1984. Association of simian virus 40 T antigen with the nuclear matrix of infected and transformed monkey cells. *Mol. Cell. Biol.* 4: 1384–92

Crute, J. J., Mocarski, E. S., Lehman, I. R. 1988. A DNA helicase induced by herpes simplex virus type I. *Nucleic Acids Res.* 16: 6585–96

Dean, F. B., Borowiec, J. A., Ishimi, Y., Deb, S., Tegtmeyer, P., Hurwitz, J. 1987a. Simian virus 40 large tumor antigen requires three core replication origin domains for DNA unwinding and replication in vitro. *Proc. Natl. Acad. Sci. USA* 84: 8276–71

Dean, F. B., Bullock, P., Murakami, Y., Wobbe, C. R., Weissback, L., Hurwitz, J. 1987b. Simian virus 40 (SV40) DNA replication: SV40 large T antigen unwinds DNA containing the SV40 origin of replication. *Proc. Natl. Acad. Sci. USA* 84: 16–20

Dean, F. B., Dodson, M., Echols, H., Hurwitz, J. 1987c. ATP-dependent formation of a specialized nucleoprotein structure by simian virus 40 (SV40) large tumor antigen at the SV40 replication origin. *Proc. Natl. Acad. Sci. USA* 84: 8981–85

Deb, S., DeLucia, A. L., Baur, C., Koff, A., Tegtmeyer, P. 1986a. Domain structure of the simian virus 40 core origin of replication. *Mol. Cell. Biol.* 6: 1663–70

Deb, S., DeLucia, A. L., Koff, A., Tsui, S., Tegtmeyer, P. 1986b. The adeneine-thymine domain of the simian virus 40 core origin detects DNA bending and coordinately regulates DNA replication. *Mol. Cell. Biol.* 6: 4578–84

Deb, S., Tsui, S., Koff, A., DeLucia, A. L., Parsons, R., Tegtmeyer, P. 1987. The T-antigen-binding domain of the simian virus 40 core origin of replication. *J. Virol.* 61: 2143–49

Deb, S. P., Tegtmeyer, P. 1987. ATP enhances the binding of simian virus 40 large T antigen to the origin of replication. *J. Virol.* 61: 3649–54

deCaprio, J. A., Ludlow, J. W., Figge, J., Shew, J., Huang, C., Lee, W., Marsilio, E., Paucha, E., Livingston, D. M. 1988. SV40 large tumor antigen forms a specific complex with the product of the retinoblastoma susceptibility gene. *Cell* 54: 275–83

Decker, R. S., Yamaguchi, M., Possenti, R., Bradley, M. K., DePamphilis, M. L. 1987. In vitro initiation of DNA replication in Simian virus 40 chromosomes. *J. Biol. Chem.* 262: 10863–72

DeLucia, A. L., Deb, S., Partin, K., Tegtmeyer, P. 1986. Functional interactions of the simian virus 40 core origin of replication with flanking regulatory sequences. *J. Virol.* 57: 138–44

DeLucia, A. L., Lewton, B. A., Tjean, R., Tegtmeyer, P. 1983. Topography of simian virus 40 A protein-DNA complexes: arrangement of pentanucleotide interaction sites at the origin of replication. *J. Virol.* 46: 143–50

DePamphilis, M. L. 1988. Transcriptional elements as components of eukaryotic origins of DNA replication. *Cell* 52: 635–38

DePamphilis, M. L., Bradley, M. K. 1986. Replication of SV40 and polyoma virus chromosomes. In *The Papovaviridae*, ed. N. P. Salzman, pp. 99–246. New York: Plenum

Desiderio, S. V., Kelly, T. J. Jr. 1981. Structure of the linkage between adenovirus DNA and the 55,000 molecular weight terminal protein. *J. Mol. Biol.* 145: 319–37

DeVilliers, J., Schaffner, W., Tyndall, C., Lupton, S., Kamen, R. 1984. Polyoma virus DNA replication requires an enhancer. *Nature* 312: 242–46

de Vries, E., Bloemers, S. M., van der Vliet, P. C. 1987. Incorporation of 5-bromodeoxycytidine in the adenovirus 2 replication origin interferes with nuclear factor 1 binding. *Nucleic Acids Res.* 15: 7223–34

deVries, E., van Driel, W., Tromp, M., van Boom, J., van der Vliet, P. C. 1985. Adenovirus DNA replication in vitro: site-directed mutagenesis of the nuclear factor I binding site of the Ad2 origin. *Nucleic Acids Res.* 13: 4935–52

Diffley, J. F. X., Stillman, B. 1986. Purification of a cellular, double-stranded DNA-binding protein required for initiation of adenovirus DNA replication by using a rapid filter-binding assay. *Mol. Cell. Biol.* 6: 1363–73

DiMaio, D., Nathans, D. 1980. Cold-sensitive regulatory mutants of simian virus 40. *J. Mol. Biol.* 140: 129–42

DiNardo, S., Voelkel, K. A., Sternglanz, R. 1984. DNA topoisomerase II mutant of *Saccharomyces cerevisiae*: Topoisomerase II is required for segregation of daughter molecules at the termination of DNA replication. *Proc. Natl. Acad. Sci. USA* 81: 2616–20

Dixon, R. A. F., Nathans, D. 1985. Purification of simian virus 40 large T antigen by immunoaffinity chromatography. *J. Virol.* 53: 1001–4

Dodson, M., Dean, F. B., Bullock, P. Echols, H., Hurwitz, J. 1987. Unwinding of duplex DNA from the SV40 origin of replication by T antigen. *Science* 238: 964–67

Dorn, A., Bollekens, J., Staub, A., Benoist, C., Mathis, D. 1987. A multiplicity of CCAAT box-binding proteins. *Cell* 50: 863–72

Dröge, P., Sogo, J. M., Stahl, H. 1985. Inhibition of DNA synthesis by aphidicolin induces supercoiling in simian virus 40 replicative intermediates. *EMBO J.* 4: 3241–46

Elias, P., Lehman, I. R. 1988. Interaction of origin binding protein with an origin of replication of herpes simplex virus 1. *Proc. Natl. Acad. Sci. USA* 85: 2959–63

Enns, R. E., Challberg, M. D., Ahern, K. G., Chow, K.-C., Mathews, C. Z., Astell, C. R., Pearson, G. D. 1983. Mutational mapping of a cloned adenovirus origin. *Gene* 23: 307–13

Enomoto, T., Lichy, J. H., Ikeda, J. E., Hurwitz, J. 1981. Adenovirus DNA replication in vitro: Purification of the terminal protein in a functional form. *Proc. Natl. Acad. Sci. USA* 78: 6779–83

Ensinger, M. G., Ginsberg, H. S. 1972. Selection and preliminary characterisation of temperature sensitive mutants of type 5 adenovirus. *J. Virol.* 10: 328–39

Fairman, M. P., Prelich, G., Stillman, B. 1987. Identification of multiple cellular factors required for SV40 replication in vitro. *Philos. Trans. R. Soc. London Ser. B* 317: 495–505

Fairman, M. P., Prelich, G., Tsurimoto, T., Stillman, B. 1988. Characterization of cellular proteins required for SV40 DNA replication in vitro. See Kelly & Stillman 1988, pp. 143–51

Fairman, M. P., Stillman, B. 1988. Cellular factors required for multiple stages of SV40 replication in vitro. *EMBO J.* 7: 1211–18

Fanning, E., Traut, W., Dornreiter, I., Dehde, S., Alliger, P., Posch, B. 1988. Sequence-specific binding of a cellular protein associated with DNA polymerase α to the SV40 core origin of DNA replication. See Kelly & Stillman 1988, pp. 177–81

Farber, J. M., Peden, K. W. C., Nathans, D. 1987. Trans-dominant defective mutants of simian virus 40 T antigen. *J. Virol.* 61: 436–45

Field, J., Gronostajski, R. M., Hurwitz, J. 1984. Properties of the adenovirus DNA polymerase. *J. Biol. Chem.* 259: 9487–95

Fowlkes, D. M., Lord, S. T., Linné, T., Pettersson, V., Philipson, L. 1979. Interactions between the adenovirus DNA-binding protein and double-stranded DNA. *J. Mol. Biol.* 132: 163–80

Friefeld, B. R., Krevolin, M. D., Horwitz, M. S. 1983a. Effects of the adenovirus H5ts125 and H5ts107 DNA binding proteins on DNA replication in vitro. *Virology* 124: 380–89

Friefeld, B. R., Lichy, J. H., Hurwitz, J., Horwitz, M. S. 1983b. Evidence for an altered adenovirus DNA polymerase in cells infected with the mutant Hsts149. *Proc. Natl. Acad. Sci. USA* 80: 1589–93

Gannon, J. V., Lane, D. P. 1987. p53 and DNA polymerase-α compete for binding to SV40 T-antigen. *Nature* 329: 456–58

Gerard, R., Gluzman, Y. 1986. Functional analysis of the role of the A + T-rich region and upstream flanking sequences in simian virus 40 DNA replication. *Mol. Cell. Biol.* 6: 4570–77

Giacherio, D., Hager, L. P. 1979. A poly (dT)-stimulated ATPase activity associated with Simian virus 40 large T antigen. *J. Biol. Chem.* 254: 8113–16

Giedroc, D. P., Keating, K. M., Williams, K. R., Coleman, J. E. 1987. The function of zinc in gene 32 protein from T4. *Biochemistry* 26: 5251–59

Giedroc, D. P., Keating, K. M., Williams, K. R., Konigsberg, W. H., Coleman, J. E. 1986. Gene 32 protein, the single-stranded DNA binding protein from bacteriophage T4, is a zinc metalloprotein. *Proc. Natl. Acad. Sci. USA* 83: 8452–56

Ginsberg, H. S., Lundholm, V., Linne, T. 1977. Adenovirus DNA-binding protein in cells infected with wild-type 5 adenovirus and two DNA-minus, temperature sensitive mutants, H5ts125 and H5ts149. *J. Virol.* 23: 142–51

Gish, W. R., Botchan, M. R. 1987. Simian virus 40–transformed human cells that express large T antigens defective for viral DNA replication. *J. Virol.* 61: 2864–76

Gluzman, Y. 1981. SV40 transformed simian cells support the replication of early SV40 mutants. *Cell* 23: 175–82

Gluzman, Y., Ahrens, B. 1982. SV40 early mutants that are defective for viral DNA synthesis but competent for transformation of cultured rat and simian cells. *Virology* 123: 78–92

Goetz, G. S., Dean, F. B., Hurwitz, J., Matson, S. W. 1988. The unwinding of duplex regions in DNA by the simian virus 40 large tumor antigen-associated DNA helicase activity. *J. Biol. Chem.* 263: 383–92

Gottlieb, P., Nasoff, M. S., Fisher, E. F., Walsh, A. M., Caruthers, M. H. 1985. Binding sites of SV40 T-antigen to SV40 binding site II. *Nucleic Acids Res.* 3: 6621–34

Grässer, F. A., Mann, K., Walter, G. 1987. Removal of serine phosphates from Simian virus 40 large T antigen increases its ability to stimulate DNA replication in vitro but has no effect on ATPase and DNA binding. *J. Virol.* 61: 3373–80

Green, M., Symington, J., Brackman, K. H., Cartas, M. A., Thornton, H., Young, L. 1981. Immunological and chemical identification of intracellular forms of adenovirus type-2 terminal protein. *J. Virol.* 40: 541–50

Gronostajski, R. M. 1987. Site specific DNA binding of nuclear factor 1: effect of the spacer region. *Nucleic Acids Res.* 15: 5545–59

Gronostajski, R. M. 1986. Analysis of nuclear factor 1 binding to DNA using degenerate oligonucleotides. *Nucleic Acids Res.* 14: 9117–32

Gronostajski, R. M., Knox, J., Berry, D., Miyamoto, N. G. 1988. Stimulation of transcription in vitro by binding sites for nuclear factor 1. *Nucleic Acids Res.* 16: 2087–98

Gronostajski, R. M., Nagata, K., Hurwitz, J. 1984. Isolation of human DNA sequences that bind to nuclear factor-I, a host protein involved in adenovirus DNA replication. *Proc. Natl. Acad. Sci. USA* 81: 4013–17

Guggenheimer, R. A., Nagata, K., Kenny, M., Hurwitz, J. 1983a. Protein-primed replication of plasmids containing the terminus of the adenovirus genome. II. Purification and characterization of a host protein required for the replication of DNA templates devoid of the terminal protein. *J. Biol. Chem.* 259: 7815–25

Guggenheimer, R. A., Nagata, K., Lindenbaum, J., Hurwitz, J. 1983b. Protein-primed replication of plasmids containing the terminus of the adenovirus genome. I. Characterization of an in vitro DNA replication system dependent on adenoviral DNA sequences. *J. Biol. Chem.* 259: 7807–14

Guggenheimer, R. A., Stillman, B. W., Nagata, K., Tamanoi, F., Hurwitz, J. 1984. DNA sequences required for the in vitro replication of adenovirus DNA. *Proc. Natl. Acad. Sci. USA* 81: 3069–73

Gutierrez, C., Guo, Z., Burhans, W., DePamphilis, M. L., Towt, J., Ju, G. 1988. Is c-myc protein directly involved in DNA replication. *Science* 240: 1202–203

Harland, R. M., Laskey, R. A. 1980. Regulated replication of DNA microinjected into eggs of X. laevis. *Cell* 21: 761–71

Hauser, J., Levine, A. S., Dixon, K. 1988. Fidelity of DNA synthesis in a mammalian in vitro replication system. *Mol. Cell. Biol.* 8: 3267–71

Hay, R. T. 1985a. Origin of adenovirus DNA replication. Role of the nuclear factor 1 binding site in vivo. *J. Mol. Biol.* 186: 129–36

Hay, R. T. 1985b. The origin of adenovirus DNA replication: minimal DNA sequence requirement in vivo. *EMBO J.* 4: 421–26

Hay, R. T., DePamphilis, M. L. 1982. Initiation of SV40 DNA replication in vivo: Location and structure of 5′ ends of DNA synthesized in the *ori* region. *Cell* 28: 767–79

Hay, R. T., McDougall, I. M. 1986. Viable viruses with deletions in the left inverted terminal repeat define the adenovirus origin of DNA replication. *J. Gen. Virol.* 67: 321–32

Hay, R. T., Stow, N. D., McDougall, M. 1984. Replication of adenovirus minichromosomes. *J. Mol. Biol.* 175: 493–510

Henninghausen, L., Siebenlist, U., Danner, D., Leder, P., Rawlins, D., Rosenfeld, P., Kelly, T. Jr. 1985. High-affinity binding site for a specific nuclear protein in the human IgM gene. *Nature* 314: 289–92

Herr, W., Sturm, R. A., Clerc, R. G., Corcoran, L. M., Baltimore, D. 1988. The POV domain: a large conserved region in the mammalian pit-1, oct-1, oct-2 and *Caenorhabditis* elegans unc-86 gene products. *Genes Dev.* 2: 1513–16

Hertz, G. Z., Young, M. R., Mertz, J. E. 1987. The A+T rich sequence of the simian virus 40 origin is essential for replication and is involved in bending of the viral DNA. *J. Virol.* 61: 2322–25

Hinzpeter, M., Deppert, W. 1987. Analysis of biological and biochemical parameters fo chromatin and nuclear matrix association of SV40 large T-antigen in transformed cells. *Oncogene* 1: 119–29

Holm, C., Goto, T., Wang, J. C., Botstein,

D. 1985. DNA topoisomerase-II is required at the time of mitosis in yeast. *Cell* 41: 553–63

Holmes, A. M., Wietstock, S. M., Ruyechan, W. T. 1988. Identification and characterization of a DNA primase activity present in Herpes simplex virus type 1-infected HeLa cells. *J. Virol.* 62: 1038–45

Huber, B., Vakalopoulou, E., Burger, C., Fanning, E. 1985. Identification and biochemical analysis of DNA replication-defective large T antigens from SV40-transformed cells. *Virology* 146: 188–202

Hutchison, C. J., Cox, R., Drepaul, R. S., Comperts, M., Ford, C. C. 1987. Periodic DNA synthesis in cell-free-extracts of *Xenopus* eggs. *Embo J.* 6: 2003–10

Iguchi-Ariga, S. M. M., Itani, T., Ylamaguchi, M., Ariga, H. 1987. c-myc protein can substitute for SV40 T antigen in SV40 DNA replication. *Nucleic Acids Res.* 15: 4889–99

Ikeda, J.-E., Enomoto, T., Hurwitz, J. 1981. Replication of the adenovirus DNA-protein complex with purified proteins. *Proc. Natl. Acad. Sci. USA* 78: 884–88

Ikeda, J.-E., Enomoto, T., Hurwitz, J. 1982. Adenoviral protein-primed DNA replication in vitro. *Proc. Natl. Acad. Sci. USA* 79: 2442–46

Jaskulski, D., de Riel, J. K., Mercer, W. E., Calabretta, B., Baserga, R. 1988. Inhibition of cellular proliferation by antisense oligodeoxynucleotides to PCNA cyclin. *Science* 240: 1544–46

Jeang, K.-T., Rawlins, D. R., Rosenfeld, P. J., Shero, J. H., Kelly, T. J., Hayward, G. S. 1987. Multiple tandemly repeated binding sites for cellular nuclear factor 1 that surround the major immediate-early promoters of simian and human cytomegalovirus. *J. Virol.* 61: 1559–70

Jeng, Y., Wold, W. S. M., Sugawara, K., Gilead, Z., Green, M. 1977. Adenovirus type 2 coded single stranded DNA binding protein: in vivo phosphorylation and modification. *J. Virol.* 22: 402–11

Jones, K. A., Kadonaga, J. T., Rosenfeld, P. J., Kelly, T. J., Tjian, R. 1987. A cellular DNA-binding protein that activates eukaryotic transcription and DNA replication. *Cell* 48: 79–89

Jones, K. A., Myers, R. M., Tjian, R. 1984. Mutational analysis of simian virus 40 large T antigen DNA binding sites. *EMBO J.* 3: 3247–55

Jong, A. Y. S., Scott, J. F. 1985. DNA synthesis in yeast cell-free extracts dependent on recombinant DNA plasmids purified from *Escherichia coli. Nucleic Acids Res.* 13: 2943–58

Kalderon, D., Smith, A. E. 1984. In vitro mutagenesis of a putative DNA binding domain of SV40 large-T. *Virology* 139: 109–37

Kaplan, L. M., Ariga, H., Hurwitz, J., Horwitz, M. S. 1979. Complementation of the temperature sensitive defect in H5*ts*125 adenovirus DNA replication in vitro. *Proc. Natl. Acad. Sci. USA* 76: 5534–38

Kelly, T. J. Jr. 1984. Adenovirus DNA replication. In *The Adenoviruses*, ed. H. S. Ginsberg, pp. 271–308. New York/London: Plenum

Kelly, T. J., Stillman, B., eds. *Cancer Cells. 6. Eukaryotic DNA Replication*, New York: Cold Spring Harbor Lab. Press

Kelly, T. J., Wold, M. S., Li, J. 1988. Initiation of viral DNA replication. *Adv. Virus Res.* 34: 1–42

Kenny, M. K., Balogh, L. A., Hurwitz, J. 1988. Initiation of adenovirus DNA replication. I. Mechanism of action of a host protein required for replication of adenovirus DNA templates devoid of the terminal protein. *J. Biol. Chem.* 263: 9801–808

Kenny, M., Hurwitz, J. 1988. Initiation of adenovirus DNA replication. II. Structural requirements using synthetic oligonucleotide adenovirus templates. *J. Biol. Chem.* 263: 9809–17

Kitchingman, G. R. 1985. Sequence of the DNA-binding protein of a human subgroup-E adenovirus (type-4): Comparisons with subgroup-A (type-12), subgroup-B (type-7), and subgroup-C (type-5). *Virology* 146: 90–101

Klausing, K., Scheidtmann, K-H., Baumann, E. A., Knippers, R. 1988. Effects of in vitro dephosphorylation on DNA-binding and DNA helicase activities of simian virus 40 large tumor antigen. *J. Virol.* 62: 1258–65

Klein, H., Maltzman, W., Levine, A. J. 1979. Structure function relationships of the adenovirus DNA binding protein. *J. Biol. Chem.* 254: 11051–60

Koff, A., Tegtmeyer, P. 1988. Characterization of major recognition sequences for a Herpes Simplex virus type-1 origin binding protein. *J. Virol.* 62: 4096–103

Kornberg, A. 1982. *Supplement to DNA Replication.* San Francisco: Freeman

Kruijer, W., Nicholas, J. C., van Schaik, F. M. A., Sussenbach, J. S. 1983a. Structure and function of DNA binding proteins from revertants of adenovirus type 5 mutants with a temperature sensitive DNA replication. *Virology* 124: 425–33

Kruijer, W., van Schaik, F. M. A., Speijer, J. G., Sussenbach, J. S. 1983b. Structure and function of adenovirus DNA binding protein: comparison of the amino acid sequences of the Ad5 and Ad12 proteins derived from the nucleotide sequence of the corresponding genes. *Virology* 128: 140–53

Lally, C., Dorper, T., Groger, W., Antoine, G., Winnacker, E.-L. 1984. A size analysis of the adenovirus replicon. *EMBO J.* 3: 333–37

Lanford, R. E. 1988. Expression of simian virus 40 T antigen in insect cells using a baculovirus expression vector. *Virology* 167: 72–81

Larson, D. D., Blackburn, E. H., Yaeger, P. C., Orias, E. 1966. Control of rDNA replication in tetrahymena involves a cis-acting upstream repeat of a promoter element. *Cell* 47: 229–40

LeBowitz, J. H., McMacken, R. 1986. The *Escherichia coli dna* B replication protein is a DNA helicase. *J. Biol. Chem.* 261: 4738–48

Lee, M. Y. W. T., Tan, C.-K., Downey, K. M., So, A. G. 1984. Further studies on calf thymus DNA polymerase δ purified to homogeneity by a new procedure. *Biochemistry* 23: 1906–13

Lee, S.-H., Ishimi, Y., Kenny, M. K., Bullock, P., Dean, F. B., Hurwitz, J. 1988. An inhibitor of the in vitro elongation reaction of simian virus 40 DNA replication is overcome by proliferating cell nuclear antigen. *Proc. Natl. Acad. Sci. USA* 85: 9469–73

Leegwater, P. A. J., Romboufs, R. F. A., van der Vliet, P. C. 1988. Adenovirus DNA replication in vitro: duplication of single-stranded DNA containing a panhandle structure. *Biochim. Biophys. Acta* 951: 403–10

Leegwater, P. A. J., van Driel, W., van der Vliet, P. C. 1985. Recognition site of nuclear factor 1, a sequence-specific DNA-binding protein from HeLa cells that stimulates adenovirus DNA replication. *EMBO J.* 4: 1515–21

Leppard, K., Crawford, L. V. 1984. An oligomeric form of simian virus 40 large T-antigen is immunologically related to the cellular tumor antigen p53. *J. Virol.* 50: 457–64

Li, J. J., Kelly, T. J. 1984. Simian virus 40 DNA replication in vitro. *Proc. Natl. Acad. Sci. USA* 81: 6973–77

Li, J. J., Kelly, T. J. 1985. Simian virus 40 DNA replication in vitro: Specificity of initiation and evidence for bidirectional replication. *Mol. Cell. Biol.* 5: 1238–46

Li, J. J., Peden, K. W. C., Dixon, R. A. F., Kelly, T. 1986. Functional organization of the simian virus 40 origin of DNA replication. *Mol. Cell. Biol.* 6: 1117–28

Lichy, J. H., Field, J., Horwitz, M. S., Hurwitz, J. 1982. Separation of the adenovirus terminal protein precursor from its associated DNA polymerase: role of both proteins in the initiation of adenovirus DNA replication. *Proc. Natl. Acad. Sci. USA* 79: 5225–29

Lichy, J. H., Horwitz, M. S., Hurwitz, J. 1981. Formation of a covalent complex between the 80,000 dalton adenovirus terminal protein and 5′-dCMP in vitro. *Proc. Natl. Acad. Sci. USA* 78: 2678–82

Lindenbaum, J. O., Field, J., Hurwitz, J. 1986. The adenovirus DNA binding protein and adenovirus DNA polymerase interact to catalyze elongation of primed DNA templates. *J. Biol. Chem.* 261: 10218–27

Linné, T., Philipson, L. 1980. Further characterization of the phosphate moiety of the adenovirus type 2 DNA-binding protein. *Eur. J. Biochem.* 103: 259–70

Loeber, G., Parsons, R., Tegtmeyer, P. 1989. The zinc finger region of simian virus 40 large T antigen. *J. Virol.* 63: 94–100

Lohka, M. J., Masui, Y. 1983. Formation in vitro of sperm pronuclei and mitotic chromosomes induced by amphibian ooplasmic contents. *Science* 220: 719–21

Lohka, M. J., Masui, Y. 1984. Roles of cytosol and cytoplasmic particles in nuclear envelope assembly and sperm pronuclear formation in cell-free preparations from amphibian eggs. *J. Cell. Biol.* 98: 1222–30

Lusky, M., Botchan, M. R. 1984. Characterization of bovine papillomavirus plasmid maintenance sequences. *Cell* 36: 391–401

Lusky, M., Botchan, M. R. 1986. Transient replication of BPV-1 plasmids: *cis* and *trans* requirements. *Proc. Natl. Acad. Sci. USA* 83: 3609–13

Madsen, P., Celis, J. E. 1985. S-phase patterns of cyclin (PCNA) antigen staining resemble topographical patterns of DNA synthesis. *FEBS Lett.* 193: 5–11

Manos, M. M., Gluzman, Y. 1984. Simian virus 40 large T-antigen point mutants that are defective in viral DNA replication but competent in oncogenic transformation. *Mol. Cell. Biol.* 4: 1125–33

Manos, M. M., Gluzman, Y. 1985. Genetic and biochemical analysis of transformation-competent, replication-defective simian virus 40 large T antigen mutants. *J. Virol* 53: 120–27

Martínez-Salas, E., Cupo, D. Y., DePamphilis, M. L. 1988. The need for enhancers is acquired upon formation of a diploid nucleus during early mouse development. *Genes Dev.* 2: 1115–26

Mastrangelo, I. A., Hough, P. V. C., Wall, J. S., Dobson, M., Dean, F. B., Hurwitz, J. 1989. APT-dependent assembly of double

hexamers of SV40 T antigen at the viral origin of DNA replication. *Nature* 338: 658–62
Mastrangelo, I. A., Hough, P. V. C., Wilson, V. G., Wall, J. S., Hainfeld, J. F., Tegtmeyer, P. 1985. Monomers through trimers of large tumor antigen bind region I and monomers through tetromers bind in region II of simian virus 40 origin of replication DNA as stable structures in solution. *Proc. Natl. Acad. Sci. USA* 82: 3626–30
Mathews, M. B. 1989. The proliferating cell nuclear antigen, PCNA, a cell cycle regulated DNA replication factor. In *Growth Control During Cell Aging*, ed. H. R. Warner, E. Wang. Boca Raton, Fla: CRC Press. In press
Mathews, M. B., Bernstein, R. M., Franza, B. R. Jr., Garrels, J. I. 1984. Identity of the proliferating cell nuclear antigen and cyclin. *Nature* 309: 374–76
Matsumoto, K., Moriuchi, T., Koji, T., Nakane, P. K. 1987. Molecular cloning of cDNA coding for rat proliferating cell nuclear antigen PCNA/cyclin. *EMBO J.* 6: 637–42
McGeoch, D. J., Dalrymple, M. A., Dolan, A., McNab, D., Perry, L. J., Taylor, P., Challberg, M. D. 1988. Structures of Herpes simplex virus type 1 genes required for replication of viral DNA. *J. Virol.* 62: 444–53
Mecsas, J., Sugden, B. 1987. Replication of plasmids derived from bovine papillomavirus type 1 and Epstein-Barr virus in cells in culture. *Annu. Rev. Cell. Biol.* 3: 87–108
Méchali, M., Kearsey, S. 1984. Lack of specific sequence requirement for DNA replication in *Xenopus* eggs compared with high sequence specificity in yeast. *Cell* 38: 55–64
Mercer, W. E., Avignolo, C., Baserga, R. 1984. Role of the p53 protein in cell proliferation as studied by microinjection of monoclonal antibodies.*Mol. Cell. Biol.* 4: 276–81
Mercer, W. E., Nelson, D., DeLeo, A. B., Old, L. J., Baserga, R. 1982. Microinjection of monoclonal antibody to protein p53 inhibits serum-induced DNA synthesis in 3T3 cells. *Proc. Natl. Acad. Sci. USA* 79: 6309–12
Miksicek, R., Borgmeyer, U., Nowock, J. 1987. Interaction of the TGGCA-binding protein with upstream sequences is required for efficient transcription of mouse mammary tumor virus. *EMBO J.* 6: 1355–60
Mitchell, P. J., Wang, C., Tjian, R. 1987. Positive and negative regulation of transcription in vitro: enhancer-binding protein AP-2 is inhibited by SV40 T antigen. *Cell* 50: 847–61
Miyachi, K., Fritzler, M. J., Tan, E. M. 1978. Autoantibody to a nuclear antigen in proliferating cells. *J. Immunol.* 121: 2228–34
Mohr, I. J., Stillman, B., Gluzman, Y. 1987. Regulation of SV40 DNA replication by phosphorylation of T antigen. *EMBO J.* 6: 153–60
Mueller, C., Graessmann, A., Graessmann, M. 1978. Mapping of early SV40-specific functions by microinjection of different early viral DNA fragments. *Cell* 15: 579–85
Müller, D., Ugi, I., Ballas, K., Reisser, P., Henning, R., Montenarh, M. 1987. The AT rich sequence of the SV40 control region influences the binding of SV40 T antigen to binding sites II and III. *Virology* 161: 81–90
Murakami, Y., Eki, T., Yamada, M., Prives, C., Hurwitz, J. 1986a. Species-specific in vitro synthesis of DNA containing the polyoma virus origin of replication. *Proc. Natl. Acad. Sci. USA* 83: 6347–51
Murakami, Y., Wobbe, C. R., Weissbach, L., Dean, F. B., Hurwitz, J. 1986b. Role of DNA polymerase α and DNA primase in simian virus 40 DNA replication in vitro. *Proc. Natl. Acad. Sci. USA* 83: 2869–73
Murphy, C. I., Weiner, B., Bikel, I., Piwnica-Worms, H., Bradley, M. K., Livingston, D. M. 1988. Purification and functional properties of simian virus 40 large and small T antigens overproduced in insect cells. *J. Virol.* 62: 2951–59
Nagata, K., Guggenheimer, R. A., Enomoto, T., Lichy, J. H., Hurwitz, J. 1982. Adenovirus DNA replication in vitro: Identification of a host factor that stimulates synthesis of the pre-terminal protein-dCMP complex. *Proc. Natl. Acad. Sci. USA* 79: 6438–42
Nagata, K., Guggenheimer, R. A., Hurwitz, J. 1983a. Specific binding of a cellular DNA replication protein to the origin of replication of adenovirus DNA. *Proc. Natl. Acad. Sci. USA* 80: 6177–81
Nagata, K., Guggenheimer, R. A., Hurwitz, J. 1983b. Adenovirus DNA replication in vitro: synthesis of full-length DNA with purified proteins. *Proc. Natl. Acad. Sci. USA* 80: 4266–70
Newlon, C. S. 1988. Yeast chromosome replication and segregation. *Microbiol. Rev.* 52: 568–601
Newport, J. 1987. Nuclear reconstitution in vitro: Stages of assembly around protein-free DNA. *Cell* 48: 205–17
Newport, J., Kirschner, M. 1982. A major developmental transition in early Xen-

opus embryos. II. Control of the onset of transcription. *Cell* 30: 687–96

Ninomiya-Tsuji, J., Goto, Y., Ishibashi, S., Shiroki, K., Ide, T. 1987. Induction of cellular DNA synthesis in Go-specific *ts* mutant, *ts*JK60 following infection with SV40 and adenoviruses. *Exp. Cell. Res.* 171: 509–12

Nowock, J., Borgmeyer, V., Püschel, A. W., Rupp, R. A. W., Sippel, A. E. 1985. The TGGCA protein binds to the MMTV-LTR, the adenovirus origin of replication, and the BK virus enhancer. *Nucleic Acids Res.* 13: 2045–61

O'Donnell, M. E., Elias, P., Funnell, B. E., Lehman, I. R. 1987. Interaction between the DNA-polymerase and single-stranded DNA-binding protein (infected cell protein-8) of Herpes simplex virus 1. *J. Biol. Chem.* 262: 4260–66

Oikarinen, J., Hatamochi, A., de Crombrugghe, B. 1987. Separate binding-sites for nuclear factor-1 and a CCAAT DNA-binding factor in the mouse α^2(I) collagen promoter. *J. Biol. Chem.* 262: 11064–70

Olivo, P. D., Nelson, N. J., Challberg, M. D. 1988. Herpes simplex virus DNA replication: The UL9 gene encodes an origin-binding protein. *Proc. Natl. Acad. Sci. USA* 85: 5414–18

O'Neill, E. A., Fletcher, C., Burrow, C. R., Heintz, N., Roeder, R. G., Kelly, T. J. 1988. Transcriptional factor OTF-1 is functionally identical to the DNA replication factor NFIII. *Science* 241: 1210–13

O'Neill, E. A., Kelly, T. J. 1988. Purification and characterization of nuclear factor III (origin recognition protein C), a sequence-specific DNA binding protein required for efficient initiation of adenovirus DNA replication. *J. Biol. Chem.* 263: 931–37

O'Reilly, D. R., Miller, L. K. 1988. Expression and complex formation of simian virus 40 large T antigen and mouse p53 in insect cells. *J. Virol.* 62: 3109–19

Ostrove, J. M., Rosenfeld, P., Williams, J., Kelly, T. J. Jr. 1983. In vitro complementation as an assay for purification of adenovirus DNA replication proteins. *Proc. Natl. Acad. Sci. USA* 80: 935–39

Palzkill, T. G., Newlon, C. S. 1988. A yeast replication origin consists of multiple copies of a small conserved sequence. *Cell* 53: 441–50

Paucha, E., Kalderon, D., Harvey, R. W., Smith, A. E. 1986. Simian virus 40 origin DNA-binding domain on large T antigen. *J. Virol.* 57: 50–64

Peden, K. W., Pipas, J. M. 1985. Site-directed mutagenesis of the simian virus 40 large T-antigen gene: replication-defective amino acid substitution mutants that retain the ability to induce morphological transformation. *J. Virol.* 55: 1–9

Pettit, S. C., Horwitz, M. S., Engler, J. A. 1988. Adenovirus preterminal protein synthesized in COS cells from cloned DNA is active in DNA replication in vitro. *J. Virol.* 62: 496–500

Pincus, S., Robertson, W., Rekosh, D. M. K. 1981. Characterization of the effect of aphidicolin on adenovirus DNA replication: evidence in support of a protein primer model of initiation. *Nucleic Acids Res.* 9: 4919–38

Prelich, G., Kostura, M., Marshak, D. R., Matthews, M. B., Stillman, B. 1987a. The cell-cycle regulated proliferating cell nuclear antigen is required for SV40 DNA replication in vitro. *Nature* 326: 471–75

Prelich, G., Stillman, B. 1988. Coordinated leading and lagging strand synthesis during SV40 DNA replication in vitro requires PCNA. *Cell* 53: 117–26

Prelich, G., Stillman, B. W. 1986. Functional characterization of thermolabile DNA-binding proteins that affect adenovirus DNA replication. *J. Virol.* 57: 883–92

Prelich, G., Tan, C. K., Kostura, M., Mathews, M. B., So, A. G., Downey, K. M., Stillman, B. 1987b. Functional identity of proliferating cell nuclear antigen and a DNA polymerase-δ auxiliary protein. *Nature* 326: 517–20

Prives, C., Murakami, Y., Kern, F. G., Folk, W., Basilico, C., Hurwitz, J. 1987. DNA sequence requirements for replication of polyomavirus DNA in vivo and in vitro. *Mol. Cell. Biol.* 7: 3694–704

Pruijn, G. J. M., van Driel, W., van der Vliet, P. C. 1986. Nuclear factor III, a novel sequence-specific DNA-binding protein from HeLa cells stimulating adenovirus DNA replication. *Nature* 322: 656–59

Pruijn, G. J. M., van Driel, W., van Miltenburg, R. T., van der Vliet, P. C. 1987. Promoter and enhancer elements containing a conserved sequence motif are recognized by nuclear factor III, a protein stimulating adenovirus DNA replication. *EMBO J.* 6: 3771–78

Pruijn, G. J. M., van Miltenburg, R. T., Claessens, J. A. J., van der Vliet, P. C. 1988. Interaction between the octamer-binding protein nuclear factor III and the adenovirus origin of DNA replication. *J. Virol.* 62: 3092–102

Quinn, C. O., Kitchingman, G. R. 1984. Sequence of the DNA-binding protein gene of a human subgroup-B adenovirus (type-7). Comparisons with subgroup-C (type-5) and subgroup-A (type-12). *J. Biol. Chem.* 259: 5003–5009

Rao, P. N., Johnson, R. T. 1970. Mammalian cell fusion: studies on the regulation of DNA synthesis and mitosis. *Nature* 225: 159–64

Rawlins, D. R., Rosenfeld, P. J., Kelly, T. J. Jr., Millman, G. R., Jeang, K. T., et al. 1984a. Sequence-specific interactions of cellular nuclear factor 1 and Epstein-Barr virus nuclear antigen with herpes virus DNAs. *Cancer Cells* 4: 525–42

Rawlins, D. R., Rosenfeld, P. J., Wides, R. J., Challberg, M. D., Kelly, T. J. Jr. 1984b. Structure and function of the adenovirus origin of replication. *Cell* 37: 309–19

Reisman, D., Sugden, B. 1986. Trans activation of an Epstein-Barr viral transcriptional enhancer by the Epstein-Barr viral nuclear antigen-1. *Mol. Cell. Biol.* 6: 3838–46

Rekosh, D. M. K., Russell, W. C., Bellett, A. J. D., Robinson, A. J. 1977. Identification of a protein linked to the ends of adenovirus DNA. *Cell* 11: 283–95

Richter, A., Strausfeld, U., Knippers, R. 1987. Effects of VM26 (teniposide), a specific inhibitor of type II DNA topoisomerase, on SV40 DNA replication in vivo. *Nucleic Acids Res.* 15: 3455–68

Rigby, P. W. J., Lane, D. P. 1983. Structure and function of simian virus 40 large T-antigen. *Adv. Viral Oncol.* 3: 31–57

Rijnders, A. W. M., van Bergen, B. G. M., van der Vliet, P. C., Sussenbach, J. S. 1983. Specific binding of the adenovirus terminal protein precursor-DNA polymerase complex to the origin of DNA replication. *Nucleic Acids Res.* 11: 8777–89

Rio, D., Robbins, A., Myers, R., Tjian, R. 1980. Regulation of simian virus 40 early transcription in vitro by a purified tumor antigen. *Proc. Natl. Acad. Sci. USA* 77: 5706–10

Roberts, J. D., Kunkel, T. A. 1988. Fidelity of a human cell DNA replication complex. *Proc. Natl. Acad. Sci. USA* 85: 7064–68

Roberts, J. M., D'Urso, G. 1988. An origin unwinding activity regulates initiation of DNA replication during mammalian cell cycle. *Science* 241: 1486–89

Roberts, J. M., Weintraub, H. 1986. Negative control of DNA replication in composite SV40-bovine papilloma virus plasmids. *Cell* 46: 741–52

Robinson, A. J., Bellett, A. J. D. 1974. A circular DNA-protein complex from adenoviruses and its possible role in DNA replicaton. *Cold Spring Harbor Symp. Quant. Biol.* 39: 523–31

Robinson, A. J., Bodnar, J. W., Coombs, D. H., Pearson, G. D. 1979. Replicating adenovirus 2 DNA molecules contain terminal protein. *Virology* 96: 143–58

Robinson, A. J., Younghusband, H. B., Bellett, A. J. D. 1973. A circular DNA-protein complex from adenoviruses. *Virology* 56: 54–69

Rosenfeld, P. J., Kelly, T. J. 1986. Purification of nuclear factor 1 by DNA recognition site affinity chromatography. *J. Biol. Chem.* 261: 1398–408

Rosenfeld, P. J., O'Neill, E. A., Wides, R. J., Kelly, T. J. 1987. Sequence-specific interactions between cellular DNA-binding proteins and the adenovirus origin of DNA replication. *Mol. Cell. Biol.* 7: 875–86

Salas, M. 1983. A new mechanism for the initiation of replication of ϕ29 and adenovirus DNA: priming by the terminal protein. *Curr. Top. Microbiol. Immunol.* 109: 89–106

Santoro, C., Mermod, N., Andrews, P. C., Tjian, R. 1988. A family of human CCAAT-box-binding proteins active in transcription and DNA replication: cloning and expression of multiple cDNAs. *Nature* 334: 218–24

Sasaguri, Y., Sanford, T., Aguirre, P., Padmanabhan, R. 1987. Immunological analysis of 140-kDa adenovirus-encoded DNA polymerase in adenovirus type 2-infected HeLa cells using antibodies raised against the protein expressed in *Escherichia coli. Virology* 160: 389–99

Schechter, N. M., Davies, W., Anderson, C. W. 1980. Adenovirus coded deoxyribonucleic acid binding protein. Isolation, physical properties, and effects of proteolytic digestion. *Biochemistry* 19: 2802–10

Scheidtmann, K., Echle, B., Walter, G. 1982. Simian virus 40 large T antigen is phosphorylated at multiple sites clustered in two separate regions. *J. Virol.* 44: 116–33

Schirmbeck, R., Deppert, W. 1987. Specific interaction of simian virus 40 large T antigen with cellular chromatin and nuclear matrix during the course of infection. *J. Virol.* 61: 3561–69

Schirmbeck, R., Deppert, W. 1988. Analysis of mechanisms controlling the interactions of SV40 large T antigen with the SV40 *ori* region. *Virology* 165: 527–38

Schneider, J., Fanning, E. 1988. Mutations in the phosphorylation sites of simian virus 40 (SV40) T antigen alter its origin DNA-binding specificity for sites I or II and affect SV40 DNA replication activity. *J. Virol.* 62: 1598–605

Schnos, M., Zahn, K., Inman, R. B., Blattner, F. R. 1988. Initiation protein induced helix destabilization at the κ origin: A pre-priming step in DNA replication. *Cell* 52: 385–95

Sekimizu, K., Kornberg, A. 1988. Cardiolipin activation of dna A protein, the initiation protein of replication in *Escherichia coli. J. Biol. Chem.* 263: 71731–35

Sekimizu, K., Yung, B. Y., Kornberg, A. 1988. The dna A protein of *Escherichia coli*: abundance, improved purification and membrane binding. *J. Biol. Chem.* 263: 7136–40

Shaul, Y., Ben-Levy, R., De-Medina, T. 1986. High-affinity binding-site for nuclear factor-1 next to the hepatitis-B virus-S gene promoter. *EMBO J.* 5: 1967–71

Sheenan, M. A., Mills, A. D., Sleeman, A. M., Laskey, R. A., Blow, J. J. 1988. Steps in the assembly of replication-competent nuclei in a cell-free system from *Xenopus* eggs. *J. Cell Biol.* 106: 1–13

Shinagawa, M., Ishiyama, T., Padmanabhan, R., Fujinaga, K., Kamada, M., Sato, G. 1983. Comparative sequence analysis of the inverted terminal repetition in the genomes of animal and avian adenoviruses. *Virology* 125: 491–95

Shu, L., Horwitz, M. S., Engler, J. A. 1987. Expression of enzymatically active adenovirus DNA polymerase from cloned DNA requires sequences upstream of the main open reading frame. *Virology* 161: 520–26

Shu, L., Pettit, S. C., Engler, J. A. 1988. The precise structure and coding capacity of mRNAs from early region 2B of human adenovirus serotype 2. *Virology* 165: 348–56

Simanis, V., Lane, D. P. 1985. An immunoaffinity purification procedure for SV40 large T antigen. *Virology* 144: 88–100

Simmons, D. T. 1986. DNA-binding region of the simian virus 40 tumor antigen. *J. Virol.* 57: 776–85

Simmons, D. T. 1988. Geometry of the simian virus 40 large tumor antigen-DNA complex as probed by protease digestion. *Proc. Natl. Acad. Sci. USA* 85: 2086–90

Simmons, D. T., Chou, W., Rodgers, K. 1986. Phosphorylation downregulates the DNA-binding activity of simian virus 40 T antigen. *J. Virol.* 60: 888–94

Sinha, N. K., Morris, C. F., Alberts, B. M. 1980. Efficient in vitro replication of double-stranded DNA templates by a purified T4 bacteriophage replication system. *J. Biol. Chem.* 255: 4290–303

Sitney, K. C., Budd, M. E., Campbell, J. L. 1989. DNA polymerase III, a second essential DNA polymerase, is encoded by the *S. cerevisiae* CDC2 gene. *Cell* 56: 599–605

Smale, S. T., Tjian, R. 1986. T-antigen-DNA polymerase, a complex implicated in Simian virus 40 DNA replication. *Mol. Cell. Biol.* 6: 4077–87

Smart, J. E., Stillman, B. W. 1982. Adenovirus terminal protein precursor: Partial amino acid sequence and site of covalent linkage to virus DNA. *J. Biol. Chem.* 257: 13499–506

Snapka, R. M. 1986. Topoisomerase inhibitors can selectively interfere with different stages of simian virus 40 DNA replication. *Mol. Cell. Biol.* 6: 4221–27

Snapka, R. M., Powelson, M. A., Strayer, J. M. 1988. Swiveling and decantenation of replicating simian virus 40 genomes in vivo. *Mol. Cell. Biol.* 8: 515–21

Snyder, M., Buchman, A. R., Davis, R. W. 1986. Bent DNA at a yeast autonomously replicating sequence. *Nature* 324: 87–89

Spradling, A., Orr-Weaver, T. 1987. Regulation DNA replication during *Drosophila* development. *Annu. Rev. Genet.* 21: 373–403

Stahl, H., Dröge, P., Knippers, R. 1986. DNA helicase activity of SV40 large tumor antigen. *EMBO J.* 5: 1939–44

Stahl, H., Dröge, P., Zentgraf, H., Knippers, R. 1985. A large-tumor-antigen-specific monoclonal antibody inhibits DNA replication of simian virus 40 minichromosomes in an in vitro elongation system. *J. Virol.* 54: 473–82

Stahl, H., Knippers, R. 1983. Simian virus 40 large tumor antigen on replicating viral chromatin: tight binding and localization on the viral genome. *J. Virol.* 47: 65–76

Stenlund, A., Bream, G. L., Botchan, M. R. 1987. A promoter with an internal regulatory domain is part of the origin of replication in BPV-1. *Science* 236: 1666–71

Stillman, B. 1986. Chromatin assembly during SV40 DNA replication in vitro. *Cell* 45: 555–65

Stillman, B. W. 1983. The replication of adenovirus DNA with purified proteins. *Cell* 35: 7–9

Stillman, B. W. 1981. Adenovirus DNA-replication in vitro: A protein linked to the 5′ end of nascent DNA strands. *J. Virol.* 37: 139–47

Stillman, B. W., Bellett, A. J. D. 1979. An adenovirus protein associated with the ends of replicating DNA molecules. *Virology* 93: 69–79

Stillman, B. W., Diffley, J. F. X., Prelich, G., Guggenheimer, R. A. 1986. DNA-protein interactions at the replication origins of adenovirus and SV40. *Cancer Cells* 4: 453–63

Stillman, B., Gerard, R. D., Guggenheimer, R. A., Gluzman, Y. 1985. T antigen and template requirements for SV40 DNA replication in vitro. *EMBO J.* 4: 2933–39

Stillman, B. W., Gluzman, Y. 1985. Replication and supercoiling of simian virus 40 DNA in cell extracts from human cells. *Mol. Cell. Biol.* 5: 2051–60

Stillman, B. W., Lewis, J. B., Chow, L. T., Mathews, M. B., Smart, J. E. 1981. Identi-

fication of the gene and mRNA for the adenovirus terminal protein precursor. *Cell* 23: 497–508
Stillman, B. W., Tamanoi, F., Mathews, M. B. 1982a. Purification of an adenovirus-coded DNA polymerase that is required for initiation of DNA replication. *Cell* 31: 613–23
Stillman, B. W., Topp, W. C., Engler, J. A. 1982b. Conserved sequences at the origin of adenovirus DNA replication. *J. Virol.* 44: 530–37
Stow, N. D. 1981. The infectivity of adenovirus genomes lacking DNA sequences from their left-hand termini. *Nucleic Acids Res.* 10: 5105–19
Strauss, M., Argani, P., Mohr, I. J., Gluzman, Y. 1987. Studies on the origin-specific DNA binding domain of simian virus 40 large T antigen. *J. Virol.* 61: 3326–30
Stringer, R. 1982. Mutant of Simian virus 40 large T antigen that is defective for viral DNA synthesis, but competent for transformation of cultured rat cells. *J. Virol.* 42: 854–64
Stunnenberg, H. G., Lange, H., Philipson, L., van Miltenburg, R. T., van der Vliet, P. C. 1988. High expression of functional adenovirus DNA polymerase and precursor terminal protein using recombinant vaccinia virus. *Nucleic Acids Res.* 16: 2431–44
Sturm, R., Baumruker, T., Franza, B. R. Jr., Herr, W. 1987. A 100 kd HeLa cell octamer binding protein (OBP100) interacts differently with two separate octamer-related sequences within the SV40 enhancer. *Genes Dev.* 1: 1147–60
Sturm, R. A., Das, G., Herr, W. 1988. The ubiquitous octamer-binding protein Oct-1 contains a POV domain with a homeo box subdomain. *Genes Dev.* 2: 1582–99
Sundin, O., Varshavsky, A. 1980. Terminal stages of SV40 DNA replication proceed via multiply intertwined catenated dimers. *Cell* 21: 103–14
Sundin, O., Varshavsky, A. 1981. Arrest of segregation leads to accumulation of highly intertwined catenated dimers: dissection of the final stages of SV40 DNA replication. *Cell* 25: 659–69
Tack, L. C., DePamphilis, M. L. 1983. Analysis of simian virus 40 chromosome-T-antigen complexes: T-antigen is preferentially associated with early replicating intermediates. *J. Virol.* 48: 281–95
Tack, L. C., Proctor, G. N. 1987. Two major replicating simian virus 40 chromosome classes. Synchronous replication fork movement is associated with bound large T antigen during elongation. *J. Biol. Chem.* 262: 6339–49
Tamanoi, F., Stillman, B. W. 1982. Function of the adenovirus terminal protein in the initiation of DNA replication. *Proc. Natl. Acad. Sci. USA* 79: 2221–25
Tamanoi, F., Stillman, B. W. 1983. Initiation of adenovirus DNA replication in vitro requires a specific DNA sequence. *Proc. Natl. Acad. Sci. USA* 80: 6446–50
Tan, C.-K., Castillo, C., So, A. G., Downey, K. M. 1986. An auxiliary protein for DNA polymerase-δ from fetal calf thymus. *J. Biol. Chem.* 261: 12310–16
Tegtmeyer, P., Lewton, B. A., DeLucia, A. L., Wilson, V. G., Ryder, K. 1983. Topography of simian virus 40 A protein-DNA complexes: arrangement of protein bound to the origin of replication. *J. Virol.* 46: 151–61
Tenen, D. G., Taylor, T. S., Haines, L. L., Bradley, M. K., Martin, R. G., Livingston, D. M. 1983. Binding of Simian virus 40 large T antigen from virus-infected monkey cells to wild type and mutant viral replication origins. *J. Mol. Biol.* 168: 791–808
Thrash, C., Voelkel, K., DiNardo, S., Sternglanz, R. 1984. Identification of Saccharomyces cerevisiae mutants deficient in DNA topoisomerase I activity. *J. Biol. Chem.* 259: 1375–77
Tjian, R. 1978. The binding site on SV40 DNA for a T antigen-related protein. *Cell* 13: 165–79
Tjian, R., Fey, G., Graessmann, A. 1978. Biological activity of purified simian virus 40 T antigen proteins. *Proc. Natl. Acad. Sci. USA* 75: 1279–83
Tjian, R., Robbins, A. 1979. Enzymatic activities associated with a purified simian virus 40 T antigen-related protein. *Proc. Natl. Acad. Sci. USA* 76: 610–14
Tolun, A., Aleström, P., Pettersson, U. 1979. Sequence of inverted terminal repetitions from different adenoviruses: Demonstration of conserved sequences and homology between SA-7 termini and SV40-DNA. *Cell* 17: 705–13
Traut, W., Fanning, E. 1988. Sequence-specific interactions between a cellular DNA-binding protein and the simian virus 40 origin of DNA replication. *Mol. Cell. Biol.* 8: 903–11
Tseng, B. Y., Ahlem, C. N. 1984. Mouse primase initiation sites in the origin region of simian virus 40. *Proc. Natl. Acad. Sci. USA* 81: 2342–46
Tsernoglou, D., Tucker, A. D., van der Vliet, P. C. 1984. Crystallization of a fragment of the adenovirus DNA binding protein. *J. Mol. Biol.* 172: 237–39
Tsurimoto, T., Stillman, B. 1989. Purification of a cellular replication factor, RF-C, that is required for coordinated

synthesis of leading and lagging strands during SV40 DNA replication in vitro. *Mol. Cell. Biol.* 9: 609–19

Uemura, T., Yanagida, M. 1984. Isolation of type-I and II DNA topoisomerase mutants from fission yeast: single and double mutants show different phenotypes in cell-growth and chromatin organization. *EMBO J.* 3: 1737–44

Umek, R. M., Linskens, M. H. K., Kowalski, D., Huberman, J. A. 1989. New beginnings in studies of eukaryotic DNA replication origins. *Biochim. Biophys. Acta.* 1007: 1–14

van Amerongen, H., van Grondelle, R., van der Vliet, P. C. 1987. Interaction between adenovirus DNA-binding protein and single-stranded polynucleotides studied by circular dichroism and ultraviolet absorption. *Biochemistry* 26: 4646–52

van Bergen, B. G. M., van der Ley, P. A., van Driel, W., van Mansfeld, A. D. M., van der Vliet, P. C. 1983. Replication of origin containing adenovirus DNA fragments that do not carry the terminal protein. *Nucleic Acids Res.* 11: 1975–89

van Bergen, B. G. M., van der Vliet, P. C. 1983. Temperature sensitive initiation and elongation of adenovirus DNA replication in vitro with nuclear extracts from H5ts36, H5ts149 and H5ts125 infected HeLa cells. *J. Virol.* 42: 642–48

van der Vliet, P. C., Keegstra, W., Jansz, H. S. 1978. Complex formation between the adenovirus type 5 DNA binding protein and single-stranded DNA. *Eur. J. Biochem.* 86: 389–98

van der Vliet, P. C., Levine, A. J. 1973. DNA-binding proteins specific for cells infected by adenovirus. *Nature New Biol.* 246: 170–74

van der Vliet, P. C., Sussenbach, J. S. 1975. An adenovirus gene function required for the initiation of viral DNA replication. *Virology* 67: 415–27

Van Weilink, P. S., Naaktgeboren, N., Sussenbach, J. S. 1979. Presence of protein at the termini of intracellular adenovirus type-5 DNA. *Biochim. Biophys. Acta* 563: 89–99

Vos, H. L., van der Lee, F. M., Reemst, A. M. C. B., van Loon, A. E., Sussenbach, J. S. 1988. The genes encoding the DNA binding protein and the 23K protease of adenovirus types 40 and 41. *Virology* 163: 1–10

Wahl, A. F., Geis, A. M., Spain, B. H., Wong, S. W., Korn, D., Wang, T. S.-F. 1988. Gene expression of human DNA polymerase α during cell proliferation and the cell cycle. *Mol. Cell. Biol.* 8: 5016–25

Wang, K., Pearson, G. D. 1985. Adenovirus sequences required for replication in vivo. *Nucleic Acids Res.* 13: 5173–87

Wang, T. S-F., Wong, S. W., Korn, D. 1989. Human DNA polymerase α: predicted functional domains and relationships with DNA polymerases. *FASEB J.* 3: 14–21

Weaver, D. T., Fields-Berry, S. C., DePamphilis, M. L. 1985. The termination region for SV40 DNA replication directs the mode of separation for the two sibling molecules. *Cell* 41: 565–75

Weller, S. K., Carmichael, E. P., Goldstein, D. J., Zhu, L. 1988. Use of host range mutants to identify genes involved in DNA replication of Herpes simplex virus. See Kelly & Stillman 1988, pp. 53–59

Whyte, P., Buchkovich, K. J., Horowitz, J. M., Friend, S. H., Raybuck, M., Weinberg, R. A., Harlow, E. 1988. Association between an oncogene and an anti-oncogene: the adenovirus E1A proteins bind to the retinoblastoma gene product. *Nature* 334: 124–29

Wides, R. J., Challberg, M. D., Rawlins, D. R., Kelly, T. J. 1987. Adenovirus origin of DNA replication: sequence requirements for replication in vitro. *Mol. Cell. Biol.* 7: 864–74

Wiekowski, M., Droge, P., Stahl, H. 1987. Monoclonal antibodies as probes for a function of large T antigen during the elongation process of simian virus 40 DNA replication. *J. Virol.* 61: 411–18

Wiekowski, M., Schwarz, M. W., Stahl, H. 1988. Simian virus 40 large T antigen DNA helicase. Characterization of the ATP-ase-dependent DNA unwinding activity and its substrate requirements. *J. Biol. Chem.* 263: 436–42

Williams, J. S., Eckdahl, T. T., Anderson, J. N. 1988. Bent DNA functions as a replication enhancer in *Saccharomyces cerevisiae*. *Mol. Cell. Biol.* 8: 2763–69

Wobbe, C. R., Dean, F. B., Murakami, Y., Weissbach, L., Hurwitz, J. 1986. Simian virus 40 DNA replication in vitro: Study of events preceding elongation of chains. *Proc. Natl. Acad. Sci. USA* 83: 4612–16

Wobbe, C. R., Dean, F., Weissbach, L., Hurwitz, J. 1985. In vitro replication of duplex circular DNA containing the simian virus 40 DNA origin site. *Proc. Natl. Acad. Sci. USA* 82: 5710–14

Wobbe, C. R., Weissbach, L., Borowiec, J. A., Dean, F. B., Murakami, Y., Bullock, P., Hurwitz, J. 1987. Replication of simian virus 40 origin-containing DNA in vitro with purified proteins. *Proc. Natl. Acad. Sci. USA* 84: 1834–38

Wold, M. S., Kelly, T. 1988. Purification and characterization of replication protein A, a cellular protein required for in vitro replication of simian virus 40 DNA. *Proc. Natl. Acad. Sci. USA* 85: 2523–27

Wold, M. S., Li, J. J., Kelly, T. J. 1987. Initiation of simian virus 40 DNA replication in vitro: Large-tumor-antigen and origin-dependent unwinding of the template. *Proc. Natl. Acad. Sci. USA* 84: 3643–47

Wong, R. L., Katz, M. E., Ogata, K., Tan, E. M., Cohen, S. 1987. Inhibition of nuclear DNA synthesis by an autoantibody to proliferating cell nuclear antigen/cyclin. *Cell Immunol.* 110: 443–48

Wong, S. W., Syvaoja, J., Tan, C.-K., Downey, K. M., So, A. G., Linn, S., Wang, T. S.-F. 1989. DNA polymerases α and δ are immunologically and structurally distinct. *J. Biol. Chem.* 264: 5924–28

Wu, C. A., Nelson, N. J., McGeoch, D. J., Challberg, M. D. 1988. Identification of herpes simplex virus type 1 genes required for origin-dependent DNA synthesis. *J. Virol.* 62: 435–43

Yamaguchi, M., DePamphilis, M. L. 1986. DNA binding site for a factor(s) required to initiate simian virus 40 DNA replication. *Proc. Natl. Acad. Sci. USA* 83: 1646–50

Yamaguchi, M., Hendrickson, E. A., DePamphilis, M. L. 1985. DNA primase-DNA polymerase α from simian cells. Modulation of RNA primer synthesis by ribonucleoside triphosphates. *J. Biol. Chem.* 260: 6254–63

Yang, L., Wold, M. S., Li, J. J., Kelly, T. J., Liu, L. F. 1987. Roles of DNA topoisomerases in simian virus 40 DNA replication in vitro. *Proc. Natl. Acad. Sci. USA* 84: 950–54

Yung, B. Y., Kornberg, A. 1988. Membrane attachment activates dna A protein, the initiation protein of chromosome replication in *Escherichia coli*. *Proc. Natl. Acad. Sci. USA* 85: 7202–205

Zhao, L., Padmanabhan, R. 1988. Nuclear transport of adenovirus DNA polymerase is facilitated by interaction with preterminal protein. *Cell* 55: 1005–15

Zuber, M., Tan, E. M., Ryoji, M. 1989. Involvement of proliferating cell nuclear antigen (cyclin) in DNA replication in living cells. *Mol. Cell. Biol.* 9: 57–66

Annu. Rev. Cell Biol. 1989. 5 : 247–75

LIPID TRAFFIC IN ANIMAL CELLS

Gerrit van Meer

Department of Cell Biology, Medical School, University of Utrecht, Utrecht, The Netherlands

---

CONTENTS

SCOPE

The membranes of the various intracellular organelles possess unique lipid compositions. Yet most of the lipids are synthesized in the endoplasmic reticulum (ER).[1] This implies that lipids are transported and that specificity exists in the traffic routes to the various organelles: in other words, lipids are sorted. For various reasons cellular lipid research lags behind that of the proteins. As a consequence, a large part of the basic characterization

[1] Abbreviations: C6-NBD-, N-6[7-nitro-2,1,3-benzoxadiazol-4-yl] aminocaproyl-; CL, cardiolipin; ER, Endoplasmic Reticulum; GSL(s), glycosphingolipid(s); kd, kilodalton; LBPA, lysobisphosphatidic acid; PC, phosphatidylcholine; PE, phosphatidylethanolamine; PG, phosphatidylglycerol; PI, phosphatidylinositol; PS, phosphatidylserine; SCP-2, sterol carrier protein 2; SPH, sphingomyelin.

0743–4634/89/1115–0247$02.00

still has to be done. This review summarizes our knowledge on lipid traffic, seeks to locate the sites of sorting events, and provides some ideas on the mechanisms of lipid sorting.

Although lipids can be transported in the same way as membrane proteins, by carrier vesicles, lipid traffic is special in a number of respects. (*a*) As lipid molecules are much smaller than proteins (M_r less than 1 k), a membrane contains 10–100 times more lipid molecules than proteins. (*b*) The lateral diffusion (and intermixing) of lipids is ten times faster than that of the proteins. (*c*) The basic structure of a membrane is the lipid bilayer. Therefore, in terms of composition and transport every membrane has to be regarded as a cytoplasmic and an exoplasmic pool of lipids that may or may not communicate. (*d*) The differences in lipid composition between the organelles are not absolute. As all organelles possess more or less the same lipid classes, it is the ratio between these classes and their molecular species that is unique. (*e*) In addition to the ER, most organelles have some limited lipid biosynthesizing activity and examples of metabolic interconversion of one lipid into another are known for almost every organelle. (*f*) Proteins have been purified from cytosol that, in vitro at least, are capable of transferring most types of lipid to and from most intracellular membranes.

So the question is how the dynamic interplay between local synthesis and modification, the various modes of lipid traffic, and the inherent sorting potential of various lipids results in a stable intracellular lipid heterogeneity that is remarkably uniform between many functionally different cell types. Cellular lipid traffic will undoubtedly be more complicated than the schemes provided here. However, the complications may turn out to be variations on a common theme.

LIPID COMPOSITION

General Remarks

Phospholipids constitute the main lipid mass of eukaryotic cell membranes. In addition, membranes may contain bulk quantities of cholesterol and glycosphingolipids (GSLs). The content of the latter two lipids ranges from essentially zero as in mitochondria to equimolar amounts with the phospholipids as in the apical plasma membrane of epithelial cells (see below). Glyco*glycero*lipids, triglycerides, cholesterol esters, and intermediates in lipid metabolism like di- and monoglycerides, ceramides, lysophospholipids, and free fatty acids are minor membrane components under normal conditions. In contrast, the amount of dolichol, a hitherto neglected membrane lipid, may turn out to be significant (Chojnacki & Dallner 1988).

An evaluation of organellar lipid compositions from the literature meets a number of complications. (*a*) Compositional data are often incomplete. In order to be able to reconstruct a membrane, one needs the total lipid composition, expressed in terms of concentration, mol/surface area, or percent of total lipids. However, GSLs (and dolichol), for example, are frequently omitted from the analysis altogether, simply because of difficulties of quantitation (*b*) Compositional studies have mostly been performed on membrane fractions, with the inherent difficulties of mutual contamination and incomplete yields. As an additional complication, membranes have frequently been contaminated by the lipid containing contents of organelles, e.g. secretory or endocytosed lipoproteins and the content fraction of multivesicular endosomes or lysosomes. Finally, some organelles, like endosomes, have only recently been appreciated as structurally distinct compartments and still more organelles may await discovery. (*c*) As a further complication, some organelles are differentiated into subcompartments of different composition, for example, the smooth and rough ER. (*d*) Finally, for studying lipid traffic one would like to specify the lipid composition of each of the two leaflets of the lipid bilayer. In most cases this information is not available.

Lipid Compositions along the Endo- and Exocytic Pathways

In animal eukaryotes all membrane and secretory proteins, with the exception of those destined for the mitochondria and peroxisomes, pass through the ER and are dispatched from the ER to their destination by carrier vesicles. The elucidation of the routing of these proteins has led to a model of endo- and exocytic pathways and their interconnections as illustrated in a simplified form in Figure 1.

The lipid compositions of the ER and the plasma membrane are strikingly different (Table 1). Plasma membranes are typically enriched in cholesterol, also the phospholipids SPH and PS, whereas PC is decreased. The plasma membrane invariably contains GSLs, which are thought to be absent from the ER. Extreme levels of GSLs occur in the apical plasma membrane of epithelial cells, where the GSLs constitute up to 33 mol % (see below) of the total lipids. The composition of the Golgi complex is intermediate between that of the ER and the plasma membrane (Zambrano et al 1975). In fact, a compositional gradient may exist over the Golgi stack.[2] *Trans* Golgi cisternae appear to possess a higher concentration of

[2] The Golgi complex is taken to consist of at least two compartments, the *cis* Golgi in vesicular equilibrium with the ER (Pelham, this volume), and the "*trans* Golgi network" the *trans* most cisterna of the Golgi complex that is in vesicular equilibrium with endosomes and plasma membrane, as reviewed by Griffiths & Simons (1986). The obvious fact that the stack contains more cisternae (Farquhar 1985) does not affect the present argument.

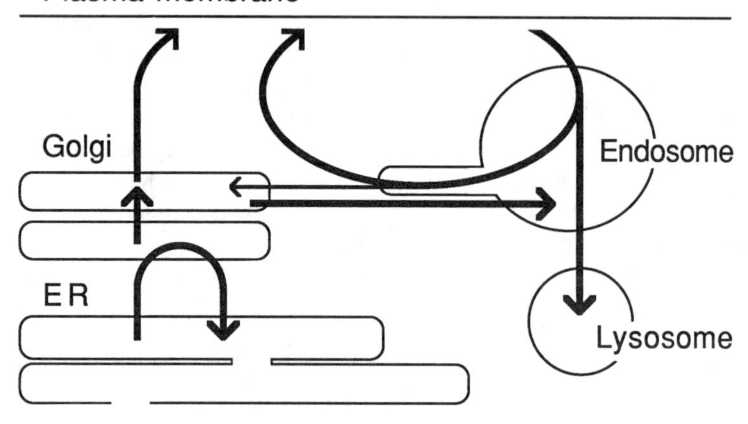

Figure 1 Pathways of vesicular traffic. Vesicles cycle between the starting point, the ER, and the *cis* Golgi (Pelham, this volume). Whether or not the intra-Golgi transport contains a unidirectional step is unclear at present. At the *trans* end of the Golgi the vesicle components are delivered to the plasmalemma and from it into the endocytic pathway. A shortcut connects the *trans* Golgi network via a prelysosomal compartment to the lysosomes; the latter present a point of no return for vesicular traffic (Kornfeld & Mellman, this volume).

cholesterol (Orci et al 1981). The composition of endocytic coated vesicles and endosomal membranes is similar to that of the plasma membrane. In virtually all cases studied, they contained high levels of cholesterol, SPH, and PS (Dickson et al 1983; Luzio & Stanley 1983; Evans & Hardison

Table 1 Typical lipid composition of some intracellular organelles of rat liver

	Mitochondrial membrane	ER	Plasma membrane	Lysosomal membrane[a]
Phospholipids:				
SPH	0.5	2.5	16.0	20.3
PC	40.3	58.4	39.3	39.7
PI	4.6	10.1	7.7	4.5
PS	0.7	2.9	9.0	1.7
PE	34.6	21.8	23.3	14.1
CL	17.8	1.1	1.0	1.0
LBPA	0.2[a]			7.0
Cholesterol/phospholipid (mol/mol):	0.03	0.08	0.40/0.76[b]	0.49[b]

[a] Data from Wherrett & Huterer 1972.
[b] Data from Colbeau et al 1971.
Data expressed as % of total phospholipid phosphorus (from Zambrano et al 1975, with permission).

1985; Helmy et al 1986; Belcher et al 1987; Urade et al 1988).[3] Lysosomal membranes have a plasma membrane-like composition, with high levels of cholesterol and SPH (Table 1) and GSLs (Henning & Stoffel 1973). However, the level of PS is very low. Moreover, the lysosome is the only organelle containing lysobisphosphatidic acid (LBPA; see Brotherus & Renkonen 1977), a phospholipid with a remarkable structure.

Lipid Composition of Organelles Not Connected by Vesicular Traffic

Mitochondria have a characteristically simple lipid composition as they are devoid of GSL, and the levels of cholesterol, SPH, and PS are very low (Table 1). The mitochondrial inner membrane mainly consists of PC and PE, and is unique in that it contains significant levels of cardiolipin (CL). The actual surface area covered by CL is twice its mol percent: each CL molecule contains two diacylglycerophosphate moieties, while the other phospholipids have one. The mitochondrial PI is localized primarily in the outer membrane, which has PC and PE as its major lipids (see Daum 1985).

PC and PE also constitute the major phospholipids of the peroxisomal membrane (73 and 18%, Crane & Masters 1986), the remainder being PI and PS. As in mitochondria, no SPH is found. In contrast, some cholesterol seems to be present (Fujiki et al 1982), which is easily explained by the fact that it can be synthesized by peroxisomes (Appelkvist 1987; Thompson et al 1987).

Quantitative Distribution of Lipids between the Various Organelles

The enrichment of a lipid in a certain organelle does not imply that this organelle contains the bulk of that lipid as the relative membrane areas of different organelles have to be considered. The subcellular distribution of a lipid can be determined from the products of the concentration of the lipid in each organelle (mol/area) and the surface area of the organelle. When the lipid compositions of the organelles in BHK cells (Brotherus & Renkonen 1977) are combined with the data on surface area (G. Griffiths,

[3] Interpretation of the differences between endosomal and plasma membrane fatty acyl compositions reported in the latter paper is difficult. The plasma membrane phospholipid composition was not plasma membrane-like, but looked like ER.

R. Back, M. Marsh, submitted for publication), the following numbers are obtained: 90% of the LBPA is found in the lysosomes, and 90% of the cellular CL is assigned to the inner mitochondrial membrane. While this probably reflects a virtually complete confinement of LBPA and CL to specific organelles, none of the major lipids displays such a biased distribution. Of the typical plasma membrane lipids SPH and PS, only 30–35% is present in the plasma membrane vs 11% of the total cellular PC and PE. Also the fraction of cellular cholesterol in the plasma membrane is remarkably low, 25–40%. Similarly, GSLs are predominantly intracellular (Tanaka & Leduc 1956; Weinstein et al 1970; Keenan et al 1972; Hansson et al 1986; Symington et al 1987). These numbers (cf van Meer 1987) are consistent with the similarity between the lipid compositions of *trans* Golgi, endosomes and lysosomes and that of the plasma membrane (see above), and with the finding that the total surface area of the former three compartments is of the same order as that of the plasma membrane (G. Griffiths, R. Back, M. Marsh, submitted for publication). The persistent idea that practically all GSL (Thompson & Tillack 1985) and 90% of both SPH and cholesterol (Lange et al 1989) are located in the plasma membrane most likely results from the fact that in the experiments essentially no attempt was made to distinguish the contribution of the endosomal membranes and the *trans* Golgi from that of the plasma membrane.

Transmembrane Lipid Asymmetry

PLASMA MEMBRANE LIPID ASYMMETRY Evidence is accumulating to suggest that plasma membranes in most cells are organized like the well-characterized erythrocyte membrane (reviewed in Op den Kamp 1979; Zachowski & Devaux 1989). Most of the choline-phospholipids SPH and PC are situated in the exoplasmic bilayer leaflet and most of the amino-phospholipids PS and PE are on the cytoplasmic surface. SPH appears to be located in the exoplasmic leaflet exclusively (Allan & Walklin 1988). Also PS is essentially unilateral, whereas significant fractions of PC and PE are present in both bilayer leaflets. GSLs are exclusively exoplasmic, while the scarce data on the distribution of cholesterol are contradictory (Zachowski & Devaux 1989). Arguments will be provided in favor of a preferential location of cholesterol in the exoplasmic leaflet.

AN AMINO-PHOSPHOLIPID TRANSLOCATOR IN THE PLASMA MEMBRANE Eukaryotic plasma membranes are stable bilayers, with low rates of transbilayer mobility (flip-flop). GSLs do not translocate at all and this appears also to be the case for SPH (Zachowski & Devaux 1989, see below). In the case of PC, two pools exist in the plasma membrane that intermix only very slowly due to $t_{1/2}$ s of flip-flop in the order of 10 h or more at 37°C (Pagano & Sleight 1985; Zachowski & Devaux 1989).[4] Conversely, the amino-phospholipids appear to be actively translocated to the cytoplasmic leaflet by an energy requiring amino-phospholipid translocator (see Devaux 1988). Also spontaneous flip-flop of the amino-phospholipids appears to be much faster than that of PC (see Pagano & Sleight 1985; Zachowski & Devaux 1989). Therefore the plasma membrane phospholipid asymmetry represents an energy consuming equilibrium state, possessing all necessary properties to be regulated. It has been suggested that since cytoskeletal elements possess the capability to interact with PS, they might stabilize the lipid asymmetry. It is, however, not obvious how such a skeleton would quantitatively trap PS in the cytoplasmic leaflet (see Zachowski & Devaux 1989).

TRANS GOLGI AND ENDOCYTIC MEMBRANES ARE PLASMA MEMBRANE-LIKE
Lipid asymmetry and a low flip-flop rate for PC are also characteristic for endocytic membranes (Sandra & Pagano 1978). Moreover, fluorescent analogs of PC and SPH did not translocate to the cytoplasmic surface when going through the endocytic pathway (Pagano & Sleight 1985; Koval & Pagano 1989). Finally, fluorescent sphingolipids did not flip-flop from their luminal location in the Golgi complex (Lipsky & Pagano 1985; van Meer et al 1987). In combination with the plasma membrane data, this suggests that membranes in the endocytic pathway (Figure 1) possess one exoplasmic pool of sphingolipids, two nonequilibrating pools of PC, and two pools of PE and of PS that readily equilibrate (at least at the plasma membrane).

RAPID PC FLIP-FLOP IN ER Studies on the lipid organization in the ER membrane have been prone to methodological artifacts (van Meer 1986). Still it has been reproducibly shown that PC, the major ER lipid, rapidly translocates across the ER membrane, possibly facilitated by a PC-specific pore, a "PC-flippase" (references in Bishop & Bell 1988). This explains how the unilateral lipid synthesis on the cytoplasmic ER surface (Bell et al 1981) can yield a lipid bilayer. Synthesis of PE, PS, and PI at the ER also occurs on the cytoplasmic surface. Simple absence of flip-flop for these phospholipids would by itself create phospholipid asymmetry, with PC being the only luminal phospholipid. However, the transmembrane organization and flip-flop rates of PE, PS, and PI in the ER are unknown. PC flip-flop must be abolished along the exocytic route. Although the

[4] Flip-flop rates with a $t_{1/2}$ of about 30 min have been reported for SPH and PC in the plasma membrane of guinea pig cells. It is not obvious why the rate of flip-flop should be so "very species-dependent" (Sune et al 1988).

mechanism remains unclear, the simplest solution would be retention of the putative PC-flippase in the ER.

LIPID TRAFFIC AND SORTING

Generation and Maintenance of the Heterogeneous Distribution of Sphingolipids

SPHINGOLIPIDS ARE SYNTHESIZED ON THE LUMINAL ASPECT OF A GOLGI MEMBRANE Any single cell expresses a limited number of GSLs, mostly one or two series each derived from ceramide by the stepwise addition of monosaccharides. Apart from ceramide synthesis in the ER, the various steps in GSL assembly occur on the luminal surface of subsequent cisternae of the Golgi complex (see Miller-Prodraza & Fishman 1984; Lipsky & Pagano 1985; Klein et al 1988). The implications of glycosyl-transferase activities at the cell surface (Pierce et al 1980) for GSL biosynthesis are unclear.

SPH is assembled by the energy independent transfer of phosphorylcholine from a PC molecule onto ceramide. In the early studies (e.g. Marggraf et al 1981; Voelker & Kennedy 1982) the plasma membrane was proposed to be a major site of SPH biosynthesis. However during biosynthetic labeling of SPH with radioactive choline, the specific activity of plasma membrane SPH notably lagged behind that of SPH in a Golgi/ER fraction (Cook et al 1988). The synthesis of a fluorescent SPH has now been shown to occur on the luminal surface of the Golgi (see Pagano 1988), and not on the plasma membrane of interphase (Lipsky & Pagano 1985; G. van Meer, submitted for publication) and mitotic cells (Kobayashi & Pagano 1989); an alternative explanation for the enrichment of SPH in the plasma membrane will be provided.

GSLs and SPH apparently do not flip-flop. Their exclusive location in the exoplasmic leaflets of the various organelles must, therefore, be established during biosynthesis. This also explains why GSLs and SPH are exclusively found in organelles connected by vesicular transport (Figure 1).[5]

SPHINGOLIPID SORTING ALONG THE EXOCYTIC PATHWAY IN EPITHELIAL CELLS A better understanding of the intracellular transport and sorting of sphingolipids has been obtained from studying their enrichment in a domain of the plasma membrane of epithelial cells.

[5] Various proteins that in vitro displayed GSL transfer activity have now been identified as activator proteins for GSL degradation in the lysosomal lumen (see Fürst et al 1988; O'Brien et al 1988). The physiologic function of a putative "SPH specific exchange protein" (Dyatlovitskaya et al 1982) is unclear.

The plasma membrane of epithelial cells is divided into an apical and a basolateral domain by tight junctions that encircle the apex of each cell. The apical domain, which a number of criteria indicate to be a specialized part of the plasma membrane (Simons & Fuller 1985), is enriched in GSLs, at the expense of PC, when compared to the basolateral plasma membrane. An apical enrichment of SPH in some kidney cells has been reported (see Venien & Le Grimellec 1988), but unfortunately not the GSL content of the corresponding membranes (see Simons & van Meer 1988; van Meer 1988). Where studied, the levels of GSLs plus SPH are high enough to completely cover the apical membrane, which implies that the glycerophospholipids must be confined to the cytoplasmic bilayer leaflet. Indeed experimental evidence indicates that only a minor fraction of glycerophospholipids is accessible to phospholipases, exchange proteins, and the chemical reagent trinitrobenzene sulfonate (Barsukov et al 1986; Venien & Le Grimellec 1988).

While current evidence indicates that some membrane proteins display a polarized distribution even in the absence of tight junctions (e.g. Salas et al 1988), the intactness of the latter is required for maintaining the compositional difference between the two plasma membrane domains in the case of lipids. The tight junctions block lipid diffusion. The barrier is exclusive to the exoplasmic leaflet of the bilayer (references in Simons & van Meer 1988). Therefore the lipid differences between the two domains reside in that outer leaflet. The cytoplasmic leaflets of the two domains share the same composition, thus lipids can freely diffuse between them. The transbilayer distribution of the various phospholipids in the basolateral membrane can now be calculated from its total phospholipid composition and that of its cytoplasmic leaflet (which is identical to that of the total apical membrane). In this way 75–90% of the PC was assigned to the exoplasmic leaflet vs 10–35% of the PE (van Meer & Simons 1986). In addition, an endogenously synthesized fluorescent SPH was exclusively localized to the exoplasmic leaflet (van Meer et al 1987). The overall phospholipid asymmetry in the basolateral plasma membrane domain of epithelial cells is thus strikingly similar to that in the erythrocyte membrane.

Lipid polarity in epithelia is generated by lipid sorting The biosynthetic transport of apical and basolateral lipids has been studied using endogenously synthesized fluorescent analogs of glucosylceramide, a typical apical lipid, and SPH, which is not polarized in the cells used (van Meer et al 1987; G. van Meer, submitted for publication). Due to the fatty acid on their backbone, the analogs, N-6[7-nitro-2,1,3-benzoxadiazol-4-yl] aminocaproyl sphingosine-, are readily exchangeable between membranes, in contrast to the natural lipids. The appearance of the analogs on the outside of the plasma membrane could then be assayed by "back-exchange", specific extraction of the fluorescent lipid from the surface by unlabeled liposomes or BSA (Lipsky & Pagano 1985). Moreover, confinement of the fluorescent lipids to discrete intracellular structures, in this type of experiment, is most easily explained by luminal localization. The conclusions from our studies were (*a*) fluorescent glucosylceramide and SPH were synthesized on the luminal, exoplasmic leaflet of an intracellular organelle membrane, and this most likely was a Golgi membrane; (*b*) the lipids did not travel to the ER; (*c*) 90% of both lipids was transported to the cell surface within 2 h at 37°C, with a $t_{1/2}$ of 20–30 min; (*d*) sidedness was completely maintained, which indicated that transport was mediated by carrier vesicles: the fluorescent lipids ended up exclusively in the exoplasmic leaflet of the plasma membrane; (*e*) two to threefold more glucosylceramide was delivered to the apical than to the basolateral surface vs equal amounts of the SPH analog. Because the experiments were set up such that redistribution by endocytosis could be excluded, lipid sorting had occurred prior to delivery to the cell surface. In analogy to the conclusion that apical plasma membrane proteins are sorted from basolateral and lysosomal proteins in the *trans* Golgi network (Simons & Fuller 1985; Matlin 1986; Griffiths & Simons 1986; Geuze et al 1987), we concluded that the *trans* Golgi network is also the site of exocytic lipid sorting (Simons & van Meer 1988).

Lipid sorting by GSL microdomain formation To generate lipid polarity in epithelial cells, apical exoplasmic GSLs must be separated from PC, which is enriched in the exoplasmic leaflet of the basolateral surface. The evidence listed above justifies the conclusion that newly synthesized GSLs and SPH are transported from the *trans* Golgi to the plasma membrane by carrier vesicles. Arguments will be presented below indicating that this must also be the case for newly synthesized exoplasmic PC. As a consequence, the lipid sorting step in the exocytic pathway will necessarily involve the lateral segregation of GSLs from PC in the luminal bilayer leaflet of the *trans* Golgi network. Immunocytochemically a lateral segregation has been visualized in this organelle for membrane proteins with different intracellular destinations (Geuze et al 1987). Interestingly, GSLs differ from PC in that they possess the capacity to self-associate by hydrogen bonding. The available data fit a model (Simons & van Meer 1988) where apical proteins and GSLs form a microdomain on the luminal surface of the *trans* Golgi membrane at the site of apical vesicle budding. The interaction between the proteins and GSLs may be direct, or mediated by a putative sorting protein. In some way such a sorting protein would also be involved in defining the apical destination of the resulting vesicle, for which interactions with proteins on the cytoplasmic surface would seem necessary (for discussions see Burgess & Kelly 1987; Pelham, Kornfeld & Mellman, Hurtley & Helenius, this volume). Few data are available on the type of bulk interactions between proteins and GSLs that would be required in such a model (Devaux & Seigneuret 1985). Still, the self-association properties of GSLs and their omnipresence in epithelia make them attractive candidates for being instrumental in the formation of microdomains. It is tempting to speculate that they are essential for generating the specialized apical plasma membrane domain. In addition, the same sorting mechanism might play a role in the polarized exocytic transport in a number of other polarized but nonepithelial cells like cytotoxic T cells, neurons, and osteoclasts.

SPHINGOLIPID SORTING IN *CIS* GOLGI The conclusion that (fluorescent) sphingolipids are not transported from their site of synthesis in the Golgi back into the ER explains the low concentration of sphingolipids in the ER. Recent evidence suggests that vesicular transport between the *cis* Golgi and the ER is bidirectional and that resident ER proteins that have entered *cis* Golgi are sorted into the return pathway (Pelham, this volume). Therefore, if GSLs and SPH are present in *cis* Golgi, which is unclear at present, they must be excluded from that return traffic by lipid sorting. In the case that sphingolipid biosynthesis would be localized in a more distal cisterna, the question whether lipid sorting occurs depends on the nature of the transport connection between *cis* Golgi and that cisterna. Various alternatives for SPH (and GSL) transport and sorting are presented in Figure 2. An alternative mechanism that would result in unidirectional transport of SPH has been proposed by Wieland et al (1987; cf Pagano

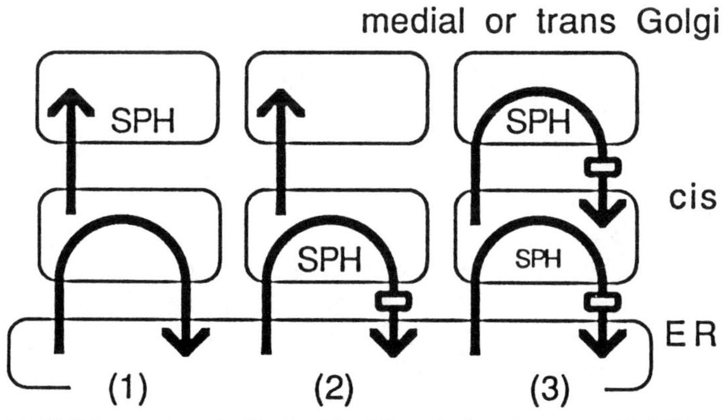

Figure 2 SPH does not reach ER. Possible (alternative) explanations: (*1*) SPH synthesis past a unidirectional vesicular transport step: no SPH sorting. (*2*) SPH synthesis before a unidirectional step: SPH excluded from return traffic. (*3*) Bidirectional vesicular traffic throughout the Golgi stack: multiple SPH sorting steps. *Arrows*: vesicular traffic, *blocks*: exclusion of SPH.

1988). Whereas import of material into *cis* Golgi is vesicular, return traffic might occur by lipid monomers. Although the data of Pelham & coworkers (this volume) make this alternative unlikely, the model has interesting consequences: traffic of retained components, proteins, SPH, and possibly also cholesterol, would be vectorial and they would be concentrated at once. Although the alternatives without sorting provide the easy explanation for the cellular distribution of sphingolipids, the sorting variants (*2*) and (*3*) (Figure 2) have attractive implications for the mechanism by which cholesterol is concentrated on its way to the plasma membrane (see below). The action of the subsequent sorting steps in option (*3*) might result in efficient sorting even if each individual sorting event, like epithelial sphingolipid sorting, were rather leaky.

Sorting, in this case, involves the exclusion of the (luminal) sphingolipids from a budding return vesicle. If phospholipid asymmetry is established in the ER by the presence of a PC-specific flippase, as discussed above, the luminal leaflet of the ER (and *cis* Golgi) membrane contains PC. In that case the Golgi sorting step segregates the sphingolipids into the exocytic vesicles and PC into the return vesicles. If asymmetry is established only later in the Golgi stack, the sphingolipids on the luminal surface would have to be sorted from a mixture of phospholipids. One may speculate that sphingolipid sorting in the Golgi stack is the more general case of the sphingolipid sorting in the *trans* Golgi of epithelial cells. Whereas in the stack, SPH would participate in hydrogen bonding with the GSLs and both GSLs and SPH would be separated from glycerophospholipids, in the *trans* Golgi network of epithelial cells, GSLs are sorted from PC whereby incorporation of SPH in the GSL microdomain seems to depend on the specific conditions in the various cell types.

UNIDIRECTIONAL VESICULAR TRANSPORT IN THE *TRANS* GOLGI? After synthesis in the Golgi, GSLs and SPH are transported to the plasma membrane with $t_{1/2}$s of 20–30 min, a process inhibited at 15°C (Miller-Prodraza & Fishman 1984; Lipsky & Pagano 1985; van Meer et al 1987). From the plasmalemma they can enter the endocytic pathway (Figure 1). Although the bulk of the lipids recycles from the endosomes straight to the plasmalemma, it has been demonstrated that a small fraction of the GSLs recycles to the plasma membrane via the Golgi (Fishman et al 1983; Klein et al 1988). It appears that, as with recycling membrane proteins (see Neefjes et al 1988), recycling GSLs may be limited to the *trans* Golgi cisternae and hence excluded from the proximal cisternae of the Golgi. After implantation into the plasma membrane, simple gangliosides were unable to reach the Golgi site where they could be converted to a more complex form (see Klein et al 1988). As argued above such exclusion could be the result of sphingolipid sorting or, alternatively, of a unidirectional transport step of GSLs in the *trans* Golgi. Although unidirectional transport in the exocytic pathway has been suggested before (Cohen & Phillips 1980), and the advantages of its location in the *trans* Golgi discussed (Rothman 1981), there is at present no direct evidence against vesicles returning from the *trans* Golgi into the Golgi stack.[6]

VESICULAR SPHINGOLIPID TRANSPORT IN THE ENDOCYTIC PATHWAY Pagano & coworkers have demonstrated (see Pagano & Sleight 1985; Koval & Pagano 1989) that fluorescent analogs of PC and SPH, after insertion into the exoplasmic leaflet of the plasma membrane of fibroblasts, are endocytosed into early peripheral endosomes and subsequently transported to perinuclear endosomes. As argued above, from the types of probes used it could be concluded that the transport was vesicular and that no translocation of the PC and SPH had occurred to the cytoplasmic bilayer leaflet in any endocytic organelle. The $t_{1/2}$ for a complete round of SPH recycling was estimated to be about 40 min (Koval & Pagano 1989).

SPHINGOLIPID SORTING IN THE ENDOSOMES Since a vesicular pathway connects the endosomes to the lysosomes (Figure 1) there must be a continuous flux of lipids into the latter. For lipids in the exoplasmic leaflet this flux appears not to be random. Fluorescent analogs of SPH present in the luminal leaflet are preferentially recycled to the plasma membrane (Koval & Pagano 1989). Cellular gangliosides have half-lives in the order of 10–50 h and therefore do not seem to reach the lysosomes efficiently either (see Miller-Prodraza & Fishman 1984). In contrast, N-Rh-PE, a fluorescent PE, was quantitatively shuttled to the lysosomes within 1 h when present in the exoplasmic leaflet of the plasma membrane (J. Davoust & M. Kail, paper in preparation; J. W. Kok, M. Ter Beest, G. Scherphof, & D. Hoekstra, paper in preparation). Lipids destined for the lysosomes can thus be sorted, like proteins, from recycling lipids in the endosomes (see Geuze et al 1987). Additional lipid sorting at the level of the coated pits of the plasma membrane may occur. Little is known about the mechanism of lipid sorting to the lysosomes; however the process apparently occurs in the luminal leaflet of the endosomal membrane and must involve lateral segregation of lipids.

For lipids the vesicular endocytic route to *trans* Golgi appears to be quantitatively of far less importance than direct recycling from the endosomes to the plasmalemma (Fishman et al 1983; Klein et al 1988; Koval

[6] In a few cases, access of cell surface markers to cisternae in the Golgi stack has been reported (see Farquhar 1985). The interpretation of these observations is unclear at present (cf Griffiths & Simons 1986; Neefjes et al 1988).

& Pagano 1989). Nothing is known concerning possible lipid sorting events along this pathway.

Preliminary experiments on transcytosis, the vesicular connection between the apical and basolateral plasma membrane domains in epithelial cells, have demonstrated that this pathway involves sphingolipid sorting similar to that in the exocytic pathway in these cells and have suggested the endosomes as the site of sorting (G. van Meer, unpublished observations).

Generation and Maintenance of the Heterogeneous Distribution of Cholesterol[7]

CHOLESTEROL BIOSYNTHESIS IN THE ER According to the general view of the past 20 years, cholesterol biosynthesis occurs on the ER (Reinhart et al 1987). In addition, a (minor) independent pathway has been shown to exist in the peroxisomal lumen (Appelkvist 1987; Thompson et al 1987). Lange & colleagues have presented cell fractionation data from which they proposed that subsequent steps in the biosynthetic process were topographically heterogeneous. It is unclear, however, whether the various peaks on their sucrose gradients represent contiguous regions of a compound structure (the ER) or entirely discrete, perhaps hitherto undescribed, organelles (Reinhart et al 1987; Lange & Muraski 1988).

VESICULAR CHOLESTEROL TRANSPORT TO THE PLASMA MEMBRANE All evidence agrees that the transport of newly synthesized cholesterol from its site of synthesis to the plasma membrane is vesicular: it is inhibited at 15°C, and influenced by energy inhibitors (Lange & Matthies 1984; Kaplan & Simoni 1985b; Slotte et al 1987). Probably due to methodological problems, however, the interpretation of the transport data is still controversial. On the one hand, the cholesterol transport appeared to be vectorial with a short (10 min; DeGrella & Simoni 1982) or long $t_{1/2}$ (1–1.5 h; Lange & Matthies 1984). On the other hand, Kaplan & Simoni (1985b) have concluded that the vesicular transport represents two-directional equilibration

[7] Compared to the methods used to localize other cellular lipids, those for cholesterol suffer from severe drawbacks. Spontaneous cholesterol exchange (see Phillips et al 1987) has a $t_{1/2}$ on the order of 2 h at 37°C, which is slow as compared to cellular transport processes. Since all cholesterol in the endocytic compartments of Figure 1 will also take part in the exchange process, the method results in an overestimate of plasma membrane cholesterol. A second method involves the use of cholesterol oxidase. The results depend on how the method is used (see Brasaemle et al 1988). It appears that more than just the plasma membrane cholesterol is accessible to externally added oxidase. The observation that newly synthesized cholesterol is not available to the enzyme (see Lange & Muraski 1988) seems insufficient as a control. Proof has to be provided that endosomal cholesterol is not oxidized. Also the interpretation of results obtained with filipin has led to serious problems (Yeagle 1985). Finally, the limitations of cell fractionation are discussed above.

of ER cholesterol with that in the plasma membrane with a $t_{1/2}$ of about 10 min. Other observations make this highly unlikely. Equilibration of exogenously inserted plasma membrane cholesterol with that in the ER, assayed as access of the cholesterol to the site of esterification, was shown to have a $t_{1/2}$ in the order of 2–5 days (see Lange & Matthies 1984; Slotte & Bierman 1987).

SORTING OF CHOLESTEROL INTO THE EXOCYTIC PATHWAY The cholesterol concentration in the plasma membrane is much higher than that in the ER (Table 1). Therefore compared to the bulk membrane lipids, the phospholipids, cholesterol must have been concentrated along the pathway. The putative SPH sorting in the Golgi suggests a mechanism for this. A preferential interaction of cholesterol with SPH in comparison with other phospholipids (the weakest interaction being with PE) has been observed in numerous studies (Demel et al 1977; Gardam et al 1989; reviewed in Barenholz & Thompson 1980; Phillips et al 1987). With the estimation by Wieland et al (1987) that about 50% of the ER lipid cycles through the *cis* Golgi every 10 min, and with SPH present in the *cis* Golgi, optimal contact of this SPH with the ER cholesterol would be guaranteed. Cholesterol may be incorporated into an SPH microdomain and subsequently be sorted as a complex to the medial Golgi and further out. In fact, the apparent cholesterol concentration gradient over the stack, measured by the density of filipin-cholesterol complexes (Orci et al 1981), would be most consistent with the model for SPH sorting effected at a series of steps, Figure 2 (*3*).

Some eukaryotes, like *Acanthamoeba*, do not contain SPH, but do enrich sterols in the plasma membrane (Dawidowicz 1987), probably during vesicle transport (Mills et al 1984). It is of possible evolutionary interest that cholesterol also displays a preference for disaturated PC, close to that for SPH (Lange et al 1979; see Phillips et al 1987): the mechanism of *cis* Golgi sorting may originally have used disaturated phospholipids. These may still be required in the sorting as there is too little SPH to complex cholesterol on a one to one basis. Plasma membranes are enriched in disaturated phospholipids (e.g. Keenan & Morré 1970; Colbeau et al 1971).

As a consequence of a preferential cholesterol-SPH interaction,[8] the

[8] The importance of SPH for maintaining the heterogeneous distribution of intracellular cholesterol is illustrated by the fact that hydrolysis of plasma membrane SPH resulted in a shift of cholesterol from the plasma membrane to intracellular membranes, while a shift out of the ER was observed after cellular uptake of exogenous SPH: the latter induced a number of effects known to be mediated by lowering the ER cholesterol concentration (see Slotte & Bierman 1988). Finally, it is interesting to note that the biosynthesis of all sphingolipids appears to be controlled by low density lipoprotein (cholesterol) uptake (Chatterjee et al 1986).

bulk of the cholesterol would be situated in the luminal, exoplasmic bilayer leaflet of membranes in the Golgi and the endocytic compartments (Figure 1)! Although X-ray studies have indeed shown this to be the case in myelin, the plasma membrane of Schwann cells and oligodendrocytes (Caspar & Kirschner 1971), in general the sparse data available on the transmembrane distribution of cholesterol in cellular membranes are contradictory, even for the well-characterized erythrocyte (Zachowski & Devaux 1989).

MONOMERIC CHOLESTEROL DIFFUSION Cholesterol equilibration between membranes by spontaneous exchange with a $t_{1/2}$ of about 2 h is slow compared to vesicular traffic (Phillips et al 1987). This value was obtained in vitro and may have been dependent on membrane collisions (Steck et al 1988). Such collisions may be limited in the cell where the process may occur more slowly. Still nonvesicular transport of cholesterol seems necessary in at least two cases (and will therefore be a common cytoplasmic process).[9] One is transport of cholesterol to the inner mitochondrial membrane of steroidogenic cells for the first step of the hormone production, the side chain cleavage, the second is the use of cholesterol derived from low density lipoproteins in the lysosomes for cholesterol ester synthesis in the ER.

CHOLESTEROL TRANSPORT INTO THE MITOCHONDRION During steroidogenesis large amounts of cholesterol are transported from lipid droplets to the inner mitochondrial membrane for the initial step of hormone production, the side chain cleavage. A large body of evidence suggests that a 13.5-kd protein, sterol carrier protein 2 (SCP-2), has a crucial role at two steps of the transport process: the cytoplasmic transport to the outer mitochondrial membrane and the transfer from the outer to the inner membrane (see Scallen et al 1985). SCP-2, which has independently been characterized as possessing a nonspecific lipid transfer activity in vitro, also stimulates a number of metabolic conversions of cholesterol. For the time being, mechanistic interpretations concerning the cytoplasmic actions of this complicated protein SCP-2 (van Amerongen et al 1989, and references therein) are frustrated by immunocytochemical studies that have

[9] The heterogeneous distribution of cholesterol could be a simple consequence of cytoplasmic exchange in combination with different affinities of the various intracellular membranes for cholesterol. Such differences were indeed observed between microsomal, mitochondrial, and plasma membrane fractions (Wattenberg & Silbert 1983), but not to an extent that could explain the level of cholesterol enrichment in the plasma membrane as compared to, for example, the ER. However, interpretation of the data was complicated by the fact that the experimental setup was based on the premise that an exchange equilibrium would exist between the various cholesterol pools. As described above, such an equilibrium appears not to apply to the cholesterol in the exoplasmic leaflet of the plasmalemma and that in the ER.

localized the protein to the peroxisomal matrix (Tsuneoka et al 1988; van Amerongen et al 1989) instead of to the cytosol (see Scallen et al 1985). In fact, this rather suggests a role for SCP-2 in peroxisomal cholesterol synthesis.

Transport of cholesterol from the lipid droplet to the site of side chain cleavage on the inner aspect of the inner membrane requires at least four discrete steps: cytoplasmic transport to the outer membrane, translocation across the outer membrane bilayer, transfer across the intermembrane space to the inner membrane (with the possible involvement of the "contact sites"), and translocation across the inner membrane. Although various inhibitors and activators of the overall transport have been characterized, the actual mechanism of each step remains obscure (reviewed in Lambeth et al 1987; van Amerongen et al 1989).

CHOLESTEROL FLIP-FLOP IN LYSOSOMES BUT NOT PLASMA MEMBRANE? Cholesterol derived from low density lipoprotein starts to exert effects at the level of the ER, i.e. stimulation of cholesterol ester synthesis and downregulation of cholesterol synthesis after just 3 h (Liscum & Faust 1987). These regulatory functions appear to depend on the actual insertion of the cholesterol into the ER membrane (Orci et al 1984; Davis & Poznansky 1987). Similar effects could be obtained by raising the plasma membrane cholesterol concentration by transfer from cholesterol-rich liposomes, but only after 24 h (Robertson & Poznansky 1985). The simplest explanation for the fact that lysosomal cholesterol does and exoplasmic plasmalemma cholesterol does not participate in cytoplasmic transport processes would be that cholesterol flip-flops to the cytoplasmic leaflet of the lysosomal membrane but not of the plasma membrane.[10] Unfortunately, the rate of cholesterol flip-flop is also a matter of controversy (see Zachowski & Devaux 1989).

In summary, in cells rapid transport of cholesterol occurs by means of carrier vesicles ($t_{1/2}$ of about 30 min), while cholesterol also redistributes by monomeric exchange ($t_{1/2}$ probably longer than 2 h). The heterogeneity in cholesterol content in vivo is best explained by cholesterol sorting in the Golgi complex (based on a differential affinity for the various phospholipid classes) in combination with preferential localization of cholesterol in the

[10] Either flip-flop in the lysosomes or the aqueous transfer of cholesterol is inhibited in type C Niemann-Pick disease, while biosynthetic (vesicular) transport is not (Liscum et al 1989). The enormous increase in lysosomal cholesterol in this disease observed by many authors (Blanchette-Mackie et al 1988; Liscum et al 1989) was accompanied by a relatively modest rise of the cholesterol level in the *trans* Golgi (Blanchette-Mackie et al 1988). Interpretation of this observation in terms of transport is difficult. Possibly, the vesicular endosome-*trans* Golgi route (Figure 1) is involved.

exoplasmic leaflet of the bilayer, and the absence of cholesterol flip-flop in Golgi membranes, plasmalemma, and endosomal membranes.

Generation and Maintenance of the Heterogeneous Distribution of Glycerophospholipids

PHOSPHATIDYLCHOLINE DISTRIBUTION AND TRANSPORT Cellular PC is synthesized in the ER and the Golgi by the CDP-choline pathway and to a small extent by the stepwise methylation of PE (Jelsema & Morré 1978; Bell et al 1981; Vance & Ridgway 1988), whereas straight incorporation of choline by a base exchange reaction appears to be a minor route (see Yaffe & Kennedy 1983; Figure 3). The creation of the ER bilayer is made possible by the presence of a PC-pore or "flippase" (see Bishop & Bell 1988; Devaux 1988), but somewhere in the Golgi stack this potential to translocate PC across the bilayer is lost: at least from the *trans* Golgi onwards throughout the endocytic pathway (Figure 1), transport has to account for two PC pools that do not rapidly intermix (see above). Kaplan & Simoni (1985a) have described how newly synthesized PC rapidly equilibrated with that in the plasma membrane [$t_{1/2}(37°C) \ll 2$ min]. However, when looking for two-pool kinetics, it is evident from the data that a significant fraction of the newly synthesized PC reached the plasma membrane with a $t_{1/2}$ of 30–60 min, a process that was influenced by energy poisons and inhibited at 15°C, all of which are characteristics of vesicular transport. The obvious prediction that this PC fraction represents the exoplasmic PC pool will have to be proven by a direct experiment.

Transport of newly synthesized PC (and PE) by a special class of phos-

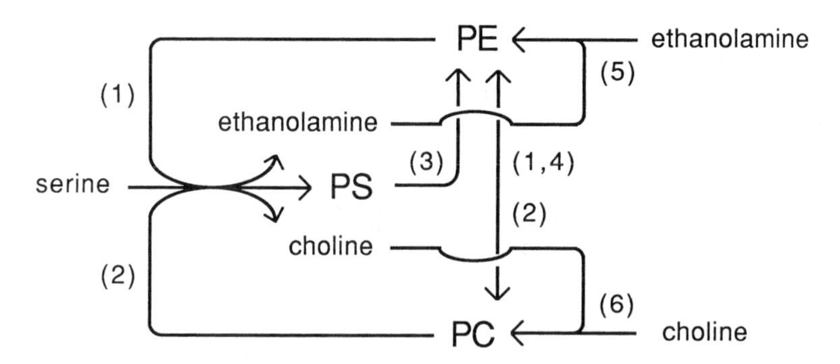

Figure 3 Import of head groups into the cellular phospholipid metabolism: (*1*) base-exchange from PE, (*2*) base-exchange from PC, (*3*) PS-decarboxylase, (*4*) PE methylation, (*5*) ethanolamine kinase; CTP-phosphoethanolamine cytidylyltransferase; CDP-ethanolamine: diacylglycerol ethanolamine phosphotransferase, (*6*) choline kinase; CTP-phosphocholine cytidylyltransferase; CDP-choline: diacylglycerol choline phosphotransferase.

pholipid-rich vesicles has been reported to be responsible for the rapid plasma membrane growth during mitosis (Bluemink et al 1983) and cell aggregation of *Dictyostelium* (De Silva & Siu 1981).

Newly synthesized PC equilibrated with the outer mitochondrial membrane PC pool with a $t_{1/2}$ in the order of 30 min (McMurray & Dawson 1969; Blok et al 1971) or faster (Yaffe & Kennedy 1983).[11] Candidates for mediating this equilibration, which according to the present view is not vesicular, are two transfer proteins that possess the capability to transport PC monomers across an aqueous barrier; one has a strict specificity for PC, the other is specific for PI and PC (reviewed in Helmkamp 1986).[12] The need for such rapid equilibration is unclear at present. However, it is clear that in order for mitochondria to grow a net transfer of PC into the mitochondria is required, since mitochondria do not synthesize PC themselves. Several investigators have demonstrated that the PC-specific transfer protein can yield such net transfer in vitro. However, since the experiments were performed under conditions where the PC concentrations of donor and acceptor membranes were far from equilibrium (discussed in Helmkamp 1986), these results cannot be simply extrapolated to the situation in the cell. How mitochondria, or for that matter peroxisomes, expand their surface, therefore remains an enigma.

Transfer of PC and PE between the outer and inner mitochondrial membranes is much slower than other intracellular PC traffic, with a $t_{1/2}$ in the order of 0.5–3 h (Bygrave 1969; McMurray & Dawson 1969; Blok et al 1971; Eggens et al 1979). In contrast with the rapid PC flip-flop in the outer membrane, translocation of a spin-labeled PC across the inner membrane was found to be extremely slow (Rousselet et al 1976).

PHOSPHATIDYLINOSITOL TRANSPORT Essentially all PI is synthesized in the ER (Jelsema & Morré 1978). It has been known for 20 years that PI equilibrates between ER and mitochondria within minutes (McMurray & Dawson 1969), and a transfer protein with a specificity for PI and PC was later characterized. Inositol lipids, as intermediates in the polyphosphoinositol signal transducing system, display a high turnover in the cytoplasmic leaflet of the plasma membrane. Therefore replenishment of plasma membrane PI is required. Although vesicles traveling from the

[11] Equilibration of the total outer mitochondrial membrane PC pool within less than an hour predicts a relatively rapid PC flip-flop in this membrane, a property recently established by a direct in vitro experiment (K. Nicolay & R. Hovius, paper in preparation).

[12] The PC and PI/PC specific transfer proteins may also be responsible for the rapid PC equilibration observed by Kaplan & Simoni (1985a). However, as for sterol carrier protein 2, the location in the cytosolic compartment remains to be demonstrated unequivocally for both transfer proteins.

ER to the plasma membrane will undoubtedly carry PI, a more rapid mechanism of PI transfer has been proposed based on the dual specificity of the PI/PC transfer protein (see Helmkamp 1986; Van Paridon et al 1987). Since the PI level is lower in the plasmalemma than in the ER membrane, the PI/PC protein has a lower probability of picking up PI as compared to PC in the plasma membrane than in the ER; this could result in a net flux of PI towards the plasmalemma and a return flow of PC.

PHOSPHATIDYLSERINE DISTRIBUTION AND TRANSPORT PS is synthesized by PS synthase, an enzyme mediating base-exchange, primarily in the ER but also in the Golgi and the plasma membrane (Figure 3, Jelsema & Morré 1978; Voelker 1985; Vance & Vance 1988). Recent evidence points at PC as the major substrate of the reaction (see Kuge et al 1986; Miller & Kent 1986; Voelker & Frazier 1986). PS is enriched at least threefold in the plasma membrane (Table 1), and probably in all membranes of the endocytic pathway as compared to the other organelles (Luzio & Stanley 1983; Urade et al 1988; cf Evans & Hardison 1985). In the plasma membrane a specific enzyme maintains its location in the cytoplasmic leaflet (see Devaux 1988).[13]

Two considerations may be relevant for understanding how PS is concentrated in the cytoplasmic leaflet of the plasma membrane and the membranes of the endocytic pathway (Figure 1). (*a*) As PS, like other phospholipids, is subject to vesicular transport, traffic back into the Golgi should be prevented; this could imply PS sorting in the cytoplasmic leaflet of the Golgi by some new mechanism (cf SPH/cholesterol sorting in the exoplasmic leaflet). Interestingly, plasma membrane PS has been reported to have a much more saturated fatty acyl composition than PS in the ER (Keenan & Morré 1970). (*b*) Significant monomeric exchange of PS occurs in the cytoplasm, but, as for cholesterol, the rate of this process appears to be much lower than that of vesicular transport (see below). As an alternative to PS sorting, the enhanced PS concentrations in the plasmalemma and the membranes of the endocytic pathway may be explained by these membranes having a higher affinity for PS. A reason for such an enhanced affinity of the former membranes as compared to the ER and

[13] The presence of the "amino-phospholipid translocator" and the fact that spontaneous flip-flop of PS across the plasma membrane is faster than that of the other phospholipids, suggest that PS is central to some regulatory mechanism. In line with the coupled bilayer hypothesis by Sheetz & Singer (1974), the mechanism may control the relative surface areas and lateral pressure of each leaflet of the plasma membrane bilayer. One could even envisage how the ATP-requiring translocator could be the driving force behind endocytic vesicle budding from the flat plasma membrane. In this view, apart from sorting considerations, clathrin might be required to translate the pressure increase in the cytoplasmic leaflet into a local site of high curvature (see also Devaux 1988).

the outer mitochondrial membrane may be the fact that they have a membrane potential that is positive on the exoplasmic side (e.g. Fuchs et al 1989): desorption of the negatively charged PS into the cytoplasm, which is against the electrical field, requires more energy than diffusion from the other organelles (independent of whether exchange is spontaneous or protein-mediated).

It has been proposed that in a number of cell types PS serves as the major, or even exclusive, precursor for PE synthesis (see Voelker 1985; Miller & Kent 1986). The responsible enzyme, PS-decarboxylase (Figure 3), has been localized to the outer leaflet of the inner mitochondrial membrane (see Daum 1985; Voelker 1985; Yorek et al 1985). A significant fraction of the cellular PS must thus be transferred from the ER, across the outer membrane to the inner membrane, where it is converted to PE. The $t_{1/2}$ of the entire process turned out to be on the order of 5–7 h (Voelker 1985; Vance & Vance 1988).[14] The transport of PS was found to be sensitive to metabolic poisons and cycloheximide (Voelker 1985); at present there is no simple explanation for these effects.

PHOSPHATIDYLETHANOLAMINE SYNTHESIS AND TRANSPORT For PE synthesized by decarboxylation of PS in the inner mitochondrial membrane (see above), transport out of the mitochondrion involves three steps: from the inner to the outer membrane (possibly involving the "contact sites"), across the outer membrane, and to the ER or other organelles. The intramitochondrial exchange has been measured to have a rate similar to that of PC, with a $t_{1/2}$ of 0.5–3 h (Bygrave 1969; McMurray & Dawson 1969; Blok et al 1971; Eggens et al 1979); the mechanism of this transport is still unclear. Transport from the outer membrane to the ER, according to the current views, occurs by monomeric exchange through the aqueous phase. This equilibration appears to be much slower than that of PI or PC, with a $t_{1/2}$ of about 70 min (McMurray & Dawson 1969; Blok et al 1971; Yaffe & Kennedy 1983). A somewhat faster time course has been reported for the equilibration[15] of newly synthesized PE with PE in the plasma membrane (Sleight & Pagano 1983; Kobayashi & Pagano 1989). Their results with various inhibitors and temperatures also are best explained by an exchange

[14] Very different time courses have been obtained from radioactive labeling and cell fractionation in yeast (Daum et al 1986) and rat liver (Vance 1988). However, the complexity of the metabolic pathways and the occurrence of multiple pools of phospholipid, the nature of which is unclear at present, complicates evaluations of transport kinetics.

[15] The authors concluded that a rapid inside-outside translocation of PE had taken place across the plasma membrane. In an independent study (see Martin & Pagano 1987) they demonstrated that C6-NBD-PE and -PS undergo a fast (protein-mediated) translocation in the opposite direction. Also in the erythrocyte membrane PE translocation in both directions is tenfold faster than that of PC (see Zachowski & Devaux 1989).

process. This conclusion was recently corroborated by the observation that, in contrast to the vesicular transport of sphingolipids, PE transport was not inhibited in mitotic cells (Kobayashi & Pagano 1989). Although PE like most phospholipids is present in all organelles along the vesicular pathway (Figure 1) and will therefore be shuttled by carrier vesicles, the evidence suggests that monomeric transfer is responsible for intracellular PE equilibration.

It has been suggested (Yorek et al 1985; see Miller & Kent 1986) that the CDP-ethanolamine pathway on the cytoplasmic surface of the ER and the Golgi (Figure 3, see Jelsema & Morré 1978; Bell et al 1981; Vance & Vance 1988) is preferentially used for the synthesis of ethanolamine plasmalogens. Plasmalogens possess a long chain alcohol in ether linkage at the C1-position of the glycerol, instead of the esterified fatty acid of the regular phospholipids. In addition, the first two carbons of the alcohol are connected by a double bond. They constitute a significant fraction of the cellular phospholipids, especially of the PE of which they routinely make up 50% or more (see Lazarow 1987; van den Bosch et al 1988). The introduction of the ether bond has been localized to the luminal leaflet of the peroxisomal membrane. All plasmalogens must, therefore, at some stage have passed through this organelle. The mechanism of transport is relatively simple: since acylation of the C2-position of the glycerol and the subsequent steps in plasmalogen biosynthesis only take place at the ER, transport occurs at the alkylglycerophosphate (or alkyl-dihydroxyacetone phosphate) stage. This should present no problem as this lysophosphatidic acid readily exchanges through the aqueous phase (see Bell et al 1981; Lazarow 1987). How the molecule initially translocates to the cytoplasmic leaflet of the peroxisomal membrane is, however, unclear.

MITOCHONDRIAL CARDIOLIPIN (CL) AND LYSOSOMAL LYSOBISPHOSPHATIDIC ACID (LBPA) CL is synthesized in the mitochondrial inner membrane, where the bulk of the CL also resides. It is unclear by what mechanism CL is retained in the inner membrane, but it appears to be leaky as some CL is present in the outer membrane (Daum 1985). Phosphatidylglycerol (PG), its precursor, is synthesized in the inner membrane. PG-synthetic activity is also present in ER (see Daum 1985). An answer to the question how this ER-produced PG is handled by the cell may provide new insights into general phospholipid traffic. However, the relative quantity of this PG is very low. Interesting systems, in this respect, are a PG overproducing cell mutant (Esko & Raetz 1980) and the lung type II cell producing surfactant PG (see Post & van Golde 1988). The selective presence of LBPA in lysosomes may be explained simply by the fact that it is synthesized there (see Bell et al 1981), as in the case of CL in the mitochondria. Interestingly, LBPA was also found in the intracellular vesicles by which Vaccinia virus surrounds its DNA in the process of obtaining its envelope (Hiller et al 1981).

SUMMARY AND PERSPECTIVES

A central question in the field of the cell biology of lipids is how cells maintain the unique lipid compositions of their organelles. At present, no coherent scheme of intracellular lipid transport can be formulated to account for these differences without numerous assumptions: there is a basic lack of reliable data on the lipid composition of purified subcellular membrane fractions and on the topography of the lipids therein. Part of the solution to this problem must come from defining the site of synthesis, the transbilayer distribution, and the translocation properties of the lipids. A relatively clear case is that of sphingolipids that are assembled somewhere in the Golgi, reside in the exoplasmic leaflet, and do not flip-flop. They can be transported by vesicles only, as they are simply not accessible for cytoplasmic exchange. The situation is more complex for the other lipids, all of which take part, as monomers, in cytoplasmic intermembrane transfer. In the case of cholesterol, a lot may be explained by a preferential interaction with sphingolipids over glycerophospholipids. Also the fact that some lipids, for instance PC, experience flip-flop in some membranes but essentially not in others imposes specificity onto overall transport. Clear areas of research will therefore be to establish the exact site of SPH synthesis, the sidedness (and the flip-flop rate) of cholesterol in cellular membranes, and the intracellular location of lipid transfer proteins. A different set of questions concerns bulk membrane traffic: does the Golgi possess a unidirectional vesicular transport step? How do cells regulate the relative surface areas of their membranes?

Maybe the most exciting direction that emerges from the recent literature is the crucial role of sphingolipids and possibly disaturated glycerophospholipids in sorting processes in the Golgi. By being shuttled exclusively from *cis* to *trans* and to the plasmalemma they may be responsible for maintaining the cholesterol gradient in the cell. They are responsible for the establishment of lipid polarity in epithelial cells. As in evolution these processes have evolved coordinately with membrane protein transport, it is expected that both membrane lipids and proteins make up indispensible parts of an intricate sorting machinery. The elucidation of this machinery will mark a breakthrough in molecular membrane biology.

ACKNOWLEDGMENTS

I apologize to all researchers to whose original contributions I could only refer via reviews by others. With thanks to Günther Daum, Philippe Devaux, Jerry Faust, Dick Hoekstra, Laura Liscum, Dick Pagano, Konrad Sandhoff, Joachim Seelig, Peter Slotte, Dennis Vance, and Dennis Voelker for opinions, reprints and preprints; to Henk van den Bosch, Arie van der Ende, Wouter van 't Hof, Arie Verkley, Karel Wirtz, and especially to Mark Marsh (London) and Klaas Nicolay for critical comments. I am grateful to them and other colleagues at the Utrecht University for their willingness to discuss what must have seemed at that time to be random issues. The research for this review was sponsored by a senior investigatorship from the Royal Netherlands Academy of Arts and Sciences.

Literature Cited

Allan, D., Walklin, C. M. 1988. Endovesiculation of human erythrocytes exposed to sphingomyelinase C: a possible explanation for the enzyme-resistant pool of sphingomyelin. *Biochim. Biophys. Acta* 938: 403–10

Appelkvist, E.-L. 1987. In vitro labeling of peroxisomal cholesterol with radioactive precursors. *Biosci. Rep.* 7: 853–58

Barenholz, Y., Thompson, T. E. 1980. Sphingomyelins in bilayers and biological membranes. *Biochim. Biophys. Acta* 604: 129–58

Barsukov, L. I., Bergelson, L. D., Spiess, M., Hauser, H., Semenza, G. 1986. Phospholipid topology and flip-flop in intestinal brush border membrane. *Biochim. Biophys. Acta* 862: 87–99

Belcher, J. D., Hamilton, R. L., Brady, S. E., Hornick, C. A., Jaeckle, S., et al. 1987. Isolation and characterization of three endosomal fractions from the liver of estradiol-treated rats. *Proc. Natl. Acad. Sci. USA* 84: 6785–89

Bell, R. M., Ballas, L. M., Coleman, R. A. 1981. Lipid topogenesis. *J. Lipid Res.* 22: 391–403

Bishop, W. R., Bell, R. M. 1988. Assembly of phospholipids into cellular membranes: Biosynthesis, transmembrane movement and intracellular translocation. *Annu. Rev. Cell Biol.* 4: 579–610

Blanchette-Mackie, E. J., Dwyer, N. K., Amende, L. M., Kruth, H. S., Butler, J. D., et al. 1988. Type-C Niemann-Pick disease: low density lipoprotein uptake is associated with premature cholesterol accumulation in the Golgi complex and excessive cholesterol storage in lysosomes. *Proc. Natl. Acad. Sci. USA* 85: 8022–26

Blok, M. C., Wirtz, K. W. A., Scherphof, G. L. 1971. Exchange of phospholipids between microsomes and inner and outer mitochondrial membranes of rat liver. *Biochim. Biophys. Acta* 233: 61–75

Bluemink, J. G., van Maurik, P. A. M., Tertoolen, L. G. J., van der Saag, P. T., de Laat, S. W. 1983. Ultrastructural aspects of rapid plasma membrane growth in mitotic neuroblastoma cells. *Eur. J. Cell. Biol.* 32: 7–16

Brasaemle, D. L., Robertson, A. D., Attie, A. D. 1988. Transbilayer movement of cholesterol in the human erythrocyte membrane. *J. Lipid Res.* 29: 481–89

Brotherus, J., Renkonen, O. 1977. Phospholipids of subcellular organelles isolated from cultured BHK cells. *Biochim. Biophys. Acta* 486: 243–53

Burgess, T. L., Kelly, R. B. 1987. Constitutive and regulated secretion of proteins. *Annu. Rev. Cell Biol.* 3: 243–93

Bygrave, F. L. 1969. Studies on the biosynthesis and turnover of the phospholipid components of the inner and outer membranes of rat liver mitochondria. *J. Biol. Chem.* 244: 4768–72

Caspar, D. L. D., Kirschner, D. A. 1971. Myelin membrane structure at 10 Å resolution. *Nature New Biol.* 231: 46–52

Chatterjee, S., Clarke, K. S., Kwiterovich, P. O. Jr. 1986. Regulation of synthesis of lactosylceramide and long chain bases in normal and familial hypercholesterolemic cultured proximal tubular cells. *J. Biol. Chem.* 261: 13474–79

Chojnacki, T., Dallner, G. 1988. The biological role of dolichol. *Biochem. J.* 251: 1–9

Cohen, B. G., Phillips, A. H. 1980. Evidence

for rapid and concerted turnover of membrane phospholipids in MOPC 41 myeloma cells and its possible relationship to secretion. *J. Biol. Chem.* 255: 3075–79

Colbeau, A., Nachbaur, J., Vignais, P. M. 1971. Enzymic characterization and lipid composition of rat liver subcellular membranes. *Biochim. Biophys. Acta* 249: 462–92

Cook, H. W., Palmer, F. B. St. C., Byers, D. M., Spence, M. W. 1988. Isolation of plasma membranes from cultured glioma cells and application to evaluation of membrane sphingomyelin turnover. *Anal. Biochem.* 174: 552–60

Crane, D. I., Masters, C. J. 1986. The effect of clofibrate on the phospholipid composition of the peroxisomal membranes in mouse liver. *Biochim. Biophys. Acta* 876: 256–63

Daum, G. 1985. Lipids of mitochondria. *Biochim. Biophys. Acta* 822: 1–42

Daum, G., Heidorn, E., Paltauf, F. 1986. Intracellular transfer of phospholipids in the yeast, *Saccharomyces cerevisiae*. *Biochim. Biophys. Acta* 878: 93–101

Davis, P. J., Poznansky, M. J. 1987. Modulation of 3-hydroxy-3-methylglutaryl-CoA reductase by changes in microsomal cholesterol content or phospholipid composition. *Proc. Natl. Acad. Sci. USA* 84: 118–21

Dawidowicz, E. A. 1987. Dynamics of membrane lipid metabolism and turnover. *Annu. Rev. Biochem.* 56: 43–61

DeGrella, R. F., Simoni, R. D. 1982. Intracellular transport of cholesterol to the plasma membrane. *J. Biol. Chem.* 257: 14256–62

Demel, R. A., Jansen, J. W. C. M., van Dijck, P. W. M., van Deenen, L. L. M. 1977. The preferential interaction of cholesterol with different classes of phospholipids. *Biochim. Biophys. Acta* 465: 1–10

De Silva, N. S., Siu, C.-H. 1981. Vesicle-mediated transfer of phospholipids to plasma membrane during cell aggregation of *Dictyostelium discoideum*. *J. Biol. Chem.* 256: 5845–50

Devaux, P. F. 1988. Phospholipid flippases. *FEBS Lett.* 234: 8–12

Devaux, P. F., Seigneuret, M. 1985. Specificity of lipid-protein interactions as determined by spectroscopic techniques. *Biochim. Biophys. Acta* 822: 63–125

Dickson, R. B., Beguinot, L., Hanover, J. A., Richert, N. D., Willingham, M. C., Pastan, I. 1983. Isolation and characterization of a highly enriched preparation of receptosomes (endosomes) from a human cell line. *Proc. Natl. Acad. Sci. USA* 80: 5335–39

Dyatlovitskaya, E. V., Timofeeva, N. G., Yakimenko, E. F., Barsukov, L. I., Muzya, G. I., Bergelson, L. D. 1982. A sphingomyelin transfer protein in rat tumors and fetal liver. *Eur. J. Biochem.* 123: 311–5

Eggens, I., Valtersson, C., Dallner, G., Ernster, L. 1979. Transfer of phospholipids between the Endoplasmic Reticulum and mitochondria in rat hepatocytes in vivo. *Biochem. Biophys. Res. Commun.* 91: 709–14

Esko, J. D., Raetz, C. R. H. 1980. Mutants of Chinese hamster ovary cells with altered membrane phospholipid composition. Replacement of phosphatidylinositol by phosphatidylglycerol in a myoinositol auxotroph. *J. Biol. Chem.* 255: 4474–80

Evans, W. H., Hardison, W. G. M. 1985. Phospholipid, cholesterol, polypeptide and glycoprotein composition of hepatic endosome subfractions. *Biochem. J.* 232: 33–36

Farquhar, M. G. 1985. Progress in unraveling pathways of Golgi traffic. *Annu. Rev. Cell Biol.* 1: 447–88

Fishman, P. H., Bradley, R. M., Hom, B. E., Moss, J. 1983. Uptake and metabolism of exogenous gangliosides by cultured cells: effect of choleragen on the turnover of G_{M1}. *J. Lipid Res.* 24: 1002–11

Fuchs, R., Mâle, P., Mellman, I. 1989. Acidification and ion permeabilities of highly purified rat liver endosomes. *J. Biol. Chem.* 264: 2212–20

Fujiki, Y., Fowler, S., Shio, H., Hubbard, A. L., Lazarow, P. B. 1982. Polypeptide and phospholipid composition of the membrane of rat liver peroxisomes: comparison with Endoplasmic Reticulum and mitochondrial membranes. *J. Cell Biol* 93: 103–10

Fürst, W., Machleidt, W., Sandhoff, K. 1988. The precursor of sulfatide activator protein is processed to three different proteins. *Biol. Chem. Hoppe-Seyler* 369: 317–28

Gardam, M. A., Itovitch, J. J., Silvius, J. R. 1989. Partitioning of exchangeable fluorescent phospholipids and sphingolipids between different lipid bilayer environments. *Biochemistry* 28: 884–93

Gueze, H. J., Slot, J. W., Schwartz, A. L. 1987. Membranes of sorting organelles display lateral heterogeneity in receptor distribution. *J. Cell Biol.* 104: 1715–23

Griffiths, G., Simons, K. 1986. The *trans* Golgi network: sorting at the exit site of the Golgi complex. *Science* 234: 438–43

Hansson, G. C., Simons, K., van Meer, G. 1986. Two strains of the Madin-Darby canine kidney (MDCK) cell line have distinct glycosphingolipid compositions. *EMBO J.* 5: 483–89

Helmkamp, G. M. Jr. 1986. Phospholipid transfer proteins: mechanism of action. *J. Bioenerg. Biomembr.* 18: 71–91
Helmy, S., Porter-Jordan, K., Dawidowicz, E. A., Pilch, P., Schwartz, A. L., Fine, R. E. 1986. Separation of endocytic from exocytic coated vesicles using a novel cholinesterase mediated density shift technique. *Cell* 44: 497–506
Henning, R., Stoffel, W. 1973. Glycosphingolipids in lysosomal membranes. *Hoppe-Seyler's Z. Physiol. Chem.* 354: 760–70
Hiller, G., Eibl, H., Weber, K. 1981. Acyl bis(monoacylglycero)phosphate, assumed to be a marker for lysosomes, is a major phospholipid of Vaccinia virions. *Virology* 113: 761–64
Jelsema, C. L., Morré, D. J. 1978. Distribution of phospholipid biosynthetic enzymes among cell components of rat liver. *J. Biol. Chem.* 253: 7960–71
Kaplan, M. R., Simoni, R. D. 1985a. Intracellular transport of phosphatidylcholine to the plasma membrane. *J. Cell Biol.* 101: 441–45
Kaplan, M. R., Simoni, R. D. 1985b. Transport of cholesterol from the Endoplasmic Reticulum to the plasma membrane. *J. Cell Biol.* 101: 446–53
Keenan, T. W., Morré, D. J. 1970. Phospholipid class and fatty acid composition of Golgi apparatus isolated from rat liver and comparison with other cell fractions. *Biochemistry* 9: 19–25
Keenan, T. W., Huang, C. M., Morré, D. J. 1972. Gangliosides: Non-specific localization in the surface membranes of bovine mammary gland and rat liver. *Biochem. Biophys. Res. Comm.* 47: 1277–83
Klein, D., Leinekugel, P., Pohlentz, G., Schwarzmann, G., Sandhoff, K. 1988. Metabolism and intracellular transport of gangliosides in cultured fibroblasts. In *New Trends in Ganglioside Research: Neurochemical and Neuroregenerative Aspects*. eds. R. W. Leeden, E. L. Hogan, G. Tettamanti, A. J. Yates, R. K. Yu, pp. 247–58 Padova: Liviana
Kobayashi, T., Pagano, R. E. 1989. Lipid transport during mitosis: alternative pathways for delivery of newly synthesized lipids to the cell surface. *J. Biol. Chem.* 264: 5966–73
Koval, M., Pagano, R. E. 1989. Lipid recycling between the plasma membrane and intracellular compartments: transport and metabolism of fluorescent sphingomyelin analogs in cultured fibroblasts. *J. Cell Biol.* In press
Kuge, O., Nishijima, M., Akamatsu, Y. 1986. Phosphatidylserine biosynthesis in cultured Chinese hamster ovary cells. III. Genetic evidence for utilization of phosphatidylcholine and phosphatidylethanolamine as precursors. *J. Biol. Chem.* 261: 5795–98
Lambeth, J. D., Xu, X. X., Glover, M. 1987. Cholesterol sulfate inhibits adrenal mitochondrial cholesterol side chain cleavage at a site distinct from cytochrome $P\text{-}450_{scc}$. Evidence for an intramitochondrial cholesterol translocator. *J. Biol. Chem.* 262: 9181–88
Lange, Y., D'Alessandro, J. S., Small, D. M. 1979. The affinity of cholesterol for phosphatidylcholine and sphingomyelin. *Biochim. Biophys. Acta* 556: 388–98
Lange, Y., Matthies, H. J. G. 1984. Transfer of cholesterol from its site of synthesis to the plasma membrane. *J. Biol. Chem.* 259: 14624–30
Lange, Y., Muraski, M. F. 1988. Topographic heterogeneity in cholesterol biosynthesis. *J. Biol. Chem.* 263: 9366–73
Lange, Y., Swaisgood, M. H., Ramos, B. V., Steck, T. L. 1989. Plasma membranes contain half the phospholipid and 90% of the cholesterol and sphingomyelin in cultured human fibroblasts. *J. Biol. Chem.* 264: 3786–93
Lazarow, P. B. 1987. The role of peroxisomes in mammalian cellular metabolism. *J. Inher. Metab. Dis.* (10 Suppl.) 1: 11–22
Lipsky, N. G., Pagano, R. E. 1985. Intracellular translocation of fluorescent sphingolipids in cultured fibroblasts: endogenously synthesized sphingomyelin and glucocerebroside analogues pass through the Golgi apparatus en route to the plasma membrane. *J. Cell Biol.* 100: 27–34
Liscum, L., Faust, J. R. 1987. Low density lipoprotein (LDL)-mediated suppression of cholesterol synthesis and LDL uptake is defective in Niemann-Pick type C fibroblasts. *J. Biol. Chem.* 262: 17002–8
Liscum, L., Ruggiero, R. M., Faust, J. R. 1989. The intracellular transport of LDL-derived cholesterol is defective in Niemann-Pick, type C fibroblasts. *J. Cell Biol.* 108: 1625–36
Luzio, J. P., Stanley, K. K. 1983. The isolation of endosome-derived vesicles from rat hepatocytes. *Biochem. J.* 216: 27–36
Marggraf, W. D., Anderer, F. A., Kanfer, J. N. 1981. The formation of sphingomyelin from phosphatidylcholine in plasma membrane preparations from mouse fibroblasts. *Biochim. Biophys. Acta* 664: 61–73
Martin, O. C., Pagano, R. E. 1987. Transbilayer movement of fluorescent analogs of phosphatidylserine and phosphatidylethanolamine at the plasma membrane of cultured cells. Evidence for a

protein-mediated and ATP-dependent process(es). *J. Biol. Chem.* 262: 5890–98
Matlin, K. S. 1986. The sorting of proteins to the plasma membrane in epithelial cells. *J. Cell Biol.* 103: 2565–68
McMurray, W. C., Dawson, R. M. C. 1969. Phospholipid exchange reactions within the liver cell. *Biochem. J.* 112: 91–108
Miller, M. A., Kent, C. 1986. Characterization of the pathways for phosphatidylethanolamine biosynthesis in Chinese hamster ovary mutant and parental cell lines. *J. Biol. Chem.* 261: 9753–61
Miller-Prodraza, H., Fishman, P. H. 1984. Effect of drugs and temperature on biosynthesis and transport of glycosphingolipids in cultured neurotumor cells. *Biochim. Biophys. Acta* 804: 44–51
Mills, J. T., Furlong, S. T., Dawidowicz, E. A. 1984. Plasma membrane biogenesis in eukaryotic cells: translocation of newly synthesized lipid. *Proc. Natl. Acad. Sci. USA* 81: 1385–88
Neefjes, J. J., Verkerk, J. M. H., Broxterman, H. J. G., van der Marel, G. A., van Boom, J. H., Ploegh, H. L. 1988. Recycling glycoproteins do not return to the *cis*-Golgi. *J. Cell Biol.* 107: 79–87
O'Brien, J. S., Kretz, K. A., Dewji, N., Wenger, D. A., Esch, F., Fluharty, A. L. 1988. Coding of two sphingolipid activator proteins (SAP-1 and SAP-2) by the same genetic locus. *Science* 241: 1098–101
Op den Kamp, J. A. F. 1979. Lipid asymmetry in membranes. *Annu. Rev. Biochem.* 48: 47–71
Orci, L., Montesano, R., Meda, P., Malaisse-Lagae, F., Brown, D., et al. 1981. Heterogeneous distribution of filipin-cholesterol complexes across the cysternae of the Golgi apparatus. *Proc. Natl. Acad. Sci. USA* 78: 293–7
Orci, L., Brown, M. S., Goldstein, J. L., Garcia-Segura, L. M., Anderson, R. G. W. 1984. Increase in membrane cholesterol: a possible trigger for degradation of HMG CoA reductase and crystalloid Endoplasmic Reticulum in UT-1 cells. *Cell* 36: 835–45
Pagano, R. E. 1988. What is the fate of diacylglycerol produced at the Golgi apparatus? *TIBS* 13: 202–5
Pagano, R. E., Sleight, R. G. 1985. Defining lipid transport pathways in animal cells. *Science* 229: 1051–57
Phillips, M. C., Johnson, W. J., Rothblat, G. H. 1987. Mechanisms and consequences of cellular cholesterol exchange and transfer. *Biochim. Biophys. Acta* 906: 223–76
Pierce, M., Turley, E. A., Roth, S. 1980. Cell surface glycosyltransferase activities. *Int. Rev. Cytol.* 65: 1–47
Post, M., van Golde, L. M. G. 1988. Metabolic and developmental aspects of the pulmonary surfactant system. *Biochim. Biophys. Acta* 947: 249–86
Reinhart, M. P., Billheimer, J. T., Faust, J. R., Gaylor, J. L. 1987. Subcellular localization of the enzymes of cholesterol biosynthesis and metabolism in rat liver. *J. Biol. Chem.* 262: 9649–55
Robertson, D. L., Poznansky, M. J. 1985. The effect of non-receptor-mediated uptake of cholesterol on intracellular cholesterol metabolism in human skin fibroblasts. *Biochem. J.* 232: 553–57
Rothman, J. E. 1981. The Golgi apparatus: Two organelles in tandem. *Science* 213: 1212–19
Rousselet, A., Colbeau, A., Vignais, P. M., Devaux, P. F. 1976. Study on the transverse diffusion of spin-labeled phospholipids in biological membranes. II. Inner mitochondrial membrane of rat liver: use of phosphatidylcholine exchange protein. *Biochim. Biophys. Acta* 426: 372–84
Salas, P. J. I., Vega-Salas, D. E., Hochman, J., Rodriguez-Boulan, E., Edidin, M. 1988. Selective anchoring in the specific plasma membrane domain: a role in epithelial cell polarity. *J. Cell Biol.* 107: 2363–76
Sandra, A., Pagano, R. E. 1978. Phospholipid asymmetry in LM cell plasma membrane derivatives: polar head group and acyl chain distributions. *Biochemistry* 17: 332–38
Scallen, T. J., Pastuszyn, A., Noland, B. J., Chanderbhan, R., Kharroubi, A., Vahouny, G. V. 1985. Sterol carrier and lipid transfer proteins. *Chem. Phys. Lip.* 38: 239–61
Sheetz, M. P., Singer, S. J. 1974. Biological membranes as bilayer couples. A molecular mechanism of drug-erythrocyte interactions. *Proc. Natl. Acad. Sci. USA* 71: 4457–61
Simons, K., Fuller, S. D. 1985. Cell surface polarity in epithelia. *Annu. Rev. Cell Biol.* 1: 243–88.
Simons, K., van Meer, G. 1988. Lipid sorting in epithelial cells. *Biochemistry* 27: 6197–202
Sleight, R. G., Pagano, R. E. 1983. Rapid appearance of newly synthesized phosphatidylethanolamine at the plasma membrane. *J. Biol. Chem.* 258: 9050–58
Slotte, J. P., Bierman, E. L. 1987. Movement of plasma-membrane sterols to the endoplasmic reticulum in cultured cells. *Biochem. J.* 248: 237–42
Slotte, J. P., Bierman, E. L. 1988. Depletion of plasma membrane-sphingomyelin rapidly alters the distribution of cholesterol

between plasma membranes and intracellular cholesterol pools in cultured fibroblasts. *Biochem. J.* 250: 653–58

Slotte, J. P., Oram, J. F., Bierman, E. L. 1987. Binding of high density lipoproteins to cell receptors promotes translocation of cholesterol from intracellular membranes to the cell surface. *J. Biol. Chem.* 262: 12904–7

Steck, T. L., Kezdy, F. J., Lange, Y. 1988. An activation-collision mechanism for cholesterol transfer between membranes. *J. Biol. Chem.* 263: 13023–31

Sune, A., Vidal, M., Morin, P., Sainte-Marie, J., Bienvenue, A. 1988. Evidence for bidirectional transverse diffusion of spin-labeled phospholipids in the plasma membrane of guinea pig blood cells. *Biochim. Biophys. Acta* 946: 315–27

Symington, F. W., Murray, W. A., Bearman, S. I., Hakomori, S.-i. 1987. Intracellular localization of lactosylceramide, the major human neutrophil glycosphingolipid. *J. Biol. Chem.* 262: 11356–63

Tanaka, N., Leduc, E. H. 1956. A study of the cellular distribution of Forssman antigen in various species. *J. Immunol.* 77: 198–212

Thompson, S. L., Burrows, R., Laub, R. J., Krisans, S. K. 1987. Cholesterol synthesis in rat liver peroxisomes. *J. Biol. Chem.* 262: 17420–25

Thompson, T. E., Tillack, T. W. 1985. Organization of glycosphingolipids in bilayers and plasma membranes of mammalian cells. *Annu. Rev. Biophys. Biophys. Chem.* 14: 361–86

Tsuneoka, M., Yamamoto, A., Fujiki, Y., Tashiro, Y. 1988. Nonspecific lipid transfer protein (sterol carrier protein-2) is located in rat liver peroxisomes. *J. Biochem.* 104: 560–64

Urade, R., Hayashi, Y., Kito, M. 1988. Endosomes differ from plasma membranes in the phospholipid molecular species composition. *Biochim. Biophys. Acta* 946: 151–63

van Amerongen, A., van Noort, M., van Beckhoven, J. R. C. M., Rommerts, F. F. G., Orly, J., Wirtz, K. W. A. 1989. The subcellular distribution of the non-specific lipid transfer protein (sterol carrier protein 2) in rat liver and adrenal gland. *Biochim. Biophys. Acta* 1001: 243–48

Vance, J. E. 1988. Compartmentalization and differential labeling of phospholipids of rat liver subcellular membranes. *Biochim. Biophys. Acta* 963: 10–20

Vance, D. E., Ridgway, N. D. 1988. The methylation of phosphatidylethanolamine. *Prog. Lipid Res.* 27: 61–79

Vance, J. E., Vance, D. E. 1988. Does rat liver Golgi have the capacity to synthesize phospholipids for lipoprotein secretion. *J. Biol. Chem.* 263: 5898–909

van den Bosch, H., Schalkwijk, C. G., Schrakamp, G., Wanders, R. J. A., Schutgens, R. B. H., et al. 1988. Aberration in de novo ether lipid biosynthesis in peroxisomal disorders. In *Biological Membranes: Aberrations in Membrane Structure and Function* eds. M. L. Karnovsky, A. Leaf, L. C. Bolis, pp. 139–50, New York: Liss

van Meer, G. 1986. The lipid bilayer of the ER. *TIBS* 11: 194–95; 401

van Meer, G. 1987. Plasma membrane cholesterol pools. *TIBS* 12: 375–76

van Meer, G. 1988. How epithelial cells grease their microvilli. *TIBS* 13: 242–43

van Meer, G., Simons, K. 1986. The function of tight junctions in maintaining differences in lipid composition between the apical and the basolateral cell surface domains of MDCK cells. *EMBO J.* 5: 1455–64

van Meer, G., Stelzer, E. H. K., Wijnaendts-van-Resandt, R. W., Simons, K. 1987. Sorting of sphingolipids in epithelial (Madin-Darby canine kidney) cells. *J. Cell Biol.* 105: 1623–35

Van Paridon, P. A., Gadella, T. W. J. Jr., Somerharju, P. J., Wirtz, K. W. A. 1987. On the relationship between the dual specificity of the bovine brain phosphatidylinositol transfer protein and membrane phosphatidylinositol levels. *Biochim. Biophys. Acta* 903: 68–77

Venien, C., Le Grimellec, C. 1988. Phospholipid asymmetry in renal brush-border membranes. *Biochim. Biophys. Acta* 942: 159–68

Voelker, D. R. 1985. Disruption of phosphatidylserine translocation to the mitochondria in baby hamster kidney cells. *J. Biol. Chem.* 260: 14671–76

Voelker, D. R., Frazier, J. L. 1986. Isolation and characterization of a Chinese hamster ovary cell line requiring ethanolamine or phosphatidylserine for growth and exhibiting defective phosphatidylserine synthase activity. *J. Biol. Chem.* 261: 1002–8

Voelker, D. R., Kennedy, E. P. 1982. Cellular and enzymic synthesis of sphingomyelin. *Biochemistry* 21: 2753–9

Wattenberg, B. W., Silbert, D. F. 1983. Sterol partitioning among intracellular membranes. Testing a model for cellular sterol disribution. *J. Biol. Chem.* 258: 2284–89

Weinstein, D. B., Marsh, J. B., Glick, M. C., Warren, L. 1970. Membranes of animal cells. VI. The glycolipids of the L cell and its surface membrane. *J. Biol. Chem.* 245: 3928–37

Wherrett, J. R., Huterer, S. 1972. Enrich-

ment of bis-(monoacylglyceryl) phosphate in lysosomes from rat liver. *J. Biol. Chem.* 247: 4114–20
Wieland, F. T., Gleason, M. L., Serafini, T. A., Rothman, J. E. 1987. The rate of bulk flow from the Endoplasmic Reticulum to the cell surface. *Cell* 50: 289–300
Yaffe, M. P., Kennedy, E. P. 1983. Intracellular phospholipid movement and the role of phospholipid transfer proteins in animal cells. *Biochemistry* 22: 1497–507
Yeagle, P. L. 1985. Cholesterol and the cell membrane. *Biochim. Biophys. Acta* 822: 267–87
Yorek, M. A., Rosario, R. T., Dudley, D. T., Spector, A. A. 1985. The utilization of ethanolamine and serine for ethanolamine phosphoglyceride synthesis by human Y79 retinoblastoma cells. *J. Biol. Chem.* 260: 2930–36
Zachowski, A., Devaux, P. F. 1989. Bilayer asymmetry and lipid transport across biomembranes. *Comm. Mol. Cell. Biophys.* In press
Zambrano, F., Fleischer, S., Fleischer, B. 1975. Lipid composition of the Golgi apparatus of rat kidney and liver in comparison with other subcellular organelles. *Biochim. Biophys. Acta* 380: 357–69

Annu. Rev. Cell Biol. 1989. 5: 277–307

PROTEIN OLIGOMERIZATION IN THE ENDOPLASMIC RETICULUM

Stella M. Hurtley and Ari Helenius

Department of Cell Biology, Yale University School of Medicine, New Haven, Connecticut 06510

CONTENTS

INTRODUCTION

Membrane, secretory and lysosomal proteins enter the secretory pathway of eukaryotic cells at the level of the endoplasmic reticulum (ER). In

0743–4634/89/1115–0277$02.00

the ER lumen, they undergo covalent modifications and acquire their secondary and tertiary structure. While still in this compartment, many of them also proceed to assemble into homo- or hetero-oligomers. It is increasingly clear from a variety of systems that a protein's three-dimensional structure and oligomeric state controls not only its functional properties but also its intracellular transport, ultimate localization in the cell and overall life span. Misfolded, misassembled, and unassembled polypeptides are generally retained in the ER and specifically degraded. This property of the ER provides an inherent quality control that is both refined and efficient. This is probably important to the cell so that structurally defective proteins are prevented from reaching their target compartments. Any harmful effects that they might have are thus limited. It also provides another level at which the expression of proteins—and hence cellular functions—can be regulated.

The quality control takes at least two forms. Misfolded polypeptides are retained as aggregates frequently associated with a heat shock related protein BiP (binding protein, also known as glucose regulated protein, GRP 78). This often occurs when conditional mutants or genetically engineered proteins are expressed, and when N-linked glycosylation is inhibited. Misfolded proteins also occur as side-products during normal protein synthesis. While correct folding is probably a general requirement for all proteins, many oligomeric proteins are subject to a further requirement: they must oligomerize to become transport competent. Like misfolded proteins, unassembled subunits and partially assembled forms of oligomeric proteins are retained in the ER, but the retention mechanism appears to be different. They are not aggregated and retain their capacity to assemble into oligomers for long periods of time. This form of transport restriction is more stringent for membrane-bound than for soluble protein subunits, and it does not generally seem to involve strong interactions with BiP.

In this review we have collected some of the information currently available on the processes of protein folding and oligomerization in the ER. Our main focus is the fate of misfolded, defective, and incompletely assembled proteins. While there is a large body of literature describing the assembly of individual oligomeric proteins, only few attempts have been made to discuss the process from a general cell biological point of view. The reviews by Carlin & Merlie (1986) and Rose & Doms (1988) provide excellent introductions to the topic. Reviews by H. Pelham (this volume), Burgess & Kelly (1987), Lodish (1988), and Pfeffer & Rothman (1987) are also relevant.

QUATERNARY INTERACTIONS IN SECRETORY AND MEMBRANE PROTEINS

Many Membrane and Secretory Proteins Are Oligomeric

As a rule, plasma membrane glycoproteins tend to be oligomeric and they generally oligomerize in the ER (Table I). Extracellular matrix components also usually leave the ER as oligomers and many continue to assemble into higher order complexes in the Golgi and the extracellular space (Kivirikko & Myllylä 1986). Although it is easy to name exceptions, the majority of soluble secretory proteins and lysosomal hydrolases seem to be monomeric.

Since most proteins interact with other proteins during their lifetime, it is not always easy to define precisely what constitutes a bona fide oligomer. Table I provides a list of secretory and plasma membrane proteins with multiple subunits that fulfill one or more of four operational criteria: (*a*) the main function and/or stability critically depends on the quaternary structure, (*b*) intracellular transport requires oligomerization, (*c*) isolation using various methods routinely yields an oligomeric complex, and (*d*) the subunits are permanently associated, and are handled and degraded by the cell as a unit.

The subunit composition of the listed proteins (Table I) ranges from simple homo-dimers to complex hetero-oligomers. Though many have cyclic, dihedral, or cubic symmetries, no fixed rules for the number of subunits or for the geometry are apparent. The polypeptide subunits that make up these proteins arise by two distinct mechanisms: they are either synthesized separately, or they are generated by posttranslational proteolytic cleavages from precursor polypeptides (the sign $\wedge$ in Table I indicates wherever a proteolytic cleavage is involved or suspected). Sometimes the final quaternary structure is derived through a combination of both mechanisms (e.g. influenza hemagglutinin (HA) and the insulin receptor). Though most oligomers are stabilized exclusively by noncovalent interactions, some have inter-chain disulfide bonds (as indicated with a dash in Table I). These are formed either after association of two subunits with each other (e.g. PDGF), or they may be formed initially as intra-chain disulfides in a precursor protein, which is cleaved to yield disulfide bonded subunits (e.g. influenza HA and the insulin receptor).

The list in Table I is obviously far from complete. Many known oligomers are not included. In some cases, the exact subunit composition remains ambiguous. Indeed, some proteins occur in alternative oligomeric states. Sometimes alternative forms of the oligomer may be expressed concomitantly in the same cell (Östman et al 1988). The detailed quaternary

Table 1 Oligomeric structure of some membrane and secretory proteins

Protein	Structure	Probable assembly site	Reference
Membrane proteins			
Receptors			
Transferrin receptor	a-^{a}a	ER	Goding & Burns 1981
LDL receptor	$(aa)_x$	—	Esser & Russell 1988; van Driel et al 1987
Asialoagalactoglycoprotein receptor	a_6	—	Loeb & Drickamer 1987
Integrins	ab	ER	Hynes 1987
Glycine receptor	a_3b_2	—	Langosch et al 1988
Asialoglycorpotein receptor	a_4b_2	ER	Bischoff et al 1988
Insulin receptor	$(a^{\wedge b}$-b)-(b^-a)	ER	Olson et al 1988
Acetylcholine receptor	a_2bcd	ER	Carlin & Merlie 1986
T-cell receptor	a-b(cde)f-f(or-g)	ER	Samelson et al 1985; Clevers et al 1988
Ectoenzymes			
Aminopeptidase	aa	—	Hussain et al 1981
Dipeptidylpeptidase IV	aa	—	Hussain 1985
Sucrase isomaltase	(a^b)(a^b)	—	Cowell et al 1986
Maltase glucoamylase	(a^b)(a^b)	—	Norén et al 1986
Acetylcholinesterase			
Torpedo, mammalian erythrocyte	aa	—	Sikorav et al 1988
Quail muscle	$(a_4)_3$-b_3	ER and Golgi	Rotundo 1984
Viral Antigens			
Herpes virus gB	aa	—	Claesson-Welsh & Spear 1986
Mumps virus HN	a-a	ER	Waxham et al 1986; Yamada et al 1988
Vesicular stomatitis virus G	a_3	ER	Rose & Doms 1988
Influenza neuraminidase	a_4	—	Varghese et al 1983
Sendai virus F protein	$(a^\wedge b)_4$	—	Sechoy et al 1987
Sendai virus HN protein	(a-a)(a-a)	—	Markwell & Fox 1980
Influenza hemagglutinin	$(a^\wedge$-$b)_3$	ER	Wilson et al 1981; Copeland et al 1986
Rous sarcoma virus envelope protein	$(a^\wedge$-$b)_3$	ER	Wills et al 1984; Einfeld & Hunter 1988

Semliki Forest virus spike	$(ab^\wedge c)_3$	ER	Vogel et al 1986; Ziemicki & Garoff 1978
Others			
Glycophorin	aa	—	Silverberg & Marchesi 1978
Variant surface glycoprotein (*Tryp. b*)	aa	—	Metcalf et al 1987
Na^+/K^+ATPase	ab	ER	Tamkun & Fambrough 1986
MHC Class I	ab	ER	Carlin & Merlin 1986
MHC Class II	ab(c)	ER	Carlin & Merlin 1986
Voltage sensitive Na channel	a-bc	—	Schmidt & Catterall 1987; Catterall 1986
mIgM	a-b-b-a(c)	ER	Hombach et al 1988
Secretory and lysosomal proteins			
Retinol binding protein	aa	ER	Rask et al 1983
von Willebrand's factor	$(a\text{-}a)_n$	ER	Titani & Walsh 1988
Concanavalin A	a_4	—	Reeke et al 1975
Yeast invertase	a_8	ER	Esmon et al 1987
TSH, FSH, LH, hCG	ab	ER	Peters et al 1984; Corless et al 1987; Magner & Weintraub 1982
Lysosomal neuraminidase	ab	—	Verheijen et al 1985
Fibroin	a-b	ER	Takei et al 1987
IgG	a-b-b-a	ER	Bergman & Kuehl 1979a
PDGF	a-a, or a-b, or b-b	ER	Östman et al 1988
β-hexosaminidase A or B	$ab^\wedge c$ or $(a^\wedge b)(a^\wedge b)$	—	Mahuran et al 1988
Fibrinogen	(a-b-c)-(a-b-c)	ER	Yu et al 1983
IgM	$(a\text{-}b\text{-}b\text{-}a)_5c$	ER	Tartakoff & Vassalli 1979; Mains & Sibley 1983
Extracellular matrix components			
Fibronectin	a-b	—	Gardner & Fambrough 1983
Laminin	a-b-c	ER	Peters et al 1985
Procollagen I, II, IV	a-a-a, or a-a-b, or a-b-c	ER	Kivirikko & Myllylä 1986

[a] Most oligomers are stabilized exclusively by noncovalent interactions, though some have inter-chain disulfide bonds as indicated by a hyphen.
[b] The carat indicates wherever a proteolytic cleavage is involved or suspected in the generation of subunits.

structure may vary with cell type and developmental stage (Tsim et al 1988). To complicate matters further, many proteins have transient precursor forms and intracellular transport forms, which are different in their subunit composition from the final products.

While proteins such as the T-cell receptor with six or seven different subunits including a disulfide bonded homodimer may seem extremely complex, it is worth noting that protein complexes in energy transducing membranes of mitochondria and choroplasts can have even larger numbers of subunits. Bacterial membranes also contain numerous large assemblies. The assembly of these complexes—while an intriguing topic in itself—is outside the scope of this review. They do not traverse the ER, and, unlike the proteins in the plasma membrane, they are generally assembled from individually transported subunits at the site of final residence.

The Structure of Oligomers

The X-ray structure is known for a large number of oligomeric proteins including two of the soluble proteins, and the ectodomains of four of the membrane proteins in Table I: IgG (Silverton et al 1977); Concanavalin A (Reeke et al 1975); influenza hemagglutinin (HA) (Wilson et al 1981); influenza neuraminidase (Varghese et al 1983); MHC Class I (Bjorkman et al 1987); and (at lower resolution) the variant glycoprotein of *Trypanosoma brucei* (Metcalf et al 1987). The structure of the photosynthetic reaction center in *Rhodopseudomonas viridis*, though unrelated to the proteins in Table I, is also interesting because it is the only oligomeric membrane protein for which the entire three-dimensional structure, including the membrane spanning region, is known (Deisenhofer et al 1985).

The oligomeric structure of influenza HA, influenza neuraminidase and the major histocompatibility Class I antigen, HLA-A2, are shown in Figure 1. In each case the structure is that of a proteolytic ectodomain fragment devoid of transmembrane and cytoplasmic domains. These well-studied examples illustrate several important structural features. They show that subunits generally occur as independently folded entities. Although influenced by quaternary interactions, the basic folding motif is intrinsic to each subunit. Only in unusual cases, such as collagen (Kivirikko & Myllylä 1986), is the secondary and tertiary structure of the monomer dictated primarily by the quaternary structure.

The area of contact between two stably associated proteins usually exceeds 600 $Å^2$ in size (Chothia & Janin 1975). Association depends on the overall complementarity of these surfaces as well as the presence of key amino acid side chains. Single mutations in the contact sites are often sufficient to prevent association (Wu et al 1983). Therefore subunits of the same protein but from different species or strains sometimes cannot

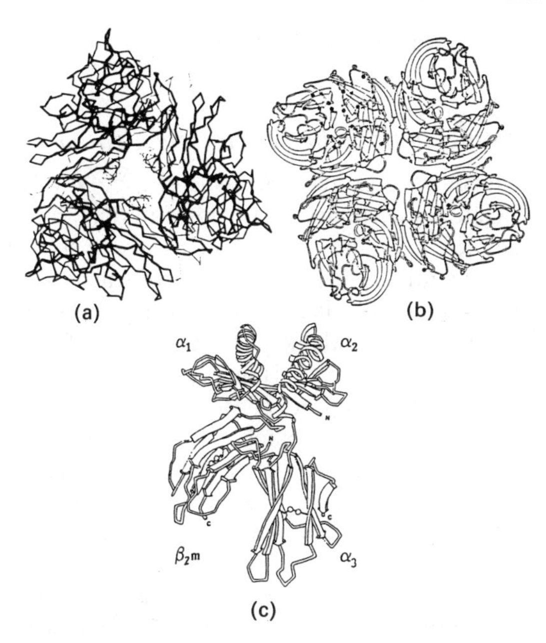

Figure 1 The ectodomain structure of three oligomeric membrane glycoproteins. (a) *Influenza hemagglutinin*—influenza HA is a homotrimer in which each 84-kd subunit consists of two disulfide-bonded polypeptide chains, HA1 (*dark lines*) and HA2 (*light lines*) derived by proteolytic cleavage from a common precursor HAO (Wiley & Skehel 1987). (Figure courtesy of W. Weis). (b) *Influenza neuraminidase*—in the influenza neuraminidase tetramer each 50-kd subunit in the protease-related catalytic ectodomain forms an independent globular domain. The interaction, in this case, involves the continuation of a single beta sheet across the interface, a structure not uncommon in subunit and domain interfaces (Varghese et al 1983). (Reproduced with permission from Varghese et al 1983). (c) *HLA-A2*-HLA-A2 (an MHC Class I antigen) consists of a transmembrane heavy chain (45 kd) in a complex with the β_2-microglobulin subunit (12 kd) that is proximal and peripheral to the membrane (Bjorkman et al 1987). The complex has a tertiary structure similar to the immunoglobulins, but the interfaces of β_2-microglobulin with the three domains of the heavy chain are different from those found in a typical immunoglobulin. (Reproduced with permission from Bjorkman et al 1987).

assemble with each other to form stable oligomers (Boulay et al 1988; Sklyanskaya et al 1988). Bound water molecules and cations are occasionally seen in the interface between proteins, and their presence may be of importance for stability. While N-linked carbohydrate side chains are not found in the interface proper, they sometimes cover an area of the protein surface that spans over two subunits, and may thereby contribute to the stability of the subunit interaction (Varghese et al 1983). When present, inter-subunit disulfide bonds greatly enhance the stability of the quaternary structure.

Topology of Interactions in Membrane Proteins

In the case of membrane proteins an issue of topology needs to be considered: In which topological domain(s) do quaternary interactions occur? The three examples in Figure 1 are proteins where subunit interactions occur primarily, or entirely, via the ectodomains. Anchor-free ectodomains of HA, when expressed in cells, can form trimers (Doyle et al 1986; I. Singh et al, in preparation) and anchor-free mutants of HLA heavy chains combine with β_2-microglobulin (Krangel et al 1984). Similar findings have been reported for insulin receptor ectodomains (Johnson et al 1988). It is also known that proteins that are anchored exclusively through a lipid tail can oligomerize even though they lack transmembrane and cytoplasmic peptide domains (Sikorav et al 1988; Metcalf et al 1987; B. Crise, P. Zagouras, J. Rose, personal communication; M. Davis, personal communication).

In other cases, it is apparent that the transmembrane and cytosolic domains determine the quaternary structure. Proteolytic cleavage experiments have shown that self-association of chicken asialoglycoprotein receptors is mediated by a sequence in the transmembrane region (Loeb & Drickamer 1987). Recent studies suggest that specific interactions occur between the two transmembrane alpha helices in a glycophorin dimer (Bormann et al 1989). Even the influenza HA trimer seems to derive additional stability from intramembrane interactions (Doms & Helenius 1986). The X-ray structure of the *Rhodopseudomonas viridis* photosynthetic center (Deisenhofer et al 1985) shows direct contacts inside the bilayer between transmembrane alpha helices in the L and M subunits. Finally, crosslinking studies by van Driel et al (1987) indicate that the thirty terminal residues in the cytoplasmic domain are required for formation and stability of LDL receptor dimers.

It is apparent that interactions can—and do—occur in all three topological domains of membrane proteins. The stability and specificity of association probably depend on contributions from them all. How specific interactions occur within the hydrophobic interior of the bilayer provides an interesting challenge for future structural studies.

METHODS USED TO STUDY OLIGOMERIZATION

To monitor protein oligomerization in cells, a variety of techniques can be used. These include chemical crosslinking, hydrodynamic analysis, coimmunoprecipitation, coelectrophoresis, cofractionation, and acquisition of function (Markwell & Fox 1980; Doms & Helenius 1986; Samelson et al 1985; Östman et al 1988; Smith et al 1987; Slieker et al 1988). Several indirect methods such as increased resistance to proteolysis and denaturation (Dulis 1983; Doms & Helenius 1986), as well as loss and acquisition of antigenic epitopes, can also be exploited (Copeland et al 1988). Combined with radioactive pulse labeling and biochemical analysis of posttranslational changes such as carbohydrate side chain processing, these methods allow determination of assembly kinetics and efficiency, identification of assembly intermediates, and assignment of the cellular location of assembly.

While straightforward in principle, such analysis is not always easy. High levels of expression and radioactive incorporation of label are needed to permit radioactive pulses short enough for the analysis of rapid co- and post-translational assembly events. Pulses as short as one min are required to detect early folding intermediates. The methodology frequently hinges on the availability of antibodies specific for the antigen at different stages of folding and assembly. The antibodies used should not cause the dissociation of noncovalent associations, a complication sometimes encountered (Ziemiecki & Garoff 1978). The ease by which some noncovalent protein-protein interactions are dissociated by nonionic detergents and by the unavoidable dilution during cell lysis also needs to be considered. To prevent artificial dissociation, phospholipid-detergent mixtures can sometimes be used to lower the chemical potential of the detergent (Helenius & Simons 1975). Detergents such as digitonin that are particularly mild with respect to protein-protein interactions have proved useful in other cases (Oettgen et al 1986; Hombach et al 1988).

INITIAL FOLDING AND ASSEMBLY

Translation of most membrane, secretory, and lysosomal proteins in mammalian cells occurs on membrane bound ribosomes. Due to the spacing of the ribosomes on the RER membrane (50 nm), nascent chains cannot interact and oligomerize cotranslationally. However, nothing prevents nascent chains from associating with completed subunits already in the lumen of the ER. Such early assembly has been reported in at least two cases; IgG (Bergman & Kuehl 1979a) and fibrinogen (Yu et al 1983). Bergman & Kuehl (1979a) showed that nascent heavy chains of immunoglobulins can be disulfide bonded to completed light chains after approximately 38 kd has been translated. To place the kinetics of these events in context, synthesis of a 500 amino acid protein (molecular weight about 55 kd) takes approximately one min and a polysome completes a new chain every 5–10 sec (see Alberts et al 1983).

Polypeptides are translocated through the membrane in unfolded form. Once in the lumen of the ER, the nascent chains begin to fold immediately. The folding process, which yields compact protein subunits, takes 3–4 min or less. In many cases, folding proceeds vectorially domain by domain from the N-terminus towards the C-terminus. Intra-domain disulfide bonds are usually formed early and help to stabilize domain structure (Bergman & Kuehl 1979b).

While there can be no doubt that the primary sequence determines the ultimate three-dimensional structure of a protein, it is apparent that the process of folding in the ER is facilitated by a set of resident enzymes and protein factors. In fact, it appears that the nascent polypeptide chain is greeted on the lumenal side of the ER membrane by a "welcoming committee" of enzymes and factors that facilitate and accelerate folding in several ways:

1. Removal of the signal sequence by signal peptidase may be important for correct, rapid folding (Randall & Hardy 1986).
2. Reshuffling of disulfides by protein disulfide isomerase (PDI) accelerates folding (Freedman 1987).
3. Addition of N-linked carbohydrate side chains by oligosaccharide transferase allows correct folding of many proteins (Schlesinger & Schlesinger 1987).
4. Catalyzed isomeration of prolines by *cis-trans* proline isomerase may accelerate folding by increasing the rate of isomerization (Freedman 1987).
5. Association with other ER factors, such as BiP, may accelerate the folding process and help to prevent aggregation of folding intermediates (see Pelham, this volume).
6. In special cases, such as procollagen synthesis, amino acid side chain modifications by enzymes such as proline hydroxylase and lysyl hydroxylase are required for correct folding and assembly (Kivirikko & Myllylä 1986).
7. In some cases, early after synthesis and translocation, the C-terminus is replaced by a glycolipid anchor (Ferguson & Williams 1988).

There is reason to believe that at least part of this machinery forms multicomponent enzyme complexes at sites of translocation. It has been demonstrated that PDI, in addition to its role in disulfide formation, serves as the β subunit of proline hydroxylase (Koivu et al 1987). Recent studies also indicate that it may be responsible for recognizing the consensus sequences for N-linked glycosylation on the nascent chains (Geetha-Habib et al 1988), and perhaps presenting it to the oligosaccharide transferases. Furthermore, it is increasingly clear that many of the resident proteins of the rough ER, including some of the proteins mentioned above, are anchored to a complex, interconnected protein network that is not disrupted by treatment with nonionic detergents (Kreibich et al 1978; Hortsch et al 1987). The rough ER may thus contain a proteinaceous network that provides a scaffold for the translocation components on the cytoplasmic side and the processing factors on the lumenal side. This network may also play a role in the retention of defective protein products as discussed below.

In most membrane proteins that have a single membrane spanning sequence, lumenal and cytoplasmic domains fold independently of each other and remain structurally independent of each other. It is therefore not surprising that topological domains can often be eliminated or exchanged between proteins without causing folding problems. How proteins, which span the membrane several times, manage to fold is not yet clear, but the process is apparently controlled by start and stop transfer signals in the sequence (see Sabatini et al 1982).

OLIGOMERIC ASSEMBLY

As indicated in Table 1, oligomerization usually occurs in the ER. It may begin when one of the oligomerization partners is still a nascent chain (Bergman & Kuehl 1979a; Takei et al 1987; Yu et al 1983). More commonly, however, it takes place posttranslationally. Pulse-chase experiments show half-maximal assembly of different proteins ranging between five min and about two hr post synthesis (Doms et al 1987; Olson & Lane 1987; Carlin & Merlie 1986). Why the kinetics differ so much between proteins is unclear, but oligomer complexity and subunit concentration may be important factors. While the ER is the primary site of oligomerization, it is not uncommon to find further assembly reactions occurring in the Golgi complex and even later in the secretory pathway (Rotundo 1984; Sporn et al 1986).

Initial folding of at least one domain of a subunit is required for correct oligomeric assembly. The reason is simple; subunits must display distinct surface features in order to recognize and bind to each other specifically and stably. The assembly reactions are, indeed, very specific: for example, related influenza HA molecules from different subtypes fail to recognize each other as assembly partners (Boulay et al 1988; Sklyanskaya et al 1988). The requirement for initial folding does not mean that the entire subunit must be folded into a compact structure. This point is illustrated by procollagen, where monomers first interact via the N-terminal domains.

This leads to the cooperative folding of the central triple-helical domain. The N-terminal domains are subsequently removed by proteolysis, which suggests that their function is primarily to ensure trimerization (see Kivirikko & Myllylä 1986).

Assembly of complex oligomers occurs step by step so that monomers associate to form dimers, which proceed to associate with additional subunits or partial complexes. Such a stepwise assembly pathway is particularly clear for complex hetero-oligomers, such as the T-cell receptor, where some intermediate subcomplexes have been identified (Minami et al 1987; see Clevers et al 1988). Whenever subunits are synthesized in non-stoichiometric amounts, the subunit made in the lowest amount is a candidate for determining the level of expression of functional, transport-competent complexes (Peters et al 1985; Minami et al 1987; Plant & Greininger 1986). In order to control the expression of a protein complex, the cell thus only needs to regulate the synthesis of one subunit.

It is far from clear how assembly actually takes place at the molecular level. The simplest model involves self-assembly via random collisions either in the fluid volume of the ER or in the lateral plane of the ER membrane. There are several indirect arguments that tend to support such a mechanism. In vitro studies have shown that certain oligomers will form without added assembly factors (see Ghelis & Yon 1982). In the case of influenza HA in double-infected cells, it has been shown that subunits are recruited randomly from a mixed pool (Boulay et al 1988). The assembly of both vesicular stomatitis virus G-protein and influenza HA trimers have been shown to be independent of metabolic energy (Doms et al 1987; Copeland et al 1988). Their trimerization proceeds in the absence of ATP, and it occurs at 4°C, albeit at a reduced rate. We have, moreover, observed that the rate of trimerization of HAO and G protein increases with increasing levels of expression (K. Wagner, I. Braakman, A. Helenius, unpublished results).

Prior to oligomerization, folded subunits generally behave like free monomers on sucrose gradients. Usually, they do not coprecipitate with other proteins, and they do not seem to be tightly associated with any large sequestering structures (Copeland et al 1986, 1988; Doms et al 1987; Yewdell et al 1988; Samelson et al 1985). Immunocytochemistry using trimer specific monoclonal antibodies to HA has, moreover, suggested that HA trimerization is not restricted to a specific area of the ER (Copeland et al 1988).

Other data can be interpreted as evidence against unassisted, random assembly. Bergman & Kuehl (1979a) reported that most newly made light chains of IgG associate preferentially with heavy chains. This suggests the existence of distinct pools of light chains. We have found, moreover, that influenza HA trimers in double-infected cells are not totally randomly assembled into mixed trimers when expressed at low levels, in contrast to the results found at high expression levels (Boulay et al 1988, T. Stegmann & A. Helenius, unpublished observations). This suggests that there is incomplete mixing of subunits and thus a preference for subunits from the same polysome to oligomerize with each other.

In an increasing number of cases, it has been reported that protein factors are present in partially assembled forms of oligomeric proteins in the ER but not in the final oligomers. Such proteins may serve a role as assembly factors. A 28-kd polypeptide, ω or TRAP, interacts transiently with assembly intermediates of the T-cell receptor (see Clevers et al 1988; Bonifacino et al 1988). Certain unassembled proteins and assembly intermediates bind the heat shock related protein BiP in the ER (see Pelham, this volume). The association of BiP with intermediates of immunoglobulin assembly provides the best studied case (Bole et al 1986). Unassembled heavy chains, and immunoglobulin assembly intermediates, are associated with BiP until the final light chain is added. In the absence of light chains, heavy chains with intact CH1 domains remain in the ER in association with BiP and are slowly degraded (Bole et al 1986; Hendershot et al 1987). BiP can also be found associated with VSV G-protein during initial folding, but it dissociates well before trimerization (C. E. Machamer et al, in preparation). Evidence for transient association of BiP with influenza HA during the first 15 min after HA synthesis has been reported (Gething et al 1986), but in another strain of virus such association was not observed even after a very short (two min) pulse (Hurtley et al 1989). Aside from the immunoglobulin example, there is at present little published data to indicate a role for BiP in the assembly of oligomeric proteins, but as discussed by Pelham (this volume) such a function remains a distinct possibility.

EXAMPLES OF OLIGOMERIC ASSEMBLY AND QUALITY CONTROL IN THE ER

We have selected eight specific examples to illustrate how different secretory and membrane proteins fold and oligomerize. They include membrane and soluble proteins of varying complexity.

T-Cell Receptor

The T-cell receptor complex is one of the most complicated cell surface receptors known (see Clevers et al 1988; Table 1). The seven subunits are all transmembrane polypeptides encoded by at least six separate genes. Assembly occurs through a series of intermediates in the ER. Correct assembly is required for receptor expression on the cell surface (Samelson et al 1985; Minami et al 1987; Berkhout et al 1988). In the case of the murine receptor, one subunit, zeta, is made in lower copy number than the others. Its concentration effectively determines the amount of receptor expressed (Minami et al 1987; Berkhout et al 1988).

Because the zeta subunits are limiting for assembly, 70–95% of the other subunits remain unassembled and are degraded without reaching the cell surface (Minami et al 1987; Sussman et al 1988). Zeta-free partial complexes seem to leave the ER and pass through the Golgi complex from which they are routed to the lysosomes and degraded (Minami et al 1987; Sussman et al 1988). Complexes lacking additional subunits are, in contrast, retained in the ER, and degraded there or in some closely associated compartment, without passage to lysosomes (Chen et al 1988; Lippincott-Schwartz et al 1988). The degradation is not sensitive to inhibitors that prevent lysosomal degradation; it is very temperature dependent and for most of the subunits it begins with a definite, but variable, lag-time (Lippincott-Schwartz et al 1988). While one of the subunits, delta, is degraded within an hr, the others are more stable and require several hr before significant turnover can be observed (Chen et al 1988). No BiP association has been observed with retained T-cell receptor subunits (Alarcon et al 1988; Bonifacino et al 1988). However, a protein already mentioned, called ω or TRAP, has been found to associate transiently with receptor subunits prior to assembly (Alarcon et al 1988; Bonifacino et al 1988).

Very recent studies have shown that alpha and beta subunits can exit the ER without the full complement of subunits when modified at their C-terminus so that they acquire a glycolipid anchor. Not only are they now expressed on the cell surface, but they can be released as a soluble, disulfide-linked dimer using phosphotidylinositol phospholipase (M. Davis, personal communication). This result suggests that the membrane anchor of these subunits plays a role in their retention prior to assembly, but is not required for alpha-beta dimerization.

The T-cell receptor system illustrates many general points concerning oligomerization, retention, and degradation of protein subunits. It provides a possible example of an assembly factor that is not part of the final complex; it allows a glimpse at how cells may regulate and coordinate the expression of subunits for complex proteins; it highlights the presence of a degradation system in connection with the ER; and it suggests that a final quality control step may, for some proteins, occur at the level of the Golgi complex.

VSV G-protein

The glycoprotein of VSV, like influenza HA, forms homotrimers in the ER with a half-time of about 6–10 min after translocation (Doms et al 1987, 1988; Copeland et al 1988; Kreis & Lodish 1986). During the first 2–3 min after synthesis, a large fraction of the monomers associates with BiP, but this interaction is dissociated as the subunit folds into a form which has a full complement of disulfide bonds and antigenic epitopes (C. E. Machamer et al, in preparation; A. De Silva et al, in preparation). Since BiP is no longer associated by the time trimers form, it is unlikely that it serves as an assembly factor. It may, however, help the protein during initial folding.

Evidence from various mutant G-proteins suggests that trimerization, though necessary, is not sufficient for transport out of the ER (Doms et al 1988). Transport mutants of G-protein with defects in the lumenal domain are generally unable to undergo normal initial folding and do not trimerize (Doms et al 1988). They are found as aggregates (Doms et al 1988) in permanent association with BiP (C. E. Machamer et al, in preparation). Mutants with a defect in the cytosolic tail domain can, in contrast, generally fold and trimerize correctly. However, many of them are transported slowly and inefficiently from the ER. These mutant proteins are not associated with BiP (Doms et al 1988; C. E. Machamer et al, in preparation). Thus the efficiency of transport from the ER can depend on the structure of both lumenal and cytoplasmic domains, which fold (or misfold) independently of each other. Both domains must be correctly structured for efficient transport. This aspect of sorting may be quite important for the transport of membrane proteins that have domains on both sides of the membrane.

A temperature sensitive folding mutant of G, *tsO45*, has been extensively studied. This protein occurs as a misfolded aggregate after synthesis at 39°C. At least one disulfide bond is missing in the lumenal domain of the protein, antigenic epitopes are absent, and the protein is associated with BiP (Doms et al 1987; C. E. Machamer et al, in preparation; A. De Silva et al, in preparation). After shift-down to 32°C, the protein emerges within 30 sec as a monomer. It continues to fold acquiring the correct intrachain disulfides and antigenic epitopes within 2–3 min. BiP dissociates, the protein trimerizes, and transport to the cell surface and incorporation into functional virus particles follows (Kreis & Lodish 1986; Doms et al 1987; C. E. Machamer et al, in preparation; A. De Silva et al, in preparation). Interestingly, the rescue of the mutant G-protein from the aggregate after shift-down depends on the presence of ATP (Balch & Keller 1986; Balch et al 1986; Doms et al 1987). This may reflect the fact that BiP association with misfolded proteins is ATP sensitive, but it is not yet known whether the ER lumen actually contains ATP.

The fate of the correctly folded *tsO45*-G-protein after a shift-up in temperature to 39°C is interesting: if still in the ER, it misfolds, aggregates, associates with BiP, and loses some of the native disulfide bonds that have already formed (A. De Silva et al, in preparation). However, if already transported to the *cis*-Golgi or beyond, it undergoes a conformational change but does not aggregate. It continues to the surface (Doms et al 1987, A. De Silva et al, in preparation). These findings indicate that the main quality control step occurs at the level of ER to Golgi transport. It has already been well documented that the rate limiting step in transport of most proteins is, indeed, at this level (Lodish et al 1983; Fries et al 1984; Rask et al 1983).

Class I MHC Antigens

Major histocompatibility complex antigens that belong to Class I are dimers consisting of a transmembrane heavy chain, μ and a peripheral subunit, β_2-microglobulin (see Figure 1 and Table 1). They provide an interesting example of the different levels of stringency in quality control for export of membrane bound vs soluble polypeptides. The subunits are assembled in the ER within 5 to 10 min post synthesis (see Carlin & Merlie 1986). The secretion of μ depends on its assembly with β_2-M (Krangel et al 1979; Owen et al 1980). In contrast, soluble β_2-M can be secreted in the absence of μ chain expression (Severinsson & Peterson 1984). Constructed mutants of the heavy chain, which lack the cytoplasmic tail and transmembrane portion, can also assemble with β_2-M and be secreted, thus indicating that the transmembrane domain is not important for the conformation of the ectodomain (Zuniga et al 1983; Krangel et al 1984). Blocking glycosylation with tunicamycin does not affect the efficiency of assembly or transport of Class I molecules (Owen et al 1980).

Williams et al (1988), studying cells expressing a variant μ chain in which only 20% of assembled Class I molecules are secreted, showed that the other 80% is retained in the ER. The secreted and the nonsecreted forms exhibit differential sensitivity to trypsin, which indicates that the two forms are conformationally distinct. The retained fraction is quite stable in the cell, does not appear to be aggregated, nor is any BiP association obvious. Miyazaki et al (1986a,b) examined engineered Class I molecule lacking either the μ chain glycosylation site or an internal disulfide bond. In both cases, assembly with β_2-M intracellularly is observed, but secretion is inhibited by 90%, and the retained fraction is not rapidly degraded. In the case of the glycosylation mutant, the fraction found on the surface is functionally active. Thus, as for VSV G-protein, assembly is required but not sufficient for the transport of the μ chains.

Class II MHC Antigens

Class II MHC antigens assemble in the ER to form a trimer consisting of α, β and the invariant chain (for review see Carlin & Merlie 1986). The assembly of α with β is required for their export from the ER (Rabourdin-Combe & Mach 1983; Malissen et al 1983, 1984). The invariant chain, however, is not required for $\alpha\beta$ transport (Malissen et al 1984). The invariant chain itself is made in excess and retained in the ER until assembled with α and β (Kvist et al 1982, Kaufman et al 1984). Before the transport competent trimeric complex reaches the plasma membrane, but after transport to the *trans*-Golgi, the invariant chain is removed from the complex (Owen et al 1981).

Glycoprotein Hormones

The assembly and secretion of glycoprotein hormones demonstrates several points about the stringency of quality control. Lutropin (LH), chorionic gonadotropin (CG), follicle stimulating hormone (FSH), and thyrotropin have a common α subunit and different β subunits. Hetero-dimer assembly occurs in the ER (Magner & Weintraub 1982; Peters et al 1984). In human choriocarcinoma cells, which synthesize CG, free α monomers, free β monomers, and $\alpha\beta$ dimers are secreted (Peters et al 1984). However, the secretion of free β monomers is inefficient, with a majority being degraded intracellularly. The rate by which the beta chain folds and acquires intrachain disulfide bonds is the rate-limiting step in alpha-beta dimerization (Ruddon et al 1987).

Corless et al (1987) compared the efficiency of secretion of CG and LH subunits, when expressed individually or together. Alpha subunits were secreted in the absence of β subunits provided that they were correctly glycosylated (Corless et al 1987; Matzuk & Boime 1988). The CG and LH β subunits, which are 80% homologous, experience different levels of quality control: whereas the β subunit of CG is secreted quantitatively, only one-tenth of the β subunits of LH are secreted when expressed alone. Ninety percent of the LH β chains are retained in the ER and slowly degraded (half-time, 8 h) (Corless et al 1987). Degradation of nonglycosylated, retained α subunits was characterized by an hour lag followed by rapid degradation within the next hour (Matzuk & Boime 1988)—a kinetic pattern common to ER degradation of many proteins.

The assembly of platelet-derived growth factor (PDGF) provides another example of concurrent, alternative assembly of subunits. Cells can produce and secrete three types of disulfide-linked PDGF dimers, A-A,

A-B, and B-B. Free subunits are not secreted. Interestingly the cell can also vary the efficiency of secretion of the oligomers; while A-B is efficiently secreted from transfected cells, B-B is partially retained (Östman et al 1988).

The extensive information available on polypeptide hormones, only briefly outlined here, illustrates a variety of features shared with other assembly systems. It shows that alternative assembly states are possible in the same cell, that soluble subunits can sometimes be secreted without assembly, that the glycosylation state can affect the assembly process, and that retention—when it takes place—occurs in the ER followed by delayed degradation.

Plasma Retinol Binding Protein

The retinol binding protein (RBP) of blood plasma is a soluble 21-kd protein that binds a single retinol molecule. It is synthesized in liver parenchymal cells. After synthesis it associates with retinol in the ER and is secreted, complexed with transerythrin (Goodman 1984). Several studies have shown that apo-RBP, synthesized in retinol-depleted cells or animals, is specifically retained in the ER (Rask et al 1983; Ronne et al 1983). When retinol is added, the presynthesized apo-proteins associate with it and secretion commences. Structural analysis, including X-ray crystallography of RBP (Newcomer et al 1984), suggests that retinol binding may induce a conformational change in the protein. This change is presumably needed to allow transport from the ER to the Golgi complex. It is not known what mechanisms are responsible for the retention of the apo-RBP in the ER.

There are several other proteins whose folding, assembly, and transport have been extensively characterized and deserve detailed discussion. Among the most informative and best studied are the immunoglobulins, the acetylcholine receptor, various procollagens, and the influenza virus HA. Readers are referred to recent reviews describing their assembly (Carlin & Merlie 1986; Merlie & Smith 1986; Kivirikko & Myllylä 1986; Rose & Doms 1988).

OLIGOMERIZATION ALLOWS TRANSPORT

The previous examples illustrate that proteins can generally proceed into the secretory pathway only after they have acquired a completely folded three-dimensional structure. If they are oligomers, they usually have to assemble to become transport competent. The overall efficiency of oligomeric assembly can determine the level at which a protein is expressed.

The most common scenario is that oligomers are rapidly and selectively transported from the ER, whereas unassembled subunits, assembly intermediates, and misassembled proteins are retained. The amount of completed oligomers at any given time in the ER may therefore be low (Copeland et al 1988; Yewdell et al 1988). Export restrictions apply to both soluble and membrane bound subunits, but the stringency varies from protein to protein. Membrane proteins and membrane bound subunits are generally more efficiently retained, whereas soluble subunits (particularly if they have a small molecular weight such as β_2-microglobulin and IgG light chains) are sometimes secreted without assembly (Severinsson & Peterson 1984; Tartakoff & Vassalli 1977). The requirement for oligomerization is obviously not universal; there are examples of monomeric proteins that are efficiently transported (Peters & Davidson 1982). The p62 precursor protein of Semliki Forest virus is efficiently transported in the absence of the complementary second subunit, E1 (Kondor-Koch et al 1983). While necessary for the transport of many proteins, assembly alone does not guarantee exit from the ER. This point is clearly shown in studies of transport mutants of influenza HA, VSV G-protein, and MHC Class I antigens (Gething et al 1986; Doms et al 1988; Miyazaki et al 1986 a,b).

In some exceptional cases assembly with other proteins actually leads to retention. For example, when a fraction of lysosomal β-galactosidase binds to the ER protein egasyn, the bound β-galactosidase is retained in the ER (Medda et al 1987). In a similar example E19 of adenovirus resides in the ER where it binds to Class I MHC antigens. It blocks surface expression of the antigen and prevents recognition of the infected cell by the immune system (Kvist et al 1978; Anderson et al 1985).

MISFOLDED PROTEINS ARE RETAINED

From what has been said above, it is clear that the cell imposes conformational transport restraints not only on unassembled proteins but also on misfolded proteins. Abnormal products retained include many proteins with deletions, insertions and point mutations (Adams & Rose 1985; Gething et al 1986), chimeric proteins (Rizzolo et al 1985), proteins produced in the presence of amino acid analogues, and proteins synthesized in the presence of inhibitors of N-linked glycosylation. The role of N-linked carbohydrate side chains in folding and transport has been examined for many proteins (Reckhow & Enns 1988; Duronio et al 1988; Machamer & Rose 1988a,b; Schlesinger & Schlesinger 1987; Slieker et al 1986; Sibley & Wagner 1981; Prives & Olden 1980; Olden et al 1978; Housley et al 1980; Tamkun & Fambrough 1986). Some are unaffected by the inhibition of glycosylation, others are rendered nonfunctional and misfolded. In the latter cases, it seems that the presence of carbohydrate side chains in strategic positions on the polypeptide chain are required for the correct folding of one or more domains.

Cellular proteins synthesized in the absence of any inhibitors may also exhibit incomplete folding and defective oligomeric assembly. While some soluble secretory proteins (such as the zymogens in the pancreas) can be very efficiently drained from the ER and secreted (Palade 1975), others, such as the β subunit of CG, do not acquire a mature conformation with great efficiency and the incorrectly folded fraction is retained or degraded (Ruddon et al 1987). Five to ten percent of influenza virus HA expressed in CV-1 cells is misfolded, retained, and degraded in the ER (Copeland et al 1986; Gething et al 1986; Hurtley et al 1989). The fraction of misfolded protein often increases with elevated temperature (Gallagher et al 1988; Machamer et al 1988b; Gibson et al 1979).

The fact that retained proteins are actually misfolded can be determined in several ways. They usually lack a full complement of antigenic epitopes, they are excessively sensitive to proteases, and they lack native disulfide bonds. As a rule misfolded proteins are retained as aggregates (Schlesinger & Schlesinger 1987; Gibson et al 1979; Machamer et al 1988a,b; Hurtley et al 1989). In this respect they clearly differ from correctly folded but unassembled subunits, which, though retained, seem to be essentially unaggregated. In some cases, the misfolded and aggregated proteins have been found to be crosslinked by interchain disulfides (Machamer & Rose 1988b; Hurtley et al 1989; Tooze et al 1989). In these cases, the interchain disulfides are clearly abnormal because the native protein subunits are not covalently associated with each other.

The aggregates range in size from dimers to large granules of several million daltons. They are often associated with BiP, which binds permanently and noncovalently in an ATP- and temperature-sensitive fashion (Hurtley et al 1989; Machamer & Rose, 1988 a,b; Kassenbrock et al 1988; Munro & Pelham 1986). In some cases secretory proteins accumulate in the ER as large granules in which the proteins may be crosslinked by disulfide bonds. Such granules have been described, for example, in pancreatic exocrine cells and in thyrotropes after thyroidectomy (Palade 1956; Tooze et al 1989).

Our recent studies on influenza HA show that aggregation and disulfide crosslinking of nonglycosylated and incorrectly folded HA occurs during or immediately after synthesis. It is followed by BiP association and the expression of aberrant antibody binding epitopes within two min (Hurtley et al 1989). The most surprising finding was that only one domain—the stem region of the spike protein—displayed signs of misfolding. This suggests that the molecular defect resulting in retention may be a relatively subtle one. Sucrose gradient centrifugation after solubilization with nonionic detergent indicated that the aggregates were heterogeneous in size, and that they were not very large (9-25S, Hurtley et al 1989; I. Singh et al, in preparation). Their actual size in situ could, however, be larger than observed after detergent treatment, which could explain, at least in part, why the complexes are retained.

When misfolded, soluble and membrane proteins seem to be handled in a similar way. Recent studies on misfolded forms of both membrane-anchored and anchor-free molecules of influenza HA (Hurtley et al 1989; I. Singh et al, in preparation) show that retention is equally efficient and that the aggregates formed are virtually identical in size and composition. The main differences are that aggregation, BiP binding, and antigenic changes occur considerably later for the anchor-free HA than for the anchored form ($t_{1/2}$ 15 vs 1–3 min).

Whether the sequestration of misfolded proteins is an active process driven by cellular machinery, or a passive phenomenon caused by the properties of the proteins themselves, is not known. Aggregation of folding intermediates is, however, often encountered during in vitro folding experiments (Ghelis & Yon 1982), and it has been suggested that aggregation competes with the correct folding process in living cells (King et al 1986). The tendency of incompletely folded polypeptides to form irreversible aggregates may thus simply reflect basic physical properties of folding. The magnitude of this effect, which may be intrinsic to all proteins, may be aggravated by mutations and cotranslational processing defects to a point where it leads to quantitative, irreversible misfolding and aggregation. It is interesting that many proteins that misfold at 37°C, can fold normally at lower temperatures (Machamer & Rose 1988a,b; Gallagher et al 1988; Hearing et al 1989a,b).

DEGRADATION OF RETAINED PROTEINS

Since large numbers of proteins in the ER are transport incompetent, the question arises as to how the cell disposes of them. Although inherently more sensitive to proteases than the correctly folded proteins, misfolded and unassembled proteins in the ER are often relatively long-lived. Half-lives of more than 6–10 hrs are not unusual (Corless et al 1987; Hurtley et al 1989). In some instances, such as the delta subunit of the T cell receptor, degradation can, however, be much faster (Chen et al 1988; Lippincott-Schwartz et al 1988), and it seems well established that different proteins are disposed of at different rates. Frequently, a lag period with little degradation is followed by a rapid degradation phase. Such a lag could suggest that the proteins need to be moved into a specific compartment. It could furthermore benefit those subunits that assemble slowly because it would allow time for assembly reactions to be completed. To accommodate a large amount of aggregated and unassembled proteins arising during extreme conditions of misfolding, the ER can actually expand in volume (Sifers et al 1988; Iozzo & Pacifici 1976). This effect is dramatically illustrated in cells whose proline hydroxylase is inhibited: misfolded, untrimerized procollagen reversibly accumulates in large, extended ER cisternae (Iozzo & Pacifici 1976; Pacifici & Iozzo 1988).

The degradation of some of the unassembled and misfolded proteins probably occurs in the ER or in some closely associated compartment. The identity of the degradative enzymes involved is unknown. Studies by Lippincott-Schwartz et al (1988) on the degradation of T-cell receptor subunits indicate that the degradation is insensitive to inhibition by lysosomotropic agents and displays a sharp threshold in its temperature dependence. However, only two out of the seven subunits seem to be susceptible to this sort of degradation (Chen et al 1988). Degradation of some other misfolded and unassembled products is, however, blocked by inhibitors of lysosomal degradation, which suggests that it takes place in lysosomes or autophagic vacuoles (Davis & Hunter 1987). It is well-known that, at least under certain conditions, the proteins of the ER can be degraded by an autophagosomal route (Masaki et al 1987). Additional studies are clearly necessary to determine where and how degradation of the wide variety of retained proteins occurs and how these processes are controlled.

GENERAL FEATURES OF QUALITY CONTROL

The control of transport from the ER to the Golgi complex presents a complex picture. It is quite clear that one single mechanism cannot explain all of the sorting events observed. Given the available information described above and other data on resident ER proteins, one can at present distinguish at least three types of selective retention in the ER:

1. Retention of resident ER proteins—many of the soluble ER proteins have C-terminal KDEL or HDEL sequences, which serve as retention signals (see Pelham, this volume). They either associate with receptors in the ER, or they are continuously retrieved from a post-ER compartment (Pelham, this volume). Other ER proteins are known to have other types of sequences that serve as retention determinants (Poruchynsky et al 1985; Poruchynsky & Atkinson 1988; Pääbo et al 1987). Retention of resident proteins may rely on association with a large protein network in the ER.
2. Unassembled subunits and assembly intermediates—these proteins do not possess retention sequences but rather express common, as yet

unidentified, structural properties that lead to retention. These properties could include hydrophobicity and/or flexibility of exposed peptide segments. Unlike misfolded proteins, these polypeptides are not present as stable aggregates, nor do they, once initial folding has occurred, seem to form easily detected, stable complexes with BiP. Immunoglobulin subunits and assembly intermediates are exceptional since they do clearly associate with BiP (Bole et al 1986; Hendershot et al 1987).

To explain their retention one must invoke weak interactions, perhaps with a protein scaffold within the ER lumen (see Pfeffer & Rothman 1987). Such interactions may restrict diffusion of these proteins sufficiently to prevent their exit from the ER, but still allow enough freedom of diffusion to sustain the degree of mixing needed for subunits to find each other and oligomerize. It is important to emphasize that other mechanisms, including specific recycling from intermediate compartments between ER and Golgi, could also explain their retention. It is noteworthy that many unassembled soluble subunits, in contrast to membrane bound counterparts, are not strictly retained. This suggests that a membrane associated mechanism may be involved.

3. Misfolded proteins—these proteins generally form aggregates and the aggregates are often associated quite stably with BiP. Aggregation occurs rapidly after synthesis possibly as a consequence of the poor solubility of incompletely or incorrectly folded forms. The retention could be explained by large aggregate size that would prevent diffusion within the ER. The evidence indicates, however, that even small complexes of misfolded proteins are efficiently retained. Another explanation could be the association with BiP, which itself is retained by virtue of its KDEL sequence. The limited data suggests that soluble and membrane bound polypeptides, when misfolded, are equally well retained, and that misfolding in any topological domain suffices to retain transmembrane proteins. In the case of transmembrane proteins, when the lumenal domain is defective, aggregation and BiP binding frequently occur. The limited data on proteins with defectives in the cytoplasmic domain suggest that retention is by a different mechanism that does not seem to involve either aggregation or BiP association.

CONCLUSIONS AND FUTURE DIRECTIONS

The role of the three-dimensional structure in controlling transport of membrane and secretory proteins and lysosomal hydrolases is now well established. Quality control seems to occur at two different levels: initial folding and oligomeric assembly. While correct initial folding is required for all proteins, correct assembly is imposed as an additional requirement for many oligomeric proteins. In the case of membrane proteins, the control extends to the lumenal, transmembrane, and cytoplasmic domains.

Misfolded and misassembled proteins are produced continuously by the cell since the reactions involved in folding, covalent modification, and oligomeric assembly have a finite margin of error. Apparently, the more complex the structure, the larger the fraction of misfolded and incompletely assembled side products and the bigger the pool of flawed protein the cell has to manage. Retention in the ER prevents the expression and secretion of misfolded hormones, receptors, ion channels, and recognition markers thus limiting their harmful effects. The restrictions in transport out of the ER can be viewed as a late stage in a series of mechanisms by which the cell seeks to assure faithful expression of genetic information at the level of functional and correctly localized proteins. Posttranslational control of this type is not exclusive to the ER. Misfolded and unassembled proteins are specifically and rapidly degraded in the cytosol, mitochondria, chloroplasts, and the nucleus.

The underlying sorting reactions are specific, efficient, and remarkably subtle. The system has probably emerged through a co-evolutionary process whereby the products to be transported and the elements of the ER have undergone mutual adaptation. It is not surprising that many current studies in which genetically modified proteins are expressed in cells (or wild-type proteins are synthesized in heterologous cell types) fail to produce transport competent protein products. The proteins may simply be mismatched with respect to the ER machinery of the cell or cell type in question.

Future studies need to address four general problems. They must define what structural properties in newly synthesized proteins determine transport competence, and which properties lead to retention. Systematic studies correlating the structure of mutant proteins with the efficiency of transport may help to illuminate why proteins are rendered transport incompetent: Is it, for example, the presence of hydrophobic surface patches? Is it flexibility or lack of defined domain structure? Or is it general insolubility? Are transport competent proteins somehow specifically selected for transport, rather than carried by bulk flow?

To understand how proteins are retained in the ER, the basic aspects of ER structure and function must be elucidated as well as the mechanism of ER to Golgi traffic. Specifically, we need to know to what degree soluble and membrane-bound proteins are mobile in the ER membrane and lumen, how transport vesicles are formed, whether there are intermediate compartments between the ER and Golgi, and to what degree selectivity in transport depends on bulk flow, specific retention, accelerated transport, or selective recycling. The difficulty in explaining ER sorting is compounded by the lack of reliable estimates of the amount of material that constitutively leaves and returns to the ER, of the concentration dependence of assembly of oligomers, of the rate of mixing of subunits within the ER pool, of the overall efficiency of oligomerization, and so on. Since oligomerization is one of the most important and most common post-translational modifications in the secretory pathway, with wide consequences outside the field of cell biology, we may expect rapid progress in this area of investigation.

A separate question involves the role of BiP and other heat shock related proteins in quality control in the ER. The accumulated evidence reviewed in this volume by Pelham suggests a variety of functions for these proteins ranging from assistance during protein translocation, to folding, assembly, retention, and targeting for degradation. Heat shock proteins in other compartments and in prokaryotes carry out crucial functions related to folding and assembly.

Finally, an effort must be made to determine how the cell disposes of the defective products which accumulate in the ER. It is possible that there are several pathways for this including local proteases, autophagocytosis, and as yet undiscovered degradative compartments that are part of the ER system.

Acknowledgments

We thank Dr. Robert W. Doms for critical comments on the manuscript and all the members of the group who have provided us with data and information. Funding was provided by a National Institutes of Health grant and by a European Molecular Biology Organisation post-doctoral fellowship to S.M.H.

Literature Cited

Adams, G. A., Rose, J. K. 1985. Incorporation of a charged amino acid into the membrane-spanning domain blocks cell surface transport but not membrane anchoring of a viral glycoprotein. *Mol. Cell. Biol.* 5: 1442–48

Alarcon, B., Berkhout, B., Breitmeyer, J., Terhorst, C. 1988. Assembly of the human T cell receptor-CD3 complex takes place in the endoplasmic reticulum and involves intermediary complexes between the CD3-$\gamma.\delta.\varepsilon$ core and single T cell receptor α or β chains. *J. Biol. Chem.* 263: 2953–61

Alberts, B., Bray, D., Lewis, J., Raff, M., Roberts, K., Watson, J. D., eds 1983. In *Molecular Biology of the Cell.* Chapters 3, 6, 7. New York: Garland

Anderson, M., Pääbo, S., Nilsson, T., Peterson, P. A. 1985. Impaired intracellular transport of class I MHC antigens as a possible means for adenovirus to evade immune surveillance. *Cell* 43: 215–22

Balch, W. E., Elliot, M., Keller, D. S. 1986. ATP-coupled transport of vesicular stomatitis virus G protein between the endoplasmic reticulum and the Golgi. *J. Biol. Chem.* 261: 14681–89

Balch, W. E., Keller, D. S. 1986. ATP-coupled transport of vesicular stomatitis virus G protein. Functional boundaries of secretory compartments. *J. Biol. Chem.* 261: 14690–96

Bergman, L. W., Kuehl, W. M. 1979a. Formation of intermolecular disulfide bonds on nascent immunoglobulin polypeptides. *J. Biol. Chem.* 254: 5690–94

Bergman, L. W., Kuehl, W. M. 1979b. Formation of an intrachain disulfide bond on nascent immunoglobulin light chains. *J. Biol. Chem.* 254: 8869–76

Berkhout, B., Alarcon, B., Terhorst, C. 1988. Transfection of genes encoding the T cell receptor-associated CD3 complex into COS cells results in assembly of the macromolecular structure. *J. Biol. Chem.* 263: 8528–36

Bischoff, J., Libresco, S., Shia, M. A., Lodish, H. F. 1988. The H1 and H2 polypeptides associate to form the asialoglycoprotein receptor in human hepatoma cells. *J. Cell Biol.* 106: 1067–74

Bjorkman, P. J., Saper, M. A., Samraoui, B., Bennett, W. S., Strominger, J. L., et al. 1987. Structure of the human class I histocompatibility antigen, HLA-A2. *Nature* 329: 506–18

Bole, D. G., Hendershot, L. M., Kearney, J. F. 1986. Posttranslational association of immunoglobulin heavy chain binding protein with nascent heavy chains in nonsecreting and secreting hybridomas. *J. Cell Biol.* 102: 1558–66

Bonifacino, J. S., Lippincott-Schwartz, J., Chen, C., Antusch, D., Samelson, L. E., et al. 1988. Association and dissociation of the murine T cell receptor associated protein (TRAP). *J. Biol. Chem.* 263: 8965–71

Bormann, B.-J., Knowles, W. J., Marchesi, V. T. 1989. Synthetic peptides mimic the assembly of transmembrane glycoproteins. *J. Biol. Chem.* 264: 4033–37

Boulay, F., Doms, R. W., Webster, R. G., Helenius, A. 1988. Posttranslational oligomerization and cooperative acid activation of mixed influenza hemagglutinin trimers. *J. Cell Biol.* 106: 629–39

Burgess, T. L., Kelly, R. B. 1987. Constitutive and regulated secretion of proteins. *Annu. Rev. Cell Biol.* 3: 243–93

Carlin, B. E., Merlie, J. P. 1986. Assembly of multisubunit membrane proteins. In *Protein Compartmentalization*, ed. A. W. Strauss, I. Boime, G. Kreil, pp. 71–86. New York: Springer-Verlag

Catterall, W. A. 1986. Molecular properties of voltage-sensitive sodium channels. *Annu. Rev. Biochem.* 55: 953–85

Chen, C., Bonifacino, J. S., Yuan, L. C., Klausner, R. D. 1988. Selective degradation of T cell antigen receptor chains in a pre-Golgi compartment. *J. Cell Biol.* 107: 2149–61

Chothia, C., Janin, J. 1975. Principles of protein-protein recognition. *Nature* 256: 705–8

Claesson-Welsh, L., Spear, P. G. 1986. Oligomerization of herpes simplex virus glycoprotein B. *J. Virol.* 60: 803–6

Clevers, H., Alarcon, B., Wileman, T., Terhorst, C. 1988. The T cell receptor/CD3 complex: A dynamic protein ensemble. *Annu. Rev. Immunol.* 6: 629–62

Copeland, C. S., Doms, R. W., Bolzau, E. M., Webster, R. G., Helenius, A. 1986. Assembly of influenza hemagglutinin trimers and its role in intracellular transport. *J. Cell Biol.* 103: 1179–91

Copeland, C. S., Zimmer, K.-P., Wagner, K. R., Healey, G. A., Mellman, I., et al. 1988. Folding, trimerization, and transport are sequential events in the biogenesis of influenza virus hemagglutinin. *Cell* 53: 197–209

Corless, C. L., Matzuk, M. M., Ramabhadran, T. P., Krichevsky, A., Boime, I. 1987. Gonadotropin beta subunits determine the rate of assembly and the oligosaccharide processing of hormone dimer in transfected cells. *J. Cell Biol.* 104: 1173–81

Cowell, G. M., Tranum-Jensen, J., Sjöström, H., Norén, O. 1986. Topology and quaternary structure of pro-sucrase/isomaltase and final-form sucrase/isomaltase. *Biochem. J.* 237: 455–61

Davis, G. L., Hunter, E. 1987. A charged amino acid substitution within the transmembrane anchor of the Rous sarcoma virus envelope glycoprotein affects surface expression but not intracellular transport. *J. Cell Biol.* 105: 1191–1203

Deisenhofer, J., Epp, O., Miki, K., Huber, R., Michel, H. 1985. Structure of the protein subunits in the photosynthetic reaction centre of *Rhodopseudomonas viridis* at 3 Å resolution. *Nature* 318: 618–24

Doms, R. W., Helenius, A. 1986. Quaternary structure of influenza virus hemagglutinin after acid treatment. *J. Virol.* 60: 833–39

Doms, R. W., Keller, D. S., Helenius, A., Balch, W. E. 1987. Role for adenosine triphosphate in regulating the assembly and transport of vesicular stomatitis virus G protein trimers. *J. Cell Biol.* 105: 1957–69

Doms, R. W., Ruusala, A., Machamer, C., Helenius, J., Helenius, A., et al. 1988. Differential effects of mutations in three domains on folding, quaternary structure, and intracellular transport of vesicular stomatitis virus G protein. *J. Cell Biol.* 107: 89–99

Doyle, C., Sambrook, J., Gething, M.-J. 1986. Analysis of progressive deletions of the transmembrane and cytoplasmic domains of influenza virus. *J. Cell Biol.* 103: 1193–1204

Dulis, B. H. 1983. Regulation of protein expression in differentiation by subunit assembly. *J. Biol. Chem.* 258: 2181–87

Duronio, V., Jacobs, S., Romero, P. A., Herscovics, A. 1988. Effects of inhibitors of N-linked oligosaccharide processing on the biosynthesis and function of insulin and insulin-like growth factor-I receptors. *J. Biol. Chem.* 263: 5436–45

Einfeld, D., Hunter, E. 1988. Oligomeric structure of a prototype retrovirus glycoprotein. *Proc. Natl. Acad. Sci. USA* 85: 8688–92

Esmon, P. C., Esmon, B. C., Schauer, I. E., Taylor, A., Schekman, R. 1987. Structure, assembly, and secretion of octameric invertase. *J. Biol. Chem.* 262: 4387–94

Esser, V., Russell, D. W. 1988. Transport-deficient mutations in the low density lipoprotein receptor. *J. Biol. Chem.* 263: 13276–81

Ferguson, M. A. J., Williams, A. F. 1988. Cell-surface anchoring of proteins via glycosyl-phosphatidylinositol structures. *Annu. Rev. Biochem.* 57: 285–320

Freedman, R. 1987. Folding into the right shape. *Nature* 329: 196–97

Fries, E., Gustaffson, L, Peterson, P. A. 1984. Four secretory proteins synthesized by hepatocytes are transported from the endoplasmic reticulum to the Golgi complex at different rates. *EMBO J.* 3: 147–52

Gallagher, P., Henneberry, J., Wilson, I., Sambrook, J., Gething, M.-J. 1988. Addition of carbohydrate side chains at novel sites on influenza virus hemagglutinin can modulate the folding, transport, and activity of the molecule. *J. Cell Biol.* 107: 2059–73

Gardner, J. M., Fambrough, D. M. 1983. Fibronectin expression during myogenesis. *J. Cell Biol.* 96: 474–85

Geetha-Habib, M., Nova, R., Kaplan, H., Lennarz, W. J. 1988. Glycosylation site binding protein, a component of oligosaccharyl transferase, is highly similar to three other 57 kd luminal proteins of the ER. *Cell* 54: 1053–60

Gething, M.-J., McCammon, K., Sambrook, J. 1986. Expression of wild-type and mutant forms of influenza virus hemagglutinin: The role of folding in intracellular transport. *Cell* 46: 939–50

Ghelis, C., Yon, J. 1982. *Protein Folding*. ed: B. Horecker, N. O. Kaplan, J. Marmur, H. A. Scheraga. New York: Academic, 562 pp

Gibson, R., Schlesinger, S., Kornfeld, S. 1979. The nonglycosylated glycoprotein of vesicular stomatitis virus is temperature sensitive and undergoes intracellular aggregation at elevated temperatures. *J. Biol. Chem.* 254: 3600–7

Goding, J. W., Burns, G. F. 1981. Monoclonal OKT-9 recognizes the receptor for transferrin on human acute lymphocytic leukemia cells. *J. Immunol.* 127: 1256–58

Goodman, D. S. 1984. Retinol binding proteins. In *Retinoids* 2: 41–48. eds. M. B. Sporn, A. B. Roberts, D. S. Goodman. New York: Academic

Hearing, J., Hunter, E., Rodgers, L., Gething, M.-J., Sambrook, J. 1989a. Isolation of Chinese hamster ovary cells lines temperature conditional for the cell-surface expression of integral membrane glycoproteins. *J. Cell Biol.* 108: 339–53

Hearing, J., Gething, M.-J., Sambrook, J. 1989b. Addition of truncated oligosaccharides to influenza virus hemagglutinin results in its temperature-conditional cell-surface expression. *J. Cell Biol.* 108: 355–65

Helenius, A., Simons, K. 1975. Solubilization of membranes by detergents. *Biochim. Biophys. Acta* 415: 29–79

Hendershot, L., Bole, D., Köhler, G., Kearney, J. F. 1987. Assembly and secretion of heavy chains that do not associate post-translationally with heavy chain-binding protein. *J. Cell Biol.* 104: 761–67

Hombach, J., Leclerq, L., Radbruch, A., Rajewsky, K., Reth, M. 1988. A novel 34-kd protein co-isolated with the IgM molecule in surface IgM-expressing cells. *EMBO J.* 7: 3451–56

Hortsch, M., Crimaudo, C., Meyer, D. I. 1987. Structure and function of the endoplasmic reticulum. In *Integration and control of metabolic processes: Pure and applied aspects*. ed. O. L. Kon, pp. 3–11. Miami: ICSU Miami

Housley, T. J., Rowland, F. N., Ledger, P. W., Kaplan, J., Tanzer, M. L. 1980. Effects of tunicamycin on the biosynthesis of procollagen by human fibroblasts. *J. Biol. Chem.* 255: 121–28

Hurtley, S. M., Bole, D. G., Hoover-Litty, H., Helenius, A., Copeland, C. S. 1989. Interactions of misfolded influenza virus hemagglutinin with binding protein (BiP). *J. Cell Biol.* 108: 2117–26

Hussain, M. M. 1985. Reconstitution of purified dipeptidyl peptidase IV. A comparison with aminopeptidase N with respect to morphology and influence of anchoring peptide on function. *Biochim. Biophys. Acta* 815: 306–12

Hussain, M. M., Tranum-Jensen, J., Norén, O., Sjöström, H., Christiansen, K. 1981. Reconstitution of purified amphiphilic pig intestinal microvillus aminopeptidase. *Biochem. J.* 199: 179–86

Hynes, R. O. 1987. Integrins: A family of cell surface receptors. *Cell* 48: 549–54

Iozzo, R. V., Pacifici, M. 1986. Ultrastructural localization of the major proteoglycan and Type II procollagen in organelles and extracellular matrix of cultured chondroplasts. *Histochemistry* 86: 113–22

Johnson, J. D., Wong, M. L., Rutter, W. J. 1988. Properties of the insulin receptor ectodomain. *Proc. Natl. Acad. Sci. USA* 85: 7516–20

Kassenbrock, C. K., Garcia, P. D., Walter,

P., Kelly, R. B. 1988. Heavy-chain binding protein recognizes aberrant polypeptides translocated *in vitro*. *Nature* 333: 90–93
Kaufman, J. F., Auffray, C., Korman, A. J., Shackelford, D. A., Strominger, J. 1984. The class II molecules of the human and murine major histocompatibility complex. *Cell* 36: 1–13
King, J., Yu, M.-H., Siddiqi, J., Haase, C. 1986. Genetic identification of amino acid sequences influencing protein folding. In *Protein Engineering*, eds. M. Inouye, R. Sarma. pp. 275–91. New York: Academic
Kivirikko, K. I., Myllylä, R. 1986. Post-translational processing of procollagens. *Ann. NY Acad. Sci.* 460: 187–201
Koivu, J., Myllylä, R., Helaakoski, T., Pihlajaniemi, T., Tasanen, K., et al. 1987. A single polypeptide acts both as the β subunit of prolyl 4-hydroxylase and as a protein disulfide-isomerase. *J. Biol. Chem.* 262: 6447–49
Kondor-Koch, C., Burke, B., Garoff, H. 1983. Expression of Semliki Forest virus proteins from cloned complementary DNA. I. The fusion activity of the spike glycoprotein. *J. Cell Biol.* 97: 644–51
Krangel, M. S., Orr, H. T., Strominger, J. L. 1979. Assembly and maturation of HLA-A and HLA-B antigens in vivo. *Cell* 18: 979–91
Krangel, M. S., Pious, D., Strominger, J. L. 1984. Characterization of a B lymphoblastoid cell line mutant that secretes HLA-A2. *J. Immunol.* 132: 2984–91
Kreibich, G., Ulrich, B. L., Sabatini, D. D. 1978. Proteins of rough microsomal membranes related to ribosome binding. I. Identification of ribophorins I and II, membrane proteins characteristic of rough microsomes. *J. Cell Biol.* 77: 464–87
Kreis, T. E., Lodish, H. F. 1986. Oligomerization is essential for the transport of vesicular stomatitis viral glycoprotein to the cell surface. *Cell* 46: 929–37
Kvist, S., Östberg, L., Persson, H., Philipson, L., Peterson, P. A. 1978. Molecular association between transplantation antigens and cell surface antigen in adenovirus-transformed cell line. *Proc. Natl. Acad. Sci. USA* 75: 5674–78
Kvist, S., Wiman, K., Claesson, L., Petersson, P. A., Dobberstein, B. 1982. Membrane insertion and oligomeric assembly of HLA-DR histocompatibility antigens. *Cell* 29: 61–69
Langosch, D., Thomas, L., Betz, H. 1988. Conserved quaternary structure of ligand-gated ion channels: The postsynaptic glycine receptor is a pentamer. *Proc. Natl. Acad. Sci. USA* 85: 7394–98
Lippincott-Schwartz, J., Bonifacino, J. S., Yuan, L. C., Klausner, R. D. 1988. Degradation from the endoplasmic reticulum: Disposing of newly synthesized proteins. *Cell* 54: 209–220
Lodish, H. F. 1988. Transport of secretory and membrane glycoproteins from the rough endoplasmic reticulum to the Golgi. *J. Biol. Chem.* 263: 2107–10
Lodish, H. F., Kong, N., Snider, M., Strous, G. J. A. M. 1983. Hepatoma secretory proteins migrate from the rough endoplasmic reticulum to Golgi at characteristic rates. *Nature* 304: 80–83
Loeb, J. A., Drickamer, K. 1987. The chicken receptor for endocytosis of glycoproteins contains a cluster of N-acetylglucosamine-binding sites. *J. Biol. Chem.* 262: 3022–29
Machamer, C. E., Rose, J. K. 1988a. Influence of new glycosylation sites on expression of the Vesicular Stomatitis virus G protein at the plasma membrane. *J. Biol. Chem.* 263: 5948–54
Machamer, C. E., Rose, J. K. 1988b. Vesicular Stomatitis virus G proteins with altered glycosylation sites display temperature-sensitive intracellular transport and are subject to aberrant intermolecular disulfide bonding. *J. Biol. Chem.* 263: 5955–60
Magner, J. A., Weintraub, B. D. 1982. Thyroid-stimulating hormone subunit processing and combination in microsomal subfractions of mouse pituitary tumor. *J. Biol. Chem.* 257: 6709–15
Mahuran, D. J., Neote, K., Klavins, M. H., Leung, A., Gravel, R. A. 1988. Proteolytic processing of pro-α and pro-β precursors from human β-hexosaminidase. *J. Biol. Chem.* 263: 4612–18
Mains, P. E., Sibley, C. H. 1983. The requirement of light chain for the surface deposition of the heavy chain of immunoglobulin M. *J. Biol. Chem.* 258: 5027–33
Malissen, B., Price, M. P., Goverman, J. M., McMillan, M., White, J., et al. 1984. Gene transfer of H-2 class II genes: Antigen presentation by mouse fibroblast and hamster B-cell lines. *Cell* 36: 319–27
Malissen, B., Steinmetz, M., McMillan, M., Pierres, M., Hood, L. 1983. Expression of I-A^k class II genes in mouse L cells after DNA-mediated gene transfer. *Nature* 305: 440–43
Markwell, M. A. K., Fox, C. F. 1980. Protein-protein interactions within paramyxoviruses identified by native disulfide bonding or reversible chemical cross-linking. *J. Virol.* 33: 152–66
Masaki, R., Yamamoto, A., Tashiro, Y. 1987. Cytochrome P-450 and NADPH-cytochrome P-450 reductase are degraded in the autolysosomes in rat liver. *J. Cell Biol.* 104: 1207–15
Matzuk, M. M., Boime, I. 1988. The role of

asparagine-linked oligosaccharides of the α subunit in the secretion and assembly of human chorionic gonadotropin. *J. Cell Biol.* 106: 1049–59

Medda, S., Stevens, A. M., Swank, R. T. 1987. Involvement of the esterase active site of egasyn in compartmentalization of β-glucuronidase within the endoplasmic reticulum. *Cell* 50: 301–10

Merlie, J. P., Smith, M. M. 1986. Synthesis and assembly of acetylcholine receptor, a multisubunit membrane glycoprotein. *J. Membrane Biol.* 91: 1–10

Metcalf, P., Blum, M., Freymann, D., Turner, M., Wiley, D. C. 1987. Two variant surface glycoproteins of *Trypanosoma brucei* of different sequence classes have similar 6 Å resolution X-ray structures. *Nature* 325: 84–86

Minami, Y., Weissman, A. M., Samelson, L. E., Klausner, R. D. 1987. Building a multichain receptor: Synthesis, degradation and assembly of the T-cell antigen receptor. *Proc. Natl. Acad. Sci. USA* 84: 2688–92

Miyazaki, J.-I., Appella, E., Ozato, K. 1986a. Intracellular transport blockade caused by disruption of the disulfide bridge in the third external domain of major histocompatibility complex class I antigen. *Proc. Natl. Acad. Sci. USA* 83: 757–61

Miyazaki, J.-I., Appella, E., Zhao, H., Forman, J., Ozato, K. 1986b. Expression and function of a nonglycosylated major histocompatibility class I antigen. *J. Exp. Med.* 163: 856–71

Munro, S., Pelham, H. R. B. 1986. An Hsp70-like protein in the ER: Identity with the 78 kd glucose regulated protein and immunoglobulin heavy chain binding protein. *Cell* 46: 291–300

Newcomer, M. E., Jones, T. A., Åqvist, J., Sundelin, J., Eriksson, U., et al. 1984. The three-dimensional structure of retinol-binding protein. *EMBO J.* 3: 1451–54

Norén, O., Sjöström, H., Cowell, G. M., Tranum-Jensen, J., Hansen, O. C., et al. 1986. Pig intestinal microvillar maltase-glycoamylase. *J. Biol. Chem.* 261: 12306–9

Oettgen, H. C., Pettey, C. L., Maloy, W. L., Terhorst, C. 1986. A T3-like protein complex associated with the antigen receptor on murine T cells. *Nature* 320: 272–75

Olden, K., Pratt, R. M., Yamada, K. 1978. Role of carbohydrates in protein secretion and turnover: Effects of tunicamycin on the major cell surface glycoprotein of chick embryo fibroblasts. *Cell* 13: 461–73

Olson, T. S., Bamberger, M. J., Lane, M. D. 1988. Post-translational changes in tertiary and quaternary structure of the insulin proreceptor. *J. Biol. Chem.* 263: 7342–51

Olson, T. S., Lane, M. D. 1987. Post-translational acquisition of insulin binding activity by the insulin proreceptor. *J. Biol. Chem.* 262: 6816–22

Östman, A., Rall, L., Hammacher, A., Wormstead, M. A., Coit, D., et al. 1988. Synthesis and assembly of a functionally active recombinant platelet-derived growth factor AB heterodimer. *J. Biol. Chem.* 263: 16202–8

Owen, M. J., Kissonerghis, A.-M., Lodish, H. F. 1980. Biosynthesis of HLA-A and HLA-B antigens in vivo. *J. Biol. Chem.* 255: 9678–84

Owen, M. J., Kissonerghis, A.-M., Lodish, H. F., Crumpton, M. J. 1981. Biosynthesis and maturation of HLA-DR antigens in vivo. *J. Biol. Chem.* 256: 8987–93

Pääbo, S., Bhat, B. M., Wold, W. S., Peterson, P. A. 1987. A short sequence in the COOH-terminus makes an adenovirus membrane glycoprotein a resident of the endoplasmic reticulum. *Cell* 50: 311–17

Pacifici, M., Iozzo, R. V. 1988. Remodeling of the rough endoplasmic reticulum during stimulation of procollagen secretion by ascorbic acid in cultured chondrocytes. *J. Biol. Chem.* 263: 2483–92

Palade, G. E. 1956. Intracisternal granules in the exocrine pancreas. *J. Biophys. Biochem. Cytol.* 2: 417–22

Palade, G. 1975. Intracellular aspects of the process of protein synthesis. *Science* 189: 347–58

Peters, T. Jr., Davidson, L. K. 1982. The biosynthesis of rat serum albumin. *J. Biol. Chem.* 257: 8847–53

Peters, B. P., Hartle, R. J., Krzesicki, R. F., Kroll, T. G., Perini, F., et al. 1985. The biosynthesis, processing, and secretion of laminin by human choriocarcinoma cells. *J. Biol. Chem.* 260: 14732–42

Peters, B. P., Krzesicki, R. F., Hartle, R. J., Perini, F., Ruddon, R. W. 1984. A kinetic comparison of the processing and secretion of the αβ dimer and the uncombined α and β subunits of chorionic gonadotropin synthesized by human choriocarcinoma cells. *J. Biol. Chem.* 259: 15123–30

Pfeffer, S. R., Rothman, J. E. 1987. Biosynthetic protein transport and sorting by the endoplasmic reticulum and Golgi. *Annu. Rev. Biochem.* 56: 829–52

Plant, P. W., Grieninger, G. 1986. Noncoordinate synthesis of the fibrinogen subunits in hepatocytes cultured under hormone-deficient conditions. *J. Biol. Chem.* 261: 2331–36

Poruchynsky, M. S., Atkinson, P. H. 1988. Primary sequence domains required for the retention of rotavirus VP7 in the endoplasmic reticulum. *J. Cell Biol.* 107: 1697–706

Poruchynsky, M. S., Tyndall, C., Both, G.

W., Sato, F., Bellamy, A. R., et al. 1985. Deletions into an NH_2-terminal hydrophobic domain result in secretion of rotavirus VP7, a resident endoplasmic reticulum membrane glycoprotein. *J. Cell Biol.* 101: 2199–209

Prives, J. M., Olden, K. 1980. Carbohydrate requirement for expression and stability of acetylcholine receptor on the surface of embryonic muscle cells in culture. *Proc. Natl. Acad. Sci. USA* 77: 5263–67

Rabourdin-Combe, C., Mach, B. 1983. Expression of HLA-DR antigens at the surface of mouse L cells co-transfected with cloned human genes. *Nature* 303: 670–74

Randall, L. L., Hardy, S. J. S. 1986. Correlation of competence for export with lack of tertiary structure of the mature species: A study in vivo of maltose-binding protein in E. coli. *Cell* 46: 921-28

Rask, L., Valtersson, C., Anundi, H., Kvist, S., Erikson, U., et al. 1983. Subcellular localization in normal and vitamin A-deficient rat liver of vitamin A serum transport proteins, albumin, ceruloplasmin and class I major histocompatibility antigens. *Exp. Cell Res.* 143: 91–102

Reckhow, C. L., Enns, C. A. 1988. Characterization of the transferrin receptor in tunicamycin-treated A431 cells. *J. Biol. Chem.* 263: 7297–301

Reeke, G. N. Jr., Becker, J. W., Edelman, G. M. 1975. The covalent and three-dimensional structure of Concanavalin A. *J. Biol. Chem.* 250: 1525–47

Rizzolo, L. J., Finidoro, J., Gonzalez, A., Arpin, M., Ivanov, I. E., et al. 1985. Biosynthesis and intracellular sorting of growth hormone-viral envelope glycoprotein hybrids. *J. Cell Biol.* 101: 1351–62

Ronne, H., Ocklind, C., Wiman, K., Rask, L., Öbrink, B., et al. 1983. Ligand-dependent regulation of intracellular protein transport: Effect of Vitamin A on the secretion of the retinol-binding protein. *J. Cell Biol.* 96: 907–10

Rose, J. K., Doms, R. W. 1988. Regulation of protein export from the endoplasmic reticulum. *Annu. Rev. Cell Biol.* 4: 257–88

Rotundo, R. L. 1984. Asymmetric acetylcholinesterase is assembled in the Golgi apparatus. *Proc. Natl. Acad. Sci. USA* 81: 479–83

Ruddon, R. W., Krzesicki, R. F., Norton, S. E., Beebe, J. S., Peters, B. P., et al. 1987. Detection of a glycosylated, incompletely folded form of human chorionic gonadotropin β subunit that is a precursor of hormone assembly in trophoblastic cells. *J. Biol. Chem.* 262: 12533–40

Sabatini, D. D., Kreibich, G., Morimoto, T., Adesnik, M. 1982. Mechanisms for the incorporation of proteins in membranes and organelles. *J. Cell Biol.* 92: 1–22

Samelson, L. E., Harford, J. B., Klausner, R. D. 1985. Identification of the components of the murine T cell antigen receptor complex. *Cell* 43: 223–31

Schlesinger, M. J., Schlesinger, S. 1987. Domains of virus glycoproteins. *Adv. Virus Res.* 33: 1–44

Schmidt, J., Catterall, W. A. 1987. Palmitylation, sulfation, and glycosylation of the α subunit of the sodium channel. *J. Biol. Chem.* 262: 13713–23

Sechoy, O., Phillipot, J. R., Bienvenue, A. 1987. F protein-F protein interaction within the Sendai virus identified by native bonding or chemical cross-linking. *J. Biol. Chem.* 262: 11519–23

Severinsson, L., Peterson, P. A. 1984. β_2-microglobulin induces intracellular transport of human class I transplantation antigen heavy chain in *Xenopus laevis* oocytes. *J. Cell Biol.* 99: 226–32

Sibley, C. H., Wagner, R. A. 1981. Glycosylation is not required for membrane localization or secretion of IgM in a mouse cell lymphoma. *J. Immunol.* 126: 1868–73

Sifers, R. N., Brashears-Macatee, S., Kidd, V. J., Muensch, H., Woo, S. L. C. 1988. A frameshift mutation results in α_1-antitrypsin that is retained within the rough endoplasmic reticulum. *J. Biol. Chem.* 263: 7330–35

Sikorav, J.-L., Duval, N., Anselmet, A., Bon, S., Krejci, E., et al. 1988. Complex alternative splicing of acetylcholinesterase transcripts in *Torpedo* electric organ; primary structure of the precursor of the glycolipid-anchored dimeric form. *EMBO J.* 7: 2983–93

Silverberg, M., Marchesi, V. T. 1978. The anomalous behaviour of the major sialoglycoprotein from the human erythrocyte. *J. Biol. Chem.* 253: 95–98

Silverton, E. W., Navia, M. A., Davies, D. R. 1977. Three-dimensional structure of an intact human immunoglobulin. *Proc. Natl. Acad. Sci. USA* 74: 5140–44

Sklyanskaya, E. I., Shie, M., Komarov, Y. S., Yamnikova, S. S., Kaverin, N. V. 1988. Formation of mixed hemagglutinin trimers in the course of double infection with influenza viruses belonging to different subtypes. *Virus Res.* 10: 153–65

Slieker, L. J., Martensen, T. M., Lane, M. D. 1986. Synthesis of epidermal growth factor receptor in human A431 cells. *J. Biol. Chem.* 261: 15233–41

Slieker, L. J., Martensen, T. M., Lane, M. D. 1988. Biosynthesis of the epidermal growth factor receptor: Post-translational glycosylation-independent acquisition of tyrosine kinase autophosphorylation activity. *Biochem. Biophys. Res. Comm.* 153: 96–103

Smith, M. M., Lindstrom, J., Merlie, J. P. 1987. Formation of the α-bungarotoxin binding site and assembly of the nicotinic acetylcholine receptor subunits occur in the endoplasmic reticulum. *J. Biol. Chem.* 262: 4367–76

Sporn, L. A., Marder, V. J., Wagner, D. D. 1986. Inducible secretion of large, biologically potent von Willebrand factor multimers. *Cell* 46: 185–90

Sussman, J. J., Bonifacino, J. S., Lippincott-Schwartz, J., Weissman, A. M., Saito, T., et al. 1988. Failure to synthesize the T cell CD3-ζ chain: Structure and function of a partial T cell receptor complex. *Cell* 52: 85–95

Takei, F., Kikuchi, Y., Mizuno, S., Shimura, K. 1987. Further evidence for the importance of the subunit combination of silk fibroin in its efficient secretion from the posterior silk gland cells. *J. Cell Biol.* 105: 175–80

Tamkun, M. M., Fambrough, D. M. 1986. The $(Na^+ + K^+)$-ATPase of the chick sensory neurons. *J. Biol. Chem.* 261: 1009–19

Tartakoff, A., Vassalli, P. 1977. Plasma cell immunoglobulin secretion. *J. Exp. Med.* 146: 1332–45

Tartakoff, A., Vassalli, P. 1979. Plasma cell immunoglobulin M molecules. *J Cell Biol.* 83: 284–99

Titani, K., Walsh, K. A. 1988. Human von Willebrand factor: The molecular glue of platelet plugs. *Trends Biochem. Sci.* 13: 94–97

Tooze, J., Kern, H. F., Fuller, S. D., Howell, K. E. 1989. Condensation-sorting events in the RER of exocrine pancreatic cells. *J. Cell Biol.* 109: 35–50

Tsim, K. W. K., Randall, W. R., Barnard, E. A. 1988. Synaptic acetylcholinesterase of chicken muscle changes during development from a hybrid to a homogeneous enzyme. *EMBO J.* 7: 2451–56

van Driel, I. R., Davis, C. G., Goldstein, J. L., Brown, M. S. 1987. Self-association of the low density lipoprotein receptor mediated by the cytoplasmic domain. *J. Biol. Chem.* 262: 16127–34

Varghese, J. N., Laver, W. G., Colman, P. M. 1983. Structure of the influenza virus glycoprotein antigen neuraminidase at 2.9 Å resolution. *Nature* 303: 35–40

Verheijen, F. W., Palmeri, S., Galjaard, H. 1987. Purification and partial characterization of lysosomal neuraminidase from human placenta. *Eur. J. Biochem.* 162: 63–67

Vogel, R. H., Provencher, S. W., von Bonsdorff, C.-H., Adrian, M., Dubochet, J. 1986. Envelope structure of Semliki Forest virus reconstituted from cryoelectron micrographs. *Nature* 320: 533–35

Waxham, W. N., Merz, D. C., Wolinsky, J. S. 1986. Intracellular maturation of mumps virus hemagglutinin-neuraminidase glycoprotein: Conformational changes detected with monoclonal antibodies. *J. Virol.* 59: 392–400

Wiley, D. C., Skehel, J. J. 1987. The structure and function of the hemagglutinin membrane glycoprotein of influenza virus. *Annu. Rev. Biochem.* 56: 365–94

Williams, D. B., Borriello, F., Zeff, R. A., Nathenson, S. G. 1988. Intracellular transport of class I histocompatibility molecules. *J. Biol. Chem.* 263: 4549–60

Wills, J. W., Srinivas, R. V., Hunter, E. 1984. Mutations of the Rous sarcoma virus *env* gene that affect the transport and subcellular location of the glycoprotein products. *J. Cell Biol.* 99: 2011–23

Wilson, I. A., Skehel, J. J., Wiley, D. C. 1981. Structure of the haemagglutinin membrane glycoprotein of influenza virus at 3 Å resolution. *Nature* 289: 366–73

Wu, G. E., Hozumi, N., Murialdo, H. 1983. Secretion of a λ_2 immunoglobulin chain is prevented by a single amino acid substitution in its variable region. *Cell* 33: 77–83

Yamada, A., Takeuchi, K., Hishimaya, M. 1988. Intracellular processing of mumps virus glycoproteins. *Virology* 165: 268–73

Yewdell, J. W., Yellen, A., Bächi, T. 1988. Monoclonal antibodies localize events in the folding, assembly, and intracellular transport of the influenza virus hemagglutinin glycoprotein. *Cell* 52: 843–52

Yu, S., Sher, B., Kudryk, B., Redman, C. M. 1983. Intracellular assembly of human fibrinogen. *J. Biol. Chem.* 258: 13407–10

Ziemicki, A., Garoff, H. 1978. Subunit composition of the membrane glycoprotein complex of Semliki Forest virus. *J. Mol. Biol.* 122: 259–69

Zuniga, M. C., Malissen, B., McMillan, M., Brayton, P. R., Clark, S. S., et al. 1983. Expression and function of transplantation antigens with either altered or deleted cytoplasmic domains. *Cell* 34: 535–44

Annu. Rev. Cell Biol. 1989. 5: 309–39

DROSOPHILA EXTRACELLULAR MATRIX

J. H. Fessler and L. I. Fessler

Department of Biology and the Molecular Biology Institute,
University of California, Los Angeles, Los Angeles, California 90024

CONTENTS

INTRODUCTION

Extracellular matrix has been implicated in vertebrate development (Hay 1981) and in the maintenance of the differentiated state of cells (Bissel et al 1982). The molecular interactions that are necessary for these processes occur between matrix glycoproteins and cell surface receptors. Recent characterization of the equivalent *Drosophila* macromolecules show that both key ligands and important receptors have been conserved during evolution. *Drosophila* homologues of other vertebrate molecules that mediate cell-cell adhesion are also being found. Therefore, the genetically manipulatable *Drosophila* system is now available for study of develop-

0743–4634/89/1115–0309$02.00

mental problems that involve the interactions of cells with their macromolecular environment.

Cell-matrix interactions and cell adhesion in vertebrate development have been reviewed (Edelman 1986; Ekblom et al 1986; McClay & Ettensohn 1987; Edelman 1988), as have the extracellular matrix molecules that influence neural development (Sanes 1989). There are reviews on the contribution of adhesion to *Drosophila* development (Fehon et al 1987; Semeriva et al 1989) and the potential role of matrix and cell adhesion molecules in neural development of *Drosophila* (Anderson 1988). Insect extracellular matrix was reviewed previously (Ashhurst 1982). Our subject is primarily that part of the specialized extracellular matrix that is adjacent to cells, and the cell surface macromolecules that interact with it.

Development of Drosophila *Extracellular Space*

Although the basic processes and problems of development are similar in *Drosophila* and in vertebrates, some differences need to be noted. The fertilized *Drosophila* egg undergoes nuclear cleavages as a syncytium before cell membranes form, and only afterwards can the embryonic extracellular matrix be investigated. Maternally provided mRNAs significantly influence the first 5 h of development, which is a substantial portion of the 22 h period of embryogenesis. Maternal proteins also assist the early zygote. As embryo, larva, and adult all have open circulatory systems, the extracellular hemolymph that bathes organs combines the functions of vertebrate blood and interstitial matrix. Within larvae are separate groups of cells, called imaginal discs, that are set aside for the future adult. Each of these groups of cells is enclosed by a basement membrane envelope, which thereby forms a separate extracellular compartment.

Domain Structure of Extracellular Molecules

The major components of extracellular matrix are large glycoproteins with multiple domains that are classified by function and structure (Timpl 1989). Functional domains are defined as regions of a molecule that are necessary for interaction with specific adjacent molecules of the extracellular matrix or cell surface. Structural domains may be apparent in electron micrographs of the molecule, but are mainly discerned by sequence analysis. New insights arise from investigating domains, e.g. the discovery of the RGD sequence as part of the domains of several extracellular molecules recognized by the integrin group of cell surface receptors (Ruoslahti 1988a). Comparative analyses of *Drosophila* and vertebrate molecules highlight domains that have been well conserved during evolution, with implications for maintained, interactive functions, e.g. the carboxyl terminal domain of the basement membrane collagen IV, which forms a junctional complex between adjacent collagen molecules. Domains may be evolutionary building blocks that are shared by otherwise apparently different molecules, e.g. portions of the murine laminin A chain and basement membrane proteoglycan core protein (Sasaki et al 1988).

Although each *Drosophila* extracellular component is described here as a single molecular form, some may exist as groups of related molecules, similar to the several vertebrate fibronectins that arise by differential RNA splicing (Hynes 1985). Although all extracellular matrix components are glycoproteins, the precise modulating role of carbohydrate side chains on the properties of the core proteins are poorly understood. The distinction between proteoglycans, with longer glycosaminoglycan chains, and other glycoproteins is diminishing and a start has been made in understanding their core proteins (Ruoslahti 1988b).

Basement Membranes

Little is known about the mutual arrangement of extracellular macromolecules in the vicinity of cell surfaces. Basement membrane is a special type of matrix that is deposited in a polar fashion at the basal side of epithelial cells. It encloses fat and muscle cells, and, in vertebrates, Schwann cells. Probably there are many components in a typical vertebrate basement membrane (Martin & Timpl 1987), and major constituents are type IV collagen, basement membrane heparan sulfate proteoglycan, laminin, and entactin, also known as nidogen. Immunoelectron microscopy locates these components mostly in an amorphous layer of about 100 nm thickness (Abrahamson et al 1988; Schittny et al 1988). Electron microscopy of insects shows similar thin, amorphous layers. The basement membrane collagen IV forms a covalently linked network within this layer, which is usually not fibrous, and has a complex topology (Yurchenco & Ruben 1987) that may vary between basement membranes. Laminin molecules are extracted from basement membranes with one molecule of entactin bound to a specific region (Paulsson et al 1987), and laminin molecules self-aggregate in the presence of Ca^{++} (Yurchenco et al 1985; Paulsson 1988). Sites for mutual interactions of laminin, basement membrane proteoglycan, and collagen IV have been mapped on these molecules, which also interact with receptors on cell surfaces.

COMPONENTS OF *DROSOPHILA* EXTRACELLULAR MATRIX

Collagen IV

One polypeptide folding motif, the collagen triple helix, is shared by the collagen matrix molecules and some proteins with quite different functions, such as lung surfactants. Searches of *Drosophila* genomic DNA libraries with probes coding for chicken collagen I detected about ten *Drosophila* genes that potentially code for collagen motifs (Monson et al 1982; Natzle et al 1982; Le Parco et al 1986a). Le Parco et al reported nine unique locations over the genome for these potential collagen genes, which were identified by in situ hybridization to polytene chromosomes. Transcripts of 0.7–6.3 kb appeared differentially during the *Drosophila* life cycle.

The transcribed portion of one *Drosophila* collagen gene, and its associated introns, has been completely sequenced, and the corresponding protein has been characterized (Blumberg et al 1988; Lunstrum et al 1988). This gene corresponds to clone DCg1 (Natzle et al 1982; Le Parco et al 1986a) at chromosome locus 25C. Strong sequence similarity of the carboxyl domain of this protein with the corresponding parts of mouse and human type IV collagen chains identified it as a basement membrane collagen IV (Blumberg et al 1987; Cecchini et al 1987). The three polypeptides of a triple helical collagen molecule may be derived from one, two, or three genes. Major forms of the human and mouse basement membrane collagens IV are heterotrimers, $[(\alpha 1)_2 \alpha 2]$IV (Glanville 1987). This form was suggested for a collagen that was extracted from 2-aminopropionitrile-fed *Drosophila* larvae (Murray & Leipzig 1986). Extracts of normal embryos and larvae and cultures of *Drosophila* cells yield, after reduction, primarily one 195 kd collagenous peptide, together with variable, small amounts of another collagen chain that migrate electrophoretically slightly more slowly (Lunstrum et al 1988). The peptide maps of these chains appeared identical. The chains may differ in minor modifications or represent differently spliced transcripts of one gene. Electron microscopy, circular dichroism, and sedimentation studies proved the triple helical form of the collagen in samples that contained considerably less than 33% of the minor polypeptide. Therefore the major component must be able to form homotrimeric molecules, independently of whether or not a second *Drosophila* basement membrane collagen gene exists.

The electron microscopic appearance of *Drosophila* collagen IV and its amino acid sequence have been correlated, and both indicate domains homologous to human and mouse collagens IV. The collagen thread is about 440 nm long and has a globule at one end. The globule is made up of the COOH domains of the three component chains. The amino acid sequence of each 230 residue domain has 58% or more identity with the corresponding NC1 domains of the human $\alpha 1$(IV) and mouse $\alpha 2$(IV) chains. As in these vertebrate sequences, there is a major internal duplication, and the positions of all 12 cys residues are conserved. Proteins with similar three-dimensional structure have high hydrophobic correlation coefficients (Sweet & Eisenberg 1983). The hydrophobic correlation coefficients of this *Drosophila* sequence with the corresponding sequences of mouse $\alpha 1$(IV) and $\alpha 2$(IV) are nearly as great as the hydrophobic correlation between the two mouse sequences. The NC1 peptide regions from the two mouse chains together make up the heterotrimeric globule of mouse collagen IV (Glanville 1987). One may expect that the principal function of the vertebrate collagen IV globules to pairwise join collagen IV molecules end-to-end has been maintained.

Drosophila collagen IV molecules, like their vertebrate homologues, can form disulfide-bonded junctions by overlap of the NH_2 ends of their collagen threads, in both parallel and antiparallel alignments (reviewed in Glanville 1987). However, there are differences. These suggest that the vertebrate junction, which consists of four collagen threads that overlap by about 30 nm, is a specialized form of a more general overlap, or segment junction, bounded by disulfide links between the molecules. The amino acid sequence of this part of the *Drosophila* collagen IV chain has better homologies with its vertebrate counterparts than over other regions of the collagen thread, but they are not strong. However, the overlap region of vertebrate collagen IV, also called the 7S region, is duplicated in *Drosophila* and correspondingly "double-length" overlaps may occur. Mouse collagen IV molecules strongly prefer to self-assemble into tetrameric overlap junctions, but *Drosophila* collagen IV molecules form dimers and, in decreasing amounts, higher forms. This is consistent with the homotrimeric helix model of *Drosophila* collagen IV, as association of the four threads (4×3 chains) requires that each has an asymmetrically placed hydrophobic reaction edge, and this can only arise in a heterotrimeric collagen (Siebold et al 1987).

A slightly modulated, but clearly homologous form of the double junctional region found near the NH_2 end of the *Drosophila* sequence is repeated at one-third of the thread length towards the COOH globule. This domain is also bounded by cys residues. Equivalent cys residues also occur in human and mouse IV chains, though to varying extents (Figure 1). Various side-by-side arrangements of *Drosophila* collagen IV molecules should be possible in which adjacent molecules overlap by one-third of their length and are fastened to each other by segment junctions. This could give rise to a variety of microfibrils or networks held together by disulfide-bonded segment junctions (Figure 2).

A correctly formed collagen triple helix requires that every third residue along each chain is glycine. The collagenous sequences of human, mouse, and *Drosophila* collagen IV have about 20 sites each where this rule is violated. The relative locations of about half of these imperfections are remarkably conserved between the vertebrate and *Drosophila* sequences (Figure 1). In contrast, neither the length of an imperfection, which can

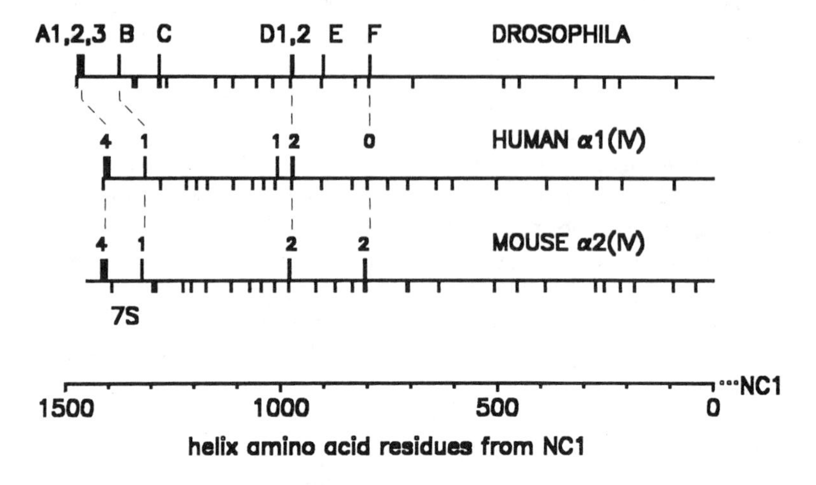

Figure 1 Relative locations of cysteine residues and imperfections of helix in *Drosophila* α1(IV), human α1(IV) and mouse α2(IV) chains. The collagen thread portions of the three chains are diagrammed as horizontal lines, aligned at their thread/NC1 domain junctions. Each junction is taken as origin for the coordinate of residue numbers. The positions of cysteine residues are indicated by the longer vertical bars, above each horizontal line. The alphanumeric designation of each thread cysteine residue of *Drosophila* α1(IV) is shown. The small numbers associated with the human α1(IV) and mouse α2(IV) chains give the number of cysteine residues in clusters that cannot be resolved at the scale of this diagram (or the lack of a cysteine residue). Dashed lines indicate suggested correspondence of locations of cysteine residues between different species. The 7S region of the vertebrate chains is indicated. The two cysteine residues indicated for mouse α2(IV) at approximately coordinate 800 are taken to form a disulfide link. The corresponding loop-out of the polypeptide between them is not shown. This is about 20 amino acids long and is allowed for in the placement of all residues of this mouse chain at the amino end. Imperfections of helical sequence are indicated by short vertical lines below the horizontal lines (with permission, *J. Biol. Chem.* 263: 18328–37).

vary from 1 to 18 residues, nor the nature of the amino acids, which make up the imperfection, seem to be conserved. Statistical analysis of the disposition of *Drosophila* collagen IV threads in eletron micrographs shows that there are locations of increased flexibility along the thread. These regions tend to correspond to clusters of sequence imperfections. Intricate topologies of the collagen IV scaffold probably require flexing of collagen threads (Yurchenco & Ruben 1987) and the imperfections may make this possible. However, the biological functions of these imperfections are presently not known.

The nucleotide sequence that codes for the 1775 amino acids of *Droso-*

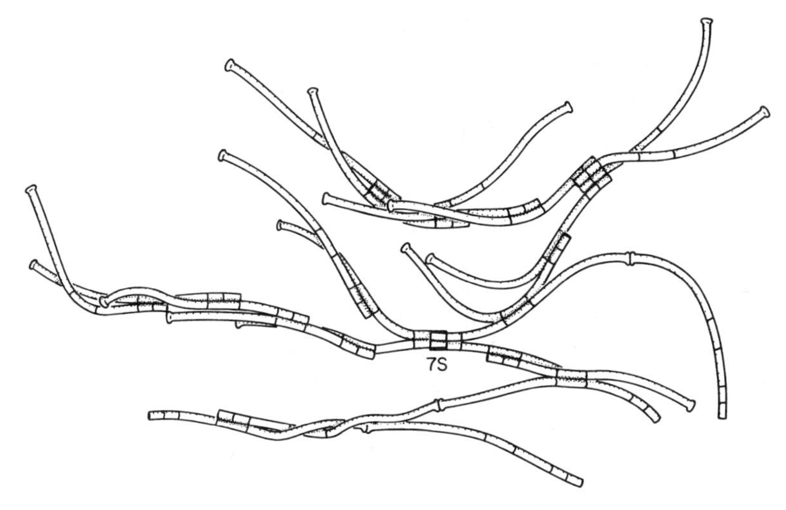

Figure 2 Diagram to illustrate the variety of possible arrangements of *Drosophila* collagen IV molecules that may be stabilized by segment junctions. Each three-stranded molecule is drawn as a flexible cylinder with a protrusion at one end to denote the carboxyl (NC1) domain. The other end of each cylinder is at the A1 cysteine residue (see Figure 1 for locations of cysteine residues). The length of each cylinder, without protrusion, denotes the length of the collagen thread portion of each molecule. Ring marks around each cylinder, from the non-knob end, denote the locations of cysteine residues B, C, D and F. Cysteine residues A1, 2, and 3 are assumed to be at the amino end. Cysteine residue E is not indicated. Segment junctions between both parallel and antiparallel arrangements of intervals of the type (A-B), (A-C) and (D-F) are illustrated. A tetrameric junction of the vertebrate 7S type is indicated, and to the left of it a portion of a potential microfibril. The complementary antiparallel portion of this microfibril has been omitted for the sake of clarity (with permission, *J. Biol. Chem.* 263: 18328–37).

phila collagen IV is interrupted by seven small introns and is preceded by an untranslated leader that contains one more intron. The exons that code for the 1453 amino acids of the *Drosophila* collagen IV thread do not obey any of the rules proposed for collagen exons. The collagenous sequence is preceded by a signal peptide and by approximately 50 additional amino acids. The sequence RGDT, which has been shown to be a cell binding domain in vertebrate collagen I (Dedhar et al 1985), also occurs in the collagen sequence of *Drosophila* and human collagen IV chains, but in different locations. The *Drosophila* collagen IV sequence also has another potential cell binding site, RGDS.

Laminin

Drosophila laminin was isolated from the media of Kc cell cultures and from embryos (Fessler et al 1987; Montell & Goodman 1988). Its electron microscopic appearance is similar to canonical, vertebrate laminin (Martin & Timpl 1987): a cross with arms of 69, 36 nm, and two arms of 30 nm. Each arm has two globules towards its end (Figure 3). The closely similar velocity sedimentation properties of *Drosophila* and mouse laminins indicate similar molecular flexibilities. *Drosophila* laminin is a weakly sulfated glycoprotein that binds to heparin-Sepharose, but only has some of the lectin affinities of vertebrate laminins. Most antibodies raised against mouse laminin react only poorly with *Drosophila* laminin. It is a disulfide-linked molecule consisting of three chains, A, B1, and B2, with apparent molecular masses of 400, 215, and 185 kd, respectively. Clones for all three chains have been obtained from cDNA expression libraries (Montell & Goodman 1988; K. Garrison et al, manuscript in preparation), and the corresponding genes were located by in situ hybridization on polytene chromosomes at 65A10-11, 28D, and 67C, respectively.

The sequences of the B1 chain (Montell & Goodman 1988) and the B2 chain (Chi & Hui 1988; Chi & Hui 1989; K. Garrison et al, manuscript in preparation) have been determined and mapped onto the corresponding sequences of mouse and human laminin chains. The functional domains of mouse and human laminin have been related to its three chains, which have been completely sequenced (Barlow et al 1984; Pikkarainen et al 1987; Sasaki & Yamada 1987; Sasaki et al 1987; Durkin et al 1988a, Pikkarainen et al 1988; Sasaki et al 1988; Timpl 1989). The long arm of mouse laminin contains all three chains in parallel orientation, each mostly folded as an α helix, with the three helices winding around each other as

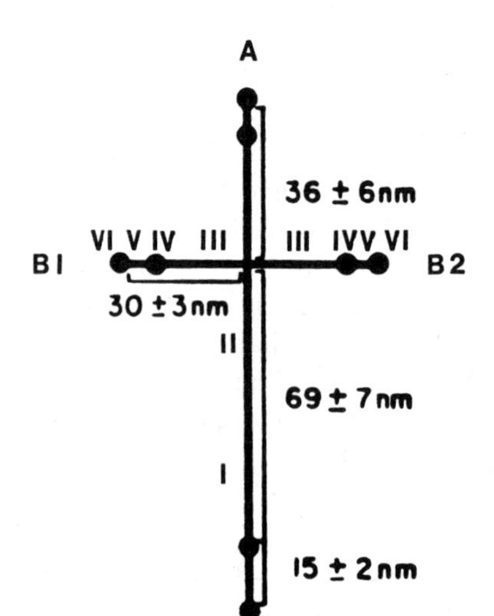

Figure 3 Diagram of the domains of *Drosophila* laminin. This diagram is based on the measurements of electron micrographs of *Drosophila* laminin. The domains of the laminin B1 and B2 chains are numbered in Roman numerals.

a coiled-coil. The COOH ends of the B1 and B2 chains are at the peripheral end of the thread portion of the long arm. The terminal globule of this arm is entirely made up of the COOH domain of the A chain. A neurite outgrowth promoting domain is in the peripheral portion of the long arm at or near the terminal globule (Timpl 1989). *Drosophila* laminin also promotes the differentiation and outgrowth of embryonic neurons in cell culture (unpublished observations in this laboratory). The neurite growth promoting fragment of vertebrate laminin has a high affinity for cells, which is not blocked by RGD peptides, and also promotes the growth of nonneuronal cells (Dillner et al 1988; Timpl 1989).

The predominantly α helix-like sequence of the B1 chain along the long arm of laminin is interrupted by a stretch of about 30 residues that are rich in gly and cys. This α domain is also present in the *Drosophila* B1 chain, but is absent from both mouse and *Drosophila* B2 chains, and from the mouse A chain. The α domain has 36% homology between *Drosophila* and mouse, which is higher than the 20–25% homologies between the remaining stretches, domains I and II, found along the long arm of both B1 and B2 chains. The amino acid sequences of domains I, and to a lesser extent of domains II, show characteristic heptad repeats that will cause residues with hydrophobic side chains to lie next to each other in the postulated three-chain coiled-coil. This is the main common characteristic of this portion of the two B chains in mouse and *Drosophila*. Several proline residues along this portion of the mouse A chain presumably interfere with continuous α helical folding. Vertebrate laminin molecules devoid of the normal A chain are secreted by some cultured cells and have neurite promoting activity (Martin & Timpl 1987; Edgar et al 1988). At this time it is unclear what the relative contributions of the A and B chains are to the neurite growth promoting activity. An adhesive protein concentrated in synaptic clefts of rat neuromuscular junctions has been named S-laminin. This chain has 40% homology with the laminin B1 chains of both mouse and *Drosophila* (Hunter et al 1989).

The COOH end-regions of the B1 and B2 chains are probably disulfide-linked to each other. At the central end of domain II all three chains are most likely held together by disulfide links. Then the chains diverge, each to form a short arm. Successive sequence domains of each B chain are denoted as III, IV, V, and VI. The globules of the short arms correspond to domains IV and VI. The intervening thread portions are made up of domains III and V. The *Drosophila* B chains show the same structural motifs. The key feature of the thread portions of domains III and V are homologous repeats of 45–60 amino acids, with eight cys residues in each, which are homologous with epidermal growth factor (EGF)-related proteins. The multiple repeats are distributed and modulated in corresponding ways along the III and V domains in mouse and *Drosophila*, with homologies of 50–55% between the two species, and clusters of identical amino acids. The intervening globular domain IV has only about half the homology between these species. Proteolytic fragmentation of laminin shows that a cell binding domain is associated with the central region of the laminin cross. A putative binding sequence, YIGSR, was identified in one of the EGF-like repeats of domain III of the mouse B1 chain by competition with synthetic peptides (Graf et al 1987). While this sequence is absent from the mouse B2 chain, the variant YSGSR occurs in the *Drosophila* B1 chain and YFGSR in the *Drosophila* B2 chain, in corresponding locations. The YIGSR sequence is probably recognized by a 67 kd laminin receptor (Martin & Timpl 1987; Timpl 1989). The NH_2 part of the A chain, which makes up the third short arm of mouse laminin, also has the EGF-like type III and V domains. Its type III domain is split into domains IIIa and IIIb by a small globular region not visible in the electron microscope. Domain IIIb contains an RGD sequence as another, and different, potential cell binding site. Twenty-two out of about 50 amino acid positions are highly conserved in the repeat motif of domains III and V of the A chain, and are of the same type as occurs in the mouse B1 and B2 chains.

Sequence homologies were found between the short arm of the mouse A chain and parts of the basement membrane heparan sulfate proteoglycan core protein. The homology covers both the EGF-like regions and their adjacent, inner globular domains (Sasaki et al 1988). This indicates a hierarchy of conserved elements: EGF-like motifs, sequences of these EGF-like motifs, and domains comprised of multiple EGF-like sequences together with adjacent polypeptide regions. The strong conservation of these regions in the vertebrate laminin A chain and the basement membrane proteoglycan suggest not only evolutionary relatedness but also some common biological functions. The terminal domain VI globules of the three short arms of mouse laminin share significant homologies and this domain may function as a collagen binding site (Martin & Timpl 1987). This domain has high sequence similarity in mouse and *Drosophila* B chains. One molecule of entactin is strongly bound to a central region of vertebrate laminin and appears like an additional arm in electron micrographs (Martin & Timpl 1987). It is not yet clear whether laminin from *Drosophila* cell cultures has a small amount of entactin attached to it, and in laminin extracted from *Drosophila* it could not yet be shown that it was associated with entactin (L. I. Fessler, unpublished observation). Extracts of sea urchin, leech, and hydra all indicate molecules with the characteristic cross-shape of laminin, some with an additional arm (McCarthy et al 1987; Beck et al 1989).

Glutactin

Glutactin is a new acidic sulfated glycoprotein of basement membranes that was purified from the media of cultured *Drosophila* Kc cells (Olson et al 1987; P. F. Olson et al, submitted for publication). Immunofluorescence microscopy of embryos located glutactin to basement membranes, demonstrated the segmentally invaginated envelope of the central nervous system, and showed mesectodermal cells that send processes to the junction of muscle apodemes at segment boundaries. Glutactin cDNA hybridized to chromosomal locus 29D. The nucleic acid sequence codes for a 1024-residue-long polypeptide that has a signal peptide, a major amino domain that is interrupted by one intron, and a highly acidic carboxyl domain. Much of the 600 residue long amino domain has homology significant to better than 20 standard deviation units with known acetyl- and butyrylcholine esterases, rat lysophospholipase, and *Drosophila* esterase-6, but lacks the catalytically critical serine residue of these serine esterases. This paradoxical homology also includes a region of thyroglobulin precursors, which also lack the critical serine residue. Presumably a substantial domain has been conserved, but no significant enzymatic activity has yet been found. The amino and carboxyl domains of glutactin are separated by 13 contiguous threonine residues. The carboxyl domain has an excess of 52 acidic residues, and glutamic acid together with glutamine account for 44% of its amino acids, which are arranged in several repeating patterns. There are four O-sulfated tyrosines. Glutactin preferentially binds Ca^{++} in the presence of excess Mg^{++}. Our previous mention of glutactin was as a putative *Drosophila* entactin because of similar electrophoretic mobility, tyrosine sulfation, and location in tissues with mammalian entactin, but sequence comparison now shows them to be different proteins (Durkin et al 1988b, P. F. Olson et al, submitted for publication).

Fibronectin

A putative *Drosophila* fibronectin was found in fly hemolymph (Gratecos et al 1988). This glycoprotein was identified by immunoprecipitation with antibodies made against human fibronectin and it bound to gelatin and heparin. Electrophoresis and immunoblotting of the reduced protein showed a single 230 kd peptide. There may be cellular, and circulating forms of fibronectin and SDS extracts of whole flies yielded much more fibronectin than hemolymph.

A potential role for fibronectin in *Drosophila* embryogenesis was indicated by injection of antibodies and by cell culture experiments (Gratecos et al 1988). Embryos injected with the total IgG fraction of an antihuman fibronectin antiserum produced two abnormal phenotypes. Either embryos did not make a blastoderm and died, or a normal blastoderm arose but gastrulation failed and only multiple ectodermal folds developed. Injection of RGD peptides also caused the latter phenotype (Naidet et al 1987). Early *Drosophila* gastrulae can be dissociated into undifferentiated cells. Fetal calf serum enables these embryonic cells to grow and differentiate in vitro (Seecof et al 1971; Seecof et al 1972; Gerson et al 1976a; Gerson et al 1976b; Furst & Mahowald 1985). Cell attachment is prevented if the calf serum has been depleted of fibronectin (Gratecos et al 1988). Addition of human fibronection to the depleted serum restores cell spreading, and human fibronectin alone also functions, though less effectively. Addition of RGDS peptide to the calf serum causes rounding-up of spread *Drosophila* cells and prevents their differentiation. These results suggest that the occurrence and interaction of fibronectins with an integrin receptor may be similar in *Drosophila* and vertebrates. Different vertebrate fibronectins arise by differential splicing and this influences their tissue distribution and function (Hynes 1985; Kornblihtt & Gutman 1988). An equivalent variety of *Drosophila* fibronectins and their receptors might exist and have subtle effects on morphogenetic processes.

Proteoglycans

Basement membranes of *Drosophila* stain with alcian blue and ruthinium red (Rizki & Rizki 1984), reagents that recognize polyanions such as sulfated proteoglycans. In vertebrates, heparan sulfate proteoglycans occur as components of cell membranes and as constituents of basement membranes (Ruoslahti 1988b). A sulfated proteoglycan-like glycoprotein called papilin, has been located in *Drosophila* basement membranes (Campbell et al 1987). A chondroitin sulfate proteoglycan is present in imaginal discs (Brower et al 1987). The 185-kd glial cell associated glycoprotein 5B12 of crickets has an epitope that is sensitive to chondroitinase ABC and is also found in basal lamina during early development (Meyer et al 1988). The gene product of the *per* locus of *Drosophila* has sequence similarity with chondroitin sulfate proteoglycan core proteins and has been reported to be a heparan sulfate proteoglycan (Jackson et al 1986; Reddy et al 1986).

Variants of *Drosophila* Kc cells secrete papilin or its approximately 400 kd core protein (Campbell et al 1987). The unique electron microscopic appearance of papilin is a 225+/−15 nm thread which is disulfide-linked into a loop with fine protruding threads (Figure 4). Oligomers are formed by disulfide linking of several loops. The mass of total carbohydrate exceeds that of the polypeptide and consists primarily of glucosamine, galactosamine, uronic acids, and neutral sugars in the approximate ratios

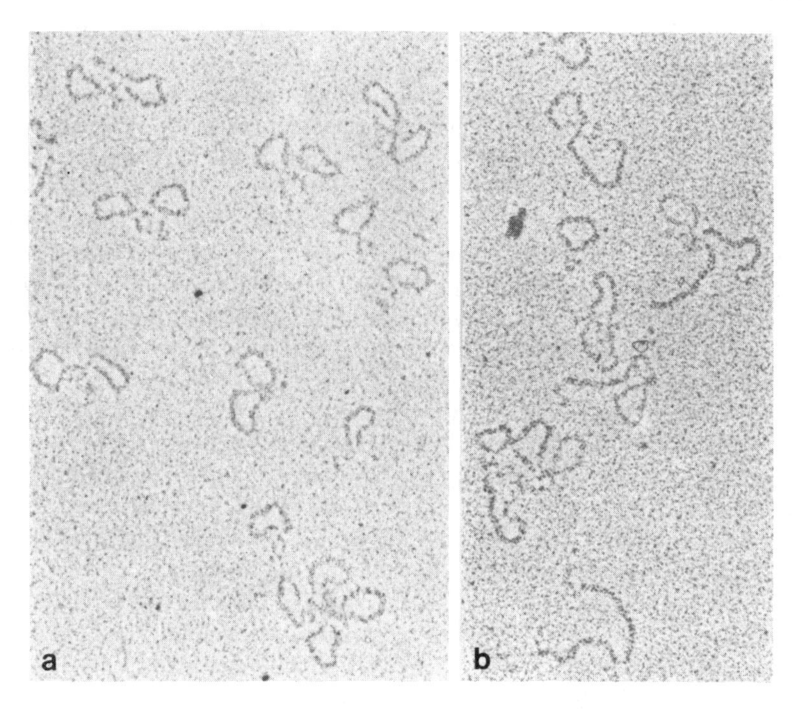

Figure 4 Electron micrographs of papilin. Purified papilin was adsorbed onto pentylamine-treated grids, stained with uranyl acetate and rotary shadowed. Many of the molecules are disulfide-linked oligomers of 225 ± 15 nm long loops closed by disulfide-bonds (*a*). Some molecules are in a linear form (*b*), and reduction converts most of them to 225-nm long threads (with permission, *J. Biol. Chem.* 262: 17605–12).

2 : 1 : 2 : 1. Approximately 80 sulfated side chains are O-linked through a neutral sugar.

A monoclonal antibody, which specifically labelled discs of late third instar larvae, was used to purify an approximately 500-kd sulfated proteoglycan (Brower et al 1987). Chondroitinase ABC facilitated extraction, and further incubation with this enzyme slightly decreased the apparent size of the products. The antigen was localized in specific regions of the epithelium as a network between the cells and the basement membrane. Brower et al postulate that this may serve as an extensible matrix during evagination of the disc. Subsequently, the antigen occurs in the matrix between the dorsal and ventral surfaces of the evaginated wing pouch.

DROSOPHILA INTEGRINS

Integrins are a family of transmembrane proteins that bind various extracellular ligands. Vertebrate and *Drosophila* integrins are implicated in cell interactions during differentiation and development. Each integrin receptor is a heterodimeric complex of an α and a β integrin subunit stabilized by Ca^{++} ions. In vertebrates at least four different β chains and about ten α subunits are distinguished (Buck & Horwitz 1987; Hynes 1987; Ruoslahti 1988a). Antibodies to integrins and synthetic peptides of the form RGD inhibit the binding of various vertebrate matrix molecules to integrins on the surface of cultured cells or lipid vesicles (Ruoslahti & Pierschbacher 1987). They can also interfere with cell attachment, spreading and migration, neurite outgrowth and differentiation (Buck & Horwitz 1987). Similar methodology demonstrated in vivo integrin-ligand interactions in the migration of neural crest cells (Bronner-Fraser 1986) and in amphibian gastrulation (Boucaut et al 1984).

Drosophila integrins were found during a search for cell-surface antigens that vary during development (Wilcox et al 1981; Brower et al 1984; Brower et al 1985) and in a quest for understanding a mutation that interferes with basement membrane function (MacKrell et al 1988). The cell-surface antigens were called position specific antigens: PS1, PS2, and PS3. Immunoprecipitation with monoclonal antibodies specific for one polypeptide antigen coisolated a second antigen. Electrophoretic analyses showed peptide pairs with the characteristics of the α and β subunits of integrins (Brower et al 1984; Wilcox et al 1984; Wilcox & Leptin 1985; Leptin et al 1987). PS1 and PS2 correspond to distinct integrin α subunits, and PS3 is a common integrin β chain.

The amino acid sequence of PS2 was determined from its cDNA and is homologous to vertebrate integrin α subunits (Bogaert et al 1987). The polypeptide of 1394 amino acids has a signal peptide and a potential transmembrane domain towards its COOH end. The glycosylated form of this chain migrates as a 160-kd peptide. Cleavage produces a 140-kd peptide disulfide linked to a 25-kd peptide. Many integrin α subunits are cleaved into a heavy and a light chain, which are disulfide linked. The cut is in an extracellular region. The heavy chain represents most of the extracellular domain and the light chain contains the transmembrane and cytoplasmic domains. The PS2-α heavy chain has about 35% homology with those of the human fibronectin, vitronectin, and IIb/IIIa platelet receptors, and cys residues and Ca^{++} binding domains are conserved. The PS2-α heavy chain differs from these vertebrate homologues near its COOH end, with an additional sequence of about 300 amino acids containing many ser residues. Presumably this exposed sequence undergoes proteolytic degradation and accounts for a smaller 125-kd PS2-α heavy chain. During extraction it was difficult to protect the *Drosophila* integrins from proteolytic modifications, and initially a greater number of electrophoretic bands was observed. The PS2-α light chain has 25–28% amino acid identity with its vertebrate equivalents, especially in the transmembrane and cytoplasmic domains. Overall, the homology of the *Drosophila* PS2-α subunit with different vertebrate integrin α subunits is almost as extensive as the homology of the vertebrate α subunits to one another. A minor variant of the *Drosophila* PS2-α subunit, differing by about 25 amino acids, may exist. The gene for PS2-α is at locus 15A on the X chromosome. The PS1-α chain may have a 145-kd precursor; it is found predominantly as a 116-kd heavy chain and its sequence is not yet determined (Leptin et al 1989).

The formation of basement membranes and attachment of muscle cells is perturbed in embryos carrying the lethal (1) myospheroid mutation [*l(1)mys*] (Wright 1960). The normal *mys* gene was cloned (Digan et al 1986) and its cDNA sequence codes for a polypeptide homologous with vertebrate integrin β chains (MacKrell et al 1988). In *mys* mutants the expression of the PS3 antigen is blocked and the protein is missing (Leptin et al 1989). The cDNA codes for a 90-kd peptide with a signal sequence and has 45% amino acid sequence identity with the integrin β chains of the human and chicken fibronectin receptors overall. Its 23 residue transmembrane domain is 60% identical with that of chicken integrin β chain. The 47 amino acid long cytoplasmic domain is 88% identical over its first 33 amino acids with chicken integrin β, and 73% identical over its whole length with platelet protein IIIa. At a conserved site is a tyrosine, which may be a substrate for protein kinases (Tamkun et al 1986), and phosphorylation of this site may have a regulatory function (Hirst et al 1986). Thus this cytoplasmic domain has more than adequate similarity to several vertebrate homologues and suggests similar functions and possibly interaction with the cytoskeleton. Antibodies made against a synthetic peptide of this part of the chick $\beta 1$ chain have cross-reacted with the *Drosophila*, human, and *Xenopus* homologues (DeSimone & Hynes 1988; Marcantonio & Hynes 1988). At the NH_2-end is a region of 100 amino acids of moderate similarity, followed by a serine-rich segment not found in vertebrate homologues. This segment is followed by a domain of 265 amino acids that has high homology with chicken integrin β, human platelet protein IIIa, and the β subunit of human leukocyte adhesion protein. The first block of 21 residues of this domain is about 80% identical. It is this region of vertebrate chains to which RGD peptides can be crosslinked after interaction with integrin receptors (D'Souza et al 1988). There are several other blocks of high similarity in this domain, with 80–95% identities over 16–21 consecutive amino acid residues. Another noteworthy domain of about 165 amino acids contains 30 cys residues in a conserved and characteristic motif. Both the number and relative positions of the 56 cys residues of the processed *Drosophila* chain are the same as in the vertebrate integrin $\beta 1$ subunits. All integrin β chains have a highly folded, disulfide-linked structure, which causes a characteristic decrease of electrophoretic mobility upon reduction of the disulfide linkages. This protein shows the same domains as its vertebrate homologue, and its blocks of amino acid identities suggest several conserved interactive sites, as exemplified by the region to which RGD peptide has been linked in vertebrate forms. The chromosome locus of this integrin β chain is 7D1-5.

EXPRESSION OF MATRIX PROTEINS AND THEIR RECEPTORS

Histochemical and cytological studies of *Drosophila* demonstrate that basement membranes and extracellular matrix surround internal organs (e.g. Rizki & Rizki 1983, 1984) and that during metamorphosis dissolution and resynthesis occur (Fristrom & Rickoll 1982; Fristrom 1989). The synthesis of this extracellular matrix can now be followed with specific antibodies and cDNA probes.

Embryos, 0–2 h old, contain small amounts of extracellular matrix proteins and the oocyte may receive a small maternal contribution of fibronectin (Gratecos et al 1987), collagen IV, laminin, glutactin (K. Garrison et al, manuscript in preparation), and integrin β chains (Leptin et al 1989). These can be detected in extracts of pregastrula embryos by Western blotting. This finding is supported by the observation that cDNA probes for collagen IV hybridize strongly in the ovary (Le Parco et al 1986b). However, discrete localization of these proteins by immunostaining is not observed until later in development. After gastrulation, greatly increased zygotic synthesis of these materials occurs in the embryo. Coordinate transcription of the three laminin genes peaks at about 6–12 h (Montell & Goodman 1988) and the maximal collagen IV message level occurs at 12–15 h according to Northern blots (Natzle et al 1982) and in situ hybridization (Mirre et al 1988). Peak synthesis of laminin (10–12 h), and of collagen IV and glutactin (12–15 h), determined by Western blotting, follows the transcription pattern (K. Garrison et al, manuscript in preparation). The peak synthesis of total embryo laminin precedes that of collagen IV and glutactin (Figure 5). In *Drosophila* we have not seen the very early appearance of only laminin B chains that has been reported in early vertebrate embryos (Ekblom et al 1986). The later appearance of the laminin A chain during renal development and its critical role in the polarization of differentiated epithelium are noteworthy (Klein et al 1988). Immunostaining of whole amounts (Figure 6) or sections of embryos with antibodies to laminin (Fessler et al 1987; K. Garrison et al, manuscript in

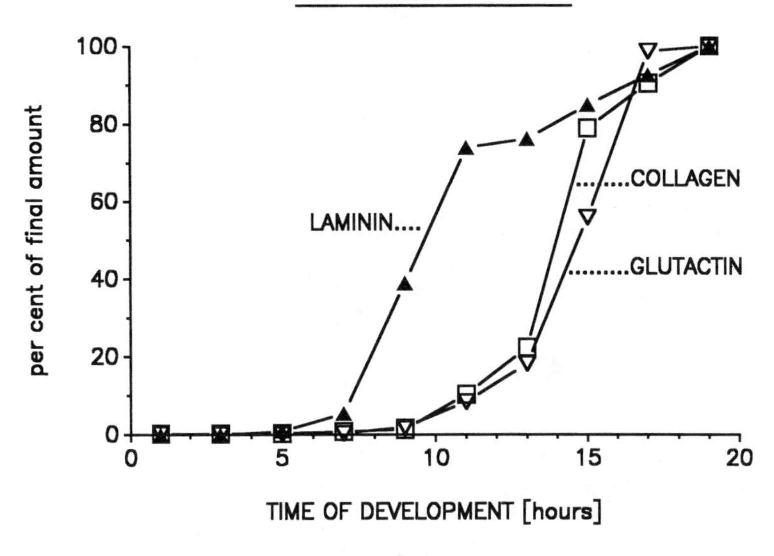

Figure 5 Accumulation of laminin, collagen IV, and glutactin during embryogenesis. Embryos were solubilized in SDS (sodium dodecyl sulfate), and after electrophoretic separation the total quantities of laminin, collagen IV, and glutactin were determined by quantitative Western immunoblot analyses. Larvae hatched at 22 h.

preparation) is first seen after gastrulation at the time of full germ band extension. A slightly segmented pattern of laminin is seen adjacent to the germ band and in the amnioserosa and gut primordia. Later there is strong staining surrounding the internal organs, e.g. gut, Malpighian tubules, and spiracles. At the site of dorsal closure the matrix adjacent to epithelia stains intensely. The nerve cord and brain are covered by a basement membrane, and as motor nerves exit from the nerve cord they are sheathed with a basement membrane. As sensory organs form, there is initial staining only for laminin. Subsequently the sensory bodies stain strongly. Pairs of mesectodermal cells (Beer et al 1987) adhering to the dorsal surface of the nerve cord at the eleven segmental borders stain (P. F. Olson et al, submitted for publication) (Figure 7). These dorsal median cells are anchored in a channel lined by basement membrane that transects the nerve cord. In a lateral view these channels are clearly seen as dorsoventral bands that traverse the nerve cord. In the abdominal segments these mesectodermal cells produce thin processes that connect to the segmental insertions, apodemes, of the lateral body wall muscles. Since these cells

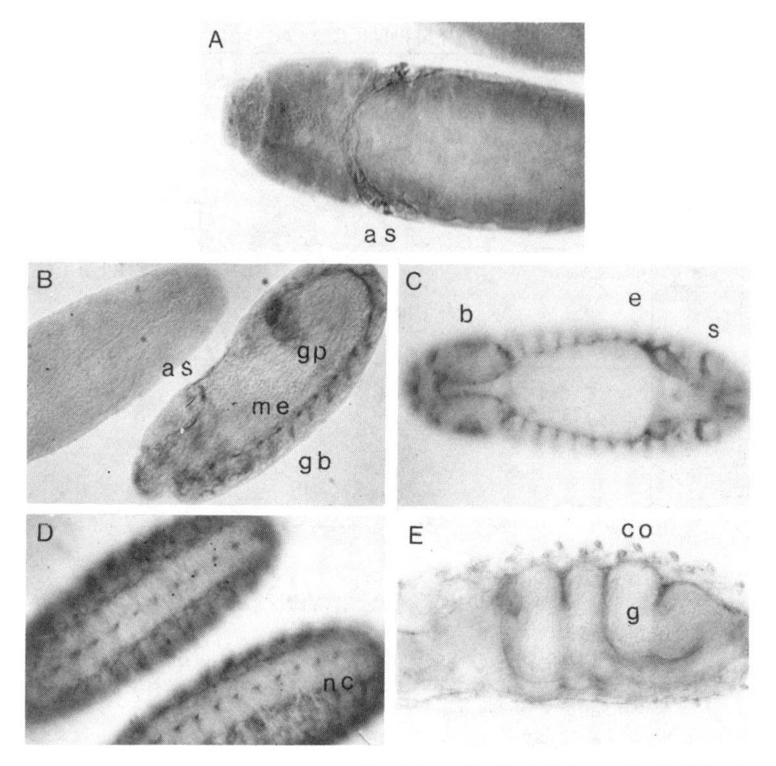

Figure 6 Whole embryos stained with anti-laminin antibodies. Embryos that had developed for 4 to 14 h were immunostained with anti-*Drosophila* laminin antibodies and a second antibody conjugated to horseradish peroxidase. The abbreviations are: as = amnioserosa, b = brain, co = chordotonal organ, e = epidermal layer, gb = germ band, gp = hind gut primordia, g = gut, me = mesoderm or mesectoderm, nc = nerve cord, s = spiracle.

are at each segment boundary, they serve as good segmental markers (Wirz et al 1986). With some variation, the above overall staining pattern for laminin is also seen with antibodies to collagen IV and glutactin (P. F. Olson et al, submitted for publication).

Antibodies raised against disulfide-linked laminin detect the above pattern of laminin that is deposited in extracellular matrix. In contrast, an antibody made against reduced B1 laminin chains primarily detected intracellular laminin, for example, in the cytoplasm of thin cell sheets at the periphery of the gut. However, this antibody most prominently stained some individual hemocytes that are scattered around the gut, and both peripheral to and alongside the nerve cord and brain. These cells were double-stained with antilaminin B1 and collagen IV antibodies (Lunstrum

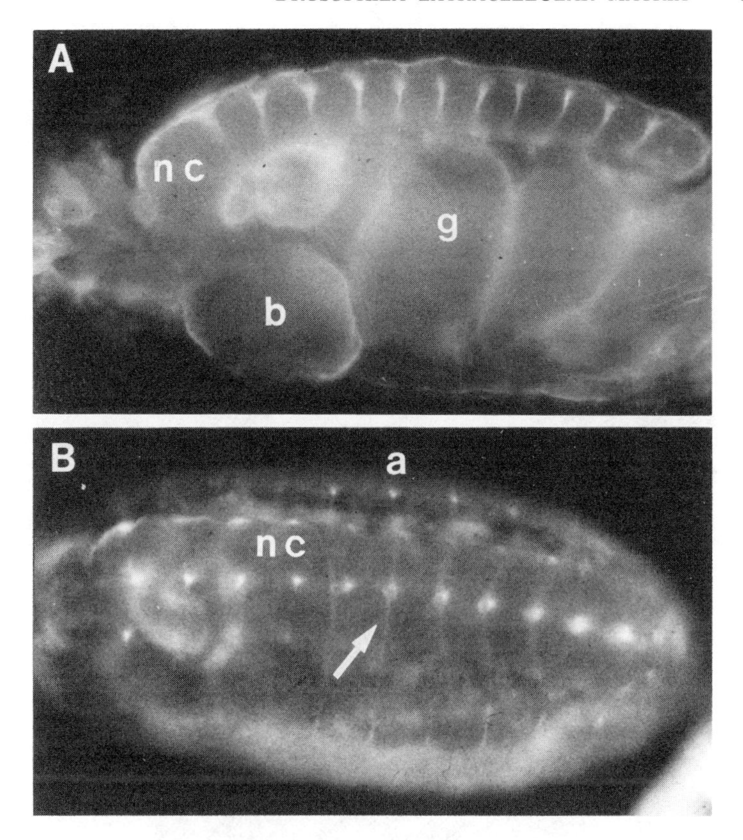

Figure 7 Whole embryos stained with anti-glutactin antibodies. Fluoresceine-labeled second antibody shows the location of glutactin in the envelopes of the nerve cord (nc), brain (b), and gut (g) of a whole-mounted embryo seen in a lateral view with the ventral region at the top (*A*). Segmental channels penetrate the nerve cord. In (*B*) the embryo is rotated through 90° and the focal plane is approximately at the interface of nerve cord (nc) and gut (g). This shows the dorsal-median cells with processes (indicated by an arrow) that extend outward to the apodemes (a). The longitudinal body wall muscles insert into these apodemes, which are at the segmental boundaries.

et al 1988; K. Garrison et al, manuscript in preparation). This suggests that these hemocytes are able to synthesize several basement membrane components.

The accumulation of transcripts for collagen IV is shown in detailed in situ hybridization analyses in embryos by Mirre & associates (1988), and by Lunstrum et al 1988 as shown in Figure 8. In the 10 h embryo, individual cells, hemocytes, were labeled around the gut in regions coinciding with visceral mesoderm where muscles form later, and at segmented intervals

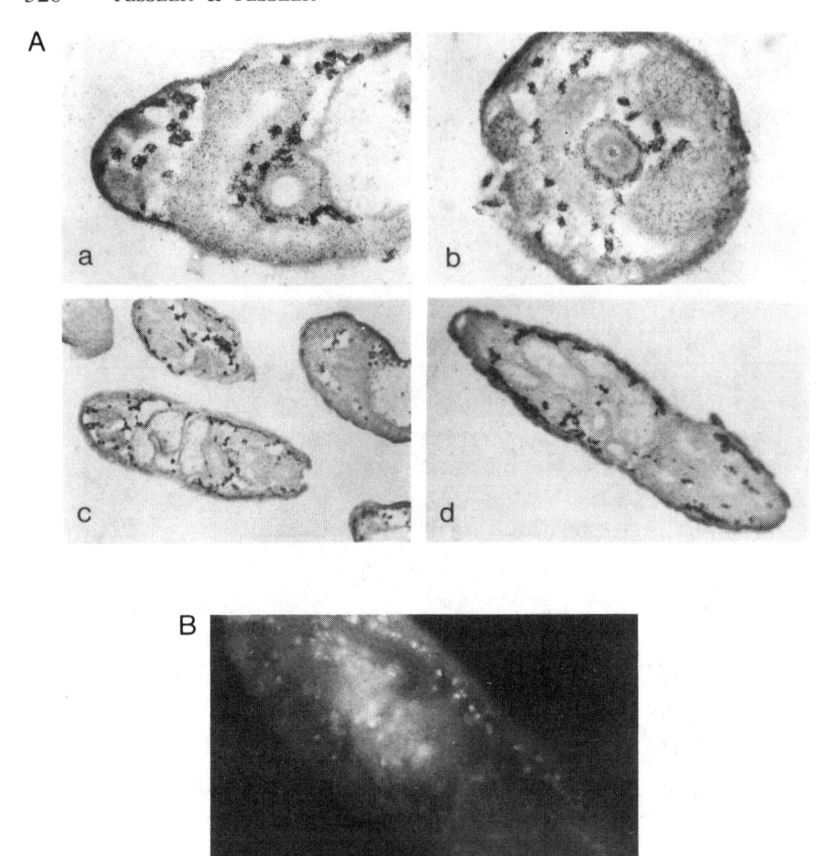

Figure 8 Localization of collagen IV synthesis in *Drosophila* embryos and larvae by (*A*) in situ hybridization of RNA transcripts and (*B*) immunofluorescence staining of cells. (*A*) Shows photomicrographs of sections of 12–18 h embryos (*a–c*) and first instar larva (*d*) that were hybridized with a DNA probe coding for the collagen helical domain. After 4 weeks' exposure the autoradiographs were developed and the sections were Giemsa counter-stained. (*B*) Shows a whole-mount of a permeabilized 12–16 h *Drosophila* embryo stained with antibodies to *Drosophila* collagen IV. The level of focus is adjusted dorsal to the internal organs to show most clearly the hemocytes (with permission, *J. Biol. Chem.* 263: 18318–27).

along the nerve cord. Transcripts were also seen in the mesoderm, which gives rise to the fat bodies. After 12 h of development all these signals were greatly amplified, coinciding with the maximum time of collagen IV synthesis. The circulating hemocytes, which make basement membrane products, may collaborate with other cells in a manner analogous to a situation in vertebrates. Cultured fibroblasts that adhere to vertebrate muscle cells cooperatively contribute to a basement membrane that is formed between them (Kühl et al 1984; Sanderson et al 1986).

During the first and second instar larval stages, collagen IV transcription is maximal (Natzle et al 1982) and is localized to fat bodies and hemocytes (Le Parco et al 1986b). Then collagen IV synthesis diminishes markedly and reappears in the late third instar larval lymph gland, fat body, cells of the hemolymph, and in adepithelial cells of imaginal discs. During early pupation the hemocytes multiply and are seen all over the pupa, invading larval tissues, especially muscles and gut, and also around imaginal discs. These hemocytes strongly hybridize with a probe to collagen IV (Knibiehler et al 1987).

The two known *Drosophila* integrin complexes $\alpha 1/\beta$ and $\alpha 2/\beta$ are widely distributed, in a complementary manner (Bogaert et al 1987; Leptin et al 1989). In the embryo the $\alpha 2/\beta$ integrin is restricted to mesodermal derivatives, somatic and visceral muscles, while the $\alpha 1/\beta$ integrin is located in endodermal and ectodermal derivatives, such as epidermis, gut, and fat bodies, but not in muscles. Both complexes are most prominent at muscle attachment sites at apposing locations; one at the epidermal and the other on the muscle side, which suggests collaboration between them in muscle attachment. Their expression is also complementary in portions of the imaginal wing disc. After evagination these two domains appose each other at the internal interface between the dorsal and ventral wing epithelia. At the earliest embryonic stages the β subunit is detected as diffuse cytoplasmic staining, and its mRNA is generally distributed (L. I. Fessler, unpublished observation). Both the mRNA and the protein may be present in the oocyte. Although mRNA for the $\alpha 2$ chain is present early, synthesis of the α chains commences only at the extended germ band stage, and only after α/β complexes can form are integrin chains found at the cell surface. Conversely, in the absence of the β chain, in *mys* mutants, the α chains are only present in the cytoplasm and not at the cell surface. When maternal contributions of β chain transcripts were prevented by recombination, the *mys* phenotype was more severe and mutants could be distinguished earlier, yet development still proceeded through gastrulation, and delays in germ band retraction were the first defects (Wieschaus & Noell 1986; Leptin et al 1989). However, embryos that lacked a maternal contribution, but received one copy of the wild type *mys*$^{+}$ from their fathers, developed normally. In contrast to these findings, injection of RGDS peptides into embryos prevented normal gastrulation (Naidet et al 1987). It is unlikely that this critical interference involved integrin complexes with the known β chain. Some other RGDS-sensitive receptor, which itself could be another integrin, might be an explanation. It was proposed (Newman & Wright

1981) that there is defective assembly of basement membranes in *mys* embryos, and this now raises the possibility of integrins facilitating an ordered accumulation of extracellular matrix molecules at cell surfaces. Conversely, ligand matrix might induce clustering of integrins.

Hormonal influences on matrix and cell surface protein synthesis were studied in vitro (Rickoll & Fristrom 1983). 20-hydroxyecdysone causes cultured wing discs to evaginate, and this is accompanied by changed expression of cell surface proteins. This agent induces similar changes of protein synthesis in cultured *Drosophila* S3 cells and alters their adhesive properties (Rickoll & Galewsky 1988). Expression of mammalian extracellular matrix synthesis is profoundly influenced by the family of transforming growth factor β (TGF-β) proteins (Sporn et al 1987). The virtual translation product of the 50 kb *Drosophila* gene *decapentaplegic* (*dpp*) has repeats homologous to TGF-β (Padgett et al 1987). The diverse influence of the gene includes effects on embryonic dorso-ventral pattern formation, and the local growth of imaginal discs that affect the resulting adult structures (Hoffmann & Goodman 1987). A portion of *dpp*, which includes TGF-β homologous sequences, can rescue some *dpp* mutants. A corresponding portion of *dpp* has been transfected into cultured *Drosophila* cells. Its controlled expression causes changes in transcription of collagen and integrin genes (F. M. Hoffmann, personal communication).

CELL SURFACE RECEPTORS AND ADHESION MOLECULES

Due to space limitations we can only outline recent characterizations of the following *Drosophila* cell surface molecules. This exciting and important start to unraveling cell-cell interactions requires a review of its own.

EGF Motif Proteins

The sequence motif of epidermal growth factor precursor (Gray et al 1983) occurs in laminin, tenascin (Erickson & Lightner 1988), some other matrix molecules, and in nerve growth factor receptor precursor (Johnson et al 1986), and TGF-α (Derynck et al 1984). It also occurs in the extracellular domains of the *lin-12* homeotic protein of *C. elegans* (Greenwald 1985) and of the *Drosophila* neurogenic proteins *Notch*, *Delta* and *slit* (Wharton et al 1985; Vassin et al 1987; Rothberg et al 1988). The *Notch* locus is involved in cell-cell interaction and is essential for the proper differentiation of *Drosophila* ectoderm (Lehmann et al 1983). It has a stronger homology to mouse laminin than to the other vertebrate proteins containing the EGF-like repeat (Sasaki et al 1988). The extracellular domain of the *Notch* protein contains 36 EGF-like tandem repeats, no two of which are identical. Genetic analysis showed that different functional classes of *Notch* mutations are correlated with amino acid substitutions in different EGF repeats and that the total number of repeats is important (Kelley et al 1987). Thus different EGF-like repeats may subserve different functions in protein-protein interactions. This suggests that different EGF-like domains within a laminin chain could serve different functions, and that the similarities between mouse and *Drosophila* laminin structures reflect closely similar biological functions. It seems less likely that modulations have been maintained solely for structural needs. The mitogenic action of EGF is mimicked by laminin and by a fragment of laminin with EGF motifs. It fails to occur in cells that lack EGF receptors, but EGF and laminin do not compete for receptors (Panayotou et al 1989).

Lethal (2) Giant Larvae

Recessive mutations at the *lethal(2)giant larvae (l(2)gl)* locus of *Drosophila* cause malignant neuroblastomas of the larval brain and tumors of the imaginal discs (Schmidt 1989). Transplantation experiments show that soon after the beginning of embryonic organogenesis the *l(2)gl*-deficient cells are committed to tumorous development. Alternative splicing leads to either a 78-kd core protein, or to a 127-kd protein that has an additional carboxyl tail. Genetic constructs show that the 78-kd core protein alone suffices for tumor suppression, though many animals do not survive pupation (Jacob et al 1987). Immunostaining shows that the 127-kd protein behaves as a cell surface protein. It is intensely expressed in normal animals in regions where, in mutants that lack this protein, tumors form (Klambt & Schmidt 1986). Sequence analysis shows a putative signal peptide, though not at the immediate amino terminus, and several repeated sequence features, particularly in the carboxyl tail. There is extended sequence similarity to the cadherin family of cell adhesion molecules of vertebrates (Schmidt 1989). The 127-kd protein also contains the potential cell-contact site RGDV (Lutzelschwab et al 1987). The RNA transcripts and proteins occur primarily during the two major terminal phases of cell proliferation, namely early embryogenesis and late third larval instar (Mechler et al 1985).

Immunoglobulin Family Proteins

AMALGAM This protein has homology with N-CAM in its three immunoglobulin-like repeats (Seeger et al 1988). Expression peaks at 6–8 h of embryogenesis, and the protein is found on the extracellular surface of mesodermal and neuronal cells and axons. Later it disappears and then reappears during pupal development. Deletion of the gene shows that amalgam is not required to maintain the integrity of the central nervous system, and no defects were noted during embryogenesis.

FASCICLIN II This protein also resembles N-CAM and has five immunoglobulin-like repeats and two fibronectin type III repeats in its extracellular domain. Its expression in grasshoppers on a subset of facsiculating axons suggests a surface recognition molecule involved in growth cone guidance (Harrelson & Goodman 1988; Snow et al 1988). Its expression pattern resembles that of the vertebrate protein L1 (Dodd & Jessel 1988).

Other Axonal Guidance Molecules

FASCICLIN I This glycoprotein is expressed in transverse commisures of the *Drosophila* and grasshopper central nervous systems (Bastiani et al 1987; Zinn et al 1988), similar to Tag-1 of vertebrates (Dodd & Jessel 1988), and on sensory axon pathways in the peripheral nervous system. It is attached to the extracellular surface of the plasma membrane and contains four homologous repeats.

FASCICLIN III This protein is a transmembrane protein with a unique extracellular domain that occurs on subsets of *Drosophila* transverse commisures (Patel et al 1987).

Eye-Specific Proteins

CHAOPTIN This membrane protein is restricted to *Drosophila* photoreceptor cells. It contains 41 tandemly arranged, leucine-rich repeats that are potentially amphipathic and presumably help to localize it to the extracellular side of the lipid bilayer (Reinke et al 1988). The microvilli of developing rhabdomeres are disorganized in *chaoptin* mutants and normal apposition of the membranes of adjacent cells is disrupted. It is suggested that chaoptin is a cell adhesion molecule that helps to align microvilli (Van Vactor et al 1988).

SEVENLESS AND *BRIDE OF SEVENLESS* The product of the *sevenless* gene is a transmembrane protein with an intracellular domain homologous to a tyrosine kinase and a large extracellular domain that is anchored at both ends in the cell membrane (Tomlinson & Ready 1986; Banerjee et al 1987; Hafen et al 1987). The protein is required by each R7 cell of ommatidia for receiving a signal needed for its normal development from the adjacent R8 cell. For expression of this signal the R8 cell requires the product of the gene *bride of sevenless* (Reinke & Zipursky 1988).

CONCLUSIONS

Until recently, the unifying concepts of cell structure and function stopped at the cell boundary. Beyond that lay an extracellular matrix that varied so much among organisms that it was mostly ignored, or left as the preserve of pioneer vertebrate embryologists and researchers of connective tissue diseases. Now it is becoming evident that key components of the macromolecular interactions of cells with their immediate environment have not been lost in evolution. This is seen in the components of basement membranes, in molecular bridges between extracellular matrix and cytoskeleton, and in a tantalizing array of extracellular macromolecules that modulate cell behavior. We understand the implications of some conservations, such as the NC1 junctional domain of collagen IV that is found in man, *Drosophila*, and the nematode, *C. elegans* (J. M. Kramer, personal communication). In other cases, such as the parallel, fine modulations of structure of laminin chains in man, mouse, and *Drosophila*, the similarities are pointers towards some common, as yet not understood, function. The biological action of fragments of such molecules can be studied in cell culture, but interpretations are complicated by multiple domains and a variety of cell receptors. That the subtle effects of laminin on neural cells are more than vagaries of cell culture is demonstrated by a *C. elegans* mutant that is defective in a laminin chain and in the extension of neuron and muscle growth processes (E. Hedgecock, personal communication). Excision of *Drosophila* genes such as *amalgam*, and even *mys*, had less dramatic effects on early embryonic development than might at first be expected. This latter result points out the biological importance of parallel, redundant interactive mechanisms of cells. They can be further investigated with double mutants. The recent technique of locating enhancers by P element mediated insertion of β-galactosidase markers (O'Kane & Gehring 1987) can also contribute to the understanding of extracellular matrix genes such as collagen IV (C. Wilson, personal communication). The biological consequences of multiple repeated motifs, such as the EGF repeats of *Notch*, can be investigated with mutants and by insertion of engineered pericellular macromolecules into the genome. Construction of a physiologically functioning pericellular matrix requires control of the expression of a whole group of genes, and this is being investigated in the TGF-β-like *decapentaplegic* system. While there will be differences between *Drosophila* and vertebrate extracellular matrices, as may be the case for some proteoglycans, it is the similarities that matter. In our context, *Drosophila* is but another tool of the cell biologist to assist in discovering how cells interact and become the integrated system that we call an organism.

ACKNOWLEDGMENTS

We thank Ms. B. Tendis for preparing this manuscript and acknowledge financial support from the Muscular Dystrophy Association and by United States Public Health Service grant AG02128.

Literature Cited

Abrahamson, D. R., Irvin, M. N., St. John, P. L., Accavitti, M. A., Heck, L. W., Couchman, J. R. 1988. Molecular orientation of laminin within basement membranes. *FASEB J.* 2: A629

Anderson, H. 1988. *Drosophila* adhesion molecules and neural development. *TINS* 11: 472–75

Ashhurst, D. 1982. The structure and development of insect connective tissues. In *Insect Ultrastructure*, ed. R. C. King, H. Akai, 1: 313–50. New York: Plenum

Banerjee, U., Renfranz, P. J., Pollack, J. A., Benzer, S. 1987. Molecular characterization and expression of *sevenless*, a gene involved in neuronal pattern formation in the *Drosophila* eye. *Cell* 49: 281–91

Barlow, D. P., Green, N. M., Kurkinen, M., Hogan, B. L. 1984. Sequencing of laminin B chain cDNAs reveals C-terminal regions of coiled-coil alpha-helix. *EMBO J.* 3: 2355–62

Bastiani, M. J., Harrelson, A. L., Snow, P. M., Goodman, C. S. 1987. Expression of fasciclin I and II glycoproteins on subsets of axon pathways during neuronal development in the grasshopper. *Cell* 48: 745–55

Beck, K., McCarthy, R. A., Chiquet, M., Masuda-Nakagawa, L., Schlage, W. K. 1989. Structure of the basement membrane protein laminin: Variations on a theme. In *Cytoskeletal and Extracellular Proteins in Biophysics*, ed. U. Aebi, J. Engel, 3: 102–5. New York/Berlin: Springer-Verlag

Beer, J., Technau, G. M., Campos-Ortega, J. A. 1987. Lineage analysis of transplanted individual cells in embryos of *Drosophila melanogaster. Roux's Arch. Dev. Biol.* 196: 222–30

Bissel, M. J., Hall, H. G., Parry, G. 1982. How does the extracellular matrix direct gene expression? *J. Theor. Biol.* 99: 31–68

Blumberg, B., MacKrell, A. J., Fessler, J. H. 1988. *Drosophila* basement membrane procollagen α1(IV). II. Complete cDNA sequence, genomic structure, and general implications for supramolecular assemblies. *J. Biol. Chem.* 263: 18328–37

Blumberg, B., MacKrell, A. J., Olson, P. F., Kurkinen, M., Monson, J. M., et al. 1987. Basement membrane procollagen IV and its specialized carboxyl domain are conserved in *Drosophila*, mouse and human. *J. Biol. Chem.* 262: 5947–50

Bogaert, T., Brown, N., Wilcox, M. 1987. The *Drosophila* PS2 antigen is an invertebrate integrin that, like the fibronectin receptor, becomes localized to muscle attachments. *Cell* 51: 929–40

Boucaut, J. C., Darribere, T., Poole, T. J., Aoyama, H., Yamada, K. M., Thiery, J. P. 1984. Biologically active synthetic peptides as probes of embryonic development: A competitive peptide inhibitor of fibronectin function inhibits gastrulation in amphibian embryos and neural crest cell migration in avian embryos. *J. Cell Biol.* 99: 1822–30

Bronner-Fraser, M. 1986. An antibody to a receptor for fibronectin and laminin perturbs cranial neural crest development in vivo. *Dev. Biol.* 117: 528–36

Brower, D. L., Piovant, M., Reger, L. A. 1985. Developmental analysis of *Drosophila* position-specific antigens. *Dev. Biol.* 108: 120–30

Brower, D. L., Piovant, M., Salatino, R., Brailey, J., Hendrix, M. J. 1987. Identification of a specialized extracellular matrix component in *Drosophila* imaginal discs. *Dev. Biol.* 119: 373–81

Brower, D. L., Wilcox, M., Piovant, M., Smith, R. J., Reger, L. A. 1984. Related cell-surface antigens expressed with positional specificity in *Drosophila* imaginal discs. *Proc. Natl. Acad. Sci. USA* 81: 7485–89

Buck, C. A., Horwitz, A. F. 1987. Cell surface receptors for extracellular matrix molecules. *Annu. Rev. Cell Biol.* 3: 179–205

Campbell, A. G., Fessler, L. I., Salo, T., Fessler, J. H. 1987. Papilin: A *Drosophila* proteoglycan-like sulfated glycoprotein from basement membranes. *J. Biol. Chem.* 262: 17605–12

Cecchini, J.-P., Knibiehler, B., Mirre, C., Le Parco, Y. 1987. Evidence for a type-IV-related collagen in *Drosophila melanogaster*. Evolutionary constancy of the carboxyl-terminal noncollagenous domain. *Eur. J. Biochem.* 165: 587–93

Chi, H. C., Hui, C. F. 1988. cDNA and amino acid sequences of *Drosophila* laminin B2 chain. *Nucl. Acids Res.* 16: 7205–6

Chi, H. C., Hui, C. F. 1989. Primary structure of the *Drosophila* laminin B2 chain and comparison with human, mouse, and *Drosophila* laminin B1 and B2 chains. *J. Biol. Chem.* 254: 1543–50

Dedhar, S., Ruoslahti, E., Pierschbacher, M. D. 1985. A cell surface receptor complex for collagen type I recognizes the Arg-Gly-Asp sequence. *J. Cell Biol.* 104: 585–93

Derynck, R., Roberts, A. B., Winkler, M. E., Chen, E. Y., Goeddel, D. V. 1984. Human transforming growth factor-alpha: precursor structure and expression in E. coli. *Cell* 38: 287–97

DeSimone, D. W., Hynes, R. O. 1988. *Xenopus laevis* integrins. Structural conservation and evolutionary divergence of integrin beta subunits. *J. Biol. Chem.* 263: 5333–40

Digan, M. E., Haynes, S. R., Mozer, B. A., Dawid, I. B., Forquignon, F., Gans, M. 1986. Genetic and molecular analysis of *fs(1)h*, a maternal effect homeotic gene in *Drosophila*. *Dev. Biol.* 114: 161–69

Dillner, L., Dickerson, K., Manthorpe, M., Ruoslahti, E., Engvall, E. 1988. The neurite-promoting domain of human laminin promotes attachment and induces characteristic morphology in non-neuronal cells. *Exp. Cell Res.* 177: 186–98

Dodd, J., Jessel, T. M. 1988. Axon guidance and the patterning of neuronal projections in vertebrates. *Science* 242: 692–99

D'Souza, S. E., Ginsberg, M. H., Burke, T. A., Lam, S. C.-T., Plow, E. F. 1988. Localization of an Arg-Gly-Asp recognition site within an integrin adhesion receptor. *Science* 242: 91–93

Durkin, M. E., Bartos, B. B., Liu, S.-H., Phillips, S. L., Chung, A. E. 1988a. Primary structure of the mouse laminin B2 chain and comparison with laminin B1. *Biochemistry* 27: 5198–5204

Durkin, M. E., Chakravarti, S., Bartos, B. B., Liu, S.-H., Friedman, R. L., Chung, A. E. 1988b. Amino acid sequence and domain structure of entactin. Homology with epidermal growth factor precursor and low density lipoprotein receptor. *J. Cell Biol.* 107: 2749–56

Edelman, G. M. 1986. Cell adhesion molecules in the regulation of animal form and tissue pattern. *Annu. Rev. Cell Biol.* 2: 81–116

Edelman, G. M. 1988. In *Topobiology—An Introduction to Molecular Embryology*. New York: Basic Books. 240 pp.

Edgar, D., Timpl, R., Thoenen, H. 1988. Structural requirements for the stimulation of neurite outgrowth by two variants of laminin and their inhibition by antibodies. *J. Cell Biol.* 106: 1299–1306

Ekblom, P., Vestweber, D., Kemler, R. 1986. Cell-matrix interactions and cell adhesion during development. *Annu. Rev. Cell Biol.* 2: 27–47

Erickson, H. P., Lightner, V. A. 1988. Hexabrachion protein (Tenascin, Cytoactin, Brachionectin) in connective tissue, embryonic brain and tumors. *Adv. Cell Biol.* 2: 55–90

Fehon, R. G., Gauger, A., Schubiger, G. 1987. Cellular recognition and adhesion in embryos and imaginal discs of *Drosophila melanogaster*. *Symp. Soc. Dev. Biol.* 44: 141–70

Fessler, L. I., Campbell, A. G., Duncan, K. G., Fessler, J. H. 1987. *Drosophila* laminin: Characterization and localization. *J. Cell Biol.* 105: 2383–91

Fristrom, D. 1988. The cellular basis of epithelial morphogenesis. *Tissue Cell* 20: 645–90

Fristrom, D. K., Rickoll, W. L. 1982. Morphogenesis of imaginal discs of *Drosophila*. In *Insect Ultrastructure*, ed. R. King, H. Akai, 1: 247. New York: Plenum

Furst, A., Mahowald, A. P. 1985. Differentiation of primary embryonic neuroblasts in purified neural cell cultures from *Drosophila*. *Dev. Biol.* 109: 184–92

Gerson, I., Seecof, R. L., Teplitz, R. L. 1976a. Ultrastructural differentiation during embryonic *Drosophila* myogenesis in vitro. *In Vitro* 12: 615–22

Gerson, I., Seecof, R. L., Teplitz, R. L. 1976b. Ultrastructural differentiation during *Drosophila* neurogenesis in vitro. *J. Neurobiol.* 7: 447–55

Glanville, R. W. 1987. Type IV collagen. In *Biology of Extracellular Matrix: Structure and Function of Collagen Types*, ed. R. Mayne, R. E. Burgeson, pp. 43–79. New York: Academic

Graf, J., Iwamoto, Y., Sasaki, M., Martin, G. R., Kleinman, H. K., et al. 1987. Identification of an amino acid sequence in laminin mediating cell attachment, chemotaxis, and receptor binding. *Cell* 48: 989–96

Gratecos, D., Astier, M., Semeriva, M. 1987. A new approach to monoclonal antibody production. In vitro immunization with antigens on nitrocellulose using *Drosophila* myosin heavy chain as an example. *J. Immuno. Meth.* 103: 169–78

Gratecos, D., Naidet, C., Astier, M., Thiery, J. P., Semeriva, M. 1988. *Drosophila* fibronectin: A protein that shares properties similar to those of its mammalian homologue. *EMBO J.* 7: 215–33

Gray, A., Dull, T. J., Ullrich, A. 1983. Nucleotide sequence of epidermal growth factor cDNA predicts a 128,000-molecular weight protein precursor. *Nature* 303: 722–25

Greenwald, I. 1985. *lin-12*, a nematode homeotic gene, is homologous to a set of mammalian proteins that includes epidermal growth factor. *Cell* 43: 583–90

Hafen, E., Basler, K., Edstroem, J. E., Rubin, G. 1987. *Sevenless*, a cell-specific homeotic gene of *Drosophila* encodes a putative transmembrane receptor with a tyrosine kinase domain. *Science* 236: 55–63

Harrelson, A. L., Goodman, C. S. 1988. Growth cone guidance in insects: Fasciclin II is a member of the immunoglobulin superfamily. *Science* 242: 700–8

Hay, E. D. 1981. In *Cell Biology of the Extracellular Matrix*, ed. E. D. Hay, pp. 379–409. New York: Plenum

Hirst, R., Horwitz, A., Buck, C., Rohrschneider, L. 1986. Phosphorylation of the fibronectin receptor complex in cells transformed by oncogenes that encode tyrosine kinases. *Proc. Natl. Acad. Sci. USA* 83: 6470–74

Hoffmann, F. M., Goodman, W. 1987. Identification in transgenic animals of the *Drosophila* decapentaplegic sequences required for embryonic dorsal pattern formation. *Genes & Dev.* 1: 615–25

Hunter, D. D., Merle, J. P., Sanes, J. R. 1989. A laminin-like adhesive protein concentrated in the synaptic cleft of the neuromuscular junction. *Nature* 338: 229–34

Hynes, R. O. 1985. Molecular biology of fibronectin. *Annu. Rev. Cell Biol.* 1: 67–90

Hynes, R. O. 1987. Integrins: A family of cell surface receptors. *Cell* 48: 549–54

Jackson, F. R., Bargiello, T. A., Yun, S. H., Young, M. W. 1986. Products of per locus of *Drosophila* shares homology with proteoglycans. *Nature* 320: 185–88

Jacob, L., Opper, M., Metzroth, B., Phannavong, B., Mechler, B. 1987. Structure of the *l(2)gl* gene of *Drosophila* and delimitation of its tumor suppressor domain. *Cell* 50: 215–25

Johnson, D., Lanahan, A., Buck, C. R., Sehgal, A., Morgan, C., et al. 1986. Expression and structure of the human NGF receptor. *Cell* 47: 545–54

Kelley, M. R., Kidd, S., Deutsch, W. A., Young, M. W. 1987. Mutations altering the structure of epidermal growth factor-like coding sequences at the *Drosophila Notch* locus. *Cell* 51: 539–48

Klambt, C., Schmidt, O. 1986. Developmental expression and tissue distribution of the *lethal(2)giant larvae* protein of *Drosophila melanogaster*. *EMBO J.* 5: 2955–61

Klein, G., Langegger, M., Timpl, R., Ekblom, P. 1988. Role of laminin A chain in the development of epithelial cell polarity. *Cell* 55: 331–41

Knibiehler, B., Mirre, C., Cecchini, J. P., Le Parco, Y. 1987. Haemocytes accumulate collagen transcripts during *Drosophila melanogaster* metamorphosis. *Roux's Arch. Dev. Biol.* 196: 243–47

Kornblihtt, A. R., Gutman, A. 1988. Molecular biology of the extracellular matrix proteins. *Biol. Rev.* 63: 465–507

Kühl, U., Ocalan, M., Timpl, R., Mayne, R., Hay, E., von der Mark, K. 1984. Role of muscle fibroblasts in the deposition of type IV collagen in the basal lamina of myotubes. *Differentiation* 28: 164–72

Lehmann, R., Jimenez, F., Dietrich, U., Campos-Ortega, J. A. 1983. On the phenotype and development of mutants of early neurogenesis in *Drosophila melanogaster*. *Roux's Arch. Dev. Biol.* 192: 62–74

Le Parco, Y., Cecchini, J. P., Knibiehler, B., Mirre, C. 1986a. Characterization and expression of collagen-like genes in *Drosophila melanogaster*. *Biol. Cell* 56: 217–26

Le Parco, Y., Knibiehler, B., Cecchini, J. P., Mirre, C. 1986b. Stage and tissue-specific expression of a collagen gene during *Drosophila melanogaster* development. *Exp. Cell Res.* 163: 405–12

Leptin, M., Aebersold, R., Wilcox, M. 1987. *Drosophila* position-specific antigens resemble the vertebrate fibronectin-receptor family. *EMBO J.* 6: 1037–43

Leptin, M., Bogaert, T., Lehmann, R., Wilcox, M. 1989. The function of PS integrins during *Drosophila* embryogenesis. *Cell* 56: 401–8

Lunstrum, G. P., Bächinger, H.-P., Fessler, L. I., Duncan, K. G., Nelson, R. E., Fessler, J. H. 1988. *Drosophila* basement membrane procollagen IV. I. Protein characterization and distribution. *J. Biol. Chem.* 263: 18318–27

Lutzelschwab, R., Klambt, C., Rossa, R., Schmidt, O. 1987. A protein product of the *Drosophila* recessive tumor gene, *l(2) giant larvae*, potentially has cell adhesion properties. *EMBO J.* 6: 1791–97

MacKrell, A. J., Blumberg, B., Haynes, S. R., Fessler, J. H. 1988. The lethal *myospheroid* gene of *Drosophila* encodes a membrane protein homologous to vertebrate integrin beta subunits. *Proc. Natl. Acad. Sci. USA* 85: 2633–37

Marcantonio, E. E., Hynes, R. O. 1988. Antibodies to the conserved cytoplasmic domain of the integrin beta-1 subunit react with proteins in vertebrates, invertebrates and fungi. *J. Cell Biol.* 106: 1765–72

Martin, G. R., Timpl, R. 1987. Laminin and other basement membrane components. *Annu. Rev. Cell Biol.* 3: 57–85

McCarthy, R. A., Beck, K., Burger, M. M. 1987. Laminin is structurally conserved in sea urchin basal lamina. *EMBO J.* 6: 1037–43

McClay, D. R., Ettensohn, C. A. 1987. Cell adhesion in morphogenesis. *Annu. Rev. Cell Biol.* 3: 319–45

Mechler, B. M., McGinnis, W., Gehring, W. J. 1985. Molecular cloning of *lethal(2)giant larvae*, a recessive oncogene of *Drosophila melanogaster*. *EMBO J.* 4: 1551–57

Meyer, M. R., Brunner, P., Edwards, J. S. 1988. Developmental modulation of a glial cell-associated glycoprotein 5B12 in an insect *acheta-domesticus*. *Dev. Biol.* 130: 374–91

Mirre, C., Cecchini, J. P., Le Parco, Y., Knibiehler, B. 1988. De novo expression of a type IV collagen gene in *Drosophila* embryos is restricted to mesodermal derivatives and occurs at germ band shortening. *Development* 102: 368–76

Monson, J. M., Natzle, J., Friedman, J., McCarthy, B. J. 1982. Expression and novel structure of a collagen gene in *Drosophila*. *Proc. Natl. Acad. Sci. USA* 79: 1761–65

Montell, D. J., Goodman, C. S. 1988. *Drosophila* substrate adhesion molecule: sequence of laminin B1 chain reveals domains of homology with mouse. *Cell* 53: 463–73

Murray, L. W., Leipzig, G. V. 1986. Collagen from *Drosophila melanogaster* larvae and adults is similar to type IV collagen. *J. Cell. Biol.* 103: 389a

Naidet, C., Semeriva, M., Yamada, K. M., Thiery, J. P. 1987. Peptides containing the cell-attachment recognition signal Arg-Gly-Asp prevent gastrulation in *Drosophila* embryos. *Nature* 325: 348–50

Natzle, J. E., Monson, J. M., McCarthy, B. J. 1982. Cytogenetic location and expression of collagen-like genes in *Drosophila*. *Nature* 296: 368–71

Newman, S. M., Wright, T. R. F. 1981. A histological and ultrastructural analysis of developmental defects produced by the mutation, *lethal(1)myospheroid* in *Drosophila melanogaster*. *Dev. Biol.* 86: 393–402

O'Kane, C. J., Gehring, W. J. 1987. Detection in situ of genomic regulatory elements in *Drosophila*. *Proc. Natl. Acad. Sci. USA* 84: 9123–27

Olson, P. F., Sterne, R., Fessler, L. I., Fessler, J. H. 1987. Entactin, a sulfated glycoprotein of *Drosophila* basement membranes. *J. Cell. Biochem. Suppl.* 11C: 26

Padgett, R. W., St. Johnston, R. D., Gelbart, W. M. 1987. A transcript from a *Drosophila* pattern gene predicts a protein homologous to the transforming growth factor-beta family. *Nature* 325: 81–84

Panayotou, G., End, P., Aumailley, M., Timpl, R., Engel, J. 1989. Domains of laminin with growth-factor activity. *Cell* 56: 93–101

Patel, N. H., Snow, P. M., Goodman, C. S. 1987. Characterization and cloning of fasciclin III: a glycoprotein expressed on a subset of neurons and axon pathways in *Drosophila*. *Cell* 48: 975–88

Paulsson, M. 1988. The role of Ca^{2+} binding in the self-aggregation of laminin-nidogen complexes. *J. Biol. Chem.* 263: 5425–30

Paulsson, M., Aumailley, M., Deutzmann, R., Timpl, R., Beck, K., Engel, J. 1987. Laminin-nidogen complex. Extraction with chelating agents and structural characterization. *Eur. J. Biochem.* 166: 11–19

Pikkarainen, T., Eddy, R. Fukushima, Y., Byers, M., Shows, T., et al. 1987. Human laminin B1 chain. A multidomain protein with gene (LAMB1) locus in the q22 region of chromosome 7. *J. Biol. Chem.* 262: 10454–62

Pikkarainen, T., Kallunki, T., Tryggvason, K. 1988. Human laminin B2 chain. Comparison of the complete amino acid sequence with the B1 chain reveals variability in sequence homology between different structural domains. *J. Biol. Chem.* 263: 6751–58

Reddy, P., Jacquier, A. C., Abovich, N., Petersen, G., Rosbash, M. 1986. The *period* clock locus of *Drosophila melanogaster* codes for a proteoglycan. *Cell* 46: 53–62

Reinke, R., Krantz, D. E., Yen, D., Zipursky, S. L. 1988. *Chaoptin*, a cell surface glycoprotein required for *Drosophila* photoreceptor cell morphogenesis, contains a repeat motif found in yeast and human. *Cell* 52: 291–301

Reinke, R., Zipursky, S. L. 1988. Cell-cell interaction in the *Drosophila* retina: The *bride of sevenless* gene is required in photoreceptor cell R8 for R7 cell development. *Cell* 55: 321–30

Rickoll, W. L., Fristrom, J. W. 1983. The effects of 20-hydroxyecdysone on the metabolic labeling of membrane proteins in *Drosophila* imaginal discs. *Dev. Biol.* 95: 275–87

Rickoll, W. L., Galewsky, S. 1988. 20-hydroxyecdysone increases the metabolic labeling of extracellular glycoproteins of *Drosophila* S3 cells. *Insect Biochem.* 18: 337–46

Rizki, T. M., Rizki, R. M. 1983. Basement membrane polarizes lectin binding sites of

Drosophila larval fat body cells. *Nature* 30: 340–42

Rizki, T. M., Rizki, R. M. 1984. The cellular defense system of *Drosophila melanogaster*. In *Insect Ultrastructure*, ed. R. C. King, H. Akai, 2: 579–604. New York: Plenum

Rothberg, J. M., Hartley, D. A., Walther, Z., Artavanis-Tsakonas, S. 1988. *slit*: An EGF-homologous locus of *D. melanogaster* involved in the development of the embryonic central nervous system. *Cell* 55: 1047–59

Ruoslahti, E. 1988a. Fibronectin and its receptors. *Annu. Rev. Biochem.* 57: 375–413

Ruoslahti, E. 1988b. Structure and biology of proteoglycans. *Annu. Rev. Cell Biol.* 4: 227–55

Ruoslahti, E., Pierschbacher, M. D. 1987. New perspectives in cell adhesion: RGD and integrins. *Science* 238: 491–97

Sanderson, R. D., Fitch, J. M., Linsenmayer, T. R., Mayne, R. 1986. Fibroblasts promote the formation of a continuous basal lamina during myogenesis in vitro. *J. Cell Biol.* 102: 740–47

Sanes, J. R. 1989. Extracellular matrix molecules that influence neural development. *Annu. Rev. Neurosci.* 12: 521–46

Sasaki, M., Kato, S., Kohno, K., Martin, G. R., Yamada, Y. 1987. Sequence of the cDNA encoding the laminin B1 chain reveals a multidomain protein containing cysteine-rich repeats. *Proc. Natl. Acad. Sci. USA* 84: 935–39

Sasaki, M., Kleinman, H. K., Huber, H., Deutzmann, R., Yamada, Y. 1988. Laminin, a multidomain protein. *J. Biol. Chem.* 263: 16536–44

Sasaki, M., Yamada, Y. 1987. The laminin B2 chain has a multidomain structure homologous to the B1 chain. *J. Biol. Chem.* 262: 17111–17

Schittny, J. C., Timpl, R., Engel, J. 1988. High resolution immunoelectron microscopic localization of functional domains of laminin, nidogen, and heparan sulfate proteoglycan in epithelial basement membrane of mouse cornea reveals different topological orientations. *J. Cell Biol.* 107: 1599–1610

Schmidt, O. 1989. A recessive tumor gene function in *Drosophila* is involved in cell adhesion. *J. Neurogen.* In press

Seecof, R. L., Alleaume, N., Teplitz, R. L., Gerson, I. 1971. Differentiation of neurons and myocytes in cell cultures made from *Drosophila* gastrulae. *Exp. Cell Res.* 69: 161–73

Seecof, R. L., Teplitz, R. L., Gerson, I., Ikeda, K., Donady, J. J. 1972. Differentiation of neuromuscular junctions in cultures of embryonic *Drosophila* cells. *Proc. Natl. Acad. Sci. USA* 69: 566–70

Seeger, M. A., Haffley, L., Kaufman, T. C. 1988. Characterization of amalgam: A member of the immunoglobulin superfamily from *Drosophila*. *Cell* 55: 589–600

Semeriva, M., Naidet, C., Krejci, E., Gratecos, D. 1989. Towards the molecular biology of cell adhesion in *Drosophila*. *Trends Genet.* 5: 24–28

Siebold, B., Qian, R. A., Glanville, R. W., Hofmann, H., Deutzmann, R., Kühn, K. 1987. Construction of a model for the aggregation and cross-linking region (7S domain) of type IV collagen based upon an evaluation of the primary structure of the α1 and α2 chains in this region. *Eur. J. Biochem.* 168: 569–75

Snow, P. M., Zinn, K., Harrelson, A. L., McAllister, L., Schilling, J., et al. 1988. Characterization and cloning of fasciclin I and fasciclin II glycoproteins in the grasshopper. *Proc. Natl. Acad. Sci. USA* 85: 5291–95

Sporn, M. F., Roberts, A. B., Wakefield, L. M., de Crombrugghe, B. 1987. Some recent advances in the chemistry and biology of transforming growth factor-beta. *J. Cell Biol.* 105: 1039–45

Sweet, R. M., Eisenberg, D. 1983. Correlation of sequence hydrophobicities measures similarity in three-dimensional protein structure. *J. Mol. Biol.* 171: 479–88

Tamkun, J. W., DeSimone, D. W., Fonda, D., Patel, R. S., Buck, C., et al. 1986. Structure of integrin, a glycoprotein involved in the transmembrane linkage between fibronectin and actin. *Cell* 46: 271–82

Timpl, R. 1989. Structure and function of extracellular matrix components. *Eur. Biochem. J.* 180: 487–502

Tomlinson, A., Ready, D. F. 1986. *Sevenless*, a cell specific homeotic mutation of the *Drosophila* eye. *Science* 231: 400–2

Van Vactor, D., Krantz, D. E., Reinke, R., Zipursky, S. L. 1988. Analysis of mutants in chaoptin, a photoreceptor cell-specific glycoprotein in *Drosophila*, reveals its role in cellular morphogenesis. *Cell* 52: 281–90

Vassin, H., Bremer, K. A., Knust, E., Campos-Ortega, J. A. 1987. The neurogenic locus *Delta* of *Drosophila melanogaster* is expressed in neurogenic territories and encodes a putative transmembrane protein with EGF-like repeats. *EMBO J.* 11: 3431–40

Wharton, K. A., Johansen, K. M., Xu, T., Artavanis-Tsakonas, S. 1985. Nucleotide sequence from the neurogenic locus *notch* implies a gene product that shares homology with proteins containing EGF-like repeats. *Cell* 43: 567–81

Wieschaus, E., Noell, E. 1986. Specificity of embryonic lethal mutations in *Drosophila* analyzed in germ line clones. *Roux's Arch. Dev. Biol.* 195: 63–73

Wilcox, M., Brower, D. L., Smith, R. J. 1981. A position-specific cell surface antigen in the *Drosophila* wing imaginal disc. *Cell* 25: 159–64

Wilcox, M., Brown, N., Piovant, M., Smith, R. J., White, R. A. 1984. The *Drosophila* position-specific antigens are a family of cell surface glycoprotein complexes. *EMBO J.* 3: 2307–13

Wilcox, M., Leptin, M. 1985. Tissue-specific modulation of a set of related cell surface antigens in *Drosophila. Nature* 316: 351–54

Wirz, J., Fessler, L. I., Gehring, W. J. 1986. Localization of the Antennapedia protein in *Drosophila* embryos and imaginal discs. *EMBO J.* 5: 3327–34

Wright, T. F. 1960. The phenogenetics of the embryonic mutant, *lethal myospheroid*, in *Drosophila melanogaster. J. Exp. Zool.* 143: 77–99

Yurchenko, P. D., Ruben, G. C. 1987. Basement membrane structure in situ: evidence for lateral associations in the type IV collagen network. *J. Cell Biol.* 105: 2559–68

Yurchenko, P. D., Tsilibary, E. C., Charonis, A. S., Furthmayr, H. 1985. Laminin polymerization in vitro. Evidence for a two-step assembly with domain specificity. *J. Biol. Chem.* 260: 7636–44

Zinn, K., McAllister, L., Goodman, C. S. 1988. Sequence analysis and neuronal expression of fasciclin I in grasshopper and *Drosophila. Cell* 53: 577–87

Annu. Rev. Cell Biol. 1989. 5 : 341–95

SIMPLE AND COMPLEX CELL CYCLES

Fred Cross, James Roberts, and Harold Weintraub

Fred Hutchinson Cancer Research Center, 1124 Columbia Street, Seattle, Washington 98104

CONTENTS

PREFACE

This review focuses on areas of the cell cycle that are coming under the scrutiny of molecular biology and genetics. We have benefited from countless reviews, particularly those of Mitchison (1971), Prescott (1976, 1987), Pardee et al (1978), Pringle & Hartwell (1981), Baserga (1984, 1985), Nurse (1985), and Kirschner et al (1985). Of necessity, we are forced to

0743–4634/89/1115–0341$02.00

omit much of the literature, however this is adequately covered in the above reviews. In some cases we have speculated rather freely; however, we have tried to confine such speculation to areas where we believe specific molecular tests are at hand.

Where possible we have attempted to connect work on cell cycle control in diverse systems and to ask what common molecular mechanisms may be involved. The organization is as follows:

1. recent results concerning regulation of the simple embryonic cycle, which lacks G_1 and G_2 phases as well as many of the regulatory controls that function in more complex cell cycles;
2. the control of DNA replication and mitosis in terms of cytoplasmic *trans*-acting factors and nuclear responsiveness to those factors;
3. the control of cell division in the yeasts *S. cerevisiae* and *S. pombe* and in tissue culture cells.

INTRODUCTION

Complex Cell Cycles Rest upon a Basic Cell Cycle

Supported largely from studies of cell division in cleavage stage frog embryos, the notion of a basic or fundamental cell cycle has emerged in recent years (Wilson 1925: see review by Kirschner et al 1985). It is reasonable to assume that such a basic cell cycle forms the framework for the more complex cell cycles exhibited by a variety of different types of cells and different types of organisms. In cleavage stage frog embryos, this basic cell cycle is represented by an S period followed by an M period and so on (i.e. an S-M cycle), with no detectable G_1 or G_2 phase.

In certain special cases, cells can carry out periodic synthesis of DNA in the absence of mitosis. For example, successive rounds of controlled (i.e. uniform doubling of DNA content once per cell generation) chromosomal DNA synthesis occurs in the absence of mitosis during formation of polytene chromosomes and, also, in certain specific mutants of *Drosophila*, [e.g. the *gnu* mutant (Freeman et al 1986; Freeman & Glover 1987)]. Therefore it is not unreasonable to suppose that, under certain defined conditions, an S cycle can occur independent of the mitotic cycle. There is also evidence that a basic M cycle can exist independently of the replication cycle. In embryos treated with aphidocolin to inhibit DNA replication (Kimelman et al 1987; Raff & Glover 1988), many aspects of the mitotic cycle continue (periodic chromosome condensation and decondensation, nuclear envelope breakdown and reformation, centrosome replication, and periodic surface contraction waves associated with mitosis); likewise in tissue culture cells treated with hydroxyurea and caffeine, periodic chromosome condensation and decondensation can occur (Schlegel & Pardee 1987; Schlegel et al 1987).

The interrelationships between the potential S cycle, the potential M cycle, and the basic cell cycle (the S-M cycle) are now coming under experimental scrutiny, since many of the principal genes and gene products have been identified. However in most cells, a variety of cellular needs impinge upon and thereby modify the basic cell cycle. For example, cells that need to achieve a certain relationship between growth (i.e. accumulation of mass) and cell size may need to impose certain constraints on the basic cell cycle that require cells not to enter S or M before they achieve a critical size, hence acquiring a G_1 or G_2 period. Tissue culture cells seem to have controls that prevent division until the environment is found to be appropriate (e.g. by receiving growth factor and nutritional signals or the proper cell contacts) or until important cellular elements such as centrioles, cytoskeleton, membrane systems, mitochondria, and so on are sufficiently accumulated. Such controls, which may be considered by the algorithm "do not go to the next step until processes x, y, and z are completed," are likely to be complex signaling systems that impinge on the basic machinery of the S-M cycle.

THE SIMPLE EMBRYONIC CELL CYCLE

MPF and Cyclin

In the past few years, we have come to realize that there is a basic cell cycle in all eukaryotes with common molecular components and strategies.

CYCLING Cleavage stage frog embryos divide every 45 min and display periodic surface contraction waves every 45 min at the metaphase-anaphase transition. They also accumulate and then lose a biochemical activity called MPF (maturation promoting factor or mitosis promoting factor) every 45 min. All of these events can occur in the absence of a cell nucleus and centrioles, and they are thought to reflect a master-oscillator that controls major features of the cell cycle such as chromosome duplication, condensation, and segregation (Hara et al 1980; Newport & Kirschner 1984).

The cell cycle in frog embryos is driven by a continuous requirement for protein synthesis. Inhibition of protein synthesis with cycloheximide arrests cells after S phase but before division. Thus all the protein components necessary for controlled replication of DNA are present and do not have to be synthesized. The cycloheximide arrest can be overcome by injection of MPF. MPF was originally identified as an activity present in frog eggs that causes maturation of frog oocytes (Masui & Markert 1971;

Smith & Ecker 1971), which involves chromosome condensation and nuclear envelope breakdown among many other changes. Subsequently, MPF was shown to be present also in mitotic mammalian cells (Sunkara et al 1979) and in yeast (Weintraub et al 1982; Tachibana et al 1987). Components of MPF activity (pre-MPF) are present throughout the cell cycle but functional MPF is only detectable as the cells approach mitosis (Dunphy & Newport 1988b; Cyert & Kirschner, 1988). The presence of pre-MPF throughout the cell cycle suggests that it is not the protein(s) whose synthesis is required to drive the basic cell cycle. However, since injection of MPF into cycloheximide-arrested embryos relieves the cycloheximide-induced cell cycle arrest (Newport & Kirschner 1984), it is likely that the cycloheximide-sensitive protein activates MPF which, in turn, is limiting for activation of the panoply of events required for mitosis. This cycloheximide-sensitive protein appears to be a cyclin (see below).

Cyclin was first discovered in clam embryos where it increases steadily in abundance throughout each embryonic cell cycle and at mitosis is rapidly degraded (Evans et al 1983; Rosenthal et al 1980). There are several related cyclin species produced by different genes (T. Hunt, personal communication, J. Ruderman, personal communication), which are likely to have different functions. When clam cyclin A mRNA is injected into frog oocytes, the oocyte nucleus breaks down and the chromosomes condense (Swenson et al 1986; see also, Pines & Hunt 1987). Since the same series of events is seen after injection of MPF, it is possible that cyclin can activate a latent form of MPF, which, in turn, directly or indirectly activates nuclear envelope breakdown, chromosome condensation, and so on. The fact that cyclin is the only protein whose synthesis is required to drive the frog cell cycle has been shown in a cell free system where multiple cycles of DNA replication and chromatin condensation occur (A. Murray, personal communication). In this system elimination of all embryonic frog mRNA with RNase prevents these cycles; however subsequent addition of RNase inhibitor and cyclin mRNA restores cycling activity. Similarly, injection of antisense cyclin oligodeoxynucleotides together with RNAse H, which specifically destroys cyclin mRNA and thus prevents cyclin synthesis, stops cell cycling. Thus the protein synthesis requirement for the frog cell cycle can be reduced to a requirement for cyclin synthesis.

ENTRY INTO M Since cyclin seems to accumulate gradually throughout interphase, a crucial aspect in understanding the role of cyclin accumulation in triggering MPF activation is how the system senses a threshold level of cyclin at the G_2/M boundary. One possibility is that an inhibitor of MPF or cyclin is present that needs to be titrated out. As discussed at length below, the p34 component of MPF in HeLa cells becomes heavily phosphorylated at G_2/M and then dephosphorylated at M. The factors controlling the phosphorylation and dephosphorylation of p34 are not known and may be a key feature in how cyclin activates MPF. It is possible that cyclin accumulation leads to the phosphorylation or dephosphorylation of MPF and/or that MPF is responsible for its own phosphorylation and dephosphorylation. It has been known for some time that injection of small amounts of crude MPF into oocytes in the presence of cycloheximide causes the activation of a latent form of MPF (Wasserman & Masui 1975; Gerhart et al 1984). This might occur because MPF activates itself. The "autocatalytic" property of MPF might be important in accentuating MPF activity once a threshold level of cyclin is achieved.

EXIT FROM M The destruction of cyclin at mitosis would explain the concurrent decline in MPF activity at mitosis if maintenance of MPF activity is dependent on cyclin. An important question is how cyclin is destroyed at mitosis, for without this event the system would not "reset", MPF would persist, and cycling would probably not occur. Possibly high levels of MPF activity activate a pathway that determines subsequent cyclin destruction (see Murray 1987), and subsequent resetting of the cycle, e.g. high levels of MPF directly phosphorylating cyclin could result in cyclin becoming a suitable substrate for a general cellular protease. The drop in MPF activity following first meiotic cleavage in starfish oocytes is inhibited by protease inhibitors (Picard et al 1985), possibly due to inhibition of cyclin degradation.

MPF activity in mammalian tissue culture cells (Rao & Smith 1981; Sunkara et al 1979; Adlakha et al 1983, 1984) is inhibited by factors called IMF (inhibitor of MPF) that arise in early G_1 and are thought to be responsible for exit from mitosis (see below) and for prevention of entry into mitosis until DNA replication is completed. In frogs, a complex pathway of MPF inactivation has been defined functionally (Gerhart et al 1984) and biochemically (Cyert & Kirschner 1988). This inactivation involves other factors; e.g. an inhibitor called INH and a complex cascade of phosphorylation and dephosphorylation reactions that cooperate with Ca^{+2} to control MPF activity. Whether the frog factors (e.g. INH) are related to the mammalian G_1 factors (IMF) is not known.

An activity called cytostatic factor (CSF) can stabilize MPF, at least in frog eggs where the chromosomes remain arrested at meiosis (Meyerhof & Masui 1979a,b). The mechanism of CSF action is not known and it has not been purified. As far as we are aware it is not known whether CSF stabilizes MPF directly, or stabilizes MPF by stabilizing cyclin. A role for CSF in cycling cells has not been described.

CONTROLS WITHIN M In frog embryos the loss of MPF activity at mitosis, possibly as a consequence of cyclin degradation, seems to allow or induce a resetting of the mitotic cycle and subsequent entry into interphase. An intriguing but as yet unanswered question is whether loss of MPF is required to remove the block to DNA rereplication (see discussion below).

In clams colchicine arrests cells in mitosis, but while one cyclin, cyclin A is degraded, another cyclin, cyclin B remains high (T. Hunt, personal communication). Thus in clams the destruction of cyclin B seems to be dependent on the completion of a mitotic event inhibited by colchicine. This suggests that some cells might monitor the effectiveness of mitosis and not continue the cycle in the absence of a successful chromosome segregation process. Direct evidence for this has been provided by Nichlas & Kubai (1985), who have mechanically moved a single chromosome off of the mitotic spindle and found that the ensuing anaphase waits until the disturbed chromosome re-aligns.

In frogs mitosis need not be completed for MPF cycles to continue; however, it is still unclear whether subsequent replicative cycles are permitted. In clams, however, inhibition of mitosis prevents cycling and also prevents cyclin B breakdown. In this respect, clam embryos seem to be more similar to tissue culture cells with respect to the degree to which various cell cycle events are coupled. In frog embryos, the usual controls that require completion of mitosis or DNA replication for further cycling seem to be absent or short-circuited in some way.

RESETTING For the potential M cycle, resetting may occur by degradation of cyclin and decline of MPF. What resets the potential S cycle is less well understood. Injection of cytostatic factor (CSF), a factor present in extracts from frog eggs, into frog embryos causes cells to arrest in mitosis, probably by stabilizing MPF. Under these conditions, injected plasmid DNA will go through one round of DNA replication (Newport & Kirschner 1984). Thus even when the cell is arrested in mitosis, all of the factors needed for replication of exogenous DNA seem to be present. How then does the embryo assure that with each cell cycle one and only one round of DNA replication occurs? This fundamental question is discussed at length below, particularly as it applies to tissue culture cells. In the frog embryo, the question was first addressed by injecting DNA into fertilized eggs and demonstrating that the injected DNA comes under the control of the normal cellular machinery for replication and rereplication control (Harland & Laskey 1980). One clue concerning how the cell regulates rereplication is the observation in CSF-arrested embryos that while a small amount of injected DNA gives regulated replication, large amounts seem to over-replicate (Newport & Kirschner 1984). If this result is corroborated by density shift experiments, it would suggest that excess DNA titrates out a component involved in replication control. Interestingly, in this situation, the endogenous nuclear DNA remains condensed and does not seem to over-replicate, which suggests that the control to rereplication occurs at the time of initial DNA duplication, although in this very complex experimental situation, it is not difficult to imagine other explanations. Also, it is unclear whether the injected DNA remains condensed when over-replication occurs in the presence of CSF.

Because injected DNA is replicated in a controlled fashion in CSF-arrested embryos where MPF is stable, one might guess that it is the loss or decrease of MPF (or cyclin) that is the final trigger for resetting the ability to reinitiate replication. In the permissive environment of the frog embryo where all factors for controlled replication seem to be in excess in both S phase and M phase, it seems that the control on rereplication is the only factor preventing massive and rapid DNA accumulation.

In this view, MPF would serve two primary functions in propelling the cell cycle: a segregation function (condensing chromosomes and breaking down the nuclear envelope) and a replication function (removing the block to rereplication) and hence could couple the two events of segregation and replication. Given that the *CDC28*$^{+}$ protein in *S. cerevisiae* clearly functions at the G_1/S boundary and the *cdc2*$^{+}$ protein in *S. pombe* functions at both G_1/S and G_2/M (see below), one might speculate that perhaps the G_1/S role of MPF is actually the removal of the block to rereplication. In frog embryos this could occur during mitosis, but in yeast perhaps it is delayed until G_1/S.

CDC Mutants, MPF, and Cyclin[1]

When purified from frog eggs, MPF contains peptides of 45 and 32 kd (Lohka et al 1988). The 32-kd peptide is roughly the same size and is antigenically related to the *S. pombe cdc2*$^{+}$ protein (Gautier et al 1988), as well as to a similarly sized peptide in human cells (Draetta et al 1987). The *S. cerevisiae CDC28*$^{+}$ protein is about 34 kd (Reed et al 1985) and its cloned gene complements *S. pombe cdc2* mutants; similarly, the *S. pombe cdc2*$^{+}$ gene complements *S. cerevisiae cdc28*$^{-}$ mutants (Beach et al 1982; Booher & Beach 1986). The human analog of the *S. pombe cdc2*$^{+}$

[1] *S. cerevisiae* gene nomenclature uses capitalized gene names (e.g. *CDC28*) to refer to the wild-type allele, and lower case names (e.g. *cdc28*) to refer to recessive (non-functional) mutant alleles. *S. pombe* nomenclature uses lower case throughout, with wild-type being referred to as "+" (e.g. *cdc2*$^{+}$) and mutants as "−" or by an allele number. We have followed these conventions here, except that for clarity we refer to wild-type as "+" for both organisms, and to non-functional mutations as "−" (e.g. for *S. cerevisiae*, *CDC28*$^{+}$ and *cdc28*$^{-}$; for *S. pombe*, *cdc2*$^{+}$ and *cdc2*$^{-}$). We hope this will reduce the inevitable typographical confusion.

gene also complements *S. pombe cdc2⁻* mutants (Lee & Nurse 1987). A specific connection between the *cdc2⁺/CDC28⁺* products and MPF activity comes from the observation that the 13-kd protein product of the *S. pombe* gene, *suc1⁺*, which interacts genetically with the *cdc2⁺* protein, will deplete frog egg extracts of MPF activity (Dunphy et al 1988). Also, overexpression of *suc1⁺* will delay mitosis (Hayles et al 1986b), which suggests a possible regulatory role of *suc1⁺* at G_2/M. (The relationship, if any, between other known MPF inhibitors such as IMF or INH and *suc1⁺* is not known.) The 34-kd proteins are associated with a kinase activity; they are themselves phosphorylated (and dephosphorylated in some cases) and, at least for the human, frog, and *S. cerevisiae* p34 analogs, they form complexes with several other defined polypeptides.

The role of the associated proteins and phosphorylation in either activating or inhibiting the cell cycle activity of this protein complex is beginning to be understood, but the details remain to be elucidated and there are surprising differences in detail between the three systems. *cdc2⁻* mutants in *S. pombe* arrest at G_1/S and also arrest at G_2/M (Nurse & Bissett 1981; see reviews by Nurse 1985; Lee & Nurse 1988; Dunphy & Newport 1988a). Hence, the *cdc2⁺* protein is involved in two decisions, possibly executing related functions for each. Overexpression of *cdc2⁺* does not drive the cells into an accelerated cycling state (Durkacz et al 1986) which suggests that the activity of the *cdc2⁺* protein is highly controlled, perhaps in a way that is analogous to the way MPF is controlled in frog embryos. However, there are specific mutants of *cdc2⁺* that cause cells to enter M prematurely (Lee & Nurse 1988; see below).

In *S. cerevisiae* most *cdc28⁻* alleles cause arrest only at G_1/S (but see Piggott et al 1982). However, common methods used to isolate *cdc28 ts* alleles may be biased in favor of alleles unable to arrest at a putative late execution point. The *cdc* screen requires a uniform terminal arrest morphology. But since cells arrested at the first *CDC28⁺* execution point (G_1/S) are unbudded and cells arrested at G_2/M are budded, mutation causing arrest at both points would not give a uniform arrest phenotype (Pringle & Hartwell 1981). In addition, the mating selection scheme of Reed (1980) required first cycle arrest at G_1. This problem does not apply to the selection of dominant negative *CDC28* alleles, one of which was characterized as a G_1 arresting mutation (Mendenhall et al 1988). Also, *ts* mutations in *CDC28* generated by random in vitro mutagenesis were reported to result in predominantly unbudded G_1/S arrest (Lorincz & Reed 1986).

It has been argued that in contrast to *S. pombe*, START in *S. cerevisiae* is, in fact, an overlapping control point where the cell commits both to S and M concurrently (Nurse 1985). This would explain the apparent differences in execution points for mutants in the two homologous genes (*cdc2* and *CDC28*). However, a clear explanation is not yet at hand and other possibilities certainly exist. An extreme possibility is that some other gene, yet to be defined, performs the equivalent G_2/M function in *S. cerevisiae* that *cdc2*$^+$ performs in *S. pombe*. This scenario seems unlikely, since *S. cerevisiae CDC28*$^+$ gene can complement the G_2/M function of *cdc2*$^+$ mutants in *S. pombe*. Alternatively, the G_1/S function of START might represent a block to rereplication (see above and below), which is removed at M in *S. pombe*, but at G_1/S in *S. cerevisiae*.

cdc13$^+$ is an *S. pombe* gene required for entry into M (Fantes 1982; Booher & Beach 1988; Hagan et al 1988). It is not needed for entry into S. A specific mutant allele of *cdc13* can suppress a specific *cdc2* mutation, which suggests that the *cdc13*$^+$ protein interacts directly with the *cdc2*$^+$ protein. *cdc13*$^+$ has recently been shown to have homology to clam and sea urchin cyclin A (Solomon et al 1988; Goebl & Byers 1988). Thus in terms of entry into M, the roles for MPF and cyclin in frog eggs have genetic counterparts in the known functions and product of *cdc2*$^+$ and *cdc13*$^+$ in yeast.

It is not clear whether there is a positive role played by MPF or its p34 component in S phase entry in higher cells resembling that for *cdc2*$^+$ in *S. pombe* and *CDC28*$^+$ in *S. cerevisiae*. Neither *cdc13*$^+$ (in *S. pombe*) nor cyclin (in frogs or clams) are known to have a role at G_1/S. Possibly, frog p34 (*cdc2*$^+$) is constitutively active in frog embryos in its G_1/S mode, but requires periodic activation by cyclin to transfer to a G_2/M mode. This simple explanation does not explain how the G_1/S mode becomes regulated in yeast. A newly characterized gene, *WHI1*$^+$/*DAF1*$^+$, in *S. cerevisiae* contains homology to cyclin A and seems to be involved in coupling START to size control (Cross 1988; Nash et al 1988). It is thus possible that different cyclins direct *cdc2*$^+$/*CDC28*$^+$ p34 to a G_1/S or a G_2/M mode, respectively. In this respect, it is noteworthy that overexpression of wild-type *WHI1*$^+$/*DAF1*$^+$ results in smaller cells, possibly because they enter S phase prematurely.

A Specific Role for RNA Transcription or Stabilization

In *Aspergillus*, two genes, *NimA* and *BimE*, are involved in mitotic control (Oakley & Morris 1983; Osmani et al 1987, 1988a,b). *NimA* is recessive and loss of function results in cell cycle arrest before mitosis at G_2/M. *BimE* mutations result in premature chromosome condensation and M phase blockade. Epistasis experiments suggest that *BimE* negatively controls the activity of *NimA*. Direct measurements of *NimA* mRNA show that in a normal cell cycle, *NimA* mRNA peaks at M phase. Induced expression of *NimA* drives interphase cells into mitosis. In *BimE*-arrested cells, *NimA* mRNA is high. *BimE* therefore acts as if it negatively controls *NimA* mRNA levels, either as a negative regulator of transcription or as a regulator of *NimA* turnover. *NimA* has no detectable homology to *cdc2*$^+$ or *CDC28*, although its sequence suggests it is a protein kinase.

Thus far *NimA* is the only defined gene whose mRNA levels seem to be correlated with cell cycle progression, although actinomycin sensitive points have been described in tissue culture cells (Schlegel et al 1987). In particular, in the presence of caffeine and hydroxyurea, premature chromosome condensation (PCC) occurs in tissue culture cells. Under these conditions, actinomycin D, an inhibitor of RNA synthesis, prevents PCC, which suggests that RNA synthesis is needed for mitosis to occur. Indirect experiments suggest that hydroxyurea may lead to stabilization of mitosis specific mRNA and caffeine to a stabilization of mitosis specific proteins. TsBN2 tissue culture cells (see below) prematurely enter mitosis at the nonpermissive temperature. Expression of this phenotype also requires new RNA and protein synthesis.

How *BimE* and *NimA* relate to the apparently ubiquitous MPF-cyclin controls is not known. Thus far, neither p34 or cyclin have been identified in *Aspergillus* by either genetic or biochemical criteria or DNA homology, although their presence is expected. From the genetic studies described above, *NimA* might be imagined to work upstream or downstream from a cyclin-MPF type of mitotic control.

Kinase Activity and the Molecular Anatomy of p34 Complexes

The molecular associations and kinase specificities of p34 complexes are just beginning to be studied in depth in at least five systems: *S. pombe*, *S. cerevisiae*, human tissue culture cells, *Xenopus laevis* embryos, and starfish embryos. In three of these systems where the data is most complete (frogs, *S. cerevisiae*, and HeLa cells), the results are surprisingly different; nevertheless, several important themes have emerged. First, in HeLa cells and in *S. cerevisiae*, p34 can change its associations with other protein factors in a cell cycle dependent manner; second, in HeLa cells and *S. cerevisiae*, the kinase activity of the p34 complex can also change, particularly with regard to its specificity for generic substrates such as casein vs more defined substrates such as histone H1 or the associated proteins in the p34 complex; third, p34 itself can be phosphorylated and this is correlated (HeLa cells and *S. pombe*) with its ability to form specific complexes and to act as a specific kinase. These common features stand out among a myriad of differences (see below); in fact, presently, one is more impressed by the apparent differences between these systems than by their similarities.

Using limited fractionation protocols combined with coimmunoprecipitation and kinase assays, the behavior of p34 has been studied in synchronized human cells (Draetta & Beach 1988; Draetta et al 1988). A complex of three proteins can be observed: (*a*) p34, corresponding to the yeast genes *cdc2*$^+$ in *S. pombe* (Simanis & Nurse 1986) and *CDC28* in *S. cerevisiae* (Reed et al 1985) and now thought to be part of MPF; (*b*) p13, corresponding to the yeast gene *suc1*$^+$, which suppresses certain *cdc2* mutants in *S. pombe* (Hayles et al 1986a,b) and forms a complex with the *cdc2*$^+$ protein (Brizuela et al 1987; Hindley et al 1987); and (*c*) p62, possibly corresponding to *S. pombe cdc13*$^+$, a gene that has homology to cyclin A and gives a mutant phenotype of G_2/M arrest (Hagan et al 1988; Booher & Beach 1988).

In G_1, p34 is present, but is only partially phosphorylated; the complex is not formed and kinase activity is low (Draetta & Beach 1988). For cells arrested in S phase with hydroxyurea, the same situation exists. In G_2/M cells, p34 is highly phosphorylated; the phosphorylated form is seen in a complex with p62 and p13; the complex, but not the free p34, is active as a kinase, phosphorylating p62 and exogenous histone H1 as well as casein. (Although p13 seems to be a part of the active kinase complex, it inhibits MPF activity in the frog microinjection assay.) In cells arrested in mitosis by nocodazole, the complex and its kinase activities persist, but p34 loses its phosphate, possibly in preparation for release of unphosphorylated p34 for the next cell cycle. It is at this time that transferable MPF activity is at its peak and consequently, the dephosphorylation of the complex may be the signal for the final conversion to active MPF (D. Beach, personal communication). It is thought that the cell cycle-dependent phosphorylation of p34 may be causally responsible for allowing associations with p13 and p62 and subsequent MPF activity. Possibly cyclin itself (p62?) directly or indirectly controls the level of p34 phosphorylation.

Recent data have shown that p34 is phosphorylated on serine, threonine, and tyrosine residues and that it is a substrate (at least in vitro) for the src tyrosine kinase (Draetta et al 1988). Interestingly, tyrosine phosphorylation seems to begin at G_1/S. A possible interaction of p34 with the *c-mos* oncogene (a serine/threonine protein kinase) is suggested by the observation that anti-sense oligonucleotides directed against *c-mos* (Sagata et al 1988) inhibit hormone induced nuclear envelope breakdown and chromosome condensation in frog oocytes, functions attributed to active MPF. Human p34 loses all of its phosphate in G_0, but becomes rapidly phosphorylated with serum stimulation (Lee et al 1988). Similarly, in *S. pombe*, p34 phosphorylation and kinase activity are lost with nutrient deprivation (Simanis & Nurse 1986). These data suggest that the activity of p34 may be controlled by a complex series of phosphorylations and dephosphorylations.

In *S. cerevisiae* there are similarities but also clear differences with human cells. Consistent with its role at START, the active kinase complex first appears in G_1 (Mendenhall et al 1987; Wittenberg & Reed 1988). This may reflect the G_1/S role of *CDC28*$^+$. In this organism, p34 (*CDC28*$^+$) is observed in association with a p40 protein, which is also a substrate for its kinase activity. The availability of p40 to join this complex may be a crucial step in controlling the kinase activity (Mendenhall et al 1987, 1988; Wittenberg & Reed 1988). Recently, attempts have been made to use genetic suppression to identify the gene products that interact with *CDC28*$^+$ p34 (S. Reed, personal communication). Suppressors of *ts cdc28* mutants have been identified by using a DNA clone library in high copy number plasmids. Two suppressors have homology to cyclin A; the other to *S. pombe suc1*$^+$. An understanding of how these genes relate to the construction of the p34 complexes will surely lead to a much better understanding of the biochemistry of *CDC28*$^+$ p34 function. The genetic identity of p40 remains unknown.

The complexes described in humans and frogs both have subunits of 34 kd, corresponding to *cdc2*$^+$ in *S. pombe* and *CDC28*$^+$ in *S. cerevisiae*. However, in humans a second subunit is 62 kd and in frogs it is 45 kd. The 45- and 62-kd proteins have not yet been identified genetically, however one might speculate that they are related to cyclins. The level of the described human complex is highest at G_2/M, consistent with its putative MPF role. If the p34 in this complex also has a G_1/S role as predicted by the *cdc2* mutations in *S. pombe*, then the biochemical fractionations have not yet succeeded in identifying a corresponding molecular species. Similarly, the 45-kd species in the frog complex is associated with functional MPF. Thus far a 13-kd protein has not been described for frog MPF; however *S. pombe* p13 (*suc1*$^+$) clearly binds and inhibits the complex. The potential role of p13 in frogs is not known although the possibility that it is absent in a system that does not seem to couple M cycles with S cycles is provocative.

For its G_2/M mode, p34 activation is thought to initiate a cascade of reactions that includes nuclear envelope breakdown and chromosome condensation, which are associated with (and possibly controlled by) lamin phosphorylation and histone H1 phosphorylation, respectively. As noted, the p34 complex itself is an effective kinase for histone H1 and appears to be identical to a previously described histone H1-kinase (Arion et al 1988; Labbe et al 1988a,b). Early evidence that H1 kinase can advance mitosis in *Physarum* has been provided (Bradbury et al 1974a,b; Inglis et al 1976). A large number of kinases that are directly or indirectly activated by MPF have been described (Erikson & Maller 1986; Lohka et al 1987; Cicirelli et al 1988). Presumably, these kinases mediate the downstream cascade of events that follow MPF activation.

Are there clues concerning the subcellular site of action of the p34 kinases? *cdc28* (and *cdc37*) *ts* alleles have a kar-phenotype: the nuclei of mating pairs of cells have an abnormally high probability of failing to fuse (Dutcher & Hartwell 1983a,b; Rose et al 1986). This phenotype requires only one of the parents to be *cdc28*$^-$. In heterokaryons one or the other of the unfused parental nuclei can be incorporated into a bud from the heterokaryon, and produce a haploid derived from mixed cytoplasm. *cdc28*$^-$ and *cdc37*$^-$ nuclei in heterokaryons are inefficiently rescued, which implies that the function of the *CDC28* protein for nuclear function (fusion or rescue into a bud) must be performed prior to G_1 (the stage at which mating occurs) (Dutcher & Hartwell 1983). It is not obvious why a protein kinase should fail to complement a defective one in mixed cytoplasm. *CDC28* p34 might have a nuclear (possibly specifically spindle-pole body) localization (Dutcher & Hartwell 1983a,b; Pringle & Hartwell 1981). In this context, it is intriguing to note that a nuclear or chromatin-associated localization has been reported for MPF from tissue culture cells (Adlakha et al 1982), and experiments in starfish embryos suggest that pre-MPF localizes to the oocyte nucleus before maturation (Kishimoto et al 1981; Picard & Doree 1984). Immunolocalization experiments show an association between *CDC28* p34 and an insoluble cytoplasmic matrix (Wittenberg et al 1987); however this observation does not address the localization of the *CDC28* p34 contained in a complex.

A simple model for coupling the S cycle to the M cycle might imagine that p34 associates with one type of cyclin (e.g. *WHI1*$^+$/*DAF1*$^+$ protein) to perform a G_1/S role, and then at G_2/M associates with a different cyclin (e.g. *cdc13*$^+$ protein). In this model, by analogy to what occurs at M, the G_1/S cyclin might be degraded as a consequence of the G_1/S (START) decision. While appealing, the model is flawed in its simplest presentation by the fact that injected DNA replicates normally in CSF-arrested embryos, where p34 would presumably be in its G_2/M state (Newport & Kirschner 1984). As discussed below, the same type of result is observed in somatic cell heterokaryons between G_1 cells and colcemid-arrested M phase cells, where the G_1 nucleus can replicate while the M phase nucleus remains arrested. In support of the model are the observations that certain mammalian cell temperature-sensitive mutants arrest in both G_1 and G_2 (Zaitsu & Kimura 1988a,b), and a specific chemical inhibitor, trichostatin, also arrests cells in G_1 and G_2 (Yoshida & Beppo 1988). A more complete account of this type of model follows where we also consider *cis*-acting changes in chromatin condensation that occur during the cell cycle.

Cell Cycle Periodicities

In yeast cells, and probably also in tissue culture cells, the timing of the cycle is governed by the requirement to double the cell's mass in each cycle, or minimally by the time to complete the functional sequences of the DNA division cycle (see below). In *Xenopus* cycles, timing is governed by MPF oscillations. However other types of timing mechanisms have been inferred in a variety of systems. Klevecz (1976) and Klevecz & King (1982) have presented evidence that mammalian cell cycle lengths are quantized in units of four hr. In *S. pombe* several cases exist of apparent oscillations of approximately one cell cycle interval that do not require the *cdc2*$^+$ component (the oscillations occur in cells arrested by *ts* mutations in this component). CO_2 production oscillates in *cdc2 ts*-arrested cells, with a periodicity slightly shorter than one cell cycle (Novak & Mitchison 1986), while step changes in nucleoside diphosphate kinase activity occur in *cdc2*-arrested cells with a periodicity equal to a normal cell cycle (Creanor & Mitchison 1986). This latter result is striking in that the normal cell cycle length is governed by the time to double cell mass (see below); does this result imply that in the absence of *cdc2* function and in the absence of progression of the DNA division cycle, cells are still able to monitor mass doublings?

Lycan et al (1987) have shown oscillations in histone RNA stability of about one cell cycle periodicity, which continue in cells blocked for DNA replication by hydroxyurea (HU). Cells were synchronized at START and then released in the presence or absence of HU. The oscillations in the stability of histone mRNA were unaffected by HU for the two cycles measured. It is important to know if the oscillations in histone mRNA stability are dependent on *CDC28*$^+$ function. An intriguing parallel can be drawn to the behavior of *cdc4* mutants: these cells bud repeatedly, at about one bud emergence per cell cycle interval, but do not replicate DNA (Hartwell 1971). This multiple budding has been shown to be dependent on *CDC28*$^+$ function (Singer et al 1984). One interpretation of this finding would be that cycles of activation of *CDC28*$^+$ p34 are continuing, and that each activation peak triggers a new bud emergence. Mutations in *CDC34*$^+$, which encode a ubiquitin-conjugating enzyme, are phenotypically similar to *cdc4*$^-$ mutations (Goebl et al 1988). One might speculate (Byers, personal communication) that the *CDC4*$^+$-*CDC34*$^+$ pathway may be required to inactivate G_1/S cyclins and allow emergence from START, much the way cyclin degradation may be required for emergence from M phase. Possibly cells blocked in START in this way can continue to reinitiate budding. While this speculation might explain the *CDC28*$^+$ requirement, it does not explain the periodicity of the multiple budding. These cases of oscillations unlinked to completion of cell cycle events are unusual in the *S. cerevisiae* cell cycle (Pringle & Hartwell 1981).

CONTROL OF DNA REPLICATION AND MITOSIS IN MAMMALIAN CELLS: REGULATION BY *CIS*- AND *TRANS*-ACTING FACTORS

In the early embryo the cell cycle is driven by oscillations in the activity of a factor that promotes the onset of mitosis. These oscillations are autonomous since they proceed independently of normal cell cycle events such as DNA replication or nuclear division. In yeast, the cell cycle may also be driven by similar oscillations, but the autonomy of the oscillator may be restricted by requirements that certain cell cycle events, such as cell growth, be completed before others commence. In fact, the available data suggest that the mechanism controlling cell cycle progression in many organisms shares common biochemical components. While many elements regulate passage of somatic cells through the cell cycle, we will discuss how the somatic cell cycle may retain as its foundation the basic embryonic S-M cycle, and how, as in yeast, the mitotic oscillator may be constrained by requirements for the successful completion of DNA replication and mitosis. In addition, we suggest that a parallel cell cycle dependent oscillation in chromosome structure also plays a central role in regulating the onset of S phase and mitosis (see also Mazia 1963).

Negative Control of DNA Replication

Usually the most obvious requirement for onset of cellular S phase is completion of the preceding mitosis. Prior to the start of DNA synthesis, during the G_1 phase of the cell cycle, the chromosome becomes competent to begin DNA replication; thus fusion of a G_1 cell to an S phase cell prematurely induces DNA synthesis in the G_1 nucleus (Rao & Johnson 1970; Brown et al 1985). In contrast, parallel experiments show that a G_2 nucleus will not rereplicate its DNA after fusion of a G_2 cell to an S phase cell (Rao & Johnson 1970). This is not due to the presence of a diffusible inhibitor in G_2 cells since replication in the S phase nucleus is unaffected. Rather, the block to rereplication must be a *cis*-acting signal that specifically marks the G_2 chromosomes or G_2 nucleus. Therefore, some event that usually is coupled temporally, if not mechanistically, to mitosis renders the genome once again capable of replication and thus is necessary for the start of the next S phase.

It is likely that the coupling between mitosis and S phase involves resetting the mechanism that limits the replication of cellular DNA to exactly once per cell cycle. It has been recognized for many years that replication of the cellular genome is rigorously controlled; all DNA is duplicated once, and only once, within each S phase. Two general mechanisms have been proposed to explain the precise duplication of the genome. One postulates that all factors necessary to replicate DNA can remain continually in excess throughout S phase, but that a *cis*-acting signal marking replicated DNA prevents its rereplication. A second control mechanism, discussed below, postulates that the limited abundance of an essential replication factor ensures that DNA sequences will be replicated only once per S phase. The first mechanism, called *cis*-acting negative control, is responsible for the once per cell cycle replication of the episomal form of Bovine papillomavirus (BPV) in mammalian cells (Roberts & Weintraub 1986, 1988). BPV *cis*-acting negative replication control requires the concerted action of three genetic elements; two of these elements are *cis*-acting DNA sequences closely linked to replication origins, the third element is a *trans*-acting regulatory protein.

The genetic components of the BPV *cis*-acting negative control system function by marking BPV genomes that have replicated (thus preventing rereplication) (Roberts & Weintraub 1988). This mark may reside on the DNA itself, such as a change in the methylation or superhelicity of the replication origin, both of which have been shown to affect replication initiation in prokaryotic systems (Baker & Kornberg 1988; Alfano & McMacken 1988; and see below). Alternatively, it was proposed that the function of these elements is to participate in a replication dependent change in chromosome structure that prevents a second round of replication. One particular version postulates replication-dependent removal of factors that prevent chromosome condensation, and this model is discussed in detail later in this review. Another specific model proposes that these replication regulatory elements act by maintaining an association between sister chromatids, and that sister chromatid pairing (which by definition can only occur on replicated DNA) prevents the re-utilization of the associated replication origin. By this mechanism, separation of sister chromatids during mitosis would inactivate (or reset) the negative controls thus rendering the genome competent to respond to replication factors.

Sister chromatid interactions have not been extensively studied, although there are some reports of an association between sister chromatids at sites other than the centromere (Goyanes & Mendez 1982; Lica et al 1986). Recent studies on double-minute chromosomes, which have no centromeres [DMs are extrachromosomal DNA elements that form during gene amplification (Balaban-Malenbaum & Gilbert 1977)] reveal a mechanism of sister chromatid pairing that acts independently of the centromere and is specifically inactivated during mitosis (Takayama & Uwaike 1988). Using BrdU labeling it was observed that sister chromatids of minute chromosomes remain paired from S until M and slowly dissociate during G_1 so that at the onset of S phase each chromatid is unpaired.

Since the sister chromatids are free to unpair during G_1, and are not free to do so during S phase and G_2, these results imply a centromere independent mechanism that actively maintains sister chromatid interactions.

There is good evidence from other organisms that during mitosis a specific mechanism that does not involve interactions between the centromere and the mitotic spindle separates sister chromatids. Sea urchin eggs treated briefly with NH_4OH undergo repeated cycles of chromosome condensation and decondensation; sister chromatid separation and a complete round of DNA replication accompany each cycle of decondensation (Mazia 1974). These events occur with no apparent mitotic spindle. Essentially the same events occur if fertilized sea urchin eggs are treated briefly with colcemid (Sluder 1979; Sluder et al 1986). Again, in the absence of a mitotic spindle, one observes cycles of chromosome condensation and decondensation accompanied by sister chromatid separation and DNA replication. The behavior of DMs also demonstrates that centromere-spindle interactions are not necessary for sister chromatid separation. Finally, there is genetic evidence in *S. pombe* supporting the notion that a specific mechanism, requiring the products of the *dis* genes, operates during mitosis to separate sister chromatids (Ohkura et al 1988).

If separation of sister chromatids, not completion of a normal mitosis, is necessary for resetting of replication controls, it should be possible to uncouple nuclear division from DNA replication. The experiments described above in sea urchin eggs are consistent with this idea. Moreover, some mutations in yeast, for example *cdc31* (Baum et al 1986), *ndc1* (Thomas & Botstein 1986), *kar1* (Rose & Fink 1987) and *esp1* (Baum et al 1988) have phenotypes consistent with the view that nuclear division, per se, is not required for the start of a new cell cycle. Polytenization of chromosomes, which in *Drosophila* salivary glands is a regular stepwise process (Hammond & Laird 1985), may also involve uncoupling nuclear division from DNA replication. It is interesting to speculate that one explanation for the absence of DNA synthesis between meiosis I and meiosis II is that in meiosis I sister chromatids do not separate.

The extent to which the cellular genome shares the specific mechanism of BPV replication control is not known. Since the BPV life cycle contains both lytic and episomal phases, BPV may have evolved specific elements to allow it to utilize the cellular replication control mechanism only during its episomal state. Except under unusual circumstances (i.e. developmentally specific gene amplification), the cellular genome does not need the option of unregulated replication. Thus whether specific DNA sequences are necessary for controlled replication of the cellular genome (possibly to focus the controlling factors at the replication origin), or whether the control mechanism simply operates in conjunction with the general DNA synthetic machinery at the replication fork, remains uncertain. In early *Xenopus* embryos specific DNA sequences are not required either to initiate DNA replication or to regulate replication to once per cell cycle (Harland & Laskey 1980), but the pattern of replication in early embryo cells may not be a good model for somatic cells.

One specific example of *cis*-acting negative replication control has been uncovered in *E. coli* (Messer et al 1985; Russell & Zinder 1987). It was demonstrated that at the onset of DNA synthesis both strands of the DNA helix were methylated at specific adenine residues within the replication origin. After one round of replication, hemimethylation of these sites prevented re-initiation at that origin, perhaps through specific association of hemimethylated DNA with the cell membrane (Ogden et al 1988), or by methylation dependent changes in DNA structure (Messer et al 1985; Russell & Zinder 1987). A second round of DNA replication could not occur until the daughter strand was remethylated.

Another mechanism that, in principle, could account for the orderly duplication of the genome has emerged from studies on the replication of nuclei in cell free extracts from activated *Xenopus* eggs (Blow & Laskey 1988). It was observed that in the presence of cycloheximide nuclei incubated with egg extract would replicate all their DNA exactly once; in the absence of protein synthesis inhibitors Blow & Laskey observed periodic nuclear envelope breakdown and reformation (presumably driven by cyclin/MPF) with one complete round of semi-conservative DNA synthesis occurring in each nuclear cycle. Again, an event occurring at mitosis was sufficient to render the genome capable of a new round of DNA replication. This event could be mimicked by artificial, chemical, or enzymatic permeabilization of the nuclear membrane, which suggested that the nuclear membrane might act as a barrier to a limiting replication initiation factor—this was called the licensing factor. If each licensing factor molecule could be used only once to start DNA synthesis, this would prevent a second round of replication until a new supply of factor was deposited on the DNA after nuclear membrane breakdown at mitosis. Certain features of this model need to be explored more completely in order to determine whether it could adequately explain replication control in somatic cells. For example, the stability of the licensing factor during prolonged cellular quiescence must be explained (e.g. during weeks of serum deprivation), since nuclear membrane breakdown is not known to occur during recovery from cell cycle arrest; however growth factor induced changes in nuclear membrane permeability and nuclear lamin phosphorylation have been reported (Jiang & Schindler 1988; Friedman & Ken 1988). One prediction might be that replicon size increases in the S phase following return from quiescence since fewer licensing factors would be present. Also, this model (and other similar models that invoke the depletion of essential replication factors during S) was formulated to explain regulated replication in a situation where initiation occurs synchronously throughout the entire genome (Blow & Watson 1987; Blow 1988). However in somatic cells it appears that initiation events occur at different times throughout S phase, which implies that all necessary replication factors remain active throughout most or all of this period. It may be difficult to reconcile the apparent need to inactivate unused licensing factor molecules with demonstrations of continuous initiation of replication throughout S phase in somatic cells. Moreover, the *Xenopus* experiments are equally consistent with the notion that artificial permeabilization of the nuclear membrane disrupts normal negative replication controls (e.g. disrupts sister chromatid association, or chromatin structure, or supercoiling) and allows a new round of DNA synthesis.

A dramatic demonstration that the availability of replication factors cannot fully explain the control of DNA replication insomatic cells was provided by Rao & Hanks (1980). In this experiment G_1 cells were fused to metaphase cells in the presence of colcemid. The mitotic cell induced G_1 nuclear membrane breakdown and premature chromosome condensation of the G_1 DNA so that the two genomes shared a common cytoplasmic environment. Remarkably, the cell entered an S phase (actually slightly more rapidly than control cells) in which the G_1 chromosomes (in their condensed state and in the absence of an intact nuclear structure) began to duplicate their DNA. Since the metaphase chromosomes in the same cytoplasm did not replicate, this experiment proves that a mechanism that prevents rereplication of DNA within one cell cycle can survive despite the breakdown of the nuclear membrane. It is not known whether anaphase chromosomes, which are unpaired chromatids, would behave differently and begin to replicate when fused to an S phase cell. In a similar experiment (discussed above) it was observed that DNA injected into CSF-arrested *Xenopus* eggs would replicate exactly once while the endogenous chromosomal DNA (which has no nuclear membrane in this high MPF environment) does not replicate (Newport & Kirschner 1984). It could be argued that the highly condensed metaphase chromosomes are especially refractory to initiating DNA replication. Thus the mechanism preventing rereplication of DNA in these experiments may differ from the mechanism that prevents rereplication of decondensed S phase DNA. Nevertheless, these experiments imply that a condensed chromosome structure may be sufficient to restrict the access of replication factors to the DNA. It should be noted that two separate control pathways preventing rereplication may be used to ensure the degree of precision required during duplication of the genome.

Periodic Accumulation of Positive Replication Factors

In most cells the completion of mitosis and erasing of rereplication controls are not sufficient for the start of S phase. Many higher eukaryotic cells have a long G_1 phase devoid of detectable DNA synthesis. During this G_1 phase the nucleus is competent to replicate its DNA, since it will rapidly initiate DNA synthesis if fused to an S phase cell. One of the major conclusions drawn from mammalian cell fusion studies is that a limiting *trans*-acting inducer(s) of S phase accumulates during G_1 (Rao & Johnson 1970; Rossow et al 1979; Pardee et al 1981; Campisi et al 1982; Croy & Pardee 1983; Rao et al 1977a; Fournier & Pardee 1975). This conclusion may be general (Gurdon 1967; de Terra 1966; Ord 1969; Johnson & Harris 1969; Gorden & Cohen 1971; Graves 1972; de Roeper et al 1977). There is no evidence in these experiments for an inhibitor of S phase in G_1 cells.

Some experiments suggest that the *trans*-acting inducer accumulates gradually throughout G_1, and induces S phase when it reaches a critical threshold level. Binucleate fibroblasts produced by cytochalasin B treatment of mitotic cells (Fournier & Pardee 1975), or by sendai virus fusion of synchronous G_1 cells (Rao et al 1977a), traversed G_1 and entered S phase significantly faster than mononucleated control cells; trinucleate fibroblasts enter S at an even greater rate (Rao et al 1977a). In every case all nuclei within a single heterokaryon entered S phase together; this is observed in many experimental situations and again strongly suggests that the start of DNA synthesis is controlled by a limiting, diffusable, *trans*-acting replication inducer (see Johnson & Rao 1971). Significantly, the kinetics of entry into S in bi- and trinucleate cells were not consistent with the simple model in which each nucleus proceeded toward S independently and where the first nucleus to start DNA synthesis initiated S in the others; rather the data suggested a cooperative interaction between the nuclei within the heterokaryon that drove them simultaneously into S earlier than any nucleus alone would have begun DNA replication. Also, older G_1 cells entered S more rapidly than young G_1 cells, and in age disparate heterokaryons, the nuclei entered S with kinetics equivalent to control mononucleate cells of the more elderly fusion partner, not at an intermediate or averaged rate (Rao et al 1977a).

These experiments describe an inducer of S phase that gradually accumulates through G_1 phase and initiates S phase above a critical threshold amount. These properties of the S phase inducer bear some similarities to those of cyclin, a mitotic inducer, and may imply that a cyclin-like activity is important in regulating the onset of DNA replication.

It seems most likely that the factor inferred in the above experiments induces the accumulation or activation of limiting replication proteins rather than being directly involved in DNA replication itself. Thus in *S. cerevisiae*, START precedes DNA replication by at least two genetically defined steps (Hereford & Hartwell 1974). Likewise, by preparing extracts from cells at defined positions in the cell cycle, it was shown that, on a per cell basis, S phase extracts were at least 20-fold more effective at initiating SV40 virus DNA replication than G_1 extracts (Roberts & D'Urso 1988). Direct biochemical analyses showed that most proteins necessary to replicate DNA were equally abundant in G_1 and S phase extracts and therefore could not account for cell cycle specific replication. However a cellular replication factor that activated the SV40 T antigen catalyzed unwinding of SV40 DNA at the replication origin increased in activity suddenly at the G_1/S boundary, and the induction of this factor was sufficient to account for the increased replication activity of S phase cell extracts. Unwinding of origin DNA is the limiting step in initiating replication at the *E. coli* and bacteriophage λ replication origins (Bramhill & Kornberg 1988; Dodson et al 1986; Schnos et al 1988). These data suggest that origin unwinding may also be the limiting reaction during entry into S phase by mammalian cells in cell culture, and that this step is controlled by a cell cycle regulated *trans*-acting replication factor.

What happens to the rate limiting unwinding activity as the cell progresses through S and completes the remainder of the cell cycle? Clearly, the abundance or activity of this factor must decrease prior to the resetting of replication controls; otherwise a new round of DNA synthesis would begin prematurely. In the SV40 cell free system, extracts prepared from G_2 cells are active, while extracts prepared from mitotic cells are inactive. These observations suggest that all factors necessary to replicate DNA might remain active throughout S and into G_2, with the limiting unwinding factor diminishing in activity during mitosis coincident with resetting negative replication controls. At mitosis both negative and positive replication controls would be removed, which results in the G_1 state in which no replication occurs.

Positive and Negative Control of Mitosis

Accumulation of a limiting inducing factor controls the onset of mitosis. Fusion of mitotic cells with G_2 cells will drive the G_2 nucleus prematurely into mitosis (Johnson & Rao 1970). Extracts from mitotic human cells injected into immature *Xenopus* oocytes induce a series of mitotic events indistinguishable from those induced by *Xenopus* MPF, including germinal vesicle breakdown, chromosome condensation, and progression through the first meiotic division (Sunkara et al 1979). This activity is absent from G_1 and S phase cells and gradually accumulates during G_2 reaching a peak in specific activity in extracts from early mitotic cells (Sunkara et al 1979).

As described above this correlates with the appearance of a multiprotein complex containing the human *cdc2*$^{+}$ equivalent (p34), which is active as a protein kinase (Draetta & Beach 1988).

When fused to mitotic cells, G_1 cells undergo nuclear membrane breakdown and premature chromosome condensation, demonstrating that sensitivity to mitosis inducing factors persists beyond M phase (Rao et al 1977b). However, there is a striking decrease in sensitivity to mitotic factors as cells traverse G_1 and enter S (Rao et al 1977b; Hittelman & Rao 1978; Rao & Hanks 1980). This is partially due to the appearance in early G_1 cells of *trans*-acting inhibitors of the mitosis inducing factors (Adlakha et al 1983). These inhibitors, called IMF, are assayed by monitoring the ability of G_1 extracts to block the maturation promoting effects of mitotic cell extracts injected into immature *Xenopus* oocytes. IMF can accumulate in the absence of protein synthesis and may form an inactive complex with mitosis inducing factors (Adlakha et al 1983). Under some conditions, notably high pH or in heterokaryons that contain multiple G_1 nuclei, fusion of G_1 and M cells causes metaphase chromosome decondensation and formation of a nuclear membrane (Obara et al 1974). This suggests that a balance between IMF and MPF can control the mitotic state of the cell. IMF are most active early in G_1 and diminish in activity as the cell progresses through G_1 and S, so that late S phase cells contain no detectable inhibitor (Adlakha et al 1984).

A Block to Condensation Can Control the Chromosome Cycle

While chromosomes from all points in the cell cycle can be condensed by mitotic factors, the degree of premature chromosome condensation (PCC) is cell cycle dependent—early G_1 PCCs are very compact while late G_1 PCCs are very extended (Rao et al 1977b; Hittelman & Rao 1978; Rao & Hanks 1980). This decline in the susceptibility of chromosomes to MPF-induced condensation must reflect a *cis*-acting change in chromosome structure as cells traverse G_1, since the concentration of *trans*-acting inhibitors of MPF diminishes during this same period (Adlakha et al 1983). Once DNA synthesis begins, the degree of S phase PCC correlates with the amount of replicated DNA. One possibility is that some event associated with DNA replication, perhaps the replication fork itself, erases the block to condensation so that the chromosomes become competent again to fully condense during mitosis. This predicts that only replicated DNA should condense in an S phase PCC. This type of mechanism would ensure that a chromosomal region would not condense until it had been replicated. It is interesting to speculate that the *cis*-acting block to rereplication may be related to such a post-replicative change in chromosome structure.

The phenotype of cells with the tsBN2 mutation, which was originally isolated as a DNA synthesis mutant of BHK cells, supports the model that DNA replication removes a *cis*-acting block to chromosome condensation (Nishimoto 1988). Mitosis is normally prevented until DNA replication has been completed. This control is lost at the non-permissive temperature in tsBN2 cells. Within two hr after shift to the non-permissive temperature, tsBN2 cells that were in S phase prematurely enter mitosis; they arrest DNA synthesis and undergo nuclear membrane breakdown and PCC (Nishimoto et al 1978). Fusion of tsBN2 cells with normal cells, at the non-permissive temperature, induced PCC and nuclear membrane breakdown in the normal cell nucleus. Thus the tsBN2 phenotype is due to the accumulation of a *trans*-acting inducer of mitosis (presumably MPF) (Hayashi et al 1982). The most striking aspect of the tsBN2 phenotype is that only cells that had entered or completed S phase undergo nuclear membrane breakdown and PCC; cells in the G_1 phase arrest with no apparent morphologic change (Nishimoto et al 1978). At the molecular level, however, changes consistent with the mitotic state did occur in these G_1 cells; histone H1 became phosphorylated (Ajiro et al 1983) and the rate of RNA and protein synthesis declined to very low levels (Nishimoto et al 1978). Mitotic H1-kinase is probably one functional activity of MPF (Labbe et al 1988a; Arion et al 1988). These observations imply that MPF became active in G_1 arrested tsBN2 cells (although this has not been shown directly), but that chromosome condensation could not occur until DNA replicated. Indeed, in S phase cells the degree of chromosome condensation depended upon the amount of DNA that had replicated prior to the temperature shift.

In contrast to the tsBN2 cells, which prematurely enter mitosis, two mammalian cell lines have been isolated that show a temperature-sensitive block to enter into mitosis (Finley et al 1984; Ciechanover et al 1984; Kulka et al 1988). Both of these lines, which arrest in the G_2 phase of the cell cycle, contain a thermolabile ubiquitin-activating enzyme E1. Ubiquitin conjugation marks proteins for rapid turnover via a specific ATP-dependent proteolytic pathway (Finley & Varshavsky 1985). It has been suggested that proteins regulating cell cycle progression might normally be labile and turn over via the ubiquitin dependent pathway; in the absence of ubiquitination these short-lived proteins can become quite stable. For example, if the BN2 protein, a putative negative regulator of mitosis inducing factors, falls into this class, its persistence in these mutants could explain the G_2 block (Nishimoto 1988).

Microscopic examination of chromosome morphology suggests that G_1 chromosomes decondenses fully within one hr after mitosis (Pederson 1972). However, a number of more sensitive approaches show that changes in chromosome structure are not restricted to entry and exit from mitosis, rather the degree of chromosome condensation continues to change throughout the cell cycle. In one series of experiments the observation that the affinity of actinomycin D for chromatin is inversely proportional to the degree of chromatin condensation was used to measure cell cycle dependent changes in chromosome structure (Pederson & Robbins 1972). Measurements of actinomycin binding to chromatin in synchronized cells demonstrated that chromosomes gradually decondense throughout G_1 and attain a maximally extended conformation early in S, followed by progressive recondensation commencing during S and proceeding until metaphase. Within S phase, newly replicated DNA bound significantly less actinomycin D than parental DNA, which suggests that some degree of condensation might begin very soon after a region of DNA has been replicated. The basic notion of a continuous cycle of chromosome condensation and decondensation paralleling the cell cycle has been confirmed by analyses employing DNase I sensitivity (Pederson 1972), quinacrine dye fluorescence (Moser et al 1975, 1981), and circular dichroism (Nicolini et al 1975) as assays for chromosome structure (reviewed in Rao & Hanks 1980).

Optimal DNA synthesis may require substantial reversal of the chromosome condensation that occurs during mitosis; however the simple hypothesis that the time between mitosis and S phase is determined solely by the rate of chromosome decondensation probably is not correct. DNA synthesis, although localized to small regions, can begin on very compact early G_1 PCC following fusion of early G_1 cells to metaphase cells in the presence of colcemid (Rao & Hanks 1980).

Mitosis Can Be Coupled to DNA Replication

Mitosis does not begin until DNA replication has finished. In heterokaryons formed by fusion of G_2 and either S or G_1 cells, the entry of the G_2 cell into mitosis is delayed until the early nucleus has completed DNA replication (Rao et al 1975; Rao & Smith 1981). It is possible that DNA damaging agents arrest cell cycle progression in G_2 (Weinert & Hartwell 1988 and references therein) (i.e. incompletely replicated DNA, perhaps the replication fork itself, might provoke the same negative signal as DNA damage). The isolation of the *rad9*$^-$ mutation in *S. cerevisiae*, which suppresses the G_2 arrest induced by DNA damage, has led to the suggestion that a negative control pathway regulates passage from S into mitosis by delaying the accumulation of mitosis inducing factors until DNA replication has been completed and DNA damage repaired (Weinert & Hartwell 1988). UV light will induce the appearance of IMF in quiescent mammalian cells (Adlakha et al 1983) and probably also in mitotic cells (Adlakha et al 1984), which supports the idea that DNA damage induces a specific signal that inhibits mitosis inducing factors. The *RAD9*$^+$ gene product may be a part of the negative signal that prevents the accumulation of mitosis promoting factors if DNA is damaged (Weinert & Hartwell 1988).

The exposure of mammalian cells to caffeine can override the effects of both unreplicated DNA and DNA damage and allow cells to enter mitosis. Exposure to caffeine can bypass the G_2 arrest caused by DNA damaging agents and thereby significantly enhance their mutagenic effects (Lau & Pardee 1982). In addition, cells arrested in S phase with hydroxyurea will exhibit nuclear membrane breakdown and premature chromosome condensation upon exposure to caffeine (Rao et al 1986; Brinkley et al 1988). These cells will also show multiple cycles of rounding and flattening suggestive of recurrent mitotic cycles, but chromosome and nuclear morphology was not described (Schlegel & Pardee 1986, 1987; Schlegel et al 1987). One experiment indicated that caffeine does not affect the timing of the mitotic events; cells exposed to caffeine and hydroxyurea showed nuclear membrane breakdown and PCC at the same time as untreated control cells entered mitosis (Musk et al 1988). This observation implies that caffeine can cause the cell to ignore negative controls on the accumulation of mitotic inducing factors but that caffeine does not directly cause the accumulation of the M factors. One interpretation is that caffeine can insulate the pathway of MPF activation from normal S and G_2 negative controls by producing an autonomous MPF cycle much like that found in early *Xenopus* embryos (see above). The autonomous MPF cycle in *Xenopus* embryos can proceed without periodic DNA replication (e.g. in hydroxyurea) and in the absence of nuclear division (e.g. in microtubule depolymerizing agents such as nocodazole). Perhaps caffeine also reverses the block to MPF oscillation present in mammalian cells arrested in mitosis (e.g. with nocodazole).

The observations discussed above suggest an interesting symmetry between the control of S phase and the control of mitosis. Both S and M are induced by the accumulation of positively acting factors to which nuclei in the preceding phase of the cell cycle are fully responsive, and nuclei in succeeding phases of the cell cycle are (at least partially) resistant. The accumulation of the S promoting factor is negatively regulated by a mechanism in G_1 that checks for adequate cell growth (see below) and the accumulation of the M promoting factor is negatively regulated by a mechanism in G_2 that checks for completion of DNA replication. Because of *cis*-modifications of the genome, once DNA replication has been completed the S inducing factors cannot induce rereplication until after mitosis, and during G_1 the genome becomes increasingly refractory to mitotic condensation until after the DNA replicates. Finally, both factors are inactivated by a *trans*-acting mechanism (during G_2 for the S phase factor and during G_1 for the M promoting factor) thus creating the G_1 and G_2 phases of the cell cycle during which times these factors are reactivated.

The symmetry between the pathways that control DNA replication and mitosis might suggest that both are responding to cellular levels of a single master oscillator and possibly that they are mechanistically coupled in ways that are not yet fully appreciated. Thus the somatic cell cycle may retain as its basis the relatively simple S-M cycle characteristic of the early embryo.

As suggested by the discovery of a mammalian homolog to the *S. pombe cdc2* gene, the notion of coupling between S and M in mammalian cells also may be supported by the isolation of genes whose products are necessary both for entry into S and entry into M. One example of this kind of gene is defined by the *tsF121* mutation in rat 3Y1 fibroblasts (Ohno et al 1984; Ohno & Kimura 1984). Cells harboring this mutation arrest at the non-permissive temperature at either the G_1/S or G_2/M boundaries. It also has been shown that rat embryo fibroblasts, which have a finite lifespan in cell culture, arrest both in G_1 and G_2 when proliferation ceases (Jat & Sharp 1989). The G_1+G_2 arrest of both *tsF121* and rat embryo fibroblasts can be suppressed by SV40 T antigen and that of the *tsF121* mutation by excess serum (Ohno & Kimura 1984; Zaitsu & Kimura 1988a,b).

A Working Model

One model discussed above proposes that the *cdc2*$^+$ or *CDC28*$^+$ p34 engages in periodic association with specific proteins forming, alternately, a complex that activates S phase specific and then mitosis specific factors. We propose that oscillations in the abundance of S and M inducing *trans*-acting factors are accompanied by parallel oscillations in chromosome structure that render the genome receptive alternately to these S and M promoting factors. In one specific model a structural element of the chromosome, perhaps a single protein, controls the degree of chromosome condensation. At the completion of mitosis and during the ensuing G_1 this element becomes modified, which results in chromosome decondensation and consequent accessibility to S phase specific replication factors. During S phase, passage of the replication fork removes this modification thereby initiating chromosome condensation specifically on replicated DNA (imparting the capacity for complete MPF-induced condensation in M phase or as PCC) and restricting access to replication factors (thus preventing rereplication). In one very simple version, MPF might be localized both to the chromosome, to participate in the control of chromosome condensation, and free in the nucleus, to induce a series of *trans*-acting events during entry into M (or S). Since this model incorporates a single *cis*-acting factor with alternately positive and negative effects on DNA replication, it unites many of the features of the *cis*-acting negative control and licensing factor models of replication control discussed above.

In conclusion, the basic cell cycle may be formulated as an oscillation of MPF specificity, alternately inducing S and then M, accompanied by parallel oscillations in chromosome structure that render the genome alternately receptive to S and M factors (and concomitantly resistant to M and then S factors). In somatic cells negative controls can interrupt these oscillations by requiring that certain conditions be met before S or M begins.

CONTROL OF CELL DIVISION

Division Control in S. cerevisiae

HOW CAN THE CELL CYCLE ARREST? Cell division cycle or *cdc* mutations, which block the cell cycle at a specific point, are considered to be mutations in genes whose products function at a specific time in the cell cycle. Such arrests can be explained by the product-substrate hypothesis (Hartwell 1976): for example, nuclear division is considered not to occur on unreplicated or partially replicated DNA because fully replicated DNA is the substrate for some critical process in nuclear division. This leads to a coherent view of the cycle as a set of sequences of dependent events, the functional sequence map. (The literature on the construction and interpretation of such a map is beyond the scope of this review; see Pringle & Hartwell 1981; Wheals 1987).

Recently it has become clear that a *cdc*$^-$ phenotype can also be explained not by an intrinsic function of the *CDC* product in the DNA-division cycle, but because of regulatory elements halting cycle progression if certain conditions do not exist. An example is provided by *RAD9*$^+$ (Weinert & Hartwell 1988). When DNA is damaged by radiation, cells respond by a G_2 delay, presumably to allow repair of damaged chromosomes using information from sister chromatids. *rad9*$^-$ mutations prevent this response, which results in increased lethality of DNA damage (Weinert & Hartwell 1988). *rad9*$^-$ mutations also suppress the *cdc*$^-$ block, but not the *ts* lethality, of a subclass of *cdc* mutations such as *cdc13*$^-$ that result in DNA damage in the G_2 phase of the cell cycle (Hartwell & Smith 1985; Weinert & Hartwell 1988; T. Weinert, personal communication). Although *cdc13*$^-$ *RAD9*$^+$ cells arrest in G_2 at the nonpermissive temperature, remaining viable while arrested, *cdc13*$^-$ *rad9*$^-$ cells go on to divide under these conditions and rapidly die as a result. Thus the *CDC13*$^+$ product does not perform a step that provides an essential substrate for performance of later steps in mitosis; it performs a step that if not performed makes mitosis ill-advised. Topoisomerase II mutations do not result in a uniform terminal arrest, despite having a defined execution point for the function (DiNardo et al 1984; Holm et al 1985). Presumably, there is no regulator analogous to *RAD9*$^{+}$ halting mitosis if topoisomerase II function has not been completed. Similarly, mutations that do not allow cytokinesis (e.g. *cdc24*; Hartwell et al 1974) allow additional rounds of DNA replication and nuclear division.

The *RAD9*$^{+}$ example suggests a possible resolution of the master oscillator view of embryonic cell cycle control (Kirschner et al 1985) with the functional sequence map. Negative regulators such as *RAD9*$^{+}$ could work by coupling the oscillator system to completion of cell cycle events. Two forms of coupling can be imagined: the oscillations themselves could be stopped by the regulator, or the oscillator could continue unimpeded, with response withheld until the negative regulator is not active, i.e. until the step monitored by the regulator is completed. Product-substrate relationships among *cdc* functions could function at the fine scale: for example, the production of dTTP by the *CDC8*$^{+}$ product (thymidylate kinase) provides a substrate for the action of the *CDC17*$^{+}$ product (DNA polymerase), which provides a substrate for the *CDC9*$^{+}$ product (DNA ligase). Globally, mutations in these genes might arrest the cycle because the end product of their joint action (completely replicated DNA) is required to inactivate a negative regulator.

COORDINATION OF GROWTH AND DIVISION AT START IN *S. cerevisiae* If the time required to double the mass of a cell is greater than the time required for the events in the DNA-division cycle [DNA replication, mitosis and cytokinesis (Mitchison 1971)], then the cell must adjust the frequency of completion of the division cycle to once per mass doubling in order to coordinate growth and division. This is a requirement probably experienced in almost all cell cycles outside of early embryogenesis and may represent a primary example of a regulatory system impinging on a more basic cell cycle. The critical-size-requirement (CSR) model proposes that there is a variable period at the initiation of the cycle that couples this initiation to mass accumulation. Cells do not pass START until they reach the critical size (Johnston et al 1977a; Hartwell & Unger 1977; Pringle & Hartwell 1981) (Figure 1*A*).

START, near the end of G_1 and before any overt steps of the DNA division cycle, is considered critical in *S. cerevisiae* cell cycle regulation. START is required for events leading to bud emergence, DNA replication, spindle pole body duplication, and transcription of the *HO* endonuclease gene (Pringle & Hartwell 1981; Breeden 1988).

Poor nutritional conditions can cause arrest of cells before START

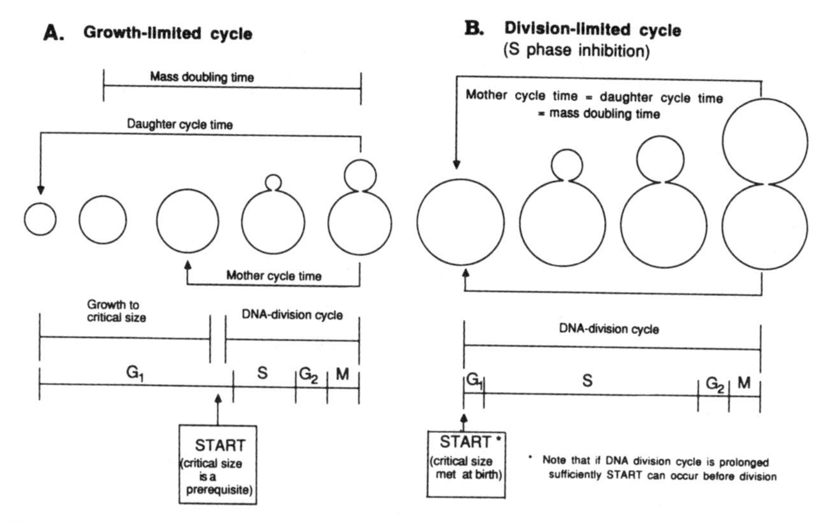

Figure 1 The relationship between growth and division in *S. cerevisiae*. (*A*) The growth-limited cycle, presumed to be the normal case, and how size control coordinates growth and division by a size-conditional restraint on the division cycle at START. (*B*) Limiting the division cycle (e.g. by hydroxyurea inhibition of DNA replication; Singer & Johnson 1981) allows enough growth in the division cycle so that size control is passed at birth. See text.

transit in a quiescent, thermotolerant state (see below). In addition, pheromones secreted by mating cells cause both partners to arrest before START transit (Bucking-Throm et al 1973). A number of *cdc* mutations also cause arrest at START (Pringle & Hartwell 1981; Reed 1980; Bedard et al 1981). START arrests have been classified as to whether arrested cells continue active growth and metabolism (e.g. arrests by mating pheromones or *cdc28*⁻ mutations) (START-I), or become quiescent and thermotolerant [arrests by nutrient deprivation, or by a number of *cdc* mutations (including *cdc25*⁻, *33*⁻, *35*⁻)] (START-II) (Reed 1980; Bedard et al 1981; Plesset et al 1987; Jahng et al 1988).

The idea that coordination of growth and division is achieved by the delay of the critical START event in a size-dependent manner (Figure 1*A*) is supported by a variety of cell physiologic experiments, and some clues as to a molecular mechanism may be emerging.

MOTHER-DAUGHTER ASYMMETRY AND GROWTH-RATE RESTRICTION Since bud emergence occurs less than a mass-doubling time before division, daughters (derived from the bud) are smaller than mothers. Since they are smaller they should take longer to reach the critical size, but should eventually pass START, thus moving into bud emergence and entry into S phase after a pre-START G_1 interval that is longer than that in mothers (Johnston et al 1977a; Hartwell & Unger 1977; Yamada & Ito 1979; Lord & Wheals 1980, 1981; Wheals 1982; Moore 1984a,b; Brewer et al 1984). Since the mother cell does not decrease in size significantly at division, a critical size requirement (CSR) model predicts no growth requirement for cells that have budded at least once previously. This is not the case however (Hartwell & Unger 1977; Johnston et al 1979; Vanoni et al 1983); mothers of increasing age increase in volume, and all cells demonstrate some growth requirement for START. If new activator proteins must be synthesized after mitosis before each START (Alberghina et al 1984; Wheals & Silverman 1982) (see below), then all cells would have some requirement for protein synthesis (and hence growth) before START.

Slow growth rates (by change of carbon source, nitrogen source, or growth in low cycloheximide) exacerbate mother-daughter asymmetry, and, in the population as a whole, expand the pre-START portion of the cell cycle (Hartwell & Unger 1977; Lord & Wheals 1980; Tyson et al 1979; Carter & Jagadish 1978a,b; Jagadish & Carter 1977, 1978; Slater et al 1977; Hartwell & Unger 1977; Johnston et al 1977a; Johnston et al 1980). (However, see Rivin & Fangman 1980 for contradictory results, discussed by Pringle & Hartwell 1981.) Small cells eventually pass START at a size similar to that at which large cells do (Johnston et al 1977a).

Cells transferred into low concentrations of cycloheximide achieve rapid coordination of the new lower growth rate to a lower rate of production of cells. The time to achieve the new rate is approximately the time required for cells past START to clear the constant DNA division cycle and enter a new pre-START period (Popolo et al 1982; Moore 1988), which suggests that the size control system responds rapidly. Although results at slow growth rates approximately confirm the CSR model, invariance of the DNA-division cycle time is not observed, especially at very slow growth rates (Hartwell & Unger 1977; Carter & Jagadish 1978a,b; Tyson et al 1979; Lord & Wheals 1980; Thompson & Wheals 1980).

EFFECT OF DIVISION LIMITATION A number of conditions (e.g. mating pheromone or hydroxyurea) slow or block the division cycle without significantly slowing mass accumulation (division-limiting conditions; Figure 1*B*). These division-limited cells execute START and later G_1 functions (*CDC4* and *CDC7* steps; Hereford & Hartwell 1974) earlier in their cycles than normal, as predicted (Johnston & Singer 1983; Singer & Johnston 1981, 1985). If the division cycle is slowed, then daughters of greater size are produced at division, and these daughters have pre-START periods closer to those of the mother cells (Yamada & Ito 1979; Singer &

Johnston 1981; Lord & Wheals 1983). If the division cycle is blocked by α-factor without blocking growth, then following release from the block, the overly large cells pass START quickly for one or several generations (Moore 1984b; Lord & Wheals 1983).

SETTING THE VALUE OF THE CRITICAL SIZE The size of the mother cell at bud emergence is dependent on carbon source (cells in carbon sources supporting a lower growth rate bud at a smaller volume). The new apparent critical size is attained within one cell cycle of shift, and most likely is attained before the first transit of START after the shift. This maps the measurement of size to the pre-START interval (Lorincz & Carter 1979; Johnston et al 1979).

BEHAVIOR OF A SIZE CONTROL MUTANT The *DAF1*-1 or *WHI1*-1 mutations (dominant mutations in the same gene) result in small cells with a shortened G_1 period. Deletions of the wild-type gene result in abnormally large cells with a longer G_1 (Sudberg et al 1980; Cross 1988; Nash et al 1988). The dominant mutations (or increases in the copy number of the wild-type gene) produce a signal that makes cells behave as if they are bigger than they really are; thus this gene may provide a clue as to how cells link growth to initiation of S phase. Interestingly, in this connection, $WHI1^+/DAF1^+$ shows homology to cyclin (Nash et al 1988).

THE QUIESCENT STATE AND ITS POSSIBLE RELATIONSHIP TO G_1 SIZE CONTROL Nutrient starvation or a variety of START-II mutational blocks causes arrest in G_1 in a quiescent state with low rates of macromolecular synthesis (Unger & Hartwell 1976; Pringle & Hartwell 1981; Plesset et al 1987; Paris & Pringle 1983; Lillie & Pringle 1980; Iida & Yahara 1984). Pringle & Hartwell (1981) proposed that these blocks give START arrest via the size control mechanism by preventing growth.

Many START-II blocks may be explained in large part by lowering of cAMP; cAMP levels may be controlled by nutritional signals via the *CDC25-RAS-CDC35* (= adenylate cyclase) pathway (reviewed by Tatchell 1986; see Broek et al 1987; Daniel et al 1987; Toda et al 1987; Cannon & Tatchell 1987; Camonis & Jacquet 1988). (There may be other pathways to quiescence; see Martegani et al 1984, 1986; Cameron et al 1988.) Low cAMP might directly inhibit START (Matsumoto et al 1983), perhaps via a requirement for cAMP-dependent phosphorylation of $CDC28^+$ p34 (Mendenhall et al 1987). $bcy1^-$ cells [lacking the regulatory subunit of cAMP-dependent protein kinase (Matsumoto et al 1983; Cannon et al 1986; Toda et al 1987)] do not show specific G_1 arrest on starvation (Matsumoto et al 1983), which suggests that without cAMP regulation the G_1 block is bypassed. However, as pointed out by Hayles & Nurse (1986), *bcy1*$^{-}$ cells may actually not bypass G_1 arrest, but rather appear to die almost immediately upon starvation, possibly due to inadequate storage of reserves compared to wild-type (Johnston 1977; Johnston et al 1977b; Sumrada & Cooper 1978; Lillie & Pringle 1980; Uno et al 1983; Toda et al 1985; C. Mann, personal communication).

The relationship of the nutrient response and the quiescent state to division control is unclear. Quiescence might keep cells from dividing simply by lowering their growth rates (Pringle & Hartwell 1981). Alternatively quiescence might entail entry into a special out-of-cycle G_0 compartment that requires specific functions for G_1 reentry (Drebot et al 1987). Adequate nutrient levels might be directly required to relieve a control on the division apparatus [e.g. starvation might lower cAMP, and cAMP-dependent phosphorylation of *CDC28*$^{+}$ p34 might be required for its activity at START (Mendenhall et al 1987)]. Similar possibilities exist for serum control of G_0/G_1 events in tissue culture cells.

THE TRANSITION PROBABILITY HYPOTHESIS Under the transition probability hypothesis (Smith & Martin 1973), cells are considered to enter a state before commitment to the DNA-division cycle; in this state no cell cycle events occur, although growth and other continuous processes may. Exit from this state is random and follows first-order decay kinetics. Rates of proliferation are governed by the rate constant for exit from this state. Since the pre-START interval is variable and environmentally modulated, the application of the transition probability model to yeast would predict random transit of START after division, without regard to the time since division or the size of the cell. This model conflicts with the data discussed above.

Nevertheless it is clear that there is considerable variability in time that cells spend before START under most conditions, and that the size of cells at START transit is quite variable. Models incorporating size control and a random transition in tandem have been proposed (Shilo et al 1976, 1977; Nurse 1980). The "sloppy size control" model (Wheals 1982) proposes that the probability of passing START is modulated from low but greater than zero for small cells to high but less than one (but see Moore 1984b) for large cells.

MOLECULAR MODELS What cells monitor in size control—protein, RNA, protein synthesis rates, level of a specific unstable protein—is certainly unclear at present.

A role for RNA production in transit of START was suggested by Singer et al (1978), Johnston & Singer (1978, 1980), and Bedard et al (1980, 1984) on the basis of the relative rates of inhibition of cell cycle transit, protein synthesis, and RNA synthesis by several relatively specific inhibitors of RNA metabolism; however a fairly rapid inhibition of protein synthesis is also observed with some of these inhibitors.

Unger & Hartwell (1976) suggested that G_1 arrest by sulfur starvation was mediated through the eventual effect of this starvation on protein synthesis. The slowing of transit of START by low cycloheximide (Hartwell & Unger 1977; Shilo et al 1979; Popolo et al 1982), even in large cells (Moore 1988), is also consistent with a high rate of protein synthesis being required for START. Similarly, certain mutations in the *PRT1*$^+$ (*CDC63*$^+$) gene, which is required for protein synthesis, cause a START-I phenotype (Bedard et al 1981; Hanic-Joyce et al 1987). Samokhin et al (1981) observed that low cycloheximide and low α-factor concentrations acted independently to inhibit START transit. While the mechanism by which low α-factor concentrations delay START (Moore 1984a, 1987) is unknown, the speculation (Cross 1988) that this occurs by a raising of the setpoint of size control is consistent with the observations of Samokhin et al (1981).

One could combine these observations in the hypothesis (Moore 1988; Pardee et al 1978; Shilo et al 1979; Popolo et al 1982; Wheals & Silverman 1982; Alberghina et al 1984; Lorincz et al 1982; Popolo et al 1986; Popolo & Alberghina 1984) that the level of an activator protein is monitored at START. If translation of this protein is highly sensitive to deficiencies in *PRT1*$^+$ function, this could explain the *cdc63* (START-I) alleles of *PRT1*. The effects of the inhibitors of RNA metabolism could be explained the same way (via an effect on protein synthesis), or by supposing an effect of these inhibitors on production of mRNA for the protein.

Mutations in the gene for such an activation protein should be START-I defective: cells should not pass START at any size. Increasing the gene dosage of the protein should decrease the setpoint of size control, and reducing the gene dosage should increase the setpoint. The *DAF1*$^+$/*WHI1*$^+$ product meets the second criterion, but not the first (Cross 1988; Nash et al 1988; F. R. Cross, unpublished research); if it is the critical element it must be one of a family of such gene products. The *CDC28*$^+$ product meets the first criterion but not the second; its levels are not affected by starvation or limiting protein synthesis (Mendenhall et al 1987), and increasing its dosage has no effect on cell size (S. I. Reed, personal communication).

Wheals & Silverman (1982) discuss a system that monitors size by the concentration of an unstable activator, which might apply to a cyclin-type model. The possible mitotic activation of *Xenopus cdc2*$^+$ p34 by cyclins (or *cdc13*$^+$ protein) could be paralleled in the START transition by activation of *CDC28*$^+$ p34 by the putative unstable protein. Nash et al (1988) have detected homology between *DAF1*$^+$/*WHI1*$^+$ and cyclin. S. I. Reed (personal communication) has isolated additional *S. cerevisiae* genes with cyclin homology, which can suppress certain *ts cdc28* alleles when present at high gene dosage. Perhaps the family of cyclin homologs function together in G_1 size control.

Division Control in Schizosaccharomyces Pombe

GROWTH PATTERN AND SIZE HOMEOSTASIS The biology and genetics of the *S. pombe* cell cycle were recently reviewed (Wheals 1987). *S. pombe* cell size increases through the cycle in a complex pattern with accelerations and decelerations of rate of growth (reviewed in Elliott & McLaughlin 1983; see also Creanor & Mitchison 1984; Mitchison & Nurse 1985). Size homeostasis is observed: large cells have shorter times to division (Fantes 1977; Myata et al 1978).

G_2 EVENTS, CDC2$^+$, AND SIZE CONTROL The G_2/M transition is size controlled. The setpoint is growth rate modulated, so nutrient down-shift gives a burst of nuclear division, and the new size is achieved within the first generation; up-shift gives transient suppression of nuclear division, and equally fast achievement of new size (Fantes & Nurse 1977).

This size control requires *wee1*$^+$, a protein kinase that inhibits mitosis in a dosage- and nutrition-dependent fashion (Fantes & Nurse 1978; Russell & Nurse 1987a). The *wee1*$^+$ protein is considered to work by inhibiting *cdc2*$^+$ p34 activity at G_2/M, based on dominant alleles of *cdc2* that result in insensitivity to *wee1*$^+$ inhibition of mitosis, short G_2, and small size (Russell & Nurse 1987a; Fantes 1981). However no change in *cdc2*$^+$ p34 phosphorylation state is observed in *wee1*$^-$ cells (Potashkin & Beach 1988), so *wee1*$^+$ regulation of *cdc2*$^+$ activity may be indirect. The dominant *wee* alleles of *cdc2* suggest that *cdc2* activity is all that is limiting in small G_2 cells. The *cdc25*$^+$ protein may act independently to activate *cdc2*$^+$ p34 (Fantes 1981; Russell & Nurse 1986, 1987a). The *nim1*$^+$ protein may inhibit *wee1*$^+$ (Russell & Nurse 1987b); a connection to nutrients has not been established. The *win1*$^+$ gene may be involved in the nutritional control (Ogden & Fantes 1986).

Novak et al (1988) showed that CO_2 removal from *S. pombe* cultures could mimic some aspects of nutritional shift-down on G_2 size control (Fantes & Nurse 1977); this was so even under conditions where the growth rate was little affected. *wee1*$^+$ was required for this effect, which suggests that the rate of CO_2 fixation may be read in some way by the *wee1*$^+$ protein. Two genes, *cdr1*$^+$ and *cdr2*$^+$, were suggested as candidates for encoding signal transducers from nutritional status to the *wee1*$^+$ protein (Young & Fantes 1987).

Reciprocal shifts with the microtubule inhibitor benomyl indicated that *cdc2*$^+$ functions before the benomyl-sensitive step, whereas *cdc13*$^+$ and *cdc25*$^+$ function interdependently with benomyl (Fantes 1982). This result is surprising since other results imply that the *cdc25*$^+$ protein is an activator of *cdc2*$^+$ (Fantes 1981; Russell & Nurse 1986, 1987a). Similarly, *cdc13*$^+$ apparently acted after *cdc2*$^+$ in this analysis, but allele-specific interactions between various forms of *cdc2* and *cdc13* (Booher & Beach 1987) suggested that the *cdc2*$^+$ protein and *cdc13*$^+$ proteins interact physically to perform their G_2/M functions. These conflicts might be explained if some of these *cdc* products have multiple functions; the specific *ts* alleles used in the reciprocal shift study (Fantes 1982) might inactivate only certain specific functions under restrictive conditions (Pringle 1978). In fact, *null* alleles of *cdc13* have an interphase (G_2) arrest phenotype (Hagan et al 1988; Booher & Beach 1988), in contrast to the partial M phase arrest observed with *cdc13.117* used by Fantes (1982) (Hagan et al 1988).

Following arrest of *cdc2 ts* cells at G_2/M, cells that were shifted to permissive temperature in the presence of cycloheximide entered mitosis (Fantes 1982). This suggests that the growth to a critical size (and any other events requiring new protein synthesis) required for entry into M can be completed before *cdc2*$^+$ p34 acts. In contrast, arrest of *S. cerevisiae* before G_1/S with α-factor cannot be reversed without protein synthesis despite the growth of the cells to a large size during arrest (Moore 1988).

G_1 SIZE CONTROL AND THE G_1/S ROLE OF CDC2$^+$ *S. pombe* growing in rich medium exhibits only a very short G_1 period (Nurse 1985). However, if very small cells are produced by nutrient limitation, or by germination of spores, or by *wee1*$^-$ mutations, cells wait a substantial portion of their cycle before initiating DNA replication. Cells in all of these situations initiate DNA replication at a similar size, without regard to their starting size or growth rate (Nurse & Thuriaux 1977; Nasmyth 1979; Nasmyth et al 1979). It is assumed that the G_1/S size control is operative in all normal cell cycles, but is cryptic because the size of cells at birth in rich medium is beyond the setpoint of G_1 size control (Nurse & Thuriaux 1977). In *wee1*$^-$ cells, G_1/S size control is the only one operative (Nurse & Thuriaux 1977; Fantes & Nurse 1978; Russell & Nurse 1987a). The *wee* alleles of *cdc2* and *wee1*$^-$ appear to make cells small only because the cells enter M at a small size; their G_1/S size control is approximately normal (Nurse & Thuriaux 1977; Nasmyth 1979).

While *cdc2*$^+$ was initially identified as having only a G_2 execution point, Nurse & Bissett (1981) concluded that it also had an execution point at G_1/S, based on two lines of evidence: (*a*) a minority of *cdc2 ts* cells could mate following arrest by high temperature (mating normally occurs between G_1 cells); (*b*) a functional *cdc2* gene was required for exit from the stationary phase, which is a G_1 arrest. While this evidence is strong, the mating assay is rather indirect (i.e. the possibility that the *cdc2 ts* cells mated in G_2 without undergoing mitosis was not excluded), and it is possible that escape from the stationary phase requires special functions (possibly including some requiring active *cdc2* p34) that are not required in cycling cells (for example, see Drebot et al 1987). *cdc10 ts* mutants arrest in G_1 and mate following arrest by high temperature (Nurse & Bissett 1981). The *cdc10* execution point is shifted to later in the cycle by strongly growth-limiting conditions in cycling cells (Nasmyth 1979; Nasmyth et al 1979), as is observed with the START execution point in *S. cerevisiae* (see Figure 1*A* and discussion above). If the G_1/S *cdc2* execution point is equivalent to START, then this execution point should be shifted to later in the cycle by growth-limitation; this would be strong confirmation that the *cdc2*$^+$ product acts at START.

Division Control in Tissue Culture Cells

The growth and division of cells in culture has been extensively studied with conflicting results. The reasons for this may include the diverse array of cell types and lines studied. Also, a bewildering array of synchronization techniques have been used in cell cycle research. Even mitotic shake-off, considered one of the least intrusive methods, can disturb the cycle substantially (Prescott 1976). Induction synchrony methods (Mitchison 1971) are potentially worse. Included in induction synchrony experiments are those in which quiescent cells are stimulated by serum growth factors. These experiments are good for studying growth factor action, but dangerous for studying the cell cycle because these experiments include events that occur only during emergence of the cells from a resting state (for example, see Stimac & Morris 1987; Mercer et al 1984).

We are unable to do justice here to the enormous literature on tissue culture cells. What follows is a summary of points of interest from our perspective. Many of these issues are well-addressed by Prescott (1976, 1987). We have tried to indicate what generalizations may hold without doing injustice to well-supported contrary views.

HOW DO TISSUE CULTURE CELLS GROW? Brooks & Shields (1985) pointed out the importance of this question for understanding size homeostasis. In their analysis of cells stimulated from quiescence, they saw a linear pattern of growth, such that large cells grew proportionately more slowly than small cells. This pattern could give rise to size homeostatis without active size monitoring. In contrast, Zetterberg (1970), Zetterberg & Larsson (1985), Ronning et al (1981), and Wheatley et al (1987) presented evidence from exponentially growing cultures that growth was autocatalytic—that is, that the rate of increase of a cell's mass was proportional to its mass. This mode of growth probably requires active size monitoring to achieve size homeostasis. This is an important point requiring further careful analysis.

IS THERE A SIZE CONTROL OVER ENTRY INTO S PHASE IN TISSUE CULTURE CELLS? In other systems [yeast, discussed above, and protozoa (Prescott 1956; Frazier 1973)] there is evidence that size regulates the division cycle. Repeated removal of amoeba cytoplasm prevented division indefinitely (Prescott 1956), and introduction of G_1 cytoplasm into G_1 Stentor cells induced S phase (Frazier 1973).

Using animals cells, Killander & Zetterberg (1965) and Darzynkiewicz et al (1982) showed a lower variability of cell protein and RNA content at the start of S phase than in postmitotic cells, which implied size regulation of entry into S. Shields et al (1978) and Hola & Riley (1987) concluded that small cells are less likely to enter S phase than large cells, although the effect was not a large one.

Slowing growth with low cycloheximide slows entry into S phase (Schneiderman et al 1971; Highfield & Dewey 1972; Brooks 1977; Zetterberg & Larsson 1985; Campisi & Pardee 1984; Ronning et al 1981; Okuda & Kimura 1988; Traganos et al 1987; Ronning & Lindmo 1983; Rossow et al 1979). The significance of these results for the critical size model can be questioned. In some cases the rest of the cycle was also elongated (Ronning et al 1981; Traganos et al 1987; Okuda & Kimura 1988), although this was not true in others (Zetterberg & Larsson 1985; Rossow et al 1979). Also, some cells can be sent into an apparent out-of-cycle G_0 state by inhibition of protein synthesis (Zetterberg & Larsson 1985). Lastly, in some of these studies cells entered S phase with little or no increase in protein content (Ronning & Lindmo 1983; Ronning & Pettersen 1984; Traganos et al 1987), which contradicts a simple critical size model. These data might fit a sloppy size control model (Wheals 1982), since these small cells were not absolutely prevented from entering S, but did so much more slowly. Baserga (1984) reviewed a number of instances where S phase could be entered without increase in cell RNA or protein. Some of these cases involved viral infection, or other treatments that might bypass normal controls. However at the least, the results summarized suggest that simple cell mass is not directly monitored.

The balance of the evidence favors at least a facilitating role for cell mass in S phase entry, although a strict size control model appears unlikely. Deeper understanding may come with analysis of the roles of p34 (and cyclin-like species?) in G_1/S regulation in these cells (Lee et al 1988).

G_2/M CONTROL Several lines of evidence suggest that tissue culture cells may have the ability to arrest in G_2 despite the apparent dominance of G_1/S control. For example, DNA damage will arrest tissue culture cells in

G_2 (see above). Several treatments can cause arrest or delay in both G_1 and G_2, thus suggesting that certain components for G_1/S and G_2/M regulation may be shared [the *tsF121* product (Zaitsu & Kimura 1988a,b), calmodulin (Rasmussen & Means 1989), and cAMP (Blomhoff et al 1988)] reminiscent of *S. pombe cdc2*$^+$ control over G_1/S and G_2/M. Treatment with trichostatin A (Yoshida & Beppu 1988) arrests cells in both G_1 and G_2; release from the G_2 block leads to nearly quantitative tetraploidization, a phenotype shared by release from certain *cdc2* blocks (P. Nurse, personal communication). Since the *tsF121* blocks can be overcome by high serum or SV40 infection, treatments that also stimulate cell growth (Zaitsu & Kimura 1988a,b), the possibility exists that some form of growth rate or size control may be operative at G_2/M in some situations. This possibility may explain some of the apparent contradictions found in the above discussion on the regulation of S phase entry by size, since in some situations control over G_2/M may dominate. Results with human p34 kinase activity and complex formation (see above) are consistent with a G_2/M role for p34 in HeLa cells.

ARE THERE G_1-SPECIFIC EVENTS? Based mainly on evidence from bacterial systems, Cooper (1979, 1981) proposed that there is no specific role for G_1 in the division cycle, since all preparation for DNA synthesis can occur in previous cell cycles. The discussion above of *S. cerevisiae* division control supports this hypothesis. The bulk of G_1 is required solely for growth and this requirement can be shifted to previous cell cycle(s). This was demonstrated in part by experiments in which division was slowed without limiting growth by a variety of DNA division cycle inhibitors; thus cells spent a disproportionate amount of time in cycle n, during which they grew larger. The result was a shortening of G_1 in cycle $n+1$. Similar experiments have been performed with mammalian cell lines. Lengthening S phase by hydroxyurea shortens the succeeding G_1 compared to untreated controls (Cress & Gerner 1977; Stancel et al 1981; Ronning & Seglen 1982; Rao et al 1984). It appears likely, however, that there is some incompressible portion of G_1, which implies that there may be events that can be performed only between M and S. One such event might be chromosome decondensation. Alternatively, there might be specific events that must occur late in G_1 in preparation for DNA synthesis. The incompressible part of G_1 has not been mapped in the cycle (i.e. do these events normally occur in early or late G_1?). The steps leading to DNA synthesis in *S. cerevisiae* after the START event (*CDC4* and *CDC7* steps), which are normally late G_1 steps, cannot occur before nuclear division; thus at least some of the incompressible G_1 in this organism is in late G_1 (Johnston & Singer 1983; Singer & Johnston 1983, 1985a).

V79–8 is a CHO cell line reported to have no G_1 phase (Liskay et al 1978, 1979, 1980) (but see Brooks et al 1983). Therefore, in this line there are no lengthy processes that are obligatory steps between M and S phase. Four mutant cell lines were derived from V79–8 that have a G_1 period (Liskay et al 1980). Three of these lines show a decreased rate of bulk protein synthesis. The fourth did not, but had a slowed generation time compared to V79–8 (as did the other three) and a slightly increased rate of protein degradation (although the specificity of this was questioned). Thus at least three of four G_{1+} derivatives of V79–8 may have a G_1 because they grow more slowly. Consistent with this idea, V79–8 cells growing in low cycloheximide increase their generation times mostly by expanding G_1 (Liskay et al 1980). V79–8 may lack G_1 because a nonphase-specific growth requirement can be shifted to previous cell cycles in this line due to its rapid rate of protein synthesis. However, Rao & Sunk (1980) have suggested that V79–8 may be able to carry out normally G_1-specific protein synthesis in M phase since the normal sharp reduction of bulk protein synthesis in M phase is less sharp in V79–8 cells.

THE G_0 STATE Another possible role for G_1 in mammalian cells is as a response point for serum growth factors. While the topic of serum regulation is beyond the scope of this review, it is clear that serum deprivation causes arrest in G_1, as well as a reduction in the rate of macromolecular synthesis. Cells appear to require a long period to emerge from this resting state into S phase compared to the normal M-S interval, so the state has been considered to be an out-of-cycle G_0 compartment (reviewed in Pardee et al 1978; Baserga 1985). The role of different growth factors in progression of G_0 or G_1 cells into S phase is complex, but there is evidence that some order of function of various factors exists in this progression (Pledger et al 1978; Campisi & Pardee 1984). Pardee et al (1978, 1981, 1986) review evidence in favor of a specific unstable activator protein that must accumulate to a sufficient level for entry into S phase (Rossow et al 1979; Campisi et al 1982; Croy & Pardee 1983). Following this accumulation (after the serum restriction point), cells are serum-independent for entry into S phase. Zetterberg & Larsson (1985) used time-lapse studies on exponentially growing 3T3 cells to locate this serum restriction point. 3T3 cells deprived of serum, when the cells were less than 3.5 hr after mitosis, underwent arrest for a period much greater than the period of serum withdrawal; cells later in G_1 or elsewhere in the cycle were not inhibited at all. Yen & Pardee (1978) obtained a similar result monitoring cell cycle progress by DNA flow cytometry. However, a twofold reduction in protein synthesis caused by serum withdrawal was not specific to any part of the cycle (Zetterberg & Larsson 1985). These results may indicate that passage of the serum restriction point is coincident with the loss of the ability to enter a G_0 state in the current cell cycle.

The ability to respond to serum withdrawal by entering into a G_0 state, as well as other aspects of restriction point control, is lost in many transformed lines (Pardee et al 1978; Campisi et al 1982). Despite the lack of restriction point control in transformed lines, transformed cells are able to coordinate growth and division (e.g. see Ronning et al 1981). Therefore the serum restriction point may have a greater relationship with the ability to enter a quiescent out-of-cycle G_0 state than with division control per se. In the discussion of *S. cerevisiae*, we offered the possibility (Pringle & Hartwell 1981; Alberghina et al 1984, 1986) that nutrient availability and cAMP regulation impinged on the cell cycle only by setting the growth rate, which then fed into division control via the size control system. Serum regulation of growth and division may occur proximally by control of growth. To refute this would require demonstration of a growth factor requirement with no effect on any parameter of general growth rate.

Supporting this idea, Zetterberg & Larsson (1985) found that low cycloheximide could mimic the effects of serum withdrawal and send early G_1 cells into a long division delay without delaying cells later in the cycle. Campisi & Pardee (1984) obtained qualitatively similar results monitoring entry into S phase rather than division. However, Larsson et al (1985a) showed that excess growth factors could block the effects of low cycloheximide on division, at least for the first cycle, without reversing the inhibition of protein synthesis by cycloheximide. They argue from these results that growth factors may differentially affect the expression of specific cell cycle regulatory proteins under some conditions.

Some of the serum regulation of division can occur in previous cell cycles. Larsson et al (1985b) examined the second-cycle consequences of the serum starvation treatments of Zetterberg & Larsson (1985) described above. Early G_1 cells that showed a long delay in the first cycle had a shortened second cycle; cells elsewhere in the first cycle that showed no delay in the first cycle had a lengthened second cycle. This lengthened cycle was mainly ascribed to a lengthened serum-sensitive period, probably therefore a longer G_1. These results are simply interpreted by a size control model: cells undergoing a lengthy division delay under conditions where the rate of protein synthesis is reduced only twofold should grow unusually large by the end of their long cycle and should be quicker to pass the size control of the next cycle. In contrast, cells that divide on schedule despite this twofold reduction in protein synthesis should be unusually small at mitosis and should require a longer G_1 to pass size control. This and other interpretations of these results in terms of the two random transitions model (Brooks et al 1980) and the restriction point model (see above)

(Pardee et al 1978) are discussed by Larsson et al (1985b). Using 3Y1 rat fibroblasts, Okuda & Kimura (1984, 1986) also observed that serum withdrawal after the serum restriction point in one cell cycle lengthened the following cell cycle by expanding G_1.

THE TRANSITION PROBABILITY HYPOTHESIS Smith & Martin (1973) observed that the generation times of cells in culture had a very curious property: they behaved as if the rate-limiting event in proliferation was a random, memory-less event, or transition in G_1. They hypothesized that cells in G_1 started in an A state, from which they would decay into a B phase of fixed length (Shields 1977, 1978; Brooks 1976, 1977). This kinetic model was later refined, but the essential element of random transitions remained (Minor & Smith 1974; Brooks et al 1980). It is a well reproduced observation that many different cell systems appear to follow random kinetics, in most cases for exit from G_1.

The transition probability model has been criticized on the grounds that other control systems could give similar kinetics, and possibly a better fit to the data (Castor 1980; Koch 1980; Smith et al 1981; Murphy et al 1984; Skehan 1988). The possibility that a deterministic system, coupled with variable segregation of mass at mitosis, could give similar kinetics has been proposed (Sennerstam 1988).

Conclusions on G_1/S and G_2/M Division Control

In *S. cerevisiae*, the predominant mode of control over division is the size control over START at G_1/S. In contrast, in *S. pombe* the predominant size control is in G_2/M, but a cryptic G_1/S control can be observed in special circumstances. It has not been ruled out that a G_2/M size control is present but cryptic in *S. cerevisiae* (this idea was proposed to explain the cell cycle of *DAF1-1/WHI1-1* cells: Cross 1988; Nash et al 1988). More specifically, could the *CDC28*$^+$ p34 product be acting at such a G_2/M size control point, as is the related *S. pombe cdc2*$^+$ product? It may be that the commitment point to M phase, and indeed the initiation of M phase itself, is concurrent with START in *S. cerevisiae*, but separated in *S. pombe* (Nurse 1985). Alternatively, commitment to *S. cerevisiae* M phase may be distinct from START. The regulatory halting of the cell cycle in G_2 in response to DNA damage by the *RAD9*$^+$ pathway (Weinert & Hartwell 1988) shows that G_2/M control can be observed; does this control work by acting on a late activity of *CDC28*$^+$ p34, by inhibiting the activity of some other as yet uncharacterized p34 species, or by inhibiting M phase progression downstream of p34 activity? IMF (inhibitors of mitotic factors) are induced by DNA damage in animal cells (Adlakha et al 1984), and caffeine can cause bypass of G_2 delay caused by irradiation. The question of whether the *RAD9*$^+$ protein, IMFs, and caffeine work by inhibiting p34 activity or by effects downstream of p34 activity should be open to direct investigation soon.

The predominant form of control in animal cells (as in *S. cerevisiae*) appears to be over G_1/S. Several lines of evidence suggest the possibility of control over both G_1/S and G_2/M, with these controls sharing some components (reminiscent of the situation in *S. pombe*). However the human p34 shows clear G_2/M regulation of its activity, but no obvious G_1/S regulation. The possibility of multiple p34 species coexisting in a single organism, with either distinct or overlapping roles, should be considered both in this case and in the case of *S. cerevisiae* (S. I. Reed, personal communication), where most evidence points only to a G_1/S role for the *CDC28*$^+$ p34 (e.g. Lorincz & Reed 1986; but see also Piggot et al 1981). Even in *S. pombe*, the evidence for G_1 arrest in cycling cells caused by *cdc2*$^-$ mutations is based on an indirect mating-competence assay for cell cycle position. However, *S. pombe cdc2*$^-$ mutations are complemented by *S. cerevisiae CDC28*$^+$ and human *cdc2*$^+$ (Beach et al 1982; Lee & Nurse 1987), and *S. cerevisiae cdc28*$^-$ mutations are complemented by *S. pombe* and human *cdc2*$^+$ (Booher & Beach 1986; S. I. Reed, personal communication). These complementation studies show that all of these genes have the capability to perform equivalent functions, probably at both G_1/S and G_2/M, but do not prove that they do so in their native environments.

The coordination of growth and division at G_1/S in *S. cerevisiae* may be attained through a family of cyclin homologs, possibly dose-dependent activators of *CDC28*$^+$ p34 (Cross 1988; Nash et al 1988; S. I. Reed, personal communication). In contrast, the coordination of growth and division at G_2/M in *S. pombe*, although likely occurring by control over the G_2/M *cdc2*$^+$ execution point, may not involve accumulation of the *cdc13*$^+$ (cyclin) product, since increasing the dosage of this gene does not advance mitosis (D. Beach, pesonal communication; P. Nurse, personal communication); rather, the *cdc25*$^+$, *nim1*$^+$ and *wee1*$^+$ products appear to be involved. The complexity of the molecular associations and dissociations involved in p34 regulation certainly leave much room for regulation at both G_1/S and G_2/M. A biochemical understanding of the activation and inactivation of p34 activities may lead to understanding of the coordination of different DNA-division cycle processes (e.g. DNA replication and nuclear division), both with each other and with the continuous process of cell growth.

Acknowledgments

F. Cross and J. Roberts are Scholars of the Lucille P. Markey Charitable

Trust. H. Weintraub was supported by a grant from the National Institutes of Health.

See note added in proof p. 395.

Literature Cited

Adlakha, R. C., Sahasrabuddhe, C. G., Wright, D. A., Lindsey, W. F., Rao, P. N. 1982. Localization of mitotic factors on metaphase chromosomes. *J. Cell Sci.* 54: 193–206

Adlakha, R. C., Sahasrabuddhe, C. G., Wright, D. A., Rao, P. N. 1983. Evidence for the presence of inhibitors of mitotic factors during G1 period in mammalian cells. *J. Cell Biol.* 97: 1707–13

Adlakha, R. C., Wang, Y. C., Wright, D. A., Sahasrabuddhe, C. G., Bigo, H., Rao, P. N. 1984. Inactivation of mitotic factors by ultraviolet irradiation of HeLa cells in mitotis. *J. Cell Sci.* 65: 279–95

Ajiro, K., Nishimoto, T., Takahashi, T. 1983. Histone H1 and H3 phosphorylation during premature chromosome condensation in a temperature sensitive mutant (*tsBN2*) of baby hamster kidney cells. *J. Biol. Chem.* 258(7): 4534–38

Alberghina, L., Mariani, L., Martegani, E. 1986. Cell cycle modelling. *BioSystems* 19: 23–44

Alberghina, L., Martegani, E., Mariani, L., Bortolan, G. 1984. A bimolecular mechanism for the cell size control of the cell cycle. *BioSystems* 16: 297–305

Alfano, C., McMacken, R. 1988. The role of template superhelicity in the initiation of bacteriophage DNA replication. *Nucleic Acids Res.* 16: 9611–30

Arion, D., Meijer, L., Brizuela, L., Beach, D. 1988. *cdc2* is a component of the M phase-specific histone H1 kinase: Evidence for identity with MPF. *Cell* 55: 371–78

Baker, T., Kornberg, A. 1988. Transcriptional activation of initiation of replication from the *E. coli* chromosomal origin: An RNA-DNA hybrid near oriC. *Cell* 55: 113–23

Balaban-Malenbaum, G., Gilbert, F. 1977. Double minute chromosomes and homogeneously staining regions in chromosomes of a human neuroblastoma cell line. *Science* 198: 739–42

Baserga, R. 1984. Growth in size and cell DNA replication. *Exp. Cell Res.* 151: 1–5

Baserga, R. 1985. *The Biology of Cell Reproduction.* Cambridge, Mass/England: Harvard Univ. 251 pp.

Baum, P., Furlong, C., Byers, B. 1986. Yeast gene required for spindle pole body duplication: homology of its product with Ca^{2+}-binding proteins. *Proc. Natl. Acad. Sci. USA* 83: 5512–16

Baum, P., Yip, C., Goetsch, L., Byers, B. 1988. A yeast gene essential for regulation of spindle pole duplication. *Mol. Cell. Biol.* 8: 5386–97

Beach, D., Durkacz, B., Nurse, P. 1982. Functionally homologous cell cycle control genes in budding and fission yeast. *Nature* 300: 706–9

Bedard, D. P., Johnston, G. C., Singer, R. A. 1981. New mutations in the yeast *Saccharomyces cerevisiae* affecting completion of the "start." *Curr. Genet.* 4: 205–14

Bedard, D. P., Li, A. W., Singer, R. A., Johnston, G. C. 1984. Mating ability during chemically induced G_1 arrest of cells of the yeast *Saccharomyces cerevisiae. J. Bacteriol.* 160: 1196–98

Bedard, D. P., Singer, R. A., Johnston, G. C. 1980. Transient cell cycle arrest of *Saccharomyces cerevisiae* by amino acid analog β-2-dl-thienylalanine. *J. Bacteriol.* 141: 100–5

Blomhoff, H. K., Blomhoff, R., Stokke, T., Davies, C. D., Brevik, K., et al. 1988. cAMP-mediated growth inhibition of a B-lymphoid precursor cell line Reh is associated with an early transient delay in G_2/M, followed by an accumulation of cells in G_1. *J. Cell. Physiol.* 137: 583–87

Blow, J. J. 1988. Eukaryotic DNA replication reconstituted outside the cell. *BioEssays* 8: 149–52

Blow, J. J., Laskey, R. A. 1988. A role for nuclear envelope in controlling DNA replication within the cell cycle. *Nature* 332: 546–48

Blow, J. J., Watson, J. V. 1987. Nuclei act as independent and integrated units of replication in a *Xenopus* cell-free DNA replication system. *EMBO J.* 6: 1997–2002

Booher, R., Beach, D. 1986. Site-specific mutagenesis of *cdc2*$^+$, a cell cycle control gene of the fission yeast *Schizosaccharomyces pombe. Mol. Cell. Biol.* 6: 3523–30

Booher, R., Beach, D. 1987. Interaction between *cdc13*$^+$ and *cdc2*$^+$ in the control of mitosis in fission yeast; dissociation of the G_1 and G_2 roles of the *cdc2*$^+$ protein kinase. *EMBO J.* 6: 3441–47

Booher, R., Beach, D. 1988. Involvement of *cdc13*$^+$ in mitotic control in *Schizo-*

saccharomyces pombe: Possible interaction of the gene product with microtubules. *EMBO J.* 7: 2321–27

Bradbury, E. M., Inglis, P. J., Matthews, H. R. 1974a. Control of cell division by very lysine rich histone (H1) phosphorylation. *Nature* 247: 257–61

Bradbury, E. M., Inglis, P. J., Matthews, H. R., Langan, T. A. 1974b. Molecular basis of control of mitotic cell division in eukaryotes. *Nature* 249: 533–56

Bramhill, D., Kornberg, A. 1988. A model for initiation at origins of DNA replication. *Cell* 54(7): 915–18

Breeden, L. 1988. Cell cycle-regulated promoters in budding yeast. *Trends Genet.* 4: 249–53

Brewer, B. J., Chlebowicz-Sledziewska, E., Fangman, W. L. 1984. Cell cycle phases in the unequal mother/daughter cell cycles of *Saccharomyces cerevisiae. Mol. Cell. Biol.* 4: 2529–31

Brinkley, B. R., Zinkowski, R. P., Mollon, W. L., Davis, F. M., Pisegna, M. A., et al. 1988. Movement and segregation of kinetochores experimentally detached from mammalian chromosomes. *Nature* 336: 251–54

Brizuela, L., Draetta, G., Beach, D. 1987. p13*suc1* acts in the fission yeast cell division cycle as a component of the p34*cdc2* protein kinase. *EMBO J.* 6: 3507–14

Broek, D., Toda, T., Michaeli, T., Levin, L., Birchmeier, C., et al. 1987. The *S. cerevisiae CDC25* gene product regulates the *RAS*/adenylate cyclase pathway. *Cell* 48: 789–99

Brooks, R. F. 1976. Regulation of the fibroblast cell cycle by serum. *Nature* 260: 248–50

Brooks, R. F. 1977. Continuous protein synthesis is required to maintain the probability of entry into S phase. *Cell* 12: 311–17

Brooks, R. F., Bennett, D. C., Smith, J. A. 1980. Mammalian cell cycles need two random transitions. *Cell* 19: 493–504

Brooks, R. F., Riddle, P. N., Richmond, F. N., Marsden, J. 1983. The G_1 distribution of "G_1-less" V79 Chinese hamster cells. *Exp. Cell Res.* 148: 127–42

Brooks, R. F., Shields, R. 1985. Cell growth, cell division and cell size homeostasis in Swiss 3T3 cells. *Exp. Cell Res.* 156: 1–6

Brown, D. B., Hanks, S. K., Murphy, E. C., Rao, P. N. 1985. Early initiation of DNA synthesis in G_1 phase HeLa cells following fusion with red cell ghosts loaded with S-phase cell extracts. *Exp. Cell Res.* 156: 251–59

Bucking-Throm, E., Duntze, W., Hartwell, L. H., Manney, T. R. 1973. Reversible arrest of haploid yeast cells at the initiation of DNA synthesis by a diffusible sex factor. *Exp. Cell Res.* 76: 99–110

Cameron, S., Levin, L., Zoller, M., Wigler, M. 1988. cAMP-independent control of sporulation, glycogen metabolism, and heat shock resistance in *S. cerevisiae. Cell* 53: 555–66

Camonis, J. H., Jacquet, M. 1988. A new *RAS* mutation that suppresses the *CDC25* gene requirement for growth of *Saccharomyces cerevisiae. Mol. Cell. Biol.* 8: 2980–83

Campisi, J., Medrano, E. E., Morreo, G., Pardee, A. B. 1982. Restriction point control of cell growth by a labile protein: Evidence for increased stability in transformed cells. *Proc. Natl. Acad. Sci. USA* 79: 436–40

Campisi, J., Pardee, A. B. 1984. Post-transcriptional control of the onset of DNA synthesis by an insulin-like growth factor. *Mol. Cell. Biol.* 4: 1807–14

Cannon, J. F., Gibbs, J. B., Tatchell, K. 1986. Suppressors of the *ras2* mutation of *Saccharomyces cerevisiae. Genetics* 113: 247–64

Cannon, J. F., Tatchell, K. 1987. Characterization of *Saccharomyces cerevisiae* genes encoding subunits of cyclic AMP-dependent protein kinase. *Mol. Cell. Biol.* 7: 2653–63

Carter, B. L. A., Jagadish, M. N. 1978a. The relationship between cell size and cell division in the yeast *Saccharomyces cerevisiae. Exp. Cell Res.* 112: 15–24

Carter, B. L. A., Jagadish, M. N. 1978b. Control of cell division in the yeast *Saccharomyces cerevisiae* cultured at different growth rates. *Exp. Cell Res.* 112: 373–83

Castor, L. N. 1980. A G_1 rate model accounts for cell-cycle kinetics attributed to "transition probability." *Nature* 287: 857–59

Cicirelli, M. F., Pelech, S. L., Krebs, E. G. 1988. Activation of multiple protein kinases during the burst in protein phosphorylation that precedes the first meiotic cell division in *Xenopus* oocytes. *J. Biol. Chem.* 263: 2009–19

Ciechanover, A., Finley, D., Varshavsky, A. 1984. Ubiquitin dependence of selective protein degradation demonstrated in the mammalian cell cycle mutant *ts85. Cell* 37: 57–66

Cooper, S. 1979. A unifying model for the G_1 period in prokaryotes and eukaryotes. *Nature* 280: 17–19

Cooper, S. 1981. The central dogma of cell biology. *Cell Biol. Int. Rep.* 5: 539–49

Creanor, J., Mitchison, J. M. 1984. Protein synthesis and its relation to the DNA-division cycle in the fission yeast *Schizosaccharomyces pombe. J. Cell Sci.* 69: 199–210

Creanor, J., Mitchison, J. M. 1986. Nucleoside diphosphokinase, an enzyme with step changes in activity during the cell cycle of the fission yeast *Schizosaccharomyces pombe*. *J. Cell Sci.* 86: 207–15

Cress, A. E., Gerner, E: W. 1977. Hydroxyurea treatment affects the G_1 phase in next generation CHO cells. *Exp. Cell Res.* 110: 347–53

Cross, F. R. 1988. *DAF1*, a mutant gene affecting size control, pheromone arrest, and cell cycle kinetics of *Saccharomyces cerevisiae*. *Mol. Cell. Biol.* 8: 4675–84

Croy, R. G., Pardee, A. B. 1983. Enhanced synthesis and stabilization of M_R 68,000 protein in transformed BALB/c-3T3 cells: Candidate for restriction point control of cell growth. *Proc. Natl. Acad. Sci. USA* 80: 4699–4703

Cyert, M. S., Kirschner, M. W. 1988. Regulation of MPF activity in vitro. *Cell* 53: 185–95

Daniel, J., Becker, J. M., Enari, E., Levitzki, A. 1987. The activation of adenylate cyclase by guanyl nucleotides in *Saccharomyces cerevisiae* is controlled by the *CDC25* start gene product. *Mol. Cell. Biol.* 7: 3857–61

Darzynkiewicz, Z., Crissman, H., Traganos, F., Steinkamp, J. 1982. Cell heterogeneity during the cell cycle. *J. Cell. Physiol.* 113: 465–74

de Roeper, A., Smith, J. A., Watt, R. A., Barry, J. M. 1977. Chromatin dispersal and DNA synthesis in G_1 and G_2 HeLa cell nuclei injected into *Xenopus* eggs. *Nature* 265: 469–70

de Terra, N. 1966. Macronuclear DNA synthesis in Stentor: Regulation by a cytoplasmic initiator. *Proc. Natl. Acad. Sci. USA* 57: 607–14

DiNardo, S., Voelkel, K., Sternglanz, R. 1984. DNA topoisomerase II mutant of *Saccharomyces cerevisiae*: topoisomerase II is required for segregation of daughter molecules at the termination of DNA replication. *Proc. Natl. Acad. Sci. USA* 81: 2616–20

Dodson, M., Echols, H., Wickner, S., Alfano, C., Mensa-Wilmot, K., et al. 1986. Specialized nucleoprotein structures at the origin of replication of bacteriophage lambda: Localized unwinding of duplex DNA by a six-protein reaction. *Proc. Natl. Acad. Sci. USA* 83(20): 7638–42

Draetta, G., Beach, D. 1988. Activation of *cdc2* protein kinase during mitosis in human cells: cell cycle-dependent phosphorylation and subunit rearrangement. *Cell* 54: 17–26

Draetta, G., Brizuela, L., Potashkin, J., Beach, D. 1987. Identification of p34 and p13, human homologs of the cell cycle regulators of fission yeast encoded by *cdc2*$^+$ and *suc1*$^+$. *Cell* 50: 319–25

Draetta, G., Worms, H. P., Morrison, D., Druker, B., Roberts, T., Beach, D. 1988. Human *cdc2* protein kinase is a major cell cycle regulated tyrosine kinase substrate. *Nature* 336: 738–44

Drebot, M. A., Johnston, G. C., Singer, R. A. 1987. A yeast mutant conditionally defective only for reentry into the mitotic cell cycle from stationary phase. *Proc. Natl. Acad. Sci. USA* 84: 7948–52

Dunphy, W. G., Brizuela, L., Beach, D., Newport, J. 1988. The *Xenopus cdc2* protein is a component of MPF, a cytoplasmic regulator of mitosis. *Cell* 54: 423–31

Dunphy, W. G., Newport, J. W. 1988a. Unraveling of mitotic control mechanisms. *Cell* 55: 925–28

Dunphy, W. G., Newport, J. W. 1988b. Mitosis-inducing factors are present in a latent form during interphase in the *Xenopus* embryo. *J. Cell Biol.* 106: 2047–56

Durkacz, B., Carr, A., Nurse, P. 1986. Transcription of the *cdc2* cell cycle control gene of the fission yeast *Schizosaccharomyces pombe*. *EMBO J.* 5: 369–73

Dutcher, S. K., Hartwell, L. H. 1983a. Test for temporal or spatial restrictions in gene product function during the cell division cycle. *Mol. Cell. Biol.* 3: 1255–65

Dutcher, S. K., Hartwell, L. H. 1983b. Genes that act before conjugation to prepare the *Saccharomyces cerevisiae* nucleus for caryogamy. *Cell* 33: 203–10

Edgar, B. A., Kiehle, C. P., Schubiger, G. 1986. Cell cycle control by the nucleo-cytoplasmic ration in early *Drosophila* development. *Cell* 44: 365–72

Elliott, S. G., McLaughlin, C. S. 1983. The yeast cell cycle: coordination of growth and division rates. *Prog. Nucleic Acid Res. Mol. Biol.* 28: 143–76

Erikson, E., Maller, J. L. 1986. Purification and characterization of a protein kinase from *Xenopus* eggs highly specific for ribosomal-protein S6. *J. Cell Biol.* 261: 350–55

Evans, T., Rosenthal, E. T., Youngblom, J., Distel, D., Hunt, T. 1983. Cyclin: a protein specified by maternal mRNA in sea-urchin eggs that is destroyed at each cleavage division. *Cell* 33: 389–96

Fantes, P. A. 1977. Control of cell size and cycle time in *Schizosaccharomyces pombe*. *J. Cell Sci.* 24: 51–67

Fantes, P. A. 1981. Isolation of cell size mutants of a fission yeast by a new selective method: characterization of mutants and implications for division control mechanisms. *J. Bacteriol.* 146: 746–54

Fantes, P. A. 1982. Dependency relations

between events in mitosis in *Schizosaccharomyces pombe*. *J. Cell Sci.* 55: 383–402

Fantes, P. A., Grant, W. D., Pritchard, R. H., Sudbery, P. E., Wheals, A. E. 1975. The regulation of cell size and the control of mitosis. *J. Theor. Biol.* 50: 213–44

Fantes, P. A., Nurse, P. 1977. Control of cell size at division in fission yeast by a growth-modulated size control over nuclear division. *Exp. Cell Res.* 107: 377–86

Fantes, P. A., Nurse, P. 1978. Control of the timing of cell division in fission yeast. *Exp. Cell Res.* 115: 317–29

Finley, D., Ciechanover, A., Varshavsky, A. 1984. Thermolability of ubiquitin-activating enzyme from the mammalian cell cycle mutant ts85. *Cell* 37: 43–55

Finley, D., Varshavsky, A. 1985. The ubiquitin system: functions and mechanisms. *Trends Biol. Sci.* 10: 343–46

Fournier, R. E., Pardee, A. B. 1975. Cell cycle studies of mononucleate and cytochalasin-B-induced binucleate fibroblasts. *Proc. Natl. Acad. Sci. USA* 72: 869–73

Frazier, E. A. J. 1973. DNA synthesis following gross alterations of the nucleocytoplasmic ratio in the ciliate *Stentor coeruleus*. *Dev. Biol.* 34(1): 77–92

Freeman, M., Glover, D. M. 1987. The *gnu* mutation of *Drosophila* causes inappropriate DNA synthesis in unfertilized and fertilized eggs. *Genes Dev.* 1: 924–30

Freeman, M., Nusslein-Volhard, C., Glover, D. M. 1986. The dissociation of nuclear and centrosomal division in *gnu*, a mutation causing giant nuclei in *Drosophila* cell. *Cell* 46: 457–68

Friedman, D. L., Ken, R. 1988. Insulin stimulates incorporation of 32-P_i into nuclear lamins A and C in quiescent BHK-21 cells. *J. Biol. Chem.* 263: 1103–6

Gautier, J., Norbury, C., Lohka, M., Nurse, P., Maller, J. 1988. Purified maturation-promoting factor contains the product of a *Xenopus* homolog of the fission yeast cell cycle control gene *cdc2*$^+$. *Cell* 54: 433–39

Gerhart, J., Wu, M., Kirschner, M. 1984. Cell cycle dynamics of an M-phase specific cytoplasmic factor in *Xenopus* oocytes and eggs. *J. Cell Biol.* 98: 1247–55

Goebl, M., Byers, B. 1988. Cyclin in fission yeast. *Cell* 54: 739–40

Goebl, M. G., Yochem, J., Jentsch, S., McGrath, J. P., Varshavsky, A., Byers, B. 1988. The yeast cell cycle gene *CDC34* encodes a ubiquitin-conjugating enzyme. *Science* 241: 1331–35

Gordon, S., Cohn, Z. 1971. Macrophage-melanoma cell heterokaryons. 3. The activation of macrophage DNA synthesis. Studies with inhibitors of protein synthesis and with synchronized melanoma cells. *J. Exp. Med.* 134(4): 935–46

Goyanes, V. J., Mendez, J. 1982. Extra-centromeric connections between sister chromatids demonstrated in human chromosomes induced to condense asymmetrically. *Hum. Genet.* 62: 324–26

Graves, J. 1972. DNA synthesis in heterokaryons formed by fusion of mammalian cells from different species. *Exp. Cell Res.* 72(2): 393–403

Gurdon, J. B. 1967. On the origin and persistence of a cytoplasmic state induring nuclear DNA synthesis in frogs' eggs. *Proc. Natl. Acad. Sci. USA* 58: 545–52

Hagan, I., Hayles, J., Nurse, P. 1988. Cloning and sequencing of the cyclin-related *cdc13*$^+$ gene and a cytological study of its role in fission yeast mitosis. *J. Cell Sci.* 91: 587–95

Hammond, M. P., Laird, C. D. 1985. Control of DNA replication and spatial distribution of defined DNA sequences in salivary gland cells of *Drosophila melanogaster*. *Chromosoma* 91: 279–86

Hanic-Joyce, P. J., Johnston, G. C., Singer, R. A. 1987. Regulated arrest of cell proliferation mediated by yeast *prt1* mutations. *Exp. Cell Res.* 172: 134–45

Hara, K., Tydeman, P., Kirschner, M. 1980. A cytoplasmic clock with the same period as the division cycle in *Xenopus* eggs. *Proc. Natl. Acad. Sci. USA* 77: 462–66

Harland, R. M., Laskey, R. A. 1980. Regulated replication of DNA microinjected into eggs of *Xenopus laevis*. *Cell* 21: 761–71

Hartwell, L. H. 1971. Genetic control of the cell division cycle in yeast II. Genes controlling DNA replication and its initiation. *J. Mol. Biol.* 59: 183–94

Hartwell, L. H. 1976. Sequential function of gene products relative to DNA synthesis in the yeast cell cycle. *J. Mol. Biol.* 104: 803–17

Hartwell, L. H., Culotti, J., Pringle, J. R., Reid, B. J. 1974. Genetic control of the cell division cycle in yeast. *Science* 183: 46–51

Hartwell, L. H., Smith, D. 1985. Altered fidelity of mitotic chromosome transmission in cell cycle mutants of *S. cerevisiae*. *Genetics* 110: 381–95

Hartwell, L. H., Unger, M. W. 1977. Unequal division in *Saccharomyces cerevisiae* and its implications for the control of cell division. *J. Cell Biol.* 75: 422–35

Hayashi, A., Yamamoto, S., Nishimoto, T., Takahashi, T. 1982. Chromosome condensing factor(s) induced in *tsBN2* cells at a nonpermissive temperature: Evidence for transferable material by cell fusion. *Cell Struct. Funct.* 7: 291–94

Hayles, J., Aves, S., Nurse, P. 1986a. *Suc1* is

an essential gene involved in both the cell cycle and growth in fission yeast. *EMBO J.* 5: 3373–79

Hayles, J., Beach, D., Durkacz, B., Nurse, P. 1986b. The fission yeast cell cycle control gene *cdc2*: isolation of a sequence *suc1* that suppresses *cdc2* mutant function. *Mol. Gen. Genet.* 202: 291–93

Hayles, J., Nurse, P. 1986. Cell cycle regulation in yeast. *J. Cell Sci. Suppl.* 4: 155–70

Hereford, L. M., Hartwell, L. H. 1974. Sequential gene function in the initiation of *Saccharomyces cerevisiae* DNA synthesis. *J. Mol. Biol.* 84: 445–61

Highfield, D. P., Dewey, W. C. 1972. Inhibition of DNA synthesis in synchronized Chinese hamster cells treated in G_1 or early S phase with cycloheximide or puromycin. *Exp. Cell Res.* 75(2): 314–20

Hindley, J., Phear, G., Stein, M., Beach, D. 1987. *Suc1*$^+$ encodes a predicted 13-kilodalton protein that is essential for cell viability and is directly involved in the division cycle of *Schizosaccharomyces pombe. Mol. Cell. Biol.* 7: 504–11

Hittelman, W. N., Rao, P. N. 1978. Mapping G_1 phase by the structural morphology of the prematurely condensed chromosomes. *J. Cell Physiol.* 95: 333–42

Hola, M., Riley, P. A. 1987. The relative significance of growth rate and interdivision time in the size control of cultured mammalian epithelial cells. *J. Cell Sci.* 88: 73–80

Holm, C., Goto, T., Wang, J. C., Botstein, D. 1985. DNA topoisomerase II is required at the time of mitosis in yeast. *Cell* 41: 553–63

Iida, H., Yahara, I. 1984. Specific early-G_1 blocks accompanied with stringent response in *Saccharomyces cerevisiae* lead to growth arrest in resting state similar to the G_0 of higher eucaryotes. *J. Cell Biol.* 98: 1185–93

Inglis, R. J., Langan, T. A., Matthews, H. R., Hardie, D. G., Bradbury, E. M. 1976. Advance of mitosis by histone phosphokinase. *Exp. Cell Res.* 97: 418–25

Jagadish, M. N., Carter, B. L. A. 1977. Genetic control of cell division in yeast cultured at different growth rates. *Nature* 269: 145–47

Jagadish, M. N., Carter, B. L. A. 1978. Effects of temperature and nutritional conditions on the mitotic cell cycle of *Saccharomyces cerevisiae. J. Cell Sci.* 31: 71–78

Jahng, K.-Y., Ferguson, J., Reed, S. I. 1988. Mutations in a gene encoding the α-subunit of a *Saccharomyces cerevisiae* G-protein indicate a role in mating pheromone signaling. *Mol. Cell. Biol.* 8: 2484–93

Jat, P. S., Sharp, P. A. 1989. Cell lines established by a temperature-sensitive SV40 large T antigen gene are growth restricted at the nonpermissive temperature. *Mol. Cell. Biol.* In press

Jiang, L.-W., Schindler, M. 1988. Nuclear transport in 3T3 fibroblasts: Effects of growth factors, transformation, and cell shape. *J. Cell Biol.* 106: 13–19

Johnson, R. T., Harris, H. 1969. DNA synthesis and mitosis in fused cells. I. HeLa homokaryons. *J. Cell Sci.* 5: 603–24

Johnson, R. T., Rao, P. N. 1970. Mammalian-cell fusion: Induction of premature chromosome condensation in interphase nuclei. *Nature* 226(247): 717–22

Johnson, R. T., Rao, P. N. 1971. Nucleocytoplasmic interactions in the achievement of nuclear synchrony in DNA synthesis and mitosis in multinucleate cells. *Biol. Rev.* 46(1): 97–155

Johnston, G. C. 1977. Cell size and budding during starvation of the yeast *Saccharomyces cerevisiae. J. Bacteriol.* 132: 738–39

Johnston, G. C., Ehrhardt, C. W., Lorincz, A., Carter, B. L. A. 1979. Regulation of cell size in the yeast *Saccharomyces cerevisiae. J. Bacteriol.* 137: 1–5

Johnston, G. C., Pringle, J. R., Hartwell, L. H. 1977a. Coordination of growth with cell division in the yeast *Saccharomyces cerevisiae. Exp. Cell Res.* 105: 79—98

Johnston, G. C., Singer, R. A. 1978. RNA synthesis and control of cell division in the yeast *S. cerevisiae. Cell* 14: 951–58

Johnston, G. C., Singer, R. A. 1980. Ribosomal precursor RNA metabolism and cell division in the yeast *Saccharomyces cerevisiae. Mol. Gen. Genet.* 178: 357–60

Johnston, G. C., Singer, R. A. 1983. Growth and the cell cycle of the yeast *Saccharomyces cerevisiae*. I. Slowing S phase or nuclear division decreases the G_1 cell cycle period. *Exp. Cell Res.* 149: 1–13

Johnston, G. C., Singer, R. A., McFarlane, E. S. 1977b. Growth and cell division during nitrogen starvation of the yeast *Saccharomyces cerevisiae. J. Bacteriol.* 132: 723–30

Johnston, G. C., Singer, R. A., Sharrow, S. O., Slater, M. L. 1980. Cell division in the yeast *Saccharomyces cerevisiae* growing at different rates. *J. Gen. Microbiol.* 118: 479–84

Killander, D., Zetterberg, A. 1965. Quantitative cytochemical studies on interphase growth. I. Determination of DNA, RNA and mass content of age determined mouse fibroblasts in vitro. *Exp. Cell Res.* 38: 272–89

Kimelman, D., Kirschner, M., Scherson, T. 1987. The events of the midblastula transition in *Xenopus* are regulated by changes in the cell cycle. *Cell* 48: 399–407

Kirschner, M., Newport, J., Gerhart, J. 1985. The timing of early developmental events in *Xenopus*. *Trends Genet.* 1: 41–47

Kishimoto, T., Hirai, S., Kanatani, H. 1981. Role of germinal vesicle material in producing maturation-promoting factor in starfish oocyte. *Dev. Biol.* 81: 177–81

Klevecz, R. R. 1976. Quantized generation time in mammalian cells as an expression of the cellular clock. *Proc. Natl. Acad. Sci. USA* 73: 4012–16

Klevecz, R. R., King, G. A. 1982. Temperature compensation in the mammalian cell cycle. *Exp. Cell Res.* 140: 307–13

Koch, A. L. 1980. Does the variability of the cell cycle result from one or many chance events? *Nature* 286: 80–82

Kulka, R. G., Raboy, B., Schuster, R., Parag, H. A., Diamond, G., et al. 1988. A Chinese hamster cell cycle mutant arrested at G_2 phase has a temperature-sensitive ubiquitin-activating enzyme, E1. *J. Biol. Chem.* 263: 15726–31

Labbe, J. C., Lee, M. G., Nurse, P., Picard, A., Doree, M. 1988a. Activation at M-phase of a protein kinase encoded by a starfish homologue of the cell cycle control gene *cdc2*$^+$. *Nature* 335: 251–54

Labbe, J. C., Picard, A., Karsenti, E., Doree, M. 1988b. An M-phase-specific protein kinase of *Xenopus* oocytes: partial purification and possible mechanism of its periodic activation. *Dev. Biol.* 127: 157–69

Larsson, O., Zetterberg, A., Engström, W. 1985a. Cell-cycle-specific induction of quiescence achieved by limited inhibition of protein synthesis: Counteractive effect of addition of purified growth factors. *J. Cell Sci.* 73: 375–87

Larsson, O., Zetterberg, A., Engström, W. 1985b. Consequences of parental exposure to serum-free medium for progeny cell division. *J. Cell Sci.* 75: 259–68

Lau, C., Pardee, A. 1982. Mechanism by which caffeine potentiates lethality of nitrogen mustard. *Proc. Natl. Acad. Sci. USA* 79: 2942–46

Lee, M., Nurse, P. 1988. Cell cycle control genes in fission yeast and mammalian cells. *Trends Genet.* 4: 287–90

Lee, M. G., Norbury, C. J., Spurr, N. K., Nurse, P. 1988. Regulated expression and phosphorylation of a possible mammalian cell-cycle control protein. *Nature* 333: 676–78

Lee, M. G., Nurse, P. 1987. Complementation used to clone a human homologue of the fission yeast cell cycle control gene *cdc2*. *Nature* 327: 31–35

Lica, L., Narayanswami, S., Hankado, B. 1986. Mouse satellite DNA, centromere structure of sister chromatid paring. *J. Cell Biol.* 103: 1145–51

Lillie, S. H., Pringle, J. R. 1980. Reserve carbohydrate metabolism in *Saccharomyces cerevisiae*: responses to nutrient limitation. *J. Bacteriol.* 143: 1384–94

Liskay, R. M., Kornfeld, B., Fullerton, P., Evans, R. 1980. Protein synthesis and the presence or absence of a measurable G_1 in cultured Chinese hamster cells. *J. Cell. Physiol.* 104: 461–67

Liskay, R. M., Leonard, K. E., Prescott, D. M. 1979. Different Chinese hamster cell lines express a G_1 period for different reasons. *Somatic Cell Genet.* 5: 615–23

Liskay, R. M., Prescott, D. M. 1978. Genetic analysis of the G_1 period: isolation of mutants (or variations) with a G_1 period from a Chinese hamster cell line lacking G_1. *Proc. Natl. Acad. Sci. USA* 75: 2873–77

Lohka, M. J., Hayes, M. K., Maller, J. L. 1988. Purification of maturation-promoting factor, an intracellular regulator of early mitotic events. *Proc. Natl. Acad. Sci. USA* 85: 3009–13

Lohka, M. J., Kyes, J. L., Maller, J. L. 1987. Metaphase protein phosphorylation in *Xenopus laevis* eggs. *Mol. Cell. Biol.* 7: 760–68

Lord, P. G., Wheals, A. E. 1980. Asymmetrical division of *Saccharomyces cerevisiae*. *J. Bacteriol.* 142: 808–18

Lord, P. G., Wheals, A. E. 1981. Variability in individual cell cycles of *Saccharomyces cerevisiae*. *J. Cell Sci.* 50: 361–76

Lord, P. G., Wheals, A. E. 1983. Rate of cell cycle initiation of yeast cells when cell size is not a rate-determining factor. *J. Cell Sci.* 59: 183–201

Lorincz, A., Carter, B. L. A. 1979. Control of cell size at bud initiation in *Saccharomyces cerevisiae*. *J. Gen. Microbiol.* 113: 287–95

Lorincz, A. T., Miller, M. J., Xuong, N.-H., Geiduschek, E. P. 1982. Identification of proteins whose synthesis is modulated during the cell cycle of *Saccharomyces cerevisiae*. *Mol. Cell. Biol.* 2: 1532–49

Lorincz, A. T., Reed, S. I. 1986. Sequence analysis of temperature-sensitive mutations in the *Saccharomyces cerevisiae* gene *CDC28*. *Mol. Cell. Biol.* 6: 4099–4103

Lycan, D. E., Osley, M. A., Hereford, L. M. 1987. Role of transcriptional and post-transcriptional regulation in expression of histone genes in *Saccharomyces cerevisiae*. *Mol. Cell. Biol.* 7: 614–21

Martegani, E., Baroni, M., Vanoni, M. 1986. Interaction of cAMP with the *CDC25*-mediated step in the cell cycle of budding yeast. *Exp. Cell Res.* 162: 544–48

Martegani, E., Vanoni, M., Baroni, M. 1984. Macromolecular syntheses in the cell cycle mutant *cdc25* of budding yeast. *Eur. J. Biochem.* 144: 205–10

Masui, Y., Markert, C. L. 1971. Cytoplasmic control of nuclear behavior during meiotic maturation of frog oocytes. *J. Exp. Zool.* 177: 129–46

Matsumoto, K., Uno, I., Ishikawa, T. 1983. Control of cell division in *Saccharomyces cerevisiae* mutants defective in a denylatecyclase and cAMP-dependent protein kinase. *Exp. Cell Res.* 146: 151–61

Mazia, D. 1963. Synthetic activities leading to mitosis. *J. Cell Comp. Physiol.* 62(Suppl. I): 123–40

Mazia, D. 1974. Chromosome cycles turned on in unfertilized sea urchin eggs exposed to NH_4OH. *Proc. Natl. Acad. Sci. USA* 71: 690–93

Mendenhall, M. D., Jones, C. A., Reed, S. I. 1987. Dual regulation of the yeast *CDC28*-p40 protein kinase complex: cell cycle, pheromone and nutrient limitation effects. *Cell* 50: 927–35

Mendenhall, M. D., Richardson, H. E., Reed, S. I. 1988. Dominant negative protein kinase mutations that confer a G_1 arrest phenotype. *Proc. Natl. Acad. Sci. USA* 85: 4426–30

Mercer, W. E., Avignolo, C., Baserga, R. 1984. Role of the p53 protein in cell proliferation as studied by microinjection of monoclonal antibodies. *Mol. Cell. Biol.* 4: 276–81

Messer, W., Bellekes, U., Lother, H. 1985. Effect of dam methylation on the activity of the *E. coli* replication origin, OriC. *EMBO J.* 4: 1327–32

Meyerhof, P. G., Masui, Y. 1979a. Chromosome condensation activity in *R. pipiens* matured in vivo and in blastomeres arrested by CS. *Exp. Cell Res.* 123: 345–53

Meyerhof, P. G., Masui, Y. 1979b. Properties of CSF from *Xenopus* eggs. *Dev. Biol.* 72: 182–87

Minor, P. D., Smith, J. A. 1974. Explanation of degree of correlation of sibling generation times in animal cells. *Nature* 248: 241–43

Mitchison, J. M. 1971. *The Biology of the Cell Cycle*. Cambridge: Univ. Press

Mitchison, J. M., Nurse, P. 1985. Growth in cell length in the fission yeast *Schizosaccharomyces pombe*. *J. Cell Sci.* 75: 357–76

Miyata, H., Miyata, M., Michio, I. 1978. The cell cycle in the fission yeast, *Schizosaccharomyces pombe*. I. Relationship between cell size and cycle time. *Cell Struct. Funct.* 3: 39–46

Moore, S. A. 1987. Alpha-factor inhibition of the rate of cell passage through the "start" step of cell division in *Saccharomyces cerevisiae* yeast: estimation of the division delay per alpha-factor receptor complex. *Exp. Cell Res.* 171: 411–25

Moore, S. A. 1984a. Yeast-cells recover from mating pheromone alpha-factor-induced division arrest by desensitization in the absence of alpha-factor destruction. *J. Biol. Chem.* 259: 1004–10

Moore, S. A. 1984b. Synchronous cell growth occurs upon synchronizing the two regulatory steps of the *Saccharomyces cerevisiae* cell cycle. *Exp. Cell Res.* 151: 542–56

Moore, S. A. 1988. Kinetic evidence for a critical rate of protein synthesis in the *Saccharomyces cerevisiae* yeast cell cycle. *J. Biol. Chem.* 263: 9674–81

Moser, G. C., Fallon, R. J., Meiss, H. K. 1981. Fluorimetric measurements and chromatin condensation patterns of nuclei from 3T3 cells throughout G_1. *J. Cell. Physiol.* 106: 293–301

Moser, G. C., Muller, H., Robbins, E. 1975. Differential nuclear fluorescence during the cell cycle. *Exp. Cell Res.* 91: 73–78

Murphy, J. S., Landsberger, F. R., Kikuchi, T., Tamm, I. 1984. Occurrence of cell division is not exponentially distributed: differences in the generation times of sister cells can be derived from the theory of survival of populations. *Proc. Natl. Acad. Sci. USA* 81: 2379–83

Murray, A. W. 1987. Cyclins in meiosis and mitosis. *Nature* 326: 542–43

Musk, S. R. R., Downes, C. S., Johnson, R. T. 1988. Caffeine induces uncoordinated expression of cell cycle functions after ultraviolet irradiation. *J. Cell Sci.* 90: 591–99

Nash, R., Tokiwa, G., Anand, S., Erickson, K., Futcher, A. B. 1988. The *WHI1*$^+$ gene of *Saccharomyces cerevisiae* tethers cell division to cell size and is a cyclin homolog. *EMBO J.* 7: 4335–46

Nasmyth, K., Nurse, P., Fraser, R. S. S. 1979. The effect of cell mass on the cell-cycle timing and duration of S-phase in fission yeast. *J. Cell Sci.* 39: 215–33

Nasmyth, K. A. 1979. A control acting over the initiation of DNA replication in the yeast *Schizosaccharomyces pombe*. *J. Cell Sci.* 36: 155–68

Newport, J., Kirschner, M. 1982. A major developmental transition in early *Xenopus* embryos: 1. characterization and timing of cellular changes at the midblastula stage. *Cell* 30: 675–86

Newport, J. W., Kirschner, M. W. 1984. Regulation of the cell cycle during early *Xenopus* development. *Cell* 37: 731–42

Nicklas, R. B., Kubai, D. F. 1985. Microtubules, chromosome movement, and reorientation after chromosomes are detached from the spindle by micromanipulation. *Chromosoma* 92: 313–24

Nicolini, C., Ajiro, K., Borun, T., Baserga, R. 1975. Chromatin changes during the cell cycle of HeLa cells. *J. Biol. Chem.* 250: 3381–85

Nishimoto, T. 1988. The "BN2" gene, a regulator for the onset of chromosome condensation. *BioEssays* 9: 121–24

Nishimoto, T., Eilen, E., Basilico, C. 1978. Premature chromosome condensation in a *ts* DNA-mutant of BHK cells. *Cell* 15: 475–83

Novak, B., Halbauer, J., Laszlo, E. 1988. The effect of CO_2 on the timing of cell cycle events in fission yeast *Schizosaccharomyces pombe*. *J. Cell Sci.* 89: 433–39

Novak, B., Mitchison, J. M. 1986. Change in the rate of CO_2 production in synchronous cultures of the fission yeast *Schizosaccharomyces pombe*: a periodic cell cycle event that persists after the DNA-division cycle has been blocked. *J. Cell Sci.* 86: 191–206

Nurse, P. 1980. Cell cycle control—both deterministic and probablistic? *Nature* 288: 9–10

Nurse, P. 1985. Cell cycle control genes in yeast. *Trends Genet.* 1: 51–55

Nurse, P., Bissett, Y. 1981. Gene required in G_1 for commitment to cell cycle and in G_2 for control of mitosis in fission yeast. *Nature* 292: 558–60

Nurse, P., Thuriaux, P. 1977. Controls over the timing of DNA replication during the cell-cycle of fission yeast. *Exp Cell Res.* 107: 376–75

Oakley, B. R., Morris, N. R. 1983. A mutation in *Aspergillus nidulans* that blocks the transition from interphase to prophase. *J. Cell Biol.* 96: 1155–58

Obara, Y., Chai, L. S., Weinfeld, H., Sandberg, A. A. 1974. Prophasing of interphase nuclei and induction of nuclear envelopes around metaphase chromosomes in HeLa and Chinese hamster homo- and heterokaryons. *J. Cell Biol.* 62: 104–13

Ogden, G., Pratt, M., Schaechter, M. 1988. The replicative origin of E. coli binds to cell membranes only when hemimethylated. *Cell* 54: 127–35

Ogden, J. E., Fantes, P. A. 1986. Isolation of a novel type of mutation in the mitotic control of *Schizosaccharomyces pombe* whose phenotypic expression is dependent on the genetic background and nutritional environment. *Curr. Genet.* 10: 509–14

Ohkura, H., Adachi, Y., Kinoshita, N., Niwa, O., Toda, T., Yanagida, M. 1988. Cold-sensitive and caffeine-supersensitive mutants of the *Schizosaccharomyces pombe* dis genes implicated in sister chromatid separation during mitosis. *EMBO J.* 7: 1465–73

Ohno, K., Kimura, G. 1984. Genetic analysis of control of proliferation in fibroblastic cells in culture. II. Alteration in proliferative and survival phenotypes in a set of temperature-sensitive mutants of rat 3Y1 cells after infection or transformation with simian virus 40. *Somat. Cell Mol. Genet.* 10(1): 29–36

Ohno, K., Okuda, A., Ohtsu, M., Kimura, G. 1984. Genetic analysis of control of proliferation in fibroblastic cells in culture. I. Isolation and characterization of mutants temperature-sensitive for proliferation or survival of untransformed diploid rat cell line 3Y1. *Somat. Cell Mol. Genet.* 10: 17–28

Okuda, A., Kimura, G. 1984. Control in previous and present generations of preparation for entry in S phase and the relationship to resting state in 3Y1 rat fibroblastic cells. *Exp. Cell Res.* 155: 24–32

Okuda, A., Kimura, G. 1986. Serum-dependent control of entry into S phase of next generation in rat 3Y1 fibroblasts. *Exp. Cell Res.* 163: 127–34

Okuda, A., Kimura, G. 1988. Non-specific elongation of cell cycle phases by cycloheximide in rat 3Y1 cells, and specific reduction of G_1 phase elongation by simian virus-40 large T-antigen. *J. Cell Sci.* 91: 296–302

Ord, M. J. 1969. Control of DNA synthesis in Amoeba proteus. *Nature* 221: 964–66

Osmani, S. A., Engle, D. B., Doonan, J. H., Morris, N. R. 1988a. Spindle formation and chromatin condensation in cells blocked at interphase by mutation of a negative cell cycle control gene. *Cell* 52: 241–51

Osmani, S. A., May, G. S., Morris, N. R. 1987. Regulation of the mRNA levels of *nimA*, a gene required for G_2-M transition in *Aspergillus nidulans*. *J. Cell Biol.* 104: 1495–1504

Osmani, S. A., Pu, R. T., Morris, N. R. 1988b. Mitotic induction and maintenance by overexpression of a G_2-specific gene that encodes a potential protein kinase. *Cell* 53: 237–44

Pardee, A. B., Coppock, D. L., Yang, H. C. 1986. Regulation of cell proliferation at the onset of DNA synthesis. *J. Cell Sci. Suppl.* 4: 171–80

Pardee, A. B., Dubrow, R., Hamlin, J. L., Kletzien, R. 1978. Animal cell cycle. *Annu. Rev. Biochem.* 47: 715–50

Pardee, A. B., Medrano, E. E., Rossow, P. W. 1981. A labile protein model for growth control of mammalian cells. In *The*

Biology of Normal Human Growth, ed. M. Ritzen, et al, pp. 59–69. New York: Raven

Paris, S., Pringle, J. R. 1983. *Saccharomyces cerevisiae*: heat and gluculase sensitivities of starved cells. *Ann. Microbiol.* 134: 379–85

Pederson, T. 1972. Chromatin structure and the cell cycle. *Proc. Natl. Acad. Sci. USA* 69: 2224–28

Pederson, T., Robbins, E. 1972. Chromatin structure and the cell division cycle. *J. Cell Biol.* 55: 322–27

Picard, A., Doree, M. 1984. The role of the germinal vesicle in producing maturation-promoting factor (MPF) as revealed by the removal and transplantation of nuclear material in starfish oocytes. *Dev. Biol.* 104: 357–65

Picard, A., Peaucellier, G., LeBouffant, F., LePeuch, C., Poree, M. 1985. Role of protein synthesis and proteases in production and inactivation of MPA during meiotic maturation of starfish oocytes. *Dev. Biol.* 109: 311

Piggott, J. R., Rai, R., Carter, B. L. A. 1982. A bifunctional gene product involved in two phases of the yeast cell cycle. *Nature* 298: 391–93

Pines, J., Hunt, T. 1987. Molecular cloning and characterization of the mRNA for cyclin from sea urchin eggs. *EMBO J.* 6: 2987–95

Pledger, W. J., Stiles, C. D., Antoniades, H. N., Scher, C. D. 1978. An ordered sequence of events is required before BALB/c-3T3 cells become committed to DNA synthesis. *Proc. Natl. Acad. Sci. USA* 75: 2839–43

Plesset, J., Ludwig, J. R., Cox, B. S., McLaughlin, C. S. 1987. Effect of cell cycle position on thermotolerance in *Saccharomyces cerevisiae*. *J. Bacteriol.* 169: 779–84

Popolo, L., Alberghina, L. 1984. Identification of a labile protein involved in the G_1 to S transition in *Saccharomyces cerevisiae*. *Proc. Natl. Acad. Sci. USA* 81: 120–24

Popolo, L., Vai, M., Alberghina, L. 1986. Identification of a glycoprotein involved in cell cycle progression in yeast. *J. Biol. Chem.* 261: 3479–82

Popolo, L., Vanoni, M., Alberghina, L. 1982. Control of the yeast cell cycle by protein synthesis. *Exp. Cell Res.* 142: 69–78

Potashkin, J. A., Beach, D. H. 1988. Multiple phosphorylated forms of the product of the fission yeast cell division cycle gene *cdc2*$^+$. *Curr. Genet.* 14: 235–40

Prescott, D. M. 1956. Relation between cell growth and cell division. II. The effect of cell size on cell growth rate and generation time in *Amoeba proteus*. *Exp. Cell Res.* 11: 86–98

Prescott, D. M. 1976. The cell cycle and the control of cellular production. *Adv. Genet.* 18: 99–177

Prescott, D. M. 1987. Cell reproduction. *Int. Rev. Cytol.* 100: 93–128

Pringle, J. R. 1978. The use of conditional lethal cell cycle mutants for temporal and functional sequence mapping of cell cycle events. *J. Cell. Physiol.* 95: 393–406

Pringle, J. R., Hartwell, L. H. 1981. The *Saccharomyces cerevisiae* cell cycle. In *The Molecular Biology of the Yeast Saccharomyces*, ed. J. N. Strathern, E. W. Jones, J. R. Broach, pp. 97–142. New York: Cold Spring Harbor Lab.

Raff, J. W., Glover, D. M. 1988. Nuclear and cytoplasmic mitotic cycles continue in *Drosophila* embryos in which DNA synthesis is inhibited with aphidicolin. *J. Cell Biol.* 107: 2009–19

Rao, P. N., Davis, F. M., Pisegna, M. A. 1986. Mitosis with unreplicated genome (MUG) induction by caffeine in CHO cells arrested in S phase. *J. Cell Biol.* 103: 169a

Rao, P. N., Hanks, S. K. 1980. Chromatin structure during the prereplicative phases in the life cycle of mammalian cells. *Cell Biophys.* 2: 327–37

Rao, P. N., Hittelman, W. N., Wilson, B. A. 1975. Mammalian-cell fusion. Regulation of mitosis in binucleate HeLA-cells. *Exp. Cell Res.* 90: 40

Rao, P. N., Johnson, R. N. 1970. Mammalian cell fusion: studies on the regulation of DNA synthesis and mitosis. *Nature* 225: 159–64

Rao, P. N., Satya-Prakash, K. L., Wang, Y. C. 1984. The role of the G_1 period in the life cycle of eukaryotic cells. *J. Cell. Physiol.* 199: 77–81

Rao, P. N., Smith, M. L. 1981. Differential response of cycling and noncycling cells to inducers of DNA synthesis and mitosis. *J. Cell Biol.* 88: 649–53

Rao, P. N., Sunk, P. S. 1980. Correlation between the high rate of protein synthesis during mitosis and the absence of a G_1 period in V79-8 cells. *Exp. Cell Res.* 125: 507–11

Rao, P. N., Sunkara, P. S., Wilson, B. A. 1977a. Regulation of DNA synthesis: Age-dependent cooperation among G_1 cells upon fusion. *Proc. Natl. Acad. Sci. USA* 74: 2869–73

Rao, P. N., Wilson, B., Puck, T. T. 1977b. Premature chromosome condensation and cell cycle analysis. *J. Cell. Physiol.* 91: 131–42

Rasmussen, C. D., Means, A. R. 1989. Calmodulin is required for cell-cycle progression during G_1 and mitosis. *EMBO J.* 8(1): 73–82

Reed, S. I. 1980. The selection of *S. cerevisiae* mutants defective in the start event of cell division. *Nature* 95: 561–77

Reed, S. I., Hadwiger, J. A., Lorincz, A. T. 1985. Protein kinase activity associated with the product of the yeast cell division cycle gene *CDC28*. *Proc. Natl. Acad. Sci. USA* 82: 4055–59

Reed, S. I., Nasmyth, K. A. 1980. Isolation of genes by complementation in yeast; molecular cloning of a cell cycle gene. *Proc. Natl. Acad. Sci. USA* 77: 2119–23

Rivin, C. J., Fangman, W. L. 1980. Cell cycle phase expansion in nitrogen-limited cultures of *Saccharomyces cerevisiae*. *J. Cell Biol.* 85: 96–107

Roberts, J. M., D'Urso, G. 1988. An origin unwinding activity regulates initiation of DNA replication during mammalian cell cycle. *Science* 241: 1486–89

Roberts, J. M., Weintraub, H. 1986. Negative control of DNA replication in composite SV40-bovine papilloma virus plasmids. *Cell* 46: 741–52

Roberts, J. M., Weintraub, H. 1988. *Cis*-acting negative control of DNA replication in eukaryotic cells. *Cell* 452: 397–404

Ronning, O. W., Lindmo, T. 1983. Progress through G_1 and S in relation to net protein accumulation of human NHIK 3025 cells. *Exp. Cell Res.* 144: 171–79

Ronning, O. W., Lindmo, T., Pettersen, E. O., Seglen, P. O. 1981. The role of protein accumulation in the cell cycle control of human NHIK 3025 cells. *J. Cell. Physiol.* 109: 411–18

Ronning, O. W., Pettersen, E. O. 1984. Doubling of cell mass is not necessary in order to achieve cell division in cultured human cells. *Exp. Cell Res.* 155: 267–72

Ronning, O. W., Seglen, P. O. 1982. The relation between protein accumulation and cell cycle traverse of human NHIK 3025 cells in unbalanced growth. *J. Cell. Physiol.* 112: 19–26

Rose, M. D., Fink, G. R. 1987. *KAR1*, a gene required for function of both intranuclear and extranuclear microtubules in yeast. *Cell* 48: 1047–60

Rose, M. D., Price, B. R., Fink, G. R. 1986. *Saccharomyces cerevisiae* nuclear fusion requires prior activation by alpha-factor. *Mol. Cell. Biol.* 6: 3490–97

Rosenthal, E. T., Hunt, T., Ruderman, J. V. 1980. Selective translation from RNA controls the pattern of protein synthesis during early development of the surf clam. *Cell* 20: 487–94

Rossow, P. W., Riddle, V. G. H., Pardee, A. B. 1979. Synthesis of labile, serum-dependent protein in early G_1 controls animal cell growth *Proc. Natl. Acad. Sci. USA* 76: 4446–50

Russell, D. W., Zinder, N. D. 1987. Hemimethylation prevents DNA replication in E coli. *Cell* 50: 1071–79

Russell, P., Nurse, P. 1986. *cdc25*$^+$ functions as an inducer in the mitotic control of fission yeast. *Cell* 45: 145–53

Russell, P., Nurse, P. 1987a. Negative regulation of mitosis by *wee1*$^+$, a gene encoding a protein kinase homolog. *Cell* 49: 559–67

Russell, P., Nurse, P. 1987b. The mitotic inducer *nim1*$^+$ functions in a regulatory network of protein kinase homologs controlling the initiation of mitosis. *Cell* 49: 569–96

Sagata, W., Oskarsson, M., Copeland, T., Brumbaugh, J., Van de Woude, G. 1988. Function of c-mos proto-oncogene product in meiotic maturation in *Xenopus* oocytes. *Nature* 335: 519–25

Samokhin, G. P., Minin, A. A., Bespalova, J. D., Titov, M. I., Smirnov, V. N. 1981. Independent action of alpha factor and cycloheximide on the rate of cell-cycle initiation in *Saccharomyces cerevisiae*. *FEMS Microbiol.* 10: 185–88

Schlegel, R., Croy, R. G., Pardee, A. B. 1987. Exposure to caffeine and suppression of DNA replication combine to stabilize the proteins and RNA required for premature mitotic events. *J. Cell Physiol.* 131(1): 85–91

Schlegel, R., Pardee, A. B. 1986. Caffeine-induced uncoupling of mitosis from the completion of DNA replication in mammalian cells. *Science* 232: 1264–66

Schlegel, R., Pardee, A. B. 1987. Periodic mitotic events induced in the absence of DNA replication. *Proc. Natl. Acad. Sci. USA* 84: 9025–29

Schneiderman, M. H., Dewey, W. C., Highfield, D. P. 1971. Inhibition of DNA synthesis in synchronized Chinese hamster cells treated in G_1 with cycloheximide. *Exp. Cell Res.* 67: 147–55

Schnos, M., Zahn, R., Inman, R., Blattner, F. 1988. Initiation protein induced helix destabilization at the lambda origin: A prepriming step in DNA replication. *Cell* 52: 385–95

Sennerstam, R. 1988. Partition of protein (mass) to sister cell pairs at mitosis: a reevaluation. *J. Cell Sci.* 90: 301–6

Sheehan, M. A., Mills, A. D., Sleeman, A. M., Laskey, R. A., Blow, J. J. 1988. Steps in the assembly of replication-competent nuclei in a cell-free system from *Xenopus* eggs. *J. Cell Biol.* 106: 1–12

Shields, R. 1977. Transition probability and the origin of variation in the cell cycle. *Nature* 267: 704–7

Shields, R. 1978. Further evidence for a random transition in the cell cycle. *Nature* 273: 755–58

Shields, R., Brooks, R. F., Riddle, P. N., Capellaro, D. F., Delia, D. 1978. Cell size, cell cycle and transition probability in mouse fibroblasts. *Cell* 15: 469–74

Shields, R., Smith, J. A. 1978. Cells regulate their proliferation through alterations in transition probability. *J. Cell. Physiol.* 91: 345–56

Shilo, B., Riddle V. G. H., Pardee, A. B. 1979. Protein turnover and cell-cycle initiation in yeast. *Exp. Cell Res.* 123: 221–27

Shilo, B., Shilo, V., Simchen, G. 1976. Cell-cycle initiation in yeast follows first-order kinetics. *Nature* 264: 767–70

Shilo, B., Shilo, V., Simchen, G. 1977. Transition-probability and cell-cycle initiation in yeast. *Nature* 267: 648–49

Simanis, V., Nurse, P. 1986. The cell cycle control gene $cdc2^+$ of fission yeast encodes a protein kinase potentially regulated by phosphorylation. *Cell* 45: 261–68

Singer, R. A., Bedard, D. P., Johnston, G. C. 1984. Bud formation by the yeast *Saccharomyces cerevisiae* is directly dependent on "start." *J. Cell Biol.* 98: 678–84

Singer, R. A., Johnston, G. C. 1981. Nature of the G_1 phase of the yeast *Saccharomyces cerevisiae*. *Proc. Natl. Acad. Sci. USA* 978: 3030–33

Singer, R. A., Johnston, G. C. 1983. Growth and the cell cycle of the yeast *Saccharomyces cerevisiae*. II. Relief of cell-cycle constraints allows accelerated cell divisions. *Exp. Cell Res.* 149: 15–26

Singer, R. A., Johnston, G. C. 1985. Growth and the DNA-division sequence in the yeast *Saccharomyces cerevisiae*. *Exp. Cell Res.* 157: 387–96

Singer, R. A., Johnston, G. C., Bedard, D. 1978. Methionine analogs and cell division regulation in the yeast *Saccharomyces cerevisiae*. *Proc. Natl. Acad. Sci. USA* 75: 6083–87

Skehan, P. 1988. Control models of cell cycle transit, exit, and arrest. *Biochem. Cell. Biol.* 66: 467–77

Slater, M. L., Sharrow, S. O., Gart, J. J. 1977. Cell cycle of *Saccharomyces cerevisiae* in populations growing at different rates. *Proc. Natl. Acad. Sci. USA* 74: 3850–54

Sluder, G. 1979. Role of spindle microtubules in the control of cell cycle timing. *J. Cell Biol.* 80: 674–91

Sluder, G., Miller, F. J., Spanjian, S. 1986. The role of spindle microtubules in the timing of the cell cycle of echinoderm eggs. *J. Exp. Zool.* 238: 325–36

Smith, J. A., Laurence, D. J. R., Rudland, P. S. 1981. Limitations of cell kinetics in distinguishing cell cycle models. *Nature* 293: 648–50

Smith, J. A., Martin, L. 1973. Do cells cycle? *Proc. Natl. Acad. Sci. USA* 70: 1263–67

Smith, L. D., Ecker, R. E. 1971. The interaction of steroids with *R, pipien* oocytes in the induction of maturation. *Dev. Biol.* 25: 233–47

Solomon, M., Booher, R., Kirschner, M., Beach, D. 1988. Cyclin in fission yeast. *Cell* 54: 738–40

Stancel, G. M., Prescott, D. M., Liskay, R. M. 1981. Most of the G_1 period in hamster cells is eliminated by lengthening the S period. *Proc. Natl. Acad. Sci. USA* 78(10): 6295–98

Standart, N., Minshull, J., Pines, J., Hunt, T. 1987. Cyclin synthesis, modification and destruction during meiotic maturation of the starfish oocyte. *Dev. Biol.* 124: 248–58

Stimac, E., Morris, D. R. 1987. Messenger RNAs coding for enzymes of polyamine biosynthesis are induced during the G_0–G_1 transition but not during traverse of the normal G_1 phase. *J. Cell. Physiol.* 133: 590–94

Sudbery, P. E., Goodey, A. R., Carter, B. L. A. 1980. Genes which control cell proliferation in the yeast *Saccharomyces cerevisiae*. *Nature* 288: 401–4

Sumrada, R., Cooper, T. G. 1978. Control of vacuole permeability and protein degradation by the cell cycle arrest signal in *Saccharomyces cerevisiae*. *J. Bacteriol.* 136: 234–46

Sunkara, P. S., Wright, D. A., Rao, P. N. 1979. Mitotic factors from mammalian cells induce germinal vesicle breakdown and chromosome condensation in amphibian oocytes. *Proc. Natl. Acad. Sci. USA* 76: 2799–2802

Swenson, K. I., Farrell, K. M., Ruderman, J. V. 1986. The clam embryo protein cyclin-A induces entry into M-phase and the resumption of meiosis in *Xenopus* oocytes. *Cell* 47: 861–70

Tachibana, K., Yanagishima, N., Kishimoto, T. 1987. Preliminary characterization of maturation-promoting factor from yeast *Saccharomyces cerevisiae*. *J. Cell Sci.* 88: 273–81

Takayama, S., Uwaike, Y. 1988. Analysis of the replication mode of double minutes using the PCC technique combined with BrdUrd labeling. *Chromosoma* 97: 198–203

Tatchell, K. 1986. *RAS* genes and growth control in *Saccharomyces cerevisiae*. *J. Bacteriol.* 166(2): 364–67

Thomas, J. H., Botstein, D. 1986. A gene required for the separation of chromosomes on the spindle apparatus in yeast. *Cell* 44: 65–76

Thompson, P. W., Wheals, A. E. 1980. Asymmetrical division of *Saccharomyces cerevisiae* in glucose-limited chemostat culture. *J. Gen. Microbiol.* 121: 401–9

Thuriaux, P., Nurse, P., Carter, B. 1978. Mutants altered in the control co-ordinating cell division with cell growth in the fission yeast *Schizosaccharomyces pombe*. *Mol. Gen. Genet.* 161: 215–20

Toda, T., Cameron, S., Sass, P., Zoller, M., Scott, J. D., et al. 1987. Cloning and characterization of BCY1, a locus encoding a regulatory subunit of the cyclic AMP-dependent protein kinase in *Saccharomyces cerevisiae*. *Mol. Cell. Biol.* 7: 1371–77

Toda, T., Cameron, S., Sass, P., Zoller, M., Wigler, M. 1987. Three different genes in *S. cerevisiae* encode the catalytic subunits of the cAMP-dependent protein kinase. *Cell* 50: 277–87

Toda, T., Uno, I., Ishikawa, T., Powers, S., Kataoka, T., et al. 1985. In yeast, *RAS* proteins are controlling elements of adenylate cyclase. *Cell* 40: 27–36

Traganos, F., Kimmel, M., Bueti, C., Darzynkiewicz, Z. 1987. Effects of inhibition of RNA or protein synthesis on CHO cell cycle progression. *J. Cell. Physiol.* 133: 277–87

Tyson, C. B., Lord, P. G., Wheals, A. E. 1979. Dependency of size of *Saccharomyces cerevisiae* cells on growth rate. *J. Bacteriol.* 138: 92–98

Unger, M. W., Hartwell, L. H. 1976. Control of cell division in *Saccharomyces cerevisiae* by methionyl-tRNA. *Proc. Natl. Acad. Sci. USA* 73: 1664–68

Uno, I., Matsumoto, K., Adachi, K., Ishikawa, T. 1983. Genetic and biochemical evidence that trehalase is a substrate of cAMP-dependent protein kinase in yeast. *J. Biol. Chem.* 258: 10867–72

Vanoni, M., Vai, M., Popolo, L., Alberghina, L. 1983. Structural heterogeneity in populations of the budding yeast *Saccharomyces cerevisiae*. *J. Bacteriol.* 156: 1282–91

Wasserman, W., Maui, Y. 1975. The cyclic behavior of a cytoplasmic factor initiating meiotic maturation in *Xenopus* oocytes. *Exp. Cell Res.* 91: 381–92

Weinert, T. A., Hartwell, L. H. 1988. The *RAD9* gene controls the cell cycle response to DNA damage in *Saccharomyces cerevisiae*. *Science* 241: 317–22

Weintraub, H., Buscaglia, M., Ferrez, M., Weiller, S., Boulet, A., et al. 1982. "MPF" activity in Saccharomyces cerevisiae. *C. R. Acad. Sci. Paris* 295: 787–90

Wheals, A. E. 1982. Size control models of *Saccharomyces cerevisiae* cell proliferation. *Mol. Cell. Biol.* 2: 361–68

Wheals, A. E. 1987. Biology of the cell cycle in yeasts. In *The Yeasts*. New York: Academic

Wheals, A., Silverman, B. 1982. Unstable activator model for size control of the cell cycle. *J. Theor. Biol.* 97: 505–10

Wheatley, D. N., Inglis, M. S., Foster, M. A., Rimington, J. E. 1987. Hydration, volume changes and nuclear magnetic resonance proton relaxation times of HeLa S-3 cells in M-phase and the subsequent cell cycle. *J. Cell Sci.* 88: 13–23

Wilson, E. B. 1925. *The Cell in Development and Heredity*. New York: Macmillan. 1232 pp.

Wittenberg, C., Reed, S. I. 1988. Control of the yeast cell cycle is associated with assembly/disassembly of the *cdc28* protein kinase complex. *Cell* 54: 1061–72

Wittenberg, C., Richardson, S. L., Reed, S. I. 1987. Subcellular localization of a protein kinase required for cell cycle initiation in *Saccharomyces cerevisiae*: evidence for an association between the *CDC28* gene product and the insoluble cytoplasmic matrix. *J. Cell Biol.* 105: 1527–38

Yamada, K., Ito, M. 1979. Simultaneous production of buds on mother and daughter cells of *Saccharomyces cerevisiae* in the presence of hydroxyurea. *Plant Cell Physiol.* 20: 1471–79

Yasuda, H., Matsumoto, Y., Mita, S., Marunouchi, T., Yamada, M. 1981. A mouse temperature-sensitive mutant defective in HI histone phosphorylation is defective in deoxyribonucleic acid synthesis and chromosome condensation. *Biochemistry* 20: 4414–19

Yen, A., Pardee, A. B. 1978. Exponential 3T3 cells escape in Mid-G_1 from their high serum requirement. *Exp. Cell Res.* 116: 103–13

Yoshida, M., Beppu, T. 1988. Reversible arrest of proliferation of rat 3Y1 fibroblasts in both the G_1 and G_2 phases by trichostatin A. *Exp. Cell Res.* 177: 122–31

Young, P. G., Fantes, P. A. 1987. *Schizosaccharomyces pombe* mutants affected in their division response to starvation. *J. Cell Sci.* 88: 295–304

Zaitsu, H., Kimura, G. 1988a. Simian virus

40 compensates a cellular mutational defect of a serum-dependent function controlling cell cycle progression in the G_2 phase. *Virology* 164: 165–70

Zaitsu, H., Kimura, G. 1988b. Serum-dependent regulation of proliferation of cultured rat fibroblasts in G_1 and G_2 phases. *Exp. Cell Res.* 174: 146–55

Zetterberg, A. 1970. Nuclear and cytoplasmic growth during interphase in mammalian cells. *Adv. Cell Biol.* 1: 211–32

Zetterberg, A., Larsson, O. 1985. Kinetic analysis of regulatory events in G_1 leading to proliferation or quiescence of Swiss 3T3 cells. *Proc. Natl. Acad. Sci. USA* 82: 5365–69

NOTE ADDED IN PROOF Below is a representative list of recent publications, some of which have been cited in the text as personal communication.

Cyert, M. S., Thorner, J. 1989. Putting it on and taking it off: Phosphoprotein phosphatase involvement in cell cycle regulation. *Cell* 57: 891–93

Doonan, J. H., Morris, N. R. 1989. The bimG gene of *Aspergillus nidulans*, required for completion of anaphase, encodes a homolog of mammalian phosphoprotein phosphatase 1. *Cell* 57: 987–96

Edgar, B. A., O'Farrell, P. H. 1989. Genetic control of cell division patterns in the *Drosophila* embryos. *Cell* 57: 177–87 (Identifies the string (*stg*) gene as a *cdc-25* homologue and suggests that *stg* and not cyclin is limiting for the embryonic cell divisions in *Drosophila*.)

Gautier, J., Matsukawa, T., Nurse, P., Maller, J. 1989. Dephosphorylation and activation of Xenopus $p34^{cdc2}$ protein kinase during the cell cycle. *Nature* 339: 626–29 (Shows that the dephosphorylated state of $p34^{cdc2}$ is a state of maximal kinase activity.)

Hadwiger, J., Wittenberg, C., Mendenhall, M., Reed, S. I. 1989. The *S. cerevisiae CKS1* gene, a homolog of the *S. pombe* $suc1^+$ gene, encodes a subunit of the $Cdc28^+$ protein kinase complex. *Mol. Cell. Biol.* 9: 2034–41

Labbe, J. C., Picard, A., Peaucellier, G., Cavadore, J. C., Nurse, P., Doree, M. 1989. Purification of MPF from Starfish: Identification as the H1 histone kinase $p34^{cdc2}$ and a possible mechanism of its periodic activation. *Cell* 57: 253–63 (The increase in protein kinase activity is associated with $p34^{cdc2}$ dephosphorylation and the decrease in protein kinase activity on leaving M phase with rephosphorylation.)

Lehner, C. F., O'Farrell, P. H. 1989. Expression and function of *Drosophila* cyclin A during embryonic cell cycle progression. *Cell* 56: 957–68 (The timing of post cellularization divisions was not governed by the rate of accumulation or level of cyclin A.)

Løbner-Olesen, A., Skarstad, K., Hansen, F. G., von Meyenburg, K., Boye, E. 1989. The DnaA protein determines the initiation mass of escherichia coli K-12. *Cell* 57: 881–89

Minshull, J., Blow, J. J., Hunt, T. 1989. Translation of cyclin mRNA is necessary for extracts of activated *Xenopus* eggs to enter mitosis. *Cell* 56: 947–56 (Cutting cyclin mRNAs with antisense oligonucleotides and endogenous RNAase H blocks entry into mitosis in a cell-free egg extract.)

Morris, N. R., Osmani, S. A., Engle, D. B., Doonan, J. H. 1989. A genetic analysis of mitosis in *Aspergillus nidulans*. *BioEssays* 10: 196–201 (A review of the elegant work on the *Aspergillus* cell cycle emphasizing the unique control of the *nim* and *bim* genes.)

Murray, A. W., Kirschner, M. W. 1989. Cyclin synthesis drives the early embryonic cell cycle. *Nature* 339: 275–80 (The addition of exogenous cyclin mRNA to an RNA depleted extract is sufficient to produce multiple cell cycles.)

Murray, A. W., Solomon, M. J., Kirschner, M. W. 1989. The role of cyclin synthesis and degradation in the control of maturation promoting factor activity. *Nature* 339: 280–86 (A proteolysis-resistant mutant of cyclin prevents the inactivation of maturation promoting factor and the exit from mitosis both in vivo and in vitro.)

North, G. 1989. Regulating the cell cycle. *Nature* 339: 97–98

Ohkura, H., Kinoshita, N., Miyatani, S., Toda, T., Yanagida, M. 1989. The fission yeast $dis2^+$ gene required for chromosome disjoining encodes one of two putative type 1 protein phosphatases. *Cell* 57: 997–1007

Riabowol, K., Draetta, G., Brizuela, L., Vandre, D., Beach, D. 1989. The *cdc2* kinase is a nuclear protein that is essential for mitosis in mammalian cells. *Cell* 57: 393–401

Russell, P., Moreno, S., Reed, S. I. 1989. Conservation of mitotic controls in fission and budding yeasts. *Cell* 57: 295–303 (*wee1*$^+$ of *S. pombe* is shown to delay M in *S. cerevisiae* and *cdc25*$^+$ is shown to have a counterpart in *S. cerevisiae*.)

Westendorf, J. M., Swenson, K. I., Ruderman, J. V. 1989. The role of cyclin B in meiosis I. *J. Cell Biol.* 108: 1431–44 (The authors suggest that the unmasking of maternal cyclin B protein allows it to interact with *cdc2* protein kinase, which is also stored in oocytes, and that the formation of this cyclin B/cdc2 complex generates active M phase-promoting factor.)

Annu. Rev. Cell Biol. 1989. 5 : 397–425

THE INTERLEUKIN 2 RECEPTOR

Kendall A. Smith

Department of Medicine, Dartmouth Medical School, Hanover, New Hampshire 03756

CONTENTS

INTRODUCTION

The idea that immune responsiveness and immune memory are determined by lymphocytotrophic hormones (interleukins) was first introduced when the T cell growth factor, interleukin 2 (IL-2), was found to bind with high affinity to sites expressed only on antigen-or lectin-activated T lymphocytes (Robb et al 1981). However, inherent in the concept that the immune system is controlled in a fashion similar to other organ systems is the necessity to demonstrate that the interleukins and their receptors actually express all of the characteristics of the more classical hormone-receptor systems. This review will focus on this point and illustrate how the T cell-IL-2-receptor system is now the system of choice for fundamental studies

0743–4634/89/1115–0397$02.00

regarding the biochemical and biophysical mechanisms responsible for ligand-receptor binding, signal transduction, and the stimulation of physiologic responses of growth and differentiation.

HORMONES AND RECEPTOR THEORY

The term hormone, first introduced by Starling (1905), is derived from a Greek root meaning "to excite" or "to arouse". Since then, classic hormones have come to be regarded as substances produced by specific glands that are secreted directly into the blood and transported to specific organs and tissues where they exert their effects. Moreover, hormones function at low concentrations, are produced at a variable rate, exert a regulatory function in response to environmental stimuli, and are under feedback regulatory control. A corollary to these characteristics is the requirement that hormones react with receptor molecules, for it is now recognized that receptors are responsible for the target cell specificity of the hormone effect. In addition, by virtue of their high affinity for the specific ligand, receptors allow the effect to be mediated by very low hormone concentrations. Of most importance, receptors transmit signals to the cell in response to ligand binding.

The concept of specific receptors was first introduced by Langley (1878) in explanation of his findings on the antagonistic effects of atropine on pilocarpine-induced salivary gland secretion in cats. Twenty-seven years later Langley (1905) elaborated on his receptor concept and defined a receptive substance as "the recipient of stimuli from drugs and hormones, which functions to transfer these stimuli to the effector organ, thereby eliciting a response." With this definition Langley first enunciated the characteristics of a true receptor that distinguish it from a binding site. To be a receptor, a binding molecule must have an affinity for the ligand that is appropriate for the concentrations that are physiologically or pharmacologically relevant; it should be detectable on those cells or tissues that are responsive; and it should show stereospecificity for active ligands and their antagonists. Inherent in these attributes is the understanding that authentic receptors transmit a message to the cell, thereby eliciting a characteristic response. Therefore, all receptors have ligand binding sites, but not all molecules expressing such sites are genuine receptors. In this regard, the term receptor was coined by Ehrlich (1913), who noted the high degree of specificity of antibodies and their relationship to the action of drugs. However, antibodies can only be considered receptors when they are expressed on the cell surface so that they can participate in signal transduction.

The first quantitative explanation of receptor function was developed by Clark (1926), who proposed that ligands and receptors obey the law of mass action. As applied to hormones and receptors, this law states that the magnitude of the biologic response is directly proportional to the number of receptors occupied, and that the maximum response is obtained only when all receptors are occupied. This serves as the foundation of the occupancy theory of receptor function. Clark recognized that the action of a drug or hormone depends upon the first of the two factors originally stated by Langley, namely binding of the ligand by the receptor. However, the occupancy theory as proposed by Clark did not address in a quantitative manner the second characteristic of a receptor, i.e. the ability of a ligand to elicit a response subsequent to binding.

Three decades later Ariëns (1954) extended the occupancy theory to explain why different ligands vary in their ability to elicit a biologic response despite equal receptor binding capacities. According to Ariëns the biologic response of a given receptor depends on two separate and independent parameters. Affinity describes the attachment or binding of the ligand to the receptor and is governed by the law of mass action. However, a ligand also has intrinsic activity, a term that describes the ability of the ligand to actually induce an effect after binding. Therefore a competitive antagonist is a ligand that possesses the appropriate structural requirements necessary to occupy a receptor (affinity), but lacks those separate and independent structural requirements necessary for eliciting a response (i.e. no intrinsic activity). The concept of intrinsic activity introduces the idea that the receptor itself can be divided into distinct components. It also suggests that certain ligands exist, or can be identified, that differ not only in their capacity to contact the binding residues of the receptor, but also in their capability of interacting productively with that distinct part of the receptor responsible for signal transduction.

An additional aspect of receptor theory is the concept of "spare receptors," introduced by Stephenson (1956) to explain several pharmacologic observations. Stephenson stated that:

1. The response to a given ligand is some unknown positive function of receptor occupancy; thus the response does not need to be directly proportional to the percentage of receptors occupied.
2. A maximum effect can be produced by an agonist when occupying only a small proportion of the receptors.
3. Different ligands can have varying capacities to initiate a response; consequently they may need to occupy different proportions of receptors, even when producing equal responses.

The concept of spare receptors implies that there may be more receptors expressed than the minimum number required to produce a maximum response, and that 100% receptor occupancy is not always necessary to elicit a maximum response. Accordingly, maximal receptor occupancy need not be the limiting factor in determining the maximum response of an agonist, nor must receptor occupancy be linearly proportional to response.

Paton (1961) proposed a theory of drug action based on the rate of drug-receptor combination in an attempt to explain phenomena such as differing intrinsic activities or relative efficacies of ligands, partial and complete antagonism, and spare receptors, all aspects of the ability of a bound ligand to produce a response. Like Clark, Paton assumed that ligand binding obeys the law of mass action. However, Paton made the additional assumption that each association of a ligand to a receptor produces a "quantum" of excitation. Moreover, he assumed that the number of excitations per unit time determined the response observed. Accordingly, the dissociation rate constant ultimately determines the difference between a full agonist, partial agonist, and antagonist. In this scheme powerful agonists will dissociate rapidly, partial agonists more slowly, and full antagonists will have very slow dissociate rates because they are essentially bound irreversibly. This theory emphasizes the quality of both the ligand and the receptor and suggests that productive interactions are like moving pictures; individual frames moving rapidly produce the appearance of a continuous signal.

After consideration of these receptor theories, several intriguing questions come to mind in relationship to the kinds of ligand-receptor systems employed by lymphocytes. Do interleukins and their receptors obey the principles of occupancy, intrinsic activity, spare receptors, and relative efficacy as proposed by standard occupancy theory? Alternatively, is the rate of IL-2-receptor interaction important in determining the response? Consideration of the IL-2-receptor phenomenon with these theories in mind may help to formulate new approaches that can be applied to the molecular mechanisms responsible for interleukin action.

IL-2 AND ITS RECEPTOR: HISTORICAL PERSPECTIVE

The first data suggesting that IL-2 might exert its mitogenic effects on activated T cells via receptors resulted from a series of experiments that comprised the first description of the distinguishing biologic characteristics of the T cell growth factor activity found in lymphocyte conditioned medium (Gillis et al 1978; Baker et al 1978; Smith et al 1979). Curiously enough, when cells are stimulated with a mitogenic lectin such as Concanavalin A (ConA), IL-2 appears in the culture medium during the first

24–48 hr, but then declines rapidly, so that by 96 hr very low levels of activity are detectable. The decline in the activity seemed to be attributable to some form of cellular consumption, because the activity by itself is stable in tissue culture-conditioned medium for months. Moreover, mixing experiments of 50% IL-2-containing medium with 50% IL-2-depleted (spent) medium failed to show any evidence of an agent such as a protease that could be responsible for extracellular degradation. Confirmation of a cell-associated removal of IL-2 from the culture medium was obtained by the experiment shown in Figure 1, where the IL-2 concentration was found to decrease progressively as the cell concentration increased. Similar experiments, performed over a shorter time interval with higher cell concentrations, indicated that only antigen/lectin-activated T cells are capable of absorbing IL-2 activity, as lipopolysaccharide (LPS)-activated spleen cells, which are comprised primarily of proliferating B cells, absorbed no IL-2 activity (Figure 2) (Smith et al 1979). Particularly noteworthy was the finding that IL-2 absorption is both time and cell concentration dependent. Moreover, lowering the temperature to 4°C does not circumvent absorption, nor does cell fixation with glutaraldehyde. Therefore, all of these findings pointed to the disappearance of IL-2 via its binding to cell surface receptors.

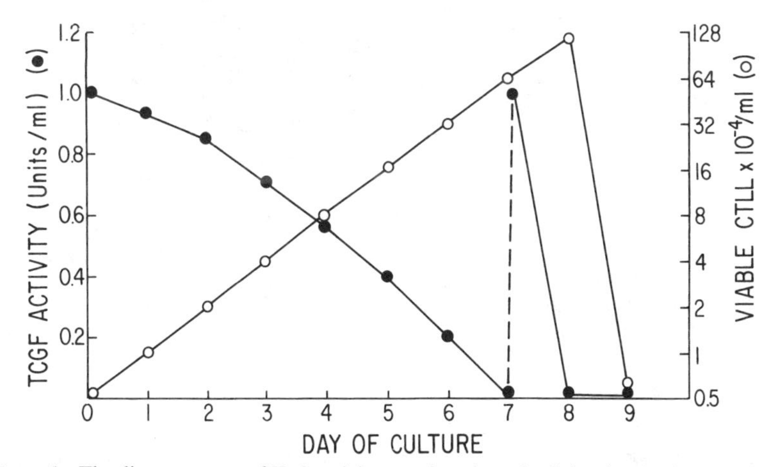

Figure 1 The disappearance of IL-2 activity as a function of cell density. At time 0 CTLL-2 cells were placed into culture with 1.0 unit/ml TCGF (equivalent to ~10 pM IL-2) (●) at a cell density of 5000 cells/ml (○). By day 7 of culture when the cell density had increased one-hundredfold, the IL-2 concentration had become undetectable. Replenishment of IL-2 at the original concentration on day 7 of culture allowed the cell density to double over the next 24 hr, but the IL-2 concentration decreased rapidly. After one more day of culture (day 9) in IL-2-depleted medium, no viable cells remained. (From Smith 1980 with permission.)

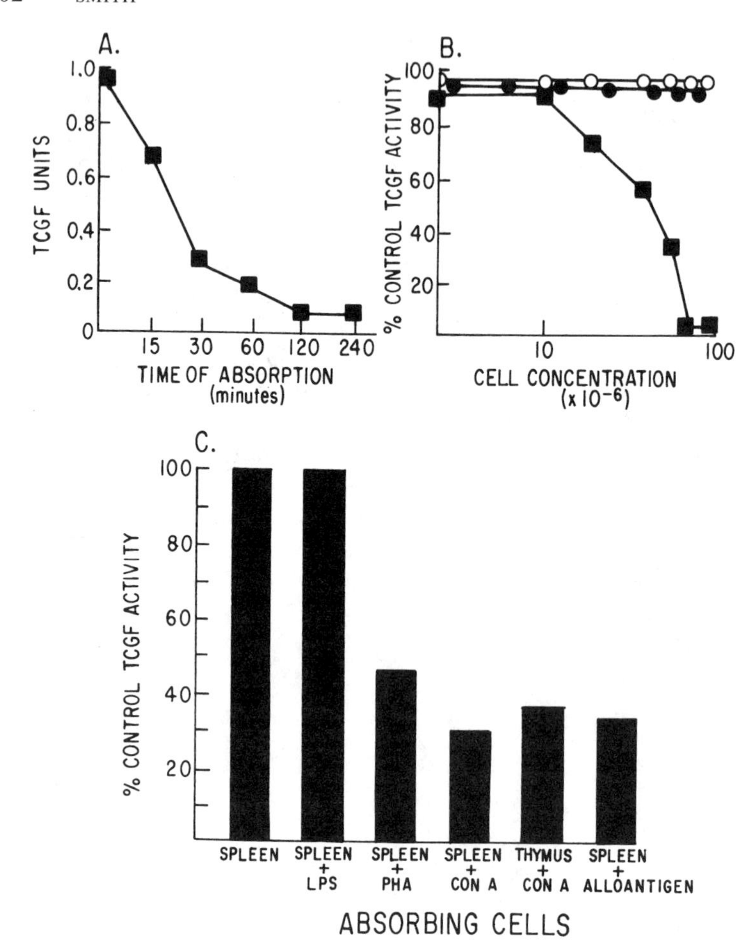

Figure 2 The absorption of IL-2. (*A*) Kinetics of IL-2 absorption by 1×10^8 Con-A-stimulated spleen cells at 37°C. (*B*) IL-2 absorption as a function of cell density during 4 hr at 37°C. Spleen cells (■); thymus cells (●), and lipopolysaccharide (LPS)-stimulated spleen cells (○). (*C*) The absorption of IL-2 for 4 hr at 37°C using different cell populations (1×10^8 cells/ml). (From Smith et al 1979.)

Cell membrane receptors for polypeptide hormones are traditionally demonstrated and characterized by radiolabeled ligand binding assays. Therefore as soon as it was technically feasible, IL-2 was radiolabeled biosynthetically using radioactive amino acids, then purified to homogeneity and used to explore its binding parameters to various target cells. The very first IL-2 binding assays were decisive: IL-2 binds specifically to a single class of high affinity sites expressed only on antigen- and lectin-activated T cells (Robb et al 1981). In addition, there is no difference when binding to whole cells is compared to binding to isolated plasma membranes (Smith 1983). Therefore cellular metabolism of the radiolabeled ligand during the short interval (ten min) required to achieve steady-state binding does not appear to play a significant role in the binding results obtained.

Comparison of the concentrations of IL-2 responsible for promoting T cell proliferation with those found to bind to high affinity IL-2 receptors was made possible by the availability of cloned IL-2-dependent cytolytic T lymphocyte lines (CTLL) (Baker et al 1979). The very first experiments revealed that IL-2 binding and biologic dose-response curves are coincident (Robb et al 1981). Thus there is a linear relationship between occupancy of high affinity IL-2 receptors and the EC_{50} (i.e. the concentration of ligand required to effect a half-maximal response). Accordingly, the EC_{50} actually is an accurate reflection of the equilibrium dissociation constant (i.e. the $EC_{50} = kd = 10$ pM), which indicates that there are no spare receptors; i.e. a maximal T cell growth response occurs at a concentraton of IL-2 that also yields 100% receptor occupancy (~ 100 pM).

Despite these initial results showing a direct relationship between receptor occupancy and response, additional findings soon pointed to a more complicated picture of IL-2 receptors. In collaboration with Leonard & Waldmann, we found that a monoclonal antibody reactive with activated T cells (anti-Tac) inhibited both radiolabeled IL-2 binding and IL-2 dependent T cell proliferation (Leonard et al 1982). Curiously enough, this antibody recognized about 10–20-fold more binding sites/cell than could be enumerated by IL-2 binding. Ultimately, studies by Robb & coworkers (1984), using 100-fold higher IL-2 concentrations, revealed a second class of IL-2 binding sites capable of binding both the antibody and IL-2. However this second class of sites was characterized by a 1000-fold lower binding affinity than had been detected previously (i.e. kd = 10 nM instead of 10 pM).

Examination of the binding and biologic response curves using IL-2 concentrations high enough to saturate both high and low affinity binding sites revealed a phenomenon that could be interpreted as consistent with the concept of spare receptors: even though there are ten-fold more low affinity binding sites compared with high affinity receptors, all of the T cell proliferative response can be accounted for by saturation solely of high affinity IL-2 receptors (Figure 3).

The phenomenon of high and low affinity IL-2 binding sites continued to be perplexing despite isolation of cDNA clones that encoded a protein

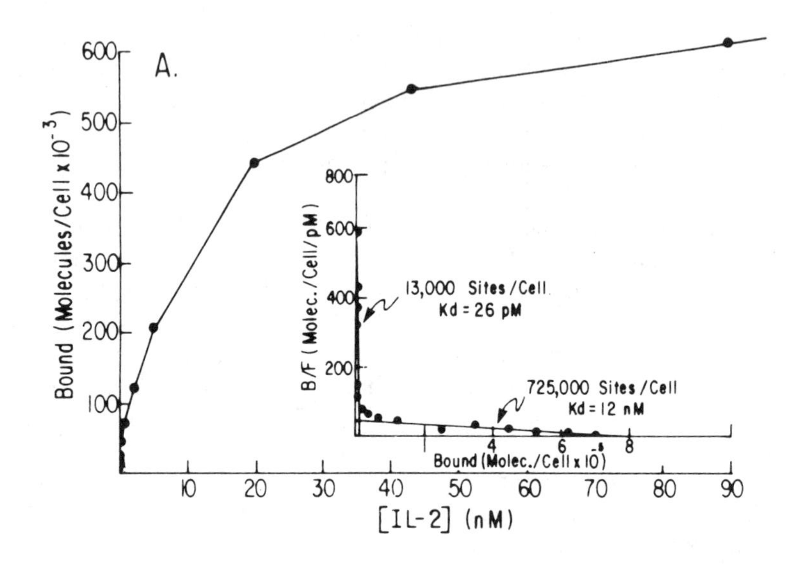

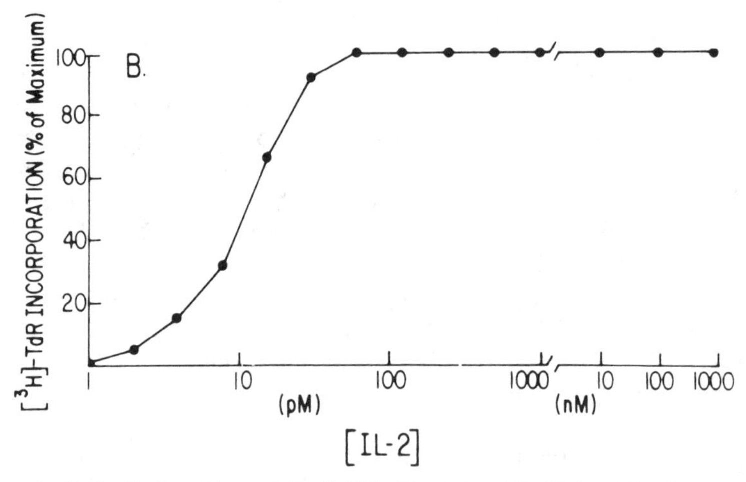

Figure 3 High affinity and low affinity IL-2 binding sites and the IL-2 proliferative response of CTLL-2 cells. (*A*) A typical IL-2 binding curve to CTLL-2 cells and IL-2-dependent murine cytolytic T lymphocyte cell line. (*Inset*) A plot of the data according to the method of Scatchard. There are 13,000 high affinity sites/cell (kd = 26 pM) and 725,000 low affinity sites/cell (kd = 12 nM). (*B*) The IL-2 concentration-dependent proliferative response. Concentrations of IL-2 that saturate the high affinity IL-2 receptors correspond with those stimulating the proliferative response ($EC_{50} \cong 20$ pM). (From Smith 1988b with permission.)

reactive with the monoclonal antibody (Leonard et al 1984; Nikaido et al 1984; Cosman et al 1984). Transfection experiments whereby this cDNA was expressed in non-T cells (mouse 3T3 cells and L cells, and human HeLa cells) revealed that the 28-kd core protein was expressed on the cell surface as a 55,000 (Mr) glycoprotein that only binds IL-2 with a low affinity (i.e. kd = 10 nM) (Sabe et al 1984). Various theories were proffered to explain the structural basis for how a single 55-kd protein could lead to two classes of binding sites differing 1000-fold in affinity (Robb 1986; Kondo et al 1986; Sharon et al 1986). However only the identification of leukemic cell lines that express an entirely distinct size of membrane protein (Mr = 75,000) and that also display an intermediate binding affinity (kd = 1 nM) finally made it possible to show that expression of two IL-2-binding proteins is required to form a high affinity IL-2 receptor (Tsudo et al 1986; Teshigawara et al 1987). Thus high affinity IL-2 receptors are constructed by cooperative binding of IL-2 to both the low affinity (55-kd chain) and intermediate affinity (75-kd chain) binding sites.

KINETIC COOPERATION: A NEW CONCEPT THAT EXPLAINS THE CONSTRUCTION OF HIGH AFFINITY RECEPTORS FROM LOW AFFINITY BINDING SITES

A reductionist approach was necessary to understand how high affinity IL-2 receptors are formed from two distinct lower affinity binding sites. Thus binding studies were performed using leukemic T cell lines that only express 75-kd chains or 55-kd chains, and the results were then compared with studies using cells expressing both chains simultaneously (Wang & Smith 1987). The results from kinetic and equilibrium binding analyses are summarized in Table 1. The equilibrium dissociation (affinity) constants can be determined from the ratio of the rate constants (i.e. $K_d = k'/k$) and from the equilibrium binding results. It is evident that both methods give values that are quite similar for each chain and that low affinity ($K_d = 10^{-8}$ M), intermediate affinity ($K_d = 10^{-9}$ M), and high affinity ($K_d = 10^{-11}$ M) IL-2 binding sites can be attributed to 55- and 75-kd chains and the formation of 75/55-kd heterodimers respectively. It is important to emphasize that finding similar values for the affinity constants using either method (i.e. kinetic vs equilibrium binding) allows one to conclude that the results of each of the three separate assays are accurate.

The veracity of the kinetic binding constants is especially important because there are such large differences between the values obtained for IL-2 binding to 75-kd chains vs 55-kd chains. For example, the rate

Table 1 Kinetic and equilibrium IL-2 binding constants

Chain[a]	Kinetic binding constants			Equilibrium dissociation constants (M)	
	Dissociation (k′)		Association (k)	Kinetic (k′/k)	Equilibrium[b]
	(S^{-1})	($t^{1/2}$)	($M^{-1}\ S^{-1}$)		
75	2.5×10^{-4}	(46 min)	3.8×10^{5}	0.7×10^{9}	$1.2(\pm 0.1) \times 10^{-9}$
55	4.0×10^{-1}	(1.7 S)	1.4×10^{7}	2.9×10^{-8}	$1.4(\pm 0.1) \times 10^{-8}$
75/55	2.3×10^{-4}	(50 min)	3.1×10^{7}	0.7×10^{-11}	$1.3(\pm 0.1) \times 10^{-11}$

[a] 75-kd chain = YT-2C2 membranes; 55-kd chain = MT-1 cells; 75/55-kd chains = induced YT membranes.
[b] Mean ± SEM from 8 separate equilibrium binding experiments. (From Wang & Smith 1987 with permission.)

constants for dissociation of the IL-2 from the two chains differ by three orders of magnitude and lead to half-times for dissociation that are striking for their disparity (i.e. 46 min for 75-kd chain-IL-2 dissociation vs <2 sec for 55-kd chain-IL-2 dissociation). Similarly, the association rate constants for the two chains differ by two orders of magnitude. Knowledge of the kinetics of IL-2 binding to these two binding sites furnishes a more dynamic picture of the intermolecular reactions and promotes a concept of the way in which IL-2 binds that cannot be obtained from equilibrium binding assays. Thus IL-2 binds to and dissociates from 55-kd chains so rapidly that it is difficult to measure. By comparison, IL-2 associates with and dissociates from 75-kd chains very slowly. Knowing the kinetics of IL-2 binding to the sites on these two chains makes it even more remarkable that at steady state their equilibrium dissociation constants differ only by a factor of ten.

Examination of the kinetic and equilibrium binding constants of the heterodimeric receptor reveals immediately how two low affinity binding sites can form a high affinity receptor: the rapid rate of IL-2 association to the site on the 55-kd chain is used together with the slow rate of dissociation of IL-2 from the site on the 75-kd chain. The combination of a rapid "on" rate and a slow "off" rate makes for an affinity 100–1000-fold greater than for IL-2 binding to either chain alone.

During evolution the expression of two protein chains capable of binding IL-2 individually as well as cooperatively has been selected for and conserved, since murine high affinity IL-2 receptors are also constructed from two chains. Although it is readily apparent how both chains form high affinity receptors given their individual binding kinetics of fast and slow on-off rates, it is curious that these two chains bind IL-2 with such distinct characteristics. However, knowing that the IL-2 receptor is constructed in this manner, and realizing the obvious advantage of 100–1000-fold higher affinity receptors, it is predictable that such a heterodimeric receptor construct may well be found for other hormone receptor systems. Actually, the nerve growth factor (NGF) receptor is another example of a heterodimeric structure: both high affinity and low affinity binding sites have been identified (Sutter et al 1979), and two chains capable of binding NGF appear operative (Hosang & Shooter 1985). Moreover, cDNA and genomic DNA have been isolated that encode one chain that binds NGF with rapid kinetics and low affinity, thereby indicating that another chain, like the 75-kd chain of the IL-2 receptor, is yet to be identified and characterized (Chao et al 1986; Johnson et al 1986; Radeke et al 1987).

BIOPHYSICAL CONSEQUENCES OF HIGH AFFINITY AND LOW AFFINITY BINDING SITES

Antigen-activated T cells always express a five to tenfold excess of low affinity 55-kd chains compared with the number of high affinity 75/55-kd heterodimers. Moreover, the ratio of low affinity binding sites to high affinity receptors is regulated by IL-2: i.e. IL-2 binding to high affinity receptors results in a decreased (by 50%) expression of this class of binding sites, but simultaneously, a marked increase (10–20-fold) in the expression of low affinity 55-kd chains (Smith & Cantrell 1985). It is intriguing that the absolute number of rapidly reacting low affinity 55-kd chains should be augmented by the interaction of IL-2 with its high affinity receptors. A simple explanation for this arrangement relates to the noncovalent interaction of the two chains. If both chains were packaged together in the Golgi apparatus and linked covalently before transportation to the cell surface, they would always be expressed in a 1 : 1 stoichiometry, as occurs, for example, with the T cell antigen receptor and cell surface immunoglobulin. However, since each chain of the IL-2 receptor is expressed and regulated independently, it follows that the formation of high affinity 75/55-kd heterodimers must occur on the cell surface. Consequently, the number of heterodimers formed will be dictated by the law of mass action (i.e. $75+55 \rightleftarrows 75\text{–}55$). Accordingly, a mechanism that guarantees the expression of an excess of one chain favors the rightward progress of the reaction, so that most of the 75-kd chains, which are in the minority, will always be occupied by 55-kd chains in the formation of high affinity 75/55-kd heterodimers.

Although the mass action hypothesis may well have operated in the selection of the arrangement of the observed excess of 55-kd chains, one wonders why the opposite configuration, i.e. an excess of 75-kd chains, is also not operative. The key to this question may reside in the difference in the kinetics of IL-2 binding by the two chains. Since the IL-2-75-kd chain binding kinetics are so sluggish in comparison with the IL-2-55-kd chain interaction, one can perceive that an excess of 75-kd chains would actually function to decrease the efficiency of IL-2 interaction with its target cells. Thus, at steady-state, even though the affinity of IL-2 binding to 75/55-kd dimers vs 75-kd chains differs 100-fold, a marked excess of 75-kd chains would still guarantee that a certain proportion of ligand would be bound essentially irreversibly to isolated 75-kd chains, given the slow IL-2 dissociation rate.

By comparison, IL-2 binding to an excess of rapidly reacting 55-kd chains could convert the actual interaction between IL-2 and 75/55-kd heterodimers from a three-dimensional search in space to a two-dimensional search in the plane of the membrane. This concept was originally proposed by Adam & Delbruck (1968) in their quest for physical principles that may govern the interaction of hormones and cell surface receptors. More recently, this concept has been invoked by Sargent & Schwyzer (1986) to account for a low affinity, rapidly reacting interaction between lipophilic peptides and the cell surface lipid bilayer. An even more concrete example of this concept has been reported for the interaction of endonucleases with nonspecific vs specific sites on DNA (Terry et al 1983; Ehbrecht et al 1985). A low affinity, high capacity reaction is operative in the interaction between endonucleases and nonspecific sites on DNA, as compared with the high affinity specific cleavage sites determined by the DNA sequence. A low affinity, rapidly reversible reaction occurs between the endonuclease and nonspecific DNA sequences, so that the search for the specific sites, where the enzyme binds with high affinity, is made linear, thereby reducing the infinitely complex three-dimensional problem to a relatively simple two-dimensional quest. One can envision the enzyme simply "walking along" a stretch of DNA until locking onto the correct sequence.

Still, it remains to be demonstrated as to whether 75- and 55-kd chains exist on the membrane as actual heterodimers, or whether IL-2 binds first to isolated 55-kd chains and then is passed, or in football jargon "handed off", either to isolated 75-kd chains or to pre-existing 75/55-kd dimers (Saito et al 1988). Whichever sequence actually occurs at the cell surface, it is evident that once IL-2 has bound either to isolated 75-kd chains or 75/55-kd dimers, its dissociation rate is essentially the same in both instances (i.e. $t^{1/2} = 45\text{–}50$ min). Recently, evidence in favor of a stable 75/55-kd-IL-2 complex was reported where IL-2 is found bound to both chains after internalization (Fung et al 1988). Presently there are data to support the idea that IL-2 interaction with 75- and 55-kd chains forms a stable trimeric complex that is internalized. Additional experiments are necessary to ascertain whether 75- and 55-kd chains exist as a stable heterodimer before binding IL-2, or whether IL-2-55-kd chain interaction occurs first prior to the formation of a stable complex. Whichever possibility is operative, the compelling concept of an excess number of rapidly reacting 55-kd chains serving to facilitate localization of the ligand to the immediate cell surface environment prior to ligand binding to functional high affinity receptors serves as a precedent for the way cell surface hormone receptors function. It is anticipated that additional examples of such a receptor system will be uncovered now that the principles have been established with the IL-2-receptor as a model.

IL-2 RECEPTOR SIGNAL TRANSDUCTION

Initially it was assumed that the IL-2 receptor and the 55-kd protein identified by monoclonal antibodies were synonymous. However when transfection of cDNA encoding the 55-kd chain failed to create a functional signal transduction system (Sabe et al 1984), attention focused on alternative explanations to account for a fully functional IL-2 receptor. As soon as it was realized that high affinity IL-2 receptors are comprised of 75/55-kd heterodimers, it was natural to question whether the structures responsible for signal transduction are contained solely within the 75-kd chain, or whether IL-2 binding to both chains is necessary to stimulate T cell growth. Presumptive evidence that solely 75-kd chains could deliver a positive growth signal to the cell was immediately suggested by the finding of Tsudo et al (1986) that MLA-144 cells express solely 75-kd chains. The MLA-144 cell line was established from tissue obtained from a gibbon ape that suffered from a spontaneous lymphoma while in captivity (Kawakami et al 1972). Several years ago it was shown that this cell line produces IL-2 constitutively (Rabin et al 1981). Since the capacity to make IL-2 is rare among leukemic cell lines, it was natural to wonder whether MLA-144 cells actually utilized IL-2 to signal growth, or whether the IL-2 production was unrelated to the mechanism responsible for autonomous growth. The initial findings were promising in that MLA-144 cells were also found to express IL-2 receptors. Capitalizing on our findings with the inhibitory effect of glucocorticoids on IL-2 production and proliferation by normal T cells (Gillis et al 1979), the MLA-144 cells were next tested for the effects of glucocorticoids. After selecting for glucocorticoid-responsive clones, it could be readily demonstrated that MLA-144 cells actually proliferate in response to the IL-2 they themselves produce (Smith 1982). Moreover, glucocorticoids regulate MLA-144 cell growth by suppressing IL-2 production: if IL-2 is supplied exogenously, glucocorticoid suppression is circumvented entirely.

Although these data certainly supported the idea that all of the signal transducing structures could be contained within the 75-kd chains, since the MLA-144 cells are transformed, it seemed prudent to interpret these results cautiously. Searching for a way to discriminate the functional capacity of IL-2 binding to 75-kd chains vs 75/55-kd dimers using normal T cells, we capitalized on the knowledge that the MoAb reactive with 55-kd chains prevents IL-2 binding to these fast reacting sites. Therefore, anti-55-kd chain should disrupt the subtle kinetic cooperation between the binding sites on the 75- and 55-kd chains and isolate the 75-kd chains so that they alone would be capable of binding IL-2. The first experiment performed to explore this possibility worked precisely as predicted (Wang & Smith 1987) (Figure 4). Normal human peripheral blood mononuclear cells cultured for three days with anti-CD3 expressed both high affinity and low affinity binding sites (Figure 4*a*). By comparison, when the binding

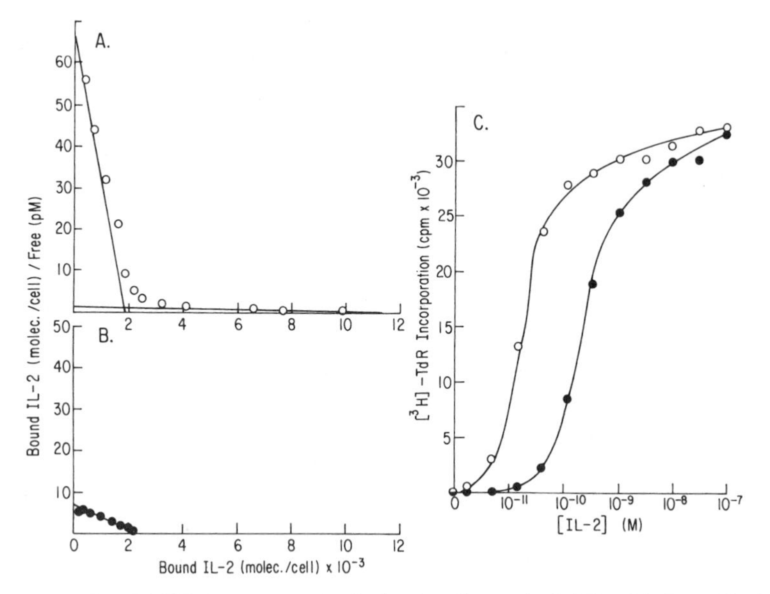

Figure 4 The 75-kd IL-2 receptor subunit signals cell growth. (*A*) IL-2 binding to human peripheral mononuclear cells (PBMC) activated with anti-cluster determinant for 3 days. There are ~2000 high affinity receptors sites/cell and ~11,000 low affinity binding sites/cell. (*B*) IL-2 binding to the same cells described in *A*, in the presence of a saturating concentration (50 nM) of a monoclonal antibody reactive with the 55-kd chains. There is a single class of intermediate affinity IL-2 binding sites equal in number to the high affinity receptors (~2000 sites/cell). (*C*) The proliferative response of day 3 activated PBMC cultured with (●) and without (○) a saturating concentration of 55-kd chain-reactive antibody. (From Wang & Smith 1987 with permission.)

assay was performed in the presence of a saturating concentration of antibody reactive with 55-kd chains, both high affinity and low affinity binding sites disappeared and left only a single class of intermediate affinity binding sites, equal in number to the high affinity receptors (Figure 4*b*). In essence the inclusion of the anti-55-kd chain served to neutralize any participation of 55-kd chains in IL-2 binding, either individually or in cooperation with 75-kd chains.

Given the ability to create normal T cells that are only able to utilize 75-kd chains for IL-2 binding, it was possible to test whether 75-kd chains alone could transduce a growth signal to the cells, or whether the complete heterodimeric high affinity IL-2 receptor was necessary. From Figure 4*c* it can be readily appreciated that the presence of anti-p55 does not prevent an IL-2-promoted growth response, a result expected if solely high affinity IL-2 binding is necessary. Instead, the antibody causes a 100-fold shift in the IL-2 dose-response curve, which is entirely consistent with the 75-kd-IL-2 interaction being responsible for signal transduction. Thus the $EC_{50} = 400$ pM, coincident with the kd for IL-2 binding to isolated p75 chains (300 pM).

Since the 75-kd chains are involved in signaling T cell proliferation, the structure of this chain, especially its cytoplasmic domain, immediately becomes relevant. Recently Hatakeyama & co-workers (1989) have isolated cDNA clones that encode the 75-kd chain, and their data reveal that the cytoplasmic domain does not contain typical kinase catalytic sequences. Therefore the IL-2 receptor may well be found to trigger a novel pathway(s).

When considering other receptors as possible models of the IL-2 receptor, it is important to distinguish between those receptors capable of stimulating detectable phosphotidylinositol (PI) hydrolysis leading to an increase in cytoplasmic free calcium and protein kinase C activation and those that do not. This distinction is especially relevant for T cells, because the T cell antigen receptor complex clearly stimulates PI turnover (Weiss et al 1986), yet triggering this receptor does not promote T cell cycle progression. Instead, the T cell antigen receptor performs a role very similar to that of the platelet-derived growth factor (PDGF) receptor; both stimulate PI turnover, and both receptors render the cells "competent" to undergo cell cycle progression. By comparison, the IL-2 receptor does not trigger detectable PI turnover (Mills et al 1988; Valge et al 1988). However it does stimulate the cells to undergo cell cycle progression to DNA replication (Stern & Smith 1986). Accordingly, the T cell antigen receptor and the IL-2 receptor are distinct, promoting separate and distinguishable biochemical changes within the cell.

IL-2 RECEPTOR METABOLISM

In other hormone systems the phenomenon of hormone-mediated "down regulation" of receptors has been traced to a ligand-mediated accelerated internalization and degradation of surface receptors. Studies using radiolabeled IL-2 reveal a similar phenomenon. In the absence of IL-2, the high affinity IL-2 receptor disappears from the cell surface with a half-life ($t^{1/2}$) of 150 min. Upon addition of ligand this disappearance time is accelerated tenfold, so that the $t^{1/2}$ is only 15 min (Smith 1988a). The chain responsible for signaling an accelerated rate of internalization is the 75-kd protein; radiolabeled IL-2 bound to either 75-kd chains or 75/55-kd high affinity heterodimers is internalized with a $t^{1/2}$ of 15 min, while IL-2 bound to 55-kd chains hardly undergoes internalization at all ($t^{1/2} \sim 10$ hr).

The consequence of the IL-2-mediated accelerated internalization of surface high affinity IL-2 receptors is down regulation of the total number of receptors. Thus the addition of IL-2 to cells results in a 50–60% decrease in the number of detectable high affinity receptors within 1–2 hr (Duprez et al 1988). Removal of IL-2 from the medium subsequently leads to the reappearance of the original number of receptors with kinetics identical to those observed during the disappearance phase. There is no evidence for receptor recycling to the cell surface. Therefore, since the number of surface receptors readjusts to a level 50% of the original number and remains at this level for 24–48 hr, it appears that the rate of synthesis and expression of new high affinity receptors is not influenced by the ligand; i.e. IL-2 does not appear to stimulate the expression of 75-kd chains in the way that it stimulates 55-kd chains. According to Duprez et al (1988) the transport of the membrane proteins after their biosynthesis takes at least 1 hr. Therefore the down regulation of surface receptors can be explained simply, considering that the rate of internalization in the presence of the ligand gives a $t^{1/2}$ of 15 min. At this rate, the endocytosis rate constant

$$\mathrm{k} = \frac{\ln 2}{15} = 4.67 \times 10^{-2}\,\mathrm{min}^{-1}, \qquad 1.$$

and the rate of receptor synthesis

$$\begin{aligned} \mathrm{R} &= \mathrm{k} \times \text{receptor number} \\ &= 4.67 \times 10^{-2}\,\mathrm{min}^{-1} \times 750\ \text{receptors/cell}^{1} \\ &= 35\ \text{receptors/cell/min}. \qquad 2. \end{aligned}$$

It is natural to question whether internalization has anything at all to

[1] Represents the number of receptors at steady-state in the presence of IL-2.

do with signal transduction. Since the 75-kd chains contain the cytoplasmic domain structures that signal cell growth and that also accelerate internalization, it could well be that the same structures are responsible for signaling both phenomena. Certainly detailed studies of the EGF receptor using truncations and site-specific mutants have shown that the tyrosine-specific kinase activity is necessary for both accelerated internalization and signal transduction (Glenney et al 1988). However signaling and internalization have recently been separated in the PDGF system (Williams 1989). In the IL-2-receptor system, ligand occupancy directly correlates with the magnitude of the proliferative response. Now that it is established that IL-2 dissociation from high affinity receptors is threefold slower than is the rate of ligand-receptor internalization (i.e. 45 min vs 15 min), it must be concluded that an occupied receptor is endocytosed before IL-2 is likely to dissociate. Accordingly, the IL-2 receptor functions like an on-off light switch. When the receptor is occupied, it is on, and when IL-2 dissociates, the receptor signaling mechanism is turned off. It follows that IL-2 receptors are only signaling when they are on the cell surface. An acid pH (<4.5) causes rapid dissociation of surface receptor-bound IL-2 (Smith & Cantrell 1985); and IL-2 is internalized via acidified endosomes that traffic and fuse with lysozomes where the ligand receptor complex is degraded by acid proteases (R. C. Budd & K. A. Smith, unpublished). Therefore as soon as the hormone-receptor complex disappears from the cell surface and the endosome becomes acidified, IL-2 dissociates from the receptor and it turns off. Accordingly, these considerations favor the occupancy theory rather than the rate theory as most likely operative in the IL-2-receptor system.

IL-2-PROMOTED T CELL G_1 PROGRESSION IS QUANTAL

Due to the ready availability of T cells in the peripheral blood of normal individuals, the development of IL-2-dependent normal T cell clones, and IL-2-dependent and independent leukemic T cell lines, studies on the determinants of T cell cycle progression have been especially illuminating and satisfying. In particular, freshly isolated normal T cells can be made to express IL-2 receptors by stimulation of the T cell antigen receptor complex. These cell populations are especially suitable for studies of cell growth, because they have not been selected for survival and growth in long-term culture. Using short-term cultures of activated normal T cells, we have established that once T cells express IL-2 receptors serum is entirely superfluous for IL-2-mediated G_1 progression to S phase (Herzberg & Smith 1987). Accordingly, the knowledge that IL-2 is the only essential factor external to the cell that is necessary to signal the entire process of G_1 progression considerably simplifies the design and interpretation of experiments to uncover the minimal parameters that participate in signalling cell division.

The development of the IL-2 radioreceptor assay, the availability of monoclonal antibodies reactive with the IL-2 receptor, and access to IL-2-responsive T cell clones together with homogeneous recombinant IL-2 finally has provided us with the cellular and molecular reagents available to dissect the parameters regulating cell cycle progression. It should be emphasized that it was most important to have established that only one ligand-receptor system is involved in T cell growth, rather than several, as is the case for murine embryonic 3T3 cell cycle progression. In addition, the ability to correlate cell growth with the number and distribution of IL-2 receptors within a given T cell population provided the methodology necessary to go beyond merely a mathematical treatment of the behavior of individual cells in a growing population.

It is relevant that all cells that have been studied during the past 50 years grow in an identical fashion (Pardee et al 1979). Prokaryotes, yeasts, protozoa, avian, and mammalian cells all proliferate with a fixed mean generation time. However, despite this invariant behavior of the cell population, individual cells comprising the population have markedly different growth rates, even among cloned, genetically identical cell populations. For example, within a given generation some cells will have a generation time of 12-hr, whereas others will require 24 hr to complete a cell cycle. Even more perplexing, cell cycle times are not passed on from one generation to the next. This is logical in that if generation times were determined genetically, the fastest growing cells would be selected, and eventually all of the cells in a population would have very fast and identical generation times. However something that contributes to cell cycle times is passed on at division, since the generation times for the daughter cells are almost identical, even though they are usually not similar to the time that the mother cell required to complete a cycle.

The key that allowed us to proceed beyond a simple mathematical treatment of cell populations to approach the molecular basis for the variability of individual cell cycle times was the ability to observe individual cells by the cytofluorograph (Cantrell & Smith 1984). A representative cytofluorographic profile of IL-2 receptor distribution on normal human T cells activated with phytohemagglutinin is shown in Figure 5*a*. A log-normal distribution of receptor densities is revealed by plotting the number of cells vs fluorescence intensity on a $\log^{10}$ scale. Thus receptor densities vary 1000-fold within the population. The similarity of the IL-2 receptor distribution profile on activated T cells by comparison with the rate-normal

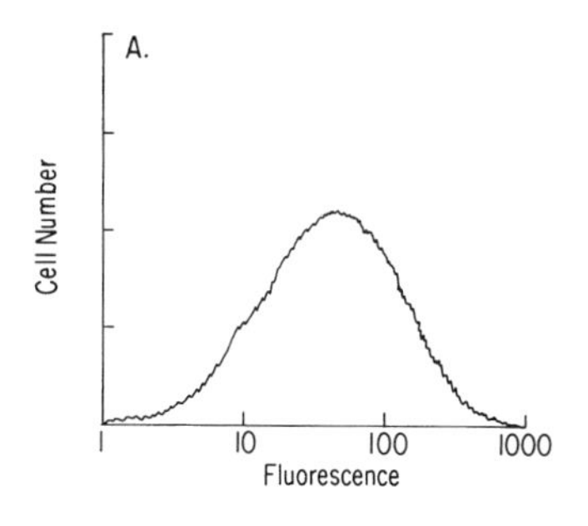

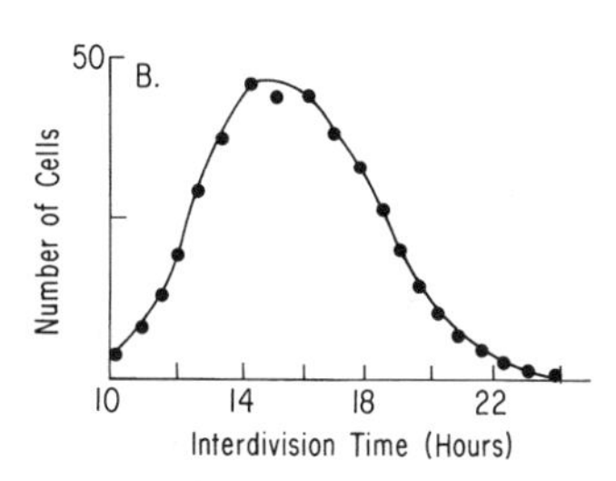

Figure 5 Profiles of IL-2 receptor density and the variability of cell cycle times. (*A*) A representative flow cytometry plot of immunofluorescence data of activated T cells reacted with monoclonal antibodies to the IL-2 receptor 55-kd subunit. (From Smith 1988b with permission.) (*B*) The cell cycle times of individual cells in an asynchronously growing population of *E. gracilis* as determined by time-lapse cinematography. (Drawn from data in Cook & Cook, 1962.)

distribution of growing *E. gracilis* depicted in Figure 5*b* is immediately apparent. These data on *E. gracilis* are from a classic paper by Cook & Cook (1962), who detailed the growth characteristics of this protozoan using time-lapse cinematography. It is readily apparent that some cells grow much faster than others, with the majority of cells distributed in a log-normal fashion about the mean. With these plots in mind, it was natural to wonder whether the T cells with a high density of IL-2 receptors might have a faster cell cycle time than those with a low density of receptors.

In a series of experiments designed to test this assumption it was found that only three parameters are responsible for determining the rate of T cell cycle progression: the IL-2 concentration, the IL-2 receptor density, and the duration during which the IL-2-receptor interaction is allowed to proceed (Figure 6) (Cantrell & Smith 1984). Moreover, the most intriguing of these findings is the last, since it appears that several hr of IL-2-receptor occupancy are necessary before any cells within the population make the irrevocable decision to proceed through to late G_1 and enter into S phase. Thus even though IL-2 binding to the receptor comes to a steady-state within 10–15 min, as long as five hr are required to stimulate the quantal (all-or-none) decision that each cell makes before initiating DNA replication. The biochemical interpretation of this phenomenon follows that there is a finite number of IL-2 receptor interactions that must occur before some critical threshold is surpassed that enables the cell to move the rest

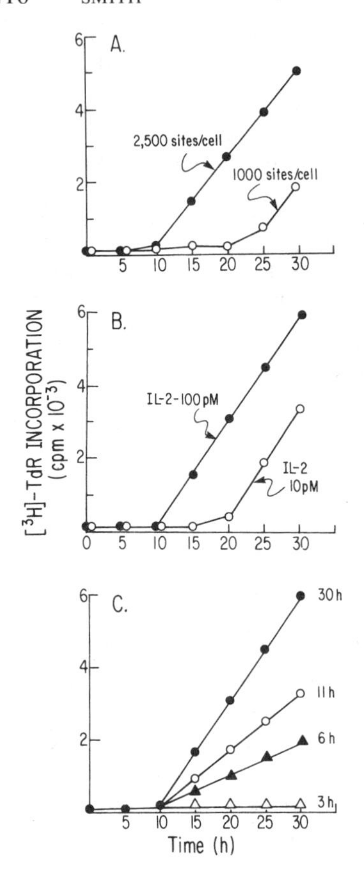

Figure 6 The variables responsible for T cell cycle progression. Synchronized T cells that expressed IL-2 receptors were exposed to IL-2 and their progression through the G_1 phase into the S phase was monitored by tritiated thymidine incorporation. (*A*) Two separate cell populations that differed in IL-2 receptor density. (*B*) The same cell population exposed to a saturating concentration of IL-2 (100 pM) vs a half-saturating IL-2 concentration (10 pM). (*C*) The effect of washing out the IL-2 after different time intervals. (Redrawn from Cantrell & Smith 1984 with permission.)

of the way through G_1. Therefore the time required for G_1 is variable, but each cell within the population has a fixed G_1 time dictated by the density of IL-2 receptors; the higher the IL-2 receptor density, the shorter the G_1 interval. Since T cells activated polyclonally have a mean high affinity IL-2 receptor density of 1500 sites/cell, the data from Figure 6 indicate that this number of receptors must be insufficient to promote cell cycle progression, since removing IL-2 after three hr prevents G_1 progression. Only an exposure time of at least five hr promotes some cells to enter the S phase. For this reason, the number of new receptors synthesized during this time interval must ultimately decide the rate of T cell division, and the $\sim$10,000 receptors synthesized and occupied during five hr (from equation 1) may be a reasonable approximation.

The most appealing molecular explanation for the threshold phenomenon just described is the idea that a critical concentration of some intracellular reactant(s) must accumulate as a consequence of receptor-mediated signal transduction. Therefore should the IL-2 receptor prove to activate the cell enzymatically, one can imagine that the concentration of the product(s) of this IL-2-directed catalysis must reach a crucial level before progression to S phase occurs. It follows that if the process should be aborted by premature removal of IL-2, the cell would return to its quiescent state rather than undergo division. Such a system would function to provide a rheostat-like form of control that would function to maintain the cell division process with exquisite fidelity.

TOWARD A MOLECULAR EXPLANATION OF IL-2 STIMULATED EVENTS

Based on our own preliminary results and those of others, the most likely explanation for the IL-2-receptor-mediated quantal decision to progress through G_1 to enter S phase is the sequential expression of a specific set of genes, the products of which both prepare the cell and signal the cell to replicate its DNA. We and others have begun to accumulate information with this hypothesis in mind, and thus far the evidence points to several genes that are expressed in a sequential manner after IL-2 stimulation. In particular, Dautry et al (1988) have used cloned murine IL-2-dependent cell lines to show that IL-2 promotes the expression of several cellular proto-oncogenes. For example, they find mRNA transcripts for c-fos and c-myc to be evident within 40 min of IL-2 stimulation. Soon, thereafter, within 1 hr, the putative cellular oncogene c-pim is expressed, followed closely by c-myb, a nuclear proto-oncogene that encodes a DNA binding protein. These data support our earlier findings (Stern & Smith 1986) with c-myb expression in IL-2-stimulated normal human T cells: maximal mRNA expression occurs at five hr, just at the midpoint of the G_1 phase and coincident with the minimum time necessary to effect the quantal decision to progress into late G_1 and S phase.

The mechanism of activation of these proto-oncogenes is of interest because Dautry et al (1988) show quite clearly that for CTLL-2 cells, IL-2 induces transcription of c-pim and c-myc. Moreover, this transcriptional activation occurs even in the presence of cycloheximide, which indicates that the expression of c-pim and c-myc does not require protein synthesis. By comparison, the mRNA transcript for c-myb does not appear to be regulated at the level of transcription. Thompson et al (1986), studying synchronized avian lymphocytes and fibroblasts, also found c-myb mRNA to be expressed at the midpoint of G_1 and its regulation to be at the level of mRNA degradation.

Additional genes that are induced by IL-2 and that may be important for cell cycle progression include the transferrin receptor and the gene encoding PCNA/cyclin. Transferrin is necessary to supply elemental iron, which serves as a co-factor for ribonucleotide reductase, a critical enzyme that functions in nucleotide synthesis (Reichard & Ehrenberg 1983). Cells deprived of transferrin transit the S phase more slowly than normal. PCNA/cyclin is an obligatory co-factor for the proper functioning of DNA polymerase-delta, which is responsible for synthesis of the leading strand during DNA replication (Downey et al 1988). The mRNA transcripts for both transferrin and PCNA first appear in mid-G_1 and peak just before the onset of DNA synthesis (A. F. Straight, D. Johnson & K. A. Smith, unpublished). Accordingly, the mechanism responsible for activating the expression of these genes may well be a part of the early G_1 threshold phenomenon that must be surpassed before the cell finally moves into the later stages of G_1 and S phase.

Obviously, much more work is necessary before it will be possible to be confident about the signaling molecules involved in G_1 progression. However it is evident that the cellular reagents are at hand for a detailed dissection of the molecular events that occur during this phase of the cell cycle: both freshly isolated and cloned normal T cells can be synchronized easily so that molecular genetic approaches can be used to detect and isolate newly expressed genes.

IL-2-STIMULATED CELLULAR DIFFERENTIATION

IL-2 was originally defined by its ability to stimulate T cell proliferation, hence its first designation as T cell growth factor (Gillis et al 1978). However IL-2 may well function to stimulate B cell proliferation and differentiation, especially during a primary immune response. Both human and murine B cells activated with polyclonal stimulants, such as staphylococcal Protein A and anti-IgM, express IL-2 binding sites (Nakanishi et al 1984; Muraguchi et al 1985). Moreover, murine B cells stimulated with LPS and anti-IgM express IL-2 binding sites at levels that are about one-third of those found on activated T cells (Lowenthal et al 1985).

Perhaps the most compelling evidence that IL-2 is important for B cell function comes from a study recently reported by Tigges et al (1989). Using a cloned murine B lymphoma cell line (BCL_1-CW13-3B3), these investigators found an average of 800 high affinity IL-2 receptors per cell and 125,000 low affinity binding sites/cell. Using concentrations of IL-2 expected to saturate the high affinity receptors, coincident dose-response curves were demonstrated for high affinity IL-2 binding, the induction of immunoglobulin J chain mRNA, and the stimulation of proliferation. For the assembly and secretion of pentamer IgM, the B cell must switch the μ

heavy chain mRNA transcript from the membranous form to the secretory form (an exon splicing event) and begin to transcribe the J chain mRNA. IL-2 induces both of these differentiative events. In a fashion similar to the induction of the transferrin receptor and PCNA/cyclin mRNA transcripts, the induction of J-chain mRNA requires at least 12 hr of IL-2 exposure. Again, there may well be a critical threshold of signals that requires several hr of IL-2 exposure for ultimate signaling. Alternatively, DNA replication could be a prerequisite for expression of J chain mRNA. In this instance, the differentiative process induced by IL-2 may be linked to the proliferative process.

It is especially noteworthy that Tigges et al (1989) found that IL-4 functions as an antagonist with respect to the IL-2-induced events in BCL_1 cells. IL-4 abrogates both IL-2-induced J chain expression and IL-2-induced BCL_1 proliferation. The mutual antagonism does not result from competition of IL-4 for IL-2 binding to its receptor. Instead, IL-4 appears to circumvent the intracellular signals generated by IL-2. These changes are reminiscent of the effects of elevated cyclic AMP, which completely prevents IL-2-mediated T cell G_1 progression (Johnson et al 1988). Obviously the mechanism of interaction of the signaling pathways of IL-2 and IL-4 warrants further investigation.

THE DEVELOPMENT OF IL-2 ANALOGUES THAT CAN SERVE AS IL-2 AGONISTS AND ANTAGONISTS

Given the understanding that there is a direct relationship between IL-2 receptor occupancy and T cell proliferation, we know that IL-2 itself requires 100% receptor occupancy to promote a maximal response. Also, on the basis of our dissection of the IL-2 receptor structure, it is evident that the receptor is complex, interacting with IL-2 via binding sites on two distinct chains, only one of which participates in signal transduction. In addition, our results suggest that the duration of receptor occupancy somehow is important for the generation of the number of signals necessary to surpass the threshold that ultimately dictates DNA replication.

This information has led us to speculate that there may be parts of the binding site on the 75-kd chain involved in signal transduction that are separate and distinct from the residues that are important for actual binding. Obviously, should this be the case, it might be possible to engineer IL-2 analogues that can occupy the receptor but not signal the cell. At the very least, it may well be possible to change the IL-2 molecule so that its interaction with 75-kd chains is impaired while the IL-2-55-kd chain binding site is left unhindered. In particular, because the duration of receptor occupancy seems to be important for signaling, it may be possible to change IL-2 so that it dissociates more rapidly from the 75-kd chain binding site. Of course this would have the effect of lowering the affinity of the interaction as well, but it could have the effect of changing the molecule into a partial agonist; i.e. a ligand that occupies the receptor but signals less efficiently. From the viewpoint of drug development this would be an especially important concept to prove or disprove since it could open the way to the development of complete antagonists. Moreover, if the signaling aspects of the receptor molecule really are different from the residues responsible for ligand binding, theoretically it should be possible to alter the IL-2 molecule in such a way as to make it an even more effective agonist than native IL-2. This would have the effect of creating spare receptors, i.e. the putative "super IL-2" would signal more effectively than IL-2 itself.

To approach these questions experimentally, in collaboration with Ciardelli & Cohen, secondary structure prediction approaches were used to generate a model of the IL-2 molecule (Cohen et al 1986). Aided by a computer-generated model, several years ago we began using a combination of peptide synthesis and recombinant DNA methods to alter IL-2, at first with the aim of testing the model structure. The model predicted the core of IL-2 to be comprised of four antiparallel α helices, a common motif for small globular proteins. Examination of the predicted helices revealed an amphipathic arrangement with hydrophobic residues comprising the faces of the helices that pack together to form the hydrophobic core of the molecule, while the hydrophilic residues were turned towards the external environment.

The model predicted that the carboxy terminus consisted of an amphipathic helix spanning residues 116 to 132 (there are 133 amino acids in the mature secreted protein). Subsequently, analysis of IL-2 crystals has revealed the actual arrangement of the peptide backbone of IL-2, which has confirmed that the molecule is primarily α helical, comprised of four major anti-parallel helices (Brandhuber et al 1987) (Figure 7). Of utmost importance for our studies, the C-terminus is an amphipathic helix spanning residues 117–133.

Inspired by studies from the late Thomas Kaiser & co-workers (1984), who engineered ideal α helical segments in small peptides, we used peptide synthesis to generate several analogues of the C-terminus, then attached the last 34 residues to the first 99 residues via the disulfide bond that occurs in the native molecule between residues 58 and 105 (Ciardelli et al 1988). Interestingly, when a heterodimeric protein is constructed, the molecule is 1000-fold less active than native monomeric IL-2, even though the natural sequence is maintained. Even so, the fact that such a heterodimeric protein

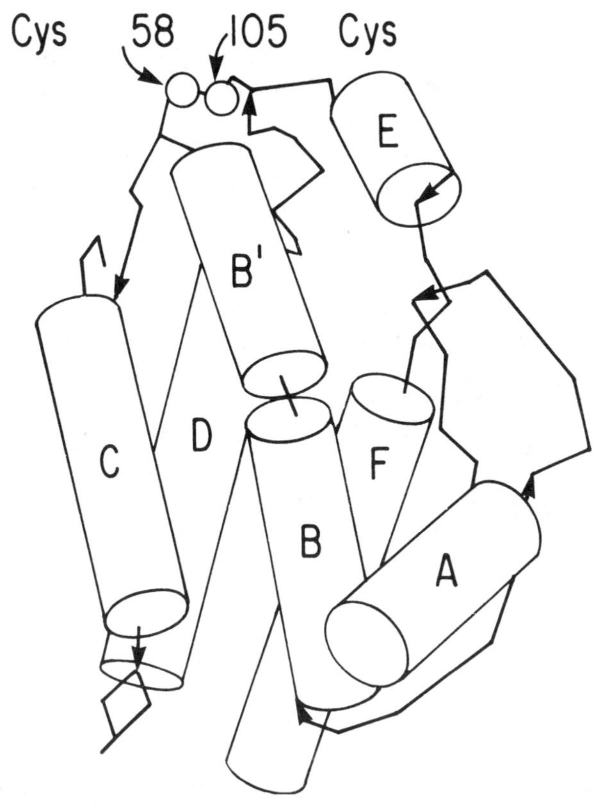

Figure 7 The tertiary configuration of IL-2 from X-ray crystallographic data. (From Brandhuber et al 1987 with permission.)

retained any activity at all allowed us to compare variants with the native sequence.

To test the significance of the amphipathic nature of the C-terminal helix we made 12 simultaneous changes in the last 19 residues, all changes chosen to enhance the propensity of helix formation. Surprisingly enough, an IL-2 hybrid with an idealized C-terminal helix has equal activity compared with the native IL-2 sequence constructed as a disulfide-linked heterodimer. However with respect to structure-activity considerations, it is most revealing that disruption of the C-terminal helix by the placement of proline residues within the helix region completely abrogates IL-2 activity, which indicates that the integrity of the secondary structure of the molecule is of primary importance for receptor binding and signal transduction.

These data serve to illustrate the importance of secondary and tertiary structures for intimate interactions between two molecules such as a hormone and its receptor, and they also reveal how difficult it is to interpret structure-activity relationships from random site-specific mutagenesis approaches: a single amino acid substitution or deletion may destroy biologic activity by disrupting crucial secondary and tertiary arrangements within the protein, but have nothing to do with the so-called active sites.

Following a protein engineering approach using peptide synthesis, cassette mutagenesis, and detailed examination of pure mutant molecules by circular dichroism to detect changes in secondary and tertiary structure, we continue to change the IL-2 molecule, searching to identify antagonists and super agonists (Landgraf et al 1989). It is worthy of emphasis that the IL-2-receptor system is now ideal for detailed SAR studies; the crystal structure of IL-2 is known, the structure and function of both chains of the receptor are known and easily quantified, and the T cell proliferative response to IL-2-receptor signals is well described.

CONCLUSIONS

The two prototypic characteristics of receptors described originally by Langley more than a century ago are now approachable for the first time in the immune system. The interleukins comprise a class of hormones that have a definite known secondary and tertiary structure under physiologic conditions. Moreover, the structure and function of the two chains comprising the IL-2 receptor are also now known. Consequently, detailed molecular engineering studies using IL-2 and its receptor are possible and open the way to a new level of considerations: those concerned with the biophysics of intermolecular reactions, energy transfer, signal transduction, and the cellular physiology of the response. These studies will surely lead to the construction and development of a new class of pharmaceuticals that either antagonize and suppress the immune response, or alternatively serve as better agonists than IL-2 itself, thereby acting as adjuvants for the immune response. Moreover, with the capacity to use cloned normal antigen-specific T cells, cloned molecules, and cloned antibodies, we now can fully exploit a reductionist approach. Consequently for the first time, answers to fundamental questions regarding exactly how cells react to environmental signals that direct and control their maturation, proliferation, and differentiation should be forthcoming.

Literature Cited

*Adam, G., Delbruck, M. 1968. Reduction of dimensionality in biological diffusion processes. In *Structural Chemistry and Molecular Biology*, eds. A. Rich, N. Davidson, pp. 198–215. San Francisco: Freeman

Ariëns, E. J. 1954. Intrinsic activity: partial agonists and partial antagonists. *J. Cardiovas. Pharmacol.* 5: S8–S15

Baker, P. E., Gillis, S., Ferm, M. M., Smith, K. A. 1978. The effect of T-cell growth factor on the generation of cytolytic T-cells. *J. Immunol.* 121: 2168–73

Baker, P. E., Gillis, S., Smith, K. A. 1979. Monoclonal cytolytic T-cell lines. *J. Exp. Med.* 149: 273–78

Brandhuber, B. J., Boone, T., Kenney, W. C., McKay, D. B. 1987. Three-dimensional structure of interleukin 2. *Science* 238: 1707–9

*Cantrell, D. A., Smith, K. A. 1984. The interleukin 2-T-cell system: a new cell growth model. *Science* 224: 1312–16

Chao, M. V., Bothwell, M. A., Ross, A. H., Koprowski, H., Lanahan, A. A., et al. 1986. Gene transfer and molecular cloning of the human NGF receptor. *Science* 232: 518–21

Ciardelli, T. L., Landgraf, B. E., Gadski, R., Strand, J., Cohen, F. E., Smith, K. A. 1988. A semisynthetic protein engineering approach to the structural analysis of Interleukin-2. *J. Molec. Recognit.* 1: 42–47

Clark, A. J. 1926. The antagonism of acetyl choline by atropine. *J. Physiol.* (*Lond.*) 61: 547–56

Cohen, F. E., Kosen, P. A., Kuntz, I. D., Epstein, L. B., Ciardelli, T. C., Smith, K. A. 1986. Structure activity studies of interleukin 2. *Science* 234: 349–52

Cook, J. R., Cook, B. 1962. Effect of nutrients on the variation of individual generations times. *Exp. Cell Res.* 28: 524–30

*Cosman, D., Cerretti, D. P., Larsen, A., Park, L., March, C., et al. 1984. Cloning, sequence and expression of human interleukin-2 receptor. *Nature* 312: 768–71

Dautry, F., Weil, D., Yu, J., Dautry-Varsat, A. 1988. Regulation of pim and myb mRNA accumulation by interleukin 2 and interleukin 3 in murine hematopoietic cell lines. *J. Biol. Chem.* 263: 17615–20

Downey, K. M., Tan, C.-K., Andrews, D. M., Li, X., So, A. G. 1988. Proposed roles for DNA polymerases α and δ at the replication fork. *Cancer Cells* 6: 403–10

Duprez, V., Cornet, V., Dautry-Varsat, A. 1988. Down-regulation of high affinity interleukin 2 receptors in a human tumor T cell line. *J. Biol. Chem.* 263: 12860–65

Ehbrecht, H.-J., Pingoud, A., Urbanke, C., Maass, G., Gualerzi, C. 1985. Linear diffusion of restriction endonucleases on DNA. *J. Biol. Chem.* 260: 6160–65

Ehrlich, P. 1913. Chemotherapeuticals: scientific principles, methods, and results. *Lancet* 4694:445–51

Fung, M. R., Ju, F., Greene, W. C. 1988. Co-internalization of the p55 and p70 subunits of the high-affinity human interleukin 2 receptor. *J. Exp. Med.* 168: 1923–28

Gillis, S., Crabtree, G. R., Smith, K. A. 1979. Glucocorticoid-induced inhibition of T-cell growth factor production. I. The effect on mitogen-induced lymphocyte proliferation. *J. Immunol.* 123: 1624–31

Gillis, S., Ferm, M. M., Ou, W., Smith, K. A. 1978. T-cell growth factor: parameters of production and a quantitative microassay for activity. *J. Immunol.* 120: 2027–32

Glenney, J. R. Jr., Chen, W. S., Lazar, C. S., Walton, G. M., Zokas, L. M., et al. 1988. Ligand-induced endocytosis of the EGF receptor is blocked by mutational inactivation and by microinjection of anti-phosphotyrosine antibodies. *Cell* 52: 675–84

Hatakeyama, M., Tsudo, M., Minamoto, S., Kono, T., Doi, T., et al. 1989. Interleukin 2 receptor β chain gene: generation of three receptor forms by cloned human α and β chain cDNAs. *Science* 244: 551–56

Herzberg, V. L., Smith, K. A. 1987. T cell growth without serum. *J. Immunol.* 139: 998–1004

*Hosang, M., Shooter, E. M. 1985. Molecular characteristics of nerve growth factor receptors on PC12 cells. *J. Biol. Chem.* 260: 655–62

Johnson, D., Lanahan, A., Buck, C. R., Sehgal, A., Morgan, C., et al. 1986. Expression and structure of the human NGF receptor. *Cell* 47: 545–54

Johnson, K. W., Davis, B. H., Smith, K. A. 1988. Cyclic AMP antagonizes interleukin-2-promoted T cell cycle progression at a discrete point in early G_1. *Proc. Natl. Acad. Sci. USA* 85: 6072–76

Kaiser, E. T., Kezdy, F. J. 1984. Amphiphilic secondary structure: design of peptide hormones. *Science* 233: 249

Kawakami, T., Huff, S., Buckley, P., Dungworth, D., Snyder, S., Gilden, R. 1972. C-type virus associated with gibbon lymphosarcoma. *Nature New Biol.* 235: 170–71

*Kondo, S., Shimizu, A., Saito, Y., Kinoshita, M., Honjo, T. 1986. Molecular basis for two different affinity states of the interleukin 2 receptor: affinity conversion

model. *Proc. Natl. Acad. Sci. USA* 83: 9026–29

Landgraf, B., Cohen, F. E., Smith, K. A., Gadski, R., Ciardelli, T. L. 1989. Structural significance of the C-terminal Amphiphilic Helix of Interleukin-2. *J. Biol. Chem.* 264: 816–22

Langley, J. N. 1878. On the physiology of the salivary secretion. *J. Physiol.* (*Lond.*) 1: 339–69

Langley, J. N. 1905. On the reaction of cells and of nerve endings to certain poisons, chiefly as regards the reaction of striated muscle to nicotine and to curari. *J. Physiol.* (*Lond.*) 33: 374–413

Leonard, W. J., Depper, J. M., Crabtree, G. R., Rudikoff, S., Pumphrey, J., et al. 1984. Molecular cloning and expression of cDNAs for the human interleukin-2 receptor. *Nature* 311: 626–35

Leonard, W. J., Depper, J. M., Uchiyama, T., Smith, K. A., Waldmann, T. A., Greene, W. C. 1982. A monoclonal antibody, anti-Tac, blocks the membrane binding and action of human T-cell growth factor. *Nature* 300: 267–69

Lowenthal, J. W., Zubler, R. H., Nabholz, M., MacDonald, H. R. 1985. Similarities between interleukin-2 receptor number and affinity on activated B and T lymphocytes. *Nature* 315: 669–72

Mills, G. B., Girard, P., Grinstein, S., Gelfand, E. W. 1988. Interleukin-2 induces proliferation of T lymphocyte mutants lacking protein kinase C. *Cell* 55: 91–100

Muraguchi, J. H., Kehrl, J. H., Longo, D. L., Volkman, D. J., Smith, K. A., Fauci, A. S. 1985. Interleukin 2 receptors on human B cells. Implication for the role of interleukin 2 in B cell function. *J. Exp. Med.* 161: 181–97

Nakanishi, K., Malek, T. R., Smith, K. A., Hamaika, T., Shevach, E. M., Paul, W. E. 1984. Both interleukin 2 and a second T-cell derived factor in EL-4 supernatant have activity as differentiation factors in IgM synthesis. *J. Exp. Med.* 160: 1605–21

Nikaido, T., Shimizu, A., Ishida, N., Sabe, H., Teshigawara, K., et al. 1984. Molecular cloning of cDNA encoding human interleukin-2 receptor. *Nature* 311: 631–35

Pardee, A. B., Shilo, B.-Z., Koch, A. L. 1979. Variability of the cell cycle. In *Hormones and Cell Culture*, eds. G. H. Sato, R. Ross, pp. 373–92. Cold Spring Harbor: Cold Spring Harbor Laboratory

Paton, W. D. M. 1961. A theory of drug action based on the rate of drug-receptor combination. *Proc. R. Soc. London Ser. B* 154: 21–69

Rabin, H., Hopkins, R. F., Ruscetti, F. W., Newbauer, R. H., Brown, R. L., Kawakami, T. G. 1981. Spontaneous release of a factor with T cell growth factor activity from a continuous line of primate tumor cells. *J. Immunol.* 127: 1852–56

Radeke, M. J., Misko, T. P., Hsu, C., Herzenberg, L. A., Shooter, E. M. 1987. Gene transfer and molecular cloning of the rat nerve growth factor receptor. *Nature* 325: 593–97

Reichard, P., Ehrenberg, A. 1983. Ribonucleotide reductase—a radical enzyme. *Science* 221: 514–19

Robb, R. J. 1986. Conversion of low-affinity interleukin 2 receptors to high-affinity state following fusion of cell membranes. *Proc. Natl. Acad. Sci. USA* 83: 3992–96

Robb, R. J., Greene, W. C., Rusk, C. M. 1984. Low and high affinity cellular receptors for interleukin 2: implications for the level of Tac antigen. *J. Exp. Med.* 160: 1126–46

Robb, R. J., Munck, A., Smith, K. A. 1981. T-cell growth factor receptors: quantitation, specificity and biological relevance. *J. Exp. Med.* 154: 1455–74

Sabe, H., Kondo, S., Shimizu, A., Tagaya, Y., Yodoi, J., et al. 1984. Properties of human interleukin 2 receptors expressed on non-lymphoid cells by cDNA transfection. *Mol. Biol. Med.* 2: 379–88

Saito, Y., Sabe, H., Suzuki, N., Kondo, S., Ogura, T., et al. 1988. A larger number of L chains (Tac) enhance the association rate of interleukin 2 to the high affinity site of the interleukin 2 receptor. *J. Exp. Med.* 168: 1563–72

Sargent, D. F., Schwyzer, R. 1986. Membrane lipid phase as catalyst for peptide-receptor interactions. *Proc. Natl. Acad. Sci. USA* 83: 5774–78

Sharon, M., Klausner, R. D., Cullen, B. R., Chizzonite, R., Leonard, W. J. 1986. Novel interleukin-2 receptor subunit detected by cross-linking under high-affinity conditions. *Science* 234: 859–63

Smith, K. A. 1980. Continuous cytotoxic T-cell lines. In *Contemporary Topics in Immunobiology*, ed. N. L. Warner, 11: 139–55. New York: Plenum

Smith, K. A. 1982. T-cell growth factor and glucocorticoids: Opposing regulatory hormones in neoplastic T-cell growth. *Immunobiology* 161: 157–73

Smith, K. A. 1983. T-cell growth factor, a lymphocytotrophic hormone. In *Genetics of the Immune Response*, eds. E. Moller, G. Moller, pp. 151–85. New York: Plenum

Smith, K. A. 1988a. Interleukin 2: inception, impact, and implications. *Science* 240: 1169–76

Smith, K. A. 1988b. Interleukin 2: a 10 year

perspective. In *Interleukin 2*, pp. 1–35. New York: Academic

Smith, K. A., Cantrell, D. A. 1985. Interleukin 2 regulates its own receptors. *Proc. Natl. Acad. Sci. USA* 82: 864–68

Smith, K. A., Gillis, S., Baker, P. E. 1979. The role of soluble factors in the regulation of T-cell immune reactivity. In *The Molecular Basis of Immune Cell Function*, ed. J. G. Kaplan, pp. 223–37. Amsterdam: Elsevier/North Holland

Starling, E. H. 1905. The chemical correlation of the functions of the body. *Lancet* 4275: 339–41

Stephenson, R. P. 1956. A modification of receptor theory. *Brit. J. Pharm.* 11: 379–93

Stern, J. B., Smith, K. A. 1986. Interleukin 2 induction of T cell G_1 progression and c-myb expression. *Science* 233: 203–6

Sutter, A., Riopelle, R. J., Harris-Warrick, R. M., Shooter, E. M. 1979. Nerve growth factor receptors: characterization of two distinct classes of binding sites on chick embryo sensory ganglia cells. *J. Biol. Chem.* 254: 5972–80

Terry, B. J., Jack, W. E., Rubin, R. A., Modrich, P. 1983. Thermodynamic parameters governing interaction of EcoRI endonuclease with specific and nonspecific DNA sequences. *J. Biol. Chem.* 258: 9820–25

Teshigawara, K., Wang, H.-M., Kato, K., Smith, K. A. 1987. Interleukin 2 high affinity receptor expression depends on two distinct binding proteins. *J. Exp. Med.* 165: 223–38

Thompson, C. B., Challoner, P. B., Neiman, P. E., Groudine, M. 1986. Expression of the c-myb proto-oncogene during cellular proliferation. *Nature* 319: 374–80

Tigges, M. A., Casey, L. S., Koshland, M. E. 1989. Mechanism of interleukin-2 signaling: mediation of different outcomes by a single receptor and transduction pathway. *Science* 243: 781–86

Tsudo, M., Kozak, R. W., Goldman, C. K., Waldmann, T. A. 1986. Demonstration of a non-Tac peptide that binds interleukin 2: a potential participant in a multichain interleukin 2 receptor complex. *Proc. Natl. Acad. Sci. USA* 83: 9694–98

Valge, V. E., Wong, J. G. P., Datlof, B. M., Sinskey, A. J., Rao, A. 1988. Protein kinase C is required for responses to T cell receptor ligands but not to interleukin-2 in T cells. *Cell* 55: 101–12

Wang, H.-M., Smith, K. A. 1987. The interleukin 2 receptor: Functional consequences of its bimolecular structure. *J. Exp. Med.* 166: 1055–69

Weiss, A., Imboden, J., Hardy, K., Manger, B., Terhorst, C., Stobo, J. 1986. The role of the T3/antigen receptor complex in T-cell activation. *Annu. Rev. Immunol.* 4: 593–619

Williams, L. T. 1989. Signal transduction by the platelet-derived growth factor receptor. *Science* 243: 1564–70

Annu. Rev. Cell Biol. 1989. 5 : 427–52

BIOGENESIS OF THE RED BLOOD CELL MEMBRANE-SKELETON AND THE CONTROL OF ERYTHROID MORPHOGENESIS

Elias Lazarides and Catherine Woods

Division of Biology 156–29, California Institute of Technology, Pasadena, California 91125

CONTENTS

INTRODUCTION

Subcellular morphogenesis can be broadly defined as the totality of regulatory events that specify the structural and functional phenotype of a cell type during the differentiation pathway of a given lineage. The success of classical studies in defining bacteriophage morphogenesis depends on a combined biochemical and genetic approach. After a detailed biochemical analysis of all the structural components of the phage particle, mutations

0743–4634/89/1115–0427$02.00

in the various components were isolated that resulted in the over accumulation of otherwise transiently unassembled intermediates. This, coupled with genetic complementation analysis and morphological studies, enabled the precise ordering of the pathway of phage assembly to be elucidated (140). In higher eukaryotes the hematopoietic lineage offers several ideal systems to unravel the mechanisms that underlie the structural morphogenesis of a given cell type. In particular, the erythroid lineages provide cell types whose different developmental histories during embryogenesis and ultimate structural phenotypes are characterized in some detail (21, 80, 114, 116). Hence the assembly of the red cell structure can be analyzed in the context of its developmental history, the sites of origin of the different erythroid lineages, and the maturational responses of the different lineages to the hormonal microenvironment of each hematopoietic site. Furthermore, a number of well defined mutations exist in various structural components of murine and human red blood cells that cause hereditary hemolytic anemia (15, 26, 106). These mutants allow genetic complementation of biochemical analysis of the assembly of erythroid structure.

In the animal kingdom erythroid cells have evolved two basic shapes. All nonmammalian erythroid cells have a biconvex ellipsoidal shape and remain nucleated throughout their mature life in the circulation. Three interconnected cytoskeletal domains, a plasma-membrane associated skeletal network, a system of intermediate filaments, and a marginal band of microtubules cooperatively contribute both to the assumption of this cell's shape during development and to its maintenance during the mature life of the circulating cell (6, 82). In contrast, during development of the majority of mammalian species, erythroid cells that originate in the embryo proper (e.g. fetal liver, embryonic spleen, and adult bone marrow) undergo several structural changes including the removal of intermediate filaments and microtubules during their later maturation stages, and the loss of their nuclei by enucleation. The resultant anuclear reticulocytes then gradually assume their characteristic biconcave discoid shape (80, 82, 114). This natural genetic divergence of mammalian and nonmammalian erythroid cells can be exploited further not only to understand the mechanisms that underlie the evolutionary divergence of cell structure, but also as a means of genetically manipulating one phenotypic background by introducing components normally expressed only in the other background. Additionally the existence of stable virally transformed lines that are blocked at various stages of lymphoid and myeloid differentiation provides the opportunity to assess how the erythroid structural phenotype becomes diversified from that of cells in other hematopoietic lineages during the ontogeny of the hematopoietic system.

In this review we concentrate on some of the mechanisms that underlie the assembly of the various components of the membrane-skeleton (MS) of erythroid cells and how these mechanisms address the problem of spatial and temporal regulation of cytoskeletal morphogenesis. We also discuss departures from simple self-assembly mechanisms and evolutionary divergence in the expression patterns of MS components that address the issue of structural divergence as part of the evolutionary change of the erythroid lineage.

THE STRUCTURE OF THE ERYTHROID MEMBRANE-SKELETON

The structural components of the erythroid MS and their interaction with each other and the plasma membrane have been well defined both by biochemical and morphological criteria. Initially these studies were carried out predominantly with human red blood cells, but were subsequently extended to avian (chicken) red blood cells, which proved to have extensive structural and functional homology with their human counterparts (9, 19, 30, 82, 90, 91). Since this aspect of the red blood cell structure has been reviewed extensively in the past, it will be reviewed here only briefly as it pertains to the discussion that follows. The principal component of the MS is the heterodimeric protein spectrin, which is crosslinked into pentagonal and hexagonal arrays by short oligomers of actin (7, 19, 23, 30, 82, 88, 90, 91, 120). The interaction of spectrin with actin is reinforced at least a million-fold by protein 4.1 (105, 109, 132). Attachment of this network to the plasma membrane is mediated by ankyrin, which interacts simultaneously with the β-subunit of spectrin and the cytoplasmic domain of the Cl^-/HCO_3^- anion transporter (AT; band 3) (7–9, 19, 23, 24, 30, 60, 82, 88, 90, 91, 120, 132, 134). Additional membrane binding sites result from the ability of protein 4.1 to interact either with the cytoplasmic domain of glycophorin C or the AT (3, 4, 107). Both sets of peripheral protein-membrane interactions may be modulated posttranslationally by various lipid metabolites. Ankyrin is fatty acid acylated with palmitic acid that may confer the ability to interact directly with the bilayer independently of its high affinity association with the AT (125). The interaction of protein 4.1 with glycophorin C is mediated by phosphoinositides that may modulate both the affinity of these two proteins for each other and of protein 4.1 directly with the bilayer (3, 4, 91). Other regulatory modifications, most notably phosphorylation mediated by Ca^{2+}/calmodulin, can modulate the affinity of the various peripheral components for each other (25, 43, 49, 87, 89, 94, 122). A number of other actin and/or spectrin binding proteins have been identified that are presumed to regulate the extent of spectrin and/or actin crosslinking (e.g. tropomyosin, adducin, protein 4.9), but because much less is known to date about their expression pattern and assembly mechanisms during erythropoiesis they will not be considered here (46, 49, 61, 94, 121, 122).

ASSEMBLY OF THE MEMBRANE-SKELETON OF ERYTHROID CELLS

General Issues

One major principle to emerge from the analysis of the biochemical properties of the individual constituents of the MS is that of simple self-assembly whereby all components carry sufficient information in their primary structure to allow them to reassemble spontaneously with each other under defined biochemical conditions in vitro. For example, purified $\alpha\beta$-spectrin subunits isolated by denaturation with chaotropic agents will spontaneously reassemble upon renaturation into heterodimers of the correct stoichiometry and antiparallel conformation (91, 135, 145). Purified spectrin heterodimers bind with high affinity to ankyrin in vitro through the β-spectrin subunit and in turn this complex can bind to the cytoplasmic domain of the AT in inverted ankyrin-depleted plasma-membrane vesicles or onto purified molecules reconstituted in lipid vesicles (7–9, 60, 105, 134). Protein 4.1 will enhance the affinity of spectrin tetramers for actin in vitro in the form of a ternary complex, and it will bind to the cytoplasmic domain of glycophorin C in the presence of phosphoinositides (7, 24, 91, 105, 132, 134).

One characteristic feature of the MS of avian and mammalian red blood cells is that at steady-state in the mature cell it is assembled exclusively on the cytoplasmic side of the plasma membrane. Thus despite the fact that all peripheral MS components can spontaneously self-assemble in vitro, in vivo nucleation of assembly must be spatially controlled during erythroid differentiation to occur only proximal to the plasma membrane without anomalous nucleation in the cytoplasm. This issue becomes more complicated if one considers that the erythroid-specific structure arises from a pre-existing structure inherent to undifferentiated progenitor cells. Since many components of the mature erythroid MS exist as multiple isoforms in nonerythroid cells (2, 12, 29, 34, 38, 41, 42, 51, 52, 55, 57, 58, 62, 66, 72, 73, 81, 85, 86, 90, 93, 129, 136), a major aspect of this problem is the change of the preexisting membrane structure of the undifferentiated erythroid cell to one characteristic of the more mature cell. Therefore, temporal control must also play an important role in the emergence of the ultimate erythroid structural phenotype. Finally, since erythroid cells of the primitive lineage of both avian and mammalian species are substantially bigger than and differ in shape to their definitive counterparts (21, 79, 80, 116), it is not a priori evident that all aspects of the control of assembly will be conserved between the primitive and definitive lineages. Therefore substantially more information than that inherent for simple self-assembly is necessary to allow for differentiation of an erythroid cell's membrane structure. These issues will be discussed here as they pertain to individual components of the MS.

Preexisting Structure in Early Erythroid Progenitor Cells and Its Remodeling during Differentiation

Early progenitor erythroid cells differentiate while attached to support cells (e.g. endothelial cells in the case of the primitive lineage and macrophages in the case of the bone-marrow derived definitive lineage) and have many characteristics in common with amoeboid cells (1, 21, 37, 80, 116). As such they have a cytoskeletal complement, which is characteristic of many other cell types including other hematopoietic lineages. For example, in the case of spectrin, a family of spectrin heterodimers exists, each molecule consisting of an α subunit and a variable subunit (7, 30, 51, 52, 83, 90, 91, 113). In avian species, a single conserved α-spectrin gene product is common to all spectrin heterodimeric species; mammals also express a homologous α-spectrin of widespread occurrence, but in addition have evolved a second erythroid-specific α-spectrin gene (12, 30, 38, 51, 52, 54, 62, 85, 86, 90, 93, 113, 136). There are three major classes of variable subunits: a γ-spectrin, which as $\alpha\gamma$-spectrin (also referred to as fodrin or brain spectrin), is expressed in a variety of undifferentiated mitotic progenitor cells as well as most fully differentiated cells, e.g. lymphocytes and neurons (30, 51, 52, 53, 62, 83, 86, 90, 113); an isoform TW_{260}, specific to the terminal web of intestinal epithelial cells (30, 51, 52, 62, 90); and a set of β-spectrin variants, the products of differential splicing of the same gene, that are characteristic of mature erythroid cells, skeletal muscle, cardiac muscle, and certain major neurons in the central nervous system (19, 30, 55, 83, 90, 97, 99, 101, 139). Coexpression of two distinct variable subunits and the accumulation of two distinct spectrin heterodimers within the same cell type are not uncommon, and can vary as a function of development along a differentiation pathway (53, 55, 83, 97, 99, 101). The ultimate structural spectrin phenotype is governed at many levels. Expression of any one spectrin isoform generally requires the obligatory coexpression of two subunits, which are the products of genes on different chromosomes (85, 93). The extent of spectrin accumulation, characteristic for a given cell type, is regulated by independent mechanisms operating on the synthesis and turnover of the individual subunits. This is even more complex in the case where different spectrins coexist in different sites within the same cell (e.g. neurons) or where switching of spectrin isoforms occurs within the same cell during its differentiation pathway. Chicken erythroid progenitors (BFUe, CFUe) provide a good example of this latter issue. Analysis of virally transformed analogues of myeloid progenitor cells (E26) and CFUe cells [avian erythroblastosis (AEV) and S13] reveals that the former express only the $\alpha\gamma$-(fodrin) spectrin isoform, whereas the latter coexpress fodrin (brain $\alpha\gamma$-spectrin) and erythroid $\alpha\beta$-spectrin at steady-state (111,141). This implies that erythroid β-spectrin expression under normal in vivo conditions may be induced after progenitor stem cells have progressed to the myeloid stage and prior to differentiation to the CFUe stage. Thus expression of erythroid β-spectrin may provide an early membrane marker of erythroid differentiation several divisions prior to the expression of the globin genes. In these retrovirally transformed erythroid progenitors both $\alpha\gamma$ and $\alpha\beta$ heterodimers assemble onto the membrane where they are unstable and are degraded (141; E. Lazarides, C. Woods, unpublished observations). Thus despite high rates of synthesis of all three subunits their steady-state levels are low in these cells due to rapid catabolism. When these cells are induced to undergo terminal differentiation the rates of synthesis of the α subunit and erythroid β subunit do not change appreciably and are comparable to what is observed in circulating mitotic embryonic hemoglobinized erythroblasts from 3 and 4 day chick embryos, but the rate of synthesis of the nonerythroid γ-spectrin subunit declines precipitously. This is accompanied by a marked decrease in the turnover rate of the erythroid $\alpha\beta$ heterodimer so that its steady-state levels increase continuously while the steady-state level of the nonerythroid heterodimer is gradually dissipated as the cells divide (141; E. Lazarides, C. Woods, unpublished observations). By the time the cells become post-mitotic they synthesize and accumulate primarily the erythroid $\alpha\beta$-spectrin heterodimer at steady-state. This pattern of spectrin subunit switching and the mechanism of heterodimer assembly observed in early erythroid differentiation is analogous to that which is observed during skeletal myogenesis in vitro (99, 101). Cardiac myocytes and certain central nervous system neurons express and accumulate simultaneously both $\alpha\gamma$- and $\alpha\beta$-spectrin heterodimers, which indicates that with cell type specific variation, preferential stabilization is a general mechanism of spectrin subunit remodeling and assembly during differentiation (55, 83, 102) (Figure 1).

In addition to the pattern of spectrin expression and accumulation described for avian erythropoiesis, mammalian erythroid development also involves the switching of a nonerythroid α-spectrin isoform to an erythroid-specific α-spectrin subunit (44, 50, 53, 108). Thus far expression

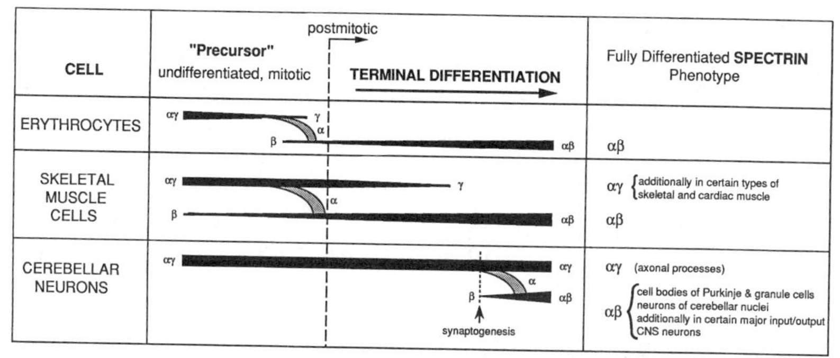

Figure 1 Proposed pathway of expression and steady-state accumulation of spectrin subunits in selected lineages of chickens. In mammals this pattern of expression is similar with the additional change of a nonerythroid α-spectrin to an erythroid α-spectrin subunit in erythropoiesis.

of the erythroid specific α-spectrin gene has been detected only in erythroid cells (7, 30, 44, 52, 53, 82, 90, 108, 113). Whether the initiation of expression of the erythroid spectrin α subunit coincides temporally with that of the erythroid-specific β subunit is unclear. However murine erythroleukemia (MEL) cells, which are arrested at the early orthochromatophilic erythroblast stage (47, 92), express all four subunits simultaneously. Down-regulation of the nonerythroid genes occurs after the cells are induced to undergo terminal differentiation similar to their chicken erythroid counterparts (53, 108).

Posttranslational Control of Spectrin Assembly

Throughout development in ovo, circulating chicken embryo erythroblasts and bone marrow and spleen-derived mammalian erythroblasts and reticulocytes synthesize α-spectrin in excess of β-spectrin during the accumulation of erythroid $\alpha\beta$-spectrin. Furthermore both subunits are synthesized in excess of the amounts that assemble onto the MS (5, 13, 59, 96). This pattern of expression and assembly is observed also in AEV- and S13-transformed avian erythroblasts and mouse splenic erythroblasts transformed with Friend virus both for $\alpha\gamma$- and $\alpha\beta$-spectrin prior to and $\alpha\beta$-spectrin subsequent to induction of terminal differentiation in culture (75–78, 141). In contrast, when MEL cells are chemically induced to undergo terminal differentiation they appear to synthesize an excess of β-spectrin over erythroid α-spectrin (53, 84, 108) as the synthesis of the nonerythroid γ subunit is down-regulated. The reasons for this discrepancy are not clear, but one possibility is that these cells fail to up-regulate the synthesis of the erythroid specific form of α-spectrin to the same extent as do the equivalent normal cells in vivo when they down-regulate the nonerythroid α subunit.

Irrespective of these differences, the above mentioned observations raise the question of how the stoichiometry and extent of spectrin subunit assembly are determined during avian and mammalian erythropoiesis since clearly they are not governed solely by the amounts of subunits synthesized and their subsequent simple self-assembly. Rather, additional post-translational events must take place to determine the extent and subunit stoichiometry of spectrin assembly.

During or immediately following synthesis, spectrin monomers combine either as $\alpha\beta$ heterodimers, which rapidly assemble onto the MS with high affinity, or as specific homo-oligomers (β_4-spectrin or α_2-spectrin). The homo-oligomeric forms are targeted for degradation by apparently distinct cytoplasmic pathways with β homo-oligomers being degraded substantially faster than α homo-oligomers (141, 142). The highly α-helical structure of the spectrin monomer may be the underlying reason for this rather inefficient kinetic pathway of spectrin assembly. All spectrin subunits appear to consist of a conserved structure of multiple repeating triple α helices, each 106 amino acids in length, which interact with a heterologous spectrin to form a large antiparallel coiled-coil molecule (123). Since the monomeric form of a long coiled-coil protein is energetically unfavorable (20), and since the assumption of an α-helical conformation occurs substantially more rapidly than the biosynthesis of the individual spectrin polypeptide (45, 130), it is reasonable to hypothesize that the formation of homo-oligomeric spectrin species during de novo synthesis results from the greater chance of coiled-coil interactions occurring between like subunits (i.e. adjacent neighbors on polysomes) rather than unlike subunits, even though the heterodimeric $\alpha\beta$ form is thermodynamically favored. Once homo-oligomers are formed, the monomers are effectively trapped in this conformation and prevented from interacting with heterologous partners because of the high energy of activation required to unwrap the coiled-coil interaction. This apparently wasteful overproduction of spectrin may therefore be necessary to ensure that adequate amounts of heterodimers assemble onto the MS. In the developing mammalian system, in CFUe, and subsequent erythroblast stages, two different α subunits and two variable (γ and β) subunits are being coexpressed (53, 84, 108). Since nonhomologous α and variable subunits can interact in vitro (29, 30, 39, 90), the situation must be even more complex; as yet the range of homo- and heterodimers formed has not been analyzed in this system. Spatial proximity of polysomes of different subunits in the cytoplasm may provide one mechanism to facilitate the accumulation of the correct doublet. Differential affinities of different heterologous spectrins for membrane binding sites (i.e. $\alpha_e\beta_e$ binds with higher affinity to ankyrin than $\alpha_{ne}\beta_e$) provides an alternative mechanism (30, 39, 40, 90).

The α_2 homo-oligomers cannot bind to the MS while the β_4 homo-oligomers are competent to bind, albeit with lower affinity than the heterodimer (143). The extreme sensitivity of β-spectrin to degradation probably ensures that only the heterodimers accumulate at the membrane under normal conditions in these cells. The extent of hetero- vs homodimer accumulation must depend on the number, degree of affinity, and degree of saturation of various membrane receptors for the two forms. For example, skeletal myotubes simultaneously accumulate erythroid $\alpha\beta$ heterodimers and erythroid β homo-oligomers. The former accumulates at the Z lines and areas of the sarcolemma where the Z lines attach, and the latter accumulates in neuromuscular junctions in association with acetylcholine receptors (14). This proposed pathway of spectrin assembly (Figure 2) predicts that whereas mutant cells deficient in β-spectrin would fail to accumulate either α- or β-spectrin, mutant cells deficient in α-spectrin biosynthesis would still accumulate some β-spectrin (β homo-oligomers) in the MS. This is what is observed in the mouse hemolytic anemia mutant ja/ja, which carries a lesion in the erythroid β-spectrin gene, and the class of sph/sph mutants, which fail to synthesize α-spectrin (5, 15). These

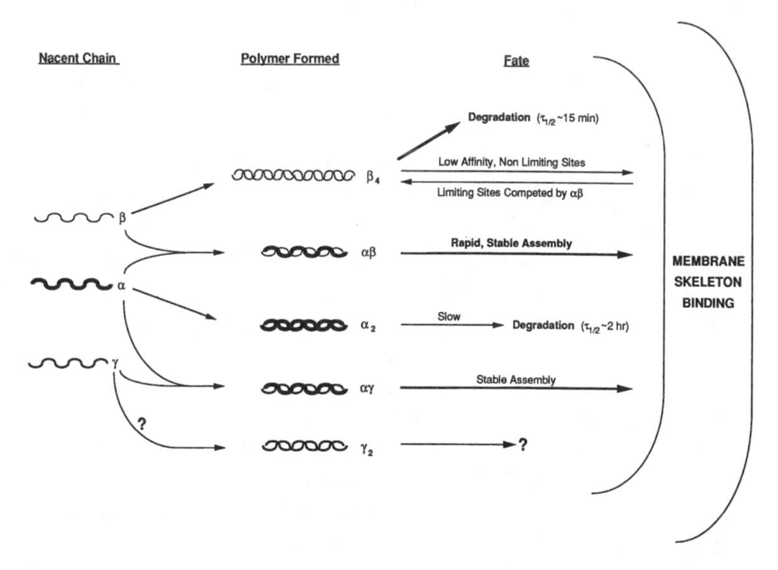

Figure 2 Proposed pathway of spectrin assembly in erythroid development and in lineages coexpressing α-, β- and γ-spectrin subunits.

observations also reveal that binding of spectrin heterodimers or homo-oligomers to a membrane site confers stability to these macromolecules against catabolism. Although neither the mechanisms that confer stability to assembled oligomers (e.g. conformational change or chemical modification) nor the mechanism that earmarks unassembled subunits for catabolism are well understood, these observations point to the importance of spectrin binding sites in determining not only the site, but also the extent and kind of spectrin accumulation. This importance of high and low affinity (primary and secondary) binding sites in determining spectrin accumulation is further revealed by the existence of the nb/nb mutant mice, which carry a lesion in the ankyrin gene. Erythroblasts from these mice fail to synthesize ankyrin and the red cells accumulate only 50% of the normal amount of spectrin (5, 15). This observation underscores the pivotal role of ankyrin in mediating spectrin binding to the membrane, but also points to the presence of additional alternative binding sites. The ability of spectrin to interact directly with phospholipid micelles may provide such an alternative (27, 95, 127). Additionally from in vitro studies, protein 4.1 may serve also as a likely candidate (4, 28, 91, 119). However the fact that red cells from patients with a form of hereditary elliptocytosis caused by a lesion in the gene for protein 4.1 lack both protein 4.1 and glycophorin C, but still possess a full complement of spectrin, suggests that the ankyrin-anion transporter association most likely, under normal conditions, provides the primary class of binding sites for spectrin at the membrane (26, 33, 106).

Expression of the Anion Transporter and Temporal Control of Peripheral Membrane-Skeleton Assembly

The spatial control of spectrin accumulation is primarily posttranslational, by differential stabilization at various membrane binding sites. Therefore the control of expression of such binding sites could prove to be a key factor in controlling the nucleation of peripheral MS component accumulation during development. In principle, two types of temporal control could be envisioned to regulate expression of primary binding sites (presumably the AT). Expression of the membrane binding sites could be initiated before expression of peripheral components. Alternatively, synthesis of all erythroid MS components could be initiated simultaneously, but assembly would instead be strictly ordered to occur only at the plasma membrane. This could occur if high affinity transmembrane receptors (e.g. the AT and glycophorin C) became competent to interact with peripheral components only after they reach their ultimate destination.

To distinguish between these two aspects of temporal control, it is necessary to analyze the initiation of the expression of the various MS components in erythroid progenitor cells prior to terminal differentiation, since hemoglobinized erythroblasts of both avian and mammalian species already coexpress all (i.e. both peripheral and transmembranous) components of the MS (13, 76–78, 138, 141, 146). To overcome the problem of the extremely low abundance of progenitor erythroid cells of both the primitive and definitive lineages, most studies on this issue have focused on progenitor cells that have been artificially amplified by transformation with RNA tumor viruses. Two such avian retroviruses, AEV and S13 virus, exhibit tropism for early BFUe progenitor cells, which then develop to the CFUe stage before becoming transformed and arrested in development (11, 56, 117). These cells can be induced to differentiate along an apparently normal pathway (by the use of a *ts* mutant of S13 or AEV virus or chemical induction with the wild-type virus) and give rise to biconvex ellipsoidal discs resembling fully mature, circulating erythrocytes (10, 65). Similarly, the anemia inducing strain of Friend virus (FVA) transforms murine proerythroblasts in vivo and these transformed cells can be induced to differentiate in response to erythropoietin into fully enucleated reticulocytes in culture (75–78). An analogue of these cells is the FVA-transformed murine erythroleukemia (MEL) cell, which is arrested at the early orthochromatophilic stage of erythroid differentiation and can be chemically induced to initiate terminal differentiation (47, 92). Studies with all four types of transformed cells have indicated that neither of the two types of temporal control discussed above are operative in governing initiation of nucleation of MS assembly. Undifferentiated AEV-, S13-, FVA-, or MEL-transformed cells express all the major peripheral components of the MS (namely α- and β-spectrin, ankyrin, and protein 4.1) as well as the nonerythroid $\alpha\gamma$ form of spectrin (fodrin). However, characteristically they do not express the mature erythroid form of the AT (44, 53, 76, 77, 92, 115, 141, 146). In these undifferentiated transformed erythroblasts, the peripheral components of the MS are synthesized at levels comparable to those in proliferative circulating cells. However these proteins assemble only transiently and inefficiently into a MS, which is then degraded; hence at steady-state only a rudimentary MS is present in these cells (77, 78, 141, 146).

Upon initiation of terminal differentiation in these transformed cells, the AT begins to be expressed concomitantly with the onset of globin gene expression and the down-regulation of nonerythroid spectrin subunit expression (76, 77, 115, 141). The accumulation of the AT is paralleled by a progressive increase in the steady-state levels of the peripheral components of the MS, similar to the level that is observed in normal maturing circulating erythroblasts during the ensuing stages of development. This increase is due primarily to an increased stability of assembled molecules

(76, 77, 96, 141, 142, 146). This asynchrony in expression between the AT and the major peripheral components does not appear to be a consequence of viral transformation, but indeed reflects the normal progression of events in erythropoiesis since such an asynchrony has been observed also in chicken yolk sac-derived progenitor cells in ovo and grown in culture (G. Cassab, C. Woods, E. Lazarides, unpublished observations). Thus the asynchronous expression of the major peripheral MS components and one of their major transmembrane receptors has been conserved in the erythroid differentiation programs of both birds and mammals. Such a pattern of expression precludes the AT serving as a nucleating site to initiate the assembly of the peripheral network. Rather the hypothesis has been advanced that the AT contributes to the stabilization of preassembled components thus protecting them from catabolism and in this manner augments their accumulation in the maturing cell (141).

The lack of AT expression in early progenitor cells still leaves open the question of how initiation of MS assembly is regulated in these cells. One possibility is that assembly is initiated by binding of peripheral MS components (e.g. ankyrin and protein 4.1) to other transmembrane receptors. Such possibilities for ankyrin include the Na^+K^+ ATPase and an 85-kd membrane glycoprotein expressed in T-lymphoma cell plasma membranes (16, 17, 63, 69, 98, 103), both of which are likely to be expressed in myeloid and early erythroid progenitors. However despite high levels of Na^+K^+ ATPase, no association between the Na^+K^+ ATPase and ankyrin has been detected in AEV cells (C. Woods, E. Lazarides, unpublished observations). Alternatively, protein 4.1 may interact directly with glycophorin C, which has been shown to be expressed at least in mouse early erythroid progenitors (64, 133). If such high affinity protein-protein interactions are indeed the main contributors to initiating peripheral MS assembly, they must be sufficiently different in nature from those occurring in terminally differentiating cells to account for the marked difference in the stability of assembled peripheral MS components between these stages of differentiation. Alternatively, assembly of peripheral components could be initiated by low affinity lipophilic interactions with the plasma membrane in early progenitor erythroid cells that are subsequently replaced by high affinity interactions when the appropriate transmembrane protein receptors are expressed later during terminal differentiation. For example, membrane-bound ankyrin is repeatedly deacylated and reacylated through rapid turnover of bound fatty acid (124, 125). Since this reaction occurs even in mature, biosynthetically inactive avian and mammalian red blood cells, both the acylating and deacylating enzymes must be plasma-membrane bound or associated and in close proximity to MS-bound ankyrin. In transformed chicken erythroid progenitors the half-life of the fatty acid bound to ankyrin turns over with the same apparent half-life as the polypeptide backbone. This is in sharp contrast to later stages of erythropoiesis where turnover of the fatty acid persists with the same apparent kinetics despite a more stable polypeptide backbone (C. Woods & E. Lazarides, unpublished observations). Because the fraction of newly synthesized ankyrin that remains unassembled is nonacylated, the possibility exists that ankyrin fatty acid acylation plays a pivotal role in controlling the rate and extent of ankyrin binding to the plasma membrane as well as the resident time of this polypeptide at the bilayer in these progenitor cells. High affinity binding sites, e.g. the AT, expressed later in development may supercede the role of fatty acid acylation in determining the resident time of ankyrin at the plasma membrane.

Likewise, spectrin has been shown to be lipophilic and capable of interacting directly with purified reconstituted erythroid lipid vesicles, specifically phosphatidylserine and phosphatidylethanolamine (27, 95, 127). Spectrin is also capable of interacting directly with an 180-kd glycoprotein expressed in T-lymphoma cells (16, 17, 63). Either mechanism could contribute to a low affinity binding of spectrin to the membrane in the absence of the high affinity binding sites provided by ankyrin binding to the AT. The binding of protein 4.1 to one of the glycophorins, glycophorin A, appears to be dependent on the presence of phosphatidylinositol 4,5-biphosphate (3, 4). Additionally erythroid protein 4.1 is capable of forming high affinity interactions with phosphatidylserine (28, 119), which indicates that protein 4.1, like spectrin and ankyrin, may belong to the class of "amphitropic proteins" capable of both protein-protein and protein-lipid interactions (22).

Regulation of Anion Transporter Expression and Assembly

In addition to its role as a transmembrane anchor to the MS, the erythroid AT serves a primary role in regulating the cytoplasmic pH of red blood cells. The electrically silent antiporter activity of this protein represents one of three major antiporter systems for regulating pH_i (48, 74, 112, 131). It has become clear that the AT expressed in red cells is the prototype member of a family of anion transporters. A small multigene family has now been described in murine (70–73), avian (34, 36, 67), and human systems (41, 129), with additional diversity being generated by use of alternative start sites and alternative splicing (2, 35, 70, 71). At least one AT gene that bears high homology in the transmembrane domain, but has diverged considerably over the cytoplasmic domains from its erythroid counterpart, is expressed in early erythroid progenitor cells. This gene is down-regulated as erythropoiesis progresses, while the transcript for the major erythroid AT polypeptide is markedly up regulated (2, 41, 72). These different AT isoforms may provide alternative membrane-binding sites in a variety of nonerythroid and early erythroid progenitor cells. This AT diversity must also contribute to the complex molecular network regulating the pH_i of a given cell type (48, 131). In this context it seems unlikely that the v-erbA oncogene of AEV, a mutated form of the thyroid hormone receptor (118, 137), could specifically act to block erythroid AT gene expression as has been proposed (146). Indeed, where such activities have been proposed as oncogene targets, it has been subsequently shown that any modulation in antiporter activity is part of a complex augmentation of pH_i regulation in anticipation of the elevated metabolic requirement of transformed cells (48, 131).

The second issue to be considered is this: given the AT is the last member of the erythroid MS complex to be expressed, by what mechanism is it assembled onto a transiently preassembled MS. The AT is cotranslationally inserted into the rough endoplasmic reticulum in the orientation found at the plasma membrane (18), i.e. with the N-terminal domain of this protein exposed to the cytoplasm. Since this domain carries the binding sites for ankyrin and protein 4.1, it is conceivable that these two proteins assemble onto the AT while the latter is still en route to the plasma membrane. The available evidence, however, suggests that the AT becomes competent to assemble onto the MS only after it reaches the plasma membrane, whereupon it is gradually recruited onto preassembled peripheral cytoskeletal binding sites (36). After the onset of its expression, the levels of synthesis and assembly of the AT are controlled primarily translationally and post-translationally since mRNA levels for this protein are maximal early in development when both its levels of synthesis and assembly are minimal. With erythroid maturation the mRNA levels for this protein decline sharply in parallel with the decline in total cytoplasmic RNA levels while both its levels of synthesis and assembly increase (36). Given the lag time for AT assembly onto the MS after its transport to the plasma membrane, it is conceivable that the protein may undergo conformational changes (e.g. oligomerization) or chemical modifications once at the plasma membrane before it becomes competent to be recruited onto its peripheral binding sites. In mature mammalian and avian red blood cells, only 10–20% of the AT polypeptides are actually bound to ankyrin even though all are competent to bind to ankyrin in vitro (8, 9, 36). Since the extent of AT recruitment onto the MS fluctuates independently of synthesis levels throughout embryonic development from minimal in early development to maximal ($\sim$ 100%) before finally settling to a final 10–30% level of MS-bound AT, additional external or internal developmental factors may influence this aspect of MS assembly. These may include hormonal fluctuations in embryonic development that may, in fact, influence the extent of chemical modification of the AT and peripheral components. This is further evidenced by the fact that AT synthesis declines well before that of ankyrin and protein 4.1 (36, 126). This aspect of MS assembly, although still poorly understood, reveals the developmental complexity of MS assembly and implies another major departure from simple self-assembly.

Ankyrin and Protein 4.1 Expression and the Final Stages of Membrane-Skeleton Assembly

In both avian and mammalian erythropoiesis, protein 4.1 and ankyrin are the last MS components to continue to be synthesized and assembled (36, 59, 78, 84, 126) (see Figure 3). In mammalian red blood cells this is also the case in reticulocytes after enucleation (59). This is at least partly due to the fact that ankyrin and protein 4.1 mRNAs persist late in erythropoiesis when levels for the majority of cytoplasmic RNAs, including those for the AT and the spectrins, have declined precipitously (36). Throughout earlier stages of erythropoiesis the amounts of ankyrin and protein 4.1 assembled are subsaturating with respect to spectrin compared to their ultimate stoichiometry in mature red blood cells. Therefore this late expression and assembly of these two interlinking components may ensure that full mechanical stability of the membrane skeleton and the red blood cell as a whole occur as one of the last steps in erythroid morphogenesis well after the cells have become postmitotic. As discussed earlier, fatty acid acylation of ankyrin may be a key controlling event in the recruitment of newly synthesized ankyrin onto the bilayer. Irrespective of the exact mechanism, ankyrin recruitment appears to be independent of newly synthesized $\alpha\beta$-spectrin. Hemolytic anemia mutations in both humans (e.g. some hereditary elliptocytosis and pyropoikilocytosis cases) and mice (sph/sph) that lead to a red cell MS deficient in spectrin do not affect levels of ankyrin (5, 15, 26, 106). Yet, similarly to spectrin, only a fraction of newly synthesized ankyrin assembles stably to the plasma membrane throughout erythropoiesis while the rest is rapidly degraded (59, 96, 142, 143). This may be because assembly of ankyrin is most likely coupled to a chemical modification (e.g. fatty acid acylation), which may be a limiting event in ankyrin assembly. Furthermore, its assembly may be a prerequisite for ankyrin to bind to spectrin since newly synthesized unassembled ankyrin appears to be unable to bind to newly synthesized unassembled β-spectrin (142, 143).

The regulation of protein 4.1 assembly differs from that of the other major peripheral components since the amounts that assemble onto the MS appear to be determined largely by the abundance of this protein's mRNA and by the efficiency of its translation (104). The majority of the protein 4.1 synthesized assembles rapidly onto the MS, irrespective of the

I. Transient Cytoskeletal Complex Assembly/Turnover

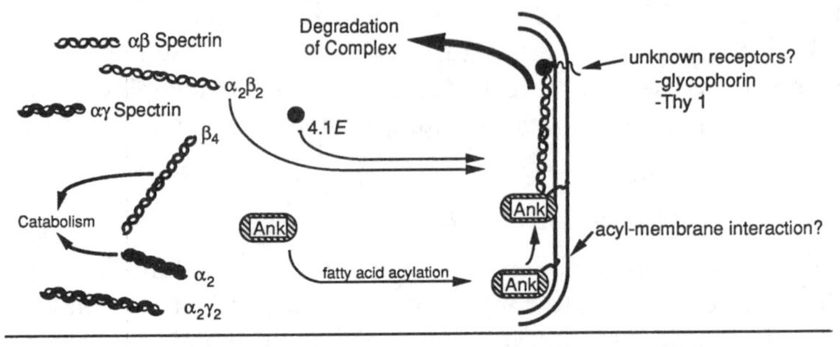

II. Anion Transporter Expression and Stabilization of Membrane-Skeleton

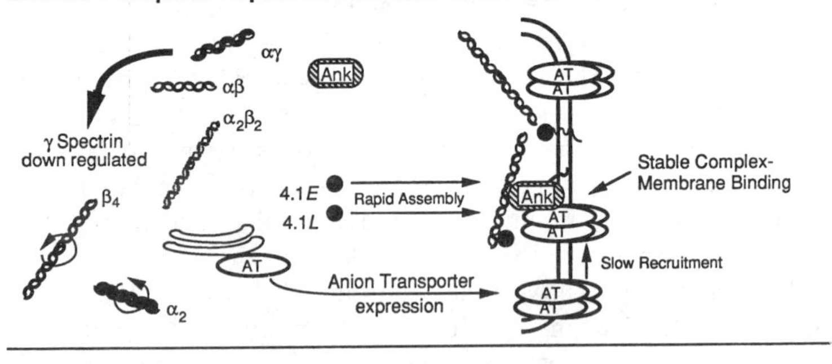

III. Final Stages of Membrane Skeleton Assembly

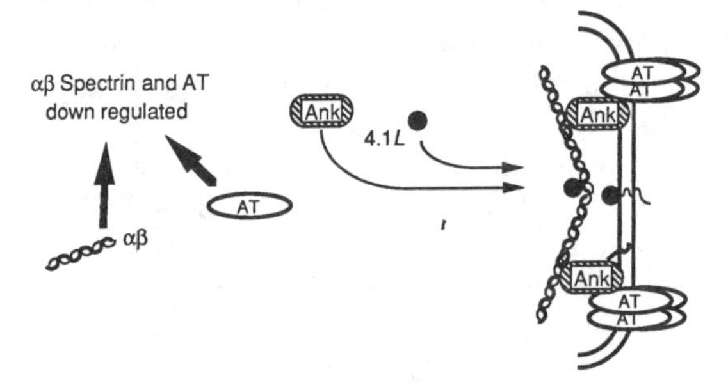

Figure 3 Proposed pathway of synthesis, assembly and turnover of major components of the erythroid membrane-skeleton during erythropoiesis. In part II, the α_2- and β_4-spectrins are metabolically turning over, as indicated by the circular arrows.

stage in development necessitating the existence of a coupling mechanism between synthesis and site of assembly to account for the extremely rapid and efficient kinetics of assembly (126). Red cells from mutant mice and humans with certain hereditary anemias related to deficiencies in ankyrin or spectrin possess normal levels of protein 4.1, which indicates that complete assembly of protein 4.1 can occur independently of ankyrin and spectrin (15, 26, 106). In both birds and mammals, protein 4.1 is encoded by a single gene at least 100 kb in size (32, 104), which gives rise to a multiplicity of isoforms through a complex pattern of differential splicing (spliceoforms) (32, 104, 128). Different variants are expressed in different tissues as well as in different hematopoietic lineages. In particular, avian and mammalian erythroid cells inhibit an intriguing divergence in the pattern of spliceoforms expressed that may be a functional prerequisite of the different structures of red blood cells between these species. In birds, the protein 4.1 gene gives rise to multiple spliceoforms during erythropoiesis whose composition varies with development and erythroid maturation (57, 58, 104, 141, 144). Although for the most part the functional differences among these variants has not been established, one of the lower molecular weight variants also appears to lack the domain responsible for crosslinking spectrin and actin (N. Burns, E. Lazarides, unpublished observations). Mammals, in contrast, have evolved a greater degree of tissue-specific expression of different spliceoforms. Only a single size class mRNA is expressed when mammalian erythroid cells undergo termination differentiation; this mRNA directs the synthesis of a major polypeptide with spectrin-actin crosslinking activity (128). Expression of the variant lacking the spectrin-actin binding domain is restricted to the lymphoid lineage and myeloid and erythroid progenitors. As erythroid cells undergo terminal differentiation, expression of this variant is down-regulated as the variant carrying the spectrin-actin crosslinking domain begins to be expressed (128). In birds the variant that lacks the spectrin-actin binding domain constitutes a substantial fraction of the total protein 4.1 variant population expressed in early erythroid progenitors. The high percentage of this protein 4.1 variant may account for the fact that assembled components of the MS in early erythroid progenitors not only are metabolically unstable but are also inefficiently crosslinked. As cells undergo terminal differentiation, the steady-state levels of this variant are diluted by the expression and assembly of variants carrying the spectrin-actin crosslinking domains (N. Burns, E. Lazarides, unpublished observations).

Actin Expression and Assembly and the Process of Enucleation

In the majority of vertebrate cells, actin filaments are supramolecular polymers composed of an actin polymer backbone and a large family of actin binding proteins, which modulate the various polymeric states of actin (110). In contrast, in adult mammalian red blood cells, actin is exclusively associated with the periphery of the plasma membrane in the form of short oligomers that crosslink spectrin tetramers in the presence of protein 4.1. Since early myeloid and erythroid progenitors are actively mitotic and of amoeboid morphology, actin filaments and their associated cytoplasmic proteins must undergo several dynamic changes before their final steady-state accumulation as short oligomers exclusively under the plasma membrane. These changes must be even more involved in nucleated red blood cells (e.g. chicken) where actin filaments accumulate not only at the plasma membrane but also at the marginal band of microtubules (68). Most of these events are poorly understood at present. In marked contrast to the other MS constituents, which gradually accumulate after the CFUe stage of development, actin steady-state levels decline drastically between the CFUe and polychromatophilic stages (61). Actin recruitment onto the MS appears to involve at least some degree of reorganization from preexisting cytoplasmic pools rather than an exclusive assembly of newly synthesized monomers; actin is almost entirely cytoplasmic in early progenitor cells (in which the MS is unstable) and gradually becomes membrane-skeletal-bound as a function of terminal differentiation and MS stabilization (61). How this process and the overall polymeric state of actin is controlled during successive cell divisions and the cyclic formation and dissociation of the contractile ring is also unclear. A number of actin binding proteins, most notably gelsolin (61), are down-regulated as a function of terminal differentiation, while others, which affect the polymeric state of actin and its ability to bind to spectrin, such as protein 4.9 and adducin (46, 49, 94, 121), are up-regulated during this time. How these various expression events cooperate positively or negatively to control the final polymeric state of actin remains to be determined. The process of enucleation in mammalian red blood cells appears to be analogous to cytokinesis in that a contractile ring of actin filaments forms at the base of the eccentrically located nucleus and appears to be responsible for the membrane constriction that develops at the base of the nucleus during this process (78). Concomitantly, components of the membrane-skeleton redistribute and become localized primarily at the emerging reticulocyte plasma membrane away from the part of the plasma membrane that ensheaths the nucleus (50). The details of this process are unclear, but the recent development of in vitro culture systems of mouse erythroblasts that can undergo differentiation including enucleation in response to erythropoiesis (75–78) should greatly aid in the study of this process.

CONCLUDING REMARKS

In many respects elucidating the mechanisms that govern the assembly of the erythroid MS and the morphogenesis of the red blood cell is at a rudimentary stage. This is especially true if one considers that for the most part conclusions are drawn without special attention to the lineage origin of the cells, both avian and mammalian species, under investigation. In fact primitive and definitive lineage cells vary substantially with respect to embryonic origin and cellular morphology, even though erythroid cells in both lineages become hemoglobinized. For example, primitive erythroid cells in mammals have a cytoplasmic structure more akin to differentiating avian (e.g. chicken) definitive erythroblasts than their mammalian definitive counterparts. Primitive mammalian erythroblasts contain cytoplasmic actin filaments and microtubules in a quasi-marginal band configuration, while their definitive counterparts lose microtubules, and actin filaments accumulate exclusively at the plasma membrane as short oligomers (23). Additionally, most studies have been carried out with transformed progenitor cells grown in tissue culture or with cells that are already differentiating in suspension. In fact all erythroid progenitors, irrespective of lineage or erythropoietic site, differentiate, at least initially, attached to various stromal cells (1, 21, 37, 116). How this mode of early differentiation affects expression and assembly of the various MS components in ways that differ from that which has been observed in unattached cells, remains to be determined.

Despite these difficulties, certain important principles that govern this aspect of membrane biogenesis in differentiation have emerged. It is evident that the erythroid cell utilizes substantially more information than simple self-assembly in directing the assembly of its MS during differentiation. Not only does it remodel its preexisting structure present in its progenitors, but it uses both temporal and spatial information in directing assembly. Gradual stabilization of unstably assembled components appears to be one of the overriding mechanisms governing assembly. Inefficient though it may be, this mechanism of assembly may have evolved to maximize assembly of a multicomponent system that has to occur over a period of several days. These principles of temporal and spatial control of assembly through controlled stabilization will probably emerge as a governing mechanism in other lineages such as myogenesis and neurogenesis where cells have to convert a rather dynamic structure in progenitor cells to a topologically organized structure in postmitotic cells. Understanding how stability of crosslinked and unassembled components is determined as a function of differentiation will provide novel insights into the mechanism of structural morphogenesis and its evolution.

ACKNOWLEDGMENTS

We are grateful to Dr. Thomas R. Coleman for his valuable comments on the manuscript and his help in the preparation of the figures. Work described from the authors' laboratory was supported by grants from the National Science Foundation and the American Cancer Society.

Literature Cited

1. Allen, T. D., Dexter, T. M. 1982. Ultrastructural aspects of erythropoietic differentiation in long-term bone marrow culture. *Differentiation* 21: 86–94
2. Alper, S. L., Kopito, R. R., Libresco, S. M., Lodish, H. F. 1988. Cloning and characterization of a murine band 3-related cDNA from kidney and from a lymphoid cell line. *J. Biol. Chem.* 263: 17092–99
3. Anderson, R. A., Lovrien, R. E. 1984. Glycophorin is linked by band 4.1 protein to the human erythrocyte membrane skeleton. *Nature* 307: 655–58
4. Anderson, R. A., Marchesi, V. T. 1985. Regulation of the association of membrane skeletal protein 4.1 with glycophorin by a phosphoinositide. *Nature* 318: 295–98
5. Barker, J. E., Bodine, D. M., Birkenmeier, C. S. 1986. Synthesis of spectrin and its assembly into the red blood cell cytoskeleton of normal and mutant mice. In *Membrane Skeletons and Cytoskeletal-membrane Associations*, UCLA Symp. New Series. eds V. Bennett, C. M. Cohen, S. E. Lux, J. Palek. 38: 313–24. New York: Liss
6. Barrett, L. A., Dawson, R. B. 1974. Avian erythrocyte development: microtubules and the formation of the disc shape. *Dev. Biol.* 36: 72–81
7. Bennett, V. 1985. The membrane skeleton of human erythrocytes and its implications for more complex cells. *Annu. Rev. Biochem.* 54: 273–304
8. Bennett, V., Stenbuck, P. J. 1979. The membrane attachment protein for spectrin is associated with band 3 in human erythrocyte membranes. *Nature* 280: 468–73
9. Bennett, V., Stenbuck, P. J. 1980. Association between ankyrin and the cytoplasmic domain of band 3 isolated from the human erythrocyte membrane. *J. Biol. Chem.* 255: 6424–32
10. Beug, H., Doederlein, G., Freudenstein, C., Graf, T. 1982. Erythroblast cell lines transformed by a temperature-sensitive mutant of avian erythroblastosis virus: a model system to study erythroid differentiation in vitro. *J. Cell Physiol. Suppl.* 1: 195–207
11. Beug, H., Hayman, M. J., Graf, T., Benedict, S. H., Wallbank, A. M., Vogt, P. K. 1985. S13—a rapidly oncogenic replication-defective avian retrovirus. *Virology* 145: 141–53
12. Birkenmeier, C. S., Bodine, D. M., Repasky, E. A., Helfman, D. M., Hughes, S. H., Barker, J. E. 1985. Remarkable homology among the internal repeats of erythroid and nonerythroid spectrin. *Proc. Natl. Acad. Sci. USA* 82: 5671–75
13. Blikstad, I., Nelson, W. J., Moon, R. T., Lazarides, E. 1983. Synthesis and assembly of spectrin during avian erythropoiesis: stoichiometric assembly but unequal synthesis of α and β spectrin. *Cell* 32: 1081–87
14. Block, R., Morrow, J. S. 1989. An unusual β spectrin associated with clustered acetylcholine receptors. *J. Cell Biol.* 108: 481–94
15. Bodine, D. M. IV, Birkenmeier, C. S., Barker, J. E. 1984. Spectrin deficient inherited hemolytic anemias in the mouse: characterization by spectrin synthesis and mRNA activity in reticulocytes. *Cell* 37: 721–29
16. Bourguignon, L. Y. W., Suchard, S. J., Nagpal, M. L., Glenney, J. R. Jr. 1985. A T-lymphoma transmembrane glycoprotein (gp180) is linked to the cytoskeletal protein fodrin. *J. Cell Biol.* 101: 477–87
17. Bourguignon, L. Y. W., Walker, G., Suchard, S. J., Balazovich, K. 1988. A lymphoma plasma membrane-associated protein with ankyrin-like properties. *J. Cell Biol.* 102: 2115–24
18. Braell, W. A., Lodish, H. F. 1981. Biosynthesis of the erythrocyte anion transport protein. *J. Biol. Chem.* 256: 11337
19. Branton, D., Cohen, C. M., Tyler, J. 1981. Interaction of cytoskeletal proteins on the human erythrocyte membrane. *Cell* 24: 24–32
20. Bronson, D. D., Schachat, F. H. 1985. Renaturation of skeletal muscle tropomyosin: implications for in vivo assembly. *Proc. Natl. Acad. Sci. USA* 82: 2359–63
21. Bruns, G. A. P., Ingram, V. M. 1973. The erythroid cells and hemoglobins of the chick embryo. *Philos. Trans. R. Soc. London Ser: B* 266: 225–69
22. Burns, P. 1988. Amphitrophic proteins:

a new class of membrane proteins. *Trends Biochem. Sci.* 13: 79–83
23. Byers, T. J., Branton, D. 1985. Visualization of the protein associations in the erythrocyte membrane skeleton. *Proc. Natl. Acad. Sci. USA* 821: 6153–57
24. Calvert, R., Bennett, P., Gratzer, W. 1980. Properties and structural role of the subunits of spectrin. *Eur. J. Biochem.* 107: 355–61
25. Cianci, C. D., Giorgi, M., Morrow, J. S. 1988. Phosphorylation of ankyrin down regulates its cooperative interaction with spectrin and protein 3. *J. Cell. Biochem.* 37: 301–15
26. Coetzer, T. L., Lawler, J., Liu, S.-C., Prchal, J. T., Gualtieri, R. J., et al. 1988. Partial ankyrin and spectrin deficiency in severe, atypical hereditary spherocytosis. *N. Engl. J. Med.* 318: 230–34
27. Cohen, A. M., Liu, S. C., Derick, L. H., Palek, J. 1986. Ultrastructural studies of the interaction of spectrin with phosphatidyl serine liposomes. *Blood* 68: 920–26
28. Cohen, A. M., Liu, S. C., Lawler, J., Derick, L., Palek, J. 1988. Identification of the protein 4.1 binding site to phosphatidyl serine vesicles. *Biochemistry* 27: 614–19
29. Coleman, T. R., Harris, A. S., Mische, S. M., Mooseker, M. S., Morrow, J. S. 1987. β-Spectrin bestows protein 4.1 sensitivity on spectrin-actin interactions. *J. Cell Biol.* 104: 519–26
30. Coleman, T. R., Fishkind, D. J., Mooseker, M. S., Morrow, J. S. 1989. Functional diversity among spectrin isoforms. In *Cell Motility and the Cytoskeleton*, 12: 225–47. New York: Liss
31. Colin, Y., Kim, C. L. V., Tsapis, A., Clerget, M., d'Auriol, L., London, J., Galibert, F., Cartron, J.-P. 1989. Human erythrocyte glycophorin C. Gene structure and rearrangement in genetic variants. *J. Biol. Chem.* 264: 3773–80
32. Conboy, J., Kan, Y. W., Shohet, S. B., Mohandas, N. 1986. Molecular cloning of protein 4.1, a major structural element of the human erythrocyte membrane skeleton. *Proc. Natl. Acad. Sci. USA* 83: 9512–16
33. Conboy, J., Mohandas, N., Tchernia, G., Kan, Y. W. 1986. Molecular basis of hereditary elliptocytosis due to protein 4.1 deficiency. *N. Engl. J. Med.* 315: 680–85
34. Cox, J. V., Moon, R. T., Lazarides, E. 1985. Anion transporter: highly cell type-specific expression of distinct polypeptides and transcripts in erythroid and nonerythroid cells. *J. Cell Biol.* 100: 1548–57
35. Cox, J. V., Lazarides, E. 1988. Alternative primary structures in the transmembrane domain of the chicken erythroid anion transporter. *J. Mol. Biochem.* 8: 1327–35
36. Cox, J. V., Stack, J. H., Lazarides, E. 1987. Erythroid anion transporter assembly is mediated by a developmentally-regulated recruitment onto a preassembled membrane cytoskeleton. *J. Cell Biol.* 105: 1405–16
37. Crocker, P. R., Morris, L., Gordon, S. 1988. Novel cell surface adhesion receptors involved in interactions between stromal macrophages and haematopoietic cells. *J. Cell Sci. Suppl.* 9: 185–206
38. Curtis, P. J., Palumbo, A., Ming, J., Fraser, P., Cioe, L., et al. 1985. Sequence comparison of human and murine α-spectrin cDNA. *Gene* 36: 357–62
39. Davis, J. Q., Bennett, V. 1984. Brain spectrin: isolation of subunits and formation of hybrid with erythrocyte spectrin subunits. *J. Biol. Chem.* 258: 7757–66
40. Davis, J. Q., Bennett, V. 1984. Brain ankyrin. A membrane-associated protein with binding sites for spectrin, tublin and the cytoplasmic domain of the erythrocyte anion channel. *J. Biol. Chem.* 259: 13550–59
41. Demuth, D. R., Showe, L. C., Ballantine, M., Palumbo, A., Fraser, P. J., et al. 1986. Cloning and structural characterization of a human nonerythroid band 3-like protein. *EMBO J.* 5: 1205–14
42. Drenckhahn, D., Schluter, K., Allen, D. P., Bennett, V. 1985. Colocalization of band 3 with ankyrin and spectrin at the basal membrane of intercalated cells in the rat kidney. *Science* 230: 1287–89
43. Eder, P. S., Soong, C.-J., Tao, M. 1986. Phosphorylation reduces the affinity of protein 4.1 for spectrin. *Biochem.* 25: 1764–70
44. Eisen, H., Bach, R., Emery, R. 1977. Induction of spectrin in erythroleukemic cells transformed by Friend virus. *Proc. Natl. Acad. Sci. USA* 74: 3898–902
45. Epstein, H. F., Schechter, A. N., Chen, R. F., Anfinsen, C. B. 1971. Folding of *Staphylococcal* nuclease: kinetic studies of two processes in acid renaturation. *J. Mol. Biol.* 60: 499–508
46. Fowler, V. M., Bennett, V. 1984. Eryth-

rocyte membrane tropomyosin. *J. Biol. Chem.* 259: 5978–89
47. Friend, C., Patuleia, C., de Harven, E. 1966. Erythrocyte maturation in vitro of murine (Friend) virus-induced leukemic cells. *Natl. Canc. Inst. Monogr.* 22: 505–22
48. Ganz, M. B., Boyarsky, A., Sterzel, R. B., Boron, W. F. 1989. Arginine vasopressin enhances pH_i regulation in the presence of HCO_3^- by stimulating three acid-base transport systems. *Nature* 337: 648–51
49. Gardner, K., Bennett, V. 1987. Modulation of spectrin-actin assembly by erythrocyte adducin. *Nature* 328: 359–62
50. Geiduschek, J. B., Singer, S. J. 1979. Molecular changes in the membranes of mouse erythroid cells accompanying differentiation. *Cell* 16: 149–63
51. Glenney, J. R. Jr., Glenney, P., Weber, K. 1982. Erythroid spectrin, brain fodrin and intestinal brush border proteins 260/240 are related molecules containing a common calmodulin-binding subunit bound to a variant cell type-specific subunit. *Proc. Natl. Acad. Sci. USA* 79: 4002–5
52. Glenney, J. R. Jr., Glenney, P. 1983. Fodrin is a general spectrin-like protein found in most cells whereas spectrin and the TW protein have restricted distribution. *Cell* 34: 503–12
53. Glenney, J. R. Jr., Glenney, P. 1984. Co-expression of spectrin and fodrin in Friend erythroleukemic cells treated with DMSO. *Exp. Cell Res.* 152: 15–21
54. Goodman, S. R., Zagon, I. S., Kulikowski, R. R. 1981. Identification of a spectrin-like protein in nonerythroid cells. *Proc. Natl. Acad. Sci. USA* 78: 7550–74
55. Goodman, S. R., Zagon, I. S. 1986. The neural cell spectrin skeleton: a review. *Am. J. Physiol.* 250: c347–60
56. Graf, T., Ade, N., Beug, H. 1978. Temperature-sensitive mutant of avian erythroblastosis virus suggests a block of differentiation as mechanism of leukaemogenesis. *Nature* 257: 496–501
57. Granger, B. L., Lazarides, E. 1984. Membrane skeletal protein 4.1 of avian erythrocytes is composed of multiple variants and exhibit tissue-specific expression. *Cell* 37: 595–607
58. Granger, B. L., Lazarides, E. 1985. Appearance of new variants of membrane skeletal protein 4.1 during terminal differentiation of avian erythroid and lenticular cells. *Nature* 313: 238–41
59. Hanspal, M., Palek, J. 1987. Synthesis and assembly of membrane skeletal proteins in mammalian red cell precursors. *J. Cell Biol.* 105: 1417–24
60. Hargreaves, W. R., Giedd, K. N., Verkleij, A., Branton, D. 1980. Reassociation of ankyrin with band 3 in erythrocyte membranes and in lipid vesicles. *J. Biol. Chem.* 255: 11965–72
61. Hinssen, H., Vandekerckhove, J., Lazarides, E. 1987. Gelsolin is expressed in early erythroid progenitor cells and negatively regulated during erythropoiesis. *J. Cell Biol.* 105: 1425–34
62. Howe, C. L., Sacramone, L. M., Mooseker, M. S., Morrow, J. S. 1985. Mechanisms of cytoskeletal regulation: modulation of membrane affinity in avian brush border and erythrocyte spectrins. *J. Cell Biol.* 101: 1379–85
63. Kalomiris, E. L., Bourguignon, L. W. W. 1988. Mouse T lymphoma cells contain a transmembrane glycoprotein (GP85) that binds ankyrin. *J. Cell Biol.* 106: 319–27
64. Katsuri, K., Harrison, P. 1985. The cell specificity and biosynthesis of mouse glycophorins studied with monoclonal antibodies. *Exp. Cell Res.* 157: 253–64
65. Keane, R. W., Lindblad, P. C., Pierik, L. T., Ingram, V. M. 1982. Isolation and transformation of primary mesenchymal cells of the chick embryo. *Cell* 17: 801–11
66. Kellokumpu, S., Neff, L., Jämsä-Kellokumpu, S., Kopito, R., Baron, R. 1988. A 115-kD polypeptide immunologically related to erythrocyte band 3 is present in Golgi membranes. *Science* 242: 1308–11
67. Kim, H. R. C., Yew, N. S., Ansorge, W., Voss, H., Schwanger, C., et al. 1988. Two different mRNAs are transcribed from a single genomic locus encoding the chicken erythrocyte anion transporter proteins (band 3). *Mol. Cell. Biol.* 8: 4416–24
68. Kim, S., Magendantz, M., Solomon, F. 1987. Development of a differentiated microtubule structure: formation of the chicken erythrocyte marginal band in vivo. *J. Cell. Biol.* 104: 51–59
69. Kobb, R., Zimmerman, M., Schoner, W., Drenckhahn, D. 1987. Localization and coprecipitation of ankyrin and Na^+,K^+-ATPase in kidney epithelial cells. *Eur. J. Cell Biol.* 45: 230–37
70. Kopito, R. R., Andersson, M. A., Lodish, H. F. 1987. Structure and organization of the murine band 3 gene. *J. Biol. Chem.* 262: 8035–40

71. Kopito, R. R., Andersson, M. A., Lodish, H. F. 1987. Multiple tissue-specific sites of transcriptional initiation of the mouse anion antiport gene in erythroid and renal cells. *Proc. Natl. Acad. Sci. USA* 84: 7149–53
72. Kopito, R. R., Lodish, H. F. 1985. Structure of the murine anion exchange protein. *J. Cell Biochem.* 29: 1–17
73. Kopito, R. R., Lodish, H. F. 1985. Primary structure and transmembrane orientation of the murine anion exchange protein. *Nature* 316: 234–38
74. Knauf, P. A. 1979. Erythrocyte anion exchange and band 3 protein. Transport kinetics and molecular structure. *Curr. Top. Membr. Transp.* 12: 249–63
75. Koury, M. J., Bondurant, M. C., Duncan, D. T., Krantz, S. B., Hankins, W. D. 1982. Specific differentiation events induced by erythropoietin in cells infected in vitro with the anemia strain of Friend virus. *Proc. Natl. Acad. Sci. USA* 79: 635–39
76. Koury, M. J., Bondurant, M. C., Mueller, T. J. 1986. The role of erythropoietin in the production of principal erythrocyte proteins other than hemoglobin during terminal erythroid differentiation. *J. Cell. Physiol.* 126: 259–65
77. Koury, M. J., Bondurant, M. C., Rana, S. S. 1987. Changes in erythroid membrane proteins during erythropoietin-mediated terminal differentiation. *J. Cell. Physiol.* 133: 438–48
78. Koury, M. J., Koury, S. T., Bondurant, M. C. 1989. Erythroid membrane and cytoskeletal development and an in vitro model of terminal mammalian erythroid differentiation. *J. Cell. Biochem. Suppl.* 13B: 207
79. Koury, S. T., Repasky, E. A., Eckert, B. S. 1987. The cytoskeleton of isolated murine primitive erythrocytes. *Cell Tissue Res.* 249: 69–77
80. Kovach, J. S., Marks, P. A., Russell, E. S., Epler, H. 1967. Erythroid cell development in fetal mice: ultrastructural characteristics and hemoglobin synthesis. *J. Mol. Biol.* 25: 131–62
81. Laury-Kleintop, L. D., Showe, L. C. 1989. Nonerythroid band 3 gene expression in uninduced and induced K562 and KMOE cell lines. *J. Cell. Biochem. Suppl.* 13B: 223
82. Lazarides, E. 1988. From genes to structural morphogenesis: the genesis and epigenesis of the red blood cell. *Cell* 51: 345–56
83. Lazarides, E., Nelson, W. J. 1985. Expression and assembly of the erythroid membrane-skeletal proteins ankyrin (goblin) and spectrin in the morphogenesis of chicken neurons. *J. Cell. Biochem.* 27: 423–41
84. Lehnert, M. E., Lodish, H. F. 1988. Unequal synthesis and differential degradation of α and β spectrin during murine erythroid differentiation. *J. Cell Biol.* 107: 413–26
85. Leto, T. L., Fortugno-Erikson, D., Barton, D., Yang-Feng, T. L., Francke, U., et al. 1988. Comparison of nonerythroid α-spectrin genes reveals strict homology among diverse species. *Mol. Cell. Biol.* 8: 1–9
86. Levine, J., Willard, M. 1981. Fodrin: axonally transported polypeptides associated with the internal periphery of many cells. *J. Cell Biol.* 90: 631–43
87. Ling, E., Danilev, Y. N., Cohen, C. M. 1988. Modulation of red cell band 4.1 function by cAMP-dependent kinase and protein kinase C. phosphorylation. *J. Biol. Chem.* 263: 2209–16
88. Liu, S. C., Derek, L. H., Palek, J. 1987. Visualization of the hexagonal lattice in the erythrocyte membrane skeleton. *J. Cell Biol.* 104: 527–36
89. Lu, P.-W., Soong, C.-J., Tao, M. 1985. Phosphorylation of ankyrin decreases its affinity for spectrin tetramer. *J. Biol. Chem.* 260: 14958–64
90. Mangeat, P.-H. 1989. Interaction of biological membranes with the cytoskeletal framework of living cells. *Biol. Cell.* 64: 261–81
91. Marchesi, V. T. 1985. Stabilizing infrastructures of membranes. *Annu. Rev. Cell Biol.* 1: 531–61
92. Marks, P. A., Rifkind, R. A. 1978. Erythroleukemic differentiation. *Annu. Rev. Biochem.* 47: 419–48
93. McMahon, A. P., Giebelhaus, D. H., Champion, J. E., Bailes, J. A., Lacey, S., et al. 1987. cDNA cloning, sequencing and chromosome mapping of a nonerythroid spectrin human α-fodrin. *Differentiation* 34: 68–78
94. Mische, S. M., Mooseker, M. S., Morrow, J. S. 1987. Erythrocyte adducin: a calmodulin-regulated actin binding protein that stimulates spectrin-actin binding. *J. Cell. Biol.* 105: 2837–45
95. Mombers, C., De Gier, J., Demel, R. A., Van Deenen, I. I. M. 1980. Spectrin-phospholipid interaction: a monolayer study. *Biochim. Biophys. Acta* 603: 52–62

96. Moon, R. T., Lazarides, E. 1984. Biogenesis of the avian erythroid membrane skeleton: receptor-mediated assembly and stabilization of ankyrin (goblin) and spectrin. *J. Cell Biol.* 98: 1899–904
97. Moon, R. T., Ngai, J., Wold, B. J., Lazarides, E. 1985. Tissue-specific expression of distinct spectrin and ankyrin transcripts in erythroid and nonerythroid cells. *J. Cell Biol.* 100: 152–60
98. Morrow, J. S., Cianci, C. D., Ardito, T., Mann, A. S., Kashgarian, M. 1989. Ankyrin links fodrin to the α subunit of $Na^{+}K^{+}$-ATPase in Madin-Darby canine kidney cells and in intact renal tubule cells. *J. Cell Biol.* 108: 455–65
99. Nelson, W. J., Lazarides, E. 1983. Switching of subunit composition of muscle spectrin during myogenesis in vitro. *Nature* 304: 364–68
100. Nelson, W. J., Lazarides, E. 1984. Goblin (ankyrin) in striated muscle: identification of the potential membrane receptor for erythroid spectrin in muscle cells. *Proc. Natl. Acad. Sci. USA* 81: 3292–96
101. Nelson, W. J., Lazarides, E. 1985. Posttranslational control of membrane skeleton (ankyrin and αβ spectrin) assembly in early myogenesis. *J. Cell Biol.* 100: 1726–35
102. Nelson, W. J., Veshnock, P. J. 1986. Dynamics of membrane skeleton (fodrin) organization during development of polarity in Madin-Darby canine kidney epithelial cells. *J. Cell. Biol.* 103: 1751–65
103. Nelson, W. J., Veshnock, P. J. 1987. Ankyrin binding to $Na^{+}K^{+}$-ATPase and implications for the organization of membrane domains in polarized cells. *Nature* 328: 533–36
104. Ngai, J., Stack, J. H., Moon, R. T., Lazarides, E. 1987. Regulated expression of multiple chicken erythroid membrane skeletal protein 4.1 variants is governed by differential RNA processing and translational control. *Proc. Natl. Acad. Sci. USA* 84: 4432–36
105. Ohanian, V., Wolfe, L. C., John, K. M., Pinder, J. C., Lux, S. E., Gratzer, W. B. 1984. Analysis of the terniary interaction of the red cell membrane skeletal proteins, spectrin, actin and 4.1. *Biochemistry* 23: 4416–20
106. Palek, J., Lux, S. E. 1983. Red cell membrane skeletal defects in hereditary and acquired hemolytic anemias. *Semin. Hematol.* 20: 189–224
107. Pasternack, G. R., Anderson, R. A., Leto, T. L., Marchesi, V. T. 1985. Interactions between protein 4.1 and band 3. *J. Biol. Chem.* 260: 3676–83
108. Pfeffer, S. R., Huima, T., Redman, C. M. 1986. Biosynthesis of spectrin and its assembly onto the cytoskeleton of Friend erythroleukemia cells. *J. Cell Biol.* 103: 103–14
109. Podgórski, A., Elbaum, D. 1985. Properties of red cell membrane proteins: mechanism of spectrin and band 4.1 interaction. *Biochemistry* 24: 7871–76
110. Pollard, T. D., Cooper, J. A. 1986. Actin and actin-binding proteins: a critical evaluation of mechanisms and functions. *Annu. Rev. Biochem.* 55: 987–1035
111. Radke, K., Beug, H., Kornfeld, S., Graf, T. 1982. Transformation of both erythroid and myeloid cells by E26, an avian leukemia virus that contains the myb gene. *Cell* 31: 643–53
112. Reinertsen, K. V., Tønnessen, T. I., Jacobsen, J., Sandvig, K., Olsnes, S. 1988. Role of chloride/bicarbonate antiport in the control of cytosolic pH: cell-line differences in activity and regulation of antiport. *J. Biol. Chem.* 263: 11117–25
113. Repasky, E., Granger, B. L., Lazarides, E. 1982. Widespread occurrence of avian spectrin in nonerythroid cells. *Cell* 29: 821–33
114. Rifkind, R. A., Chui, D., Epler, H. 1969. An ultrastructural study of early morphogenetic events during the establishment of fetal hepatic erythropoiesis. *J. Cell Biol.* 40: 343–65
115. Sabban, E. L., Sabatini, D. D., Marchesi, V. T. 1980. Biosynthesis of erythrocyte membrane protein band 3 in DMSO-induced Friend erythro-leukemia cells. *J. Cell. Physiol.* 104: 261–68
116. Sabin, F. R. 1920. Studies on the origin of blood-vessels and of red blood corpuscles as seen in the living blastoderm of chicks during the second day of incubation. *Contrib. Embryol.* 9: 215–62
117. Samarut, J., Gazzolo, L. 1982. Target cells infected by erythroblastosis virus differentiate and become transformed. *Cell* 28: 921–29
118. Sap, J., Munoz, A., Damm, K., Goldberg, Y., Ghysdael, J., et al. 1986. The c-erbA protein is a high-affinity receptor for thyroid hormone. *Nature* 224: 635–40
119. Sato, S. B., Ohnishi, S. 1983. Inter-

action of a peripheral protein of the erythrocyte membrane, band 4.1 with phosphatidylserine-containing liposomes and erythrocyte inside-out vesicles. *Eur. J. Biochem.* 130: 19–25
120. Shen, B. W., Josephs, R., Steck, T. L. 1986. Ultrastructure of unit fragments of the skeleton of the human erythrocyte membrane. *J. Cell Biol.* 102: 997–1006
121. Siegel, D. L., Branton, D. 1985. Partial purification and characterization of an actin-bundling protein, 4.9, from human erythrocytes. *J. Cell Biol.* 100: 775–85
122. Sobue, K., Muramoto, Y., Fujita, M., Kakiuchi, S. 1981. Calmodulin-binding protein of erythrocyte cytoskeleton. *Biochem. Biophys. Res. Commun.* 100: 1063–70
123. Speicher, D. W., Marchesi, V. T. 1984. Erythrocyte spectrin is comprised of many homologous triple helical segments. *Nature* 311: 177–80
124. Staufenbiel, M. 1987. Ankyrin-bound fatty acid turns over rapidly at the erythrocyte plasma membrane. *Mol. Cell. Biol.* 7: 2981–84
125. Staufenbiel, M., Lazarides, E. 1986. Ankyrin is fatty acid acylated in erythrocytes. *Proc. Natl. Acad. Sci. USA* 83: 318–22
126. Staufenbiel, M., Lazarides, E. 1986. Assembly of protein 4.1 during chicken erythroid differentiation. *J. Cell Biol.* 102: 157–63
127. Sweet, C., Zull, J. E. 1970. Interaction of the erythrocyte membrane protein, spectrin with model membrane systems. *Biochem. Biophys. Res. Commun.* 41: 135–41
128. Tang, T. K., Leto, T. L., Correas, I., Alonso, M. A., Marchesi, V. T., Benz, E. J. Jr. 1988. Selective expression of an erythroid-specific isoform of protein 4.1. *Proc. Natl. Acad. Sci. USA* 85: 3713–17
129. Tanner, M. J. A., Martin, P. G., High, S. 1988. The complete amino acid sequence of the human erythrocyte membrane anion-transport protein deduced from the cDNA sequence. *Biochem. J.* 256: 703–12
130. Teipel, J. W., Koshland, D. G. Jr. 1971. Kinetic aspects of conformational changes in proteins. II. Structural changes in renaturation of denatured proteins. *Biochemistry* 10: 798–805
131. Thomas, R. C. 1989. Bicarbonate and pH_i response. *Nature* 337: 601
132. Tyler, J. M., Reinhardt, B. N., Branton, D. 1980. Associations of erythrocyte membrane proteins. Binding of purified bands 2.1 and 4.1 to spectrin. *J. Biol. Chem.* 255: 7034–39
133. Ulmer, J. B., Dolci, E. D., Palade, G. E. 1989. Biosynthetic anomalies in the glycophorins of murine erythroleukemia cells. *J. Cell. Biochem. Suppl.* 1313: 220
134. Ungewickell, E., Bennett, P. M., Calvert, R., Ohanian, V., Gratzer, W. 1979. In vitro formation of a complex between cytoskeletal proteins of the human erythrocyte. *Nature* 280: 811–14
135. Ungewickell, G., Gratzer, W. 1978. Self-association of human spectrin: a thermodynamic and kinetic study. *Eur. J. Biochem.* 88: 379–85
136. Wasenius, V.-M., Saraste, M., Knowles, J., Vistanen, I., Lehto, V. P. 1985. Sequencing of the chicken nonerythroid spectrin cDNA reveals an internal repetitive structure homologous to the human erythrocyte spectrin. *EMBO J.* 4: 1425–30
137. Weinberger, G., Thompson, C. C., Ong, E. S., Lebo, R., Gruol, D. J., Evans, R. M. 1986. The c-erbA gene encodes a thyroid hormone receptor. *Nature* 324: 641–46
138. Weise, M. J., Chan, L.-N. L. 1978. Membrane protein synthesis in embryonic chick erythroid cells. *J. Cell Biol.* 253: 1892–97
139. Winkelmann, J. C., Leto, T. L., Watkins, P. C., Eddy, R., Shows, T. B., et al. 1988. Molecular cloning of the cDNA for human erythrocyte β spectrin. *Blood* 72: 328–34
140. Wood, W. B. 1979. Bacteriophage T4 assembly and the morphogenesis of subcellular structure. *Harvey Lect.* 73: 203–23
141. Woods, C. M., Boyer, B., Vogt, P. K., Lazarides, E. 1986. Control of erythroid differentiation: asynchronous expression of the anion transporter and the peripheral components of the membrane skeleton in AEV- and S13-transformed cells. *J. Cell Biol.* 103: 1789–98
142. Woods, C. M., Lazarides, E. 1985. Degradation of unassembled α- and β-spectrin by distinct intracellular pathways: regulation of spectrin topogenesis by β-spectrin degradation. *Cell* 40: 595–69
143. Woods, C. M., Lazarides, E. 1986. Spectrin assembly in avian erythroid development is determined by competing reactions of homo- and hetero-oligomerization. *Nature* 321: 855–89

144. Yew, N. S., Choi, H.-R., Gallarda, J. L., Engel, J. D. 1987. Expression of cytoskeletal protein 4.1 during avian erythroid cellular maturation. *Proc. Natl. Acad. Sci. USA* 84: 1035–39
145. Yoshino, H., Marchesi, V. T. 1984. Isolation of spectrin subunits and reassociation in vitro. *J. Biol. Chem.* 259: 4496–500
146. Zenke, M., Kahn, P., Disela, C., Vennström, B., Leutz, A., et al. 1988. V-erbA specifically suppresses transcription of the avian erythrocyte anion transporter (band 3) gene. *Cell* 52: 107–19

Annu. Rev. Cell Biol. 1989. 5: 453–81

MEMBRANE TRAFFIC IN ENDOCYTOSIS: INSIGHTS FROM CELL-FREE ASSAYS

*Jean Gruenberg and Kathryn E. Howell**

European Molecular Biology Laboratory, Postfach 10.2209, Heidelberg 6900, West Germany

---

CONTENTS

INTRODUCTION

Our understanding of the endocytic pathway has come largely from the studies on the receptor-mediated endocytosis of different ligands that are discussed in many reviews (for example, Steinman et al 1983; Helenius et al 1983; Goldstein et al 1985; Wileman et al 1985; Courtoy 1989). At the cell surface coated pits, containing clustered receptors, as well as other

* Present address: University of Colorado, 4200 East 9th Avenue, Denver, Colorado 80262.

0743–4634/89/1115–0453$02.00

transmembrane proteins, membrane lipids, and solutes, bud from the plasma membrane and become coated vesicles. The internalized molecules next appear in early endosomes located at the cell periphery, presumably via fusion of the coated vesicles. During the subsequent steps of membrane traffic, these molecules are either routed back to the plasma membrane, to the Golgi complex, or to the lysosomes where degradation occurs. Morphological and fractionation studies have shown that during transport toward the lysosomes internalized molecules appear sequentially in early endosomes, next in late endosomes, and eventually in lysosomes. Both late endosomes and lysosomes can often be observed in the perinuclear region of the cell. A number of studies have demonstrated that early and late endosomes are morphologically distinct and can be identified by their different content of certain proteins, including receptors. They can be separated according to their physical properties, and they exhibit specific functions.

Although early and late endosomes, as well as lysosomes, are well established stages of endocytosis, many aspects of membrane traffic in the pathway remain poorly understood and controversial. Two models for membrane traffic in endocytosis, vesicular traffic and maturation, have been considered (see review, Helenius et al 1983) and still remain under debate. Vesicular traffic, as originally proposed by Palade (1975) for the secretory pathway, predicts that early and late endosomes as well as lysosomes are preexisting organelles connected by carrier vesicles that bud from one compartment and deliver their membrane and content to the next compartment by fusion and then recycle. Early and late endosomes are then expected to contain resident proteins in addition to the molecules in transit. According to the maturation model, early endosomes are constantly being formed by the fusion of incoming vesicles with each other. This early endosome then matures while being translocated in the cell, receives Golgi components and becomes a late endosome, and eventually becomes a lysosome. After retrieval of the molecules destined to be reutilized via recycling routes, the rest of the membrane and content becomes the next stage of the pathway; no resident protein is expected to be present in early or late endosomes. Clearly the available biochemical and functional data indicate that early and late endosomes are successive stations of the pathway. Whether they contain specific resident proteins remains to be seen, however.

This review is concerned with the different steps of endocytic membrane traffic, with particular emphasis on recent studies using cell-free assays. Cell-free approaches have provided experimental systems to investigate individual steps of the pathway, in particular fusion events. In elegant morphological studies, Oates & Touster (see 1980) originally reported that

Acanthamoeba phagolysosomes can undergo fusion in vitro. More recent assays are outlined in Table 1. All cell-free assays share the fundamental and obvious feature that the step of membrane traffic reconstituted in vitro can be experimentally manipulated to an extent not possible in vivo. How cell-free each assay is varies significantly beyond this common advantage. Perforated (Simons & Virta 1987) or semi-intact (Balch et al 1987) cells have provided in vitro systems in the exocytic pathway with minimal disruption of the cellular organization. In the endocytic pathway similar systems have been employed by Goda & Pfeffer (1988) and Smythe et al (1989). Others have used postnuclear supernatants or cytosol-free membrane fractions (Davey et al 1985; Braell 1987; Woodman & Warren 1988; Diaz et al 1988; Mayorga et al 1988–1989; Mullock et al 1989) which provide simple and rapid assays. Finally, subcellular fractionation has been used to isolate different endosomal elements and to test their capacity to undergo fusion in vitro (Gruenberg & Howell 1986, 1987; Gruenberg et al 1989).

COATED PITS AND VESICLES

In vivo *Studies*

The process of coated pit/vesicle formation involves the recruitment and assembly of clathrin and specific clathrin-associated molecules at the plasma membrane to form a pit, which subsequently invaginates, buds, and produces a coated vesicle (review, Pearse 1987). While some receptors are clustered in coated pits and internalized even when unoccupied (e.g. the LDL-receptor), other receptors, for example of some growth factors and hormones, migrate into the pit and are internalized only after binding of their respective ligands (review, Goldstein et al 1985). In this latter example, protein kinase C-mediated phosphorylation and auto-phosphorylation events have been proposed as regulatory mechanisms, however these observations are controversial (reviewed in Carpenter 1987). In contrast, some requirements for the interaction of clustered receptors with the coat material are becoming apparent. Three-dimensional reconstruction of coated vesicles using samples prepared in vitreous ice have shown that clathrin-associated proteins form the inner shell of the coat, which suggests that they interact with the cytoplasmic domains of the receptors (Vigers et al 1986). This proposal has now been strengthened by molecular and biochemical observations. Naturally occurring or experimental mutations show that endocytosis of the LDL-receptor (Davis et al 1986), the large cation-independent mannose-6-phosphate receptor (CI-MPR; Lobel et al 1989), and the influenza hemagglutinin (Lazarovits & Roth 1988), which is normally not internalized, require the presence of a

Table 1 Cell-free studies of membrane traffic in endocytosis[a]

Assays	Detection systems
Binding of cytosolic clathrin to stripped plasma membrane[b]	Immuno-radiometric quantitation of clathrin binding
Conversion of coated pits from shallow to deeply invaginated and formation of coated vesicles[c]	Cell surface immuno-precipitation of ^{125}I-transferrin Morphometric analysis
Endosome-endosome fusion monitored with	
viruses (^{3}H-sialic acid-SFV and FPV)[d]	Release of ^{3}H-sialic acid by FPV neuraminidase
fluid phase markers (avidin-β-galactosidase, biotin-mouse IgG)[e]	Avidin-biotin complex quantitated with an ELISA
ligands (^{125}I-transferrin, anti-transferrin antibody)[f]	Immune complex quantitated with ^{125}I label
ligands (DNP-β-glucuronidase, mannosylated anti-DNP antibody)[g]	Immune complex quantitated with glucuronidase activity
Endosome-endosome fusion using immuno-isolated fractions monitored with	
membrane spanning G-protein of VSV and lactoperoxidase[h]	Fusion-specific iodination and immuno-precipitation
fluid phase markers (avidin and biotin-horseradish peroxidase)[i]	Avidin-biotin complex quantitated with an ELISA
Fusion of plasma membrane derived vesicles with endosomes (aggregated anti-DNP antibody, DNP-β-glucuronidase)[j]	Immune complex quantitated with glucuronidase activity
Degradation of an internalized immune complex in endosomes (aggregated anti-DNP antibody/DNP-^{125}I-BSA)[k]	Release of free ^{125}I
Recycling of the CI-MPR from endosomes to the *trans*-Golgi network[l]	Resialylation of CI-MPR quantitated by chromatography
Fusion of stripped coated vesicles containing a dye with lysosomes[m]	Hydrolysis-specific fluorescence of the dye
Transfer of a ligand (^{125}I-asialoglycoprotein) from late endosomes to lysosomes[n]	Density shift of ^{125}I-asialoglycoprotein on gradients

[a] Abbreviations: SFV, Semliki Forest virus; FPV, fowl plague virus; DNP, dinitrophenol; VSV, vesicular stomatitis virus; BSA, bovine serum albumin; CI-MPR, large cation-independent mannose-6-phosphate receptor.
[b] Moore et al 1987; Mahaffey et al 1989.
[c] Smythe et al 1989.
[d] Davey et al 1985.
[e] Braell 1987.
[f] Woodman & Warren 1988.
[g] Diaz et al 1988
[h] Gruenberg & Howell 1986, 1987.
[i] Gruenberg et al 1989.
[j] Mayorga et al 1988.
[k] Mayorga et al 1989.
[l] Goda & Pfeffer 1988.
[m] Altstiel & Branton 1983.
[n] Mullock et al 1989.

tyrosine residue (or possibly other aromatic amino acids, Davis et al 1987) in the cytoplasmic domain. Based on in vitro binding experiments, Pearse & her collaborators (see Glickman et al 1989) have proposed that tyrosine residues may specifically interact with the complex of clathrin-associated proteins (HA-II), presumably present in plasma membrane-derived coated structures (Robinson 1987). A different complex of clathrin-associated proteins (HA-I) is believed to be involved in the clathrin-coated structures observed in the Golgi region (Ahle et al 1988), which may mediate the delivery of newly synthesized lysosomal hydrolases (reviews, Farquhar 1985; Kornfeld & Mellman, this volume).

The number of coated pits and vesicles at the plasma membrane as well as the dynamics of their formation have been estimated by several investigators. Coated pits occupy 1 to 2% of the plasma membrane surface area of human fibroblasts (Anderson et al 1976) and assemble with a half-time of ≈ 5 min (Larkin et al 1986). Once formed, coated vesicles are believed to uncoat rapidly and then to fuse with each other or with early endosomes. A member of the 70-kd heat shock protein family has been purified and shown to release both heavy and light chains of clathrin in vitro (Rothman & Schmid 1986). The half-life of a coated vesicle was reported to be ≤ 1 min (Anderson et al 1977), and the fraction of the total cellular clathrin in the unassembled state varied considerably depending on the cell type (Goud et al 1985). These numbers correlate well with photobleaching studies after rhodamine-clathrin microinjection, which indicate that polymerized clathrin has a $t_{1/2} \approx 10$–15 s and represents approximately 50% of the total cellular clathrin (P. Cosson & J. Davoust, in preparation). The number of coated vesicles generated in BHK cells was estimated to be approximately 1500/min (Marsh & Helenius 1980). More recently, a slightly smaller value of 1100/min was obtained; moreover it was estimated that the coated vesicles present at any time accounted for about 6% of the volume and 8% of the surface area of the early endosome (G. Griffiths, R. Back & M. Marsh, in preparation). In both latter studies the volume contained by the given number of coated vesicles was sufficient to account for the estimated volume of fluid internalized by the cells.

Whether an alternate pathway independent of clathrin also occurs in animal cells is still debated; inhibition studies indicate that the relative contribution of this putative alternate route to the total amount of fluid internalized is unclear and controversial. Three reversible treatments have been reported to inhibit clathrin-mediated endocytosis: potassium depletion after hypotonic shock (Larkin et al 1983; Moya et al 1985; Carpentier et al 1989), hypertonic media (Daukas & Zigmond 1985; Oka & Weigel 1988; Carpentier et al 1989; Heuser & Anderson 1989), and cytoplasmic acidification (Sandvig et al 1987; Davoust et al 1987; Cosson et al 1989; Heuser 1989a). However ricin appeared to be internalized (Moya et al 1985; Sandvig et al 1987) and the internalization of fluid phase markers was not affected (Daukas & Zigmond 1985), partially reduced (Sandvig et al 1987; Carpentier et al 1989), or strongly inhibited (Davoust et al 1987; Heuser 1989a; Cosson et al 1989; Heuser & Anderson 1989). In morphological studies, coated pits appeared to be absent after potassium-depletion (Larkin et al 1983; Moya et al 1985) but were not affected by acidification (Davoust et al 1987; Sandvig et al 1987). More recently, Heuser & Anderson (1989) and Heuser (1989a) reported that any one of the three treatments produced a drastic reduction in normal coated pits and the appearance of numerous "microcages" beneath the plasma membrane, which suggests a common mechanism for the observed inhibition of endocytosis.

Few other treatments or drugs have been reported to interfere with the internalization step. Attempts to inhibit endocytosis with anti-clathrin antibodies delivered by fusion of erythrocyte ghosts have been partially successful (Doxsey et al 1987). Studies of endocytosis after inhibition of cellular metabolism in vivo have also produced conflicting results with no effect (Larkin et al 1985), only one round of internalization (Clarke & Weigel 1985), or complete inhibition (Hertel et al 1986). These studies underline the difficulty in interpreting observations obtained with treatments that interfere with essential cellular functions. Mitosis is the only physiologic condition where the internalization step is arrested without experimental manipulation. The factors present in mitotic cells that interfere with membrane traffic in endocytosis and in other pathways have not been identified (review, Warren 1985).

In vitro *Studies*

Moore et al (1987) have used an immuno-radiometric assay to monitor the binding of cytosolic clathrin at 0°C to the cytoplasmic surface of stripped plasma membranes from human skin fibroblasts immobilized via polylysine. Structures that resembled coated pits, albeit smaller, were observed by electron microscopy. It is not clear what fraction of the clathrin and clathrin-associated molecules recruited by the stripped membranes were incorporated in these structures, nor whether the reconstituted coated regions formed at random on the adsorbed membranes, or at sites where receptors may have remained clustered. In these experiments the bound clathrin was released from the membranes at 37°C in an ATP-dependent manner, and this release appeared not to depend on the presence of cytosol. In a more recent study these authors have used the same assay to reconstitute the binding of clathrin present in an extract of coat proteins prepared from purified bovine brain coated vesicles (Mahaffey et al 1989).

The authors postulate that in the assay residual clathrin present on the stripped membranes may serve as nucleation sites during subsequent binding.

Smythe et al (1989) have recently studied the budding of coated vesicles in A431 cells. After [^{125}I]-transferrin binding to the cell surface, the cells were disrupted by scraping the dish with a section of a rubber bung with a straight edge, and the broken cells were then incubated for various times at 37°C. The number of coated pits was quantitated by morphology and the amount of [^{125}I]-transferrin accessible on the cell surface was quantitated by immuno-precipitation. Conversion of coated pits from a shallow to a deeply invaginated form occurred within 5 min and only at 37°C, but required neither cytosol nor ATP. Similar findings were reported using cells broken by sonication (Heuser 1989a). Budding of coated vesicles occurred, although with a low efficiency; apparently the formed coated vesicles were stable and did not fuse with early endosomes as in vivo studies would predict. The budding step was temperature-dependent, required the presence of both ATP and cytosol, and was inhibited at acidic pH in agreement with in vivo inhibition studies. In contrast to the findings of Moore et al (1987), who observed that ATP released the bound clathrin, Smythe et al (1989) observed an increased number of pits, presumably newly formed, in the presence of both cytosol and ATP.

In these experiments some of the individual steps of coated pit/vesicle formation have been identified and manipulated in vitro. The assays now provide the means to investigate the postulated interactions between the cytoplasmic domains of transmembrane proteins, clathrin, and clathrin-associated proteins. In addition, the mechanisms controlling coated vesicle budding are more accessible to experimental manipulations. It is unclear whether coated pits form as a flat lattice, which subsequently curves as a result of coat rearrangement (see Heuser 1989a), or whether curvature is built into the clathrin lattice (see Pearse & Crowther 1987). Neither is it clear whether the budding process requires a cellular activity, as suggested by the shibire mutation in a single gene that prevents budding of coated vesicles in *Drosophila* neurons (Kosaka & Ikeda 1983) and by the ATP-dependence observed by Smythe et al (1989), or whether it is simply a consequence of self-assembly (Harrison & Kirchhausen 1983) in a process analogous to the budding of enveloped viruses (Simons & Fuller 1987).

EARLY ENDOSOMES

Recycling to the Plasma Membrane

After delivery via coated vesicles, markers are observed in early endosomal elements that exhibit a tubulo-vesicular morphology and are preferentially located at the cell periphery. Kinetic studies show that many receptors are internalized within a few min at 37°C and then rapidly recycle back to the cell surface with a $t_{1/2} \approx 5$ min (reviews, Goldstein et al 1985; Wileman et al 1985). Approximately 50% of the content of the early endosome, labeled for 5 min at 37°C with different fluid phase markers (reviews, Steinman et al 1983; Swanson & Silverstein 1988) recycles with a $t_{1/2} \approx 5$ min. Similar rapid kinetics of internalization and recycling have been observed for cell surface proteins labeled by exo-galactosylation (Haylett & Thilo 1986) or radioiodination (Draye et al 1987).

These observations indicate that many receptors, most of the internalized membrane proteins, and ≈ 50% of the fluid content recycle to the cell surface with rapid kinetics and suggest that this rapid recycling occurs from the early endosome, a proposal strengthened by two cell fractionation studies. The early endosomes, which contain the recycling receptors in transit, most likely correspond to the receptor-positive endosomal fraction isolated from rat hepatocytes after internalization of [^{125}I]-asialoorosomucoid for 2.5 min at 37°C, whereas a late endosomal fraction prepared after 14–44 min was receptor-negative (Mueller & Hubbard 1986). After [^{125}I]-transferrin internalization for 4 min and 15 min at 37°C, fractions respectively enriched and depleted in transferrin were separated by free-flow electrophoresis (Schmid et al 1988).

Receptor-Ligand Uncoupling and Sorting

Since some receptors rapidly recycle back to the cell surface from the early endosome, they are presumably sorted in this compartment from the internalized proteins destined to be degraded. In polarized cells, which have two plasma membrane domains, some internalized proteins are sorted from those destined to be degraded or recycled and are transcytosed to the opposite cell surface. Several studies indicate that this sorting process also occurs in early endosomes. In rat hepatocytes, two ligands internalized by receptor-mediated endocytosis, the asialoglycoprotein destined for the lysosomes and the polymeric IgA destined for transcytosis, colocalize in the early endosome before they rapidly (within minutes) part company (see review, Courtoy et al 1989). In Madin-Darby canine kidney (MDCK) cells, the recycling efficiency of the transferrin receptor at the basolateral surface was estimated; missorting to the apical surface via the transcytotic route was less than 0.3% (Fuller & Simons 1986). However very little is known about the mechanisms of sorting in early endosomes. Polyvalent ligands or antibodies added at the cell surface can cross-link proteins that normally recycle and change their route to the degradative pathway. This misrouting has been shown, for example, with the Fc receptor in mouse macrophages (see Ukkonen et al 1986), the implanted VSV G-protein in

BHK cells (see Gruenberg et al 1989), the CI-MPR (von Figura et al 1984), and with transferrin made polyvalent by coupling to colloidal gold in A431 cells (Fodor et al 1986). Whether these experimentally induced associations between molecules bear any similarity to the concentration of receptors observed in coated pits remains to be shown. In the absence of polyvalent ligands, increasing densities of VSV G-protein implanted in the plasma membrane reduced the fraction of internalized VSV G-protein recycling back to the cell surface. Reduced recycling may have resulted from increased lateral interactions between the G molecules themselves or possibly from the saturation of a putative sorting mechanism (Gruenberg & Howell, 1987).

The tubulo-vesicular morphology and the presence of recycling receptors in early endosomes suggest that the latter corresponds to the Compartment of Uncoupling of Receptors and Ligands or CURL (Geuze et al 1983). CURL was identified morphologically by localizing asialoglycoprotein and its receptor, using double-label immunoelectron microscopy. The fact that the early endosome and CURL correspond to the same compartment is supported by the observation that transferrin and asialoglycoprotein colocalize after internalization, but before being rapidly segregated (Stoorvogel et al 1987). Although it is now generally accepted that recycling back to the cell surface must be mediated by transport vesicles, which also return part of the fluid phase markers present in the endosome lumen, these recycling vesicles have not yet been characterized. Geuze et al (1983) have proposed that the tubular portions of tubulo-vesicular endosomes (CURL) are an intermediate in the recycling pathway, since their membrane is enriched in recycling receptors when compared to the membrane of the vesicular portions.

Comparison of Early and Late Endosomal Properties

In addition to their morphological appearance and to the distribution of rapidly recycling molecules, early endosomes exhibit other differences from late endosomes. They can be separated from one another using density gradient centrifugation (Storrie et al 1984; Kindberg et al 1984; Wall & Hubbard 1985; Branch et al 1987); diaminobenzidine induced density shift (review, Courtoy et al 1989); immuno-isolation (Mueller & Hubbard 1986; see Gruenberg et al 1989), and free-flow electrophoresis (Schmid et al 1988). Using pH-sensitive fluorescent probes, early endosomes were shown to be less acidic (pH $\approx$ 6.0–6.2) than late endosomes (pH $\approx$ 5.5–6.0) and lysosomes (pH ≤ 5) (review Mellman et al 1986). These observations were confirmed using an independent approach that involved endocytosis of either wild type or a mutant Semliki Forest virus with a lower pH threshold for fusion (Kielian et al 1986; Schmid et al 1989). Cell lines defective in acidification have been generated by mutagenesis, but they have not yet been characterized at the molecular level (reviews, Krieger 1986; Robbins & Roff 1987).

Whereas the proton gradient across the membrane of endocytic compartments is generated by a vacuolar ATPase (see reviews, Mellman et al 1986; Nelson & Taiz 1989), the moderately acidic pH of the early endosome may be due to the counter-effect of recycling (Na^+-K^+) ATPase present in early endosomes (Fuchs et al 1989; Cain et al 1989). It is generally accepted that the low pH of the endosomal lumen induces conformational changes that facilitate receptor-ligand uncoupling (review, Mellman et al 1986) or activate latent activities (e.g. fusion activity of viral spike proteins; review, Wiley & Skehel 1987). Whether the mild acidic pH of the early endosome is sufficient to account for the observed dissociation of ligands from rapidly recycling receptors or whether other mechanisms also contribute to the reaction, remains to be seen.

Recent data suggest that early and late endosomes may also contain distinct polypeptides, since the pattern of polypeptides iodinated in an early endosomal fraction of CHO cells after internalization of lactoperoxidase for 4 min at 37°C differs from that observed after 15 min (Schmid et al 1988). Preliminary studies confirmed this finding in the BHK cell (J. Gruenberg, unpublished data). Several investigators have also identified other putative endosomal proteins after gel electrophoresis of fractions, however until further characterization of these proteins is achieved little can be concluded. One endosome-specific antigen (44 kd) has been relatively well-characterized and is specific for the absorptive cells of the suckling rat ileum (Wilson et al 1987). Finally, cell-free studies of vesicle fusion show that early endosomal elements exhibit a high and specific fusion activity in vitro, which must reflect the presence of specific components in the early endosomal membrane (Gruenberg et al 1989).

In vitro *Fusion of Early Endosomal Elements*

Different groups have reconstituted in vitro the fusion events occurring between endocytic vesicles (see Table I). The diversity of markers used in these assays has proven to be useful, since a more general interpretation of some of the in vitro observations can now be made. Most assays have used a fusion-specific reaction between two external markers endocytosed separately by two cell populations (Figure 1). Once the cells have been homogenized, the endosomal vesicles derived from each cell population are free to interact with each other upon mixing. Since the selected markers are membrane-enclosed, the detection reaction can occur only if the membranes of the vesicles have fused and the product of the detection reaction

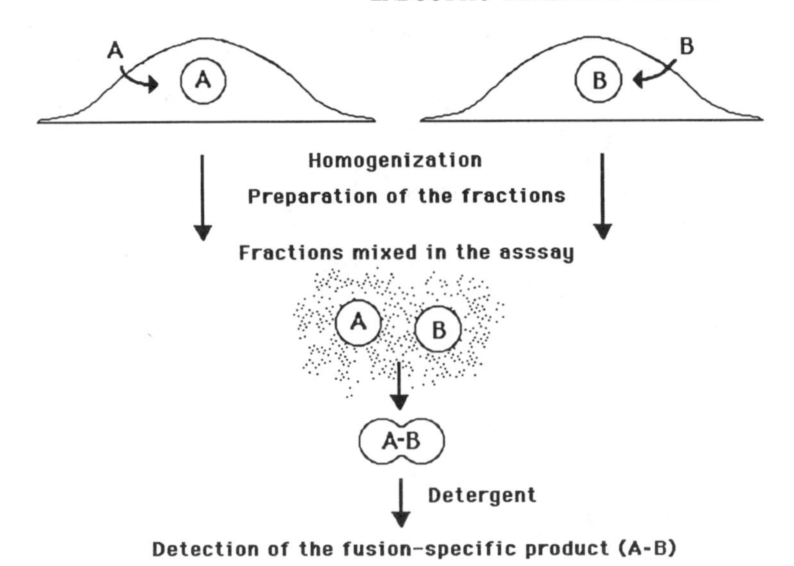

Figure 1 Outline of the cell-free assays. A and B represent a pair of markers used to detect the occurrence of fusion. They are endocytosed separately in two cell populations.

can be used to monitor the occurrence of fusion. Morphological examination of the vesicles after fusion was carried out by Gruenberg & Howell (1986) and Mayorga et al (1988) using electron-dense markers. Since the original endosome fusion assay of Davey et al (1985), most systems have used postnuclear supernatant or cytosol-free vesicle fractions. Gruenberg & Howell (1986–1987) and Gruenberg et al (1989) have used immuno-isolated endosomal fractions to introduce more defined components in the fusion reaction. The antigen was provided by the cytoplasmic domain of the VSV G-protein, which was implanted in the plasma membrane and subsequently internalized (see review, Howell et al 1989). In a subsequent fusion assay, the endosomal vesicles immobilized on a solid support via the G-protein could then be mixed with a postnuclear supernatant prepared from other cells lacking the G-protein. After the reaction, the immuno-isolated vesicles could be retrieved and washed to remove the nonreacted cellular material.

The observations made with these different in vitro assays of endosome-endosome fusion show striking similarities. The consensus finding is that an early fusion event in the endocytic pathway has been reconstituted, as defined by the time-course of markers endocytosed. Fusion activity was maximal 5 min after endocytosis of the following markers: fluid phase markers in CHO (Braell 1987) and BHK (Gruenberg et al 1989) cells, mannosylated ligands (Diaz et al 1988) and aggregated antibodies

(Mayorga et al 1988) in J774 macrophages, and the VSV G-protein in BHK cells (Gruenberg & Howell 1986, 1987; Gruenberg et al 1989). The occurrence of early endosome fusion was also shown by Woodman & Warren (1988), who used a marker that recycles from the early endosome. The vesicular partners of these early fusion events are now characterized both morphologically and biochemically as being elements of the early endosome (Gruenberg et al 1989). Fusion of early endosomes with inside-out plasma membrane vesicles, presumably via clathrin-coated regions, has also been reported (Mayorga et al 1988). The significance of this observation is difficult to assess however, since the signal was low when compared with early endosome fusion tested in the same assay. With a different assay, an immune complex was internalized for ≤ 5 min into a protease-negative endosome and subsequently degraded in vitro after incubation under conditions promoting fusion (Mayorga et al 1989). This observation was interpreted to be the result of fusion between protease-negative endosomes and either protease-containing endosomes or other vesicles containing hydrolases, possibly originating from the Golgi complex.

When tested with different assays, the fusion activity of the corresponding fractions was significantly reduced after chasing the markers in vivo to later stages of the pathway (Gruenberg & Howell 1987; Braell 1987; Diaz et al 1988; Gruenberg et al 1989). This rapid decrease occurred with a $t_{1/2} \approx 5$ min (Gruenberg & Howell 1987) and paralleled the appearance of the markers in large putative carrier vesicles (Gruenberg et al 1989). This time-course correlates well with in vivo studies that have followed the sequential appearance in early and late endosomes of a fluid phase marker in CHO cells (Storrie et al 1984), as well as EGF (Dunn & Hubbard 1984), and asialoglycoprotein in rat hepatocytes (Mueller & Hubbard 1986), and α_2-macroglobulin in 3T3 cells (Tran et al 1987; Goldenthal et al 1988). A slower and more asynchronous time-course was obtained using an SFV mutant with a low pH threshold for fusion (Kielian et al 1986).

The different assays have shown similar requirements. In vitro, fusion increased linearly for ≈ 20–40 min with no apparent time-lag and was temperature-sensitive with a minimum at 18–20°C (Braell 1987; Diaz et al 1988; Woodman & Warren 1988; Gruenberg et al 1989). A similar temperature-dependence was observed for the in vitro transfer from late endosomes to lysosomes (Mullock et al 1989). While the fusion step required ATP in all assays (and ATP hydrolysis), it did not require an acidic luminal pH of endosomes (Diaz et al 1988). Fusion also required the presence of cytosol (Woodman & Warren 1988; Diaz et al 1988; J. Gruenberg, unpublished data), which suggests that cytosolic factors are involved in this reaction. The SH alkylating agent N-ethylmaleimide (NEM) was shown to block the fusion reaction (Diaz et al 1988), and fusion was restored with fresh cytosol but not with NEM-treated cytosol (Braell 1987; Woodman & Warren 1988). In the studies of both Woodman & Warren (1988) and Diaz et al (1988), nontreated membranes reconstituted with NEM-treated cytosol exhibited some activity. Proteins present on the cytoplasmic surface of endosomal membranes are also likely to be involved, since trypsin treatment of the membranes abolishes fusion (Diaz et al 1988) and fusion cannot be restored with fresh cytosol (Woodman & Warren 1988). All the assays discussed have been carried out in the presence of KCl, and its absence from the reaction mixture reduced the fusion activity (Diaz et al 1988; J. Gruenberg, unpublished). In contrast, late endosome-lysosome fusion proceeds in the absence of KCl (Mullock et al 1989) and KCl was reported to actually inhibit *Acanthamoeba* phagolysosome fusion (see Oates & Touster 1980).

It still must be shown whether the fusion observed in these assays is mediated, as might be expected, by direct interactions between individual elements, or whether vesicular intermediates are involved. The observed fusion may also have measured the reforming of elements that vesicularized during homogenization, however this appears unlikely since gentle homogenization conditions and high latency of endosomal markers were required for optimal in vitro fusion (Gruenberg & Howell 1986, 1987; Braell 1987). In these assays it is unclear whether fission events, which are predicted to occur in vivo, paralleled the occurrence of fusion in vitro. Using early endosomal vesicles immobilized on a solid support, little release of the marker was observed during the course of the fusion reaction, which suggests either that fission events were limited or prevented by the immobilization, or that vesicles may have been formed but rapidly recaptured via the antigen present on their surface (Gruenberg et al 1989). Morphological analysis of the fusion product showed no major changes in the size of cross-sections of endosomal elements (Gruenberg & Howell 1986).

If these in vitro findings can be extrapolated to in vivo events, the individual elements of the early endosome appear to be highly dynamic, continuously exchanging membrane and content. Gruenberg et al (1989) have recently proposed that these interactions may pool the endocytosed materials destined to be degraded from more than one early endosomal elements, prior to the transfer of these materials to putative microtubule-dependent carrier vesicles. Rapid lateral interactions may also be important for the redistribution of internalized constituents of the cell surface, as was observed at the leading edge of moving and spreading cells (see

Hopkins 1985; Bretscher 1989). These interactions should guarantee the presence of effector molecules, e.g. the vacuolar ATPase, in the individual early endosomal elements. Finally one can speculate that the early endosome is organized as a dynamic network of interacting elements, which may transiently associate with each other, forming, in effect, a single functional compartment.

TRANSLOCATION OF ENDOCYTIC VESICLES IN THE CYTOPLASM

Morphological studies have shown that internalized tracers are observed sequentially in early endosomes located at the cell periphery and then in late endosomes and lysosomes, which are often clustered in the perinuclear region, particularly in polarized cells (review, Courtoy 1989). Using optical and fluorescent microscopy, the movement of endosomal vesicles containing an internalized tracer between the cell periphery and the perinuclear region has been well documented in vivo (review, Vale 1987; see Matteoni & Kreis 1987; De Brabander et al 1988). This movement depends on intact microtubules, which radiate in many cells from the microtubule-organizing center located in the perinuclear region. An intact microtubule system is also required for the clustering of late endosomes and lysosomes in this region, and depolymerization of microtubules reduces the delivery of internalized molecules to lysosomes (reviews, Vale 1987; Swanson & Silverstein 1988). Since microtubules are required both for translocation and clustering at the subsequent stages of the pathway, reduced degradation may simply result from the dispersion of the endocytic organelles.

The movement of endosomal vesicles from the periphery to the perinuclear region can be discriminated from membrane traffic to and from the early endosome by their microtubule-dependence. After microtubule depolymerization, the rate and amounts of internalization and recycling back to the plasma membrane were not affected, and the markers destined to the lysosomes appeared in distinct vesicles where they remained (Gruenberg et al 1989). While late endosomal elements as well as lysosomes clearly interact with microtubules, the factors that mediate these interactions have not been identified. Specific motor proteins are involved in vesicle translocation during axonal transport, and motor proteins have been characterized in a variety of non-neuronal cells (reviews, Vale 1987; Gelfand 1989; see Pashal & Valee 1987). It can be anticipated that motor proteins also control the interactions observed in endocytosis, although it is far from clear how the endocytic elements leaving the periphery acquire their specific capacity to interact with microtubules.

LATE ENDOSOMES

In addition to their microtubule-dependence, late endosomes also differ from early endosomes in their general appearance. Late endosomes, particularly in the perinuclear region, are often larger and exhibit a more complex organization of internal membranes than early endosomes; they have often been described as multivesicular structures (review, Courtoy 1989). Whereas the sorting of internalized proteins, in particular receptors, presumably occurs in early endosomes, the functions of late endosomes are not clear. In general, membrane traffic to late endosomes and lysosomes remains poorly understood. In most studies no attempts were made to discriminate between late endosomal elements in the perinuclear region and those presumably translocated from the peripheral early endosomes to the perinuclear region.

Recent observations show that markers internalized for 5 to 15 min at 37°C in BHK cells appear in large spherical vesicles (diameter ≈ 0.5 μm) after being observed in early peripheral endosomes, but before reaching late perinuclear endosomes, which are acid phosphatase-positive (Gruenberg et al 1989). After microtubule depolymerization with nocodazole, the markers also appeared in these vesicles, but did not reach late acid phosphatase-positive endosomes. In contrast to early endosomes, these large vesicles had little fusion activity in vitro with each other or with early endosomal elements (Gruenberg et al 1989). They contained internal membranes and appeared similar to some of the multivesicular structures that were described by many investigators. In fact, they may correspond to one of the sub-populations of multivesicular endosomes described morphologically in rat liver (Dunn et al 1986) and may also partially overlap with the late endosomal fraction prepared 15 min after internalization by Schmid et al (1988). A rather complex and structurally similar membrane organization was also observed in spherical regions of early endosomal elements, which suggests that they may give rise to the large vesicles (Gruenberg et al 1989).

These large vesicles were proposed to represent putative carrier vesicles that mediate the microtubule-dependent step of the pathway observed between early peripheral endosomes and late perinuclear endosomes (Gruenberg et al 1989). Clearly, these large carrier vesicles appear morphologically and functionally distinct from the small, short-lived vesicles traditionally considered to mediate vesicular traffic (Palade 1975). In endocytosis, these large vesicles may collect the material destined for degradation from more than one early endosomal element. (Mixing of early endosomal content occurs rapidly in vitro, hence each spherical vesicle may pool the content of different early elements). The large vesicles contained a high density of internalized VSV G-protein when compared to the early endosome, and their relatively large size could accommodate the volume internalized by 100 coated vesicles. In the presence of intact microtubules, the markers subsequently appear in late acid phosphatase-positive endosomes. Whether this process is mediated by fusion of the vesicles remains to be shown. However an indication that fusion with late endosomes may actually occur is that two electron-dense tracers internalized separately can colocalize in late perinuclear endosomes rich in cation-independent mannose-6-phosphate receptor (CI-MPR; G. Griffiths, personal communication). The role of microtubules in vivo may be to increase the frequency of encounters between these putative carrier vesicles and late endosomes, thereby increasing the efficiency of an otherwise inefficient delivery process. This suggested role of microtubules may account for the reduced fusion activity measured at this late stage of the pathway with in vitro assays, which fail to maintain structured microtubule arrays.

The observation that reduced temperature, 16–20°C, blocks degradation but not internalization has been instrumental in defining endosomes, and this temperature reduction was originally postulated to prevent fusion between endosomes and lysosomes (Dunn et al 1980). In fact, membrane traffic between early and late endosomes appears sensitive to low temperatures. An 18°C-sensitive step was observed before segregation of asialoglycoprotein and its receptor (Wolkoff et al 1984), and at 16°C the asialoglycoprotein entered the early receptor-positive endosome but not the late receptor-negative endosomes (Mueller & Hubbard 1986). More recently, markers internalized at 20°C in NRK and MDCK cells did not reach late perinuclear endosomes as identified by their high content of CI-MPR (Griffiths et al 1988; R. Parton, K. Prydz, M. Bomsel, K. Simons, G. Griffiths, in preparation). Early endosomes and structures morphologically similar to the microtubule-dependent putative carrier vesicles reported by Gruenberg et al (1989) were observed proximal to the temperature-sensitive step. These observations suggest that both drug-induced microtubule depolymerization at 37°C and reduced temperature prevent the appearance of internalized markers in late perinuclear endosomes.

A late endosome located in the perinuclear region of the cell has been described by Griffiths et al (1988) and is referred to by the authors as the prelysosomal compartment. This late endosome contains large amounts of the large cation-independent receptor for lysosomal hydrolases (CI-MPR), which is absent from bona fide lysosomes, and lgp 120, a lysosomal membrane glycoprotein (see Kornfeld & Mellman, this volume). The codistribution of CI-MPR and lgp 120 in late endosomes has also been reported in a rat hepatoma cell line (Geuze et al 1988). Late endosomes containing the CI-MPR could also be distinguished from transferrin-containing early endosomes (Woods et al 1988). The abundance of CI-MPR in this late endosome and the presence of detectable amounts of β-glucuronidase have led Griffiths et al (1988) to propose that this compartment corresponds to the site where the newly synthesized lysosomal hydrolases bound to the CI-MPR are delivered from the *trans*-Golgi network (TGN). The receptor is then believed to recycle to the TGN while the hydrolases are packaged in the lysosomes (Kornfeld & Mellman, this volume).

Although the cytochemical reaction for acid phosphatase has been traditionally used to identify lysosomes, late endosomes were also reported to be acid phosphatase-positive in CHO cells (Storrie et al 1984). More recent studies have confirmed this observation in NRK (G. Griffiths, R. Matteoni, R. Back, B. Hoflack, in preparation), MDCK (R. Parton, K. Prydz, M. Bomsel, K. Simons, G. Griffiths, in preparation), as well as BHK cells (Gruenberg et al, 1989) and have suggested that these acid phosphatase-positive structures correspond to the late endosome/prelysosomal compartment rich in CI-MPR. In fact, reexamination of the elegant separation of endosomal and lysosomal fractions from CHO cells using Percoll gradients by Sahagian & Neufeld (1983) indicates that a major portion of the acid phosphatase colocalized with the CI-MPR. However this sub-cellular distribution may not be representative of all lysosomal hydrolases, since the human acid phosphatase is synthesized as an integral membrane protein precursor, in contrast to other hydrolases (Waheed et al 1988).

In polarized cells, markers internalized from the apical and the basolateral surface had been observed in the same morphological structures (Oliver 1982; Nielsen et al 1985), but the meeting site was not identified. More recent experiments with the MDCK cell indicate that early apical and basolateral endosomes are distinct, whereas the two pathways meet after $\approx$ 15 min internalization in late CI-MPR-rich endosomes located in the perinuclear region (R. Parton, K. Prydz, M. Bomsel, K. Simons, G. Griffiths, in preparation; M. Bomsel, K. Prydz, R. Parton, J. Gruenberg, K. Simons, in preparation). As expected, both delivery to the late endosomes and meeting of the two pathways were prevented at 20°C.

IN VITRO TRANSFER FROM ENDOSOMES TO THE *TRANS*-GOLGI NETWORK (TGN)

A route connecting the plasma membrane to the Golgi complex has been observed using electron dense tracers in vivo in cells secreting high amounts of protein for export (for review, see Farquhar 1985). More recently, proteins originally present on the cell surface and endocytosed were shown to be resialylated with slow kinetics presumably in the *trans*-Golgi network (review, Kornfeld & Mellman, this volume). Since the sialyltransferase is assumed to be in the TGN, the consensus is that these proteins must pass through the TGN. Goda & Pfeffer (1988) have used the CI-MPR as a marker protein to reconstitute the transfer from endosomes to the resialylation site in a cell-free assay. Sialylation of CI-MPR was first detected after a lag time of $\approx$ 18 min and increased to a value $\approx$ 10–20% of the total after 3 hr and was ATP-dependent. AMP and ADP did not substitute for ATP, whereas 50 μM GTP-γ-S inhibited the reaction. In situ radioiodination of CI-MPR before the experiment showed that at least part of the sialylated CI-MPR was originally present in endosomes. Sialylation of the transferrin-receptor carried out in parallel was four times lower than that of CI-MPR, an observation suggesting that the transfer of CI-MPR to the TGN is selective, in agreement with the in vivo resialylation experiments of Duncan & Kornfeld (1988).

LYSOSOMES

In vivo *Transfer to Lysosomes*

The function and the biogenesis of lysosomes have been the subject of several reviews (see for example, Dingle et al 1984; Storrie 1988). Recent insights have been obtained from the observation of the subcellular distribution and the pathway followed by lysosomal glycoproteins, and lysosomal hydrolases and their receptors. These are discussed in more detail in another chapter of this volume (see Kornfeld & Mellman). The autophagic pathway, which accounts for the turnover of organelles and cytoplasm, also leads to the lysosomes and has been reviewed by Seglen (1987).

Transfer between late endosomes and lysosomes has been long thought to be mediated by the fusion between endocytic vesicles and lysosomes (review, Steinman et al 1983). The process is affected by different treatments. Low temperature prevents transfer from late endosomes to lysosomes (in vivo, Wolkoff et al 1984; Mueller & Hubbard 1986; in vitro, Mullock et al 1989), but also affects a more proximal stage of the pathway, as does microtubule depolymerization. Inhibition of luminal acidification (reviews, Steinman et al 1983; Mellman et al 1986) as well as hyperosmolarity (Oka & Weigel 1988; Park et al 1988) and acidification of the cytoplasm (Wolkof et al 1984; Samuelson et al 1988) also prevent delivery to lysosomes. However in these studies, it is not always clear what stage of the pathway is affected. Heuser (1989b) observed that cytoplasm acidification not only inhibited appearance of markers in lysosomes but also caused a redistribution of lysosomes to the cell periphery without affecting early endosome distribution. Of immediate medical importance is the finding that several pathogens can apparently survive in macrophages by preventing phagosome-lysosome fusion (see, Horwitz & Maxfield 1984; Sibley et al 1986; Hart et al 1987). Since phagosome-lysosome fusion can be reconstituted in vitro (see Oates & Touster 1980), cell-free assays may provide insights into the mechanisms controlling this fusion event and its inhibition by pathogenic agents.

In vitro *Transfer to Lysosomes*

In one of the earliest in vitro assays, Alstiel & Branton (1983) reported the fusion of coated vesicles present in a mixed population with lysosomes enriched by subcellular fractionation. Fusion was high with vesicles stripped of their coat, in contrast to coated or reconstituted vesicles, and was not affected by protease treatment of the vesicles. However low Ca concentrations or EDTA reduced the signal. These observations, together with the fact that coated vesicles in vivo had not been observed to fuse with lysosomes, led the authors to suggest that the fusion was nonspecific and reflected a Ca-mediated event of the kind observed with artificial lipid systems.

More recently Mullock et al (1989) have used the well-characterized pathway of asialoglycoproteins in rat liver to reconstitute the transfer from late endosomes to the lysosomes in vitro. For these experiments rats were first injected with [^{125}I]-asialofetuin and sacrificed after 10 min. In the transfer assay, a post-mitochondrial supernatant, prepared from the liver, was incubated at 37°C in buffered sucrose containing the reagents to be tested and then endosomes and lysosomes were separated using a Nycodenz gradient. After the in vitro incubation a large fraction of the [^{125}I] was recovered in the lysosomal peak in an ATP-, temperature-, and cytosol-dependent manner. In contrast to early endosome fusion (Braell 1987; Diaz et al 1988; Woodman & Warren 1988), NEM did not inhibit the reaction. After transfer of the marker to the lysosomes, some [^{125}I]-asialofetuin was degraded to TCA-soluble counts, as expected. Since the late, but not the early, endosomes were shown to be transfer competent, Mullock et al (1989) suggest that these transfer-competent endosomes correspond to the receptor-negative endosomes of Mueller & Hubbard (1986) and Schmid et al (1988). The authors argue that their findings do not support the maturation model for endocytosis, since the presence of lysosomes in the reaction mixture was necessary for the transfer to occur.

Properties of Lysosomes

Lysosomes are generally considered as the end station of the endocytic pathway, since nondegradable markers can be observed in the lysosomes of animal cells many days after internalization. The occurrence of a recycling route from the lysosomes back to endosomes or to the plasma membrane is poorly characterized. Some recycling has been observed with a lysosomal membrane glycoprotein in chicken cells (see Lippincott-Schwarz & Fambrough 1987). The receptors for the lysosomal hydrolases are recycled before reaching the lysosomes (Kornfeld & Mellman, this volume), and little is known about other cellular processes or functions associated with a recycling route from the lysosomes in most higher eukaryotes. However a recycling route may allow the retrieval of effector proteins delivered to lysosomes, that need to be reutilized. In contrast, some unicellular organisms clearly regurgitate their lysosomal content (e.g. paramecium; Fok & Allen 1988).

In addition to their full complement of mature hydrolases and absence of CI-MPR, lysosomes can be discriminated from late endosomes by their morphological appearance (see review Courtoy 1989), although this may be difficult in some cell types in the absence of specific markers. Their luminal pH ≈ 4.5, which corresponds to the optimal pH of many hydrolases and is about 1 pH unit lower than the late endosomes (review, Mellman et al 1986) as well as their physical properties during fractionation (de Duve 1975, see Sahagian & Neufeld 1983) are discriminating characteristics. In some fractionation studies using Percoll gradients, lysosomal hydrolases distributed at different densities (Pertoft et al 1978; Rome et al 1979; Merion & Sly 1983; Berg et al 1985; Kelly et al 1989). Although these findings were interpreted by some investigators to reflect the existence of different lysosome populations (light and heavy), light lysosomes may in fact overlap with late endosomes. The classical identification of lysosomes, which uses the histochemical reactions for acid phosphate or aryl-sulphatase, is ambiguous. Late endosomes are also acid phosphatase-positive in several cell types (Sahagian & Neufeld 1983; Storrie et al 1984; Gruenberg et al 1989; Roederer et al 1989; G. Griffiths personal communication), and aryl-sulfatase has been detected in endosomes (Kelly et al 1989) as well as other hydrolytic activities (review, Murphy 1988; see also Diment et al 1988; Gruenberg et al 1989). These observations indicate that some limited processing events, such as the processing of endocytosed antigens by antigen presenting cells (for a recent review, Lanzavecchia 1988) traditionally attributed to lysosomes, might in fact, occur in endosomes.

Recent studies indicate that lysosomes can be highly dynamic organelles. Interspecies cell fusion experiments have shown that two populations of lysosomes rapidly exchange both content and membrane in vivo, in a temperature- and microtubule-dependent manner (Ferris et al 1987; Deng & Storrie 1988). These experiments may also have monitored transfer from late endosomes to lysosomes; despite a 1–2 hr chase, some internalized tracer may still have been present in late endosomes. Further, the lysosomal membrane glycoproteins used as markers appear to label late endosomes as well as lysosomes (Brown et al 1986; Griffiths et al 1988). These experiments suggest that lysosomes, which can be present in a large number of copies (up to 1000 in a macrophage; Steinman et al 1976) may interact with each other via fusion events. These interactions may be analogous to the efficient exchange observed in vitro between early endosomes.

Several lines of evidence suggest that membrane fusion events can also occur within organelles that are apparently present as a single copy, e.g. the nuclear envelope, the endoplasmic reticulum (Louvard et al 1982; Terasaki et al 1984) and the Golgi complex (Rambourg et al 1981). In interphase cells, the polygonal lattice of the ER may be generated by localized changes and interactions between the membranes of the tubules, including tubule branching, ring closure, and fusion (Lee & Chen 1988; Dabora & Sheetz 1988). After fragmentation during mitosis, the elements of the nuclear envelope, the ER or the Golgi regenerate the daughter compartments via fusion events (Warren 1985; Gerace & Burke 1988). While these findings suggest that homotypic or lateral interactions can occur within compartments present in a single copy, the individual elements of a vesicular compartment may, in fact, form a partial, possibly transient reticulum. A particularly striking network of interconnecting tubular lysosomes was observed in J774 cells or phorbol ester-stimulated macrophages (review, Swanson & Silverstein 1988). These tubular lysosomes were up to 5 μm long, were positive for acid phosphatase cytochemistry, and contained cathepsin L by immunofluorescence. It is, however, not clear whether these tubular structures correspond to the CI-MPR-negative lysosomes that have been described in other cell types. Rapid conversion between vesicular and tubular lysosomes has been observed after cytosol acidification and subsequent recovery from acidification (Heuser 1989b).

CONCLUSIONS

The observations of many laboratories have provided different information that has been integrated in a tentative scheme of the endocytic pathway (Figure 2). Whereas the sequential stages of endocytosis have been observed in vivo (solid arrows), cell-free approaches (open arrows, see Table 1) have permitted the detailed investigation of individual steps of the pathway.

Both from in vivo and in vitro studies, the evidence is now compelling that receptor-mediated internalization occurs via clathrin-coated pits, which detach from the plasma membrane and become coated vesicles. The precise mechanism of budding has still to be explained and the subsequent

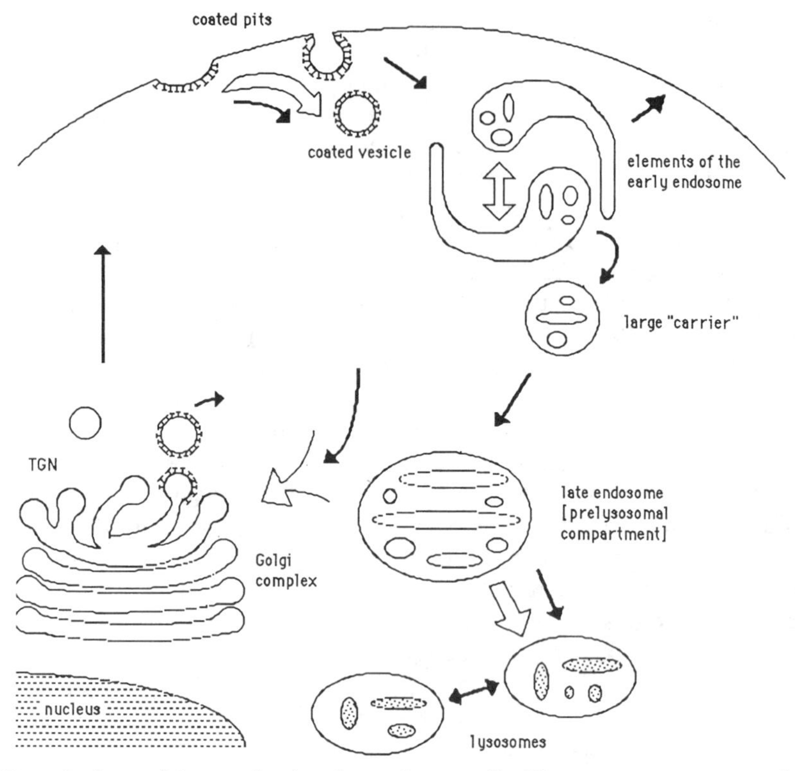

Figure 2 Sequential steps of endocytic membrane traffic. The open arrows represent the steps of the pathway that have been reconstituted with cell-free assays and the solid arrows correspond to the observations that were made in vivo. The microtubule network has not been represented, nor have other possible recycling routes between lysosomes, endosomes, and the cell surface.

fusion step of coated vesicles, presumably after uncoating, with early endosomes remains to be experimentally demonstrated. Once in the early endosomes located at the cell periphery, internalized proteins appear to be rapidly sorted and either recycled back to the cell surface, transported for degradation in the lysosomes, or transcytosed in polarized cells. Cell-free assays have revealed an unexpected activity of early endosomal elements, namely the specific and efficient exchange of membrane and content via fusion events. Homotypic or lateral interactions have also been observed within other compartments of both the secretory and the endocytic pathway. The high fusion activity of early endosomal elements suggests that these are highly dynamic and possibly form a transient network in vivo, hence it seems reasonable to consider the early endosome as a single compartment functionally. Although little is known about the mechanisms that provide the observed specificity of fusion, it can be anticipated that specific membrane components will be required as recognition signals. This agrees with the observation that early endosomes appear enriched in some polypeptides.

Early endosomes differ from later endosomes morphologically, biochemically, and functionally. Of particular significance is the observation that membrane traffic to and from the early endosome does not appear to require intact microtubules, in contrast to later stages. Whereas late endosomes clearly represent a second obligatory station for materials en route to the lysosomes, this sequence of events is less well-documented. Studies of endosomal vesicle movement in vivo as well as morphological and biochemical analysis suggest that late endosomes may encompass distinct elements, corresponding respectively to large vesicles possibly translocated on microtubules between the cell periphery and the perinuclear region (carriers) and larger perinuclear structures. These latter structures are thought to correspond to the site where lysosomal hydrolases and their receptor (CI-MPR) are delivered from the TGN. The hydrolases are then transferred to lysosomes, while CI-MPR is believed to recycle to the TGN, which suggests that, in addition to rapid recycling at the cell surface, a second and presumably slower recycling pathway connects endosomes and the TGN. CI-MPR can be transferred from endosomes to the TGN as demonstrated with a cell-free assay.

Finally, the molecules destined to degradation are routed to the lysosomes together with the lysosomal enzymes. Cell-free experiments have suggested that this transfer requires the presence of both late endosomes and lysosomes. In vivo studies have shown that lysosomes can be highly dynamic organelles and that they can exchange both membrane and content.

At present relatively little is known about the components of the fusion machinery involved either in lateral transfer between early endosomal elements or vectorial transfer between different compartments. Clearly this machinery exhibits requirements that are distinct from those of the well-described fusogenic spike proteins of enveloped viruses, which are low-pH dependent and ATP- and cytosol-independent. Whereas intracellular fusions require specific controls and targeting mechanisms to provide efficient delivery to the correct compartment, the viral mechanism has evolved to be opportunistic, acting from the extracellular or the luminal milieu. Interestingly, the fusion of early endosomal elements was inhibited by NEM treatment of the cytosol. Sensitivity of the cytosol to NEM has led Rothman and his collaborators to the purification of a factor required for the in vitro transfer from the *cis*- to medial Golgi (reviewed in Orci et al 1989). The gene coding for this factor was cloned and sequenced from CHO cells and is equivalent to sec18 in *Saccharomyces cerevisiae* (Wilson et al 1989). A recent study indicates that this factor can support the fusion of early endosomes in vitro (Diaz et al 1989). Still unclear at present is the possible role of GTP-binding proteins in endocytic membrane traffic. Based both on inhibition studies in cell-free assays and on mutagenesis experiments in yeast, GTP-binding proteins are thought to control membrane traffic in the secretory pathway (review, Bourne 1988; see Orci et al 1989). Transfer from the endosome to the TGN was inhibited by GTP-γ-S, which suggests the possible involvement of a putative GTP-binding protein.

Cell-free assays are now providing the experimental systems necessary for the dissection of these individual steps of membrane traffic at the molecular level and for the identification and characterization of the regulatory factors. Since some factors of the transfer and fusion machinery may be similar in both lateral and vectorial pathways of membrane fusion and common to the secretory pathway, the specificity of membrane traffic observed in vitro and in vivo must be provided by specific recognition signals. These are expected to be exposed on the cytoplasmic side of the membrane of each compartment and may serve as docking sites for the fusion machinery.

ACKNOWLEDGMENTS

We regret that we are unable to mention references to many original contributions due to the limited space. We are grateful to Drs. G. Griffiths, B. Hoflack, K. Simons, B. Storrie, J. Tooze, and D. Vaux for critically reading the manuscript. We are also grateful to our colleagues, who sent us papers or manuscripts in advance of publication.

Literature Cited

Ahle, S., Mann, A., Eichelsbacher, U., Ungewickell, E. 1988. Structural relationships between clathrin assembly proteins from the Golgi and the plasma membrane. *EMBO J.* 7: 919–29

Altstiel, L., Branton, D. 1983. Fusion of coated vesicles with lysosomes: measurement with a fluorescent assay. *Cell* 32: 921–29

Anderson, R. G. W., Goldstein, J. L., Brown, M. S. 1976. Location of low density lipoprotein receptors on plasma membrane of normal human fibroblasts and their absence in cells from a familial hypercholesterolemia homozygote. *Proc. Natl. Acad. Sci. USA* 73: 2434–38

Anderson, R. G. W., Brown, M. S., Goldstein, J. L. 1977. Role of the coated endocytic vesicle in the uptake of receptor-bound low density lipoprotein in human fibroblasts. *Cell* 10: 351–64

Balch, W. E., Wagner, K. R., Keller, D. S. 1987. Reconstitution of transport of vesicular stomatitis virus G protein from the endoplasmic reticulum to the Golgi complex using a cell-free system. *J. Cell Biol.* 104: 749–60

Berg, T., Kindberg, G. M., Ford, T., Blomhoff, R. 1985. Intracellular transport of asialoglycoproteins in rat hepatocytes. Evidence for two subpopulations of lysosomes. *Exptl. Cell Res.* 161: 285–96

Bourne, H. R. 1988. Do GTPases direct membrane traffic in secretion? *Cell* 53: 669–71

Braell, W. A. 1987. Fusion between endocytic vesicles in a cell-free system. *Proc. Natl. Acad. Sci. USA* 84: 1137–41

Branch, W. J., Mullock, B. M., Luzio, J. P. 1987. Rapid subcellular fractionation of the rat liver endocytic compartments involved in transcytosis of polymeric immunoglobulin A and endocytosis of asialofetuin. *Biochem J.* 244: 311–15

Bretscher, M. S. 1989. Endocytosis and recycling of the fibronectin receptor in CHO cells. *EMBO J.* 8: 1341–48

Brown, W. J., Goodhouse, J., Farquhar, M. G. 1986. Mannose-6-phosphate receptors for lysosomal enzymes cycle between the Golgi complex and endosomes. J. Cell Biol. 103: 1235–47

Cain, C. C., Sipe, D. M., Murphy, R. F. 1989. Regulation of endocytic pH by the Na^+, K^+-ATPase in living cells. *Proc. Natl. Acad. Sci. USA* 86: 544–48

Carpenter, G. 1987. Receptors for epidermal growth factor and other polypeptide mitogens. *Annu. Rev. Biochem.* 56: 881–914

Carpentier, J.-L., Sawano, F., Geiger, D., Gorden, P., Perrelet, A., Orci, L. 1989. Potassium depletion and hypertonic medium reduce "non-coated" and clathrin-coated pit formation, as well as endocytosis through these two gates. *J. Cell. Physiol.* 138: 519–26

Clarke, B. L., Weigel, P. H. 1985. Recycling of the asialoglycoprotein receptor in isolated rat hepatocytes. *J. Biol. Chem.* 260: 128–33

Cosson, P., de Curtis, I., Pouysségur, J., Griffiths, G., Davoust, J. 1989. Low cytoplasmic pH inhibits endocytosis and transport from the trans-Golgi network to the cell surface. *J. Cell Biol.* 108: 377–87

Courtoy, P. 1989. Dissection of endosomes. In *Intracellular trafficking of proteins*, ed. C. Steer, J. Hannover. New York: Cambridge Univ. In press

Dabora, S. L., Sheetz, M. P. 1988. The microtubule-dependent formation of a tubulovesicular network with characteristics of the ER from cultured cell extracts. *Cell* 54: 27–35

Daukas, G., Zigmond, S. H. 1985. Inhibition of receptor-mediated but not fluid-phase endocytosis in polymorphonuclear leukocytes. *J. Cell Biol.* 101: 1673–79

Davey, J., Hurtley, S. M., Warren, G. 1985. Reconstitution of an endocytic fusion event in a cell-free system. *Cell* 43: 643–52

Davis, C. G., Lehrman, M. A., Russell, D. W., Anderson, R. G. W., Brown, M. S., Goldstein, J. L. 1986. The J.D. mutation in familial hypercholesterolemia: amino acid substitution in cytoplasmic domain impedes internalization of LDL receptors. *Cell* 45: 15–24

Davis, C. G., van Driel, I. R., Russell, D. W., Brown, M. S., Goldstein, J. L. 1987. The low density lipoprotein receptor. *J. Biol. Chem.* 262: 4075–82

Davoust, J., Gruenberg, J., Howell, K. E. 1987. Two threshold values of low pH block endocytosis at different stages. *EMBO J.* 6: 3601–9

De Brabander, M., Nuydens, R., Geerts, H., Hopkins, C. R. 1988. Dynamic behaviour of the transferrin receptor followed in living epidermoid carcinoma (A431) cells with nanovid microscopy. *Cell Motil. Cytoskel.* 9: 30–47

De Duve, C. 1975. Exploring cells with a centrifuge. *Science* 189: 186–94

Deng, P., Storrie, B. 1988. Animal cell lysosomes rapidly exchange membrane proteins. *Proc. Natl. Acad. Sci. USA* 85: 3860–4

Diaz, R., Mayorga, L., Stahl, P. 1988. In vitro fusion of endosomes following receptor-mediated endocytosis. *J. Biol. Chem.* 263: 6093–100

Diaz, R., Mayorga, L. S., Weidman, P. J., Rothman, J. E., Stahl, P. D. 1989. Vesicle fusion following receptor-mediated endocytosis requires a protein active in Golgi transport. *Nature* 339: 398–400

Diment, S., Leech, M. S., Stahl, P. D. 1988. Cathepsin D is membrane-associated in macrophage endosomes. *J. Biol. Chem.* 263: 6901–7

Dingle, J. T., Dean, R. T., Sly, W., eds. 1984. *Lysosomes in biology and pathology.* Amsterdam/New York/Oxford: Elsevier, 479 pp.

Doxsey, S. J., Brodsky, F. M., Blank, G. S., Helenius, A. 1987. Inhibition of endocytosis by anti-clathrin antibodies. *Cell* 50: 453–63

Draye, J. P., Quintart, J., Courtoy, P. J., Baudhuin, P. 1987. Relations between plasma membrane and lysosomal membrane. I. Fate of covalently labelled plasma membrane protein. *Eur. J. Biochem.* 170: 395–403

Duncan, J. R., Kornfeld, S. 1988. Intracellular movement of two mannose 6-phosphate receptors: return to the Golgi apparatus. *J. Cell Biol.* 106: 617–28

Dunn, W. A., Hubbard, A. L., Aronson, N. N. Jr. 1980. Low temperature selectively inhibits fusion between pinocytic vesicles and lysosomes during heterphagy of 125I-

Asialofetuin by the perfused rat liver. *J. Biol. Chem.* 255: 5971–78

Dunn, W. A., Hubbard, A. L. 1984. Receptor-mediated endocytosis of epidermal growth factor by hepatocytes in the perfused rat liver: ligand and receptor dynamics. *J. Cell Biol.* 98: 2148–59

Dunn, W. A., Connolly, T. P., Hubbard, A. L. 1986. Receptor-mediated endocytosis of epidermal growth factor by rat hepatocytes: receptor pathway. *J. Cell Biol.* 102: 24–36

Farquhar, M. G. 1985. Progress in unraveling pathways of Golgi traffic. *Annu. Rev. Cell Biol.* 1: 447–88

Ferris, A. L., Brown, J. C., Park, R. D., Storrie, B. 1987. Chinese hamster ovary cell lysosomes rapidly exchange contents. *J. Cell Biol.* 105: 2703–12

Fodor, I., Egyed, A., Lelkes, G. 1986. A comparative study on the cellular processing of free and gold-conjugated transferrin. *Eur. J. Cell Biol.* 42: 74–78

Fok, A. K., Allen, R. D. 1988. The lysosome system. In *Paramecium*, ed. H.-D. Görtz, pp. 301–27. Berlin/Heidelberg: Springer-Verlag

Fuchs, R., Schmid, S., Mellman, I. 1989. A possible role for sodium potassium-ATPase in regulating ATP-dependent endosome acidification. *Proc. Natl. Acad. Sci. USA* 86: 539–43

Fuller, S. D., Simons, K. 1986. Transferrin receptor polarity and recycling accuracy in "tight" and "leaky" strains of Madin-Darby canine kidney cells. *J. Cell Biol.* 103: 1767–79

Gelfand, V. I. 1989. Cytoplasmic microtubular motors. *Curr. Opin. Cell Biol.* 1: 63–66

Geuze, H. J., Slot, J. W., Strous, J. A. M., Lodish, H. F., Schwartz, A. L. 1983. Intracellular site of asialoglycoprotein receptor-ligand uncoupling: double-label immunoelectron microscopy during receptor-mediated endocytosis. *Cell* 32: 277–87

Geuze, H. J., Stoorvogel, W., Strous, G. J., Slot, J. W., Bleekemolen, J. E., Mellman, I. 1988. Sorting of mannose 6-phosphate receptors and lysosomal membrane proteins in endocytic vesicles. *J. Cell Biol.* 107: 2491–501

Glickman, J. N., Conibear, E., Pearse, B. M. F. 1989. Specificity of binding of clathrin adaptors to signals on the mannose 6-phosphate/insulin-like growth factor II receptor. *EMBO J.* 8: 1041–47

Goda, Y., Pfeffer, S. R. 1988. Selective recycling of the mannose 6-phosphate/IGF-II receptor to the *trans* Golgi network in vitro. *Cell* 55: 309–20

Goldenthal, K. L., Hedman, K., Chen, J. W., August, J. T., Vihko, P., Pastan, I., Willingham, M. C. 1988. Pre-lysosomal divergence of alpha$_2$-macroglobulin and transferrin: A kinetic study using a monoclonal antibody against a lysosomal membrane glycoprotein (LAMP-1). *J. Histochem. Cytochem.* 36: 391–400

Goldstein, J. L., Brown, M. S., Anderson, R. G. W., Russel, D. W., Schneider, W. J. 1985. Receptor-mediated endocytosis: concepts emerging from the LDL receptor system. *Annu. Rev. Cell Biol.* 1: 1–39

Goud, B., Huet, C., Louvard, D. 1985. Assembled and unassembled pools of clathrin: quantitative study using an enzyme immunoassay. *J. Cell Biol.* 100: 521–27

Griffiths, G., Hoflak, B., Simons, K., Mellman, I., Kornfeld, S. 1988. The mannose 6-phosphate receptor and the biogenesis of lysosomes. *Cell* 52: 329–41

Gruenberg, J., Howell, K. E. 1986. Reconstitution of vesicle fusions occurring in endocytosis with a cell-free system. *EMBO J.* 5: 3091–101

Gruenberg, J., Howell, K. E. 1987. An internalized transmembrane protein resides in a fusion-competent endosome for less than 5 minutes. *Proc. Natl. Acad. Sci USA* 84: 5758–62

Gruenberg, J., Griffiths, G., Howell, K. E. 1989. Characterization of the early endosome and putative endocytic carrier vesicles in vivo and with an assay of vesicle function in vitro. *J. Cell Biol.* 108: 1301–16

Harrison, S. C., Kirchhausen, T. 1983. Clathrin, cages, and coated vesicles. *Cell* 33: 650–52

Hart D'arcy, P., Young, M. R., Gordon, A. H., Sullivan, K. H. 1987. Inhibition of phagosome-lysosome fusion in macrophages by certain mycobacteria can be explained by inhibition of lysosomal movements observed after phagocytosis. *J. Exp. Med.* 166: 933–46

Haylett, T., Thilo, L. 1986. Limited and selective transfer of plasma membrane glycoproteins to membrane of secondary lysosomes. *J. Cell Ciol.* 103: 1249–56

Helenius, A., Mellman, I., Wall, D., Hubbard, A. 1983. Endosomes. *Trends Biochem. Sci.* 8: 245–50

Hertel, C., Coulter, S. J., Perkins, J. P. 1986. The involvement of cellular ATP in receptor-mediated internalization of epidermal growth factor and hormone-induced internalization of β-adrenergic receptors. *J. Biol. Chem.* 261: 5974–80

Heuser, J. 1989a. Effects of cytoplasmic acidification on clathrin lattice morphology. *J. Cell Biol.* 108: 401–11

Heuser, J. 1989b. Changes in lysosome shape and distribution correlated with changes in cytoplasmic pH. *J. Cell Biol.* 108: 855–64

Heuser, J. E., Anderson, R. G. W. 1989. Hypertonic media inhibit receptor-mediated endocytosis by blocking clathrin-coated pit formation. *J. Cell Biol.* 108: 389–400

Hopkins, C. R. 1985. The appearance and internalization of transferrin receptors at the margins of spreading human tumor cells. *Cell* 40: 199–208

Horwitz, M. A., Maxfield, F. R. 1984. *Legionella pneumophila* inhibits acidification of its phagosome in human monocytes. *J. Cell Biol.* 99: 1936–43

Howell, K. E., Schmid, R., Ugelstad, J., Gruenberg, J. 1989. Immuno-isolation using magnetic-solid supports: subcellular fractionation for cell-free functional studies. *Meth. Cell Biol.* 31A: 264–92

Kelly, B. M., Cheng-Zhi, Y., Chang, P. L. 1989. Presence of a lysosomal enzyme, arylsulfatase-A, in the prelysosome-endosome compartments of human cultured fibroblasts. *Europ. J. Cell Biol.* 48: 71–78

Kielian, M. C., Marsh, M., Helenius, A. 1986. Kinetics of endosome acidification detected by mutant and wild-type Semliki Forest virus. *EMBO J.* 5: 3103–9

Kindberg, G. M., Ford, T., Blomhoff, R., Rickwood, D., Berg, T. 1984. Separation of endocytic vesicles in nycodenz gradients. *Anal. Biochem.* 142: 455–62

Kosaka, T., Ikeda, K. 1983. Reversible blockage of membrane retrieval and endocytosis in the garland cell of the temperature-sensitive mutant of *Drosophila melanogaster* shibire. *J. Cell Biol.* 97: 499–507

Krieger, M. 1986. Isolation of somatic cell mutants with defects in the endocytosis of low-density lipoprotein. *Meth. Enzymol.* 129: 227–53

Lanzavecchia, A. 1988. Clonal sketches of the immune response. *EMBO J.* 7: 2945–51

Larkin, J. M., Brown, M. S., Goldstein, J. L., Anderson, R. G. W. 1983. Depletion of intracellular potassium arrests coated pit function and receptor-mediated endocytosis in fibroblasts. *Cell* 33: 273–85

Larkin, J. M., Donzell, W. C., Anderson, R. G. W. 1985. Modulation of intracellular potassium and ATP: Effects on coated pit function in fibroblasts and hepatocytes. *J. Cell. Physiol.* 124: 372–78

Larkin, J. M., Donzell, W. C., Anderson, R. G. W. 1986. Potassium dependent assembly of coated pits: new coated pits form as planar clathrin lattices. *J. Cell Biol.* 103: 2619–27

Lazarovits, J., Roth, M. 1988. A single amino-acid change in the cytoplasmic domain allows the influenza-virus hemagglutinin to be endocytosed through coated pits. *Cell* 53: 743–52

Lee, C., Chen, L. B. 1988. Dynamic behavior of endoplasmic reticulum in living cells. *Cell* 54: 37–46

Lippincott-Schwartz, J., Fambrough, D. M. 1987. Cycling of the integral membrane glycoprotein, LEP100, between plasma membrane and lysosomes: kinetic and morphological analysis. *Cell* 49: 669–77

Lobel, P., Fujimoto, K., Ye, R. D., Griffiths, G., Kornfeld, S. 1989. Mutations in the cytoplasmic domain of the cation-independent mannose 6-phosphate receptor have differential effects on lysosomal enzyme sorting and endocytosis. *Cell* 57: 787–96

Louvard, D., Reggio, H., Warren, G. 1982. Antibodies to the Golgi complex and the rough endoplasmic reticulum. *J. Cell Biol.* 92: 92–107

Mahaffey, D. T., Moore, M. S., Brodsky, F. M., Anderson, R. G. W. 1989. Coat proteins isolated from clathrin coated vesicles can assemble into coated pits. *J. Cell Biol.* 108: 1615–24

Marsh, M., Helenius, A. 1980. Adsorptive endocytosis of Semliki Forest virus. *J. Mol. Biol.* 142: 439–54

Matteoni, R., Kreis, T. E. 1987. Translocation and clustering of endosomes and lysosomes depends on microtubules. *J. Cell Biol.* 105: 1253–65

Mayorga, L. S., Diaz, R., Stahl, P. D. 1988. Plasma membrane derived vesicles containing receptor-ligand complexes are fusogenic with early endosomes in a cell-free system. *J. Biol. Chem.* 263: 17213–16

Mayorga, L. S., Diaz, R., Stahl, P. D. 1989. Reconstitution of endosomal proteolysis in a cell-free system. Transfer of immune complexes internalized via Fc receptors to an endosomal proteolytic compartment. *J. Biol. Chem.* 264: 5392–99

Mellman, I., Fuchs, R., Helenius, A. 1986. Acidification of the endocytic and exocytic pathways. *Annu. Rev. Biochem.* 55: 663–700

Merion, M., Sly, W. S. 1983. The role of intermediate vesicles in the adsorptive endocytosis and transport of ligand to lysosomes by human fibroblasts. *J. Cell Biol.* 96: 644–50

Moore, M. S., Mahaffey, D. T., Brodsky, F. M., Anderson, R. G. W. 1987. Assembly of clathrin-coated pits onto purified plasma membrane. *Science* 236: 558–63

Moya, M., Dautry-Varsat, A., Goud, B., Louvard, D., Boquet, P. 1985. Inhibition of coated pit formation in Hep2 cells blocks the cytotoxicity of diphtheria toxin but not that of ricin toxin. *J. Cell Biol.* 101: 548–59

Mueller, S. C., Hubbard, A. 1986. Receptor-mediated endocytosis of asialoglycoproteins by rat hepatocytes: receptor-positive and receptor-negative endosomes. *J. Cell Biol.* 102: 932–42

Mullock, B. M., Branch, W. J., van Schaik, M., Gilbert, L. K., Luzio, J. P. 1989. Reconstitution of an endosome-lysosome interaction in a cell free system. *J. Cell Biol.* 108: 2093–2100

Murphy, R. F. 1988. Processing of endocytosed material. *Adv. in Cell Biol.* 2: 159–80

Nelson, N., Taiz, L. 1989. The Evolution of H^+-ATPases. *Trends Biochem. Sci.* 41: 113–16

Nielsen, J. T., Nielsen, S., Christensen, E. I. 1985. Transtubular transport of proteins in rabbit proximal tubules. *J. Ultrastruct. Res.* 92: 133–45

Oates, P. J., Touster, O. 1980. In vitro fusion of *Acanthamoeba* phagolysosomes. III. Evidence that cyclic nucleotides and vacuole subpopulations respectively control the rate and the extent of vacuole fusion in *Acanthamoeba* homogenates. *J. Cell Biol.* 85: 804–10

Oka, J. A., Weigel, P. H. 1988. Effects of hyperosmolarity on ligand processing and receptor recycling in the hepatic galactosyl receptor system. *J. Cell Biochem.* 36: 169–83

Oliver, C. 1982. Endocytic pathways at the lateral and basal cell surfaces of exocrine acinar cells. *J. Cell Biol.* 95: 154–61

Orci, L, Malhotra, V., Amherdt, M., Serafini, T., Rothman, J. E. 1989. Dissection of a single round of vesicular transport: Sequential intermediates for intercisternal movement in the Golgi stack. *Cell* 56: 357–68

Palade, G. 1975. Intracellular aspects of the process of protein secretion. *Science* 189: 347–85

Park, R. D., Sullivan, P. C., Storrie, B. 1988. Hypertonic sucrose inhibition of endocytic transport suggests multiple early endocytic compartments. *J. Cell. Physiol.* 135: 443–50

Paschal, B. M., Vallee, R. B. 1987. Retrograde transport by the microtubule-associated protein MAP 1C. *Nature* 330: 181–83

Pearse, B. M. F. 1987. Clathrin and coated vesicles. *EMBO J.* 6: 2507–12

Pearse, B. M. F., Crowther, R. A. 1987. Structure and assembly of coated vesicles. *Annu. Rev. Biophys. Chem.* 16: 49–68

Pertoft, H., Wärmegård, B., Höök, M. 1978. Heterogeneity of lysosomes originating from rat liver parenchymal cells. *Biochem. J.* 174: 309–17

Rambourg, A., Clermont, Y., Hermo, L. 1981. Three-dimensional structure of the Golgi apparatus. *Methods Cell Biol.* 23: 155–66

Robbins, A. R., Roff, C. F. 1987. Isolation of mutant Chinese hamster ovary cells defective in endocytosis. *Methods Enzymol.* 138: 458–70

Robinson, M. S. 1987. 100-kD coated vesicle proteins: molecular heterogeneity and intracellular distribution studied with monoclonal antibodies. *J. Cell Biol.* 104: 887–95

Roederer, M., Mays, R. W., Murphy, R. F. 1989. Effect of confluence on endocytosis by 3T3 fibroblasts: increased rate of pinocytosis and accumulation of residual bodies. *Europ. J. Cell Biol.* 48: 37–44

Rome, L. H., Garvin, J., Allietta, M. M., Neufeld, E. F. 1979. Two species of lysosomal organelles in cultured human fibroblasts. *Cell* 17: 143–53

Rothman, J. E., Schmid, S. L. 1986. Enzymatic recycling of clathrin from coated vesicles. *Cell* 46: 5–9

Sahagian, G. G., Neufeld, E. F. 1983. Biosynthesis and turnover of the mannose 6-phosphate receptor in cultured Chinese hamster ovary cells. *J. Biol. Chem.* 258: 7121–28

Samuelson, A. C., Stockert, R. J., Novikoff, A. B., Novikoff, P. M., Saez, J. C., et al. 1988. Influence of cytosolic pH on receptor-mediated endocytosis of asialo-orosomucoid. *Am. J. Physiol.* 254: 829–38

Sandvig, K., Olsnes, S., Petersen, O. W., van Deurs, B. 1987. Acidification of the cytosol inhibits endocytosis from coated pits. *J. Cell Biol.* 105: 679–89

Schmid, S. L., Fuchs, R., Male, P., Mellman, I. 1988. Two distinct subpopulations of endosomes involved in membrane recycling and transport to the lysosomes. *Cell* 52: 73–83

Schmid, S. L., Fuchs, R., Kielian, M., Helenius, A., Mellman, I. 1989. Acidification of endosome subpopulations in wild-type Chinese hamster ovary cells and temperature-sensitive acidification-defective mutants. *J. Cell Biol.* 108: 1291–300

Seglen, P. O. 1987. Regulation of autophagic protein degradation in isolated liver cells. In *Lysosomes: Their role in protein breakdown*, pp. 371–414, London: Academic

Sibley, L. D., Krahenbuhl, J. L., Adams, G. M. W., Weidner, E. 1986. Toxoplasma modifies macrophage phagosomes by secretion of a vesicular network rich in surface proteins. *J. Cell Biol.* 103: 867–74

Simons, K., Fuller, S. 1987. The budding of enveloped viruses: A paradigm for membrane sorting? In *Biological organization: Macromolecular interactions at high res-*

olution, ed. R. M. Burnett, H. J. Vogel, pp. 139–50, New York: Academic
Simons, K., Virta, H. 1987. Perforated MDCK cells support intracellular transport. *EMBO J.* 6: 2241–47
Smythe, E., Pypaert, M., Lucocq, J., Warren, G. 1989. Formation of coated vesicles from coated pits in broken A431 cells. *J. Cell Biol.* 108: 843–53
Steinman, R. M., Brodie, S. E., Cohn, Z. A. 1976. Membrane flow during pinocytosis. A stereologic analysis. *J. Cell Biol.* 68: 665–87
Steinman, R. M., Mellman, I. S., Muller, W. A., Cohn, Z. A. 1983. Endocytosis and the recycling of plasma membrane. *J. Cell Biol.* 96: 1–27
Stoorvogel, W., Geuze, H. J., Strous, G. J. 1987. Sorting of endocytosed transferrin and asialoglycoprotein occurs immediately after internalization in HepG2 cells. *J. Cell Biol.* 104: 1261–68
Storrie, B., Pool, R. R. Jr., Sachdeva, M., Maurey, K. M., Oliver, C. 1984. Evidence for both prelysosomal and lysosomal intermediates in endocytic pathways. *J. Cell Biol.* 98: 108–15
Storrie, B. 1988. Assembly of lysosomes: Perspectives from comparative molecular cell biology. *Int. Rev. Cytology* III: 53–105
Swanson, J. A., Silverstein, S. C. 1988. Pinocytic flow through macrophages. In *Processing and Presentation of Antigens*, ed. B. Pernis, S. Silverstein, H. Vogel, pp. 15–27, London: Academic
Terasaki, M., Song, J., Wong, J. R., Weiss, M. J., Chen, L. B. 1984. Localization of endoplasmic reticulum in living and glutaraldehyde-fixed cells with fluorescent dyes. *Cell* 38: 101–8
Tran, D., Carpentier, J.-L., Sawano, F., Gorden, P., Orci, L. 1987. Ligands internalized through coated or noncoated invaginations follow a common intracellular pathway. *Proc. Natl. Acad. Sci. USA* 84: 7957–61
Ukkonen, P., Lewis, V., Marsh, M., Helenius, A., Mellman, I. 1986. Transport of macrophage Fc receptors and Fc receptor-bound ligands to lysosomes. *J. Exp. Med.* 163: 952–71
Vale, R. D. 1987. Intracellular transport using microtubule-based motors. *Annu. Rev. Cell Biol.* 3: 347–78
Vigers, G. P. A., Crowther, R. A., Pearse, B. M. F. 1986. Location of the 100 kd–50 kd accessory proteins in clathrin coats. *EMBO J.* 5: 2079–85
von Figura, K., Gieselmann, V., Hasilik, A. 1984. Antibody to mannose 6-phosphate receptors induces receptor deficiency in human fibroblasts. *EMBO J.* 3: 1281–86
Waheed, A., Gottschalk, S., Hille, A., Krentler, C., Pohlman, R., et al. 1988. Human lysosomal acid phosphatase is transported as a transmembrane protein to lysosomes in transfected baby hamster kidney cells. *EMBO J.* 7: 2351–58
Wall, D. A., Hubbard, A. L. 1985. Receptor-mediated endocytosis of asialoglycoproteins by rat liver hepatocytes: Biochemical characterization of the endosomal compartments. *J. Cell Biol.* 101: 2104–12
Warren, G. 1985. Membrane traffic and organelle division. *Trends Biochem. Sci.* 502: 439–43
Wileman, T., Harding, C., Stahl, P. 1985. Receptor-mediated endocytosis. *Biochem. J.* 232: 1–14
Wiley, D. C., Skehel, J. J. 1987. The structure and function of the hemagglutinin membrane glycoprotein of influenza virus. *Annu. Rev. Biochem.* 56: 365–94
Wilson, J. M., Whitney, J. A., Neutra, M. R. 1987. Identification of an endosomal antigen specific to absorptive cells of suckling rat ileum. *J. Cell Biol.* 105: 691–703
Wilson, W. W., Wilcox, C. A., Flynn, G. C., Chen, E., Kuang, W.-J., et al. 1989. A fusion protein required for vesicle-mediated transport in both mammalian cells and yeast. *Nature* 339: 355–59
Wolkoff, A. D., Klausner, R. D., Ashwell, G., Harford, J. 1984. Intracellular segregation of asialoglycoproteins and their receptors: a prelysosomal event subsequent to dissociation of the ligand-receptor complex. *J. Cell Biol.* 98: 375–81
Woodman, P. G., Warren, G. 1988. Fusion between vesicles from the pathway of receptor-mediated endocytosis in a cell-free system. *Eur. J. Biochem.* 173: 101–8
Woods, J. W., Goodhouse, J., Farquhar, M. G. 1988. Endocytosed transferrin (Tf) and mannose-6-phosphate receptor (M6PR) antibodies label two different endosome populations. *J. Cell Biol.* 107: 767a

NOTE ADDED IN PROOF

Gerace, L., Burke, B. 1988. Functional organization of the nuclear envelope. *Annu. Rev. Cell Biol.* 4: 335–74

Annu. Rev. Cell Biol. 1989. 5 : 483–525

THE BIOGENESIS OF LYSOSOMES

Stuart Kornfeld

Divisions of Hematology & Oncology, Washington University School of Medicine, P.O. Box 8125, St. Louis, Missouri 63130

Ira Mellman

Department of Cell Biology, Yale University School of Medicine, P.O. Box 3333, New Haven, Connecticut 06510

CONTENTS

0743–4634/89/1115–0483$02.00

INTRODUCTION

Lysosomes are acidic, hydrolase-rich vacuoles capable of degrading most biological macromolecules. In spite of this digestive capacity, the acid hydrolases and associated proteins found in the lysosome's interior, as well as the proteins that comprise the lysosome membrane, are relatively long-lived. Nevertheless endogenous lysosomal constituents do turn over and are continuously replaced with newly synthesized components. In addition, dividing cells must be able to form new lysosomes. During the past several years much has been learned about many aspects of lysosome biogenesis. The selective phosphorylation of mannose residues on lysosomal enzymes, in conjunction with specific receptors for the mannose 6-phosphate recognition marker, is found to be largely responsible for the targeting or "sorting" of newly synthesized lysosomal enzymes. Several lysosomal membrane glycoproteins have been isolated and their structures determined by cDNA cloning. The routing of these various molecules to lysosomes has been studied, and the structural determinants on the proteins that direct their intracellular trafficking are beginning to be defined. In addition, the organelles involved in mediating the transport of newly synthesized lysosomal enzymes to lysosomes have been generally identified. This information has been derived from a combination of biochemistry, immunocytochemistry, molecular biology, and classical human genetics. The goal of this review is to summarize our current understanding of the biogenesis of lysosomes. The reader is referred to reviews by Robbins (1987) and von Figura and Hasilik (1986) and the recent monograph by Holtzman (1989) for more complete summaries of earlier work.

A FUNCTIONAL DEFINITION OF LYSOSOMES

The early efforts of de Duve, Straus, Novikoff, and others established that lysosomes are most easily characterized as dense vacuoles containing a variety of acid-dependent hydrolases that are responsible for degrading internalized and endogenous macromolecules (de Duve 1963; Straus 1963; Novikoff 1963). This simple definition has some limitations, however, when describing pathways of lysosome biogenesis. First, lysosomes receive input from both the endocytic and biosynthetic pathways and share a number of characteristics with the vesicles involved in their formation. In addition, lysosomes can be morphologically heterogeneous due to variations in their content of internalized and/or partially degraded material. This situation has resulted in a rich and complex terminology for lysosomes, including terms such as residual bodies, multivesicular bodies, autophagosomes, GERL, and primary and secondary lysosomes. Many of these structures are not readily distinguished from elements of the endocytic or biosynthetic pathways, which makes it difficult to identify lysosomes solely on the basis of morphology. In addition, Novikoff introduced the term GERL to define acid phosphatase-containing organelles (Golgi, endoplasmic reticulum, and lysosomes) believed to be specifically interconnected on the pathway of lysosome biogenesis.

Since it is necessary to have a clear definition of a lysosome before discussing its biogenesis, we feel that it is useful to review the thoughts of de Duve on the subject (de Duve 1963). In his early studies, de Duve recognized the difficulty in defining lysosomes. Accordingly, he stressed what he considered to be their most important aspect, namely that lysosomes represent "the association within a special group of cytoplasmic particles of a number of soluble acid hydrolases of widely differing specificity, in such a manner as to restrict to a considerable extent the accessibility of these enzymes to surrounding substrates, both the association and the latency of the enzymes being dependent on the structural integrity of the particles." de Duve noted that lysosomes were part of a complex intracellular digestive tract, and, as such, were functionally related to phagosomes [i.e. endocytic vesicles, first described by Straus (1963)], which contain newly internalized material, but at least in their earliest stages, were devoid of most acid hydrolases. He also recognized that organelles other than lysosomes may contain one or another of the acid hydrolases, such as acid phosphatase, but he felt that these organelles should not be considered lysosomes because they lack the full complement of lysosomal enzymes. de Duve proposed that lysosomes have but one function, that of acid digestion, implying that lysosomes are also the final repository for macromolecules intended for degradation, be they derived from the extracellular space (via endocytosis) or from within the cell (via autophagy). This feature of lysosomes as the terminal degradative compartment serves as a useful functional definition to distinguish these organelles from the other components of the biosynthetic and the endocytic pathway, and it is this definition that we will use in our review.

Lysosomes as a Terminal Degradative Compartment

There is ample evidence that lysosomes represent the final destination for a significant fraction of all intracellular traffic. For example, extracellular macromolecules that are accumulated or degraded by cells are typically internalized via clathrin-coated pits and coated vesicles, and then transferred to endosomes, a collection of uncoated vesicles and tubules that are acidic, of light buoyant density ($\rho = < 1.03 \text{ g ml}^{-1}$), and contain a variable (but usually low) concentration of hydrolytic enzymes (Helenius et al

1983). Internalized material actually passes sequentially through at least two distinct endosome subpopulations (corresponding to early and late) (Schmid et al 1988) before reaching the hydrolase-rich, heavier ($\rho = 1.10$ g ml^{-1}) lysosomes. Once in lysosomes, endocytic tracers are either degraded or reside indefinitely; they are generally not released intact into the extracellular medium, nor are they transferred to other organelles (Cohn et al 1966; Cohn & Benson 1965; Swanson et al 1985; Silverstein et al 1977; Steinman et al 1976; Steinman et al 1983).

Lysosomes also serve as the terminal compartment for the intracellular transport of newly synthesized lysosomal enzymes. As during endocytosis, the pathway of enzyme biosynthesis involves transport through multiple compartments prior to reaching lysosomes: transit through the Golgi, exit via clathrin-coated vesicles (primary lysosomes) (Bainton & Farquhar 1970; Schulze-Lohoff et al 1985; Lemansky et al 1987; Marquand et al 1987), and delivery to post-Golgi, acidic vacuoles that probably correspond to endosomes (Brown et al 1986) (see below). Upon reaching the heavy density lysosomes, many acid hydrolases are proteolytically processed (Brown & Swank 1983; Gieselmann et al 1985; Gabel & Foster 1987; Hasilik & von Figura 1984). Experimental manipulations, such as treatment of cells with acidophilic weak bases (e.g. chloroquine, NH_4Cl), lead to secretion of newly synthesized enzyme precursors leaving the Golgi, but have little or no effect on the retention of the previously synthesized, mature forms of the same enzymes residing in lysosomes (Gonzalez-Noriega et al 1980; Rosenfeld et al 1982). Thus neither lysosomal enzymes nor undigested lysosomal contents are released from most cells, which demonstrates the largely unidirectional nature of the transport of soluble macromolecules to this compartment. It should be pointed out that under certain conditions some cells do release lysosomal contents extracellularly (e.g. release from hepatocytes into the bile) (de Duve 1963).

A functional definition of lysosomes as the final intracellular destination for soluble proteins internalized by endocytosis (and newly synthesized lysosomal enzymes) provides a simple and consistent criterion with which to establish the identity of lysosomes. It avoids confusion due to morphological heterogeneity typical of lysosomes in different cell types, or between lysosomes and endosomes in a single cell. It is also helpful in understanding the precise pathways and mechanisms involved in the targeting and sorting of soluble lysosomal enzymes and lysosomal membrane proteins in both the biosynthetic and endocytic pathways. Thus it takes into account the expected transient presence of classical lysosomal markers in organelles and vesicles involved in lysosome formation. This is a particular problem in cell types such as the osteoclast, in which lysosomal enzymes and membrane proteins are so abundant as to be easily demonstrable in virtually all endocytic and biosynthetic organelles (Baron et al 1988).

TARGETING OF SOLUBLE LYSOSOMAL ENZYMES

In many mammalian cells, the transport of newly synthesized or externally added lysosomal enzymes to lysosomes is dependent on their specific recognition by receptors for mannose 6-phosphate (Man-6-P) (von Figura & Hasilik 1986). This was first demonstrated by biochemical studies that documented the presence of a Man-6-P recognition marker on soluble lysosomal enzymes (Kaplan et al 1977). The physiologic importance of this pathway was confirmed by the finding that fibroblasts from patients with mucolipidosis type II (ML-II; I-cell disease) fail to phosphorylate mannose residues on their newly synthesized lysosomal enzymes, which results in the secretion of a large percentage of their acid hydrolases into the culture medium. There are three key elements to this recognition system: first, the selective modification of lysosomal hydrolases by enzymes that synthesize the phosphomannosyl recognition marker; second, the recognition of this modification by receptors for Man-6-P; and third, the recognition of Man-6-P receptors by cellular components that mediate the selective transport of the bound enzyme to lysosomes. Each of these steps will be discussed in turn.

Biosynthesis and Phosphorylation

The initial steps in the biosynthesis of soluble lysosomal enzymes are shared with secretory proteins: insertion into the lumen of the rough endoplasmic reticulum (RER), signal sequence cleavage, core glycosylation of selected asparagine residues on the nascent protein with a preformed oligosaccharide (three glucose, nine mannose and two N-acetylglucosamine residues), and removal of the glucose residues and one mannose from the oligosaccharide. The first reaction unique to the synthesis of lysosomal enzymes, the acquisition of the Man-6-P recognition marker, appears to occur shortly after export from the RER. Phosphomannosyl residues are generated through the concerted action of two distinct enzymes. First, UDP-N-acetylglucosamine : lysosomal enzyme N-acetylglucosamine-1-phosphotransferase (EC 2.7.8.17) (phosphotransferase) transfers N-acetylglucosamine 1-phosphate from UDP-GlcNAc to one or more mannose residues on lysosomal enzymes to give rise to a phosphodiester intermediate (Reitman & Kornfeld 1981a,b; Waheed et al 1982). Then, N-acetylglucosamine-1-phosphodiester α-N-acetylglucosaminidase (EC 3.1.4.45) removes the N-acetylglucosamine residue to generate the active phosphomonoester (Varki & Kornfeld 1981; Waheed et al 1981).

Studies of the kinetics of β-glucuronidase phosphorylation in a murine macrophage cell line revealed that phosphorylation begins 15–20 min after the protein is synthesized and continues for the next 40–80 min (Goldberg & Kornfeld 1983). The earliest detectable phosphorylated molecules, each containing a single phosphodiester, were devoid of glucose residues, consistent with phosphorylation occurring after the action of the RER glucose trimming enzymes, glucosidase I and II. The first mannose to be phosphorylated is generally on the α1,6 branch linked to the core β-linked mannose (Figure 1, residue i), while phosphorylated oligosaccharides formed at later times contain one or two Man-6-P residues located at five different positions of the oligosaccharide, including both the α1,3 and the α1,6 branches (Figure 1) (Goldberg & Kornfeld 1981; Natowicz et al 1982;

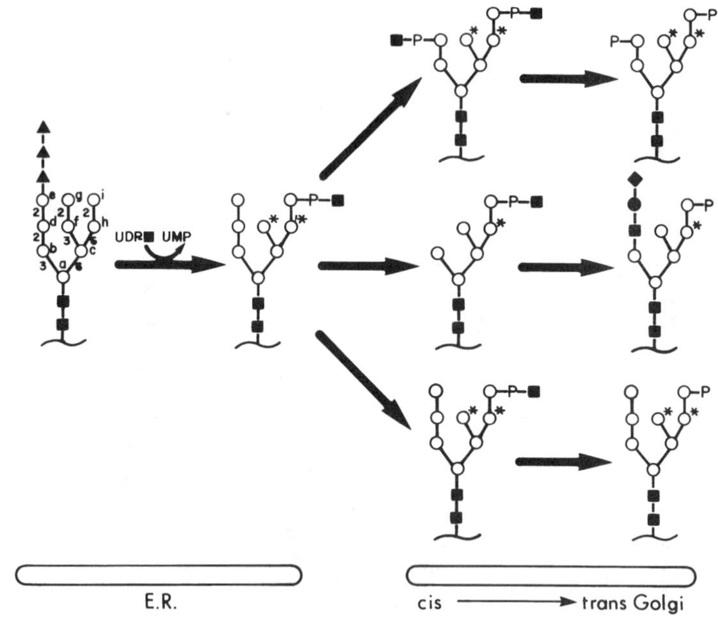

Figure 1 Schematic pathway of oligosaccharide processing on newly synthesized lysosomal enzymes. See text for discussion of the precise intracellular sites thought to be involved in the events leading to mannose 6-phosphate addition. The symbols represent: ■, N-acetylglucosamine; ○, mannose; ▲, glucose; ●, galacose; ◆, sialic acid. All mannose residues are linked α with the exception of residue a, which is linked β. The residues marked with the * are also phosphorylated in one or more species when penultimate to an α1,2-linked mannose. ER = endoplasmic reticulum.

Varki & Kornfeld 1983). Almost all diphosphorylated oligosaccharides lack the terminal mannose residue found on the α1,3 branch of the precursor oligosaccharide (Figure 1, residue e).

When the movement of lysosomal enzymes from the ER to the Golgi was blocked by low temperature or *m*-chlorocarboxylcyanide phenylhydrazone (CCCP), only the species with one phosphodiester was synthesized (Lazzarino & Gabel 1988). Reversal of these transport blocks allowed the formation of the diphosphorylated species and the conversion of the phosphodiesters to phosphomonoesters. Taken together, these data indicate that the initial phosphorylation of lysosomal enzymes occurs at a pre-Golgi site, while the conversion of monophosphorylated species to diphosphorylated forms and the hydrolysis of the diesters occurs within the Golgi. The explanation for this compartmentalization in the formation of the diphosphorylated species appears to be the requirement for the selective removal of a single mannose residue from the α1,3 branch of the oligosaccharide (Figure 1, residue e) in order for phosphotransferase to act on that branch. This particular mannose residue is not removed by the ER α-mannosidase, but is cleaved by Golgi α-mannosidase-I (Bischoff et al 1986). Consequently the formation of the diphosphorylated oligosaccharides can only occur after the lysosomal enzymes have arrived in the Golgi. Thus, when the Golgi α-mannosidase-I is selectively inhibited with deoxymannojirimycin, the treated cells produce lysosomal enzymes containing monophosphorylated oligosaccharides with the single Man-6-P restricted to the α1,6 branch (Lazzarino & Gabel 1989). While the activity of the phosphotransferase is restricted by the presence of the terminal α1,2-linked mannose residue on the α1,3 branch, it is also clear that this enzyme has a preference for transferring GlcNAc-P to Man α1,2 Man sequences (Couso et al 1986). An examination of the structure of the high mannose oligosaccharide precursor reveals, however, that these two features are consistent with each other, since removal of the blocking mannose residue (Figure 1, residue e) leaves a terminal mannose (Figure 1, residue d) linked α1,2 to its penultimate mannose (Figure 1, residue b).

Subcellular Localization of Mannose Phosphorylation

The precise site where the pre-Golgi phosphorylation occurs has not been established, but it seems unlikely to be the RER. First, membranes containing phosphotransferase have been separated from membranes with the RER markers in several cell types (Pohlmann et al 1982; Goldberg & Kornfeld 1983; Deutscher et al 1983; Minnifield et al 1986). Secondly, as discussed above, both the kinetics of phosphorylation and the failure to detect phosphorylated oligosaccharides with residual glucose residues favor a site beyond the RER. Candidates for the initial phosphorylation compartment are the transitional elements of the ER (Palade 1975; Rizzolo et al 1985; Rizzolo & Kornfeld 1988) and the smooth membrane compartment located between the transitional elements of the ER and the Golgi complex (Saraste & Kuismanen 1984; Tooze et al 1988). The latter compartment appears to represent an intermediate organelle in the transport of proteins from the ER to the Golgi and is known to accumulate viral glycoproteins at low temperatures.

Additional evidence for the compartmentalization of lysosomal enzyme phosphorylation has come from the study of Pelham (1988), who attached the ER retention signal (KDEL) to the lysosomal enzyme cathepsin D with the result that this modified lysosomal enzyme accumulated within the ER. The oligosaccharides of the retained cathepsin D were shown to be phosphorylated with almost all of the residues being phosphodiesters. The phosphorylation was inhibited at low temperatures, with the strongest inhibition occurring below 20°C. This effect of low temperature on phosphorylation is consistent with phosphorylation occurring in a post-RER compartment, while the lack of conversion of the phosphodiesters to phosphomonoesters indicates that the modified cathepsin D may not have reached the Golgi. Taken together, these data are consistent with the view that the initial phosphorylation event occurs in a post-RER, pre-Golgi compartment.

UDP-N-Acetylglucosamine: Lysosomal Enzyme N-Acetylglucosamine-1-Phosphotransferase

Phosphotransferase mediates a critical step in the Man-6-P-dependent sorting system since it selectively recognizes and phosphorylates lysosomal enzymes (Reitman & Kornfeld 1981b; Waheed et al 1982). The basis for this specificity is the enzyme's ability to recognize a protein domain that is common to all lysosomal enzymes, but is absent in non-lysosomal glycoproteins (Lang et al 1984; Little et al 1986, 1987). This recognition domain is not a simple linear sequence of amino acids since the numerous lysosomal enzymes that have been cloned do not share any significant sequence identity. Furthermore, since heat-denatured lysosomal enzymes or proteolytic fragments of lysosomal enzymes do not serve as substrates of the phosphorylating enzyme, it appears that the conformation of the protein is important for the expression of the recognition marker.

Some insight into the nature of the recognition domain has come from studies of chimeric proteins derived from human pepsinogen and cathepsin D (T. Baranski et al, in preparation). Although these two aspartyl proteases share 50% identity in amino acid sequence, pepsinogen is a secretory protein while cathepsin D is a lysosomal enzyme. When pepsinogen, engineered to contain sites for Asn-linked glycosylation at the same positions as cathepsin D, was expressed in *Xenopus* oocytes, it was glycosylated and secreted, but not phosphorylated. Cathepsin D, on the other hand, was efficiently phosphorylated when expressed in this system (Faust et al 1987). Phosphorylation also occurred using chimeric proteins containing only two regions of cathepsin D (residues 188–230 and 265–319) spliced into the pepsinogen backbone. Chimeras containing either of these regions alone were not phosphorylated. Further analysis of these sequences by site-directed mutagenesis indicated that their lysine residues were critical in permitting phosphorylation of the chimera. When the positions of these residues are localized on the three-dimensional model of pepsin (James & Siedecki 1986), it is apparent that they are in close apposition to each other on the surface of the molecule. The implication is that the recognition marker is formed by a patch of lysine residues derived from different regions of cathepsin D. This conformational nature of the recognition domain would explain the failure to detect amino acid sequence identity between lysosomal enzymes.

In addition to recognizing a protein conformation presumably shared by all lysosomal enzymes, phosphotransferase also exhibits specificity for the α1,2-linked mannose units on Asn-linked oligosaccharides, as previously noted. The oligosaccharide recognition, however, appears to be of relatively low affinity. Even the best D-mannosyl disaccharide acceptors [e.g. Man α1,2 Man (1αOMe)] exhibit a poor affinity for the phosphotransferase (20 mM) relative to the affinity exhibited by many lysosomal enzymes (10–20 μM) (Madiyalakan et al 1986, 1987). The importance of α1,2-linked mannose residues for lysosomal enzyme phosphorylation was demonstrated by showing that treatment of an intact lysosomal enzyme with an α1,2-specific mannosidase greatly decreased the ability of the molecule to be phosphorylated by phosphotransferase even though it still contained oligosaccharides with the composition $Man_5GLcNAc_2$ (Couso et al 1986). While these experiments establish the importance of oligosaccharide structure for efficient phosphorylation, they also confirm that the high affinity binding of lysosomal enzymes to phosphotransferase is determined largely by protein-protein interactions rather than by carbohydrate-protein interactions.

Phosphotransferase Function in Mucolipidosis Type II and Type III

Interesting insights into the functioning of phosphotransferase have come from studies of patients with I-cell disease (ML-II) and pseudo-Hurler polydystrophy (ML-III). Fibroblasts from patients with ML-II have extremely low or undetectable phosphotransferase levels, whereas fibroblasts from patients with ML-III have some residual phosphorylating activity consistent with their milder clinical course (Varki et al 1981; Hasilik et al 1981). Okada et al (1987) found that 5 out of 9 ML-II fibroblast lines had restoration of phosphotransferase activity to almost normal levels when the cells were grown in the presence of 88-mM sucrose. β-hexosaminidase activity in sucrose-treated cells was also restored to normal. The basis for this effect is unknown, but one possible explanation is that the sucrose somehow stabilizes the defective phosphotransferase in the responding cells.

Several complementation groups have been defined among various ML-II and ML-III fibroblast lines based on cell fusion experiments (Honey et al 1982; Mueller et al 1983; Ben-Yoseph et al 1986). The ML-III patients have been placed into two complementation groups, with the possible existence of a third. The fibroblasts of the group IIIA patients contain a heat-labile form of phosphotransferase that is inactive at 37°C, the usual assay temperature, but functional at 23°C (Little et al 1986). Interestingly, the enzyme is also stabilized in vitro by the presence of 88-mM sucrose, similar to the findings with the subset of ML-II patients' fibroblasts. These data are consistent with a mutation in the IIIA complementation group, which leads to an unstable form of phosphotransferase.

The phosphotransferase of the group IIIC patients has normal or near normal activity toward free α-methylmannoside, a substrate that is phosphorylated independent of protein recognition. In contrast, the enzyme from these mutant cells exhibits poor activity towards lysosomal enzymes, coincidental with a striking decrease in its affinity for these substrates (Varki et al 1981; Lang et al 1985; Ben-Yoseph et al 1986; Little et al 1986). The mutant enzyme is, therefore, defective in its protein recognition function. Taken together these data indicate that phosphotransferase is a protein that contains a recognition site (or subunit) and a catalytic site (or subunit) that interact to specifically recognize and phosphorylate lysosomal enzymes.

While all patients with the clinical diagnosis of ML-II or ML-III analyzed to date have been found to be deficient in phosphotransferase, a normal Lebanese individual with elevated plasma lysosomal enzymes has been shown to have a partial deficiency of the phosphodiester α-N-acetylglucosaminidase (that removes the N-acetylglucosamine residue to expose Man-6-P) (Alexander et al 1986). Fibroblasts from this individual secreted lysosomal enzymes at a greater rate than fibroblasts from two ML-II heterozygotes. This raises the possibility that a complete deficiency of this enzyme could result in the development of a clinical picture similar to ML-II or ML-III.

Two Receptors for Mannose 6-Phosphate

The next step in the targeting of newly synthesized lysosomal enzymes is their binding to mannose 6-phosphate receptors (MPRs) in the Golgi. Two distinct MPRs have been isolated and characterized. The first is an integral membrane glycoprotein with an apparent molecular weight of 215,000 or greater (Sahagian et al 1981). The other MPR is also an integral membrane glycoprotein, but it has a subunit molecular weight of approximately 46,000 (Hoflack & Kornfeld 1985). Both receptors have similar, but not identical, binding specificities toward various types of phosphorylated oligosaccharides (Hoflack & Kornfeld 1985; Hoflack et al 1987; Tong et al 1989; Tong & Kornfeld 1989). The large receptor binds ligands independent of divalent cations, whereas the small receptor exhibits enhanced ligand binding affinity in the presence of divalent cations (Hoflack & Kornfeld 1985). On this basis, the large receptor has been referred to as the cation-independent (CI) MPR and the small receptor as the cation-dependent (CD) MPR. However, the CD-MPR can bind to high affinity ligands even in the absence of divalent cations (Junghans et al 1988; Baba et al 1988). Recently cDNAs for the CI-MPR have been cloned from bovine (Lobel et al 1987, 1988), human (Morgan et al 1987; Oshima et al 1988), and rat (MacDonald et al 1988) while cDNAs for the CD-MPR have been cloned from bovine (Dahms et al 1987) and human (Pohlmann et al 1987) sources. The deduced amino acid sequences of these clones have provided valuable insights into the structures of the receptors and have revealed that the two receptors are related proteins.

The bovine CI-MPR precursor is composed of four structural domains: a 44-residue amino-terminal signal sequence, a 2269-residue extra-cytoplasmic domain, a single 23-residue transmembrane region, and a 163-residue carboxyl-terminal cytoplasmic domain (Figure 2). The extra-cytoplasmic domain contains 19 potential Asn-linked glycosylation sites, a number of which are utilized yielding a mature receptor of 275–300 kd. The CI-MPR ectodomain also contains 15 contiguous repeating segments of approximately 147 amino acids each. Each repeat shares sequence identities with all the other repeats, with the percent of identical residues ranging from 16 to 38%. In addition, there are many amino acid substitutions that are conservative in nature. The cysteine residues in the repeating segments are also regularly spaced.

While the sequence of the CI-MPR does not bear close relationship to other proteins, the 13th repeat from the amino terminus contains a 43-residue insertion that is similar to sequences found in fibronectin, factor XII, and a bovine seminal fluid protein (Hynes 1985). This segment forms part of a collagen-binding domain in fibronectin. The cytoplasmic domain

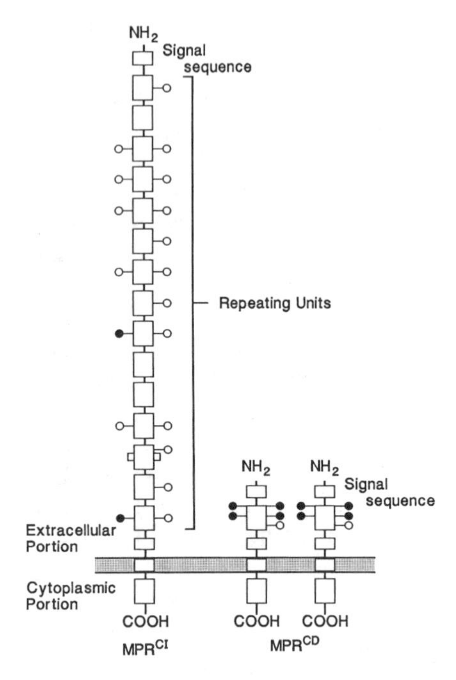

Figure 2 Schematic representation of the MPRs. The closed circles represent N-linked glycosylation sites known to be used and the open circles represent potential glycosylation sites. Modified from Dahms et al 1987.

contains sequences that are known to be potential substrates for various protein kinases, including protein kinase C, cAMP-dependent protein kinase and casein kinases I and II (MacDonald et al 1988). Moreover, the receptor has been shown to be phosphorylated at a number of these sites (Sahagian & Neufeld 1983; Corvera & Czech 1985).

The bovine CD-MPR contains a 28-residue amino-terminal signal sequence, a 159-residue extracytoplasmic domain, a single 25-residue transmembrane region, and a 67-residue carboxyl-terminal cytoplasmic domain (Figure 2). This receptor has five potential Asn-linked glycosylation sites, of which four are used (Dahms et al 1987). When the sequences of the two receptors are compared, it is evident that the entire extracytoplasmic domain of the CD-MPR is similar to each of the repeating units of the CI-MPR. The actual sequence identities range from 14 to 28%. This means that the extracytoplasmic domain of the CD-MPR is almost as similar to the various repeating units of the CI-MPR as the repeating units are to each other. The implication is that the two receptors are derived from a common ancestor with the CI-MPR arising from the duplication of a single ancestral gene. In contrast to these homologies, the signal sequences, transmembrane regions, and cytoplasmic domains of the two receptors have no apparent sequence similarities.

MPR Oligomeric Structure and Ligand Binding Activities

Chemical cross linking experiments indicate that the CD-MPR is a dimer in the membrane, and either a dimer or a tetramer in solution (Stein et al 1987a; Dahms & Kornfeld 1989). Equilibrium dialysis experiments have revealed that this receptor binds one mole of the monovalent ligand Man-6-P and 0.5 mole of a di-phosphorylated oligosaccharide per monomeric subunit (Tong & Kornfeld 1989). Therefore, each functional dimer has two Man-6-P binding sites, both of which can be occupied by a single oligosaccharide containing two Man-6-P residues. Evidence that each polypeptide monomer can fold into an independent ligand binding unit has been obtained by demonstrating that a truncated, anchor-minus form of the bovine CD-MPR, which behaves as a soluble monomer in solution, is capable of binding Man-6-P (Dahms & Kornfeld 1989). Expression of the truncated form of the human CD-MPR gave rise to a soluble dimer (Wendland et al 1989).

The quaternary structure of the CI-MPR has not been analyzed as extensively, but the available evidence suggests that the receptor could be a monomer (Perdue et al 1983) or an oligomer (Stein et al 1987b). The CI-MPR binds two moles of Man-6-P or one mole of a divalent phosphorylated oligosaccharide per monomer (Tong & Kornfeld 1989). This suggests that only two of the 15 repeating segments of this receptor function in the binding of Man-6-P. Both receptors bind Man-6-P with essentially the same affinity ($7–8 \times 10^{-6}$ M), but the CI-MPR binds a di-phosphorylated oligosaccharide with a much higher affinity than does the CD-MPR (2×10^{-9} M vs 2×10^{-7} M, respectively). Since an oligosaccharide with two phosphomonoesters binds to the MPRs with an affinity similar to that of a lysosomal enzyme, the high affinity binding of lysosomal enzymes can be explained by a two-site model in which two phosphomannosyl residues on the lysosomal enzyme interact with the receptors (Tong et al 1989). In the case of the CD-MPR, the two Man-6-P binding sites would reside on distinct polypeptide subunits rather than on a single polypeptide as occurs with the CI-MPR.

While phosphomannosyl residues are the major determinant of receptor binding, other features of the oligosaccharide influence the interaction. This is clearly illustrated by the finding that high-mannose oligosaccharides with one phosphomonoester interact with the receptors with higher affinity than do Man-6-P or pentamannose phosphate (Distler et al 1979; Creek & Sly 1982; Fischer et al 1982; Natowicz et al 1983; Tong et al 1989; Tong

& Kornfeld 1989). The major difference between these ligands is that in the high-mannose oligosaccharide, the Man-6-P residue is most commonly linked α1,2 to the underlying mannose (Natowicz et al 1982; Varki & Kornfeld 1983), whereas in pentamannose phosphate, the Man-6-P is linked α1,3 to the penultimate mannose (Bretthauer et al 1973), and in Man-6-P there is no underlying residue. These differences in binding affinities suggest that the MPRs actually recognize and bind to an extended oligosaccharide structure that includes the Man α1,2 Man sequence. In this regard, the receptors would be similar to phosphotransferase, which also recognizes the Man α1,2 Man sequence.

The CI-MPR and the IGF-II Receptor Are the Same Protein

In 1987 Morgan et al reported that the sequence of the human insulin-like growth factor II (IGF-II) receptor corresponds to that of the bovine CI-MPR, thus establishing that this receptor is a multifunctional protein (Morgan et al 1987). Subsequent biochemical studies showed that the binding sites for Man-6-P and IGF-II, a nonglycosylated polypeptide hormone, are distinct and that the receptor can bind both ligands simultaneously (Roth et al 1987; Tong et al 1988; Waheed et al 1988; MacDonald et al 1988; Kiess et al 1988). However IGF-II can inhibit lysosomal enzyme binding (in contrast to its lack of effect on Man-6-P binding), and lysosomal enzymes can impair IGF-II binding (MacDonald et al 1988; Kiess et al 1988, 1989). The latter finding may explain the observations that Man-6-P actually stimulates IGF-II binding by causing the release of prebound lysosomal enzymes and thus facilitating IGF-II binding (Roth et al 1987; Polychronakos et al 1988). Studies of the stoichiometry of IGF-II binding have revealed that the bovine receptor binds one mole of IGF-II per polypeptide chain with a high affinity (0.2 nM) (Tong et al 1988). The human and rat CI-MPRs also bind IGF-II with high affinity, but the chicken CI-MPR lacks the high affinity IGF-II binding site (Canfield & Kornfeld 1989). The CD-MPR does not bind this ligand (Tong et al 1988; Kiess et al 1988).

The biological significance of the finding that this receptor binds both IGF-II and lysosomal enzymes is uncertain. Three functions have been suggested. One is that the CI-MPR may participate in the clearance of IGF-II from the circulation, since the receptor can mediate the internalization and lysosomal degradation of IGF-II (Oka et al 1985; Kiess et al 1987). This could serve as a mechanism for regulating the turnover of IGF-II present in the plasma. A second possible consequence of IGF-II binding to the receptor is the modulation of lysosomal enzyme trafficking. In cells that synthesize IGF-II, the hormone might inhibit lysosomal enzyme binding to the receptor in the Golgi, thereby enhancing the secretion of acid hydrolases. In a similar fashion, extracellular IGF-II could inhibit lysosomal enzyme binding to cell surface receptors, and therefore prevent the recapture of secreted lysosomal enzymes (Kiess et al 1989). These complementing effects would serve to increase the levels of extracellular lysosomal enzymes, which could be beneficial for processes such as tissue and bone remodeling. IGF-II may also modulate lysosomal enzyme trafficking by altering the subcellular distribution of the CI-MPR. In a number of cell types, IGF-II and other hormones can induce the movement of receptor from intracellular compartments to the plasma membrane (Cushman & Wardzala 1980; Oppenheimer et al 1983; Wardzala et al 1984; Oka et al 1985; Lonnroth et al 1987; Appel et al 1988; Corvera et al 1988a,b; Braulke et al 1989). This redistribution of receptor could potentially impair the efficiency of lysosomal enzyme sorting in the Golgi and/or enhance the binding and internalization of extracellular ligands for the CI-MPR.

Finally, a third possible function of IGF-II binding to the CI-MPR is in signal transduction. This function has been difficult to study because IGF-II also binds to the IGF-I receptor, a member of the tyrosine kinase family of receptors that transmit signals across the plasma membrane (Rechler & Nissley 1985; Froesch et al 1985). Nevertheless, there is a growing number of reports that indicate that IGF-II can mediate a response through its own receptor (Roth 1988). These responses include the stimulation of glycogen synthesis in rat hepatoma cells (Hari et al 1987), the promotion of cell proliferation in K562 cells (Tally et al 1987), the stimulation of Ca^{2+} influx and DNA synthesis in BaLB/C 3T3 cells (Nishimoto et al 1987), and the stimulation of Na^{+}/H^{+} exchange in canine kidney proximal tubular cells (Mellas et al 1986). IGF-II also stimulates the production of inositol triphosphate in the cell membranes of the renal proximal tubules, and this effect is potentiated by Man-6-P (Rogers & Hammerman 1989). In some of these instances, the signal transduction appears to be coupled to a pertussis toxin-sensitive G protein (Nishimoto et al 1987; Kojima et al 1988; Braulke et al 1989). Clearly, this is a fascinating aspect of CI-MPR biology and one that is being intensely investigated.

Role of the MPRs in Lysosomal Enzyme Sorting and Endocytosis

Several approaches have been used to define the relative roles of the two MPRs in the sorting of newly synthesized lysosomal enzymes and in the endocytosis of extracellular enzymes. Cells deficient in the CI-MPR still sort 30–40% of their newly synthesized enzymes to lysosomes (Gabel et al 1983). When antibodies that block ligand binding to the CD-MPR are added to such cells, the secretion of acid hydrolases is increased, consistent with a role for the CD-MPR in lysosomal enzyme sorting (Stein et al 1987c). However cells that express both MPRs do not hypersecrete lysosomal enzymes when anti-CD-MPR antibodies are added. On the other hand, addition of anti-CI-MPR antibodies induces lysosomal enzyme secretion, even in the presence of functional CD-MPR (Stein et al 1987c; Nolan et al 1987; Braulke et al 1989). These data indicate that both receptors participate in lysosomal enzyme sorting, and that the CI-MPR is the dominant receptor in this process as it is able to compensate for the loss of the CD-MPR. Further insights into the role of the individual MPRs in lysosomal enzyme sorting have come from cDNA transfection experiments. Transfection of CI-MPR deficient cell lines with CI-MPR cDNA results in receptor expression and complete or almost complete correction of the hypersecretion defect (Kyle et al 1989; Lobel et al 1989). When the same cell type is transfected with the CD-MPR cDNA, the hypersecretion of lysosomal enzymes is only partially corrected (K. Johnson et al, unpublished), which reinforces the conclusion that the CD-MPR is less efficient than the CI-MPR in lysosomal enzyme sorting.

An even more striking difference between the two receptors was identified when they were tested for their ability to internalize extracellular lysosomal enzymes. These studies revealed that only the CI-MPR is efficient at receptor-mediated endocytosis, with the CD-MPR being almost entirely ineffective (Stein et al 1987c). The inability of the CD-MPR to function in endocytosis reflects its poor binding of ligand at the cell surface rather than its failure to recycle to the plasma membrane (Stein et al 1987b,c; Duncan & Kornfeld 1988).

Intracellular Trafficking of Mannose 6-Phosphate Receptors

It is clear from the above discussion that the CI-MPR can mediate the targeting of lysosomal enzymes to lysosomes by both the biosynthetic and endocytic pathways. In the biosynthetic pathway, the generation of the active phosphomannosyl monoesters on lysosomal enzymes occurs in an early Golgi compartment, which raises the possibility that the enzymes might bind to the MPR at this site and either pass through the Golgi as a complex or exit the Golgi at the *cis* side of the stack. However most biochemical evidence suggests that the major site for lysosomal enzyme sorting is the last Golgi compartment, variously referred to as the *trans* Golgi network (TGN), the *trans* Golgi reticulum, the *trans* tubular network, and GERL (Griffiths & Simons 1986). The strongest evidence for this comes from the finding that several of the hydrolases residing in lysosomes contain terminally processed Asn-linked oligosaccharides in addition to their phosphorylated high-mannose-type units (Vladutiu 1983;

Fedde & Sly 1985). Because the glycosyltransferases responsible for terminal processing are located in *trans* Golgi cisternae and the TGN, at least some lysosomal enzymes must pass through the entire Golgi apparatus before reaching lysosomes. The presence of terminally glycosylated enzymes in lysosomes cannot be explained by endocytosis of secreted enzymes (Fedde & Sly 1985). Kinetic studies of receptor trafficking in several cell types have also shown that both MPRs routinely recycle to the Golgi compartments that contain sialyltransferase (*trans* cisternae and TGN) while returning to the earlier regions of the Golgi much less frequently (Duncan & Kornfeld 1988; Goda & Pfeffer 1988; Jin et al 1989). Finally, lysosomal enzyme precursors and MPRs have been identified in clathrin-coated vesicles that appear to be derived from the TGN (Campbell et al 1983; Geuze et al 1985; Lemansky et al 1987). Taken together, these data indicate that lysosomal enzymes most likely encounter the MPRs in the TGN.

The results of the CI-MPR immunolocalization experiments are more difficult to interpret. In some cell types, such as pancreatic, hepatic, and epididymal cells, the receptor is most concentrated in the *cis* Golgi (Brown & Farquhar 1984, 1987; Brown et al 1984) while in other cell types it is found primarily in the TGN with lesser amounts detected throughout the Golgi stack (Geuze et al 1985; Brown & Farquhar 1987; Griffiths et al 1988). This difference in steady-state location of the CI-MPR in different cell types could indicate that the sorting site varies among cells. However in the absence of studies of the kinetics of the movement of the MPRs to the different compartments of the Golgi, it is difficult to draw such conclusions from these data.

The immunolocalization studies have also revealed that at steady-state most of the receptor is present in one or more populations of endosomes, with only small amounts in the Golgi and on the cell surface or in coated vesicles, and very low, or undetectable amounts in structures identified as lysosomes (Willingham et al 1981; Geuze et al 1985; Brown et al 1986; Griffiths et al 1988; Geuze et al 1988). The absence of receptor from dense lysosomes has been confirmed by direct biochemical analysis (Sahagian & Neufeld 1983). The high level of receptor in endosomes, in contrast to the low level in lysosomes, suggests that newly synthesized lysosomal enzymes are delivered to one or more endosomal compartments where the low pH induces the lysosomal enzymes to be discharged from the MPRs and allows the receptors to then recycle either back to the Golgi and/or to the plasma membrane (Sahagian 1984; Brown et al 1986; Griffiths et al 1988; Geuze et al 1988). Since significant dissociation of ligand from the CI-MPR requires pH < 5.5, it is likely that the sorting of enzyme and receptor occurs in a late endosomal compartment of pH 5.5 or less (Schmid et al 1989). Discharged lysosomal enzymes would then be transported to lysosomes along with enzymes (and other dissociated ligands) internalized via endocytosis.

Mannose 6-Phosphate Receptors Belong to a Common Pool

Several lines of evidence are consistent with there being a single pool of CI-MPR molecules that participate in lysosomal enzyme sorting in the Golgi and endocytosis at the cell surface. When anti-CI-MPR antibodies are added to the medium of cultured cells, the endocytosis of exogenous and the sorting of endogenous lysosomal enzymes is inhibited (Sahagian 1984; Gartung et al 1985; Nolan et al 1987). This result demonstrates that intracellular MRPs are accessible to the anti-MPR antibodies. One explanation for this is that the total pool of receptors is in equilibrium even though only a small percentage of MPRs are on the cell surface at any one time. However, since these experiments were carried out at 37°C, it is possible that the antibodies were internalized by fluid phase endocytosis and reached a subset of receptors that recycle between endosomes and the Golgi without actually reaching the cell surface. More direct evidence in favor of a single pool of receptors was obtained by labeling cell surface MPRs with [^{3}H]galactose at 4°C and showing that the labeled molecules were sialylated after the cells were warmed to 37°C (Duncan & Kornfeld 1988). Since sialic acid addition occurs in late (*trans*) Golgi compartments, these findings demonstrate that cell surface receptors reach the Golgi, presumably by passing through endosomes. In similar experiments cell surface CI-MPRs that were desialylated by neuraminidase digestion at 4°C were shown to be resialylated after warming (Duncan & Kornfeld 1988; Jin et al 1989). A fraction of the resialylated receptors then reappeared on the plasma membrane (Jin et al 1989), which strongly supports the concept that there is only a single pool of receptor molecules that recycle between all compartments containing MPR.

The recycling of receptor through the sorting and endocytic pathways appears to occur constitutively (Oka & Czech 1986; Braulke et al 1987; Pfeffer 1987; Duncan & Kornfeld 1988; Jin et al 1989) although the rate of movement of receptor might be modulated by ligand binding. This could explain the alterations in steady-state distribution of the CI-MPR that have been found in some (Brown et al 1984, 1986; Braulke et al 1987), but not all (Geuze et al 1984; Pfeffer 1987), studies where lysosomal enzyme synthesis was inhibited or where the dissociation of the receptor-ligand complexes was prevented. Insulin treatment of rat adipocytes or H-35 hepatoma cells causes a redistribution of IGF-II receptors (CI-MPRs) from internal membranes to the cell surface (Oppenheimer et al 1983; Wardzala et al 1984; Oka et al 1984). This is associated with a decrease in phosphorylation of the receptor molecules present in the plasma membrane (Corvera & Czech 1985; Corvera et al 1988a).

Indeed, it appears that there is a relationship between receptor phosphorylation and endocytosis. Corvera et al have shown that a highly phosphorylated form of the CI-MPR is localized to a plasma membrane subfraction that is enriched in clathrin, whereas the majority of the receptor in the plasma membrane is poorly phosphorylated (Corvera et al 1988a). Insulin treatment increases the receptor content of the plasma membrane fraction devoid of clathrin while decreasing the phosphorylation of the receptor molecules localized to the clathrin-enriched membrane fraction. Based on these data, it has been proposed that insulin-mediated dephosphorylation of the receptor impairs its concentration in clathrin-coated regions of the membrane, thereby decreasing its rate of internalization and increasing its steady-state level at the cell surface (Corvera & Czech 1985; Corvera et al 1988b).

Structural Determinants of MPR Traffic

The cloning of the cDNA for the CI-MPR has made it possible to begin identifying regions of the receptor needed for rapid endocytosis from the cell surface and efficient lysosomal enzyme sorting in the Golgi (Lobel et al 1989; Watanabe et al 1988). The approach has been to transfect CI-MPR deficient cells with normal CI-MPR cDNA or cDNAs mutated in the cytoplasmic domain and then test for the ability of these cells to sort and endocytose lysosomal enzymes. Relative to transfected wild-type receptor, mutant receptors with 40 and 89 residues deleted from the carboxyl terminus of the 163 amino acid cytoplasmic tail functioned normally in endocytosis, but were impaired in sorting of newly synthesized enzyme. These receptors were presumed to be defective in the structural determinants required for the return to or the departure from the Golgi while retaining determinants needed for rapid endocytosis. Mutant receptors with 20 or fewer residues of the cytoplasmic tail were defective in both functions, as was a mutant receptor in which the tyrosines at positions 24 and 26 were replaced by alanines. Unlike the wild-type receptor, the endocytosis-defective mutants accumulate at the cell surface, presumably unable to enter coated pits (Lobel et al 1989).

The requirement for tyrosine residues in endocytosis is analogous to results obtained with the LDL receptor (Davis et al 1986). When the single tyrosine in the cytoplasmic tail of this receptor was changed to a cysteine or to other charged or uncharged aliphatic residues, the resultant receptor was incapable of entering coated pits efficiently (Davis et al 1987). A tyrosine residue is also essential for the rapid internalization of influenza virus hemagglutinin (Lazarovits & Roth 1988). The rate-limiting step that leads to impaired lysosomal enzyme sorting by the receptors lacking the outer portion of the cytosolic tail is unknown.

Some insight into how the CI-MPR may be concentrated in clathrin-coated pits has come from the work of Pearse & colleagues who have analyzed the interaction of this receptor with coat proteins termed adaptins (Pearse 1985; Glickman et al 1989). The adaptor proteins in the plasma membrane coated pits consist of a heterodimer of 100-kd polypeptides (HA-II adaptins) plus two smaller polypeptides of 50 kd and 16 kd (Brodsky 1988). The Golgi clathrin-coated pits contain related but distinct 100-kd polypeptides (HA-I adaptins) as well as two different associated polypeptides of 47 kd and 19 kd (Brodsky 1988; Ahle et al 1988). Using affinity chromatography, Pearse has presented evidence that the plasma membrane-derived HA-II adaptors may exhibit some ability to bind to the cytoplasmic tails of the CI-MPR and several other receptors (including the LDL receptor), and that this association is dependent on the presence of tyrosine residues in the cytoplasmic domains of the receptor. In contrast, the Golgi-associated HA-I adaptors only bind to the cytoplasmic tail of the CI-MPR, and this interaction is independent of the tyrosine residues. These data suggest that the plasma membrane-associated HA-II adaptors interact with the tyrosine-containing portion of the CI-MPR cytoplasmic tail to allow rapid endocytosis of the receptor, while the Golgi-associated HA-I adaptors interact with other determinants on the cytoplasmic tail to allow diversion to the prelysosomal sorting compartment.

MANNOSE 6-PHOSPHATE-INDEPENDENT TRANSPORT OF LYSOSOMAL ENZYMES

Enzyme Targeting in Mammalian Cells

While MPRs play a major role in the intracellular transport of newly synthesized lysosomal enzymes, several considerations suggest that there may also be additional or alternative mechanisms for lysosomal enzyme targeting. In ML-II fibroblasts, for example, not all lysosomal enzymes are aberrantly secreted. Enzymes that are membrane-associated (β-glucocerebrosidase) or are synthesized as integral membrane protein precursors (human lysosomal acid phosphatase) (Erickson et al 1985; Pohlmann et al 1988; Waheed et al 1988) appear to be retained intracellularly and transported to lysosomes (van Dongen et al 1984; Lemansky et al 1985; Neufeld & McKusick 1983). Thus the mechanisms by which membrane proteins or membrane-bound components are selectively targeted to lysosomes are independent of the Man-6-P recognition marker (see below). More interestingly, hepatocytes, certain leukocytes, and other cell types in patients with ML-II do not manifest a storage disorder and have normal levels of lysosomal enzymes (Neufeld & McKusick 1983; Kornfeld 1986). The lysosomal enzymes expressed by these cell types are otherwise identical to those expressed in cells that are affected by the ML-II defect. Although these observations suggest that certain cell types have an alternative mechanism(s) for targeting their lysosomal enzymes, or for accumulating exogenous enzymes, the basis for the lysosomal enzyme transport in these cells is unknown.

Enzyme Targeting in Dictyostelium discoideum *and Yeast*

An examination of lysosomal enzyme targeting in lower eukaryotes indicates the existence of mechanims of enzyme transport to lysosomes that are independent of Man-6-P. In *Dictyostelium discoideum*, the lysosomal enzymes acquire Man-6-P methyl esters (Freeze & Wolgast 1986; Gabel et al 1984). While this modification is recognized with high affinity by the mammalian CI-MPR, no MPR activity has been detected in *Dictyostelium* membranes (Cardelli et al 1986), which makes it unlikely that phosphorylated mannose residues play a central role in targeting. Neverthcless, the fact that immature enzyme precursors are membrane bound strongly suggests that they bind to some receptor, albeit one that is not specific for Man-6-P (Cardelli et al 1986; Mierendorf ct al 1985). The actual recognition marker specifying transport to lysosomes is unknown, but is likely to be contained within the polypeptide portion of *Dictyostelium* lysosomal enzymes since inhibition of an early post-Golgi proteolytic cleavage by leupeptin or antipain leads to missorting of newly synthesized enzyme (Richardson et al 1989).

In the yeast *Saccharomyces cerevisiae*, the involvement of a protein determinant in the targeting of hydrolytic enzymes to the vacuole (the lysosome equivalent) has been directly established using a genetic approach. The propeptide of carboxypeptidase Y (CPY), a soluble vacuole protein, has been shown by site-directed mutagenesis to contain a domain necessary for vacuole targeting (Valls et al 1987; Johnson et al 1987; Klionsky et al 1988). This domain appears to represent the entire sorting signal since fusion of only the 30 amino-terminal amino acids of the propeptide to a secretory protein (invertase) is sufficient to direct delivery of the CPY-invertase chimera to the vacuole. Although it is likely that there is a receptor for this sorting signal, no candidate molecule has been isolated as yet. Like lysosomal enzyme sorting via MPR in animal cells, targeting of enzymes to the yeast vacuole is dependent on acidic pH, which presumably facilitates the dissociation of newly synthesized vacuolar hydrolases from the putative sorting receptor. Treatment of cells with the proton ATPase inhibitor bafilomycin A1 or with other agents that dissipate the pH gradient across the vacuole membrane results in the secretion of at least two vacuolar enzymes (Banta et al 1988). In addition, yeast mutants defective in vacuole acidification missort newly synthesized enzymes (T. Stevens, personal communication), just as mammalian cell mutants defective in endosome acidification (Robbins et al 1984; Roff et al 1986).

Overexpression of CYP leads to missorting and secretion of newly synthesized enzyme, consistent with saturation of a putative hydrolase sorting receptor (Rothman et al 1986; Stevens et al 1986). It is interesting, however, that this is not accompanied by the missorting of other vacuole enzymes, which suggests that multiple vacuolar enzyme receptors may exist or that vacuole targeting may be dependent on other factors apart from receptor binding (e.g. self-association of enzyme molecules). The identification of multiple mutations leading to vacuolar protein missorting (Rothman & Stevens 1986; Banta et al 1988; Robinson et al 1988), in addition to those resulting in defective vacuole acidification, is likely to provide insight into this problem and possibly into the basis for MPR-independent transport to lysosomes in mammalian cells.

STRUCTURE AND BIOGENESIS OF THE LYSOSOMAL MEMBRANE

Lysosomal membranes have long been suspected to be rich in carbohydrates (Lloyd & Foster 1986) as well as containing cholesterol and a unique phospholipid content (Bleistein et al 1980). It has also been recognized that a number of protein components must exist to provide a variety of specific functions such as ATP-dependent acidification (Mellman et al 1986) and the transmembrane transport of amino acids (Schneider et al 1984; Bernar et al 1986), fatty acids (Rome et al 1983), carbohydrates (Maguire et al 1983; Renlund et al 1986; Cohn & Ehrenreich 1969; Ehrenreich & Cohn 1969), and nutrients [such as cholesterol and cobalamin (vitamin B_{12}) (Rosenblatt et al 1985)], generated as products of lysosomal catabolism. The actual structure of these transporters, however, remains unknown.

During the past few years, much information has been obtained concerning the major glycoprotein components of the lysosomal membrane. In general, identification and characterization of these proteins was made possible by the production of polyclonal and monoclonal antibodies against isolated lysosomes (Reggio et al 1984; Lewis et al 1985) or other cellular components [intact cells (Chen et al 1985); coated vesicles (Lippincott-Schwartz & Fambrough 1986); isolated polylactosamine-containing glycoproteins (Carlsson et al 1988)]. The antibodies have defined a series of antigens, typically integral membrane glycoproteins, that are enriched in lysosomes as indicated by indirect immunofluorescence, EM-immunocytochemistry, or by their cofractionation with lysosomal enzymes on density gradients. While glycoproteins ranging in molecular mass from 20 kd to >150 kd (Barriocanal et al 1986; Marsh et al 1987; I. Mellman, unpublished) have been identified, the most commonly recognized antigens in rodent, human, and avian lysosomes exhibit molecular masses of 100–120 kd. It is likely that these antigens represent major structural components of the lysosomal membrane since glycoproteins of ~ 100–120 kd comprise ~ 50% of the total integral membrane proteins found in highly purified lysosomes (Marsh et al 1987).

Structure of the Major Lysosomal Membrane Glycoproteins

Based on their biochemical features and amino acid sequences deduced from rat, mouse, human, and chicken cDNA clones, it is clear that the major components of the lysosomal membrane comprise a closely related family of acidic, highly glycosylated membrane proteins (Howe et al 1988). Although their functions are unknown, two distinct but highly homologous proteins have been distinguished thus far. The first group of proteins, designated here as lgp-A (i.e. lysosomal membrane glycoprotein-1), is defined by rat and mouse lgp120 (Howe et al 1988; Granger et al 1989), mouse LAMP-1 (Chen et al 1988), human lamp-1 (Fukuda et al 1988), and chicken LEP-100 (Fambrough et al 1988); the second group, lgp-B, contains rat and mouse lgp110 (mouse LAMP-2) (Howe et al 1988; Granger et al 1989) and human lamp-2 (Fukuda et al 1988). While members within each group are likely to represent species-specific versions of the same protein, lgp-A and lgp-B within a single species are immunologically distinct. This has made it possible to show that individual lysosomes can contain both lgp-A and lgp-B (Green et al 1987).

The overall domain structure and several aspects of the primary sequence are highly conserved between lgp-A and lgp-B. As illustrated in Figure 3, both lgp-A and lgp-B are membrane proteins with large lumenal domains of 380–396 amino acids containing 16–20 sites for Asn-linked oligosaccharide addition, most or all of which are used (Lewis et al 1985; Fambrough et al 1988; Granger et al 1989). There are 4 pairs of cysteine residues, the relative positions of which are conserved. Adjacent cysteines are thought to be disulfide-bonded to each other (Arterburn et al 1988). All the lgps have a single hydrophobic membrane spanning segment and short cytoplasmic tails that are 10–11 residues in length. Overall, lgp-A and lgp-B are 65–70% similar irrespective of species (Howe et al 1988; Granger et al 1989). The lumenal domains appear to be partitioned into two sub-domains of approximately equal length by a ~ 30 amino acid

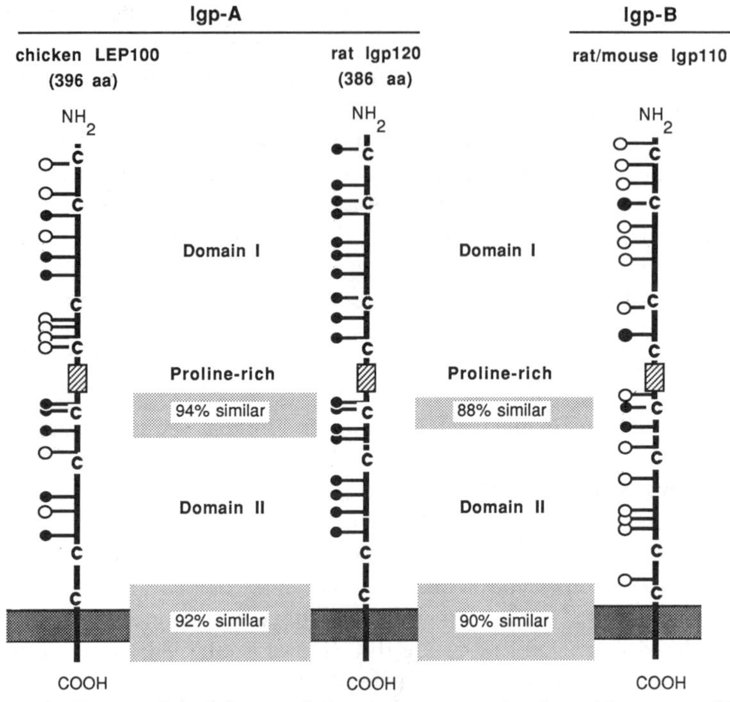

Figure 3 Conservation of the overall domain structure and amino acid sequence of lgp-A and lgp-B from avian and rodent species. Two members of the lgp-A family [chicken LEP100 (Fambrough et al 1988) and rat lgp 120 (Howe et al 1988)] are compared with each other and with a member of the closely related lgp-B family [rat/mouse lgp110 (Howe et al 1988; Granger et al 1989)]. Each protein is arbitrarily divided into two domains (I and II) separated by a 25–30 amino acid region rich in proline, serine, and threonine (boxes with diagonal lines). Each also contains 8 cysteine residues (*C*) whose positions are conserved. Circles show the positions of sites for N-linked oligosaccharide addition; glycosylation sites whose positions are conserved relative to lgp120 are indicated by filled circles. Two regions of particularly high amino acid sequence similarity (~ 90%) are shown by the shaded boxes.

segment rich in proline, threonine, and serine. Although the primary sequence of the proline-rich region is not conserved, it is directly followed (i.e. on the C-terminal side) by a stretch of ~ 20 amino acids that is highly conserved (up to 92% identical) among lgps of diverse species (Howe et al 1988).

An even more striking homology occurs in the membrane spanning and cytoplasmic domains. The 11 amino acid cytoplasmic tail sequences of rodent, human, and chicken lgp-A are completely conserved, as are 8 of the 10 amino acids in the cytoplasmic tail sequences of rodent and human lgp-B. When compared to each other, the cytoplasmic domains of lgp-A and lgp-B are ~50% identical (Figure 4). This high degree of sequence similarity found in the membrane spanning and cytoplasmic domains of lgp-A and lgp-B suggest that these regions may be critical to the functions or intracellular transport of the lgps.

The oligosaccharides of the extensively glycosylated lgps have been characterized to some extent. The Asn-linked units of lgp-A and lgp-B isolated from the murine J774 macrophage cell line are tetra-antennary with (on average) 3 of the 4 antennae being terminally sialylated (Howe et al 1988; Granger et al 1989). In both human and mouse lgps, at least one antenna contains a repeating lactosamine unit (N-acetylglucosamine-galactose) (Howe et al 1988; Carlsson et al 1988). Human lgp-A and lgp-B, but only lgp-B in the mouse, contain O-linked sugars (Carlsson et al 1988; Granger et al 1989). The high content of sialic acid accounts for the acidic pIs of these proteins (pI 2–4) (Lewis et al 1985). Importantly, lgps do not appear to be phosphorylated or to contain the Man-6-P recognition marker (Howe et al 1988; Granger et al 1989), which makes it unlikely that they are transported to lysosomes by interacting with the MRPs.

Human Lysosomal Acid Phosphatase

Although the mature form of human lysosomal acid phosphatase (LAP) is a soluble protein (and thus not officially a lgp), recent biochemical and cDNA cloning studies have demonstrated that this enzyme is synthesized

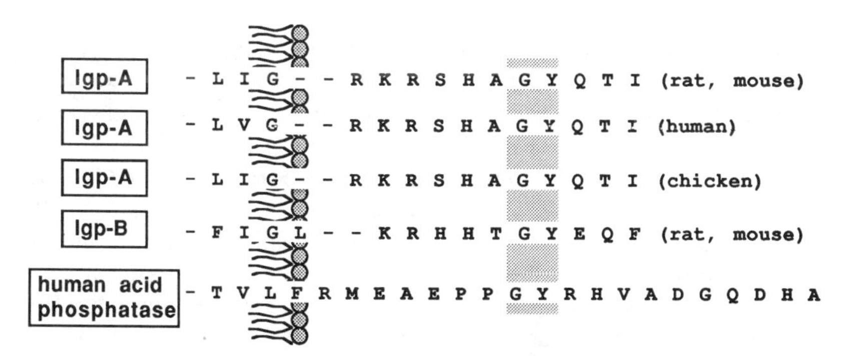

Figure 4 Cytoplasmic tail sequences of membrane proteins transported to lysosomes. While the cytoplasmic tail sequences from mouse, rat, human, and chicken lgp-A are identical, they are only 50% identical to the rodent lgp-B cytoplasmic tail sequence. This variation represents a major divergence between members of the lgp-A and lgp-B groups. Little obvious similarity exists between lgp and the cytoplasmic tail of the human lysosomal acid phosphatase precursor. However alignment of the tail sequences suggests the presence of a conserved glycine-tyrosine motif.

as an integral membrane protein precursor (Pohlmann et al 1988; Waheed et al 1988). On the basis of pulse-chase and cell fractionation experiments, it is now known that the LAP precursor is transported through the Golgi and then to dense lysosomes (see below) where it is proteolytically cleaved and accumulated as the soluble mature enzyme. The fate of the membrane anchor and remaining cytosolic tail is unknown. As previously mentioned, the fact that LAP is synthesized as a membrane protein probably accounts for its ability to be transported to lysosomes in a fashion independent of the Man-6-P recognition marker.

The deduced amino acid sequence of LAP is completely distinct from the sequences of lgp-A or lgp-B, as is its apparent domain organization. The cytoplasmic domain of the LAP precursor is 18 amino acids in length, with no obvious sequence similarities to the cytoplasmic tails of the lgps. However alignment of the LAP cytoplasmic tail sequence with the lgp tails does reveal a conserved glycine-tyrosine in approximately the same relative position, with the surrounding residues being mostly hydrophilic (Figure 4). Given the suspected role of tyrosine residues in coated pit localization of endocytic receptors, and the likely role of the LAP tyrosine in lysosomal transport (see below), it is conceivable that this conserved motif (however short) forms part of a determinant needed for targeting newly synthesized membrane proteins to lysosomes.

Biosynthesis and Intracellular Transport

Since the lgps are selectively targeted to an organelle avoided by all other newly synthesized membrane proteins, there has been a considerable interest in elucidating the intracellular pathway of lgp biogenesis, as well as in identifying the structural determinants that direct the transport of lgps to lysosomes. In broad terms, two distinct transport pathways are possible. The first involves the sorting of lgps from membrane and soluble proteins destined for the cell surface at the level of the TGN. In this pathway, newly synthesized lgps would be transported from the Golgi to endosomes and then to lysosomes together with newly synthesized, MPR-bound lysosomal enzymes. In the alternative pathway, lgps would be transported to the cell surface via the constitutive secretory pathway and then be sorted from plasma membrane proteins by selective internalization followed by transport to dense lysosomes.

Studies of lgp-A and lgp-B biosynthesis in mouse macrophages and rat fibroblasts support the first model (Green et al 1987; D'Souza & August 1986). These experiments have shown that both proteins are synthesized in the RER and transported through the Golgi stack at the same rate as proteins destined for the plasma membrane. The lgps (and newly synthesized lysosomal enzymes) then appear in dense lysosomes at essentially the same time as the newly synthesized plasma membrane proteins arrive at the cell surface. If the lgps traveled to the cell surface before going to lysosomes, their arrival in lysosomes should have been delayed relative to the arrival of the plasma membrane proteins at the cell surface. Further evidence against the transient appearance of the lgps at the cell surface is the failure to detect any lgp-A and lgp-B on the surface of J774 cells, 3T3 cells, and NRK cells and the lack of anti-lgp monoclonal antibody uptake when added to the medium of these cells (Lewis et al 1985; Green et al 1987; I. Mellman unpublished results). In addition, lgps could not be detected by immunocytochemistry in early endosomes (as defined by labeling NRK cells with endocytic tracers at 20°C) as would be expected if these proteins were targeted to lysosomes via endocytosis from the cell surface. On the other hand, the lgps were readily detected by immunocytochemistry in acidic, late endosomal vacuoles that also contained the CI-MPR and lysosomal enzymes (Griffiths et al 1988). These data are all consistent with a targeting pathway that involves movement of the lgps from the Golgi to late endosomes to dense lysosomes together with newly synthesized lysosomal enzymes. However it remains possible that transport to and from the cell surface en route to lysosomes is so rapid that it is not detected by these approaches.

A number of recent experiments have shown, however, that a small fraction ($<5\%$) of chicken lgp-A is present on the plasma membrane of some cells, such as chick fibroblasts (Lippincott-Schwartz & Fambrough 1986, 1987). These molecules can enter coated pits and be rapidly internalized, as indicated by the observation that anti-chicken lgp-A antibodies added to the media are specifically internalized and delivered to lysosomes. When chicken fibroblasts are treated with chloroquine, the fraction of lgp-A on the cell surface increases. This has led to the suggestion that at least a fraction of the chicken lgp-A may be continuously recycling between the lysosomes and the plasma membrane.

The basis for these different findings between chicken lgp-A and its rat and mouse counterparts is unclear. It is not likely to be explained by cell type differences since the distinct behaviors of the two lgps has been documented in the same cell type, namely primary osteoclasts. In chick osteoclasts, lgp-A is expressed in lysosomes and on the "ruffled border", the specialized region of the plasma membrane directly opposed to the bone surface that functions as an externalized lysosome for the digestion of bone. In contrast, in primary rodent osteoclasts, neither lgp-A nor lgp-B is found at the ruffled border; rather these proteins are only present in intracellular lysosomes (Baron et al 1985a,b). Also, chloroquine treatment of rodent fibroblasts has no effect on lgp-A or lgp-B distribution (I. Mellman, unpublished results).

Studies of the biosynthesis of human lysosomal acid phosphatase (LAP) have revealed some similarities with chicken lgp-A (Waheed et al 1988). This protein is transported to the TGN and dense lysosomes with $t_{1/2}$s of 30 min and 4 h, respectively. It then undergoes a slow proteolytic processing ($t_{1/2}$ of 4–5 h) within the dense lysosome to give rise to the mature, soluble form of acid phosphatase. While en route to lysosomes, the membrane precursor accumulates in endosomes and may recycle to the plasma membrane a number of times (K. von Figura, unpublished). When the cytoplasmic tail of the membrane-bound LAP precursor is deleted, or when its single tyrosine residue is changed to phenylalanine, the expressed mutant protein accumulates on the cell surface, presumably being unable to enter the clathrin-coated pits (K. von Figura, personal communication).

The results obtained for chicken lgp-A and human LAP may indicate that these proteins, and perhaps other lgps, are first transported to the cell surface via the constitutive secretory pathway and then internalized by endocytosis for delivery to lysosomes. Possibly the kinetics of their transport are sufficiently slow relative to rodent lgps so as to facilitate the detection of this pathway. However the presence of chicken lgp-A and human LAP on the cell surface can also be explained in the context of a transport pathway that involves intracellular sorting without obligatory passage via the cell surface. For example, human LAP and chicken lgp-A may be sorted less efficiently in the Golgi than either rodent lgp-A or lgp-B, thus leading to some leakage of these proteins to the cell surface. Alternatively, all lgps may be efficiently sorted in the TGN with their paths diverging somewhat after reaching post-Golgi endosomes. Thus rodent lgp-A and lgp-B may be targeted from endosomes to lysosomes with a high degree of fidelity while chicken lgp-A and human LAP may undergo several rounds of cycling to the plasma membrane before being diverted to lysosomes. This tendency to recycle might also explain the reported ability of chicken lgp-A and human LAP to recycle between lysosomes and the plasma membrane (Lippincott-Schwartz & Fambrough 1987). It is also consistent with the intracellular pathway taken by the CI- and CD-MPRs (Duncan & Kornfeld 1988).

It is clear that more information will be needed to distinguish among the various possible routes for lgp transport to lysosomes. The clearest indication that transport proceeds via the same pathway as the MPR and lysosomal enzymes would be the demonstration that newly synthesized lgps exit the Golgi via the same clathrin-coated vesicles, as are thought to be involved in the transport of newly synthesized lysosomal enzymes bound to MPR (see above).

Role of lgp Cytoplasmic Domains in Intracellular Transport

The high degree of conservation among lgp cytoplasmic tails and the preliminary results obtained using site-directed mutations in the human LAP precursor cytoplasmic tail strongly suggest that this domain contains a determinant necessary for the targeting of lysosomal membrane proteins to lysosomes. Using lgp-A mutants, preliminary results similar to those obtained for human LAP have been obtained (D. Fambrough, unpublished; C. Howe, I. Mellman et al, unpublished). Chimeric proteins bearing the LAP or lgp cytoplasmic domain and the extracytoplasmic domain of a protein not normally found in lysosomes are transported to the lysosome (K. von Figura, personal communication). However, this result must be interpreted cautiously since it could reflect a pathway involved in the degradation of certain types of denatured membrane proteins.

The possible role of the conserved tyrosine residue in the cytoplasmic domain of these proteins is being studied. When this residue is changed to a phenylalanine in LAP, the expressed mutant protein accumulates on the cell surface, presumably being unable to enter clathrin-coated pits (K. von Figura, personal communication). This finding differs from that obtained with the LDL receptor where this particular substitution in the cytoplasmic tail has no effect on coated pit localization (Davis et al 1987). If lgps exit the Golgi via clathrin-coated pits, their cytoplasmic tails might interact with the HA-1 class of Golgi adaptor proteins. In in vitro binding studies, these adaptor proteins appear to interact with the cytoplasmic tail of the CI-MPR (Glickman et al 1989). Furthermore this interaction is not altered when the two tyrosine residues in the CI-MPR cytoplasmic tail are mutated to alanines. It will be of interest, therefore, to determine whether or not the tyrosine residue in the cytoplasmic tail of the lgps are important in the exit of these proteins from the Golgi.

The Functions of the lgps are Unknown

Although the extent of sequence conservation among lgps strongly suggests that these proteins play a critical role in lysosome function, no specific activities have been associated with any lgp or other lysosomal membrane glycoprotein identified thus far. The general structure of lgps makes it unlikely that they are involved in any ion or metabolite transport processes. In addition, their extensive glycosylation is apparently not necessary for normal targeting, stability, or lysosome function. This has been shown using CHO glycosylation mutants that fail to add terminal sialic acid residues or any terminal sugars. In these cell lines, the lysosomal membrane proteins do not undergo terminal glycosylation reactions; nevertheless the proteins are targeted correctly to lysosomes where their survival is normal.

The lysosomes also function normally with respect to hydrolytic activity and ATP-dependent acidification (S. Schmid, R. Fuchs, H. Plutner, I. Mellman, in preparation). On the other hand, core glycosylation of the lgps is necessary (presumably for correct initial folding) since tunicamycin treatment results in the rapid degradation of newly synthesized lgps (Barriocanal et al 1986).

While lgps are sensitive to proteolysis when in detergent solution, they are stable to a variety of proteases when inserted in microsomal and lysosomal membranes (Green et al 1987; Howe et al 1988). They also appear to be present in lysosomes at exceedingly high concentrations, perhaps approaching the concentration of spike glycoproteins in enveloped viruses. This dense concentration of membrane proteins with highly conserved cytoplasmic tails might create a unique recognition marker on the cytoplasmic face of lysosomes that could specify their interaction with other organelles. Alternatively, it is possible that the high concentration of lgps serves to stabilize the lysosomal membrane against degradation by its own hydrolases, perhaps by creating a dense protease-insensitive matrix. As such, lgps may in effect define the lysosomal membrane by their resistance to auto-digestion. For example, fusion of an endosome with a lysosome would result in a dilution of the lgp concentration or introduction of lgp-poor regions of membrane. These newly incorporated non-lysosomal components would then be subject to degradation until the original lgp concentration was reached.

PATHWAYS OF LYSOSOME BIOGENESIS

A key question in the biogenesis of lysosomes is the identification of the site(s) where the biosynthetic and the endocytic pathways converge. Recent insights concerning the identity and functions of the organelles involved in intracellular sorting make it possible to begin to draw a number of preliminary conclusions concerning the cellular pathways leading to lysosomes.

The Endocytic Pathway to Lysosomes

While the transport of exogenous lysosomal enzymes to lysosomes via the endocytic pathway represents a minor pathway (accounting for only 5–10% of lysosomal enzyme delivery), it is better characterized at the morphologic level than the biosynthetic pathway that originates in the Golgi. This targeting pathway exhibits many features of the standard pathway of receptor-mediated endocytosis established for other cell surface receptors: internalization in coated pits/coated vesicles, dissociation of the ligand-receptor complex in acidic endosomes, and accumulation of the dissociated ligand in a terminal endocytic compartment, i.e. dense lysosomes. However the pathway followed by the MPRs differs from that of the LDL receptor and the other recycling receptors in that the MPRs routinely cycle to the Golgi as well as to the plasma membrane. The intracellular distribution of the MPR may reflect the underlying special organization of the endocytic pathway in many cells (Figure 5).

After internalization, the ligand-MPR complexes are delivered first to early endosomes (Schmid et al 1988, 1989), a distinct subpopulation of endosomes that appear (at least in tissue culture cells) to be distributed largely in the peripheral cytoplasm (Herman & Albertini, 1984; Gruenberg et al 1989; de Brabander et al 1988; I. Mellman, unpublished). Early endosomes have an internal pH of $\sim$ 6–6.3 (Schmid et al 1989; Sipe & Murphy 1987) that is sufficiently acidic to cause the dissociation of ligands from receptors, which rapidly recycle to the cell surface (e.g. LDL receptor, α_2-macroglobulin receptor). Thus early endosomes are likely to be the primary site from which this class of plasma membrane receptors return to the surface (Schmid et al 1988). However the MPRs would not discharge their ligands at this pH. Consequently, the lysosomal enzyme-MPR complex would either recycle to the cell surface (a type of futile cycle) or be transferred to late endosomes, a second endosome subpopulation that has a lower internal pH (Schmid et al 1988). Since late endosomes may be concentrated more in the perinuclear cytoplasm (Gruenberg et al 1989; I. Mellman, unpublished), translocation of early endosomes (or transport vesicles derived from early endosomes) from the peripheral to the perinuclear cytoplasm may be required. A variety of considerations predict that such vesicle translocations indeed occur and are mediated by microtubules (perhaps in conjunction with cytoplasmic dynein) (Gruenberg et al 1989; Vale et al 1985). In late endosomes, the lysosomal enzymes can dissociate from the MPR when the pH decreases < 5.5. The MPRs may then recycle to the Golgi, which is now in close proximity, or to the plasma membrane.

Late endosomes containing the internalized hydrolases and any remaining MPR may have several fates. First, they may fuse directly with dense, hydrolase-rich lysosomes, as suggested by light microscopic analysis (Hirsch et al 1968; Cohn et al 1966; Cohn & Benson 1965). This would result in the degradation of digestable contents and endosomal membrane components that fail to recycle rapidly after fusion. In addition, some of the late endosomes may fuse with Golgi-derived vesicles carrying newly synthesized lysosomal enzymes and lgps (Figure 5). This possibility is consistent with the observation that late endosomes contain unique membrane proteins not found in early endosomes or the plasma membrane (Schmid et al 1988). Finally, late endosomes may fuse with each other.

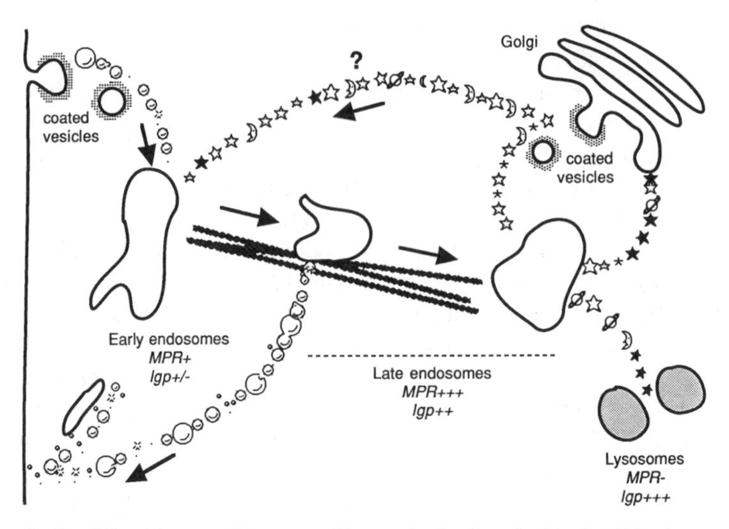

Figure 5 Possible pathways of lysosome biogenesis. As described in detail in the text, the diagram summarizes possible pathways of membrane and organelle transport that play a role in the biogenesis of lysosomes. While no attempt is made to use individual names to distinguish the various endosomal intermediates or transport vesicles that must be involved, their likely compositions—at least with respect to MPR and lgp—are shown at each stage. The suggested patterns of vesicle transport and/or interactions are indicated by the bubbles (denoting pathways between endosomes and the plasma membrane) and stars (pathways between endosomes, the Golgi complex, and lysosomes). In the proposed scheme, early endosomes in the peripheral cytoplasm are involved in a rapid constitutive pathway of endocytosis and recycling with the cell surface. As such, they would be expected to contain any MPR (and lgps) internalized from the cell surface. After a short period, the early endosomes would then begin to translocate via microtubules towards the perinuclear cytoplasm. The translocation may occur concomitantly (or be responsible for) the generation of the biochemically distinct late endosomes. First, translocation of endosomes from the peripheral cytoplasm might reduce their probability of fusion with incoming coated vesicles simply by spatial consideration. If the endosomes remained capable of participating in membrane recycling, this would result in a net alteration or simplification of their membrane composition, as observed (Schmid et al 1988) as their input of internalized membrane decreases. Second, upon reaching the perinuclear region, the late endosomes would now begin to receive new input from Golgi-derived coated vesicles (presumably containing MPR, newly synthesized lysosomal enzymes, and possibly newly synthesized lgps). The late endosomes might also fuse directly with existing late endosomes, with lysosomes, or with each other, predicting a dynamic and complex series of interactions. Irrespective of their precise origin, each subpopulation would be involved in the most critical function of late endosomes, namely the sorting of incoming ligands, newly synthesized enzymes and lgps (destined for lgp-positive/MPR-negative lysosomes) from the Golgi-derived or internalized MPR (which recycles to the Golgi and/or early endosomes).

Lysosomes, then, would be recognized as the final population of vesicles to which endocytic tracers are delivered, and from which all non-lysosomal components have either recycled or been degraded. As discussed below, it may now be possible to identify these terminal vesicles on the basis of their distinctive membrane composition.

Biochemical Definition of Lysosomes

Further information on the identity of the vesicles through which endocytic tracers and lysosomal enzymes are transported has come from recent immunocytochemical experiments. Cationized ferritin, added to the media of rat H_4S hepatoma cells, was traced kinetically through three distinct compartments. First, the tracer entered MPR-positive endosomes, then vesicles containing both MPR and lgps, and finally it accumulated in structures devoid of MPR but rich in lgps (Geuze et al 1989). Since the latter compartment represented the final destination, it can be defined as a lysosome. Based on this finding and the various biochemical observations discussed above, it would appear that lysosomes can be identified operationally as lgp-positive, MPR-negative organelles. Similar compartments were identified in NRK cells using α_2-macroglobulin as tracer (Griffiths et al 1988).

The identity and function of the MPR-positive/lgp-positive structures is less certain. While it is not yet established that they constitute an obligatory step on either the endocytic or biosynthetic pathways to lysosomes, they are of interest since they possess the features expected of an organelle responsible for sorting the various components involved in lysosome formation. They contain both lysosomal enzymes and MPRs, are acidic (allowing for enzyme dissociation from the receptor), and contain lgps that are presumably en route to lysosomes. However there is as yet insufficient information to establish them as a sorting site. Since these structures appear to be intermediates in the latter stages of the endocytic pathway to lysosomes, they can be defined as late endosomes. However, they differ from the usual operational definition of endosomes in several ways: (*a*) endocytic tracers do not gain access to MPR-positive/lgp-positive vacuoles in cells incubated at temperatures $< 20°C$ (Griffiths et al 1988), conditions that have been used to operationally limit transfer of material from low density, hydrolase-poor endosomes to heavy density, hydrolase-rich lysosomes (Dunn et al 1980; Helenius et al 1983); (*b*) they are rich in acid hydrolases, unlike most endosomes; (*c*) endosomes labeled by exposure of cells to endocytic tracers at temperatures $< 20°C$ are generally devoid of lgps (Green et al 1987).

For simplicity, however, we feel that it is preferable to refer to these vacuoles as late endosomes and not to distinguish them as a distinct organelle. This reflects our inability to follow carefully the various fusion and fission events that must be continuously occuring at rapid rates along the endocytic pathway. It considers the possibility that the 20°C block actually inhibits transfer of endocytic tracers from early to late endosomes, as opposed to inhibiting transfer from late endosomes to lysosomes; and it takes into account the possibility that late endosomes serve a function in the sorting and recycling of components involved in lysosome formation that is directly analogous to the function served by early endosomes in the sorting and recycling of receptors and ligands during endocytosis.

While transport vesicles that leave the TGN carrying a cargo of MPR-bound lysosomal enzymes and possibly lgps could potentially fuse with dense lysosomes, this seems less likely since MPRs are rarely detected in heavy density structures. It is also possible that some (or all) of these vesicles fuse with early endosomes, thereby entering the endocytic pathway at this point. If this occurs, the MPR-ligand complex would presumably migrate to late endosomes where the ligand would dissociate.

While it is clear that much additional work is needed to clarify the functions and formation of the organelles involved in lysosome biogenesis, it should be apparent that sufficient information is now available to allow some general conclusions and predictions, as illustrated in Figure 5. Many aspects of the proposed pathways can be tested with available systems and methodologies; accordingly, we should soon know how well present conceptualizations reflect reality.

ACKNOWLEDGMENTS

The authors would like to thank the past and present members of their respective laboratories (and colleagues elsewhere) for their contributions to the work discussed in this review. We would also like to give special thanks to Gareth Griffiths, Bruce Granger, Ari Helenius, Kai Simons, Hans Geuze, Kurt von Figura, George Palade, and Marilyn Farquhar for numerous helpful and enlightening discussions. Debbie Sinak also deserves thanks for her help in preparing the manuscript. Our laboratories have been supported by research grants from the National Institutes of Health (to SK & IM) and from the International Immunology Research Institute (to IM). This paper is dedicated to the memory of Martin Louis Mellman (1919–1989).

Literature Cited

Ahle, S., Mann, A., Eichelsbacher, U., Ungewickell, E. 1988. Structural relationships between clathrin assembly proteins from the Golgi and the plasma membrane. *EMBO J.* 7: 919–29

Alexander, D., Deeb, M., Talj, F. 1986. Heterozygosity for phosphodiester glycosidase deficiency: a novel human mutation of lysosomal enzyme processing. *Hum. Genet.* 73: 53–59

Appell, K. C., Simpson, I. A., Cushman, S. W. 1988. Characteristics of the stimulatory action of insulin on insulin-like growth factor II binding to rat adipose cells. *J. Biol. Chem.* 263: 10824–29

Arterburn, L. M., Chen, J. W., August, J. T. 1988. The structure and surface expression of lysosomal membrane glycoprotein LAMP-1. *J. Cell Biol.* 107: 343a (Abst.)

Baba, T., Watanabe, K., Arai, Y. 1988. Isolation and characterization of the 36-kDa D-mannose 6-phosphate receptor from porcine testis. *Carbohydr. Res.* 177: 153–61

Bainton, D., Farquhar, M. G. 1970. Segregation and packaging of granule enzymes in eosinophilik leukocytes. *J. Cell Biol.* 45: 54–73

Banta, L. M., Robinson, J. S., Klionsky, D. J., Emr, S. D. 1988. Organelle assembly in yeast: characterization of yeast mutants defective in vacuolar biogenesis and protein sorting. *J. Cell Biol.* 107: 1369–83

Baron, R., Neff, L., Brown, W., Courtoy, P. J., Louvard, D., Farquhar, M. G. 1988. Polarized secretion of lysosomal enzymes: Co-distribution of cation-independent mannose 6-phosphate receptors and lysosomal enzymes along the osteoclast exocytic pathway. *J. Cell Biol.* 106: 1863–72

Baron, R., Neff, L., Lippincott-Schwartz, J., Louvard, D., Mellman, I., Helenius, A., Marsh, M. 1985a. Distribution of lysosomal membrane proteins in the osteoclast and their relationship to acidic compartments. *J. Cell Biol.* 101: 53a (Abst.)

Baron, R., Neff, L., Louvard, D., Courtoy, P. J. 1985b. Cell-mediated extracellular acidification and bone resorption: evidence for a low pH in resorbing lacunae and localization of a 100 kD membrane protein at the osteoclast ruffled border. *J. Cell Biol.* 101: 2210–22

Barriocanal, J. G., Bonifacino, J. S., Yuan, L., Sandoval, I. V. 1986. Biosynthesis, glycosylation, movement through the Golgi system and transport to lysosomes by a N-linked carbohydrate independent mechanism of three lysosomal integral membrane proteins. *J. Biol. Chem.* 261: 1604–7

Ben-Yoseph, K., Pack, B. A., Mitchell, D. A., Elwell, D. G., Potier, M., et al. 1986. Characterization of the mutant N-acetylglucosaminylphosphotransferase in I-cell disease and pseudo-Hurler polydystrophy: complementation analysis and kinetic studies. *Enzyme* 35: 106–16

Bernar, J., Tietze, F., Kohn, L. D., Bernardini, I., Harper, G. S., et al. 1986. Characteristics of a lysosomal membrane transport system for tyrosine and other neutral amino acids in rat thyroid cells. *J. Biol. Chem.* 261: 17107–12

Bischoff, J., Luscum, L., Kornfeld, R. 1986. The effect of 1-deoxymannojirimycin on rat liver α-mannosidases. *J. Biol. Chem.* 261: 4766–74

Bleistein, J., Heidrich, H. G., Debuch, H. 1980. The phospholipids of liver lysosomes from untreated rats. *Hoppe-Seylers Z Physiol. Chem.* 361: 595–97

Braulke, T., Gartung, C., Hasilik, A., von Figura, K. 1987. Is movement of mannose-6-phosphate specific receptor triggered by binding of lysosomal enzymes? *J. Cell Biol.* 104: 1735–42

Braulke, T., Tippmer, S., Neher, E., von Figura, K. 1989. Regulation of mannose 6-phosphate/IGF-II receptor expression at the cell surface by mannose 6-phosphate, insulin-like growth factors, and epidermal growth factor. *EMBO J.* 8: 681–86

Bretthauer, R. K., Kaczorowski, G. J., Weise, M. J. 1973. Characterization of a phosphorylated pentasaccharide isolated from *Hansenula holstii* NRRL Y-2448 phosphomannan. *Biochemistry* 12: 1251–56

Brodsky, F. M. 1988. Living with clathrin: its role in intracellular membrane traffic. *Science* 242: 1396–1402

Brown, J. A., Swank, R. T. 1983. Subcellular redistribution of newly synthesized macrophage lysosomal enzymes. *J. Biol. Chem.* 258: 1604–7

Brown, W. J., Constantinescu, E., Farquhar, M. G. 1984. Redistribution of mannose 6-phosphate receptors induced by tunicamycin and chloroquine. *J. Cell Biol.* 99: 320–26

Brown, W. J., Farquhar, M. G. 1984. The mannose 6-phosphate receptor for lysosomal enzymes is concentrated in *cis* Golgi cisternae. *Cell* 36: 295–307

Brown, W. J., Farquhar, M. G. 1987. The distribution of 215 kD mannose 6-phosphate receptor within *cis* (heavy) and *trans* (light) Golgi subfractions varies in different cell types. *Proc. Natl. Acad. Sci. USA* 84: 9001–5

Brown, W. J., Goodhouse, J., Farquhar, M. G. 1986. Mannose 6-phosphate receptors for lysosomal enzymes cycle between the Golgi complex and endosomes. *J. Cell Biol.* 103: 1235–47

Campbell, C. H., Fine, R. E., Squicciarini, J., Rome, L. H. 1983. Coated vesicles from rat liver and calf brain contain cryptic mannose 6-phosphate receptors. *J. Biol. Chem.* 258: 2628–33

Canfield, W., Kornfeld, S. 1989. The chicken liver cation-independent mannose 6-phosphate receptor lacks the high affinity binding site for insulin-like growth factor II. *J. Biol. Chem.* 264: 7100–3

Cardelli, J. A., Golumbeski, G. S., Dimond, R. L. 1986. Lysosomal enzymes in *Dictyostelium discoideum* are transported at distinctly different rates. *J. Cell Biol.* 102: 1264–70

Carlsson, S. R., Roth, J., Piller, F., Fukuda, M. 1988. Isolation and characterization of human lysosomal membrane glycoproteins, h-lamp-1 and h-lamp-2. *J. Biol. Chem.* 263: 18911–19

Chen, J. W., Cha, Y., Yuksel, K. U., Gracy, R. W., August, J. T. 1988. Isolation and sequencing of a cDNA clone encoding lysosomal membrane glycoprotein mouse LAMP-1. *J. Biol. Chem.* 263: 8754–58

Chen, J. W., Murphy, T. L., Willingham, M. C., Pastan, I., August, J. T. 1985. Identification of two lysosomal membrane glycoproteins. *J. Cell Biol.* 101: 85–95

Cohn, Z. A., Benson, B. 1965. The in vitro differentiation of mononuclear phagocytes. III. The reversibility of granule and hydrolytic enzyme formation and turnover of granule constituents. *J. Exp. Med.* 122: 455–66

Cohn, Z. A., Ehrenreich, B. A. 1969. The uptake, storage, and intracellular hydrolysis of carbohydrates by macrophages. *J. Exp. Med.* 129: 201–25

Cohn, Z. A., Hirsch, J. G., Fedorko, M. E. 1966. The in vitro differentiation of mononuclear phagocytes. V. The formation of macrophage lysosomes. *J. Exp. Med.* 123: 757–66

Corvera, S., Czech, M. P. 1985. Mechanism of insulin action on membrane protein recycling: a selective decrease in the phosphorylation state of insulin-like growth factor II receptors in the cell surface membrane. *Proc. Natl. Acad. Sci. USA* 82: 7314–18

Corvera, S., Folander, K., Clairmont, K. B., Czezh, M. P. 1988a. A highly phosphorylated subpopulation of insulin-like growth factor II/mannose 6-phosphate receptors is concentrated in a clathrin-enriched plasma membrane fraction. *Proc. Natl. Acad. Sci. USA* 85: 7567–71

Corvera, S., Roach, P. J., DePaoli-Roach, A. A., Czech, M. P. 1988b. Insulin action inhibits insulin-like growth factor-II (IGF-II) receptor phosphorylation in H-35 hepatoma cells. *J. Biol. Chem.* 263: 3116–22

Couso, R., Lang, L., Roberts, R. M., Kornfeld, S. 1986. Phosphorylation of the oligosaccharide of uteroferrin by UDP-GlcNAc : glycoprotein N-acetylglucosamine-1-phosphotransferases from rat liver, *Acanthamoeba castellani* and *Dictyostelium discoideum* requires α-1,2-linked mannose residues. *J. Biol. Chem.* 261: 6326–31

Creek, K. E., Sly, W. S. 1982. Adsorptive pinocytosis of phosphorylated oligosaccharides by human fibroblasts. *J. Biol. Chem.* 257: 9931–37

Cushman, S. W., Wardzala, L. J. 1980. Potential mechanism of insulin action on glucose transport in the isolated rat adipose cell. *J. Biol. Chem.* 255: 4758–62

Dahms, N., Kornfeld, S. 1989. The cation-dependent mannose 6-phosphate receptor: structural requirements for mannose 6-phosphate binding and oligomerization. *J. Biol. Chem.* In press

Dahms, N. M., Lobel, P., Breitmeyer, J., Chirgwin, J. M., Kornfeld, S. 1987. 46 kD mannose 6-phosphate receptor: cloning, expression, and homology to the 215 kD mannose 6-phosphate receptor. *Cell* 50: 181–92

Davis, C. G., Lehrman, M. A., Russell, D. W., Anderson, R. G. W., Brown, M. S., Goldstein, J. L. 1986. The J.D. mutation in familial hypercholesterolemia: amino acid substitution in cytoplasmic domain impedes internalization of LDL receptors. *Cell* 45: 15–24

Davis, C. G., van Driel, I. R., Russell, D. W., Brown, M. S., Goldstein, J. L. 1987. The low density lipoprotein receptor. Identification of amino acids in cytoplasmic domain required for rapid endocytosis. *J. Biol. Chem.* 262: 4075–82

de Brabander, M., Nuydens, R., Geerts, H., Hopkins, C. R. 1988. Dynamic behavior of the transferrin receptor followed in living epidermoid carcinoma (A431) cells with nanovid microscopy. *Cell Motil. Cytoskel.* 9: 30–47

de Duve, C. 1963. The lysosome concept. In *Lysosomes*. Ciba Found. Symp. ed. A. V. S. de Reuck, M. P. Cameron. pp. 1–35. London: Churchill

D'Souza, M. P., August, J. T. 1986. A kinetic analysis of biosynthesis and localization of a lysosome-associated membrane glycoprotein. *Arch. Biochem. Biophys.* 249: 522–32

Deutscher, S. L., Creek, K. E., Merion, M., Hirschberg, C. B. 1983. Subfractionation of rat liver Golgi apparatus separation of enzyme activities involved in the biosynthesis of the phosphomannosyl recognition marker in lysosomal enzymes. *Proc. Natl. Acad. Sci. USA* 80: 3938–42

Distler, J., Heiber, V., Sahagian, G. G., Schmickel, R., Jourdian, G. W. 1979. Identification of mannose 6-phosphate in glycoproteins that inhibit the assimilation of β-galactosidase by fibroblasts. *Proc. Natl. Acad. Sci. USA* 76: 4235–39

Duncan, J. R., Kornfeld, S. 1988. Intracellular movement of two mannose 6-phosphate receptors: return to the Golgi apparatus. *J. Cell Biol.* 106: 617–28

Dunn, W. A., Hubbard, A. L., Aronson, N. N. 1980. Low temperature selectively inhibits fusion between pinocytic vesicles and lysosomes during heterophagy of ^{125}I-asialofetuin by the perfused rat liver. *J. Biol. Chem.* 255: 5971–78

Ehrenreich, B. A., Cohn, Z. A. 1969. The fate of peptides pinocytosed by macrophages in vitro. *J. Exp. Med.* 129: 227–45

Erickson, A. H., Ginns, E. I., Barranger, J. A. 1985. Biosynthesis of the lysosomal enzyme, glucocerebrosidase. *J. Biol. Chem.* 260: 14319–24

Fambrough, D. M., Takeyasu, K., Lippincott-Schwartz, J., Siegel, N. R., Somerville, D. 1988. Structure of LEP100, a glycoprotein that shuttles between lysosomes and the plasma membrane, deduced from the nucleotide sequence of the encoding cDNA. *J. Cell Biol.* 196: 61–67

Faust, P. L., Wall, D. A., Perara, E., Lingappa, V. R., Kornfeld, S. 1987. Expression of human cathepsin D in *Xenopus* oocytes: phosphorylation and intracellular targeting. *J. Cell Biol.* 105: 1937–45

Fedde, K. N., Sly, W. S. 1985. Ricin-binding properties of acid hydrolases from isolated lysosomes implies prior processing by terminal transferases of the *trans*-Golgi apparatus. *Biochem. Biophys. Res. Commun.* 133: 614–20

Fischer, H. D., Creek, K. E., Sly, W. S. 1982. Binding of phosphorylated oligosaccharides to immobilized phosphomannosyl receptors. *J. Biol. Chem.* 257: 9938–43

Freeze, H. H., Wolgast, D. 1986. Biosynthesis of methyl-phosphomannosyl residues in the oligosaccharides of *Dictyostelium discoideum. J. Biol. Chem.* 261: 135–41

Froesch, E. R., Schmid, C., Schwander, J., Zapf, J. 1985. Actions of insulin-like growth factors. *Annu. Rev. Physiol.* 47: 443–76

Fukuda, M., Viitala, J., Matteson, J., Carlsson, S. R. 1988. Cloning of cDNAs encoding human lysosomal membrane glycoproteins, h-lamp-1 and h-lamp-2. Comparison of their deduced amino acid sequences. *J. Biol. Chem.* 263: 18920–28

Gabel, C. A., Costello, C. E., Reinhold, V. N., Kurtz, L., Kornfeld, S. 1984. Identification of methylphosphomannosyl residues as components of the high mannose oligosaccharides of *Dictyostelium discoideum* glycoproteins. *J. Biol. Chem.* 259: 13762–69

Gabel, C. A., Foster, S. A. 1987. Post-endocytic maturation of acid hydrolases: evidence of prelysosomal processing. *J. Cell Biol.* 105: 1561–70

Gabel, C. A., Goldberg, D. E., Kornfeld, S. 1983. Identification and characterization of cells deficient in the mannose 6-phosphate receptor: evidence for an alternate pathway for lysosomal enzyme targeting. *Proc. Natl. Acad. Sci. USA* 80: 775–79

Gartung, C., Braulke, T., Hasilik, A., von Figura, K. 1985. Internalization of blocking antibodies against mannose 6-phosphate specific receptors. *EMBO J.* 4: 1725–30

Geuze, H. J., Slot, J. W., Strous, G. J. A. M., Hasilik, A., von Figura, K. 1985. Possible pathways for lysosomal enzyme delivery. *J. Cell Biol.* 101: 2253–62

Geuze, H. J., Slot, J. W., Strous, G. J., Luzio, J. P., Schwartz, A. L. 1984. A cycloheximide-resistant pool of receptors for asialoglycoproteins and mannose 6-phosphate residues in the Golgi complex of hepatocytes. *EMBO J.* 3: 2677–85

Geuze, H. J., Stoorvogel, W., Strous, G. J., Slot, J. W., Zijderhand-Bleekemolen, J., Mellman, I. 1989. Sorting of mannose 6-phosphate receptors and lysosomal membrane proteins in endocytic vesicles. *J. Cell Biol.* 107: 2491–2501

Gieselmann, V., Hasilik, A., von Figura, K. 1985. Processing of human cathepsin D in lysosomes in vitro. *J. Biol. Chem.* 260: 3215–20

Glickman, J. N., Conibear, E., Pearse, B. M. F. 1989. Specificity of binding of clathrin adaptors to signals on the mannose 6-phosphate/insulin-like growth factor II receptor. *EMBO J.* 8: 1041–47

Goda, Y., Pfeffer, S. R. 1988. Selective recycling of the mannose 6-phosphate, insulin-like growth factor II receptor to the *trans* Golgi network in vitro. *Cell* 55: 309–20

Goldberg, D. E., Kornfeld, S. 1981. The phosphorylation of β-glucuronidase oligosaccharides in mouse P388D$_1$ Cells. *J. Biol. Chem.* 256: 13060–67

Goldberg, D. E., Kornfeld, S. 1983. Evidence for extensive subcellular organization of asparagine-linked oligosaccharide processing and lysosomal enzyme phosphorylation. *J. Biol. Chem.* 258: 3159–65

Gonzales-Noriega, A., Grubb, J. H., Talkad, V., Sly, W. S. 1980. Chloroquine inhibits lysosomal enzyme pinocytosis and enhances enzyme secretion by impairing receptor recycling. *J. Cell Biol.* 85: 839–52

Granger, B. L., Howe, C. L., Gabel, C., Helenius, A., Mellman, I. 1989. Characterization and cDNA cloning of lgp110, a lysosomal membrane protein from rat and mouse cells. *J. Biol. Chem.* Submitted

Green, S. A., Zimmer, K.-P., Griffiths, G., Mellman, I. 1987. Kinetics of intracellular transport and sorting of lysosomal membrane and plasma membrane proteins. *J. Cell Biol.* 105: 1227–40

Griffiths, G., Hoflack, B., Simons, K., Mellman, I., Kornfeld, S. 1988. The mannose 6-phosphate receptor and the biogenesis of lysosomes. *Cell* 52: 329–41

Griffiths, G., Simons, K. 1986. The *trans* Golgi network: sorting at the exit site of the Golgi complex. *Science* 234: 430–43

Gruenberg, J., Griffiths, G., Howell, K. E. 1989. Characterization of the early endosome and putative endocytic carrier vesicles in vivo and with an assay of vesicle fusion in vitro. *J. Cell Biol.* In press

Hari, J., Pierce, S. B., Morgan, D. O., Sara, V., Smith, M. C., Roth, R. A. 1987. The receptor for insulin-like growth factor II mediates an insulin-like response. *EMBO J.* 6: 3367–71

Hasilik, A., Waheed, A., von Figura, K. 1981. Enzymatic phosphorylation of lysosomal enzymes in the presence of UDP-N-acetylglucosamine: absence of the activity in I-cell fibroblasts. *Biochem. Biophys. Res. Comm.* 98: 761–67

Hasilik, A., von Figura, K. 1984. Processing of lysosomal enzymes in fibroblasts. In *Lysosomes in Biology and Pathology*, eds. J. T. Dingle, R. T. Dean, W. S. Sly. New York: Elsevier/North Holland. pp. 3–17

Helenius, A., Mellman, I., Wall, D., Hubbard, A. L. 1983. Endosomes. *Trends Biochem. Chem.* 8: 245–50

Herman, B., Albertini, D. F. 1984. A time lapse video image intensification analysis of cytoplasmic organelle movements during endosome translocation. *J. Cell Biol.* 98: 565–76

Hirsch, J. G., Fedorko, M. E., Cohn, Z. A. 1968. Vesicle fusion and formation at the surface of pinocytic vesicles in macrophages. *J. Cell Biol.* 38: 629–32

Hoflack, B., Fujimoto, K., Kornfeld, S. 1987. The interaction of phosphorylated oligosaccharides and lysosomal enzymes with bovine liver cation-dependent mannose 6-phosphate receptor. *J. Biol. Chem.* 262: 123–29

Hoflack, B., Kornfeld, S. 1985. Purification and characterization of a cation-dependent mannose 6-phosphate receptor from murine $P388D_1$ macrophages and bovine liver. *J. Biol. Chem.* 260: 12008–14

Holtzman, E. 1989. *Lysosomes*. New York: Plenum. 439 pp.

Honey, N. K., Mueller, O. T., Little, L. E., Miller, A. L., Shows, T. B. 1982. Mucolipidosis II and III. The genetic relationships between two disorders of lysosomal enzyme biosynthesis. *Proc. Natl. Acad. Sci. USA* 79: 7420–24

Howe, C. L., Granger, B. L., Hull, M., Green, S. A., Gabel, C. A., et al. 1988. Derived protein sequence, oligosaccharides, and membrane insertion of the 120-kDa lysosomal membrane glycoprotein (lgp 120): Identification of a highly conserved family of lysosomal membrane glycoproteins. *Proc. Natl. Acad. Sci. USA* 85: 7577–81

Hughes, E. N., August, J. T. 1982. Murine cell surface glycoproteins. Identification, purification, and characterization of a major glycosylated component of 110,000 daltons by use of a monoclonal antibody. *J. Biol. Chem.* 257: 3970–77

Hynes, R. 1985. Molecular biology of fibronectin. *Annu. Rev. Cell Biol.* 1: 67–90

James, M. N. G., Siedecki, A. R. 1986. Molecular structure of an aspartic proteinase zymogen porcine pepsinogen, at 1.8 A resolution. *Nature* 319: 33–38

Jin, M., Sahagian, G. G., Snider, M. D. 1989. Transport of surface mannose 6-phosphate receptor to the Golgi complex in cultured human cells. *J. Biol. Chem.* 264: 7675–80

Johnson, L. M., Bankaitis, V. A., Emr, S. D. 1987. Distinct sequence determinants direct intracellular sorting and modification of a yeast vacuolar protease. *Cell* 48: 875–85

Junghans, U., Waheed, A., von Figura, K. 1988. The 'cation-dependent' mannose 6-phosphate receptor binds ligands in the absence of divalent cations. *FEBS Letters* 237: 81–84

Kaplan, A., Achord, D. T., Sly, W. S. 1977. Phosphohexosyl components of a lysosomal enzyme are recognized by pinocytosis receptors on human fibroblasts. *Proc. Natl. Acad. Sci. USA* 74: 2026–30

Kiess, W., Blickenstaff, G. D., Sklar, M. M., Thomas, C. L., Nissley, S. P., Sahagian, G. G. 1988. Biochemical evidence that the type II insulin-like growth factor receptor is identical to the cation-independent mannose 6-phosphate receptor. *J. Biol. Chem.* 263: 9339–44

Kiess, W., Haskel, J. F., Lee, L., Greenstein, L. A., Miller, B. E., et al. 1987. An antibody that blocks insulin-like growth factor (IGF) binding to the type II IGF receptor is neither an agonist nor an inhibitor of IGF-stimulated biologic responses in L6 myoblasts. *J. Biol. Chem.* 262: 12745–51

Kiess, W., Thomas, C. L., Greenstein, L. A., Lee, L., Sklar, M. M., et al. 1989. Insulin-like growth factor II inhibits both the cellular uptake of β-galactosidase and the binding of β-galactosidase to purified IGFII/mannose 6-phosphate receptor. *J. Biol. Chem.* 264: 4710–14

Klionsky, D. J., Banta, L. M., Emr, S. D. 1988. Intracellular sorting and processing of a yeast vacuolar hydrolase: the proteinase A propeptide contains vacuolar targeting information. *Mol. Cell. Biol.* 8: 2105–16

Kojima, I., Nishimoto, I., Iiri, T., Ogata, E., Rosenfeld, R. 1988. Evidence that type II insulin-like growth factor receptor is coupled to calcium gating system. *Biochem. Biophys. Res. Comm.* 154: 9–19

Kornfeld, S. 1986. Trafficking of lysosomal enzymes in normal and disease states. *J. Clin. Invest.* 77: 1–6

Kyle, J. W., Nolan, C. M., Oshima, A., Sly, W. S. 1989 Expression of human cation-independent mannose 6-phosphate receptor cDNA in receptor-negative mouse $P388D_1$ cells following gene transfer. *J. Biol. Chem.* 263: 16230–35

Lang, L., Reitman, M., Tang, J., Roberts, R. M., Kornfeld, S. 1984. Lysosomal enzyme phosphorylation: recognition of a protein dependent determinants allows specific phosphorylation of oligosaccharides present on lysosomal enzymes. *J. Biol. Chem.* 259: 14663–71

Lang, L., Takahashi, T., Tang, J., Kornfeld, S. 1985. Lysosomal enzyme phosphorylation in human fibroblasts: kinetic parameters offer a biochemical rationale for two distinct defects in N-acetylglucosaminylphosphotransferase. *J. Clin. Invest.* 76: 2191–95

Lazarovits, J., Roth, M. 1988. A single amino acid change in the cytoplasmic domain allows the influenza virus hemagglutinin to be endocytosed through coated pits. *Cell* 53: 743–52

Lazzarino, D. A., Gabel, C. A. 1988. Biosynthesis of the mannose 6-phosphate recognition marker in transport-impaired mouse lymphoma cells. *J. Biol. Chem.* 263: 10118–26

Lazzarino, D. A., Gabel, C. A. 1989. Mannose processing is an important determinant in the assembly of phosphorylated high mannose-type oligosaccharides. *J. Biol. Chem.* 264: 5015–23

Lemansky, P., Gieselmann, V., Hasilik, A., von Figura, K. 1985. Synthesis and transport of lysosomal acid phosphatase in normal and I-cell fibroblasts. *J. Biol. Chem.* 260: 9023–30

Lemansky, P., Hasilik, A., von Figura, K., Helmy, S., Fishman, J., et al. 1987. Lysosomal enzyme precursors in coated vesicles derived from the exocytic and endocytic pathways. *J. Cell Biol.* 104: 1743–48

Lewis, V. A., Green, S. A., Marsh, M., Vihko, P., Helenius, A., Mellman, I. 1985. Glycoproteins of the lysosomal membrane. *J. Cell Biol.* 100: 1839–47

Lippincott-Schwartz, J., Fambrough, D. M. 1986. Lysosomal membrane dynamics: structure and interorganellar movement of a major lysosomal membrane glycoprotein. *J. Cell Biol.* 102: 1593–1605

Lippincott-Schwartz, J., Fambrough, D. M. 1987. Cycling of the integral glycoprotein LEP100 between plasma membrane and lysosomes: kinetic and morphological analysis. *Cell* 49: 669–77

Little, L. E., Mueller, O. T., Honey, N. K., Shows, T. B., Miller, A. L. 1986. Heterogeneity of N-acetylglucosamine 1-phosphotransferase within mucolipidosis III. *J. Biol. Chem.* 261: 733–38

Little, L. E., Alcouloumre, M., Drotar, A. M., Herman, S., Robertson, R., et al. 1987. Properties of N-acetylglucosamine 1-phosphotransferase from human lymphoblasts. *Biochem. J.* 248: 151–59

Lloyd, J. B., Foster, S. 1986. The lysosome membrane. *Trends Biochem. Sci.* 11: 365–68

Lobel, P., Dahms, N. M., Breitmeyer, J., Chirgwin, J. M., Kornfeld, S. 1987. Cloning of the bovine 215-kDa cation-independent mannose 6-phosphate receptor. *Proc. Natl. Acad. Sci USA* 84: 2233–37

Lobel, P., Dahms, N. M., Kornfeld, S. 1988. Cloning and sequence analysis of the cation-independent mannose 6-phosphate receptor. *J. Biol. Chem.* 263: 2563–70

Lobel, P., Fujimoto, K., Ye, R. D., Griffiths, G., Kornfeld, S. 1989. Mutations in the cytoplasmic domain of the 275-kD mannose 6-phosphate receptor differentially alter lysosomal enzyme sorting and endocytosis. *Cell* In press

Lonnroth, P., Assmundsson, K., Eden, S., Enberg, G., Gause, I., et al. 1987. Regulation of insulin-like growth factor II receptors by growth hormone and insulin in rat adipocytes. *Proc. Natl. Acad. Sci. USA* 84: 3619–22

MacDonald, R. G., Pfeffer, S. R., Coussen, L., Tepper, M. A., Brocklebank, C. M., et al. 1988. A single receptor binds insulin-like growth factor II and mannose 6-phosphate. *Science* 239: 1134–37

Madiyalakan, K., Chowdhary, M. S., Rana, S., Matta, K. L. 1986. Lysosomal enzyme targeting: the phosphorylation of synthetic D-mannosyl saccharides by UDP-N-acetylglucosamine : lysosomal enzyme N-acetylglucosamine phosphotransferase from rat liver microsomes and fibroblasts. *Carbohydr. Res.* 152: 183–94

Madiyalakan, R., Jain, F. K., Matta, K. L. 1987. Phosphorylation of the α1,2-linked mannosylsaccharide by N-acetylglucosamine-1-phosphotransferase from fibroblasts occurs at the terminal mannose. *Biochem. Biophys. Res. Comm.* 142: 354–58

Maguire, G. A., Docherty, K., Hales, C. N. 1983. Sugar transport in lysosomes. *Biochem. J.* 212: 211–18

Marquardt, T., Braulke, T., Hasilik, A., von Figura, K. 1987. Association of the precursor of cathepsin D with coated membranes. *Eur. J. Biochem.* 168: 37–42

Marsh, M., Schmid, S., Kern, H., Harms, E., Mâle, P., et al. 1987. Rapid analytical and preparative isolation of functional endosomes by free-flow electrophoresis. *J. Cell Biol.* 104: 875–86

Mellas, J., Gavin, J. B., Hammerman, M. R. 1986. Multiplication-stimulating activity-induced alkalinization of canine renal proximal tubular cells. *J. Biol. Chem.* 261: 14437–42

Mellman, I., Fuchs, R., Helenius, A. 1986. Acidification of the endocytic and exocytic pathways. *Annu. Rev. Biochem.* 55: 663–700

Mierendorf, R. C., Cardelli, J. A., Dimond, R. L. 1985. Pathways involved in targeting and secretion of a lysosomal enzyme in *Dictyostelium discoideum*. *J. Cell Biol.* 100: 1777–87

Minnifeld, N., Creek, K. E., Navas, P., Morre, D. J. 1986. Involvement of *cis* and *trans* Golgi apparatus elements in the intracellular sorting and targeting of acid hydrolases to lysosomes. *Eur. J. Cell. Biol.* 42: 92–100

Morgan, D. O., Edman, J. C., Standing, D. N., Fried, V. A., Smith, M. C., 1987. Insulin-like growth factor II receptor as a multifunctional binding protein. *Nature* 329: 301–7

Mueller, O. T., Honey, N. K., Little, L. E., Miller, A. L., Shows, T. B. 1983. Mucolipidosis II and III. The genetic relationships between two discorders of lysosomal enzyme biosynthesis. *J. Clin. Invest.* 72: 1016–23

Natowicz, M., Baenziger, J. U., Sly, W. S. 1982. Structural studies of the phosphorylated high mannose-type oligosaccharides on human β-glucuronidase. *J. Biol. Chem.* 257: 4412–20

Natowicz, M., Hallet, D. W., Frier, C., Chi, M., Schlesinger, P. H., Baenziger, J. U. 1983. Recognition and receptor-mediated uptake of phosphorylated high mannose-type oligosaccharides by cultured human fibroblasts. *J. Cell. Biol.* 96: 915–19

Neufeld, E. F., McKusik, V. A. 1983. Disorders of lysosomal enzyme synthesis and localization: I-cell disease and pseudo-Hurler polydistrophy. In *The Metabolic Basis of Inherited Disease*, ed. J. B. Stanbury, J. B. Wyngaarden, D. S. Fredrickson, J. L. Goldstein, M. S. Brown. pp. 778–87. New York: McGraw–Hill. 5th ed.

Nishimoto, I., Hata, K., Ogata, E., Kojima, I. 1987. Insulin-like growth factor II stimulates calcium influx in competent BALB/c 3T3 cells primed with epidermal growth factor. Characteristics of calcium influx and involvement of GTP-binding protein. *J. Biol. Chem.* 262: 12120–26

Nolan, C. M., Creek, K. E., Grubb, J., Sly, W. S. 1987. Antibody to the phosphomannosyl receptor inhibits receptor recycling in fibroblasts. *J. Cell Biochem.* 35: 137–51

Novikoff, A. 1963. Lysosomes in the physiology and pathology of cells: contributions of staining methods. In *Lysosomes*, Ciba Found. Symp., ed. A. V. S. de Reuck, M. P. Cameron. pp. 36–77. London: Churchill

Oka, Y., Czech, M. P. 1986. The type II insulin-like growth factor receptor is internalized and recycles in the absence of ligand. *J. Biol. Chem.* 261: 9090–93

Oka, Y., Mottola, C., Oppenheimer, C. L., Czech, M. P. 1984. Insulin activates the appearance of insulin-like growth factor II receptors on the adipocyte cell surface. *Proc. Natl. Acad. Sci. USA* 81: 4028–32

Oka, Y., Rozek, L. M., Czech, M. P. 1985. Direct demonstration of rapid insulin-like growth factor II receptor internalization and recycling in rat adipocytes. *J. Biol. Chem.* 260: 9435–42

Okada, S., Inui, K., Furukawa, M., Midorikawa, M., Nishimoto, J., et al. 1987. Biochemical heterogeneity in I-cell disease. Sucrose-loading test classifies two distinct subtypes. *Enzyme* 38: 267–72

Oppenheimer, C. L., Pessin, J. E., Massagúe, J., Gitomer, W., Czech, M. P. 1983. Insulin action rapidly modulates the apparent affinity of the insulin-like growth factor II receptor. *J. Biol. Chem.* 258: 4824–30

Oshima, A., Nolan, C. M., Kyle, J. W., Grubb, J. H., Sly, W. S. 1988. The human cation-independent mannose 6-phosphate receptor. *J. Biol. Chem.* 263: 2553–62

Palade, G. E. 1975. Intracellular aspects of the process of protein synthesis. *Science* 189: 347–58

Pearse, B. M. F. 1985. Assembly of the mannose 6-phosphate receptor into reconstituted clathrin coats. *EMBO J.* 4: 2457–60

Pelham, H. R. B. 1988. Evidence that luminal ER proteins are sorted from secreted proteins in a post-ER compartment. *EMBO J.* 7: 913–18

Perdue, J. F., Chan, J. K., Thibault, C., Radaj, P., Mills, B., Daughaday, W. H. 1983. The biochemical characteristics of detergent-solubilized insulin-like growth factor II receptors from rat placenta. *J. Biol. Chem.* 258: 7800–11

Pfeffer, S. R. 1987. The endosomal concentration of a mannose 6-phosphate receptor is unchanged in the absence of ligand synthesis. *J. Cell Biol.* 105: 229–34

Pohlmann, R., Krentler, C., Schmidt, B., Schroeder, W., Lorkowski, G., et al. 1988. Human lysosomal acid phosphatase: cloning, expression and chromosomal assignment. *EMBO J.* 7: 2343–50

Pohlmann, R., Nagel, G., Schmidt, B., Stein, M., Lorkowski, G., et al. 1987. Cloning of a cDNA encoding the human cation-dependent mannose 6-phosphate-specific receptor. *Proc. Natl. Acad. Sci. USA* 84: 5575–79

Pohlmann, R., Waheed, A., Hasilik, A., von Figura, K. 1982. Synthesis of phosphorylated recognition marker in lysosomal enzymes is located in the *cis* part of Golgi apparatus. *J. Biol. Chem.* 257: 5323–25

Polychronakos, C., Guyda, H. J., Posner, B. I. 1988. Mannose 6-phosphate increases the affinity of its cation-independent receptor for insulin-like growth factor II by displacing inhibitory endogenous ligands. *Biochem. Biosphys. Res. Comm.* 157: 632–38

Rechler, M. M., Nissley, S. P. 1985. The nature and regulation of the receptors for insulin-like growth factors. *Annu. Rev. Physiol.* 47: 425–62

Reggio, H., Bainton, D., Harms, E., Courdrier, E., Louvard, D. 1984. Antibodies against lysosomal membranes reveal a 100,000 mol. wt. protein that cross-reacts with H^+-K^+ ATPase from gastric mucosa. *J. Cell Biol.* 99: 1511–26

Reitman, M. L., Kornfeld, S. 1981a. UDP-N-acetylglucosamine: glycoprotein N-acetylglucosamine-1-phosphotransferase, proposed enzyme for the phosphorylation of the high mannose oligosaccharide units of lysosomal enzymes. *J. Biol. Chem.* 256: 4275–81

Reitman, M. L., Kornfeld, S. 1981b. Lysosomal enzyme targeting: N-acetylglucosaminyl-phosphotransferase selectively phosphorylates native lysosomal enzymes. *J. Biol. Chem.* 256: 11977–80

Renlund, M., Tietze, F., Gahl, W. A. 1986. Defective sialic acid egress from isolated fibroblast lysosomes of patients with Sallas disease. *Science* 232: 759–62

Richardson, J. M., Wyochik, N. A., Ebert, D. L., Dimond, R. L., Cardelli, J. A. 1989. Inhibition of early but not late proteolytic processing events leads to the missorting and oversecretion of precursor forms of lysosomal enzymes in Dictyostelium discoideum. *J. Cell Biol.* In press

Rizzolo, L. J., Finidori, J., Gonzalez, A., Arpin, M., Ivanov, I. E., et al. 1985. Biosynthesis and intracellular sorting of growth hormone-viral envelope glycoprotein hybrids. *J. Cell Biol.* 101: 1351–62

Rizzolo, L. J., Kornfeld, R. 1988. Post-translational protein modification in the endoplasmic reticulum. *J. Biol. Chem.* 263: 9520–25

Robbins, A. R. 1988. Endocytosis and compartmentalization of lysosomal enzymes in normal and mutant mammalian cells: Mannose 6-phosphate pathways. In *Protein Transfer and Organelle Biogenesis*, pp. 463–520. Orlando: Academic

Robbins, A. R., Oliver, C., Bateman, J. L., Krag, S. S., Galloway, C. J., Mellman, I. 1984. A single mutation in Chinese hamster ovary cells impairs both Golgi and endosomal functions. *J. Cell Biol.* 99: 1296–1308

Robinson, J. S., Klionsky, D. J., Banta, L. M., Emr, S. D. 1988. Protein sorting in yeast: isolation of mutants defective in the processing and sorting of multiple vacuolar hydrolases. *Mol. Cell. Biol.* In press

Roff, C. F., Fuchs, R., Mellman, I., Robbins, A. R. 1986. Chinese hamster ovary cells with temperature-sensitive defects in endocytosis. I. Loss of functions on shifting to the non-permissive temperature. *J. Cell Biol.* 103: 2283–97

Rogers, S. A., Hammerman, M. R. 1989. Mannose 6-phosphate potentiates insulin-like growth factor II-stimulated inositol triphosphate production in proximal tubular basolateral membranes. *J. Biol. Chem.* 264: 4273–76

Rome, L. H., Hill, D. F., Bame, K. J., Crain, L. J. 1983. Utilization of exogenously added aceyl coenzyme A by intact isolated lysosomes. *J. Biol. Chem.* 258: 3006–11

Rosenblatt, D. S., Hosack, A., Matiaszuk, N. V., Cooper, B. A., LaFramboise, R. 1985. Defect in vitamin B_{12} release from lysosomes: newly described inborn error of vitamin B_{12} metabolism. *Science* 228: 1319–21

Rosenfeld, M. G., Kreibich, G., Popov, D., Kato, K., Sabaitini, D. D. 1982. Biosynthesis of lysosomal hydrolases: their synthesis in bound polysomes and the role of co- and posttranslational processing in determining their subcellular distribution. *J. Cell Biol.* 93: 135–43

Roth, R. A. 1988. Structure of the receptor for insulin-like growth factor II: the puzzle amplified. *Science* 239: 1269–71

Roth, R. A., Stover, C., Hari, J., Morgan, D. O., Smith, M. C., et al. 1987. Interactions of the receptor for insulin-like growth factor II with mannose 6-phosphate and antibodies to the mannose 6-phosphate receptor. *Biochem. Biophys. Res. Comm.* 149: 600–6

Rothman, J. H., Stevens, T. H. 1986. Protein sorting in yeast: mutants defective in vacuole biogenesis mislocalize vacuolar proteins into the late secretory pathway. *Cell* 47: 1041–51

Rothman, J. H., Hunter, C. P., Valls, L. A., Stevens, T. H. 1986. Overproduction-induced mislocalization of a yeast vacuolar protein allows isolation of its structural gene. *Proc. Natl. Acad. Sci. USA* 83: 3248–52

Sahagian, G. G. 1984. The mannose 6-phosphate receptor: function, biosynthesis and translocation. *Biol. Cell* 51: 207–14

Sahagian, G. G., Distler, J., Jourdian, G. W. 1981. Characterization of a membrane-associated receptor from bovine liver that binds phosphomannosyl residues of bovine testicular β-galactosidase. *Proc. Natl. Acad. Sci. USA* 78: 4289–93

Sahagian, G. G., Neufeld, E. F. 1983. Biosynthesis and turnover of the mannose 6-phosphate receptor in cultured Chinese hamster ovary cells. *J. Biol. Chem.* 258: 7121–28

Saraste, J., Kuismanen, E. 1984. Pre- and post-Golgi vacuoles operate in the transport of Semliki Forest Virus membrane glycoproteins to the cell surface. *Cell* 38: 535–49

Schmid, S., Fuchs, R., Mâle, P., Mellman, I. 1988. Two distinct subpopulations of endosomes involved in membrane recycling and transport to lysosomes. *Cell* 52: 73–83

Schmid, S., Fuchs, R., Kielian, M., Helenius, A., Mellman, I. 1989. Acidification of endosome subpopulations in wild-type Chinese hamster ovary cells and temperature-sensitive acidification-defective mutants. *J. Cell Biol.* 108: 1291–1300

Schulze-Lohoff, E., Hasilik, A., von Figura, K. 1985. Cathepsin D precursors in clathrin-coated organelles from human fibroblasts. *J. Cell Biol.* 101: 824–29

Schneider, J. A., Jonas, A. J., Smith, M. L., Greene, A. A. 1984. Lysosomal transport of cystine and other small molecules. *Biochem. Soc. Trans.* 12: 908–10

Silverstein, S. C., Steinman, R. M., Cohn, Z. A. 1977. Endocytosis. *Annu. Rev. Biochem.* 46: 669–772

Sipe, D. M., Murphy, R. F. 1987. High resolution kinetics of transferrin acidification in Balb/c 3T32 cells: exposure to pH 6 followed by temperature-sensitive alkalinization during recycling. *Proc. Natl. Acad. Sci. USA* 84: 7119–23

Stein, M., Braulke, T., Krentler, C., Hasilik, A., von Figura, K. 1987a. 46-kDa mannose 6-phosphate-specific receptor: purification, subunit composition, chemical modification. *Biol. Chem. Hoppe-Seyler* 368: 927–36

Stein, M., Braulke, T., Krentler, C., Hasilik, A., von Figura, K. 1987b. 46-kDa mannose 6-phosphate-specific receptor: biosynthesis, processing, subcellular location and topology. *Biol. Chem. Hoppe-Seyler* 368: 937–47

Stein, M., Zijderhand-Bleekemolen, J. E., Geuze, H., Hasilik, A., von Figura, K. 1987. Mr 46,000 mannose 6-phosphate specific receptor: its role in targeting of lysosomal enzymes. *EMBO J.* 6: 2677–81

Steinman, R. M., Brodie, S. E., Cohn, Z. A. 1976. Membrane flow during pinocytosis. A stereological analysis. *J. Cell Biol.* 55: 665–87

Steinman, R. M., Mellman, I., Muller, W. A., Cohn, Z. A. 1983. Endocytosis and the recycling of plasma membrane. *J. Cell Biol.* 96: 1–27

Stevens, T. H., Rothman, J. H., Payne, G. S., Schekman, R. 1986. Gene dosage-dependent secretion of yeast vacuolar carboxypeptidase Y. *J. Cell. Biol.* 102: 1551–57

Straus, W. 1963. Comparative observations on lysosomes and phagosomes in kidney and liver of rats after administration of horseradish peroxidase. In *Lysosomes* Ciba Found. Symp., ed. A. V. S. de Reuck, M. P. Cameron. pp. 151–75. London: Churchill

Swanson, J. A., Yiriniec, B. D., Silverstein, S. C. 1985. Phorbol esters and horseradish peroxidase stimulate pinocytosis and redirect the flow of pinocytosed fluid in macrophages. *J. Cell Biol.* 100: 1217–22

Tally, M., Li, C. H., Hall, K. 1987. IGF-2 stimulated growth mediated by the somatomedin type 2 receptor. *Biochem. Biophys. Res. Commun.* 148: 811–16

Tong, P., Gregory, W., Kornfeld, S. 1989. Ligand interactions of the cation-independent mannose 6-phosphate receptor. The stoichiometry of mannose 6-phosphatase binding. *J. Biol. Chem.* 264: 7962–69

Tong, P., Kornfeld, S. 1989. Ligand interactions of the cation-dependent mannose 6-phosphate receptor. Comparison with the cation-independent mannose 6-phosphate receptor. *J. Biol. Chem.* 264: 7970–75

Tong, P., Tollefsen, S. E., Kornfeld, S. 1988. The cation-independent mannose 6-phosphate receptor binds insulin-like growth factor II. *J. Biol. Chem.* 263: 2585–88

Tooze, S. A., Tooze, J., Warren, G. 1988.

Site of addition of N-acetyl galactosamine to the E1 glycoprotein of mouse hepatitis virus A59. *J. Cell Biol.* 106: 1475–87

Vale, R. D., Schnapp, B. J., Mitchinson, T., Steuer, E., Reese, T. S., Sheetz, M. P. 1985. Different axoplasmic proteins generate movement in opposite directions along microtubules in vitro. *Cell* 4: 623–32

Valls, L. A., Hunter, C. P., Rothman, J. H., Stevens, T. H. 1987. Protein sorting in yeast: the localization determinant of yeast vacuolar carboxypeptidase Y resides in the propeptide. *Cell* 30: 887–97

Varki, A., Kornfeld, S. 1981. Purification and characterization of rat liver α-N-acetyl-glucosaminyl phosphodiesterase. *J. Biol. Chem.* 256: 9937–43

Varki, A., Kornfeld, S. 1983. The spectrum of anionic oligosaccharides released by endo-β-N-acetylglucosaminidase H from glycoproteins. *J. Biol. Chem.* 258: 2808–18

Varki, A., Reitman, M. L., Kornfeld, S. 1981. Identification of a variant of mucolipidosis III (pseudo-Hurler polydystrophy): a catalytically active N-acetyl-gluco-saminylphosphotransferase that fails to phosphorylate lysosomal enzymes. *Proc. Natl. Acad. Sci. USA* 78: 7773–77

Vladutiu, G. D. 1983. Effect of the co-existence of galactosyl and phosphomannosyl residues of β-hexosaminidase on the processing and transport of the enzyme in MLI fibroblasts. *Biochem. Biophys. Acta* 760: 363–70

van Dongen, J. M., Barneveld, R. A., Geuze, H. J., Galjaard, H. 1984. Immunocytochemistry of lysosomal hydrolases and their precursor forms in normal and mutant human cells. *Histochem. J.* 16: 941–54

von Figura, K., Hasilik, A. 1986. Lysosomal enzymes and their receptors. *Annu. Rev. Biochem.* 55: 167–93

Waheed, A., Gottschalk, S., Hille, A., Krentler, C., Pohlmann, R., et al. 1988. Human lysosomal acid phosphatase is transported as a transmembrane protein to lysosomes in transfected baby hamster kidney cells. *EMBO J.* 7: 2351–58

Waheed, A., Hasilik, A., von Figura, K. 1982. UDP-N-acetylglucosamine: lysosomal enzyme precursor N-acetylglucosamine-1-phosphotransferase. *J. Biol. Chem.* 257: 12322–31

Waheed, A., Braulke, T., Junghans, U., von Figura, K. 1988. Mannose 6-phosphate/insulin-like growth factor II receptor: the two types of ligands bind simultaneously to one receptor at different sites. *Biochem. Biophys. Res. Comm.* 152: 1248–54

Waheed, A., Hasilik, A., von Figura, K. 1981. Processing of the phosphorylated recognition marker in lysosomal enzymes. *J. Biol. Chem.* 256: 5717–21

Wardzala, L. J., Simpson, I. A., Rechler, M. M., Cushman, S. W. 1984. Potential mechanisms of the stimulatory action of insulin on insulin-like growth factor II binding to the isolated rat adipose cell. *J. Biol. Chem.* 259: 8378–83

Watanabe, H., Kyle, J. W., Nolan, C. M., Sly, W. S. 1988. The human mannose 6-phosphate/IGF II receptor: structure/function analysis of the mutationally altered cytoplasmic domain following gene transfer and amplification in receptor-negative mouse L-cells. *J. Cell Biol.* 107: 550a.

Wendland, M., Hille, A., Nagel, G., Waheed, A., von Figura, K., Pohlmann, R. 1989. Synthesis of a truncated 46 kDa mannose 6-phosphate receptor that is secreted and retains ligand binding. *Biochem. J.* 260: In press

Willingham, M. C., Pastan, I. H., Sahagian, G. G., Jourdian, G. W., Neufeld, E. F. 1981. Morphologic study of the internalization of a lysosomal enzyme by the mannose 6-phosphate receptor in cultured Chinese hamster ovary cells. *Proc. Natl. Acad. Sci. USA* 78: 6967–71

SUBJECT INDEX

R

CUMULATIVE INDEXES

CONTRIBUTING AUTHORS, VOLUMES 1–5

CHAPTER TITLES, VOLUMES 1–5